MATRIX THEORY AND
FINITE MATHEMATICS

INTERNATIONAL SERIES IN
PURE AND APPLIED MATHEMATICS

William Ted Martin, E. H. Spanier,
G. Springer, and P. J. Davis
CONSULTING EDITORS

McGraw-Hill
Book
Company

New York
St. Louis
San Francisco
Düsseldorf
Johannesburg
Kuala Lumpur
London
Mexico
Montreal
New Delhi
Panama
Rio de Janeiro
Singapore
Sydney
Toronto

MARTIN PEARL

Professor of Mathematics
University of Maryland

Mathematician, National Bureau of Standards

Matrix Theory and Finite Mathematics

This book was set in Times New Roman.
The editors were Jack L. Farnsworth and Norma Frankel;
the designer was Rolando Morales;
and the production supervisor was Thomas J. Lo Pinto.
The drawings were done by Reproduction Drawings Ltd.
The printer and binder was R. R. Donnelley & Sons Company.

209407

23932

MATRIX THEORY AND FINITE MATHEMATICS

1234567890 DODO 79876543

72006823

Library of Congress Cataloging in Publication Data

Pearl, Martin, 1928—
  Matrix theory and finite mathematics.

  (International series in pure and applied
mathematics)
  1. Matrices. I. Title.
QA263.P34   512.9'43   72-6823
ISBN 0-07-049027-9

(16/2/94)

CONTENTS

PREFACE

The term "applied mathematics" stands for a conglomeration of ordinary and partial differential equations, Fourier series, Laplace transforms, and other similar topics from the domain of classical analysis. Until recently applied mathematics (without quotation marks) consisted almost exclusively of "applied mathematics" and numerical analysis. However, since the end of World War II two factors have combined to change the situation radically. Foremost is the advent of the computer. Whereas it had long been practically possible to solve a great many continuous, and hence infinite, problems when only hand computation was available, it usually happened that the amount of computation necessary to solve a meaningful finite problem was prohibitively large. At about the same time that the computer became an indispensable tool in applied mathematics, enormous contributions were being made to our knowledge of finite mathematics. It is interesting to note that of the vast quantity of mathematics developed in the last quarter of a century, virtually all the material that can be presented at the undergraduate level comes from the realm of finite mathematics.

One consequence of finite mathematics becoming an integral part of applied mathematics is the recent resurgence of interest in matrices, as a subject both for research and for undergraduate courses. The basic course in matrix theory was greatly changed after World War II by shifting the emphasis to finite-dimensional linear transformations. Today there is a growing feeling that both points of view are necessary. The aim of this book is to provide a unified treatment of matrix algebra coupled with an introduction to three of the most important modern applications of matrices.

Chapter 1 contains most of the material usually encountered in a one-semester undergraduate course in vectors and matrices. After much soul-searching I finally decided not to add a section dealing with the orthogonal and unitary reduction of normal matrices. I was also greatly tempted to include some discussion of the generalized inverse of a matrix. The sole reason for excluding both these topics is that they are not used elsewhere in the book. Thus, as things now stand, virtually all the material covered in Chap. 1 is needed for at least one of the three later chapters. It is interesting to note just how much matrix theory is required in the later chapters.

The vectors and matrices of Chap. 1 take their entries from an arbitrary field. In the application of Chap. 2 (The Theory of Games) this field is specialized to be the field of real numbers. When discussing Markov chains in Chap. 3 it is more convenient to use the field of complex numbers. In Chap. 4 (The Theory of Graphs) the theorems of Chap. 1 are applied both to matrices with real entries and to matrices whose entries are from the field with two elements, $\mathbb{Z}_2$. This provides an unusually striking example, especially to the reader who is not a professional mathematician, of the power and beauty of abstract mathematics.

After Chap. 1 the remaining chapters may be covered in any order. The following dependency diagram should prove useful:

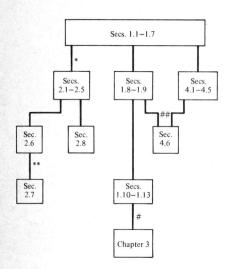

* Section 1.10 is used in the proof of Theorem 2.4.10 (Farkas' theorem) and in the preceding lemma. If Farkas' theorem is stated without proof, only Secs. 1.1 to 1.7 are needed for Chap. 2 (but see next note).

** Theorem 1.9.32 is required for Lemma 2.7.18 and Theorem 2.7.21. However, not much is lost by leaving them out.

The Jordan canonical form is used in Sec. 3.2 and is basic to all of Chap. 3.

Both Secs. 1.8 to 1.9 and Secs. 4.1 to 4.5 are needed for Sec. 4.6.

In order to achieve this flexibility it was occasionally necessary to repeat a definition or the discussion of a particular point. It seems a very small price to pay.

In the numerous times I have taught courses from the various preliminary versions of this book (at Imperial College of the University of London and at the University of Maryland) I have found that Chap. 1 requires slightly more than a semester's time. With the current trend of introducing more and more linear algebra into the standard calculus sequence it will undoubtedly soon be possible to cover the first chapter in a significantly shorter time. Another means for decreasing the time needed for Chap. 1 is to assign Secs. 1.8 and 1.11 as independent reading at the very beginning of the semester. Neither of these sections depends on any previous material in the course and the student can study them at his leisure.

Chapter 4 fits conveniently into a semester, but I have repeatedly found that somewhat more than a semester is needed to complete either Chap. 2 or Chap. 3. On the other hand, colleagues who have used my notes in their courses have been able to construct a very satisfactory semester course by taking roughly the first third of each of Chaps. 2, 3, and 4.

I was faced with a notational dilemma in Chap. 3. The standard matrix notation for a Markov chain calls for the probability vector to be written as a row vector so that the entry a_{ij} of the transition matrix is the probability of going from state S_i to state S_j. Unfortunately, this leads to the stationary vector being the characteristic vector of A^T. In addition, in Sec. 1.10, when a linear transformation acts on a vector, the corresponding matrix appears on the left and the vector is written as a column vector. For these reasons, I decided to express the probability vectors as column vectors and to have a_{ij} represent the probability of going from state S_j to state S_i.

Many of the results in Chaps. 2 to 4 are often obtained without relying on the machinery of matrix theory. My purpose here, however, is to demonstrate how matrices can be used and, therefore, whenever possible I have presented proofs employing matrices. On the other hand, I have also tried to provide the reader with a reference to an alternate approach.

I am deeply indebted to Professor George A. Barnard, who originally suggested that I teach a course at Imperial College along the lines of this book. At the same time I would like to thank Professor Walter Hayman of Imperial College, whose generous hospitality to me as a visiting lecturer in 1964–1965 provided me with the facilities for preparing a preliminary edition. Professors Richard Good, Howard Egan, and William Anderson taught several courses from subsequent preliminary editions and made a number of very valuable criticisms. Also, many ideas and suggestions came to me through numerous discussions with Dr. Alan Goldman, Jack Edmonds, Lambert Joel, and Joel Levy of the National Bureau of Standards. In addition, many kind students, both at Imperial College and at the University of Maryland, pointed out errors and commented on my lapses of various sorts. I would especially like to thank Betty Vanderslice for the care and patience she showed in typing endless versions of the manuscript.   Finally, I wish to express my appreciation to the McGraw-Hill Book Company for several years of benign neglect.

MARTIN PEARL

MATRIX THEORY AND
FINITE MATHEMATICS

0

FIELDS

0.1 FIELDS

In Chap. 1 we will develop in great detail the basic facts concerning matrices whose entries are taken from a field. In the three subsequent chapters several quite different problems will be discussed that make heavy use of the theory of matrices. A striking feature of this treatment is that it will be appropriate to use a different field for each of the applications we study. Since we will not restrict ourselves to a specific field in Chap. 1, all the results we obtain there can be utilized equally well no matter which field is needed for a given application. Thus, by introducing the concept of an arbitrary field at this point, before launching into the study of matrices, we can develop at one time all the matrix tools which will be essential in later chapters.

Definition 0.1.1 A set $\mathbb{F}$, together with two operations denoted by $+$ and $\cdot$, called *addition* and *multiplication* respectively, is called a *field* provided the following 11 conditions hold:

1 *Closure under addition:* $a + b$ belongs to $\mathbb{F}$ whenever a and b belong to $\mathbb{F}$.

2 *Associativity of addition:*

$$a + (b + c) = (a + b) + c$$

for all a, b, and c belonging to $\mathbb{F}$.

3 *Commutativity of addition:*

$$a + b = b + a$$

for all a and b belonging to $\mathbb{F}$.

4 *Additive identity:* there exists an element of $\mathbb{F}$, denoted by 0 and called *zero*, such that

$$a + 0 = 0 + a = a$$

for all a belonging to $\mathbb{F}$.

5 *Additive inverse:* For each element a of $\mathbb{F}$ there exists an element of $\mathbb{F}$, denoted by $-a$, such that

$$a + (-a) = (-a) + a = 0$$

6 *Closure under multiplication:* $a \cdot b$ belongs to $\mathbb{F}$ whenever a and b belong to $\mathbb{F}$.

7 *Associativity of multiplication:*

$$a \cdot (b \cdot c) = (a \cdot b) \cdot c$$

for all a, b, and c belonging to $\mathbb{F}$.

8 *Commutativity of multiplication:*

$$a \cdot b = b \cdot a$$

for all a and b belonging to $\mathbb{F}$.

9 *Multiplicative identity:* There exists an element of $\mathbb{F}$, distinct from 0, denoted by 1 and called *one*, such that

$$a \cdot 1 = 1 \cdot a = a$$

for all a belonging to $\mathbb{F}$.

10 *Multiplicative inverse:* For each element a of $\mathbb{F}$, other than 0, there exists an element of $\mathbb{F}$, denoted by a^{-1} and called the *inverse* of a, such that

$$a \cdot a^{-1} = a^{-1} \cdot a = 1$$

11 *The distributive laws:*

$$a \cdot (b + c) = a \cdot b + a \cdot c$$
$$(a + b) \cdot c = a \cdot c + b \cdot c$$

for all a, b, and c belonging to $\mathbb{F}$.

It is customary to write *ab* instead of $a \cdot b$.   Also we occasionally write a/b instead of ab^{-1}.

Roughly speaking, a field is a set in which the four basic arithmetic operations (addition, subtraction, multiplication, and division by a nonzero element) can be performed in a "reasonable" way.   Operations which have not been mentioned, e.g., finding a square root, may or may not be possible.   Note that parts 4 and 9 of the definition require that a field contain at least two elements.

The most frequently encountered fields are:

1   The real numbers
2   The complex numbers
3   The rational numbers

Other examples of fields are given in the exercises below.   We will see that one of them is used extensively in Chap. 4.

EXERCISES

1   Which of the following sets are fields (with the usual addition and multiplication)?   For those sets which are not fields determine which of the 11 parts of Definition 0.1.1 are violated.

(*a*) The numbers of the form $a + bi$, where a and b are rational numbers and $i = \sqrt{-1}$.

(*b*) The numbers of the form $a + b\sqrt{2}$, where a and b are rational numbers.

(*c*) The integers.

(*d*) The set of all polynomials in x with real coefficients.

(*e*) The set of all quotients of polynomials in x with real coefficients, $f(x)/g(x)$, $g(x) \neq 0$.

(*f*) The subsets of the plane, with addition and multiplication defined to be union and intersection respectively.

2   Show that a field has exactly one additive identity and exactly one multiplicative identity.

3   Show that every element of a field has exactly one additive inverse.

4   Show that every element of a field other than 0 has exactly one multiplicative inverse.   Show that 0 does not have a multiplicative inverse.

5   Prove that if a and b are elements of a field such that $ab = 0$ and $a \neq 0$, then $b = 0$.

6 Let $\mathbb{Z}$ denote the set of integers and let p be a prime in $\mathbb{Z}$. We partition $\mathbb{Z}$ into p subsets

$$0' = \{0, \quad p, \quad 2p, \ldots; \quad -p, \quad -2p, \ldots\}$$
$$1' = \{1, 1 + p, 1 + 2p, \ldots; \quad 1 - p, 1 - 2p, \ldots\}$$
$$2' = \{2, 2 + p, 2 + 2p, \ldots; \quad 2 - p, 2 - 2p, \ldots\}$$

$$\cdots\cdots\cdots\cdots\cdots\cdots\cdots\cdots\cdots\cdots\cdots\cdots\cdots$$

$$(p - 1)' = \{-1 + p, -1 + 2p, -1 + 3p, \ldots; \quad -1 - p, -1 - 2p, \ldots\}$$

that is, x' is the set of all integers which when divided by p leave a remainder of x. The set consisting of the p elements $0', 1', 2', \ldots, (p - 1)'$ is denoted by $\mathbb{Z}_p$. Show that if a and c are integers which belong to the same element of $\mathbb{Z}_p$ and b and d are integers which belong to the same element of $\mathbb{Z}_p$, then $a + b$ and $c + d$ belong to the same element of $\mathbb{Z}_p$ and ab and cd belong to the same element of $\mathbb{Z}_p$.

We define the operations of addition and multiplication in $\mathbb{Z}_p$ as follows: $a' + b'$ is defined to be the element of $\mathbb{Z}_p$ which contains $a + b$, and $a' \cdot b'$ is defined to be the element of $\mathbb{Z}_p$ which contains ab. Show that with these definitions $\mathbb{Z}_p$ is a field. [*Hint:* Show that if a is any integer which does not belong to $0'$, then no two of the integers $a, 2a, 3a, \ldots, (p - 1)a, pa$ belong to the same element of $\mathbb{Z}_p$. Where is the hypothesis that p is a prime used?] Show that if p is not a prime, then $\mathbb{Z}_p$ is not a field.

Write out the addition and multiplication tables for $\mathbb{Z}_2$, $\mathbb{Z}_3$, $\mathbb{Z}_5$, and $\mathbb{Z}_6$.

1

MATRICES

1.1 INTRODUCTION AND NOTATION

Several quite different topics will be treated in these pages, but throughout there will be a single unifying concept—that of a matrix. This chapter is devoted to matrices themselves.[1] In the subsequent chapters we will introduce a few of the many subjects to which the material of this chapter can be applied, namely, zero-sum two-person games, linear programming, finite Markov chains, and directed and undirected graphs. Perhaps the most widely known and historically most important use for matrices is as a device for solving a system of simultaneous linear equations. A brief treatment of this problem is contained in Sec. 1.9.

---

[1] During the course of this chapter two digressions will be necessary. The first presents just enough information about finite permutations to enable us to define the determinant of a matrix. This material will be found in the first part of Sec. 1.8, ending with Definition 1.8.11. The second digression, devoted to the greatest common divisor and the least common multiple of polynomials in one variable, occupies all of Sec. 1.11. Since the discussion of each of these topics is self-contained and independent of the remainder of the text, a good deal of time can be saved if the reader is already familiar with the contents of these sections by the time they arise.

We will begin by defining a matrix with entries from a field $\mathbb{F}$ and continue with a study, in some depth, of the general theory of matrices with entries from $\mathbb{F}$. Although the treatment of the material throughout this chapter, including the proofs in the exercises, is in terms of an unspecified field $\mathbb{F}$ whose elements are referred to as *numbers* or *scalars*, readers who are unfamiliar with the concept of a field will lose very little (at least until Chap. 4) by always interpreting $\mathbb{F}$ as being either the set of all real numbers or the set of all complex numbers. However, in all exercises and examples of a numerical nature in this chapter, $\mathbb{F}$ is assumed to be the field of real numbers.

Definition 1.1.1   Let m and n be positive integers. A rectangular array of $m \cdot n$ numbers belonging to $\mathbb{F}$ arranged in m rows and n columns is called an $m \times n$ *matrix* with *entries* from $\mathbb{F}$. We call m and n the *row* and *column dimensions* of the matrix.

Matrices will be denoted by capital letters of the Latin alphabet and will be displayed in several ways. The simplest situation occurs when all the entries of the matrix are known. In such a case we write down the matrix explicitly. For example,

$$A = \begin{bmatrix} 1 & 3 & -4 \\ 0 & -6 & 2 \end{bmatrix}$$

is a 2×3 matrix all of whose entries are specified.

Often, especially when proving theorems about matrices, we do not wish (or are unable) to state exactly which number is in each position of the matrix. Indeed, we may even not wish to specify the dimensions. In such cases we will discuss an $m \times n$ matrix A and write

$$A = \begin{bmatrix} a_{11} & a_{12} & \cdots & a_{1n} \\ a_{21} & a_{22} & \cdots & a_{2n} \\ \cdots\cdots\cdots\cdots\cdots\cdots \\ a_{m1} & a_{m2} & \cdots & a_{mn} \end{bmatrix} \tag{1.1.2}$$

Elements of $\mathbb{F}$ (in particular, entries of a matrix) will be denoted either by lowercase Latin or lowercase Greek letters. Thus a_{21} stands for the number in the second row and the first column of A. When it is not necessary to illustrate all the entries of A, we will use the more condensed notation

$$A = [a_{ij}] \qquad \begin{matrix} i = 1, 2, \ldots, m \\ j = 1, 2, \ldots, n \end{matrix}$$

Either way we have used a *double-subscript notation*. The first subscript, i, of a_{ij} denotes the row in which the entry a_{ij} is found. The second subscript, j, denotes the column. Thus we say that a_{ij} is the *entry in the (i, j) position* of A. In order to characterize a matrix, it is necessary to give not only the numbers of which it is composed but also the positions within the matrix in which the numbers are located.

There is another important way of illustrating a matrix. A matrix can be *partitioned* into matrices of smaller dimensions. For example, the matrix A of (1.1.2) can be expressed as

$$A = \begin{bmatrix} A_{11} & A_{12} \\ A_{21} & A_{22} \end{bmatrix}$$

where

$$A_{11} = \begin{bmatrix} a_{11} & a_{12} & \cdots & a_{1t} \\ a_{21} & a_{22} & \cdots & a_{2t} \\ \cdots\cdots\cdots\cdots\cdots \\ a_{s1} & a_{s2} & \cdots & a_{st} \end{bmatrix} \quad A_{12} = \begin{bmatrix} a_{1,t+1} & a_{1,t+2} & \cdots & a_{1n} \\ a_{2,t+1} & a_{2,t+2} & \cdots & a_{2n} \\ \cdots\cdots\cdots\cdots\cdots\cdots \\ a_{s,t+1} & a_{s,t+2} & \cdots & a_{sn} \end{bmatrix}$$

$$A_{21} = \begin{bmatrix} a_{s+1,1} & a_{s+1,2} & \cdots & a_{s+1,t} \\ a_{s+2,1} & a_{s+2,2} & \cdots & a_{s+2,t} \\ \cdots\cdots\cdots\cdots\cdots\cdots \\ a_{m1} & a_{m2} & \cdots & a_{mt} \end{bmatrix}$$

$$A_{22} = \begin{bmatrix} a_{s+1,t+1} & a_{s+1,t+2} & \cdots & a_{s+1,n} \\ a_{s+2,t+1} & a_{s+2,t+2} & \cdots & a_{s+2,n} \\ \cdots\cdots\cdots\cdots\cdots\cdots \\ a_{m,t+1} & a_{m,t+2} & \cdots & a_{mn} \end{bmatrix}$$

Again we have used a double-subscript notation, this time for the *submatrices* into which the original matrix A was partitioned. We say that A_{ij} is the submatrix in the *(i, j) block position* of A (relative to this partition of A).

Definition 1.1.3   Two matrices are *equal* only if they are identical, i.e., if they have the same dimensions and if the entries in corresponding positions are equal. For

$$A = [a_{ij}] \qquad \begin{array}{l} i = 1, 2, \ldots, m \\ j = 1, 2, \ldots, n \end{array}$$

$$B = [b_{kl}] \qquad \begin{array}{l} k = 1, 2, \ldots, p \\ l = 1, 2, \ldots, q \end{array}$$

we understand $A = B$ to mean that:

1 $m = p$ and $n = q$

2 $a_{ij} = b_{ij}$ $\begin{array}{l} i = 1, 2, \ldots, m \\ j = 1, 2, \ldots, n \end{array}$

There are two kinds of matrices which have special roles in the general theory, and we use them often enough to warrant giving them special names.

Definition 1.1.4   Let A be an $m \times n$ matrix.   If $m = n$, then A is said to be a *square matrix* of *order n*.

An $n \times 1$ matrix is called an *n-dimensional column vector*.   A $1 \times n$ matrix is called an *n-dimensional row vector*.

Thus, in the context of this book, a vector is a particular type of matrix.

A matrix (and, in particular, a vector) all of whose entries are the zero element of $\mathbb{F}$ is called a *zero matrix* (or a *zero vector*) and is denoted by 0, irrespective of its dimensions.   The dimensions of a zero matrix will usually be evident from the context.   When this is not the case, we will specify the dimensions.

1.2   THE ARITHMETIC OF MATRICES

In the last section we defined a matrix and introduced a simple and efficient notation.   Now we define the three basic arithmetic operations performed on matrices and present some of their elementary properties.   These operations are:

1   Addition of matrices
2   Multiplication of a matrix by a scalar
3   Multiplication of matrices

Definition 1.2.1 Addition of matrices   Let A and B be two $m \times n$ matrices,

$$A = [a_{ij}] \qquad B = [b_{ij}] \qquad \begin{array}{l} i = 1, 2, \ldots, m \\ j = 1, 2, \ldots, n \end{array}$$

We define $A + B$ to be the $m \times n$ matrix which has $a_{ij} + b_{ij}$ in the (i, j) position.

Corollary 1.2.2   Let A, B, and C be $m \times n$ matrices.   Then

1 $A + (B + C) = (A + B) + C$

i.e., addition is defined and is *associative*,

2 $A + B = B + A$

i.e., addition is *commutative*.

PROOF   (1) Since A, B, and C are $m \times n$ matrices, $B + C$ and $A + B$ are both defined by Definition 1.2.1 and are also $m \times n$ matrices.   In the same way, $A + (B + C)$ and $(A + B) + C$ are both defined by Definition 1.2.1 and are $m \times n$ matrices.

In order to show that $A + (B + C) = (A + B) + C$ we must show that for any choice of i and j $(1 \leq i \leq m,\ 1 \leq j \leq n)$ the entry in the (i, j) position of $A + (B + C)$ is equal to the entry in the (i, j) position of $(A + B) + C$.   For each choice of i and j, the entry in the (i, j) position of $A + (B + C)$ is $a_{ij} + (b_{ij} + c_{ij})$, where a_{ij}, b_{ij}, and c_{ij} are the entries in the (i, j) positions of A, B and C, respectively.   Similarly, the entry in the (i, j) position of $(A + B) + C$ is $(a_{ij} + b_{ij}) + c_{ij}$.   Since addition is associative in $\mathbb{F}$, both $A + (B + C)$ and $(A + B) + C$ have $a_{ij} + b_{ij} + c_{ij}$ in the (i, j) position.   Hence $A + (B + C) = (A + B) + C$.

(2) Left as an exercise.   ////

Since the addition of matrices is associative whenever it is defined, we may omit all the parentheses and write $A + B + C$ instead of $A + (B + C)$ or $(A + B) + C$.   It follows from repeated applications of Definition 1.2.1 and the part 1 of Corollary 1.2.2 that the entry in the (i, j) position of any finite sum of $m \times n$ matrices is the sum of the entries in the (i, j) positions of the matrices.   On the other hand, if not all matrices of the collection have the same dimensions, their sum is not defined.

Definition 1.2.3 Multiplication of a matrix by a scalar   Let A be an $m \times n$ matrix, $A = [a_{ij}]$, and let α be an element of $\mathbb{F}$.   Then αA and $A\alpha$ are both defined to be the $m \times n$ matrix which has αa_{ij} in the (i, j) position.

Thus multiplication of a matrix by a scalar is defined for all matrices A and all elements α of $\mathbb{F}$.   Observe that multiplication of a matrix by a scalar does not change the dimensions of the matrix.

According to Definition 1.2.3, we have $1A = A$ for every matrix A and $0A = 0$, where the first zero is an element of $\mathbb{F}$ and the second zero is the zero matrix whose dimensions are the same as those of A.

Definition 1.2.4 Multiplication of matrices   Let A be an $m \times n$ matrix and B be an $n \times p$ matrix

$$A = [a_{ij}] \qquad B = [b_{jk}] \qquad \begin{aligned} i &= 1, 2, \ldots, m \\ j &= 1, 2, \ldots, n \\ k &= 1, 2, \ldots, p \end{aligned}$$

AB is defined to be the $m \times p$ matrix which has

$$\sum_{j=1}^{n} a_{ij} b_{jk} = a_{i1} b_{1k} + a_{i2} b_{2k} + \cdots + a_{in} b_{nk}$$

in the (i, k) position.

Definition 1.2.4 requires that the number of columns of A be equal to the number of rows of B. If the number of columns of A is not equal to the number of rows of B, the product AB is not defined. Thus, for a pair of matrices A, B, it is possible that none, one, or both of the products AB and BA are defined. The product $A^2 = AA$ is defined only for those matrices A for which $m = n$, that is, only for square matrices. Then A^2 is also a square matrix and has the same order as A.

Considerable care must be exercised when operating with matrices. As we have seen (e.g., in Corollary 1.2.2), many of the common properties of arithmetic operations remain valid when applied to matrices of the proper dimensions. However, there are numerous exceptions, especially in dealing with the multiplication of matrices. Some of these are illustrated in the following example.

EXAMPLE 1.2.5 Let

$$A = \begin{bmatrix} 1 & 2 \\ 2 & 4 \end{bmatrix} \quad B = \begin{bmatrix} -2 & 2 \\ 1 & -1 \end{bmatrix} \quad C = \begin{bmatrix} 1 \\ 1 \end{bmatrix} \quad D = \begin{bmatrix} -1 \\ 2 \end{bmatrix} \quad \alpha = 4$$

Then

1 $A + B = \begin{bmatrix} -1 & 4 \\ 3 & 3 \end{bmatrix} \quad C + D = \begin{bmatrix} 0 \\ 3 \end{bmatrix}$

2 $\alpha A = \begin{bmatrix} 4 & 8 \\ 8 & 16 \end{bmatrix} \quad \alpha C = \begin{bmatrix} 4 \\ 4 \end{bmatrix}$

3 $AB = 0 \quad BA = \begin{bmatrix} 2 & 4 \\ -1 & -2 \end{bmatrix} \quad AC = \begin{bmatrix} 3 \\ 6 \end{bmatrix} = AD$

4 $A + C$ and CA are not defined ////

Example 1.2.5 demonstrates several peculiarities of matrix multiplication. Although $A \neq 0$, $B \neq 0$, a simple calculation shows that $AB = 0$. Also, $AC = AD$, $A \neq 0$, and $C \neq D$. Thus the usual cancellation is generally not permitted. Later we will determine under what circumstances cancellation can be performed. Notice also that although AB and BA are both defined and have the same dimensions, they are not equal.

Let A, B, and C be $m \times n$, $p \times q$, and $s \times t$ matrices respectively.   According to Definition 1.2.4, $A(BC)$ is defined if and only if both the conditions $n = p$ and $q = s$ are satisfied, and then $A(BC)$ is an $m \times t$ matrix.   Exactly the same two conditions must be satisfied in order that $(AB)C$ be defined, and then $(AB)C$ is also an $m \times t$ matrix.   Thus, if one of the products $A(BC)$ and $(AB)C$ is defined, both are defined and they have the same dimensions.   In fact, we will show that the two products of A, B, and C, taken in that order, are equal.

Corollary 1.2.6   Let

$$i = 1, 1, \ldots, m$$
$$j = 1, 2, \ldots, n$$
$$A = [a_{ij}] \qquad B = [b_{jk}] \qquad C = [c_{kl}] \qquad k = 1, 2, \ldots, p$$
$$l = 1, 2, \ldots, q$$

be $m \times n$, $n \times p$, $p \times q$ matrices respectively.   Then

$$A(BC) = (AB)C$$

i.e., multiplication of matrices is defined and *associative*.

PROOF   To prove the corollary we must show that for each choice of i and l ($1 \leq i \leq m$, $1 \leq l \leq q$), the entry in the (i, l) position of $A(BC)$ is equal to the entry in the (i, l) position of $(AB)C$.   Set

$$D = AB = [d_{ik}] \qquad \begin{matrix} i = 1, 2, \ldots, m \\ k = 1, 2, \ldots, p \end{matrix}$$

$$F = BC = [f_{jl}] \qquad \begin{matrix} j = 1, 2, \ldots, n \\ l = 1, 2, \ldots, q \end{matrix}$$

According to Definition 1.2.4,

$$d_{ik} = \sum_{j=1}^{n} a_{ij} b_{jk} \qquad f_{jl} = \sum_{k=1}^{n} b_{jk} c_{kl}$$

The entry in the (i, l) position of $A(BC) = AF$ is

$$\sum_{j=1}^{n} a_{ij} f_{jl} = \sum_{j=1}^{n} a_{ij} \left(\sum_{k=1}^{p} b_{jk} c_{kl} \right) \qquad (1.2.7)$$

and the entry in the (i, l) position of $(AB)C = DC$ is

$$\sum_{k=1}^{p} d_{ik} c_{kl} = \sum_{k=1}^{p} \left(\sum_{j=1}^{n} a_{ij} b_{jk} \right) c_{kl} \qquad (1.2.8)$$

By expanding the right-hand sides of equations (1.2.7) and (1.2.8) and rearranging the terms it is easily seen that they are equal.   Each expression consists of all terms of the form

$$a_{ij}\,b_{jk}\,c_{kl} \qquad (1.2.9)$$

where i and l are fixed and are determined by the position we are considering, i.e., the (i, l) position, and where j takes on the values $1, 2, \ldots, n$ and k takes on the values $1, 2, \ldots, p$.   There are the same np terms of the form (1.2.9) in each of the sums (1.2.7) and (1.2.8).   Thus, each entry of $A(BC)$ is equal to the entry of $(AB)C$ in the same position, and consequently $A(BC) = (AB)C$.   ////

Since matrix multiplication is always associative when it is defined, we may omit the parentheses and write ABC instead of $A(BC)$ or $(AB)C$.

At first glance Definition 1.2.4 may appear rather arbitrary and artificial. However, the following example shows one way in which this definition arises quite naturally.   Let us suppose that $x_1, x_2, \ldots, x_p$ and $y_1, y_2, \ldots, y_n$ are related by the system of equations

$$\begin{aligned}
y_1 &= b_{11}x_1 + b_{12}x_2 + \cdots + b_{1p}x_p \\
y_2 &= b_{21}x_1 + b_{22}x_2 + \cdots + b_{2p}x_p \\
&\cdots\cdots\cdots\cdots\cdots \\
y_n &= b_{n1}x_1 + b_{n2}x_2 + \cdots + b_{np}x_p
\end{aligned} \qquad (1.2.10)$$

Suppose further that $y_1, y_2, \ldots, y_n$ and $z_1, z_2, \ldots, z_m$ are related by the equations

$$\begin{aligned}
z_1 &= a_{11}y_1 + a_{12}y_2 + \cdots + a_{1n}y_n \\
z_2 &= a_{21}y_1 + a_{22}y_2 + \cdots + a_{2n}y_n \\
&\cdots\cdots\cdots\cdots\cdots \\
z_m &= a_{m1}y_1 + a_{m2}y_2 + \cdots + a_{mn}y_n
\end{aligned} \qquad (1.2.11)$$

We can eliminate the y's completely by substituting equations (1.2.10) into equations (1.2.11).   This expresses each of the z's in terms of the x's.   Carrying out this computation, we get

$$\begin{aligned}
z_1 &= c_{11}x_1 + c_{12}x_2 + \cdots + c_{1p}x_p \\
z_2 &= c_{21}x_1 + c_{22}x_2 + \cdots + c_{2p}x_p \\
&\cdots\cdots\cdots\cdots\cdots \\
z_m &= c_{m1}x_1 + c_{m2}x_2 + \cdots + c_{mp}x_p
\end{aligned} \qquad (1.2.12)$$

where

$$c_{ik} = a_{i1}b_{1k} + a_{i2}b_{2k} + \cdots + a_{in}b_{nk} = \sum_{j=1}^{n} a_{ij}a_{jk} \qquad \begin{aligned} i &= 1, 2, \ldots, m \\ k &= 1, 2, \ldots, p \end{aligned} \qquad (1.2.13)$$

Let us form matrices from the coefficients in the systems of equations (1.2.10), (1.2.11), and (1.2.12). Set

$$A = [a_{ij}] \quad B = [b_{jk}] \quad C = [c_{ik}] \quad \begin{matrix} i = 1, 2, \ldots, m \\ j = 1, 2, \ldots, n \\ k = 1, 2, \ldots, p \end{matrix}$$

Then A, B, C are $m \times n$, $n \times p$, $m \times p$ matrices respectively. The adoption of Definition 1.2.4 allows us to write the systems of equations (1.2.10) to (1.2.12) in a very simple and compact way:

$$Y = BX \quad Z = AY \quad Z = CX \quad (1.2.14)$$

where X, Y, and Z are the column vectors

$$X = \begin{bmatrix} x_1 \\ x_2 \\ \vdots \\ x_p \end{bmatrix} \quad Y = \begin{bmatrix} y_1 \\ y_2 \\ \vdots \\ y_n \end{bmatrix} \quad Z = \begin{bmatrix} z_1 \\ z_2 \\ \vdots \\ z_m \end{bmatrix}$$

Substitution of the first equation of (1.2.14) into the second equation of (1.2.14) yields

$$Z = A(BX) \quad (1.2.15)$$

Then Definition 1.2.4 and equation (1.2.13) permit us to write $C = AB$, and hence the equations of (1.2.12) can be expressed as

$$Z = (AB)X$$

Throughout this section no use has been made of the fact that division is possible within $\mathbb{F}$. Later (specifically in Secs. 1.10, 1.12, and 1.13) we will need to consider matrices whose entries come from a set having all the properties of a field except those related to division.[1] It is clear that all the results of this section remain valid for such matrices.

EXERCISES

1 Find all 2×2 matrices which commute under the operation of multiplication with the matrix

(a) $\begin{bmatrix} 1 & 2 \\ 0 & 1 \end{bmatrix}$ (b) $\begin{bmatrix} 1 & 2 \\ 0 & -1 \end{bmatrix}$

[1] Such a system is called a *ring* in modern algebra. Division by a nonzero number in a field is equivalent to part 10 of Definition 0.1.1.

2   Find all 3×3 matrices which commute under the operation of multiplication with the matrix

$$\begin{bmatrix} 1 & 0 & 1 \\ 0 & 1 & -2 \\ 0 & 0 & 2 \end{bmatrix}$$

3   Let

$$A = \begin{bmatrix} 1 & 2 \\ -3 & 6 \end{bmatrix}$$

Find
(a) All 2×2 matrices X such that $AX = 0$.
(b) All 2×2 matrices X such that $XA = 0$.
(c) All 2×2 matrices X such that $AX = A$.
(d) All 2×2 matrices X such that $XA = A$.
(e) All 2×2 matrices X such that $XA = \begin{bmatrix} 1 & 0 \\ 0 & 1 \end{bmatrix}$.

4   Repeat Exercise 3 using the matrix $\begin{bmatrix} 1 & 2 \\ 3 & 6 \end{bmatrix}$ for A.

5   Find all 2×2 matrices X such that $X^2 = 0$.

6   Verify the computation leading to (1.2.13).

7   Prove part 2 of Corollary 1.2.2.

8   Let α, β be scalars, A, B be $m \times n$ matrices, and C be an $n \times p$ matrix. Prove each of the following:

$$(\alpha + \beta)A = \alpha A + \beta A \quad (1.2.16a)$$
$$\alpha(A + B) = \alpha A + \alpha B \quad (1.2.16b)$$
$$\alpha(AC) = (\alpha A)C = A(\alpha C) \quad (1.2.16c)$$
$$\alpha(\beta A) = (\alpha \beta)A \quad (1.2.16d)$$

9   Let A be an $m \times n$ matrix and B and C be $n \times p$ matrices. Prove that

$$A(B + C) = AB + AC \quad (1.2.17)$$

i.e., all the indicated operations are defined, and matrix multiplication is *left-distributive* with respect to matrix addition.

10   State and prove the right-distributive law of matrix multiplication with respect to matrix addition.

11   Let $A = [a_{ij}]$ and $B = [b_{jk}]$ be $m \times n$ and $n \times p$ matrices respectively and let $AB = C = [c_{ik}]$. We denote the ith *row* of A by $A_{(i)}$ and the kth *column* of B by $B^{(k)}$, that is,

$$A_{(i)} = [a_{i1} \quad a_{i2} \quad \cdots \quad a_{in}] \qquad B^{(k)} = \begin{bmatrix} b_{1k} \\ b_{2k} \\ \vdots \\ b_{nk} \end{bmatrix}$$

Prove:

$$c_{ik} = A_{(i)}B^{(k)} \qquad (1.2.18a)$$
$$C_{(i)} = a_{i1}B_{(1)} + a_{i2}B_{(2)} + \cdots + a_{in}B_{(n)} \qquad (1.2.18b)$$
$$C^{(k)} = A^{(1)}b_{1k} + A^{(2)}b_{2k} + \cdots + A^{(n)}b_{nk} \qquad (1.2.18c)$$

12 Show that the set of matrices of the form

$$\begin{bmatrix} a & b \\ -b & a \end{bmatrix}$$

where a and b are real numbers, form a field. Show that if a and b are permitted to be complex numbers, this set of matrices is not a field.

13 Let $A = [a_{ij}]$ be a square matrix of order n. A is called a *diagonal matrix* if $a_{ij} = 0$ whenever $i \neq j$. Show that multiplication of diagonal matrices of order n is commutative.

14 A diagonal matrix $A = [a_{ij}]$ of order n is called a *scalar matrix* if $a_{11} = a_{22} = \cdots = a_{nn}$. Show that a scalar matrix of order n commutes with every matrix of order n under multiplication.

15 The scalar matrix $A = [a_{ij}]$ of order n in which $a_{11} = a_{22} = \cdots = a_{nn} = 1$ is called the *identity matrix* of order n and is denoted by I_n (or simply by I when the order is clear from the context).[1] Show that for any $m \times n$ matrix A

$$I_m A = AI_n = A \qquad (1.2.19)$$

16 Prove that a matrix which commutes under multiplication with every square matrix of order n is a scalar matrix of order n.

17 Let $m_1, m_2, n_1, n_2, p_1, p_2$ be positive integers. For $i, j = 1, 2$, let A_{ij} be an $m_i \times n_j$ matrix and B_{ij} be an $n_i \times p_j$ matrix. Show that

$$\begin{bmatrix} A_{11} & A_{12} \\ A_{21} & A_{22} \end{bmatrix} \begin{bmatrix} B_{11} & B_{12} \\ B_{21} & B_{22} \end{bmatrix} = \begin{bmatrix} A_{11}B_{11} + A_{12}B_{21} & A_{11}B_{12} + A_{12}B_{22} \\ A_{21}B_{11} + A_{22}B_{21} & A_{21}B_{12} + A_{22}B_{22} \end{bmatrix}$$

$$(1.2.20)$$

---

[1] Note that the identity matrix of order n is denoted by I_n whereas $I_{(n)}$ stands for the nth row of some identity matrix I. According to these conventions, the nth row of the $m \times m$ identity matrix would be denoted by $I_{m(n)}$. However, this notation is quite cumbersome and will be avoided.

Generalize (1.2.20) to a product of two matrices A, B, where A is partitioned into r row blocks and s column blocks and B is partitioned into s row blocks and t column blocks.

1.3 VECTOR SPACES

As we will see, the language of this section has a strong geometric flavor and, in fact, had its beginnings in the analytic approach to the study of euclidean spaces. As every reader who has encountered plane or multidimensional analytic geometry knows, vectors in a plane can be identified with ordered pairs of real numbers, and vectors in a three-dimensional space can be identified with ordered triples of real numbers. Although it is possible to give a geometric interpretation to most of the theorems and definitions stated in this section in algebraic language, a detailed discussion of this material would carry us much too far afield. However, the reader can profit by keeping in mind that this interpretation exists. In fact, it will often happen that restating an algebraic definition or theorem in geometric terms and drawing a two- or three-dimensional diagram to illustrate it will help the reader see what is really going on.

Definition 1.3.1   Let $\mathbb{V}$ be a nonempty set of $m \times n$ matrices with entries from the field $\mathbb{F}$. $\mathbb{V}$ is called a *vector space* (over $\mathbb{F}$) provided the following two conditions are satisfied:

1   Closure under addition: $X + Y$ belongs to $\mathbb{V}$ whenever X and Y belong to $\mathbb{V}$.
2   Closure under multiplication by a scalar: αX belongs to $\mathbb{V}$ whenever X belongs to $\mathbb{V}$ and α belongs to $\mathbb{F}$.

If $\mathbb{U}$ and $\mathbb{V}$ are vector spaces over $\mathbb{F}$ and every element of $\mathbb{U}$ is also an element of $\mathbb{V}$, we say that $\mathbb{U}$ is a *subspace* of $\mathbb{V}$.

A vector space, then, is a set of matrices which is closed under the operations of addition and scalar multiplication.[1] Vector spaces arise in many

---

[1] In more abstract treatments than the one given here the general topic of vector spaces is sharply divided into two parts. A *vector space* over $\mathbb{F}$ is defined to be any set $\mathbb{V}$ of elements which are closed under addition and multiplication by a scalar and which satisfy certain combining axioms such as parts 1 and 2 of Corollary 1.2.2 and equations (1.2.16a), (1.2.16b), and (1.2.16d). If the vector spaces are permitted to be infinite-dimensional, the subject falls in the realm of analysis (see, for example, Taylor [9]). When the vector spaces are assumed to be finite-dimensional, many difficulties disappear and they can be studied algebraically. The vector spaces of this chapter all have finite dimension. For a more abstract treatment of finite-dimensional vector spaces see Halmos [6] or Greub [4]. The elements of the vector spaces we will discuss will always be matrices and usually be vectors (as defined in Definition 1.1.4). Note that the multiplication of matrices does not enter into the definition of a vector space.

different ways.   In Examples 1.3.2 to 1.3.6 we describe some of the more important and more commonly encountered vector spaces.   Several others will be found in the exercises.

EXAMPLE 1.3.2   For each choice of m and n the set of all $m \times n$ matrices with entries in $\mathbb{F}$ is a vector space which we denote by $\mathbb{F}_n{}^m$.   Two important special cases of this example are:

1   The set of all m-dimensional column vectors with entries in $\mathbb{F}$ (denoted by $\mathbb{F}_1{}^m$ or by $\mathbb{F}^m$)
2   The set of all n-dimensional row vectors with entries from $\mathbb{F}$ (denoted by $\mathbb{F}_n{}^1$ or by $\mathbb{F}_n$)   ////

EXAMPLE 1.3.3   The set consisting only of a zero matrix is a vector space. We call this vector space a *zero space* and denote it by 0.   ////

EXAMPLE 1.3.4   For a fixed matrix A, the set of all multiples of A by elements of $\mathbb{F}$ is a vector space   For example, let

$$A = \begin{bmatrix} 1 & 2 & -1 \\ 0 & 1 & 1 \end{bmatrix}$$

Then $V = \left\{ \alpha \begin{bmatrix} 1 & 2 & -1 \\ 0 & 1 & 1 \end{bmatrix} \middle| \alpha \in \mathbb{F} \right\} = \left\{ \begin{bmatrix} \alpha & 2\alpha & -\alpha \\ 0 & \alpha & \alpha \end{bmatrix} \middle| \alpha \in \mathbb{F} \right\}$

is a vector space.   ////

EXAMPLE 1.3.5   For a fixed $m \times n$ matrix A and a fixed positive integer p, the set of all $n \times p$ matrices X satisfying the equation

$$AX = 0$$

is a vector space.   In the special case where X is a column vector (that is, $p = 1$) this vector space is called the *column null space* (or simply *null space*) of A and is denoted by $\mathbb{N}(A)$.

For example, suppose A is the matrix of Example 1.3.4 and we set $p = 1$. $\mathbb{N}(A)$ consists of all three-dimensional column vectors

$$X = \begin{bmatrix} x_1 \\ x_2 \\ x_3 \end{bmatrix}$$

such that

$$AX = \begin{bmatrix} 1 & 2 & -1 \\ 0 & 1 & 1 \end{bmatrix} \begin{bmatrix} x_1 \\ x_2 \\ x_3 \end{bmatrix} = \begin{bmatrix} 0 \\ 0 \end{bmatrix} = 0$$

The single matrix equation $AX = 0$ is equivalent to the pair of linear equations

$$x_1 + 2x_2 - x_3 = 0$$
$$x_2 + x_3 = 0$$

We can solve for two of these unknowns in terms of the third. For example,

$$x_2 = -x_3$$
$$x_1 = -2x_2 + x_3 = 3x_3$$

expresses x_1 and x_2 in terms of x_3. Substituting for x_1 and x_2, we have

$$\mathbb{N}(A) = \left\{ \begin{bmatrix} 3x_3 \\ -x_3 \\ x_3 \end{bmatrix} \middle| x_3 \in \mathbb{F} \right\} = \left\{ x_3 \begin{bmatrix} 3 \\ -1 \\ 1 \end{bmatrix} \middle| x_3 \in \mathbb{F} \right\} \qquad ////$$

EXAMPLE 1.3.6 For k fixed $m \times n$ matrices $A_1, A_2, \ldots, A_k$ the set of all matrices which can be expressed as

$$\alpha_1 A_1 + \alpha_2 A_2 + \cdots + \alpha_k A_k$$

where $\alpha_1, \alpha_2, \ldots, \alpha_k \in \mathbb{F}$, is a vector space of $m \times n$ matrices.[1]
For example, let

$$A_1 = \begin{bmatrix} 1 & 2 \\ 0 & 1 \end{bmatrix} \qquad A_2 = \begin{bmatrix} 0 & 0 \\ 1 & 1 \end{bmatrix}$$

Then

$$\mathbb{V} = \left\{ \alpha_1 \begin{bmatrix} 1 & 2 \\ 0 & 1 \end{bmatrix} + \alpha_2 \begin{bmatrix} 0 & 0 \\ 1 & 1 \end{bmatrix} \middle| \alpha_1, \alpha_2 \in \mathbb{F} \right\} = \left\{ \begin{bmatrix} \alpha_1 & 2\alpha_1 \\ \alpha_2 & \alpha_1 + \alpha_2 \end{bmatrix} \middle| \alpha_1, \alpha_2 \in \mathbb{F} \right\}$$

is a vector space. $\qquad ////$

The process used in Example 1.3.6 to obtain a vector space from a given finite set of matrices all of which have the same dimensions is sufficiently important to warrant giving it a name.

---

[1] What is the geometrical interpretation of this vector space when the matrices $A_1, A_2, \ldots, A_k$, are three-dimensional vectors considered to be points in three-dimensional euclidean space?

Definition 1.3.7 Let $A_1, A_2, \ldots, A_k, B$ be $m \times n$ matrices. We say that B is a *linear combination* of $A_1, A_2, \ldots, A_k$ if there exist scalars $\alpha_1, \alpha_2, \ldots, \alpha_k$ such that

$$B = \alpha_1 A_1 + \alpha_2 A_2 + \cdots + \alpha_k A_k$$

Example 1.3.6 states that the set of all linear combinations of any finite set of matrices of the same dimensions is a vector space.

Definition 1.3.8 Let V be a vector space and let $A_1, A_2, \ldots, A_k$ be matrices. If:

1 $A_1, A_2, \ldots, A_k$ belong to V
2 Every element of V can be expressed as a linear combination of A_1, $A_2, \ldots, A_k$

then the set

$$S = \{A_1, A_2, \ldots, A_k\}$$

is said to *span* (or form a *spanning set* for) V.

Every finite set of matrices of the same dimensions spans exactly **one** vector space, namely, the one constructed in Example 1.3.6. In Examples 1.3.4 and 1.3.5 we have vector spaces which are spanned by a single matrix. The vector space V of Example 1.3.6 is spanned by the pair of matrices

$$S_1 = \left\{ \begin{bmatrix} 1 & 2 \\ 0 & 1 \end{bmatrix}, \begin{bmatrix} 0 & 0 \\ 1 & 1 \end{bmatrix} \right\} \qquad (1.3.9)$$

A nonzero vector space has many spanning sets. For example, it is not difficult to show that each of the sets

$$S_2 = \left\{ \begin{bmatrix} 2 & 4 \\ 0 & 2 \end{bmatrix}, \begin{bmatrix} 0 & 0 \\ 1 & 1 \end{bmatrix} \right\}$$

$$S_3 = \left\{ \begin{bmatrix} -4 & -8 \\ 0 & -4 \end{bmatrix}, \begin{bmatrix} 0 & 0 \\ 2 & 2 \end{bmatrix}, \begin{bmatrix} 0 & 0 \\ 1 & 1 \end{bmatrix} \right\} \qquad (1.3.10)$$

$$S_4 = \left\{ \begin{bmatrix} 1 & 2 \\ 1 & 2 \end{bmatrix}, \begin{bmatrix} 2 & 4 \\ 3 & 5 \end{bmatrix}, \begin{bmatrix} -1 & -2 \\ 1 & 0 \end{bmatrix} \right\}$$

is a spanning set for the vector space V of Example 1.3.6.

Any finite set of matrices of a vector space V which contains a spanning set for V is itself a spanning set. Thus, spanning sets may contain more matrices than are necessary. It is left as an exercise to show that S_3 and S_4 above contain unnecessary matrices and that $\mathbb{S}_1$ and $\mathbb{S}_2$ do not.   Note also that if $A_1, A_2, \ldots, A_k$ are matrices belonging to the vector space $\mathbb{U}$ and $\mathbb{V}$ is the vector space spanned by $A_1, A_2, \ldots, A_k$, then $\mathbb{V}$ is a subspace of $\mathbb{U}$.

Let us pause here to point out a danger.   We have spoken about vector spaces which are obtained from a given spanning set.   Conversely, should we be given a vector space, we would like to be able to find some spanning set for it. However, it is not until Sec. 1.5 that we are able to prove that every subspace of $\mathbb{F}_n{}^m$ has at least one spanning set.   One vector space for which we can furnish a particularly simple spanning set is $\mathbb{F}^m$, the set of all m-dimensional column vectors.   Clearly the m m-dimensional *unit column vectors*

$$I^{(1)} = \begin{bmatrix} 1 \\ 0 \\ \vdots \\ 0 \end{bmatrix}, \quad I^{(2)} = \begin{bmatrix} 0 \\ 1 \\ \vdots \\ 0 \end{bmatrix}, \quad \ldots, \quad I^{(m)} = \begin{bmatrix} 0 \\ 0 \\ \vdots \\ 1 \end{bmatrix} \qquad (1.3.11)$$

belong to $\mathbb{F}^m$.   In order to show that they span $\mathbb{F}^m$ we must prove that every element of $\mathbb{F}^m$ can be expressed as a linear combination of them.   Let

$$X = \begin{bmatrix} x_1 \\ x_2 \\ \vdots \\ x_m \end{bmatrix}$$

be any element of $\mathbb{F}^m$.   Then

$$X = x_1 I^{(1)} + x_2 I^{(2)} + \cdots + x_m I^{(m)}$$

as required.

There are other ways in which vector spaces arise.   Specifically, new vector spaces can be formed by combining old ones.

Definition 1.3.12   Let $V_1, V_2, \ldots, V_k$ be vector spaces of $m \times n$ matrices. Then:

1 The set of matrices which are contained in all the vector spaces $\mathbb{V}_1$, $\mathbb{V}_2, \ldots, \mathbb{V}_k$ form a vector space called the *intersection* of $\mathbb{V}_1, \mathbb{V}_2, \ldots, \mathbb{V}_k$ and denoted by

$$\mathbb{V}_1 \cap \mathbb{V}_2 \cap \cdots \cap \mathbb{V}_k$$

2 The set of all $m \times n$ matrices which can be expressed in the form

$$V_1 + V_2 + \cdots + V_k$$

where $V_1 \in \mathbb{V}_1, V_2 \in \mathbb{V}_2, \ldots, V_k \in \mathbb{V}_k$, form a vector space called the *sum* of $\mathbb{V}_1, \mathbb{V}_2, \ldots, \mathbb{V}_k$ and denoted by

$$\mathbb{V}_1 + \mathbb{V}_2 + \cdots + \mathbb{V}_k$$

We have already used part 2 of Definition 1.3.12. If $\mathbb{V}_1, \mathbb{V}_2, \ldots, \mathbb{V}_k$ are vector spaces of the type of Example 1.3.4 and each $\mathbb{V}_i$ is spanned by a single matrix A_i, then Example 1.3.6 can be considered as applying part 2 of Definition 1.3.12 to $\mathbb{V}_1, \mathbb{V}_2, \ldots, \mathbb{V}_k$.

EXAMPLE 1.3.13   Let $\mathbb{V}$ be the vector space spanned by

$$\left\{ V_1 = \begin{bmatrix} 2 \\ -1 \\ 0 \\ 1 \end{bmatrix}, \quad V_2 = \begin{bmatrix} 1 \\ 0 \\ 0 \\ 1 \end{bmatrix}, \quad V_3 = \begin{bmatrix} 1 \\ 1 \\ 0 \\ 1 \end{bmatrix} \right\}$$

and let

$$A = \begin{bmatrix} 1 & 2 & 1 & 0 \\ -1 & -2 & -1 & 0 \end{bmatrix}$$

Find a spanning set for $\mathbb{V} \cap \mathbb{N}(A)$.

The vector

$$X = \begin{bmatrix} x_1 \\ x_2 \\ x_3 \\ x_4 \end{bmatrix}$$

is contained in $\mathbb{V}$ if and only if it can be represented as

$$X = \alpha_1 V_1 + \alpha_2 V_2 + \alpha_3 V_3 = \begin{bmatrix} 2\alpha_1 + \alpha_2 + \alpha_3 \\ -\alpha_1 + \alpha_3 \\ 0 \\ \alpha_1 + \alpha_2 + \alpha_3 \end{bmatrix} \qquad (1.3.14)$$

for some $\alpha_1, \alpha_2, \alpha_3 \in \mathbb{F}$.   Similarly X is contained in $\mathbb{N}(A)$ if and only if

$$AX = 0 \qquad (1.3.15)$$

Substituting (1.3.14) into (1.3.15), we see that X belongs to $\mathbb{V} \cap \mathbb{N}(A)$ if and only if

$$1(2\alpha_1 + \alpha_2 + \alpha_3) + 2(-\alpha_1 + \alpha_3) = 0$$
$$-1(2\alpha_1 + \alpha_2 + \alpha_3) - 2(-\alpha_1 + \alpha_3) = 0$$

which reduces to the single condition

$$\alpha_2 = -3\alpha_3$$

Thus

$$\mathbb{V} \cap \mathbb{N}(A) = \{\alpha_1 V_1 - 3\alpha_3 V_2 + \alpha_3 V_3 \,|\, \alpha_1, \alpha_3 \in \mathbb{F}\}$$

$$= \left\{ \begin{bmatrix} 2\alpha_1 - 2\alpha_3 \\ -\alpha_1 + \alpha_3 \\ 0 \\ \alpha_1 - 2\alpha_3 \end{bmatrix} \middle| \alpha_1, \alpha_3 \in \mathbb{F} \right\} = \left\{ \alpha_1 \begin{bmatrix} 2 \\ -1 \\ 0 \\ 1 \end{bmatrix} + \alpha_3 \begin{bmatrix} -2 \\ 1 \\ 0 \\ -2 \end{bmatrix} \middle| \alpha_1, \alpha_3 \in \mathbb{F} \right\}$$

and

$$\left\{ \begin{bmatrix} 2 \\ -1 \\ 0 \\ 1 \end{bmatrix}, \begin{bmatrix} -2 \\ 1 \\ 0 \\ -2 \end{bmatrix} \right\}$$

is a spanning set for $\mathbb{V} \cap \mathbb{N}(A)$.   ////

It seems intuitively desirable to get rid of the unnecessary matrices in a spanning set. A means of accomplishing this is contained in:

Lemma 1.3.16   Let $\mathbb{V}$ be a vector space spanned by $\mathbb{S} = \{A_1, A_2, \ldots, A_k\}$ and let $\mathbb{U}$ be a vector space spanned by $k - 1$ of the elements of $\mathbb{S}$. Then $\mathbb{U} \subseteq \mathbb{V}$. Moreover, $\mathbb{U} = \mathbb{V}$ if and only if the matrix removed from $\mathbb{S}$ can be expressed as a linear combination of the remaining elements of $\mathbb{S}$.

PROOF   It will be convenient for us to assume that the elements of $\mathbb{S}$ are ordered in such a way that $\mathbb{U}$ is spanned by $A_1, A_2, \ldots, A_{k-1}$. Each element of $\mathbb{U}$ is a linear combination of the matrices $A_1, A_2, \ldots, A_{k-1}$ and hence may be considered as a linear combination of $A_1, A_2, \ldots, A_{k-1}, A_k$, where the coefficient of A_k is zero. Thus every element of $\mathbb{U}$ belongs to $\mathbb{V}$, and we have shown that $\mathbb{U} \subseteq \mathbb{V}$.

If $\mathbb{U} = \mathbb{V}$, then A_k belongs to $\mathbb{U}$ and hence can be expressed as a linear combination of $A_1, A_2, \ldots, A_{k-1}$. Conversely, suppose A_k can be expressed as

$$A_k = \alpha_1 A_1 + \alpha_2 A_2 + \cdots + \alpha_{k-1} A_{k-1} \qquad (1.3.17)$$

for some $\alpha_1, \alpha_2, \ldots, \alpha_{k-1} \in \mathbb{F}$. We wish to show that $\mathbb{U} = \mathbb{V}$. By the preceding paragraph we have $\mathbb{U} \subseteq \mathbb{V}$, and it only remains to show that $\mathbb{V} \subseteq \mathbb{U}$. For each $V \in \mathbb{V}$ there exist scalars $\beta_1, \beta_2, \ldots, \beta_{k-1}, \beta_k$ in $\mathbb{F}$ such that

$$V = \beta_1 A_1 + \beta_2 A_2 + \cdots + \beta_{k-1} A_{k-1} + \beta_k A_k \qquad (1.3.18)$$

Substituting equation (1.3.17) into equation (1.3.18) yields

$$V = \beta_1 A_1 + \beta_2 A_2 + \cdots + \beta_{k-1} A_{k-1} + \beta_k(\alpha_1 A_1 + \alpha_2 A_2 + \cdots + \alpha_{k-1} A_{k-1})$$
$$= (\beta_1 + \beta_k \alpha_1)A_1 + (\beta_2 + \beta_k \alpha_2)A_2 + \cdots + (\beta_{k-1} + \beta_k \alpha_{k-1})A_{k-1}$$

which shows that $V \in \mathbb{U}$. Hence $\mathbb{V} \subseteq \mathbb{U}$, and it follows that $\mathbb{U} = \mathbb{V}$ as required.   ////

Definition 1.3.19   A finite set $A_1, A_2, \ldots, A_k$ of elements of a vector space $\mathbb{V}$ is *linearly dependent* if there exist scalars $\alpha_1, \alpha_2, \ldots, \alpha_k$ in $\mathbb{F}$, not all of which are zero, such that

$$\alpha_1 A_1 + \alpha_2 A_2 + \cdots + \alpha_k A_k = 0 \qquad (1.3.20)$$

On the other hand, if the only instance in which equation (1.3.20) is satisfied is when $\alpha_1 = \alpha_2 = \cdots = \alpha_k = 0$, then $A_1, A_2, \ldots, A_k$ are *linearly independent*.

Definitions 1.3.7 and 1.3.19 and the process described in Lemma 1.3.16 are closely related. Both definitions are applicable to a set of matrices only when all the matrices concerned have the same dimensions. It is clear that if one of the matrices $A_1, A_2, \ldots, A_k$ can be expressed as a linear combination of the others, then a relation such as equation (1.3.20) exists in which not all of the scalars $\alpha_1, \alpha_2, \ldots, \alpha_k$ are zeros. Thus $A_1, A_2, \ldots, A_k$ are linearly dependent. A little care must be exercised in examining the converse, however. A finite set of matrices, all of the same dimensions and containing at least two matrices, in which it is not possible to express one of them as a linear combination of the others is easily shown to be linearly independent. On the other hand, a set consisting of a single matrix is a linearly independent set if the matrix is not a zero matrix, and it is a linearly dependent set if the matrix is a zero matrix.

EXAMPLE 1.3.21   Let

$$A_1 = \begin{bmatrix} 1 & 0 \\ 2 & -1 \end{bmatrix} \qquad A_2 = \begin{bmatrix} 0 & 1 \\ -1 & 1 \end{bmatrix} \qquad A_3 = \begin{bmatrix} 1 & 1 \\ 1 & 0 \end{bmatrix} \qquad A_4 = \begin{bmatrix} 1 & 2 \\ 0 & 1 \end{bmatrix}$$

In order to determine the linear dependence or independence of A_1, A_2, A_3, A_4 we need to know whether the equation

$$\alpha_1 A_1 + \alpha_2 A_2 + \alpha_3 A_3 + \alpha_4 A_4 = 0$$

has any nonzero solutions. This matrix equation is equivalent to the system of linear equations

$$\alpha_1 + \alpha_3 + \alpha_4 = 0 \qquad \alpha_2 + \alpha_3 + 2\alpha_4 = 0$$
$$2\alpha_1 - \alpha_2 + \alpha_3 = 0 \qquad -\alpha_1 + \alpha_2 + \alpha_4 = 0$$

and it is easily seen that

$$\alpha_1 = 0 \qquad \alpha_2 = 1 \qquad \alpha_3 = 1 \qquad \alpha_4 = -1$$

is a solution.   Thus A_1, A_2, A_3, A_4 are linearly dependent.   ////

According to Lemma 1.3.16, if a set of matrices span V and one of them can be expressed as a linear combination of the others, it is superfluous and can be discarded.   The remaining matrices also span V.   This process can be repeated until we have a spanning set for V in which none of the matrices can be expressed as a linear combination of the others.   If V is not the zero space, the matrices of this spanning set are linearly independent.

Definition 1.3.22   A set of matrices $S = \{A_1, A_2, \ldots, A_k\}$ of a vector space V is a *basis* for V if:

1 S spans V.
2 S is a linearly independent set.

EXAMPLE 1.3.23   Let W be the vector space spanned by the matrices A_1, A_2, A_3, A_4 of Example 1.3.21.   We wish to find a basis for W.   According to Example 1.3.21, A_4 can be expressed as a linear combination of A_1, A_2 and A_3 (specifically, $A_4 = A_2 + A_3$) and, by Lemma 1.3.16, A_1, A_2, A_3 also span W.
In order to determine whether A_1, A_2, A_3 form a basis for W we repeat the procedure of Example 1.3.21 and find that

$$A_1 + A_2 - A_3 = 0$$

Thus A_1, A_2, A_3 are linearly dependent.   Since A_3 can be expressed as a linear combination of A_1 and A_2, it follows that A_1, A_2 also span W.
In order to determine whether A_1, A_2 form a basis for W we need to know whether the equation

$$\beta_1 A_1 + \beta_2 A_2 = 0$$

has any solutions other than $\beta_1 = \beta_2 = 0$.   The same procedure that we used before shows that there are no other solutions.   Thus A_1, A_2 form a basis.
In exactly the same way it can be shown that any two of the matrices A_1, A_2, A_3, A_4 form a basis for W.   ////

In speaking of a basis for a vector space we have had to treat the zero space separately.   It will simplify matters greatly if we now define an empty set of matrices to be linearly independent and to be a basis for a zero space.

It is not difficult to show that the unit column vectors are linearly independent and consequently form a basis for $\mathbb{F}^m$. We leave it as an exercise to show that in Example 1.3.6 the sets $\mathbb{S}_1$ and $\mathbb{S}_2$ of (1.3.9) and (1.3.10) are bases for $\mathbb{V}$ but $\mathbb{S}_3$ and $\mathbb{S}_4$ are not bases.

We stated earlier that it will not be until Sec. 1.5 that we can prove that every subspace of $\mathbb{F}_n{}^m$ has a spanning set. Since a basis is a particular type of spanning set, we cannot hope to show now that every subspace of $\mathbb{F}_n{}^m$ has a basis. However, the problem of finding a basis once a spanning set is obtained is relatively simple.

Theorem 1.3.24   If $\mathbb{S}$ is a spanning set for a vector space $\mathbb{V}$, then some subset of $\mathbb{S}$ is a basis for $\mathbb{V}$.

PROOF   The proof consists of repeated use of Lemma 1.3.16 and makes use of the process described in Example 1.3.23.   If $\mathbb{V} = 0$, then the empty set is a subset of $\mathbb{S}$ which is a basis for $\mathbb{V}$.   Now suppose $\mathbb{V} \neq 0$.   Then some of the matrices of $\mathbb{S}$ are nonzero.   If none of the matrices of $\mathbb{S}$ can be expressed as a linear combination of the others, then $\mathbb{S}$ is a basis for $\mathbb{V}$. On the other hand, if one of the elements of $\mathbb{S}$ can be expressed as a linear combination of the others, then that element can be discarded and, by Lemma 1.3.16, the resulting set, say $\mathbb{S}_1$, again spans $\mathbb{V}$.   If none of the matrices of $\mathbb{S}_1$ can be expressed as a linear combination of the others, then $\mathbb{S}_1$ is a basis for $\mathbb{V}$.   Otherwise Lemma 1.3.16 can be applied again. Since $\mathbb{S}$ initially contains only a finite number of elements, after a finite number of steps the process ends and the resulting subset of $\mathbb{S}$ is a basis of $\mathbb{V}$.   ////

There are many characterizations of a basis for a vector space.   Probably the one most often used is:

Theorem 1.3.25   Let $\mathbb{S}$ span $\mathbb{V}$.   Then $\mathbb{S}$ is a basis for $\mathbb{V}$ if and only if every element of $\mathbb{V}$ has a unique expression as a linear combination of the elements of $\mathbb{S}$.

PROOF   Let $\mathbb{S} = \{A_1, A_2, \ldots, A_k\}$ span $\mathbb{V}$ and let each element of $\mathbb{V}$ have a unique representation as a linear combination of $A_1, A_2, \ldots, A_k$.   We wish to show that $\mathbb{S}$ is a linearly independent set.   Suppose the zero element of $\mathbb{V}$ can be expressed as

$$0 = \alpha_1 A_1 + \alpha_2 A_2 + \cdots + \alpha_k A_k$$

for some $\alpha_1, \alpha_2, \ldots, \alpha_k \in \mathbb{F}$.   Clearly

$$0 = 0A_1 + 0A_2 + \cdots + 0A_k$$

Since the representation of each element of $\mathbb{V}$ is unique, it follows that

$$\alpha_1 = \alpha_2 = \cdots = \alpha_k = 0$$

and hence $\mathbb{S}$ is a linearly independent set.

Conversely let $\mathbb{S}$ be a basis for $\mathbb{V}$ and let

$$X = \alpha_1 A_1 + \alpha_2 A_2 + \cdots + \alpha_k A_k$$
$$X = \beta_1 A_1 + \beta_2 A_2 + \cdots + \beta_k A_k$$

be representations of an element X of $\mathbb{V}$. Subtracting one equation from the other, we have

$$0 = X - X = (\alpha_1 - \beta_1)A_1 + (\alpha_2 - \beta_2)A_2 + \cdots + (\alpha_k - \beta_k)A_k$$

Since $A_1, A_2, \ldots, A_k$ are linearly independent, it follows that

$$\alpha_1 - \beta_1 = \alpha_2 - \beta_2 = \cdots = \alpha_k - \beta_k = 0$$

Thus the representation for X is unique.   ////

We close this section by pointing out a dilemma that our notation forces upon us. In Lemma 1.3.16, in Definitions 1.3.7, 1.3.8, and 1.3.19, and in numerous definitions and theorems to come the order implicit in writing down a set of elements of a vector space, for example, $\mathbb{S} = \{A_1, A_2, \ldots, A_k\}$, is of no importance. (Remember, however, that in defining a matrix it was stressed that the order in which the entries appear within the matrix is of extreme importance.) When speaking of spanning sets or of linearly independent sets, for example, the order in which the matrices are given is completely immaterial. The difficulty would disappear if only there were a simple way of exhibiting such sets without giving their elements some order. This problem becomes especially acute in the next section when the rows or columns of an $m \times n$ matrix are considered to be elements of some subspace of $\mathbb{F}^m$ or $\mathbb{F}_n$.

EXERCISES

1 Show that

$$\begin{bmatrix} 1 \\ 1 \\ 0 \end{bmatrix} \quad \begin{bmatrix} 0 \\ 1 \\ 1 \end{bmatrix} \quad \begin{bmatrix} 1 \\ 2 \\ -1 \end{bmatrix}$$

form a basis for $\mathbb{F}^3$.

2   Show that

$$A_1 = \begin{bmatrix} 1 \\ 2 \\ 3 \end{bmatrix} \qquad A_2 = \begin{bmatrix} -1 \\ -1 \\ 2 \end{bmatrix} \qquad A_3 = \begin{bmatrix} -1 \\ 1 \\ 12 \end{bmatrix}$$

and

$$B_1 = \begin{bmatrix} 0 \\ 1 \\ 5 \end{bmatrix} \qquad B_2 = \begin{bmatrix} 1 \\ 3 \\ 8 \end{bmatrix}$$

span the same vector space V.   Show that $\{B_1, B_2\}$ is a basis for V.   Is
$S = \{A_1, A_2, A_3\}$ a basis for V?   If not, find a subset of S which is a basis.

3   Let V and W be the vector spaces defined in Examples 1.3.6 and 1.3.23.
Find bases for $V + W$ and $V \cap W$.

4   Let the vector space spanned by

$$C_1 = \begin{bmatrix} 1 \\ -1 \\ 1 \end{bmatrix} \qquad C_2 = \begin{bmatrix} 6 \\ 6 \\ 1 \end{bmatrix} \qquad C_3 = \begin{bmatrix} 8 \\ 4 \\ 3 \end{bmatrix}$$

be denoted by U.   Find a basis for U.   Find a basis for the intersection
of U and the vector space V of Exercise 2.   Show that $U + V = F^3$.

5   Interpret the vectors of Exercises 2 and 4 as points in three-dimensional
euclidean space.   Sketch U, V, and $U \cap V$.   Explain why each of the
sets $\{A_1, A_2, A_3\}$ and $\{B_1, B_2\}$ spans V.

6   Find a basis for the vector space X spanned by the set of matrices

$$\begin{bmatrix} 1 & 0 \\ 1 & 0 \end{bmatrix} \quad \begin{bmatrix} 1 & 1 \\ 0 & 0 \end{bmatrix} \quad \begin{bmatrix} 0 & 1 \\ 0 & 1 \end{bmatrix} \quad \begin{bmatrix} 0 & 0 \\ 1 & 1 \end{bmatrix} \quad \begin{bmatrix} 1 & 0 \\ 0 & 1 \end{bmatrix} \quad \begin{bmatrix} 0 & 1 \\ 1 & 0 \end{bmatrix}$$

Find all subsets of this set of matrices which are bases for X.

7   Show that $F_n{}^m$ has a basis consisting of mn matrices each of which has
exactly one nonzero entry.

8   Verify that the sets of matrices described in Examples 1.3.4 to 1.3.6 are
vector spaces.

9   Verify that S_1 of (1.3.9) and S_2 of (1.3.10) are bases for the vector space
V of Example 1.3.6 and that S_3 and S_4 of (1.3.10) are not bases for V.

10   Which of the following sets of matrices form vector spaces?
   (a) For a fixed square matrix A, the set of all matrices which commute
       with A under multiplication.
   (b) For fixed $m \times n$ and $n \times m$ matrices A and B, the set of all $n \times n$
       matrices X such that $AX = XB$.
   (c) For a fixed integer n, the set of all square matrices X of order n such
       that $X^2 = 0$.
   (d) For fixed matrices A and B, the set of all matrices X such that $AXB = 0$.

(e) The set of all vectors

$$X = [x_1 \quad x_2 \quad x_3]$$

such that $x_1 + x_2 + x_3 = 1$.

(f) The set of all vectors

$$X = [x_1 \quad x_2 \quad x_3]$$

such that $x_1 + x_2 + x_3 = 0$.

11 Verify that the sum and intersection of vector spaces as defined in Definition 1.3.12 are vector spaces.

12 Let U_1, U_2, and U_3 be subspaces of the vector space V.   Prove that
(a) $U_1 + U_2 = U_2 + U_1$.
(b) $U_1 + U_2 = U_1$ if and only if $U_2 \subseteq U_1$.
(c) $U_1 + (U_2 + U_3) = (U_1 + U_2) + U_3 = U_1 + U_2 + U_3$.
[Henceforth we write $U_1 + U_2 + U_3$ instead of either $U_1 + (U_2 + U_3)$ or $(U_1 + U_2) + U_3$.]
(d) $(U_1 \cap U_2) + (U_1 \cap U_3) \subseteq U_1 \cap (U_2 + U_3)$
Construct examples to show that both equality and inequality are possible in (d).

13 Let V be a vector space of $m \times n$ matrices.   For fixed matrices B and C of dimensions $p \times m$ and $n \times q$ respectively, we denote the set of all $p \times n$ matrices which can be expressed as BV for some $V \in V$ by BV and the set of all $m \times q$ matrices which can be expressed as VC for some $V \in V$ by VC, that is,

$$BV = \{BV \mid V \in V\} \qquad VC = \{VC \mid V \in V\}$$

Prove that BV and VC are vector spaces.

14 Let

$$A = \begin{bmatrix} 3 & -3 & 1 \\ 3 & 0 & -1 \\ 0 & 3 & -2 \\ 0 & 0 & 0 \end{bmatrix}$$

Find bases for AF^3 and $F_4 A$.   Find bases for the vector spaces AV and AU, where V and U are the vector spaces of Exercises 2 and 4 respectively.

15 Let $S = \{V_1, V_2, \ldots, V_k\}$ be a basis for a vector space V of $m \times n$ matrices and let B be a $p \times m$ matrix.   Prove that

$$BS = \{BV_1, BV_2, \ldots, BV_k\}$$

spans BV.   Prove that BS is a basis for BV if and only if $N(B) \cap V = 0$. State and prove the corresponding results for VC.

16 The *row null space*, denoted by $N_R(A)$, of the $m \times n$ matrix A is defined by

$$\mathbb{N}_R(A) = \{X \mid XA = 0, \; X \in \mathbb{F}_m\}$$

Prove that $\mathbb{N}_R(A)$ is a vector space.   Find a basis for $\mathbb{N}_R(A)$, where A is the matrix of Exercise 14.

17 Let $\mathbb{U}$ and $\mathbb{V}$ be vector spaces and let $\mathbb{S}$ be a spanning set for $\mathbb{V}$.   Prove that $\mathbb{V} \subseteq \mathbb{U}$ if and only if $\mathbb{S} \subseteq \mathbb{U}$.

18 Let A and B be $m \times n$ and $p \times n$ matrices respectively.   Show that if $\mathbb{N}(A) = \mathbb{N}(B)$, then $\mathbb{N}(AC) = \mathbb{N}(BC)$ for every $n \times q$ matrix C.

19 Let $\mathbb{S}$ be a linearly independent subset of a vector space $\mathbb{V}$.   Prove that $\mathbb{S}$ is a basis for $\mathbb{V}$ if and only if every finite subset of $\mathbb{V}$ which contains $\mathbb{S}$ properly is linearly dependent.

20 Let $\mathbb{S}$ be a spanning set for the vector space $\mathbb{V}$.   Prove that if there exists an element of $\mathbb{V}$ which can be expressed in exactly one way as a linear combination of the elements of $\mathbb{S}$, then every element of $\mathbb{V}$ can be expressed in exactly one way as a linear combination of the elements of $\mathbb{S}$.

21 Let $\mathbb{S}_1$ and $\mathbb{S}_2$ be bases for the subspaces $\mathbb{U}_1$ and $\mathbb{U}_2$ of the vector space $\mathbb{V}$ respectively.   Show that $\mathbb{S}_1 \cup \mathbb{S}_2$ spans $\mathbb{U}_1 + \mathbb{U}_2$.   Show that $\mathbb{S}_1 \cup \mathbb{S}_2$ is a basis for $\mathbb{U}_1 + \mathbb{U}_2$ if and only if $\mathbb{U}_1 \cap \mathbb{U}_2 = 0$.[1]

22 Let $\mathbb{U}_1$ and $\mathbb{U}_2$ be subspaces of the vector space $\mathbb{V}$.   Prove that $\mathbb{U}_1 \cup \mathbb{U}_2$ is a vector space if and only if either $\mathbb{U}_1 \subseteq \mathbb{U}_2$ or $\mathbb{U}_2 \subseteq \mathbb{U}_1$.

23 Prove that the simultaneous linear homogeneous equations

$$a_{11}x_1 + a_{12}x_2 + \cdots + a_{1n}x_n = 0$$
$$a_{21}x_2 + a_{22}x_2 + \cdots + a_{2n}x_n = 0$$
$$\cdots\cdots\cdots\cdots\cdots\cdots\cdots\cdots\cdots\cdots$$
$$a_{m1}x_1 + a_{m2}x_2 + \cdots + a_{mn}x_n = 0$$

have a nonzero solution if and only if the n vectors

$$A^{(1)} = \begin{bmatrix} a_{11} \\ a_{21} \\ \vdots \\ a_{m1} \end{bmatrix}, \quad A^{(2)} = \begin{bmatrix} a_{12} \\ a_{22} \\ \vdots \\ a_{m2} \end{bmatrix}, \quad \cdots, \quad A^{(n)} = \begin{bmatrix} a_{1n} \\ a_{2n} \\ \vdots \\ a_{mn} \end{bmatrix}$$

are linearly dependent.

24 Show that the set of all solutions

$$\begin{bmatrix} x_1 \\ x_2 \\ \vdots \\ x_n \end{bmatrix}$$

of the equations of Exercise 23 form a vector space.

---

[1] $\mathbb{S}_1 \cup \mathbb{S}_2$ is the *union* of $\mathbb{S}_1$ and $\mathbb{S}_2$ and is the set which consists of all elements which belong to at least one of the sets $\mathbb{S}_1$ and $\mathbb{S}_2$.

1.4 THE ROW SPACE AND THE COLUMN SPACE OF A MATRIX

We will see that the columns and rows, rather than the individual elements, are the true building blocks of a matrix. It is often essential to think of an $m \times n$ matrix as being composed of an ordered set of m n-dimensional columns or, in the same way, as an ordered set of m n-dimensional rows. For an $m \times n$ matrix A we have denoted the n columns of A by $A^{(1)}, A^{(2)}, \ldots, A^{(n)}$ and the m rows of A by $A_{(1)}, A_{(2)}, \ldots, A_{(m)}$. We then consider A as being partitioned into the n $m \times 1$ submatrices, that is, m-dimensional column vectors, $A^{(1)}$, $A^{(2)}, \ldots, A^{(n)}$.[1] Consequently we may say that $A^{(1)}, A^{(2)}, \ldots, A^{(n)}$ belong to $\mathbb{F}^m$ and thus the vector space that they span is a subspace of $\mathbb{F}^m$. We call this vector space the *column space* of A and denote it by $\mathbb{C}(A)$. In the same way, we consider A as being partitioned into m $1 \times n$ matrices, that is, n-dimensional row vectors, $A_{(1)}, A_{(2)}, \ldots, A_{(m)}$. These vectors belong to $\mathbb{F}_n$. The vector space spanned by them is a subspace of $\mathbb{F}_n$ called the *row space* of A and denoted by $\mathbb{R}(A)$.

Because most of our discussion of matrices in the remainder of this chapter is in terms of $\mathbb{C}(A)$ and $\mathbb{R}(A)$, row and column vectors (or simply vectors) will be studied separately in some detail. Although our treatment is specifically, at any one moment, in terms either of column or row vectors, it should be remembered that these methods and results are equally applicable to column vectors and row vectors and, in fact, often to vector spaces of matrices of arbitrary but fixed dimensions. In the future, when we have need for some result about column vectors which in an earlier section is stated and proved in terms of row vectors (or vice versa), we will simply refer to the row-vector (or column-vector) result.

EXAMPLE 1.4.1 Find bases for the row and column spaces of the matrix

$$B = \begin{bmatrix} -3 & 0 & 6 & 0 \\ 1 & 0 & -2 & 0 \\ -1 & 0 & 2 & 0 \end{bmatrix}$$

The three row vectors

$$B_{(1)} = [-3 \quad 0 \quad 6 \quad 0] \qquad B_{(2)} = [1 \quad 0 \quad -2 \quad 0] \qquad B_{(3)} = [-1 \quad 0 \quad 2 \quad 0]$$

span $\mathbb{R}(B)$ but do not form a linearly independent set since

$$B_{(1)} + 2B_{(2)} - B_{(3)} = 0$$

---

[1] We are purposely confusing the jth column of A with the m-dimensional column vector having the same entries. Both are denoted by $A^{(j)}$.

Since $B_{(3)}$ is expressible as a linear combination of $B_{(1)}$ and $B_{(2)}$, it is superfluous and the two vectors $B_{(1)}$ and $B_{(2)}$ also span $\mathbb{R}(B)$. However,

$$B_{(1)} + 3B_{(2)} = 0$$

Thus we can eliminate $B_{(2)}$ and conclude that $\{B_{(1)}\}$ is a basis for $\mathbb{R}(B)$.

In the same way, the four column vectors

$$B^{(1)} = \begin{bmatrix} -3 \\ 1 \\ -1 \end{bmatrix} \quad B^{(2)} = \begin{bmatrix} 0 \\ 0 \\ 0 \end{bmatrix} \quad B^{(3)} = \begin{bmatrix} 6 \\ -2 \\ 2 \end{bmatrix} \quad B^{(4)} = \begin{bmatrix} 0 \\ 0 \\ 0 \end{bmatrix}$$

span $\mathbb{C}(B)$ but do not form a linearly independent set since

$$2B^{(1)} + \alpha B^{(2)} + B^{(3)} + \beta B^{(4)} = 0$$

for every choice of α and β. As before, $B^{(2)}$, $B^{(3)}$, and $B^{(4)}$ can be eliminated, and $\{B^{(1)}\}$ is a basis for $\mathbb{C}(B)$.

Note that $\mathbb{C}(B)$ is the same vector space as $\mathbb{N}(A)$ of Example 1.3.5.   ////

Because of the way multiplication of matrices is defined, the row and column spaces of a product are closely related to the row and column spaces of the factors. Specifically:

Lemma 1.4.2 Let A and B be $m \times n$ and $n \times p$ matrices respectively. Then

1 $\mathbb{R}(AB) \subseteq \mathbb{R}(B)$.

2 $\mathbb{C}(AB) \subseteq \mathbb{C}(A)$.

PROOF Set $AB = C$. A careful consideration of Definition 1.2.4 leads to the following relations between the rows of B and C and between the columns of A and C:

$$\begin{aligned} C_{(i)} &= a_{i1}B_{(1)} + a_{i2}B_{(2)} + \cdots + a_{in}B_{(n)} & i &= 1, 2, \ldots, m \\ C^{(j)} &= A^{(1)}b_{1j} + A^{(2)}b_{2j} + \cdots + A^{(n)}b_{nj} & j &= 1, 2, \ldots, p \end{aligned} \quad (1.4.3)$$

(See Exercise 11 of Sec. 1.2.)

Now let $X \in \mathbb{R}(AB) = \mathbb{R}(C)$. Since $\mathbb{R}(C)$ is the vector space spanned by the m rows of C, there exist scalars $\alpha_1, \alpha_2, \ldots, \alpha_m$ such that

$$X = \alpha_1 C_{(1)} + \alpha_2 C_{(2)} + \cdots + \alpha_m C_{(m)}$$

Substituting for each $C_{(i)}$ the expression of equation (1.4.3) and rearranging terms yields

$$X = (\alpha_1 a_{11} + \alpha_2 a_{21} + \cdots + \alpha_m a_{m1})B_{(1)}$$
$$+ (\alpha_1 a_{12} + \alpha_2 a_{22} + \cdots + \alpha_m a_{m2})B_{(2)}$$
$$+ \cdots + (\alpha_1 a_{1n} + \alpha_2 a_{2n} + \cdots + \alpha_m a_{mn})B_{(n)}$$

Thus X has been expressed as a linear combination of the rows of B, and it follows that X belongs to the vector space spanned by the rows of B; that is, $X \in \mathbb{R}(B)$. Consequently $\mathbb{R}(AB) = \mathbb{R}(C) \subseteq \mathbb{R}(B)$.

In the same way we can show that $\mathbb{C}(AB) = \mathbb{C}(C) \subseteq \mathbb{C}(A)$.   ////

Consideration of the row and column spaces of a matrix permits us to determine when matrix equations of the form $XB = C$ and $BX = C$ have solutions. We have already pointed out that a set of simultaneous linear equations can be written in this form. Unlike the situation which occurs when X, B, and C represent numbers, it is possible that there are no solutions, even when $B \neq 0$. Likewise, it is also possible that there are infinitely many solutions, even when $B \neq 0$. Finally, it will become evident that one of these equations may have a solution while the other does not. (See Exercises 3 and 4 of Sec. 1.2.)

Theorem 1.4.4 (a) Let B and C be $m \times n$ and $p \times n$ matrices respectively. A necessary and sufficient condition that there exist a $p \times m$ matrix A such that $AB = C$ is that $\mathbb{R}(C) \subseteq \mathbb{R}(B)$. Moreover, when such a matrix A does exist, A is unique if and only if the rows of B are linearly independent.

(b) Let B and C be $m \times n$ and $m \times p$ matrices respectively. A necessary and sufficient condition that there exist an $n \times p$ matrix A such that $BA = C$ is that $\mathbb{C}(C) \subseteq \mathbb{C}(B)$. Moreover, when such a matrix A does exist, A is unique if and only if the columns of B are linearly independent.

PROOF It is sufficient to present a proof for (a) since the proof of (b) is essentially identical.

First let $\mathbb{R}(C) \subseteq \mathbb{R}(B)$. We wish to show that the equation $AB = C$ has a solution. Since each of the p rows of C is contained in the row space of B, it can be expressed as a linear combination of the rows of B. Thus, for each i, $i = 1, 2, \ldots, p$, there exist m scalars, a_{ij}, $j = 1, 2, \ldots, m$, such that

$$C_{(i)} = a_{i1}B_{(1)} + a_{i2}B_{(2)} + \cdots + a_{im}B_{(m)} \qquad (1.4.5)$$

The p equations of (1.4.5) require pm numbers a_{ij}, $i = 1, 2, \ldots, p$; $j = 1, 2, \ldots, m$. Define the $p \times m$ matrix A by setting $A = [a_{ij}]$. Then it follows from (1.4.5) that $C = AB$.

Conversely, suppose that there is a $p \times m$ matrix $A = [a_{ij}]$ such that $C = AB$. Then the p equations of (1.4.5) are satisfied, and, as in Lemma 1.4.2, we have $\mathbb{R}(AB) = \mathbb{R}(C) \subseteq \mathbb{R}(B)$. We have now shown that $C = AB$ has a solution if and only if $\mathbb{R}(C) \subseteq \mathbb{R}(B)$.

Finally, let us determine when the solution A is unique. Thus we assume that a matrix $A = [a_{ij}]$ exists such that $AB = C$. It follows from Theorem 1.3.25 that for each of the p equations of (1.4.5) the m numbers a_{ij}, $j = 1, 2, \ldots, m$, satisfying the equation are unique if and only if $B_{(1)}, B_{(2)}, \ldots, B_{(m)}$ form a basis for $\mathbb{R}(B)$, that is, if and only if the rows of B are linearly independent.   ////

Corollary 1.4.6   Let B and C be $m \times n$ and $p \times n$ matrices respectively. If $\mathbb{R}(C) \subseteq \mathbb{R}(B)$, then $\mathbb{N}(C) \supseteq \mathbb{N}(B)$. Hence, if $\mathbb{R}(C) = \mathbb{R}(B)$, then $\mathbb{N}(B) = \mathbb{N}(C)$.

PROOF   Since $\mathbb{R}(C) \subseteq \mathbb{R}(B)$, there is a matrix A such that $AB = C$. Let $X \in \mathbb{N}(B)$. Then

$$CX = ABX = A0 = 0$$

and $X \in \mathbb{N}(C)$. Consequently $\mathbb{N}(C) \supseteq \mathbb{N}(B)$.   ////

In Sec. 1.7 we will prove the converse of Corollary 1.4.6; i.e., if $\mathbb{N}(C) \supseteq \mathbb{N}(B)$, then $\mathbb{R}(C) \subseteq \mathbb{R}(B)$. Surprisingly, this is a much deeper result.[1]

An important special case of Theorem 1.4.4 occurs when C is the identity matrix I. The n columns of an identity matrix of order n are the n unit column vectors (1.3.11) and hence form a basis for $\mathbb{F}^n$. Then we have:

Corollary 1.4.7   Let B be an $m \times n$ matrix.

1   A necessary and sufficient condition that there exist a matrix A such that $AB = I$ (where I is necessarily of order n) is that $\mathbb{R}(B) = \mathbb{F}_n$.

2   A necessary and sufficient condition that there exist a matrix A such that $BA = I$ (where I is necessarily of order m) is that $\mathbb{C}(B) = \mathbb{F}^m$.

PROOF   Let us apply Theorem 1.4.4 to the equation $AB = I$. A solution exists if and only if $\mathbb{F}_n = \mathbb{R}(I) \subseteq \mathbb{R}(B)$. However, $\mathbb{R}(B) \subseteq \mathbb{F}_n$ for every $m \times n$ matrix B, and hence the condition $\mathbb{F}_n \subseteq \mathbb{R}(B)$ is equivalent to the condition $\mathbb{F}_n = \mathbb{R}(B)$.   ////

---

[1] In fact, it is false for arbitrary vector spaces; see, for example, Taylor [9].

EXERCISES

1 Let

$$A = \begin{bmatrix} 1 & 2 & -1 & 1 & 4 \\ 0 & 1 & 2 & -1 & -3 \\ 2 & 7 & 4 & -1 & -1 \end{bmatrix}$$

Find a basis for $\mathbb{R}(A)$ and a basis for $\mathbb{C}(A)$.

2 Let

$$B = \begin{bmatrix} 1 & 2 & -1 \\ 0 & 0 & 1 \\ 1 & 0 & 1 \end{bmatrix} \qquad C = \begin{bmatrix} 1 & -1 & 0 \\ 2 & 0 & 1 \end{bmatrix}$$

Find a matrix A such that $AB = C$. Is A unique? Does there exist a matrix A such that $AB = I_3$? Does there exist a matrix A such that $BA = I_3$?

3 Does $[3 \quad -2 \quad -3 \quad 1]$ belong to the row space of

$$\begin{bmatrix} 1 & -1 & -1 & 2 \\ 0 & 1 & 0 & 1 \\ 1 & 0 & -1 & 0 \end{bmatrix}$$

4 Let

$$B = \begin{bmatrix} 1 & 2 & -1 \\ 1 & 1 & 1 \\ 0 & 1 & -2 \end{bmatrix} \qquad C = \begin{bmatrix} 1 & 0 & 1 \\ 2 & 1 & 4 \end{bmatrix}$$

Show that there does not exist a matrix A such that $AB = C$ (*a*) by a direct computation and
(*b*) by showing that $\mathbb{R}(C) \nsubseteq \mathbb{R}(B)$ and applying Theorem 1.4.4.

5 Find a basis for $\mathbb{C}(A) \cap \mathbb{C}(B)$, where A and B are given in Exercises 1 and 4 respectively.

6 For what values of x is $\mathbb{F}^3$ the column space of the matrix

$$A = \begin{bmatrix} 1 & 0 & 1 \\ -1 & 1 & 2 \\ 0 & 3 & x \end{bmatrix}$$

For what values of x is $\mathbb{F}_3$ the row space of A?

7 Let

$$B = \begin{bmatrix} 1 & 1 & 3 \\ -1 & 2 & 0 \\ 1 & 0 & 2 \end{bmatrix} \qquad C = \begin{bmatrix} x & -1 & 1 \\ y & 0 & 1 \end{bmatrix}$$

For what values of x and y do there exist matrices A such that $AB = C$?

8 Let A be an $m \times n$ matrix. Prove that $\mathbb{C}(A) = A\mathbb{F}^n$ and $\mathbb{R}(A) = \mathbb{F}_m A$.

9   Prove Theorem 1.4.4(*b*).

10   Let A and B be $p \times n$ and $q \times n$ matrices respectively. Show that if $\mathbb{R}(A) = \mathbb{R}(B)$, then $\mathbb{R}(AC) = \mathbb{R}(BC)$ for every $n \times s$ matrix C.

11   Let A and B be $m \times n$ and $p \times n$ matrices respectively and let C be the $(m + p) \times n$ matrix

$$C = \begin{bmatrix} A \\ B \end{bmatrix}$$

Prove:

(*a*) $\mathbb{R}(A) + \mathbb{R}(B) = \mathbb{R}(C)$.

(*b*) $\mathbb{R}(A) = \mathbb{R}(C)$ if and only if $\mathbb{R}(B) \subseteq \mathbb{R}(A)$.

(*c*) $\mathbb{N}(A) \cap \mathbb{N}(B) = \mathbb{N}(C)$.

12   Let A and B be $m \times n$ and $n \times p$ matrices respectively. Prove:

(*a*) $\mathbb{N}(AB) \supseteq \mathbb{N}(B)$.

(*b*) $\mathbb{N}(AB) = \mathbb{N}(B)$ if and only if $\mathbb{N}(A) \cap \mathbb{C}(B) = 0$.

13   Let A and B be $m \times n$ and $n \times p$ matrices respectively. Prove that

$$\mathbb{N}(A) \supseteq \mathbb{C}(B) \qquad \text{if and only if} \qquad \mathbb{R}(A) \subseteq \mathbb{N}_R(B)$$

14   Show that the set of simultaneous linear equations

$$\begin{aligned}
a_{11}x_1 + a_{12}x_2 + \cdots + a_{1n}x_n &= d_1 \\
a_{21}x_1 + a_{22}x_2 + \cdots + a_{2n}x_n &= d_2 \\
&\cdots\cdots\cdots\cdots\cdots\cdots\cdots \\
a_{m1}x_1 + a_{m2}x_2 + \cdots + a_{mn}x_n &= d_m
\end{aligned} \qquad (1.4.8)$$

has a solution if and only if the vector

$$\begin{bmatrix} d_1 \\ d_2 \\ \vdots \\ d_m \end{bmatrix}$$

is contained in the column space of the $m \times n$ matrix $[a_{ij}]$ formed from the coefficients in the equations.

15   Show that if the rows of the $m \times n$ matrix A are linearly dependent, then $\mathbb{C}(A) \neq \mathbb{F}^m$. [*Hint:* Show that when the rows of A are linearly dependent, the system of equations (1.4.8) does not have a solution for some choice of $d_1, d_2, \ldots, d_m$.]

1.5 THE STEINITZ REPLACEMENT THEOREM

The replacement theorem of Steinitz (Theorem 1.5.1) is of fundamental importance in the study of vector spaces. Although the replacement theorem by itself does not establish the existence of a basis, it will permit us to show (in Theorem

1.5.6) not only that every subspace of $\mathbb{F}_n{}^m$ has a basis but also that all bases of a given vector space contain the same number of elements. A corollary to the replacement theorem provides a particularly simple means of identifying bases and leads to a method of constructing them. In addition, it will be clear that the matrices which belong to a basis are in no way "special" matrices, as one might think the unit vectors are. In fact, it will follow from the replacement theorem that for any linearly independent set of matrices of a vector space there is some basis which contains it.

Theorem 1.5.1 The Steinitz replacement theorem[1] Let $\mathbb{A} = \{A_1, A_2, \ldots, A_k\}$ be a basis for the vector space $\mathbb{V}$ and let $\mathbb{B} = \{B_1, B_2, \ldots, B_l\}$ be a set of linearly independent matrices belonging to $\mathbb{V}$. Then $l \leq k$. Moreover, there exist $k - l$ of the matrices of $\mathbb{A}$ which, together with the matrices of $\mathbb{B}$, form a basis for $\mathbb{V}$. In particular, if $l = k$, then $\mathbb{B}$ is a basis for $\mathbb{V}$.

PROOF   The basic idea underlying the proof is that there is a process for repeatedly replacing one of the elements of $\mathbb{A}$ by one of the elements of $\mathbb{B}$ in such a way that the set which is constructed at each stage spans $\mathbb{V}$. Because we always replace one element of $\mathbb{A}$ with one element of $\mathbb{B}$, the total number of matrices in the set at any time remains unchanged, namely, k.

Since $\mathbb{A}$ spans $\mathbb{V}$, it follows that the $k + 1$ matrices $B_l, A_1, A_2, \ldots, A_k$ also span $\mathbb{V}$. Moreover, this set is linearly dependent since B_l can be expressed as a linear combination of the elements of the spanning set $\mathbb{A}$. Thus there exist numbers $\beta_l, \alpha_1, \alpha_2, \ldots, \alpha_k$, not all of which are zero, such that

$$\beta_l B_l + \alpha_1 A_1 + \alpha_2 A_2 + \cdots + \alpha_k A_k = 0 \qquad (1.5.2)$$

Furthermore, at least one of the numbers $\alpha_1, \alpha_2, \ldots, \alpha_k$ is not zero, for if $\alpha_1 = \alpha_2 = \cdots = \alpha_k = 0$, then $\beta_l \neq 0$ and $\beta_l B_l = 0$, from which it follows that $B_l = 0$. Hence $\{B_l\}$ is a linearly dependent set. However, we know that $\{B_l\}$ is a linearly independent set since it is a subset of the linearly independent set $\mathbb{B}$. To prevent our notation from becoming terribly cumbersome we will assume that the matrices constituting $\mathbb{A}$ are numbered in such a way that we may choose $\alpha_k \neq 0$ in equation (1.5.2) (see the last paragraph of Sec. 1.3). Then A_k can be expressed as a linear combi-

---

[1] Steinitz published this theorem in 1913, undoubtedly not aware that the same result had appeared in a book by Grassmann in 1862.

nation of B_l, $A_1, A_2, \ldots, A_{k-1}$. According to Lemma 1.3.16, A_k can be discarded, and the set

$$\mathbb{A}_1 = \{B_l, A_1, A_2, \ldots, A_{k-1}\}$$

spans $\mathbb{V}$. In other words, A_k has been "replaced" by B_l.

We repeat this process, beginning now with $\mathbb{A}_1$. Since $\mathbb{A}_1$ spans $\mathbb{V}$, it is clear that the $k + 1$ matrices $B_{l-1}, B_l, A_1, A_2, \ldots, A_{k-1}$ also span $\mathbb{V}$. This set is linearly dependent since B_{l-1} can be expressed as a linear combination of the elements of the spanning set $\mathbb{A}_1$. As before, there exist numbers $\delta_{l-1}, \delta_l, \gamma_1, \gamma_2, \ldots, \gamma_{k-1}$, not all zero, such that

$$\delta_{l-1}B_{l-1} + \delta_l B_l + \gamma_1 A_1 + \gamma_2 A_2 + \cdots + \gamma_{k-1}A_{k-1} = 0$$

Moreover, at least one of the numbers $\gamma_1, \gamma_2, \ldots, \gamma_{k-1}$ is not zero, for if $\gamma_1 = \gamma_2 = \cdots = \gamma_{k-1} = 0$, then at least one of δ_{l-1} and δ_l is not zero and we have

$$\delta_{l-1}B_{l-1} + \delta_l B_l = 0$$

which contradicts the linear independence of the set $\{B_{l-1}, B_l\}$. Again, for simplicity of notation, we will assume that the elements of $\mathbb{A}$ are numbered in such a way that we may now choose $\gamma_{k-1} \neq 0$. Then A_{k-1} can be expressed as a linear combination of $B_{l-1}, B_l, A_1, A_2, \ldots, A_{k-2}$. As before, it follows from Lemma 1.3.16 that

$$\mathbb{A}_2 = \{B_{l-1}, B_l, A_1, A_2, \ldots, A_{k-2}\}$$

spans $\mathbb{V}$.

We continue in this way, repeating the process as long as we are able. At each repetition one of the elements of $\mathbb{A}$ is replaced by one of the elements of $\mathbb{B}$. The process can no longer be carried out only when either (1) all the elements of $\mathbb{A}$ have been discarded or (2) all the elements of $\mathbb{B}$ have been brought into the spanning set. The first part of Theorem 1.5.1 merely states that the process can come to an end either by (1) or (2) above happening simultaneously (in which case $l = k$) or by (2) alone occurring (in which case $l < k$). In other words, we wish to show that the third possibility, namely, that (1) occurs before (2) (in which case we would have $l > k$), cannot occur. Let us suppose then that $l > k$ is possible. This means that after k steps of the process we will have shown that

$$\mathbb{A}_k = \{B_{l-k+1}, B_{l-k+2}, \ldots, B_{l-1}, B_l\}$$

spans $\mathbb{V}$. Furthermore, $l > k$ implies that $l - k + 1 \geq 2$, that is, B_1 does not belong to $\mathbb{A}_k$. However, B_1 belongs to $\mathbb{V}$, and thus we must be able to express B_1 as a linear combination of $B_{l-k+1}, B_{l-k+2}, \ldots, B_{l-1}, B_l$. But this is impossible since, by hypothesis, $\mathbb{B}$ is a linearly independent set. This contradiction arose from the supposition that $l > k$, which we now discard. Hence we have shown that $l \leq k$, as required.

The last spanning set we construct using the replacement process described above is $\mathbb{A}_l$. $\mathbb{A}_l$ consists of the l vectors which belonged to $\mathbb{B}$ and $k - l$ of the vectors of $\mathbb{A}$. To prove that the k elements of $\mathbb{A}_l$ form a basis for $\mathbb{V}$ it remains only to show that $\mathbb{A}_l$ is a linearly independent set. By Theorem 1.3.24, some subset of $\mathbb{A}_l$, say $\mathbb{S}$, is a basis for $\mathbb{V}$. Let $\mathbb{S}$ have m elements. Since $\mathbb{S}$ is a subset of $\mathbb{A}_l$, we have $m \leq k$. On the other hand, considering $\mathbb{S}$ as a basis for $\mathbb{V}$ and $\mathbb{A}$ as a linearly independent set of k matrices of $\mathbb{V}$, it follows from the first part of the theorem that $k \leq m$. Combining these results, we have

$$k \leqq m \leqq k$$

proving that $k = m$ and that $\mathbb{A}_l = \mathbb{S}$ is indeed a basis for $\mathbb{V}$. The last sentence of Theorem 1.5.1 merely points out the special case in which $k - l = 0$. ////

Note that the replacement theorem does not guarantee that a given vector space actually has a spanning set. We will not prove this until Theorem 1.5.6. However, if we assume the existence of bases, the replacement theorem does imply the very important:

Corollary 1.5.3 Any two bases for a vector space contain the same number of elements.

PROOF Let $\mathbb{A} = \{A_1, A_2, \ldots, A_k\}$ and $\mathbb{B} = \{B_1, B_2, \ldots, B_l\}$ be two bases for the same vector space $\mathbb{V}$ and suppose that $l \leq k$. By the replacement theorem there exist $k - l$ matrices of $\mathbb{A}$ which, together with the matrices of $\mathbb{B}$, form a basis for $\mathbb{V}$. However, $\mathbb{B}$ is already a basis for $\mathbb{V}$, and consequently a set consisting of the matrices of $\mathbb{B}$ and any further matrices of $\mathbb{V}$ is linearly dependent. Thus $k - l = 0$. ////

Corollary 1.5.3 gives us a most useful characterization of a basis, namely:

Corollary 1.5.4 Let $\mathbb{V}$ be a vector space which has a basis consisting of k elements. A set of necessary and sufficient conditions that the matrices $B_1, B_2, \ldots, B_l$ of $\mathbb{V}$ form a basis for $\mathbb{V}$ is:

1 $k = l$.

2 $B_1, B_2, \ldots, B_l$ are linearly independent.

In the special case where $\mathbb{V} = \mathbb{F}^m$ we have already found a basis consisting of m elements, namely, the m unit column vectors [see (1.3.11)]. It is the existence of a basis for $\mathbb{F}^m$ which will permit us to prove that every subspace of $\mathbb{F}^m$ has a basis. In order to apply the same methods to $\mathbb{F}_n{}^m$, we need a basis for $\mathbb{F}_n{}^m$. Any basis will do. The set of matrices of $\mathbb{F}_n{}^m$ which can most easily be shown to be a basis is the *mn unit m × n matrices*

$$\{I(i, j) \,|\, i = 1, 2, \ldots, m; j = 1, 2, \ldots, n\}$$

where $I(i, j)$ is the $m \times n$ matrix which has a 1 in the (i, j) position and 0s elsewhere. Note that when $n = 1$, these matrices become the m unit column vectors.

According to the replacement theorem, since the mn unit matrices form a basis for $\mathbb{F}_n{}^m$, every linearly independent subset of $\mathbb{F}_n{}^m$ contains at most mn elements. In other words:

Lemma 1.5.5 Every set of $mn + 1$ matrices of $\mathbb{F}_n{}^m$ is linearly dependent.

It is this lemma which provides the means of showing that every subspace of $\mathbb{F}_n{}^m$ actually possesses a basis.

Theorem 1.5.6 Every subspace of $\mathbb{F}_n{}^m$ has a basis consisting of not more than mn matrices.

PROOF For $\mathbb{V} = 0$ we have defined the empty set to be a basis for $\mathbb{V}$. Now assume that $\mathbb{V} \neq 0$. Then $\mathbb{V}$ contains linearly independent sets of matrices. In particular, every set consisting of one nonzero matrix belonging to $\mathbb{V}$ is a linearly independent subset of $\mathbb{V}$. Let $\mathbb{A}$ be a linearly independent set of elements of $\mathbb{V}$ which is not contained in any larger linearly independent set of elements of $\mathbb{V}$. The existence of such a set follows from Lemma 1.5.5, and in fact $\mathbb{A}$ contains at most mn matrices. Thus we may set $\mathbb{A} = \{A_1, A_2, \ldots, A_k\}$. In order to show that $\mathbb{A}$ is a basis for $\mathbb{V}$ it remains only to prove that $A_1, A_2, \ldots, A_k$ span $\mathbb{V}$. Let $X \in \mathbb{V}$. By our method of selection of the set $\mathbb{A}$, the matrices $X, A_1, A_2, \ldots, A_k$ are linearly dependent. Hence there exist scalars, $\beta, \alpha_1, \alpha_2, \ldots, \alpha_k$, not all zero, such that

$$\beta X + \alpha_1 A_1 + \alpha_2 A_2 + \cdots + \alpha_k A_k = 0$$

Moreover, $\beta \neq 0$, for if $\beta = 0$, this equation would state that $A_1, A_2, \ldots,$

A_k are linearly dependent, which would be a contradiction. Hence

$$X = -\frac{\alpha_1}{\beta} A_1 - \frac{\alpha_2}{\beta} A_2 - \cdots - \frac{\alpha_k}{\beta} A_k$$

Thus $A_1, A_2, \ldots, A_k$ span $\mathbb{V}$. ////

Definition 1.5.7 Let $\mathbb{V}$ be a vector space. The number of elements in any basis for $\mathbb{V}$ is called the *dimension*[1] of $\mathbb{V}$ and is written dim $\mathbb{V}$.

Let A be an $m \times n$ matrix. The dimension of the column space of A is called the *column rank* of A and is denoted by $r_C(A)$. Similarly, the dimension of the row space of A is called the *row rank* of A and is denoted by $r_R(A)$. The dimension of the null space of A is the *(column) nullity* of A and is denoted by $v(A)$.

By Theorem 1.5.6, every subspace $\mathbb{V}$ of $\mathbb{F}_n{}^m$ has a dimension, and by Corollary 1.5.3, the dimension of $\mathbb{V}$ is well defined; i.e., the dimension of $\mathbb{V}$ is independent of which basis for $\mathbb{V}$ is used.

EXAMPLE 1.5.8 For the 3×4 matrix

$$B = \begin{bmatrix} -3 & 0 & 6 & 0 \\ 1 & 0 & -2 & 0 \\ -1 & 0 & 2 & 0 \end{bmatrix}$$

of Example 1.4.1 we showed that $\{B_{(1)}\}$ and $\{B^{(1)}\}$ are bases for $\mathbb{R}(B)$ and $\mathbb{C}(B)$ respectively. Thus

$$r_R(B) = r_C(B) = 1$$

It was also pointed out that $\mathbb{C}(B) = \mathbb{N}(A)$, where

$$A = \begin{bmatrix} 1 & 2 & -1 \\ 0 & 1 & 1 \end{bmatrix}$$

is the matrix of Example 1.3.4 Thus

$$v(A) = 1 \qquad ////$$

---

[1] The use of the word "dimension" in this context may cause some confusion   When speaking of a single matrix (and, in particular, of a single vector) *dimension* refers to its physical size, i.e., to the number of rows and columns it has. When speaking about a vector space, dimension refers to the number of elements in a basis. Thus we may have a two-dimensional vector space consisting of three-dimensional column vectors. If we were to emphasize the geometric interpretation of vector spaces, we would call any one-dimensional vector space a *line* and any two-dimensional vector space a *plane*, even if they are subspaces of a three-dimensional vector space.

If U is a nonzero subspace of the vector space V, then a basis S for U is a linearly independent set of matrices which belong to V and, by Theorem 1.5.1, can be extended to a basis T for V. Hence dim $U \leq$ dim V. Moreover, according to Corollary 1.5.4, S is also a basis for V if and only if the dimension of V is equal to the dimension of U. Thus, when U is a subspace of V, we have $U = V$ if and only if dim $U =$ dim V.

There is an immediate application of these results to Lemma 1.4.2.

Corollary 1.5.9   Let A and B be $m \times n$ and $n \times p$ matrices respectively. Then:

  1 $r_R(AB) \leq r_R(B)$.
  2 $r_C(AB) \leq r_C(A)$.

At this point we have no way of comparing the column rank of AB with the column rank of B or the row rank of AB with the row rank of A. These difficulties will be resolved later.

In the remainder of this section we will consider the dimension of a sum of vector spaces. Specifically, let $U_1, U_2, \ldots, U_l$ be vector spaces of $m \times n$ matrices and let

$$A_1 = \{A_{11}, A_{12}, \ldots, A_{1k_1}\}$$
$$A_2 = \{A_{21}, A_{22}, \ldots, A_{2k_2}\}$$
$$\cdots\cdots\cdots\cdots\cdots\cdots\cdots\cdots\cdots$$
$$A_l = \{A_{l1}, A_{l2}, \ldots, A_{lk_l}\}$$

be bases for $U_1, U_2, \ldots, U_l$ respectively. Clearly the $k_1 + k_2 + \cdots + k_l$ (not necessarily distinct) matrices of the set

$$A = \{A_{11}, A_{12}, \ldots, A_{1k_1}, A_{21}, \ldots, A_{lk_l}\} \qquad (1.5.10)$$

span the vector space $U_1 + U_2 + \cdots + U_l$. By Theorem 1.3.24 some subset of A is a basis for $U_1 + U_2 + \cdots + U_l$. Thus we have proved

Lemma 1.5.11   Let $U_1, U_2, \ldots, U_l$ be vector spaces of $m \times n$ matrices. Then

$$\dim U_1 + U_2 + \cdots + U_l \leq \dim U_1 + \dim U_2 + \cdots + \dim U_l \qquad (1.5.12)$$

Moreover, we have equality in (1.5.12) if and only if A is a linearly independent set. Let us give this a name.

Definition 1.5.13   Let $U_1, U_2, \ldots, U_l$ be vector spaces of $m \times n$ matrices and set $U = U_1 + U_2 + \cdots + U_l$. If

$$\dim U = \dim U_1 + \dim U_2 + \cdots + \dim U_l$$

we say that $\mathbb{U}$ is a *direct sum* of $\mathbb{U}_1, \mathbb{U}_2, \ldots, \mathbb{U}_l$ and write

$$\mathbb{U} = \mathbb{U}_1 \oplus \mathbb{U}_2 \oplus \cdots \oplus \mathbb{U}_l$$

A natural example of a direct sum occurs when a vector space is identified by means of some basis. In that case the vector space is immediately expressable as a direct sum of one-dimensional subspaces. Specifically, let $\mathbb{A} = \{A_1, A_2, \ldots, A_k\}$ be any basis for the k-dimensional vector space $\mathbb{V}$. For each i, $i = 1, 2, \ldots, k$, denote by $\mathbb{V}_i$ the one-dimensional vector space spanned by the single matrix A_i. Then

$$\dim \mathbb{V} = k = \dim \mathbb{V}_1 + \dim \mathbb{V}_2 + \cdots + \dim \mathbb{V}_k$$

and therefore $\mathbb{V} = \mathbb{V}_1 \oplus \mathbb{V}_2 \oplus \cdots \oplus \mathbb{V}_k$. Note that Theorem 1.3.25 states that every element X of $\mathbb{V}$ has a unique expression of the form

$$X = X_1 + X_2 + \cdots + X_k$$

where $X_i \in \mathbb{V}_i$. This is, in fact, an important property of the direct sum.

Theorem 1.5.14   Let $\mathbb{U}_1, \mathbb{U}_2, \ldots, \mathbb{U}_l$ be vector spaces of $m \times n$ matrices and let $\mathbb{U} = \mathbb{U}_1 + \mathbb{U}_2 + \cdots + \mathbb{U}_l$. Then

$$\mathbb{U} = \mathbb{U}_1 \oplus \mathbb{U}_2 \oplus \cdots \oplus \mathbb{U}_l \qquad (1.5.15)$$

if and only if every element X belonging to $\mathbb{U}$ has a unique representation as

$$X = X_1 + X_2 + \cdots + X_l \qquad (1.5.16)$$

where $X_i \in \mathbb{U}_i$, $i = 1, 2, \ldots, l$.

PROOF   For each i, $i = 1, 2, \ldots, l$, select a basis

$$\mathbb{A}_i = \{A_{i1}, A_{i2}, \ldots, A_{ik_i}\}$$

for $\mathbb{U}_i$ and let $\mathbb{A}$ be the set of $k_1 + k_2 + \cdots + k_l$ (not necessarily distinct) matrices

$$\mathbb{A} = \{A_{11}, A_{12}, \ldots, A_{1k_1}, A_{21}, \ldots, A_{lk_l}\}$$

Clearly $\mathbb{A}$ spans $\mathbb{U}$. According to the discussion preceding Definition 1.5.13, equation (1.5.15) holds if and only if the elements of $\mathbb{A}$ are (distinct and) linearly independent and hence form a basis for $\mathbb{U}$. Thus, in order to prove the theorem we will show that $\mathbb{A}$ is a linearly independent set if and only if every matrix X belonging to $\mathbb{U}$ has a unique representation of the form of equation (1.5.16).

First, let us assume that $U = U_1 \oplus U_2 \oplus \cdots \oplus U_l$. Then the elements of A are linearly independent. Let $X \in U$ and suppose that

$$X = X_1 + X_2 + \cdots + X_l = Y_1 + Y_2 + \cdots + Y_l \qquad (1.5.17)$$

where X_i, $Y_i \in U_i$, $i = 1, 2, \ldots, l$. We wish to show that the two representations for X given in equation (1.5.17) are in fact the same representation. It follows from equation (1.5.17) that

$$0 = X - X = (X_1 - Y_1) + (X_2 - Y_2) + \cdots + (X_l - Y_l) \qquad (1.5.18)$$

Setting $X_i - Y_i = Z_i$, equation (1.5.18) becomes

$$0 = Z_1 + Z_2 + \cdots + Z_l \qquad (1.5.19)$$

and $Z_i \in U_i$. Since A_i is a basis for U_i, there exist scalars $\alpha_{i1}, \alpha_{i2}, \ldots,$ α_{ik_i} such that

$$Z_i = \alpha_{i1} A_{i1} + \alpha_{i2} A_{i2} + \cdots + \alpha_{ik_i} A_{ik_i} \qquad (1.5.20)$$

Substituting each of the l equations of (1.5.20) into equation (1.5.19), we have

$$0 = \alpha_{11} A_{11} + \alpha_{12} A_{12} + \cdots + \alpha_{1k_1} A_{1k_1} + \alpha_{21} A_{21} + \cdots + \alpha_{lk_l} A_{lk_l}$$

But we previously assumed that A is a linearly independent set. Therefore

$$\alpha_{11} = \alpha_{12} = \cdots = \alpha_{1k_1} = \alpha_{21} = \cdots = \alpha_{lk_l} = 0$$

from which it immediately follows that

$$Z_1 = Z_2 = \cdots = Z_l = 0$$

Consequently $X_i = Y_i$, $i = 1, 2, \ldots, l$, as required.

Conversely, let us now assume that the representation (1.5.16) is unique for every element X of U. In particular, it is unique for $X = 0$. We wish to show that A is a linearly independent set. Let

$$0 = \alpha_{11} A_{11} + \alpha_{12} A_{12} + \cdots + \alpha_{1k_1} A_{1k_1} + \alpha_{21} A_{21} + \cdots + \alpha_{lk_l} A_{lk_l}$$

and set

$$Z_i = \alpha_{i1} A_{i1} + \alpha_{i2} A_{i2} + \cdots + \alpha_{ik_i} A_{ik_i} \qquad i = 1, 2, \ldots, l$$

Then we have

$$0 = Z_1 + Z_2 + \cdots + Z_l$$

But certainly

$$0 = 0 + 0 + \cdots + 0$$

is one representation for 0 in the form of (1.5.16). Since we have assumed that the representation of 0 is unique, it follows that

$$Z_1 = Z_2 = \cdots = Z_l = 0$$

However, $A_i = \{A_{i1}, A_{i2}, \ldots, A_{ik_i}\}$, being a basis for U_i, is a linearly independent set. Consequently $Z_i = 0$ implies that $\alpha_{i1} = \alpha_{i2} = \cdots = \alpha_{ik_i} = 0$. We can now conclude that $\mathbb{A}$ is a linearly independent set. ////

EXERCISES

1 Let

(a) $A = \begin{bmatrix} 1 & 3 & -2 \\ 2 & 6 & -4 \\ 0 & 0 & 0 \end{bmatrix}$ (b) $A = \begin{bmatrix} 1 & 2 & -1 \\ 0 & 1 & 1 \\ 2 & 1 & 0 \end{bmatrix}$

(c) $A = \begin{bmatrix} 1 & 2 & -1 \\ 0 & 1 & 1 \end{bmatrix}$

Find bases for $\mathbb{R}(A)$, $\mathbb{C}(A)$, and $\mathbb{N}(A)$.

2 Find the row rank, the column rank, and the nullity of:

(a) $\begin{bmatrix} 1 & 2 & 0 & -1 & -1 & 1 \\ -1 & 0 & 1 & 1 & 0 & -2 \\ 0 & 2 & 1 & 0 & -1 & -1 \end{bmatrix}$

(b) $\begin{bmatrix} 1 & 2 & 0 & -1 & -1 & 1 \\ -1 & 0 & 1 & 1 & 0 & -2 \end{bmatrix}$

3 Extend the linearly independent set of vectors

$$\begin{bmatrix} 0 \\ 2 \\ -1 \\ 0 \end{bmatrix} \quad \begin{bmatrix} 0 \\ 1 \\ 1 \\ 0 \end{bmatrix}$$

to a basis for $\mathbb{F}^4$ using the method of the replacement theorem and the basis for $\mathbb{F}^4$ consisting of the unit column vectors.

4 Extend the basis for each $\mathbb{N}(A)$ of Exercise 1 to a basis for $\mathbb{F}^3$ using the method of the replacement theorem and the basis for $\mathbb{F}^3$ consisting of the unit column vectors.

5 Let

$$A = \begin{bmatrix} 2 & 1 \\ 4 & 2 \end{bmatrix} \quad B = \begin{bmatrix} 3 & 0 \\ 0 & 4 \end{bmatrix}$$

Find bases for each of the following vector spaces:
(a) $\{X \mid XA = AX\}$ (b) $\{X \mid XA = BX\}$ (c) $\{X \mid AX = XB\}$

6 Let

$$A = \begin{bmatrix} 1 & 1 & 0 \\ 0 & 0 & 0 \\ 0 & 1 & 0 \end{bmatrix} \qquad B = \begin{bmatrix} 1 & 0 & 1 \\ 0 & 2 & 1 \end{bmatrix}$$

Find bases for:
(a) $\mathbb{R}(A) \cap \mathbb{R}(B)$.
(b) $\mathbb{R}(A) + \mathbb{R}(B)$.   Is the sum direct?
(c) $\mathbb{N}(A) \cap \mathbb{N}(B)$.
(d) $\mathbb{N}(A) + \mathbb{N}(B)$.   Is the sum direct?
(e) $\mathbb{N}(B) + \mathbb{C}(A)$.   Is the sum direct?

7 Let $\mathbb{V}$ and $\mathbb{U}$ be the vector spaces of Exercises 2 and 4 of Sec. 1.3.   Find
a basis $\mathbb{S}$ for $\mathbb{V} \cap \mathbb{U}$.   Extend $\mathbb{S}$ to a basis for $\mathbb{V}$ using the method of the
replacement theorem and the basis $\{B_1, B_2\}$ for $\mathbb{V}$.   Extend $\mathbb{S}$ to a basis
for $\mathbb{U}$ using the basis $\{C_2, C_3\}$ for $\mathbb{U}$.

8 Let $\{X_1, X_2, X_3\}$ be a basis for the vector space $\mathbb{V}$.   Show that
$\{X_1 + X_2, X_2 + X_3, X_3\}$ is also a basis for $\mathbb{V}$.

9 Show that if $m < n$, then m simultaneous linear homogeneous equations
in n unknowns have a nonzero solution.   (*Hint:* Prove that the columns
of the matrix of coefficients are linearly dependent.)

10 Let $\mathbb{U}$ and $\mathbb{V}$ be vector spaces, $\mathbb{U} \subseteq \mathbb{V}$.   Show that there exists a subspace
$\mathbb{W}$ of $\mathbb{V}$ such that $\mathbb{V} = \mathbb{U} \oplus \mathbb{W}$.

11 Let $\mathbb{V}_1$, $\mathbb{V}_2$ be subspaces of $\mathbb{V}$ for which $\mathbb{V} = \mathbb{V}_1 + \mathbb{V}_2$.   Prove that
$\mathbb{V} = \mathbb{V}_1 \oplus \mathbb{V}_2$ if and only if $\mathbb{V}_1 \cap \mathbb{V}_2 = 0$.   (This is a very important
characterization of a direct sum.)

12 Let $\mathbb{V}_1$, $\mathbb{V}_2$ be vector spaces of $m \times n$ matrices.   Prove that

$$\dim(\mathbb{V}_1 + \mathbb{V}_2) = \dim \mathbb{V}_1 + \dim \mathbb{V}_2 - \dim(\mathbb{V}_1 \cap \mathbb{V}_2)$$

Show that Exercise 11 is a direct consequence of this formula.   (*Hint:*
Begin with a basis for $\mathbb{V}_1 \cap \mathbb{V}_2$.   Apply Exercise 10 with $\mathbb{U} = \mathbb{V}_1 \cap \mathbb{V}_2$,
$\mathbb{V} = \mathbb{V}_1$ and also with $\mathbb{U} = \mathbb{V}_1 \cap \mathbb{V}_2$, $\mathbb{V} = \mathbb{V}_2$.)

13 Let $\mathbb{V}_1$, $\mathbb{V}_2$ be subspaces of F_n^m.   Prove that

$$\dim(\mathbb{V}_1 \cap \mathbb{V}_2) \geqq \dim \mathbb{V}_1 + \dim \mathbb{V}_2 - mn$$

14 Let $\mathbb{V}_1$, $\mathbb{V}_2, \ldots, \mathbb{V}_l$ be subspaces of the vector space $\mathbb{V}$ such that
$\mathbb{V} = \mathbb{V}_1 + \mathbb{V}_2 + \cdots + \mathbb{V}_l$.   For each j, $j = 1, 2, \ldots, l$, set

$$\mathbb{W}_j = \mathbb{V}_1 + \mathbb{V}_2 + \cdots + \mathbb{V}_{j-1} + \mathbb{V}_{j+1} + \cdots + \mathbb{V}_l$$

that is, $\mathbb{W}_j$ is the sum of all of the $\mathbb{V}_i$'s except $\mathbb{V}_j$.   Show that

$V = V_1 \oplus V_2 \oplus \cdots \oplus V_l$ if and only if

$$W_j \cap V_j = 0 \qquad j = 1, 2, \ldots, l$$

Show that Exercise 11 is a special case of this result.

15   Let A be a square matrix of order n. Prove that there exists a positive integer p and $p + 1$ scalars $\alpha_p, \alpha_{p-1}, \ldots, \alpha_1, \alpha_0$ belonging to $\mathbb{F}$, not all zero, such that

$$\alpha_p A^p + \alpha_{p-1} A^{p-1} + \cdots + \alpha_1 A + \alpha_0 I = 0$$

Prove that p can always be chosen to be less than $n^2 + 1$.

16   Let A and B be $m \times n$ and $p \times q$ matrices, respectively, and denote by C the $(m + p) \times (n + q)$ matrix

$$C = \begin{bmatrix} A & 0 \\ 0 & B \end{bmatrix}$$

Prove that

(a) $\mathbb{R}(C) = \mathbb{R}([A \quad 0]) \oplus \mathbb{R}([0 \quad B])$.

(b) $\mathbb{N}(C) = \mathbb{N}([A \quad 0]) \cap \mathbb{N}([0 \quad B])$.

(c) $\mathbb{C}(C) = \mathbb{C}\left(\begin{bmatrix} A \\ 0 \end{bmatrix}\right) \oplus \mathbb{C}\left(\begin{bmatrix} 0 \\ B \end{bmatrix}\right)$.

Use these results to give relations between the row ranks of A, B, C, between the column ranks of A, B, C, and between the nullities of A, B, C.

17   Let V_1, V_2, and V be subspaces of $\mathbb{F}^n$ such that $V = V_1 \oplus V_2$ and let A be an $m \times n$ matrix.   Show that if $\mathbb{N}(A) \cap V = 0$, then

$$AV = AV_1 \oplus AV_2$$

18   Let V_1 and V_2 be subspaces of $\mathbb{F}^n$ and let A be an $m \times n$ matrix.   Prove:

(a) $A(V_1 + V_2) = AV_1 + AV_2$ (b) $A(V_1 \cap V_2) \subseteq AV_1 \cap AV_2$

Give an example to show that we do not always have equality in (b). Further, prove that if $\mathbb{N}(A) \cap (V_1 + V_2) = 0$, then

$$A(V_1 \cap V_2) = AV_1 \cap AV_2$$

19   Let V be a subspace of $\mathbb{F}^n$ and let A be an $m \times n$ matrix.   Prove that

$$\dim AV = \dim V - \dim[\mathbb{N}(A) \cap V]$$

[*Hint:* Extend a basis for $\mathbb{N}(A) \cap V$ to a basis for V.   Multiply the elements of this basis by A.]

20   Let V_1, V_2, and V be subspaces of $\mathbb{F}^n$ such that $V = V_1 \oplus V_2$ and let A be an $m \times n$ matrix.   Prove that $AV = AV_1 \oplus AV_2$ if and only if

$$\mathbb{N}(A) \cap V = [\mathbb{N}(A) \cap V_1] \oplus [\mathbb{N}(A) \cap V_2]$$

Show that Exercise 17 follows directly from this result. (*Hint:* It is always true that $N(A) \cap V \supseteq [N(A) \cap V_1] \oplus [N(A) \cap V_2]$. Thus we are concerned only with the inclusion in the opposite direction.)

1.6 NONSINGULAR MATRICES

We have associated three vector spaces with a given matrix A, namely, the column space $C(A)$, the row space $R(A)$, and the column null space $N(A)$. [The row null space $N_R(A)$ has also been introduced, but it does not play a prominent role in our treatment of the subject.]   There are two relations between the dimensions of these vector spaces which are basic to all the work that follows. First, we will show, in Theorem 1.6.1, that for any matrix the sum of the column rank and the nullity is equal to the number of columns.   The second relation which we will prove between the dimensions of these vector spaces is contained in Theorem 1.7.11 and states that for any matrix the row rank and the column rank are equal.   As a result it will then not be necessary to distinguish the row rank from the column rank, and we will speak simply of the rank of a matrix.

For an $m \times n$ matrix the dimensions of the vectors belonging to the column space and the null space are generally different; that is, $C(A) \subseteq F^m$, $N(A) \subseteq F^n$. Thus, unless $m = n$, it is impossible to speak of the sum of these vector spaces. However, the dimensions of these vector spaces can be added.

Theorem 1.6.1   For any $m \times n$ matrix A

$$r_C(A) + v(A) = n$$

PROOF   We begin by choosing any basis S_1 for $N(A)$.   According to the definition of the nullity, S_1 contains $v(A)$ elements, say

$$S_1 = \{X_1, X_2, \ldots, X_{v(A)}\}$$

The elements of S_1 are n-dimensional column vectors.   By Theorem 1.5.1, S_1 can be extended to a basis for F^n; that is, there exist $n - v(A)$ vectors in F^n, say $X_{v(A)+1}, X_{v(A)+2}, \ldots, X_n$ such that

$$S = \{X_1, X_2, \ldots, X_{v(A)}, X_{v(A)+1}, X_{v(A)+2}, \ldots, X_n\}$$

is a basis for F^n.

In order to prove that $r_C(A) = n - v(A)$, we will show that the $n - v(A)$ m-dimensional column vectors $AX_{v(A)+1}, AX_{v(A)+2}, \ldots, AX_n$ form a basis for $C(A) = AF^n \subseteq F^m$.   Set

$$S_2 = \{X_{v(A)+1}, X_{v(A)+2}, \ldots, X_n\}$$

so that $S = S_1 \cup S_2$. Since S is a basis for F^n, it follows that AF^n is spanned by

$$AS = \{AX_1, AX_2, \ldots, AX_{v(A)}, AX_{v(A)+1}, AX_{v(A)+2}, \ldots, AX_n\}$$

But

$$AX_1 = AX_2 = \cdots = AX_{v(A)} = 0$$

because $X_1, X_2, \ldots, X_{v(A)}$ belong to $N(A)$. Thus

$$AS_2 = \{AX_{v(A)+1}, AX_{v(A)+2}, \ldots, AX_n\}$$

spans AF^n and contains $n - v(A)$ elements. It remains only to show that AS_2 is a linearly independent set.

Denote the vector space which has S_2 as a basis by V. Then it is easily seen that

$$F^n = N(A) \oplus V$$

and consequently (see Exercise 11 of Sec. 1.5)

$$N(A) \cap V = 0$$

Now let $\alpha_{v(A)+1}, \alpha_{v(A)+2}, \ldots, \alpha_n$ be scalars for which

$$\alpha_{v(A)+1}AX_{v(A)+1} + \alpha_{v(A)+2}AX_{v(A)+2} + \cdots + \alpha_n AX_n = 0$$

and set $\quad Y = \alpha_{v(A)+1}X_{v(A)+1} + \alpha_{v(A)+2}X_{v(A)+2} + \cdots + \alpha_n X_n$

Then $\quad AY = \alpha_{v(A)+1}AX_{v(A)+1} + \alpha_{v(A)+2}AX_{v(A)+2} + \cdots + \alpha_n AX_n = 0$

and consequently $\quad\quad\quad\quad\quad\quad Y \in N(A) \cap V = 0$

But $X_{v(A)+1}, X_{v(A)+2}, \ldots, X_n$ are linearly dependent, and therefore

$$\alpha_{v(A)+1} = \alpha_{v(A)+2} = \cdots = \alpha_n = 0$$

as required. $\quad\quad\quad\quad\quad\quad\quad\quad\quad\quad\quad\quad\quad\quad\quad$ ////

There is a theorem similar to Theorem 1.6.1 connecting the row rank, the row nullity, and the number of rows of a matrix, the statement and proof of which are left as an exercise.

In Sec. 1.7 we will prove that the row rank and the column rank of any matrix are equal. As a preliminary we prove a lemma which is, in fact, a special case of the general theorem. As we will see, the machinery needed to prove the general theorem depends heavily on this lemma.

Lemma 1.6.2 Let A be a square matrix. The rows of A are linearly independent if and only if the columns of A are linearly independent.

PROOF It is sufficient to show that when the rows of A are linearly independent, the columns of A are also linearly independent. Hence, assume that $A_{(1)}, A_{(2)}, \ldots, A_{(n)}$ are linearly independent and let $\alpha_1, \alpha_2, \ldots, \alpha_n$ be numbers for which

$$\alpha_1 A^{(1)} + \alpha_2 A^{(2)} + \cdots + \alpha_n A^{(n)} = 0 \qquad (1.6.3)$$

If we denote the vector

$$\begin{bmatrix} \alpha_1 \\ \alpha_2 \\ \vdots \\ \alpha_n \end{bmatrix}$$

by X, then equation (1.6.3) can be written as

$$AX = 0$$

According to Corollary 1.5.4, $A_{(1)}, A_{(2)}, \ldots, A_{(n)}$ form a basis for $\mathbb{F}_n$, and it follows from part 1 of Corollary 1.4.7 that there exists a square matrix of order n, say C, such that $CA = I$. Then

$$X = I \cdot X = CAX = C \cdot 0 = 0$$

and we have shown that $A^{(1)}, A^{(2)}, \ldots, A^{(n)}$ are linearly independent, as required.   ////

Square matrices with the property that their columns (and therefore, their rows) are linearly independent will play a special role in studying general rectangular matrices. Let us gather together some of the characterizations for such matrices which we have established up to this point. In later sections we will find still other characterizations for these matrices and will add them to the list.

Theorem 1.6.4 For a square matrix A of order n the following conditions are equivalent:

1 The columns of A are linearly independent.
2 The rows of A are linearly independent.
3 $C(A) = \mathbb{F}^n$.
4 $R(A) = \mathbb{F}_n$.
5 $N(A) = 0$.
6 There exists a square matrix B of order n such that

$$AB = BA = I \qquad (1.6.5)$$

PROOF

$$
\begin{array}{ll}
1 \Leftrightarrow 2 & \text{Lemma 1.6.2} \\
1 \Leftrightarrow 3 & \text{Corollary 1.5.4} \\
2 \Leftrightarrow 4 & \text{Corollary 1.5.4} \\
3 \Leftrightarrow 5 & \text{Theorem 1.6.1} \\
6 \Rightarrow 3 & \text{Corollary 1.4.7}
\end{array}
$$

It remains only to prove that $3 \Rightarrow 6$. By the two parts of Corollary 1.4.7, there exist matrices B and C such that $AB = CA = I$. Then

$$C = CI = CAB = IB = B \qquad ////$$

In proving that $3 \Rightarrow 6$ above, we showed that two matrices, B and C, which satisfy the equations

$$AB = I \qquad CA = I$$

must be equal. In particular, this implies that for a given matrix A, there is at most one solution to (1.6.5) [see part (e) of Exercises 3 and 4 of Sec. 1.2].

Definition 1.6.6 A square matrix satisfying any one (and hence all) of the six conditions of Theorem 1.6.4 is said to be *nonsingular*. For a nonsingular matrix A, the (necessarily unique) solution B of

$$AB = BA = I$$

is called the *inverse* of A and is denoted by A^{-1}.

Note that Definition 1.6.6 does not deal at all with matrices which are not square. A matrix which is not nonsingular (and this includes all nonsquare matrices) is called *singular*.

Nonsingular matrices are closely analogous to nonzero numbers, just as the zero matrix is closely analogous to the number zero. We shall see that singular matrices occupy a position somewhere between the two and might be thought of as "semizero" matrices. For example, although we have given examples to demonstrate that the cancellation law for multiplication of matrices is not valid in all cases, e.g., Example 1.2.5, it is valid when the matrix to be canceled is nonsingular. For if $AB = AC$ and A is nonsingular, multiplying on the left by A^{-1}, we have

$$B = IB = A^{-1}AB = A^{-1}AC = IC = C$$

As an important special case, note that if A is nonsingular and $AB = 0 \ (= A0)$, then $B = 0$.

But we must be careful.   Since multiplication of matrices is generally not commutative, it does not follow from $AB = CA$ that $B = C$, even when A is nonsingular.   In this case we shall say that B and C are *similar*.   In Sec. 1.10 this situation is studied in great detail.

Nonsingular matrices have another very important property:

Theorem 1.6.7   Let A be an $m \times n$ matrix and let P and Q be nonsingular matrices of orders m and n respectively.   Then A and PAQ have the same row rank and the same column rank.

PROOF   It will be sufficient for us to prove that multiplying a matrix on the left by a nonsingular matrix does not change either the row rank or the column rank; that is, PA and A have the same row rank and the same column rank.   In exactly the same way it can be shown that multiplication on the right by a nonsingular matrix does not affect either the row rank or the column rank.   We can then conclude that PAQ and PA (and hence A) have the same row rank and the same column rank.

Applying part 1 of Lemma 1.4.2 twice, we have

$$\mathbb{R}(PA) \subseteq \mathbb{R}(A)$$

and
$$\mathbb{R}(A) = \mathbb{R}(IA) = \mathbb{R}(P^{-1}PA) \subseteq \mathbb{R}(PA)$$

Thus $\mathbb{R}(A) = \mathbb{R}(PA)$ and consequently A and PA have the same row rank.

Now let us prove that A and PA have the same column rank.   Since P is a square matrix of order m, PA and A have the same dimensions.   Thus, making use of Theorem 1.6.1, in order to show that PA and A have the same column rank it is sufficient to show that they have the same nullity.   Clearly

$$\mathbb{N}(A) \subseteq \mathbb{N}(PA)$$

since if $AX = 0$, then $PAX = 0$.   In the same way,

$$\mathbb{N}(PA) \subseteq \mathbb{N}(P^{-1}PA) = \mathbb{N}(A)$$

Combining these relations, we have

$$\mathbb{N}(A) = \mathbb{N}(PA)$$

and consequently PA and A have the same nullity.   ////

The proof of Theorem 1.6.7 provides a typical example of the special properties of nonsingular matrices.   The existence of the inverse gives us much more freedom in moving matrices from one side of an equation to the other.

In the next section we will prove that the row rank and the column rank of any matrix are equal. The proof consists of little more than repeated applications of Theorem 1.6.7. Beginning with an arbitrary $m \times n$ matrix, we multiply it on the left and right by carefully chosen nonsingular matrices. This, of course, does not change either the row rank or the column rank. Finally, we arrive at an $m \times n$ matrix which has a particularly simple form and in which the row rank and column rank are equal.

EXERCISES

1 Show that the matrix

$$A = \begin{bmatrix} 1 & 0 & 1 \\ 1 & 1 & 0 \\ 1 & 1 & 1 \end{bmatrix}$$

is nonsingular by showing that:
(a) The rows of A form a basis for $\mathbb{F}_3$,
(b) The columns of A form a basis for $\mathbb{F}^3$,
(c) $N(A) = 0$.

2 Find the inverse of

$$A = \begin{bmatrix} 1 & 3 \\ -1 & 0 \end{bmatrix}$$

by solving for B in the equation:

(a) $AB = I_2$ (b) $BA = I_2$

3 Repeat Exercise 2 using the matrix of Exercise 1 and replacing I_2 by I_3.
4 Verify Theorem 1.6.1 for the matrices:

(a) $\begin{bmatrix} 1 & 2 & 1 & 1 & 0 & 1 \\ 2 & 0 & 1 & 1 & 0 & -1 \end{bmatrix}$ (b) $\begin{bmatrix} 1 & 2 \\ 2 & 0 \\ 1 & 1 \\ 1 & 1 \\ 0 & 0 \\ 1 & -1 \end{bmatrix}$

5 Let

$$A = \begin{bmatrix} 1 & 1 & 2 \\ 0 & -1 & 0 \end{bmatrix}$$

Show that $AB = I_2$ has a solution but that $CA = I_3$ does not. Is B unique?

6 Find the inverse of the matrix of Exercise 6 of Sec. 1.4.   For which values of x does the inverse exist?

7 Prove that the $(m + n) \times (m + n)$ matrix

$$\begin{bmatrix} I_n & 0 \\ A & I_m \end{bmatrix}$$

is nonsingular for every $m \times n$ matrix A.   Find the inverse.

8 Prove that the $n \times n$ matrix

$$\begin{bmatrix} 1 & 0 & 0 & \cdots & 0 \\ a_{21} & 1 & 0 & \cdots & 0 \\ a_{31} & a_{32} & 1 & \cdots & 0 \\ \multicolumn{5}{c}{\cdots\cdots\cdots\cdots\cdots\cdots} \\ a_{n1} & a_{n2} & a_{n3} & \cdots & 1 \end{bmatrix}$$

is nonsingular for every choice of $a_{21}, a_{31}, a_{32}, \ldots, a_{n1}, a_{n2}, \ldots$.

9 State and prove the row analog of Theorem 1.6.1 and apply it to the two matrices of Exercise 4.

10 Show that if A is nonsingular, then A^{-1} is also nonsingular and its inverse is given by

$$(A^{-1})^{-1} = A$$

11 Prove that if A and B are nonsingular matrices of the same order, then AB is nonsingular and its inverse is given by

$$(AB)^{-1} = B^{-1}A^{-1}$$

12 Show that if A and B are square matrices of the same order and either A or B is singular, then AB is singular.

13 Show that if A and B are $m \times n$ and $n \times m$ matrices and if either A or B is singular, then at least one of AB and BA is singular.   Find an example to show that one of AB and BA may be nonsingular.

14 Let A be an $m \times n$ matrix $(m < n)$ whose rows are linearly independent. Prove that there exists an $(n - m) \times n$ matrix B such that the $n \times n$ matrix

$$\begin{bmatrix} A \\ B \end{bmatrix}$$

is nonsingular.

15 Show that if A and B are square matrices and $AB = I$, then $BA = I$.

16 Let A be an $m \times n$ matrix and let P and Q be nonsingular matrices of order m and n respectively.   Prove that

$$\mathbb{N}(PAQ) = Q^{-1}\mathbb{N}(A)$$

Use this relation, Theorem 1.6.1, and Exercise 9 to give a different proof of Theorem 1.6.7.

17 Let A be a square matrix of order n. Prove that

$$\mathbb{F}^n = \mathbb{N}(A) \oplus \mathbb{C}(A)$$

if and only if $\mathbb{N}(A^2) = \mathbb{N}(A)$. (*Hint:* Use Exercise 11 of Sec. 1.5.)

18 Let A be a square matrix such that $A^2 = A$. Prove that $A = I$ or A is singular.

19 Let A be a square matrix of order n. Prove that a necessary and sufficient condition for there to exist numbers $\alpha_p, \alpha_{p-1}, \ldots, \alpha_1, \alpha_0$ belonging to $\mathbb{F}$ for which $\alpha_p A^2 + \alpha_{p-1} A^{p-1} + \cdots + \alpha_1 A + \alpha_0 I = 0$ and $\alpha_0 \neq 0$ is that A be nonsingular. (See Exercise 15 of Sec. 1.5.)

1.7   THE ELEMENTARY OPERATIONS

When a matrix is multiplied on either side by a nonsingular matrix, neither its row rank nor its column rank is changed.   This section is devoted to showing how a given matrix can be transformed by a sequence of multiplications by non-singular matrices into a matrix having a particularly simple form (1.7.9).   It will be clear that the row rank and the column rank of the resulting matrix, and hence of the original matrix, are equal.   This sequence of multiplications can best be described in terms of *operations* on the rows and columns of the matrix.   We will also show how each of these operations can be carried out by means of a mul-tiplication by a properly chosen nonsingular matrix.

In this section we define two sets of operations on a matrix.   One set consists of three column operations, and the other consists of the corresponding three row operations.   These six operations will be called the *elementary opera-tions*.   We will also determine the nonsingular matrices which effect these opera-tions.   They will be called the *elementary matrices*.

The First Elementary Operation

Let A and B be $m \times n$ matrices and let r and s be distinct integers, $1 \leq r < s \leq m$. Further suppose that the rows of A and B are identical except for their rth and sth rows.   Specifically, the rth row of A is identical with the sth row of B, and the sth row of A is identical with the rth row of B; that is,

$$B_{(i)} = A_{(i)} \qquad i \neq r, s$$
$$B_{(r)} = A_{(s)}$$
$$B_{(s)} = A_{(r)}$$

We say that B has been obtained from A by the operation of interchanging the *r*th and *s*th rows of A.   We denote the operation of interchanging the *r*th and *s*th rows of a matrix by $\mathscr{R}_{rs}$ and write

$$\mathscr{R}_{rs} : A \rightarrow B$$

$\mathscr{R}_{rs}$ is called the *elementary row operation of the first kind*.   Clearly we also have

$$\mathscr{R}_{rs} : B \rightarrow A$$

and thus the operation $\mathscr{R}_{rs}$ when performed twice on a matrix leaves it unchanged.

We have said that the elementary operations can be carried out by multiplying the given matrix by properly chosen nonsingular matrices.   Let us construct a nonsingular matrix corresponding to the operation $\mathscr{R}_{rs}$.   For distinct integers r and s, $1 \leqq r < s \leqq m$, we define the square matrix of order m, which we call R_{rs}, by

that is, R_{rs} has 1 in all the diagonal positions except the (r, r) and (s, s), positions, 1 in the (r, s) and (s, r) positions, and 0 elsewhere.   Note that

$$\mathscr{R}_{rs} : I \rightarrow R_{rs}$$

R_{rs} is called the *elementary row matrix of the first kind*, corresponding to the elementary row operation $\mathscr{R}_{rs}$ In order to show that this definition is justified we must prove that the operation $\mathscr{R}_{rs}$ can be carried out by means of a multiplication on the left by R_{rs}, that is:

Lemma 1.7.1   Let $\mathscr{R}_{rs}$ be an elementary row operation of the first kind and let R_{rs} be the corresponding elementary row matrix of the first kind of order m.   Then

$$\mathscr{R}_{rs} : A \rightarrow R_{rs} A$$

for every $m \times n$ matrix A.

PROOF   The proof consists of comparing the entries of the matrix product $R_{rs}A$ with the result obtained by performing the operation $\mathscr{R}_{rs}$ on A. The details are left as an exercise.   ////

EXAMPLE 1.7.2   Let

$$A = \begin{bmatrix} 1 & 2 \\ 3 & 4 \\ 5 & 6 \end{bmatrix} \qquad \begin{matrix} r = 2 \\ s = 3 \end{matrix}$$

Then

$$R_{23} = \begin{bmatrix} 1 & 0 & 0 \\ 0 & 0 & 1 \\ 0 & 1 & 0 \end{bmatrix}$$

and it can easily be verified that

$$\mathscr{R}_{23}: \quad \begin{bmatrix} 1 & 2 \\ 3 & 4 \\ 5 & 6 \end{bmatrix} \to \begin{bmatrix} 1 & 2 \\ 5 & 6 \\ 2 & 3 \end{bmatrix} = \begin{bmatrix} 1 & 0 & 0 \\ 0 & 0 & 1 \\ 0 & 1 & 0 \end{bmatrix} \begin{bmatrix} 1 & 2 \\ 3 & 4 \\ 5 & 6 \end{bmatrix} \qquad ////$$

The Second Elementary Operation

Let A and B be $m \times n$ matrices, r an integer, $1 \leq r \leq m$, and λ a nonzero number. Suppose that the rows of A and B are related as follows:

$$B_{(i)} = A_{(i)} \qquad i \neq r$$
$$B_{(r)} = \lambda A_{(r)}$$

B has been obtained from A by the operation of multiplying all the entries of the rth row of A by λ.   We denote this operation by $\mathscr{R}_r(\lambda)$ and write

$$\mathscr{R}_r(\lambda) : A \to B$$

$\mathscr{R}_r(\lambda)$ is called the *elementary row operation of the second kind*.

Let us construct a nonsingular matrix corresponding to the operation $\mathscr{R}_r(\lambda)$.   For an integer r, $1 \leq r \leq m$, and a nonzero number λ, we define the square matrix of order m, which we call $R_r(\lambda)$, by

$$R_r(\lambda) = \begin{bmatrix} 1 & & & \vdots & & \\ & \ddots & & \vdots & & \\ & & 1 & \vdots & & \\ \cdots & \cdots & \cdots & \lambda & \cdots & \cdots \\ & & & \vdots & 1 & \\ & & & \vdots & & \ddots \\ & & & \vdots & & & 1 \end{bmatrix} \begin{matrix} \\ \\ \\ r \\ \\ \\ \end{matrix}$$

that is, $R_r(\lambda)$ has 1 in all the diagonal positions except the (r, r) position, λ in the (r, r) position, and 0 elsewhere.   Note that

$$\mathcal{R}_r(\lambda) : I \to R_r(\lambda)$$

$R_r(\lambda)$ is the *elementary row matrix of the second kind*, corresponding to the elementary row operation $\mathcal{R}_r(\lambda)$.   It is important to remember that λ must be a nonzero number.

The row operation $\mathcal{R}_r(\lambda)$ can be carried out by means of a multiplication on the left by $R_r(\lambda)$, that is:

Lemma 1.7.3   Let $\mathcal{R}_r(\lambda)$ be an elementary row operation of the second kind and let $R_r(\lambda)$ be the corresponding elementary row matrix of the second kind of order m.   Then

$$\mathcal{R}_r(\lambda) : A \to R_r(\lambda)A$$

for every $m \times n$ matrix A.

PROOF   The proof is left as an exercise.   ////

EXAMPLE 1.7.4   Let

$$A = \begin{bmatrix} 1 & 2 \\ 3 & 4 \\ 5 & 6 \end{bmatrix} \qquad \begin{matrix} r = 2 \\ \lambda = -7 \end{matrix}$$

Then

$$R_2(-7) = \begin{bmatrix} 1 & 0 & 0 \\ 0 & -7 & 0 \\ 0 & 0 & 1 \end{bmatrix}$$

and it can easily be verified that

$$\mathcal{R}_2(-7): \quad \begin{bmatrix} 1 & 2 \\ 3 & 4 \\ 5 & 6 \end{bmatrix} \to \begin{bmatrix} 1 & 2 \\ -21 & -28 \\ 5 & 6 \end{bmatrix} = \begin{bmatrix} 1 & 0 & 0 \\ 0 & -7 & 0 \\ 0 & 0 & 1 \end{bmatrix}\begin{bmatrix} 1 & 2 \\ 3 & 4 \\ 5 & 6 \end{bmatrix} \qquad ////$$

The Third Elementary Operation

Let A and B be $m \times n$ matrices, r and s distinct integers, $1 \le r, s \le m$, and α any number.   Suppose that the rows of A and B are related as follows:

$$B_{(i)} = A_{(i)} \qquad i \ne r$$
$$B_{(r)} = A_{(r)} + \alpha A_{(s)}$$

Note that we require r and s to be distinct. B has been obtained from A by the operation of adding the sth row of A, multiplied by α, to the rth row of A. We denote this operation by $\mathscr{R}_{rs}(\alpha)$ and write

$$\mathscr{R}_{rs}(\alpha) : A \to B$$

$\mathscr{R}_{rs}(\alpha)$ is called the *elementary row operation of the third kind.*

For distinct integers r and s, $1 \leq r, s \leq m$, and a number α we define the square matrix of order m, which we call $R_{rs}(\alpha)$, by

$$
R_{rs}(\alpha) =
\begin{bmatrix}
1 & & & & \vdots & & \\
 & \ddots & & & \vdots & & \\
 & & 1 & & \vdots & & \\
\cdots\cdots & & 1 & \cdots & \alpha & \cdots\cdots & \Big| r \\
 & & & \ddots & \vdots & & \\
 & & & & 1 & & \\
 & & & & \vdots & \ddots & \\
 & & & & \vdots & & 1
\end{bmatrix}
$$

that is, $R_{rs}(\alpha)$ has 1 in all the diagonal positions, α in the (r, s) position, and 0 elsewhere. Note that

$$\mathscr{R}_{rs}(\alpha) : I \to R_{rs}(\alpha)$$

$R_{rs}(\alpha)$ is the *elementary row matrix of the third kind*, corresponding to the elementary row operation $\mathscr{R}_{rs}(\alpha)$.

The row operation $\mathscr{R}_{rs}(\alpha)$ can be carried out by means of a multiplication on the left by $R_{rs}(\alpha)$, that is:

Lemma 1.7.5   Let $\mathscr{R}_{rs}(\alpha)$ be an elementary row operation of the third kind and let $R_{rs}(\alpha)$ be the corresponding elementary row matrix of the third kind of order m.   Then

$$\mathscr{R}_{rs}(\alpha) : A \to R_{rs}(\alpha)A$$

for every $m \times n$ matrix A.

PROOF.   The proof is left as an exercise.   ////

EXAMPLE 1.7.6   Let

$$
A = \begin{bmatrix} 1 & 2 \\ 3 & 4 \\ 5 & 6 \end{bmatrix}
\qquad
\begin{array}{l} r = 2 \\ s = 3 \\ \alpha = -2 \end{array}
$$

Then
$$R_{23}(-2) = \begin{bmatrix} 1 & 0 & 0 \\ 0 & 1 & -2 \\ 0 & 0 & 1 \end{bmatrix}$$

and it can easily be verified that

$$\mathscr{R}_{23}(-2): \quad \begin{bmatrix} 1 & 2 \\ 3 & 4 \\ 5 & 6 \end{bmatrix} \rightarrow \begin{bmatrix} 1 & 2 \\ -7 & -8 \\ 5 & 6 \end{bmatrix} = \begin{bmatrix} 1 & 0 & 0 \\ 0 & 1 & -2 \\ 0 & 0 & 1 \end{bmatrix} \begin{bmatrix} 1 & 2 \\ 3 & 4 \\ 5 & 6 \end{bmatrix} \qquad ////$$

We have now defined and illustrated the three elementary row operations $\mathscr{R}_{rs}$, $\mathscr{R}_r(\lambda)$, and $\mathscr{R}_{rs}(\alpha)$. The second set of three operations on a matrix consists of the *elementary column operations*:

1 $\mathscr{C}_{rs}$, interchange the rth and sth columns of the matrix
2 $\mathscr{C}_r(\lambda)$, multiply all the entries of the rth column of the matrix by λ
3 $\mathscr{C}_{rs}(\alpha)$, add the sth column of the matrix, multiplied by α, to the rth column

To these operations there correspond the elementary column matrices, C_{rs}, $C_r(\lambda)$, and $C_{rs}(\alpha)$. It is left to the reader to construct these matrices and show that the elementary column operations can be carried out by multiplying the matrix on the right by the corresponding elementary matrices.

When the elementary operations and matrices were introduced, it was claimed that any matrix could be reduced to a simple form by means of them and without changing either the row rank or the column rank of the matrix. As we will see, this process depends on Theorem 1.6.7 and hence we must first show that:

Lemma 1.7.7   The elementary matrices are nonsingular.

PROOF.   It is sufficient to show that each elementary matrix has an inverse.

$$[R_{rs}]^{-1} = R_{rs} \qquad [C_{rs}]^{-1} = C_{rs}$$

$$[R_r(\lambda)]^{-1} = R_r\left(\frac{1}{\lambda}\right)^* \qquad [C_r(\lambda)]^{-1} = C_r\left(\frac{1}{\lambda}\right)$$

$$[R_{rs}(\alpha)]^{-1} = R_{rs}(-\alpha) \qquad [C_{rs}(\alpha)]^{-1} = C_{rs}(-\alpha) \qquad ////$$

Thus an elementary operation performed on a matrix changes neither the row rank nor the column rank. Let us now demonstrate how a given matrix can be reduced to one having a simpler form, namely, (1.7.9), by means of elementary operations.

* It is for this reason that we insist that $\lambda \neq 0$.

Theorem 1.7.8   Let A be an $m \times n$ matrix.   There is a finite sequence of elementary row and column operations which transforms A into the $m \times n$ matrix.[1]

$$\begin{bmatrix} I & 0 \\ 0 & 0 \end{bmatrix} \qquad (1.7.9)$$

PROOF   The proof consists of describing the process of reducing A to the desired form.   Since we would quickly run out of letters if we changed the notation for the matrix we are operating on after each operation, it will be convenient to refer to it as A throughout the reduction process.

If $A = 0$, then A already has the form (1.7.9) and there is nothing to prove.   Henceforth we assume $A \neq 0$.   First we wish to obtain a nonzero number in the $(1, 1)$ position of A.   If a_{11} is already nonzero, this step is unnecessary.   Since $A \neq 0$, the entry in some position of A, say the entry in the (r, s) position, is not zero.   This number can be brought into the $(1, 1)$ position by at most two elementary operations of the first kind.   If $r \neq 1$, we interchange the first and rth rows of A.   If $s \neq 1$, we interchange the first and sth columns of A.   In this way we arrive at a matrix (still called A) for which $a_{11} \neq 0$.   Now multiply the entries of the first row of A by $1/a_{11}$.   This is an elementary row operation of the second kind, and in the resulting matrix A we have $a_{11} = 1$.

By at most $m + n - 2$ elementary operations of the third kind we can obtain zeros in all positions of the first row and first column of A except for the $(1, 1)$ position and thereby reduce A to a matrix of the form

$$A = \begin{bmatrix} 1 & 0 \\ 0 & \tilde{A} \end{bmatrix} \qquad (1.7.10)$$

where $\tilde{A}$ is an $(m - 1) \times (n - 1)$ matrix.   This is accomplished in the following way:

1   For $i = 2, 3, \ldots, m$, add the first row of A, multiplied by $-a_{i1}$, to the ith row of A,

2   For $j = 2, 3, \ldots, n$, add the first column of A, multiplied by $-a_{1j}$, to the jth column of A.

The resulting matrix has the form (1.7.10).

If $\tilde{A} = 0$, we are finished.   If $\tilde{A} \neq 0$, we repeat the entire process.

---

[1] If the row rank of A is equal to the number of rows of A, the row of zero matrices of (1.7.9) does not actually appear.   Similarly, if the column rank of A is equal to the number of columns of A, the column of zero matrices of (1.7.9) does not actually appear.

All the elementary operations on the matrix A of (1.7.10) needed to reduce the submatrix $\tilde{A}$ can easily be chosen in such a way that at no time are the first row and first column of A disturbed. In this way we reduce A to the form

$$A = \begin{bmatrix} 1 & 0 & 0 \\ 0 & 1 & 0 \\ 0 & 0 & \tilde{A} \end{bmatrix}$$

where $\tilde{A}$ is now an $(m-2) \times (n-2)$ matrix.

Continuing this process, after a finite number of elementary operations we arrive at the required form

$$\begin{bmatrix} I & 0 \\ 0 & 0 \end{bmatrix} \qquad ////$$

Both the row rank and the column rank of a matrix of the form of (1.7.9) are equal to the order of I. Applying Theorem 1.6.7, we have now proved:

Theorem 1.7.11 The row rank and the column rank of a matrix are equal.

It is no longer necessary to distinguish between the row rank and the column rank of a matrix, and henceforth we use the term *rank* to refer to either one. The rank of a matrix A will be denoted by $r(A)$.

When the elementary row and column operations are carried out by means of elementary row and column matrices, Theorem 1.7.8 becomes:

Corollary 1.7.12 Let A be an $m \times n$ matrix of rank r. There exist elementary matrices $R_1, R_2, \ldots, R_k$ and $C_1, C_2, \ldots, C_l$ which transform A into the form (1.7.9), that is,

$$R_k \cdots R_2 R_1 A C_1 C_2 \cdots C_l = \begin{bmatrix} I_r & 0 \\ 0 & 0 \end{bmatrix} \qquad (1.7.13)$$

Let us multiply both sides of equation (1.7.13) on the left by $R_1^{-1} R_2^{-1} \cdots R_k^{-1}$ and on the right by $C_l^{-1} \cdots C_2^{-1} C_1^{-1}$. Then

$$A = R_1^{-1} R_2^{-1} \cdots R_k^{-1} \begin{bmatrix} I_r & 0 \\ 0 & 0 \end{bmatrix} C_l^{-1} \cdots C_2^{-1} C_1^{-1}$$

However, we have shown that the inverse of an elementary matrix is again an elementary matrix, and so we have proved:

Corollary 1.7.14 Let A be an $m \times n$ matrix of rank r. There exist elementary matrices $R_1, R_2, \ldots, R_k$ and $C_1, C_2, \ldots, C_l$ such that

$$A = R_1 R_2 \cdots R_k \begin{bmatrix} I_r & 0 \\ 0 & 0 \end{bmatrix} C_l \cdots C_2 C_1 \qquad (1.7.15)$$

Note that although we use the same notation in both cases, the elementary matrices of equation (1.7.15) are the inverses of the elementary matrices of equation (1.7.13) and are in the opposite order. It is in the form of equation (1.7.15) that the reduction of a matrix by elementary matrices is most often used.

Theorem 1.7.11 is fundamental to all of matrix theory. The rank of a matrix is one of its most important attributes. In Sec. 1.9 we will see another way in which the rank can be defined. Now, however, let us show how Theorem 1.7.11 leads to a proof of the converse of Corollary 1.4.6. If B and C are $m \times n$ and $p \times n$ matrices respectively, then $\mathbb{R}(B)$ and $\mathbb{R}(C)$ are both subspaces of $\mathbb{F}_n$ while $\mathbb{N}(B)$ and $\mathbb{N}(C)$ are subspaces of $\mathbb{F}^n$. In Corollary 1.4.6 we showed that if $\mathbb{R}(C) \subseteq \mathbb{R}(B)$, then $\mathbb{N}(C) \supseteq \mathbb{N}(B)$. It follows from Theorem 1.6.1 (without Theorem 1.7.11) that if $\mathbb{N}(C) \supseteq \mathbb{N}(B)$, then $r_C(C) \leq r_C(B)$. With Theorem 1.7.11 we have:

Theorem 1.7.16 Let B and C be $m \times n$ and $p \times n$ matrices respectively. If $\mathbb{N}(C) \supseteq \mathbb{N}(B)$, then $\mathbb{R}(C) \subseteq \mathbb{R}(B)$. Hence, if $\mathbb{N}(C) = \mathbb{N}(B)$, then $\mathbb{R}(C) = \mathbb{R}(B)$.

PROOF Let $\mathbb{N}(C) \supseteq \mathbb{N}(B)$ and set

$$D = \begin{bmatrix} B \\ C \end{bmatrix}$$

According to part (c) of Exercise 11 in Sec. 1.4,

$$\mathbb{N}(B) \cap \mathbb{N}(C) = \mathbb{N}(D) \qquad (1.7.17)$$

Also, it is clear that

$$\mathbb{R}(B) \subseteq \mathbb{R}(D) \qquad \mathbb{R}(C) \subseteq \mathbb{R}(D) \qquad (1.7.18)$$

However, since $\mathbb{N}(C) \supseteq \mathbb{N}(B)$, we have

$$\mathbb{N}(B) \cap \mathbb{N}(C) = \mathbb{N}(B)$$

Comparing this equation with equation (1.7.17) yields

$$\mathbb{N}(B) = \mathbb{N}(D)$$

Thus $v(B) = v(D)$, and it follows from Theorems 1.6.1 and 1.7.11 that $r(B) = r(D)$.   Combining this with the first part of (1.7.18), we have shown that $\mathbb{R}(B)$ is a subspace of $\mathbb{R}(D)$ whose dimension is equal to the dimension of $\mathbb{R}(D)$.   Applying Corollary 1.5.4, we conclude that

$$\mathbb{R}(B) = \mathbb{R}(D)$$

The second part of (1.7.18) now yields

$$\mathbb{R}(C) \subseteq \mathbb{R}(B) \qquad ////$$

Thus the row space and the null space of a matrix determine each other, though neither one determines the matrix itself.

EXERCISES

1 Reduce the matrices

$$(a)\quad A = \begin{bmatrix} 0 & 0 & 1 & 1 \\ 2 & -4 & 2 & 0 \\ -1 & 2 & 0 & 1 \end{bmatrix} \qquad (b)\ B = \begin{bmatrix} 0 & 1 & 1 & 0 \\ 2 & 0 & 2 & -4 \\ -2 & 3 & 1 & 4 \end{bmatrix}$$

to the form (1.7.9) by means of elementary operations.

2 Transform one of the matrices of Exercise 1 into the other by means of elementary operations by first reducing both matrices to form (1.7.9).   Show that it is possible to transform one matrix into the other without transforming either into form (1.7.9).

3 Find the rank of the matrix

$$A = \begin{bmatrix} 1 & 2 & -1 & 0 & 1 \\ 2 & 0 & 0 & 1 & 1 \\ 1 & -2 & 1 & 1 & 0 \end{bmatrix}$$

(*a*) By finding a basis for $\mathbb{R}(A)$.
(*b*) By finding a basis for $\mathbb{C}(A)$.
(*c*) By finding a basis for $\mathbb{N}(A)$ and applying Theorem 1.6.1.
(*d*) By transforming A to form (1.7.9).

4 Reduce the matrix A of the examples of this section to the form (1.7.9).

5 Let

$$A = \begin{bmatrix} 0 & 1 & 2 \\ -1 & 1 & 0 \\ 1 & 0 & 0 \end{bmatrix}$$

Show that A can be reduced to the form (1.7.9) using only elementary row operations. Show also that A can be reduced to the form (1.7.9) using only elementary column operations. Can this be done with the matrices of Exercise 1? Why?

6 Describe each of the elementary column matrices C_{rs}, $C_r(\lambda)$, and $C_{rs}(\alpha)$. Show that every elementary column matrix is equal to an elementary row matrix of the same kind.

7 Let A be a $4 \times n$ matrix. Find a matrix R such that RA can be obtained from A by performing all the following operations:
(a) Interchanging the first and third rows of A.
(b) Adding a fifth row to A which is equal to the fourth row of A.
(c) Multiplying the fourth row of A by 3.
(d) Deleting the second row of A.

8 Prove that a matrix is nonsingular if and only if it can be expressed as a product of elementary matrices.

9 We wish to perform one elementary row operation $\mathscr{R}$ and one elementary column operation $\mathscr{C}$ on a given matrix. Show that the same matrix is obtained irrespective of whether $\mathscr{R}$ or $\mathscr{C}$ is performed first. Show that this result is no longer true if both operations are row operations or both are column operations. (*Hint:* Use the elementary matrices.)

10 Supply proofs for Lemmas 1.7.1, 1.7.3, and 1.7.5 and Corollary 1.7.14.

11 Let A and B be $m \times n$ and $p \times q$ matrices respectively and let C be the $(m + p) \times (n + q)$ matrix

$$C = \begin{bmatrix} A & 0 \\ 0 & B \end{bmatrix}$$

Prove that $r(C) = r(A) + r(B)$ and $v(C) = v(A) + v(B)$.

12 Two $m \times n$ matrices A and B are said to be *equivalent* if there exist non-singular matrices P and Q such that $B = PAQ$. Prove that equivalence of matrices is an equivalence relation.

13 Prove that two matrices are equivalent if and only if one of them can be transformed into the other by a finite number of elementary operations.

14 Prove that two $m \times n$ matrices are equivalent if and only if they have the same rank.

15 Prove that the first elementary row operation can be expressed in terms of the second and third elementary row operations.

16 Prove that a nonsingular matrix can be reduced to the identity matrix using only elementary row operations or only elementary column operations (see Exercise 5).

17 Let A and B be $m \times n$ matrices for which there exist matrices C and D such that $A = CB$, $B = AD$. Prove that there exist matrices F and G such that $A = BF$, $B = GA$.

18 Let B and C be $m \times n$ and $p \times n$ matrices respectively. Prove that a necessary and sufficient condition that there exist a $p \times m$ matrix A such that $AB = C$ is that $\mathbb{N}(C) \supseteq \mathbb{N}(B)$.

19 Let A be an $m \times n$ matrix. Show that A can be expressed as

$$A = XY$$

where $X \in \mathbb{F}^m$, $Y \in \mathbb{F}_n$, if and only if $r(A) \leq 1$. Use the methods of this exercise to give an independent proof of the fact that if the row rank of a matrix is 1, then the column rank of the matrix is 1.

20 Let $A = [a_{ij}]$ be an $m \times n$ matrix. We define the $n \times m$ matrix $C = [c_{ij}]$ by

$$a_{ij} = c_{ji} \qquad \begin{aligned} i &= 1, 2, \ldots, m \\ j &= 1, 2, \ldots, n \end{aligned}$$

that is, the entry in the (i, j) position of A is equal to the entry in the (j, i) position of C. Then C is called the *transpose* of A and is denoted by A^T. Prove:

(a) $(A + B)^T = A^T + B^T$.
(b) $(\alpha A)^T = \alpha A^T$.
(c) $(AB)^T = B^T A^T$.
(d) $(A^T)^T = A$.
(e) $r(A) = r(A^T)$.
(f) If A is nonsingular, then A^T is nonsingular and $(A^T)^{-1} = (A^{-1})^T$.

21 Prove that if A and C are $m \times n$ matrices such that $\mathbb{R}(A) = \mathbb{R}(C)$, there is a nonsingular matrix B of order m such that $A = BC$. [*Hint:* Show that $\mathbb{R}(PAQ) = \mathbb{R}(PCQ)$ for all nonsingular matrices P and Q of orders m and n respectively. Choose P and Q so that PCQ has the form (1.7.9) and use Exercise 14 of Sec. 1.6.]

22 Let A be a square singular matrix. Prove that there exists a nonzero matrix B such that $AB = BA = 0$. [*Hint:* First prove the result for matrices which have the form (1.7.9).]

23 Let A be a square matrix. Prove that $r(A) = r(A^2)$ if and only if

$$\mathbb{F}^n = \mathbb{C}(A) \oplus \mathbb{N}(A)$$

24 Show that when $\mathbb{F}$ is the real field,

$$\mathbb{F}^n = \mathbb{C}(A^T) \oplus \mathbb{N}(A)$$

for every $m \times n$ matrix A. Show that this equation is no longer valid when $\mathbb{F}$ is the complex field.

25 Let A and B be $m \times n$ and $n \times p$ matrices respectively. Prove:
(a) $r(AB) \leq \min\{r(A), r(B)\}$.
(b) $v(AB) \leq v(A) + v(B)$. [*Hint:* Extend a basis for $\mathbb{N}(B)$ to a basis for $\mathbb{N}(AB)$.]
(c) $r(A) + r(B) - n \leq r(AB)$.

26 Let A and B be $m \times n$ matrices. Prove that

$$r(A + B) \leq r(A) + r(B)$$

27 Let A and B be $m \times n$ and $n \times p$ matrices respectively. Prove that $\mathbb{N}(A) = \mathbb{C}(B)$ if and only if $\mathbb{R}(A) = \mathbb{N}_R(B)$. (*Hint:* See Exercise 13 of Sec. 1.4. Use an argument based on ranks and nullities.)

28 Let A be an $m \times n$ matrix of rank r in which the first r rows and the first r columns are linearly independent sets. Let $\tilde{A}$ be the $r \times r$ submatrix of A formed by deleting all rows and all columns of A except the first r. Prove that $\tilde{A}$ is nonsingular. Construct an example to show that if the rank of A is greater than the number of linearly independent rows and columns used to form $\tilde{A}$, then $\tilde{A}$ need not be nonsingular. (*Hint:* Instead of deleting these rows and columns from A, change them into rows and columns of zeros by multiplying A on the left and on the right by properly chosen nonsingular matrices.)

29 Prove that a set of necessary and sufficient conditions that a matrix have rank r is:
(a) There is a nonsingular submatrix of order r.
(b) For each s, $s > r$, every submatrix of order s is singular.

30 Prove that a set of necessary and sufficient conditions that a matrix A have rank r is:
(a) A has a nonsingular submatrix $\tilde{A}$ of rank r.
(b) Every square submatrix of A of order $r + 1$ which has $\tilde{A}$ as a submatrix is singular.
(*Hint:* Show that each row of A which does not meet $\tilde{A}$ can be expressed uniquely as a linear combination of the rows of A which do meet $\tilde{A}$.)

31 Let A, X, Y, and α be $n \times n$, $n \times 1$, $1 \times n$, and 1×1 matrices respectively and let A be singular. Prove that if the $(n + 1) \times (n + 1)$ matrix

$$\begin{bmatrix} A & X \\ Y & \alpha \end{bmatrix}$$

is nonsingular for some value of α, then it is nonsingular for all values of α.

1.8 PERMUTATIONS[1]

In the next section we will introduce the determinant of a square matrix. The definition of the determinant we will use requires a knowledge of some of the basic facts about the set of permutations of the set $\mathbb{N} = \{1, 2, \ldots, n\}$. Let us begin with a definition.

Definition 1.8.1 A *permutation* of a set $\mathbb{N}$ is a one-to-one mapping of $\mathbb{N}$ onto itself. The set of all permutations of $\mathbb{N}$ is denoted by $\mathbb{S}_n$.

We will be concerned only with permutations of a set $\mathbb{N}$ consisting of the first n positive integers. In order to avoid having to make exceptions for the special case when $n = 1$, we assume throughout this section (and only this section) that $n \geq 2$. A permutation σ which maps 1 onto a, 2 onto b, 3 onto $c, \ldots, n$ onto d is denoted by

$$\sigma = \begin{pmatrix} 1 & 2 & 3 & \cdots & n \\ a & b & c & \cdots & d \end{pmatrix} \qquad (1.8.2)$$

The mapping which maps every element of $\mathbb{N}$ onto itself is a permutation and is called the *identity permutation*. It is denoted either by I or by

$$\begin{pmatrix} 1 & 2 & 3 & \cdots & n \\ 1 & 2 & 3 & \cdots & n \end{pmatrix}$$

There are n different possibilities for the entry a of (1.8.2); that is, a can be any one of the numbers $1, 2, 3, \ldots, n$. Once a has been chosen, there are $n - 1$ choices for b. Continuing in this way, we have:

Lemma 1.8.3 $\mathbb{S}_n$ contains $n!$ elements.

Since the domain and range of a permutation are identical, for any two elements of $\mathbb{S}_n$, say σ_1 and σ_2, we may define the product $\sigma_1 \sigma_2$ by

$$(\sigma_1 \sigma_2)(x) = \sigma_1(\sigma_2(x)) \qquad x \in \mathbb{N}$$

Then $\sigma_1 \sigma_2$ is again a permutation of $\mathbb{N}$. It is easy to show that with this definition, multiplication of permutations of $\mathbb{N}$ is associative. It will soon become apparent that, in general, multiplication of permutations is not commutative.

---

[1] Most of this section forms one of the two digressions from matrix theory in this chapter. The other one is Sec. 1.11. Each is self-contained and independent of the remainder of the book.

EXAMPLE 1.8.4   Let $\mathbb{N} = \{1, 2, 3, 4, 5\}$ (so that $n = 5$) and set

$$\sigma_1 = \begin{pmatrix} 1 & 2 & 3 & 4 & 5 \\ 2 & 1 & 5 & 3 & 4 \end{pmatrix} \qquad \sigma_2 = \begin{pmatrix} 1 & 2 & 3 & 4 & 5 \\ 3 & 5 & 2 & 4 & 1 \end{pmatrix}$$

According to the definition of a product of permutations, we have

$$\sigma_1\sigma_2: \quad \begin{aligned} 1 &\rightarrow (\sigma_1\sigma_2)(1) = \sigma_1(\sigma_2(1)) = \sigma_1(3) = 5 \\ 2 &\rightarrow (\sigma_1\sigma_2)(2) = \sigma_1(\sigma_2(2)) = \sigma_1(5) = 4 \\ 3 &\rightarrow (\sigma_1\sigma_2)(3) = \sigma_1(\sigma_2(3)) = \sigma_1(2) = 1 \\ 4 &\rightarrow (\sigma_1\sigma_2)(4) = \sigma_1(\sigma_2(4)) = \sigma_1(4) = 3 \\ 5 &\rightarrow (\sigma_1\sigma_2)(5) = \sigma_1(\sigma_2(5)) = \sigma_1(1) = 2 \end{aligned}$$

and hence
$$\sigma_1\sigma_2 = \begin{pmatrix} 1 & 2 & 3 & 4 & 5 \\ 5 & 4 & 1 & 3 & 2 \end{pmatrix} \qquad ////$$

It will be convenient for us to identify the permutation τ of $\mathbb{N}' = \{1, 2, \ldots, n - 1\}$ with the permutation σ of $\mathbb{N} = \{1, 2, \ldots, n\}$ when

$$\tau(x) = \sigma(x) \qquad x = 1, 2, \ldots, n - 1$$

and
$$\sigma(n) = n \qquad (1.8.5)$$

Henceforth we will consider τ and σ satisfying (1.8.5) to be the same permutation. In particular, all identity permutations are the same permutation.

Definition 1.8.6   Let σ be a permutation of $\mathbb{N} = \{1, 2, \ldots, n\}$. If there exist $a, b \in \mathbb{N}$, $a \neq b$, such that

$$\begin{aligned} \sigma(a) &= b \\ \sigma(b) &= a \\ \sigma(x) &= x \qquad x \neq a, b \end{aligned}$$

then σ is a *transposition* and is denoted by $(a \quad b)$.

In other words, a transposition is a permutation of $\mathbb{N}$ which interchanges two elements of $\mathbb{N}$ and leaves all other elements of $\mathbb{N}$ fixed.

Theorem 1.8.7   Every permutation can be expressed as a product of transpositions.

PROOF   The proof is by an induction on n, beginning with $n = 2$. Clearly $\mathbb{S}_2$ consists of the two elements

$$I = \begin{pmatrix} 1 & 2 \\ 1 & 2 \end{pmatrix} \qquad (1 \quad 2) = \begin{pmatrix} 1 & 2 \\ 2 & 1 \end{pmatrix}$$

Since (1 2) is already the product of one transposition and

$$I = (1 \quad 2)(1 \quad 2)$$

the theorem is true for $n = 2$.

Now suppose that every permutation of $\mathbb{N}' = \{1, 2, \ldots, n - 1\}$ is expressible as a product of transposition and let σ be a permutation of $\mathbb{N} = \{1, 2, \ldots, n\}$. If $\sigma(n) = n$, we may consider σ to be a permutation of $\mathbb{N}'$ and hence σ can be expressed as a product of transpositions. On the other hand, suppose

$$\sigma(n) = a \neq n$$

In this case we set

$$\tau = (a \quad n)\sigma$$

and then

$$\tau(n) = (a \quad n)\sigma(n) = (a \quad n)(a) = n$$

that is, τ maps n onto itself. Consequently τ can be considered as a permutation of N' and thus can be expressed as a product of transpositions, say

$$\tau = \rho_1 \rho_2 \cdots \rho_k$$

Clearly

$$\sigma = (a \quad n)(a \quad n)\sigma = (a \quad n)\tau = (a \quad n)\rho_1 \rho_2 \cdots \rho_k$$

completing the induction.   ////

The number of transpositions into which a permutation can be factored is not unique. Indeed, to any factorization we can always add two more (identical) transpositions. We can show, however, that for a given permutation the number of transpositions in its various factorizations is either always odd or always even. The proof is most easily described using the following polynomial in the n variables $x_1, x_2, \ldots, x_n$:

$$f(x_1, x_2, \ldots, x_n) = \prod_{i<j}^{n} (x_i - x_j)$$

$$= (x_1 - x_2)(x_1 - x_3) \cdots (x_1 - x_n)(x_2 - x_3) \cdots (x_{n-1} - x_n)$$

For any permutation σ of $\mathbb{N}$ we define the polynomial σf in the same n variables, $x_1, x_2, \ldots, x_n$, by

$$(\sigma f)(x_1, x_2, \ldots, x_n) = \prod_{i<j}^{n} (x_{\sigma(i)} - x_{\sigma(j)})$$

i.e., the permutation σ acts on the subscripts of the variables. When $\sigma(i) < \sigma(j)$, the term $x_{\sigma(i)} - x_{\sigma(j)}$ is one of the factors of $f(x_1, x_2, \ldots, x_n)$. On the other hand, when $\sigma(i) > \sigma(j)$, then $x_{\sigma(i)} - x_{\sigma(j)} = -(x_{\sigma(j)} - x_{\sigma(i)})$ and $x_{\sigma(j)} - x_{\sigma(i)}$ is one of the factors of $f(x_1, x_2, \ldots, x_n)$. Thus, aside from possible differences of sign, the factors of $f(x_1, x_2, \ldots, x_n)$ and $(\sigma f)(x_1, x_2, \ldots, x_n)$ are identical. The exact relation between f and σf will follow from:

Lemma 1.8.8 Let $\rho = (a \ b)$ be a transposition of $\mathbb{N}$. Then

$$(\rho f)(x_1, x_2, \ldots, x_n) = -f(x_1, x_2, \ldots, x_n)$$

PROOF In order to see more easily the effect of ρ on f we must first express $f(x_1, x_2, \ldots, x_n)$ in the form

$$f(x_1, x_2, \ldots, x_n) = \Theta(x_a - x_b) \prod_{\substack{c \neq a, b \\ d \neq a, b \\ c < d}} (x_c - x_d) \prod_{c \neq a, b} (x_a - x_c) \prod_{c \neq a, b} (x_b - x_c)$$

$$(1.8.9)$$

where $\Theta = +1$ or $\Theta = -1$. That this can be done follows from the discussion which precedes the lemma. Clearly Θ is uniquely determined by the choice of a and b. For a given pair of distinct integers, a and b, it is possible (but not at all necessary) to determine whether $\Theta = +1$ or $\Theta = -1$.

Now let us investigate the effect of ρ on each of the terms on the right-hand side of (1.8.9). Clearly ρ leaves Θ and $\prod_{\substack{c \neq a, b \\ d \neq a, b \\ c < d}} (x_c - x_d)$ unchanged and has the effect of interchanging the two terms $\prod_{c \neq a, b} (x_a - x_c)$ and $\prod_{c \neq a, b} (x_b - x_c)$. Finally, $\rho(x_a - x_b) = x_b - x_a = -(x_a - x_b)$. Thus the total effect of ρ is to change the sign of $f(x_1, x_2, \ldots, x_n)$, that is,

$$(\rho f)(x_1, x_2, \ldots, x_n) = -f(x_1, x_2, \ldots, x_n) \qquad ////$$

Let σ be a permutation of $\mathbb{N} = \{1, 2, \ldots, n\}$ for which some factorization into transpositions contains an odd number of transpositions. It follows from repeated applications of Lemma 1.8.8 that

$$(\sigma f)(x_1, x_2, \ldots, x_n) = -f(x_1, x_2, \ldots, x_n)$$

On the other hand, if some factorization of σ were to contain an even number of transpositions, then, in the same way, we would have

$$(\sigma f)(x_1, x_2, \ldots, x_n) = f(x_1, x_2, \ldots, x_n)$$

Clearly this would be a contradiction, and consequently we have proved:

Theorem 1.8.10   Every permutation of $\mathbb{N} = \{1, 2, \ldots, n\}$, $n \geq 2$, can be expressed as a product of a finite number of transpositions.   If the number of transpositions in one such factorization of the permutation σ is odd (even), then the number of transpositions in every factorization of σ is odd (even).   In this case we call σ an *odd* (*even*) *permutation*.

Because of Theorem 1.8.10 we are able to make the following definition:

Definition 1.8.11   The *signature* of a permutation σ (written sgn σ) is defined by

$$\text{sgn } \sigma = \begin{cases} +1 & \text{if } \sigma \text{ is an even permutation} \\ -1 & \text{if } \sigma \text{ is an odd permutation} \end{cases}$$

The operation $\mathscr{R}_{rs}$ is a transposition of the rows of a matrix, interchanging the rth and sth rows.   Similarly $\mathscr{C}_{rs}$ is a transposition of the columns of a matrix. We can generalize this to any permutation of $\mathbb{N} = \{1, 2, \ldots, n\}$.   Let A be a matrix with n rows and let $\mathscr{R}_\sigma$ be the operation of performing the permutation σ on the rows of A.   As in Sec. 1.7, the operation $\mathscr{R}_\sigma$ can be carried out by multiplying the given matrix on the left by a properly chosen matrix nonsingular R_σ.   As before, R_σ is obtained by performing the operation $\mathscr{R}_\sigma$ on the identity matrix:

$$\mathscr{R}_\sigma : I \to R_\sigma$$

where R_σ has 1 in the $(\sigma(1), 1)$, $(\sigma(2), 2)$, $\ldots$, $(\sigma(n), n)$ positions and 0 elsewhere.

Theorem 1.8.12   Let σ be a permutation of $\mathbb{N} = \{1, 2, \ldots, n\}$.   Then

$$\mathscr{R}_\sigma : A \to R_\sigma A$$

for every matrix A with n rows.

PROOF   According to Theorem 1.8.7, there exist transpositions $\rho_1 = (r_1 \quad s_1)$, $\rho_2 = (r_2 \quad s_2)$, $\ldots$, $\rho_k = (r_k \quad s_k)$ such that

$$\sigma = \rho_1 \rho_2 \cdots \rho_k$$

Then $\qquad\qquad \mathscr{R}_\sigma = \mathscr{R}_{\rho_1} \mathscr{R}_{\rho_2} \cdots \mathscr{R}_{\rho_k} = \mathscr{R}_{r_1 s_1} \mathscr{R}_{r_2 s_2} \cdots \mathscr{R}_{r_k s_k}$

Set $\qquad\qquad R_\sigma = R_{r_1 s_1} R_{r_2 s_2} \cdots R_{r_k s_k}$

It now follows by repeated applications of Lemma 1.7.1 that

$$\mathscr{R}_\sigma(A) = \mathscr{R}_{r_1 s_1} \mathscr{R}_{r_2 s_2} \cdots \mathscr{R}_{r_k s_k}(A)$$
$$= R_{r_1 s_1} R_{r_2 s_2} \cdots R_{r_k s_k} A$$
$$= R_\sigma A$$

for every matrix A with n rows.   ////

We call R_σ the *permutation matrix* corresponding to the permutation σ. Thus there is a one-to-one correspondence between permutations of $\mathbb{N} = \{1, 2, \ldots, n\}$ and permutation matrices of order n.   Moreover,

$$R_\sigma R_\tau = R_{\sigma\tau}$$

for all $\sigma, \tau \in \mathbb{S}_n$.

Because we have defined multiplication of permutation by

$$(\sigma_1 \sigma_2)(x) = \sigma_1[\sigma_2(x)]$$

rather than by

$$(x)(\sigma_1 \sigma_2) = [(x)\sigma_1]\sigma_2$$

the treatment of permutations of the columns of a matrix is not as straight-forward as the discussion above (see Exercises 18 and 19).

EXERCISES

1   For the permutations σ_1 and σ_2 of Example 1.8.4, find:

(a) $\sigma_2 \sigma_1$ (b) $(\sigma_1)^2$ (c) $(\sigma_1)^3$

(d) $(\sigma_1)^4$ (e) $(\sigma_1)^5$ (f) $(\sigma_2)^2$

2   Express

$$\sigma = \begin{pmatrix} 1 & 2 & 3 & 4 & 5 & 6 & 7 & 8 & 9 \\ 7 & 9 & 1 & 6 & 5 & 2 & 3 & 4 & 8 \end{pmatrix}$$

as a product of transpositions.   Is σ odd or even?

3   Let σ be a permutation of $\mathbb{N}$.   Prove that there is a unique permutation $\tilde{\sigma}$ of $\mathbb{N}$ such that

$$\tilde{\sigma}\sigma = \sigma\tilde{\sigma} = I$$

$\tilde{\sigma}$ is called the *inverse* of σ and is denoted by σ^{-1}.

4   For the permutations σ_1 and σ_2 of Example 1.8.4, find:

(a) σ_1^{-1} (b) σ_2^{-1}

(c) $\sigma_1 \sigma_2 \sigma_1^{-1}$ (d) $\sigma_2 \sigma_1 \sigma_2^{-1}$

5   Find the inverse of the permutation σ of Exercise 2.   Express σ^{-1} as a product of transpositions.   Is σ^{-1} odd or even?

6   Show that the inverse of the permutation σ of (1.8.2) can be found by interchanging the two rows of σ (and then rearranging the columns so that the first row is again $1, 2, \ldots, n$).

7 Prove that if a permutation is expressed as a product of transpositions, then the inverse of the permutation is the product of the transpositions in the opposite order.

8 Prove that exactly half of the permutations of $\mathbb{N}$ are even.

9 Prove the assertion that the multiplication of permutations is associative.

10 Let σ and τ be permutations of $\mathbb{N}$. Prove that $(\sigma\tau)^{-1} = \tau^{-1}\sigma^{-1}$.

11 Let σ be a permutation of $\mathbb{N}$. Show that

$$\{(i, \sigma(i)) \mid i \in \mathbb{N}\} = \{(\sigma^{-1}(j), j) \mid j \in \mathbb{N}\}$$

12 Show that $\mathbb{S}_n = \{\sigma^{-1} \mid \sigma \in \mathbb{S}_n\}$.

13 Let σ and τ be permutations of $\mathbb{N}$. Prove:
   (a) $\operatorname{sgn}(\sigma\tau) = (\operatorname{sgn} \sigma)(\operatorname{sgn} \tau)$ (b) $\operatorname{sgn} \sigma = \operatorname{sgn} \sigma^{-1}$

14 Prove that

$$\begin{pmatrix} 1 & 2 & \cdots & a-1 & a & a+1 & \cdots & n-1 & n \\ 1 & 2 & \cdots & a-1 & n & a & \cdots & n-2 & n-1 \end{pmatrix}$$
$$= (a \quad a+1)(a \quad a+2) \quad \cdots \quad (a \quad n-1)(a \quad n)$$

15 Let $\mathbb{N}' = \{1, 2, \ldots, n-1\}$ and $\mathbb{N} = \{1, 2, \ldots, n\}$. A permutation σ of $\mathbb{N}$ which leaves n fixed is considered to be a permutation of $\mathbb{N}'$. Denote the set of permutations of $\mathbb{N}'$ by $\mathbb{S}_{n-1}$ and let $a \in \mathbb{N}$, $a \neq n$. Prove that

$$\mathbb{S}_{n-1}(a \quad n) = \{\sigma(a \quad n) \mid \sigma \in \mathbb{S}_{n-1}\}$$

is the set of all permutations of $\mathbb{N}$ which map a onto n and that

$$(a \quad n)\mathbb{S}_{n-1} = \{(a \quad n)\sigma \mid \sigma \in \mathbb{S}_{n-1}\}$$

is the set of all permutations of $\mathbb{N}$ which map n onto a. How many permutations are there in each set?

16 For the permutations σ_1 and σ_2 of Example 1.8.4, find:

(a) R_{σ_1} (b) R_{σ_2}
(c) $R_{\sigma_1 \sigma_2}$ (d) $R_{\sigma_2 \sigma_1}$

17 Let $\mathscr{C}_\sigma$ be the operation of performing the permutation σ on the columns of a matrix and let

$$\mathscr{C}_\sigma : I \to C_\sigma$$

Find:

(a) C_{σ_1} (b) C_{σ_2}
(c) $C_{\sigma_1 \sigma_2}$ (d) $C_{\sigma_2 \sigma_1}$

where σ_1 and σ_2 are the permutations of Example 1.8.4.

18 Prove:

(a) $C_{\sigma_1} C_{\sigma_2} = C_{\sigma_2 \sigma_1}$ (b) $C_\sigma = R_\sigma^{-1}$

(c) $R_{\sigma^{-1}} = R_\sigma^{-1}$ (d) $C_{\sigma^{-1}} = C_\sigma^{-1}$

(e) $R_\sigma^{-1} = R_\sigma^{T}$ (f) $C_\sigma^{-1} = C_\sigma^{T}$

19 Formulate and prove a theorem corresponding to Theorem 1.8.12 for column operations. Verify that the matrices C_{σ_1} and C_{σ_2} of Exercise 17 conform with these results.

20 Let $\mathbb{S}_1$ be a subset of $\mathbb{N} = \{1, 2, \ldots, n\}$ consisting of m integers, $\mathbb{S}_1 = \{s_1, s_2, \ldots, s_m\}$, $s_1 < s_2 < \cdots < s_m$. Let $\mathbb{S}_2$ be the set of $n - m$ elements of $\mathbb{N}$ which do not belong to $\mathbb{S}_1$, $\mathbb{S}_2 = \{s_{m+1}, s_{m+2}, \ldots, s_n\}$, $s_{m+1} < s_{m+2} < \cdots < s_n$. Set

$$\sigma_{\mathbb{S}} = \begin{pmatrix} 1 & 2 & \cdots & m & m+1 & m+2 & \cdots & n \\ s_1 & s_2 & \cdots & s_m & s_{m+1} & s_{m+2} & \cdots & s_n \end{pmatrix}$$

Show that

$$\text{sgn } \sigma_{\mathbb{S}} = (-1) \exp\left(\sum_{j=1}^{m} s_j - j \right) = (-1) \exp\left(\sum_{j=m+1}^{n} s_j - j \right)$$

(*Hint:* Use an induction on m.) Can this result be extended to three subsets $\mathbb{S}_1, \mathbb{S}_2, \mathbb{S}_3$ of $\mathbb{N}$?

1.9 DETERMINANTS[1]

In this section we will define the determinant of a square matrix and investigate many of its properties. Actually, we will present two quite different definitions of the determinant (Definition 1.9.1 and Theorem 1.9.20) and show that they are equivalent. The determinant function can be developed not only for matrices whose entries are from a field but also for matrices with entries from any set in which there are "reasonable" operations of addition, subtraction, and multiplication. This can be done since the operation of division is never mentioned.

---

[1] In his four-volume treatise, "The Theory of Determinants in the Historical Order of Development," (Macmillan, London, 1906 et seq.) Thomas Muir quotes extensively from a letter written by Leibnitz in 1693 to de L'Hospital. In this letter Leibnitz announced a new method for displaying and solving a system of linear equations in several unknowns. Essentially he had formulated a double-subscript notation similar to the one we use today, and he used it to obtain what is now known as Cramer's rule (see Exercise 19 of Sec. 1.9). Cramer discovered the same results independently, and his book, which was published in 1750 (unlike Leibnitz' letters, which were not published until 1850), attracted considerable attention. The concept of a matrix not merely as an array of numbers but as an object which could be added and multiplied is due primarily to Hamilton (1853) and Cayley (1858).

By reasonable operations we will mean that all the properties of a field as given in Definition 0.1.1, except for item 10, are satisfied.[2]   In order to keep our notation simple, *in this section* we will denote a set having reasonable operations of addition, subtraction, and multiplication by $\mathbb{F}$.   The reader will lose very little (at least until Sec. 1.10) by continuing to think of $\mathbb{F}$ as a field.

The determinant is a function whose domain is the set of all square matrices with entries from $\mathbb{F}$ and whose range is $\mathbb{F}$ itself.   The determinant is not defined for matrices which are not square.

Definition 1.9.1   Let $A = [a_{ij}]$ be a square matrix of order n.   We define the *determinant* of A to be

$$\sum_{\sigma \in \mathbb{S}_n} \text{sgn } \sigma \, a_{1\sigma(1)} \, a_{2\sigma(2)} \cdots a_{n\sigma(n)} \qquad (1.9.2)$$

and denote it by *det A*.

For any positive integer n we have denoted the set of all permutations of $\mathbb{N} = \{1, 2, \ldots, n\}$ by $\mathbb{S}_n$ and showed that $\mathbb{S}_n$ contains $n!$ elements.   For $n = 1$, $\mathbb{S}_1$ consists of the identity permutation I alone.   For $n \geq 2$, exactly half of the elements of $\mathbb{S}_n$ are even permutations, and exactly half are odd permutations. (See Exercise 8 of Sec. 1.8.)

Thus, for $n \geq 2$, sgn $\sigma = +1$ for $\frac{1}{2}n!$ permutations σ belonging to $\mathbb{S}_n$ and sgn $\sigma = -1$ for the remaining $\frac{1}{2}n!$ permutations.

A careful consideration of Definition 1.9.1 will show that the determinant of a matrix of order n is a homogeneous polynomial of degree n in the entries of the matrix, consisting of $n!$ terms, each with a coefficient of either $+1$ or -1.   Moreover, each of these terms contains exactly one entry from each row and exactly one entry from each column of the matrix.   Conversely, because Definition 1.9.1 calls for us to use all permutations of $\mathbb{N} = \{1, 2, \ldots, n\}$, every product of n entries of the matrix which consists of exactly one entry from each row and exactly one entry from each column is contained in (1.9.2).

EXAMPLE 1.9.3   Set $n = 1$ and $A = [a_{11}]$.   Then $\mathbb{S}_1 = \{I\}$, and hence

$$\det A = \text{sgn } I \, a_{1, \, I(1)} = a_{11} \qquad \qquad ////$$

---

[2] We have already pointed out that such a system is called a *ring*.   In Sec. 1.10 we will apply the results of this section to matrices whose entries are from the ring of polynomials in a single variable x with coefficients from a field.

EXAMPLE 1.9.4 Set $n = 2$ and $A = [a_{ij}]$, $i, j = 1, 2$. Then $\mathbb{S}_2 = \{I, (1\ \ 2)\}$, and hence

$$\det A = \operatorname{sgn} I\, a_{1, I(1)} a_{2, I(2)} + \operatorname{sgn}(1\ \ 2)\, a_{1,(1\ \ 2)(1)} a_{2,(1\ \ 2)(2)}$$
$$= a_{11} a_{22} - a_{12} a_{21} \qquad\qquad ////$$

EXAMPLE 1.9.5 For $n = 3$ we have

$$\mathbb{S}_3 = \left\{ I, (1\ \ 2), (1\ \ 3), (2\ \ 3), \begin{pmatrix} 1 & 2 & 3 \\ 2 & 3 & 1 \end{pmatrix}, \begin{pmatrix} 1 & 2 & 3 \\ 3 & 1 & 2 \end{pmatrix} \right\}$$

Since

$$\begin{pmatrix} 1 & 2 & 3 \\ 2 & 3 & 1 \end{pmatrix} = (1\ \ 2)(2\ \ 3) \qquad \begin{pmatrix} 1 & 2 & 3 \\ 3 & 1 & 2 \end{pmatrix} = (2\ \ 3)(1\ \ 2)$$

it follows that

$$\operatorname{sgn}\begin{pmatrix} 1 & 2 & 3 \\ 2 & 3 & 1 \end{pmatrix} = \operatorname{sgn}\begin{pmatrix} 1 & 2 & 3 \\ 3 & 1 & 2 \end{pmatrix} = +1$$

and therefore

$$\det A = \operatorname{sgn} I\, a_{11} a_{22} a_{22} + \operatorname{sgn}(1\ \ 2)\, a_{12} a_{21} a_{33} + \operatorname{sgn}(1\ \ 3)\, a_{13} a_{22} a_{31}$$
$$+ \operatorname{sgn}(2\ \ 3)\, a_{11} a_{23} a_{32} + \operatorname{sgn}\begin{pmatrix} 1 & 2 & 3 \\ 2 & 3 & 1 \end{pmatrix} a_{12} a_{23} a_{31}$$
$$+ \operatorname{sgn}\begin{pmatrix} 1 & 2 & 3 \\ 3 & 1 & 2 \end{pmatrix} a_{13} a_{21} a_{32}$$
$$= a_{11} a_{22} a_{33} - a_{12} a_{21} a_{33} - a_{13} a_{22} a_{31} - a_{11} a_{23} a_{32}$$
$$+ a_{12} a_{23} a_{31} + a_{13} a_{21} a_{32} \qquad\qquad ////$$

These formulas agree with the definition of the determinant for 1×1, 2×2, and 3×3 matrices that the reader has met before. The determinant of larger matrices can be evaluated in the same way, but the number of terms becomes unpleasantly large very soon. In this section we will find several other ways to evaluate the determinant of a matrix.

In Definition 1.9.1 the row and column subscripts of each entry of A appear to play different roles. Let us demonstrate that in fact this is not the case.

Lemma 1.9.6 Let $A = [a_{ij}]$ be a matrix of order n. Then

$$\det A = \sum_{\tau \in \mathbb{S}_n} \operatorname{sgn} \tau\, a_{\tau(1)1} a_{\tau(2)2} \cdots a_{\tau(n)n} \qquad (1.9.7)$$

PROOF. We wish to show that each of the $n!$ terms of (1.9.2) is identical with one of the $n!$ terms on the right-hand side of equation (1.9.7). More specifically, we wish to show that the term in (1.9.2) corresponding to the permutation σ is equal to the term of (1.9.7) corresponding to the permutation $\tau = \sigma^{-1}$; that is,

$$\text{sgn } \sigma \, a_{1\sigma(1)} a_{2\sigma(2)} \cdots a_{n\sigma(n)} = \text{sgn } \sigma^{-1} a_{\sigma^{-1}(1)1} a_{\sigma^{-1}(2)2} \cdots a_{\sigma^{-1}(n)n}$$

for every $\sigma \in \mathbb{S}_n$. Let j be one of the integers of $\mathbb{N} = \{1, 2, \ldots, n\}$. There exists $i \in \mathbb{N}$ such that $\sigma(i) = j$. In fact, $i = \sigma^{-1}(j)$. Then

$$a_{i\sigma(i)} = a_{ij} = a_{\sigma^{-1}(j)j}$$

Since

$$\{(i, \sigma(i)) \mid i \in \mathbb{N}\} = \{(\sigma^{-1}(j), j) \mid j \in \mathbb{N}\}$$

(see Exercise 11 of Sec. 1.8) it follows that

$$\text{sgn } \sigma \, a_{1\sigma(1)} a_{2\sigma(2)} \cdots a_{n\sigma(n)} = \text{sgn } \sigma \, a_{\sigma^{-1}(1)1} a_{\sigma^{-1}(2)2} \cdots a_{\sigma^{-1}(n)n}$$

But $\text{sgn } \sigma = \text{sgn } \sigma^{-1}$, and thus

$$\det A = \sum_{\sigma \in \mathbb{S}_n} \text{sgn } \sigma^{-1} a_{\sigma^{-1}(1)1} a_{\sigma^{-1}(2)2} \cdots a_{\sigma^{-1}(n)n} \qquad (1.9.8)$$

The set of elements of $\mathbb{S}_n$ is exactly the same as the set of inverses of the elements of $\mathbb{S}_n$ (see Exercise 12 of Sec. 1.8). Consequently we may take the summation in equation (1.9.8) for all σ^{-1} belonging to $\mathbb{S}_n$. Thus equation (1.9.8) becomes

$$\det A = \sum_{\sigma^{-1} \in \mathbb{S}_n} \text{sgn } \sigma^{-1} a_{\sigma^{-1}(1)1} a_{\sigma^{-1}(2)2} \cdots a_{\sigma^{-1}(n)n} \qquad (1.9.9)$$

Set $\sigma^{-1} = \tau$. Then equation (1.9.9) becomes equation (1.9.7).   ////

A very important property of the determinant function is proved in Theorem 1.9.18, namely, that the determinant of a product of matrices of order n is equal to the product of the determinants of the matrices. The main tool used in the proof of this result is the expression of a matrix as a product of elementary matrices and a matrix of the form (1.7.9), as given in Corollary 1.7.14. Thus we first investigate the effect each of the elementary row and column operations has on the determinant. The three elementary operations are considered in Lemmas 1.9.10, 1.9.14, and 1.9.15. The second part of each of these lemmas shows that Theorem 1.9.18 holds in the special case when one of the matrices of the product is an elementary matrix. Because of Lemma 1.9.6 it will be sufficient to consider only the three row operations.

Lemma 1.9.10 (a) det $R_{rs} = -1$.
  (b) det $R_{rs} A = -\det A = (\det R_{rs})(\det A)$
for every square matrix A.

PROOF (a) R_{rs} has exactly n entries which are not zero. These occur in the (r, s) and (s, r) positions and in all the diagonal positions except the (r, r) and (s, s) positions, and they are all equal to 1. Thus, only one term of the expression (1.9.2) for det R_{rs} is not zero, namely, the term containing exactly the n nonzero entries of R_{rs}. This is the term of the expression (1.9.2) corresponding to the transposition $\rho = (r \ \ s)$ of $\mathbb{S}_n$. Thus

$$\det R_{rs} = \operatorname{sgn} \rho \, 1 \cdot 1 \cdots 1 = -1$$

(b) Let $A = [a_{ij}]$ and set $R_{rs} A = B = [b_{ij}]$. Then A and B are related by

$$
\begin{aligned}
b_{ij} &= a_{ij} && i \neq r, s && j = 1, 2, \ldots, n && (1.9.11)\\
b_{rj} &= a_{sj} \\
b_{sj} &= a_{rj}
\end{aligned}
$$

We can express equations (1.9.11) more simply by using permutations. Let σ be any element of $\mathbb{S}_n$ and set $(r \ \ s)\sigma = \tau$. Then equations (1.9.11) reduce to the single equation

$$b_{\tau(j)j} = a_{\sigma(j)j} \qquad j = 1, 2, \ldots, n$$

For if $\sigma(j) = i \neq r, s$, then

$$\tau(j) = (r \ \ s)[\sigma(j)] = (r \ \ s)(i) = i$$

and hence

$$b_{\tau(j)j} = b_{ij} = a_{ij} = a_{\sigma(j)j}$$

while if $\sigma(j) = r$, then

$$\tau(j) = (r \ \ s)[\sigma(j)] = (r \ \ s)(r) = s$$

and hence

$$b_{\tau(j)j} = b_{sj} = a_{rj} = a_{\sigma(j)j}$$

and if $\sigma(j) = s$, then

$$\tau(j) = (r \ \ s)[\sigma(j)] = (r \ \ s)(s) = r$$

and hence

$$b_{\tau(j)j} = b_{rj} = a_{sj} = a_{\sigma(j)j}$$

Thus, using Lemma 1.9.6, we have

$$
\begin{aligned}
\det B &= \sum_{\tau \in \mathbb{S}_n} \operatorname{sgn} \tau \, b_{\tau(1)1} b_{\tau(2)2} \cdots b_{\tau(n)n}\\
&= \sum_{(r \ \ s)\sigma \in \mathbb{S}_n} \operatorname{sgn}(r \ \ s)\sigma \, a_{\sigma(1)1} a_{\sigma(2)2} \cdots a_{\sigma(n)n}\\
&= - \sum_{(r \ \ s)\sigma \in \mathbb{S}_n} \operatorname{sgn} \sigma \, a_{\sigma(1)1} a_{\sigma(2)2} \cdots a_{\sigma(n)n}
\end{aligned}
$$

since sgn$(r\ s)\sigma = -\text{sgn}\ \sigma$ for every $\sigma \in \mathbb{S}_n$.   It is clear that the set of $n!$ permutation

$$\{(r\ s)\sigma \mid \sigma \in \mathbb{S}_n\}$$

is exactly the set of $n!$ elements of $\mathbb{S}_n$.   Thus the last summation above is simply the sum for all permutations contained in $\mathbb{S}_n$.   Consequently we have shown that det $B = -\det A$.   ////

Part (a) of Lemma 1.9.10 can be stated as

$$\det R_{rs} = \text{sgn}(r\ s)$$

Since any permutation can be expressed as a product of transpositions, it follows that:

Corollary 1.9.12   For every permutation σ

$$\det R_\sigma = \text{sgn}\ \sigma = \det C_\sigma$$

Lemma 1.9.10 has another very useful corollary.

Corollary 1.9.13   If two rows (or two columns) of a square matrix are identical, then the determinant of the matrix is zero.[1]

PROOF   Let the rth and sth rows of A be identical, $r \neq s$.   Then

$$A = R_{rs} A$$

and hence, by Lemma 1.9.10,

$$\det A = \det(R_{rs} A) = -\det A$$

Thus det $A = 0$.   ////

Corollary 1.9.13 is a special case of Corollary 1.9.17, which states that the determinant of any square matrix whose rows or columns are linearly dependent is zero.

We return to the investigation of the effect on the determinant of the elementary operations.

Lemma 1.9.14   (a)   det $R_{rs}(\alpha) = 1$.
          (b)   det $R_{rs}(\alpha)A = \det A = [\det R_{rs}(\alpha)](\det A)$

for every square matrix A.

---

[1] This proof fails when $\mathbb{F}$ has the property $1 + 1 = 0$ (as, for example, when $\mathbb{F} = \mathbb{Z}_2$). The corollary is still true in this case, but a different proof is necessary.   See Exercise 23 and set $m = 2$.

PROOF   (a) $R_{rs}(\alpha)$ contains exactly $n + 1$ nonzero entries.   These occur in the n diagonal positions, where we have 1s, and in the (r, s) position, where we have α.   First, using Definition 1.9.1 for the determinant of $R_{rs}(\alpha)$, we wish to show that every term of expression (1.9.2) which contains α as a factor is equal to zero.   Clearly α appears in the term of (1.9.2) corresponding to the permutation σ if and only if $\sigma(r) = s$.   In fact, such a term is equal to sgn σ multiplied by all the entries in the $(j, \sigma(j))$ positions of $R_{rs}(\alpha), j = 1, 2, \ldots, n$.   In particular, let us look at the entry in the $(s, \sigma(s))$ position of $R_{rs}(\alpha)$.   Set $\sigma(s) = t$.   Since we already have $\sigma(r) = s$ and $r \neq s$, it follows that $\sigma(s) \neq s$, that is, $t \neq s$.   But the only nonzero entry in the sth row of $R_{rs}(\alpha)$ is the 1 in the (s, s) position.   Thus the entry in the $(s, t) = (s, \sigma(s))$ position of $R_{rs}(\alpha)$ is zero, and consequently the term of (1.9.2) corresponding to σ is also zero.

With the elimination of the possibility that α is a factor of some term of (1.9.2), it becomes clear that the only nonzero term in (1.9.2) is the one which corresponds to the permutation I of $\mathbb{S}_n$.   Hence

$$\det R_{rs}(\alpha) = (\text{sgn } I)\, 1 \cdot 1 \cdots 1 = 1$$

(b) Let $A = [a_{ij}]$ and set $R_{rs}(\alpha)A = B = [b_{ij}]$.   Then A and B are related by

$$b_{ij} = a_{ij} \qquad i \neq r \qquad j = 1, 2, \ldots, n$$
$$b_{rj} = a_{rj} + \alpha a_{sj}$$

As before, it will be useful to express this relation in terms of permutations. Let σ be any element of $\mathbb{S}_n$.   Then we have

$$b_{i\sigma(i)} = a_{i\sigma(i)} \qquad i \neq r$$
$$b_{r\sigma(r)} = a_{r\sigma(r)} + \alpha a_{s\sigma(r)}$$

Thus, by Definition 1.9.1, we have

$$
\begin{aligned}
\det B &= \sum_{\sigma \in \mathbb{S}_n} \text{sgn } \sigma\, b_{1\sigma(1)} b_{2\sigma(2)} \cdots b_{r\sigma(r)} \cdots b_{n\sigma(n)} \\
&= \sum_{\sigma \in \mathbb{S}_n} \text{sgn } \sigma\, a_{1\sigma(1)} a_{2\sigma(2)} \cdots (a_{r\sigma(r)} + \alpha a_{s\sigma(r)}) \cdots a_{n\sigma(n)} \\
&= \sum_{\sigma \in \mathbb{S}_n} \text{sgn } \sigma\, a_{1\sigma(1)} a_{2\sigma(2)} \cdots a_{r\sigma(r)} \cdots a_{n\sigma(n)} \\
&\quad + \alpha \sum_{\sigma \in \mathbb{S}_n} \text{sgn } \sigma\, a_{1\sigma(1)} a_{2\sigma(2)} \cdots a_{s\sigma(r)} \cdots a_{n\sigma(n)} \\
&= \det A + \alpha \det C
\end{aligned}
$$

where C is a matrix in which the rth and sth rows are equal.   In fact,

$$C_{(i)} = A_{(i)} \qquad i \neq r$$
$$C_{(r)} = A_{(s)} = C_{(s)}$$

According to Corollary 1.9.13, det $C = 0$ and thus det $B = $ det A.   ////

Lemma 1.9.15 (*a*)   det $R_r(\lambda) = \lambda$.
   (*b*)   det $R_r(\lambda)A = \lambda$ det $A = [$det $R_r(\lambda)](\text{det } A)$
for every square matrix A.

PROOF   The proof is left as an exercise.   ////

In defining the third elementary operation we were careful to specify that $\lambda \neq 0$. According to the first parts of Lemmas 1.9.10, 1.9.14, and 1.9.15, the determinant of an elementary matrix is never zero.

Corollary 1.7.14, together with Lemmas 1.9.10, 1.9.14, and 1.9.15, leads to an expression, (1.9.16), for the determinant of any square matrix in terms of determinants of elementary matrices and a matrix of the form (1.7.9). Since the determinant of each of these matrices is readily obtainable, equation (1.9.16) describes a method for evaluating the determinant which is often more practical than using Definition 1.9.1 directly. For a given matrix A, Corollary 1.7.14 states that there exist elementary matrices $R_1, R_2, \ldots, R_k$ and $C_1, C_2, \ldots, C_l$ such that

$$A = R_1 R_2 \cdots R_k \begin{bmatrix} I_r & 0 \\ 0 & 0 \end{bmatrix} C_l C_{l-1} \cdots C_1$$

where I_r is an identity matrix whose order is equal to the rank of A. By repeated use of Lemmas 1.9.10, 1.9.14, and 1.9.15 and their column-matrix counterparts we have

$$\text{det } A = \det\left(R_1 R_2 \cdots R_k \begin{bmatrix} I_r & 0 \\ 0 & 0 \end{bmatrix} C_l C_{l-1} \cdots C_1 \right)$$

$$= (\text{det } R_1)\left[\det\left(R_2 \cdots R_k \begin{bmatrix} I_r & 0 \\ 0 & 0 \end{bmatrix} C_l C_{l-1} \cdots C_1 \right) \right]$$

$$= (\text{det } R_1)(\text{det } R_2)\left[\det\left(R_3 \cdots R_k \begin{bmatrix} I_r & 0 \\ 0 & 0 \end{bmatrix} C_l C_{l-1} \cdots C_1 \right) \right]$$

. . . .

$$= (\text{det } R_1)(\text{det } R_2) \cdots (\text{det } R_k)\left[\det\left(\begin{bmatrix} I_r & 0 \\ 0 & 0 \end{bmatrix} C_l C_{l-1} \cdots C_1 \right) \right]$$

$$= (\text{det } R_1)(\text{det } R_2) \cdots (\text{det } R_k)\left[\det\left(\begin{bmatrix} I_r & 0 \\ 0 & 0 \end{bmatrix} C_l C_{l-1} \cdots C_2 \right) \right](\text{det } C_1)$$

. . . .

$$= (\text{det } R_1)(\text{det } R_2) \cdots (\text{det } R_k)\left(\det \begin{bmatrix} I_r & 0 \\ 0 & 0 \end{bmatrix} \right)(\text{det } C_l)(\text{det } C_{l-1}) \cdots (\text{det } C_1)$$

$$\text{(1.9.16)}$$

Since the determinant of each elementary matrix is nonzero, it follows that $\det A = 0$ if and only if $\det\begin{bmatrix} I_r & 0 \\ 0 & 0 \end{bmatrix} = 0$. But it is clear from Definition 1.9.1 that

$$\det\begin{bmatrix} I_r & 0 \\ 0 & 0 \end{bmatrix} = \begin{cases} 1 & \text{if the rank of } A \text{ is } n \text{ (in which case the zero} \\ & \text{matrices do not actually appear)} \\ 0 & \text{if the rank of } A \text{ is less than } n \text{ (in which} \\ & \text{case the zero matrices do actually appear)} \end{cases}$$

Thus we have another characterization of nonsingular matrices to add to those of Theorem 1.6.4.

Corollary 1.9.17 Let A be a square matrix. Then A is nonsingular if and only if $\det A \neq 0$.

Corollary 1.7.14 shows that a nonsingular matrix A can be expressed as a product of elementary matrices; for when A is nonsingular, the matrix of (1.7.9) becomes an identity matrix. Since each elementary row matrix is also an elementary column matrix, and conversely, we can consider A to be expressed either as a product of elementary row matrices or as a product of elementary column matrices. By applying the same method that was used to obtain equation (1.9.16) to a product of two matrices of the same order we can prove the main theorem of this section, namely:

Theorem 1.9.18 For any two matrices, A and B, of the same order,

$$\det AB = (\det A)(\det B)$$

PROOF If either A or B is singular, then AB is also singular (see Exercise 12 of Sec. 1.6), and, by Corollary 1.9.17, we have

$$\det AB = 0 = (\det A)(\det B)$$

Now suppose that A and B are both nonsingular. Then the theorem follows from the second parts of Lemmas 1.9.10, 1.9.14, and 1.9.15. Since A is nonsingular, A can be expressed as a product of elementary row matrices, say

$$A = R_1 R_2 \cdots R_k$$

As before,

$$\begin{aligned} \det A &= \det(R_1 R_2 \cdots R_k B) \\ &= (\det R_1)(\det R_2) \cdots (\det R_k)(\det B) \\ &= (\det R_1 R_2 \cdots R_k)(\det B) \\ &= (\det A)(\det B) \end{aligned}$$

////

There is an alternative definition for the determinant of a matrix which emphasizes the functional rather than the computational aspects of determinants. This definition is in terms of three elementary properties of the determinant function.   One drawback of this type of definition is that it is often necessary to prove that such a function actually exists.   By presenting Definition 1.9.1 first we avoid this step.   However, in Theorem 1.9.20 we will prove that these properties define a unique function whose domain is the set of square matrices with entries from $\mathbb{F}$ and whose range is $\mathbb{F}$.   Clearly this function must be the determinant.

Two of the three properties of the determinant function that we will use have already been explicitly stated.   We will also need:

Lemma 1.9.19   The determinant is an *n-linear* function of its rows, that is, if A, B and C are $n \times n$ matrices which are identical except for the rth row, and if

$$C_{(r)} = \alpha A_{(r)} + \beta B_{(r)}$$

for some α and β in $\mathbb{F}$ then

$$\det C = \alpha \det A + \beta \det B$$

PROOF   The proof is left as an exercise.   ////

Theorem 1.9.20   There is exactly one function D whose domain is the set of square matrices with entries from $\mathbb{F}$ and whose range is $\mathbb{F}$ which satisfies the three conditions:

1   If two rows of A are equal, then $D(A) = 0$.
2 D is an n-linear function of the rows of A.
3 $D(I) = 1$.

Consequently, $D(A) = \det A$.

PROOF   We have already shown that the determinant, as defined in Definition 1.9.1, has properties 1, 2, and 3.

Conversely, let D be a mapping which satisfies conditions 1, 2, and 3.   First, we wish to show that if B is a matrix obtained from A by interchanging the rth and sth rows of A, then

$$D(B) = -D(A) \qquad (1.9.21)$$

Let C be defined by

$$C_{(r)} = C_{(s)} = A_{(r)} + A_{(s)} = B_{(r)} + B_{(s)}$$
$$C_{(i)} = A_{(i)} = B_{(i)} \qquad i \neq r, s$$

By condition 1,

$$D(C) = 0$$

However, applying condition 2 to both the rth and sth rows of C, we have

$$D(C) = D(A) + D(B) + D(X) + D(Y)$$

where
$$X_{(r)} = X_{(s)} = A_{(r)}$$
$$Y_{(r)} = Y_{(s)} = A_{(s)}$$
$$X_{(i)} = Y_{(i)} = A_{(i)} \qquad i \neq r, s$$

Thus $D(X) = D(Y) = 0$, and we have established (1.9.21).

The analog of Corollary 1.9.12 follows directly from (1.9.21), that is,

$$D(R_\sigma) = \text{sgn } \sigma \qquad (1.9.22)$$

for every permutation σ.

Now let $A = [a_{ij}]$ be an $n \times n$ matrix. By repeated use of condition 2 we have

$$D(A) = \sum_\alpha a_{1\alpha(1)} a_{2\alpha(2)} \cdots a_{n\alpha(n)} D(I_{\alpha(1), \alpha(2), \cdots, \alpha(n)}) \qquad (1.9.23)$$

where α is any map of $\mathbb{N} = \{1, 2, \ldots, n\}$ into itself and $I_{\alpha(1), \alpha(2), \cdots, \alpha(n)}$ represents the $n \times n$ matrix whose rows are $I_{\alpha(1)}, I_{\alpha(2)}, \ldots, I_{\alpha(n)}$. (Note that there are n^n such maps, of which only $n!$ are permutations.) Making use of condition 1, we see that

$$D(I_{\alpha(1), \alpha(2), \cdots, \alpha(n)}) = 0$$

unless α is a permutation of $\mathbb{N}$. When α is a permutation, $I_{\alpha(1), \alpha(2), \cdots, \alpha(n)}$ has 1 in the $(1, \alpha(1)), (2, \alpha(2)), \ldots, (n, \alpha(n))$ positions and 0 elsewhere, that is,

$$I_{\alpha(1), \alpha(2), \cdots, \alpha(n)} = C_\alpha = R_{\alpha^{-1}}$$

[see part (b) of Exercise 18 in Sec. 1.8]. Since $\text{sgn } \alpha = \text{sgn } \alpha^{-1}$ for every permutation α, the theorem follows by substituting (1.9.22) into (1.9.23).

////

As we have pointed out, the direct use of Definition 1.9.1 is not a convenient way of computing the determinant of a given matrix except when its order is very small. The determinant can also be obtained by applying Theorem 1.9.18 to (1.7.15). This method is not much better. However, several practical schemes have been devised for computing the determinant. We will present one which, in addition, leads to a very useful expression for the inverse of a matrix.

Let $A = [a_{ij}]$ be a square matrix of order n. According to Definition 1.9.1, the determinant of A is a linear homogeneous polynomial in the entries of any one row of A each of whose coefficients is a homogeneous polynomial of degree $n - 1$ in the entries of the remaining $n - 1$ rows. More specifically, if we select one row of A, say the pth row, and group together those terms of the determinant which contain a_{p1}, those which contain a_{p2}, etc., we have

$$\det A = a_{p1} A_{p1} + a_{p2} A_{p2} + \cdots + a_{pn} A_{pn} \qquad (1.9.24)$$

where $\displaystyle A_{pq} = \sum_{\substack{\sigma \in S_n \\ \sigma(p) = q}} \operatorname{sgn} \sigma \, a_{1\sigma(1)} a_{2\sigma(2)} \cdots a_{p-1,\sigma(p-1)} a_{p+1,\sigma(p+1)} \cdots a_{n\sigma(n)}$

$$= \sum_{\substack{\sigma \in S_n \\ \sigma(p) = q}} \operatorname{sgn} \sigma \prod_{\substack{r \in \mathbb{N} \\ r \neq p}} a_{r\sigma(r)} \qquad (1.9.25)$$

Equation (1.9.24) is called the pth *row expansion* of $\det A$, and A_{pq} is the *cofactor* of a_{pq} in A. The form of A_{pq} in equation (1.9.25) is strongly suggestive of Definition 1.9.1 for the determinant. Theorem 1.9.26 shows that A_{pq} is, up to a possible difference in sign, the determinant of a submatrix of A of order $n - 1$.

Theorem 1.9.26   Let $A = [a_{ij}]$ be a matrix of order n and let B_{pq} be the submatrix of A formed by deleting the pth row and the qth column of A. Then

$$A_{pq} = (-1)^{p+q} \det B_{pq}$$

PROOF   Let

$$B_{pq} = [b_{rs}] \qquad r, s = 1, 2, \ldots, n - 1$$

Then

$$B_{pq} = \begin{bmatrix}
a_{11} & a_{12} & \cdots & a_{1,q-1} & a_{1,q+1} & \cdots & a_{1n} \\
a_{21} & a_{22} & \cdots & a_{2,q-1} & a_{2,q+1} & \cdots & a_{2n} \\
\vdots & & & & & & \vdots \\
a_{p-1,1} & a_{p-1,2} & \cdots & a_{p-1,q-1} & a_{p-1,q+1} & \cdots & a_{p-1,n} \\
a_{p+1,1} & a_{p+1,2} & \cdots & a_{p+1,q-1} & a_{p+1,q+1} & \cdots & a_{p+1,n} \\
\vdots & & & & & & \vdots \\
a_{n1} & a_{n2} & \cdots & a_{n,q-1} & a_{n,q+1} & \cdots & a_{nn}
\end{bmatrix}$$

Hence, for $1 \leq r, s \leq n, r \neq p, s \neq q$,

$$a_{rs} = \begin{cases}
b_{rs} & r \leq p - 1,\ s \leq q - 1 \\
b_{r-1,s} & r \geq p + 1,\ s \leq q - 1 \\
b_{r,s-1} & r \leq p - 1,\ s \geq q + 1 \\
b_{r-1,s-1} & r \geq p + 1,\ s \geq q + 1
\end{cases} \qquad (1.9.27)$$

The four parts of equation (1.9.27) can be expressed in a single equation using permutations. Define the permutations ρ and τ by

$$\rho = \begin{pmatrix} 1 & 2 & \cdots & p-1 & p & p+1 & \cdots & n-1 & n \\ 1 & 2 & \cdots & p-1 & n & p & \cdots & n-2 & n-1 \end{pmatrix}$$

$$= (p \ \ p+1)(p \ \ p+2)\cdots(p \ \ n-1)(p \ \ n)$$

$$\tau = \begin{pmatrix} 1 & 2 & \cdots & q-1 & q & q+1 & \cdots & n-1 & n \\ 1 & 2 & \cdots & q-1 & n & q & \cdots & n-2 & n-1 \end{pmatrix}$$

$$= (q \ \ q+1)(q \ \ q+2)\cdots(q \ \ n-1)(q \ \ n)$$

Then equation (1.9.27) becomes

$$a_{rs} = b_{\rho(r)\tau(s)} \qquad \begin{matrix} r \neq p \\ s \neq q \end{matrix} \qquad (1.9.28)$$

For if $r \leq p-1$, then $\rho(r) = r$, while if $r \geq p+1$, then $\rho(r) = r-1$. Note that we have already excluded the possibility that $r = p$. It then follows that $\rho(r)$ is never equal to n. A similar argument shows that $\tau(s)$ is the correct second subscript for b in equation (1.9.28).

Let σ be any permutation of $\mathbb{N} = \{1, 2, \ldots, n\}$ such that $\sigma(p) = q$. By equation (1.9.28), we have

$$a_{r\sigma(r)} = b_{\rho(r)\tau\sigma(r)} \qquad r \neq p$$

Taking the product of these terms for all $r \in \mathbb{N}$, $r \neq p$, yields

$$\prod_{\substack{r \in \mathbb{N} \\ r \neq p}} a_{r\sigma(r)} = \prod_{\rho(r) \in \mathbb{N}'} b_{\rho(r)\tau\sigma(r)} \qquad (1.9.29)$$

where $\mathbb{N}' = \{1, 2, \ldots, n-1\}$. We now set $\rho(r) = t$. Then $r = \rho^{-1}(t)$ and

$$\tau\sigma(r) = \tau\sigma\rho^{-1}(t)$$

If we set $\tau\sigma\rho^{-1} = \omega$, then equation (1.9.29) becomes

$$\prod_{\substack{r \in \mathbb{N} \\ r \neq p}} a_{r\sigma(r)} = \prod_{t \in \mathbb{N}'} b_{t\omega(t)} = b_{1\omega(1)}b_{2\omega(2)}\cdots b_{n-1,\,\omega(n-1)}$$

Moreover, since ρ and τ are fixed permutations of $\mathbb{N}$, it follows that as σ runs through the $(n-1)!$ permutations of $\mathbb{N}$ for which $\sigma(p) = q$, then $\omega = \tau\sigma\rho^{-1}$ runs through the $(n-1)!$ distinct permutations of $\mathbb{N}'$. Hence

$$\det B_{pq} = \sum_{\omega \in \mathbb{S}_{n-1}} \operatorname{sgn} \omega \ b_{1\omega(1)}b_{2\omega(2)}\cdots b_{n-1,\,\omega(n-1)}$$

$$= \sum_{\substack{\sigma \in \mathbb{S}_n \\ \sigma(p)=q}} \operatorname{sgn} \tau\sigma\rho^{-1} \prod_{\substack{r \in \mathbb{N} \\ r \neq p}} a_{r\sigma(r)}$$

$$= \operatorname{sgn} \tau \operatorname{sgn} \rho^{-1} \sum_{\substack{\sigma \in \mathbb{S}_n \\ \sigma(p)=q}} \operatorname{sgn} \sigma \prod_{\substack{r \in \mathbb{N} \\ r \neq p}} a_{r\sigma(r)}$$

$$= \operatorname{sgn} \tau \operatorname{sgn} \rho^{-1} A_{pq} = (-1)^{2n-p-q}A_{pq} = (-1)^{p+q}A_{pq}$$

since $\qquad \operatorname{sgn} \tau = (-1)^{n-q} \qquad \operatorname{sgn} \rho^{-1} = \operatorname{sgn} \rho = (-1)^{n-q}$ ////

Thus we have shown that in the expression for the determinant of a matrix the coefficient of each entry is, up to sign, the determinant of some $(n-1) \times (n-1)$ submatrix of the matrix. Let us give these submatrices and their determinants names.

Definition 1.9.30 Let $A = [a_{ij}]$ be an $n \times n$ matrix. The submatrix B_{pq} of order $n-1$ formed from A by deleting the pth row and the qth column of A is called the *minor* of a_{pq} in A. The number

$$A_{pq} = (-1)^{p+q} \det B_{pq}$$

is called the *cofactor* of a_{pq}. Finally, the matrix of order n which has the cofactor A_{pq} of a_{pq} in the (q, p) position, $p, q, = 1, 2, \ldots, n$, is called the *adjoint*[1] of A and is denoted[2] by *adj* A.

According to equation (1.9.24), the cofactor of a_{pq} is the coefficient of a_{pq} in the pth row expansion for the determinant of A. This entire discussion can be repeated expressing the determinant of A as a linear homogeneous polynomial in the entries of the qth column of A. The coefficient of a_{pq} is the same in both expressions for det A, namely, the cofactor A_{pq} of a_{pq} in A. Thus, in addition to equation (1.9.24) we have

$$\det A = a_{1q} A_{1q} + a_{2q} A_{2q} + \cdots + a_{nq} A_{nq} \qquad (1.9.31)$$

which we call the qth *column expansion* of det A.

The adjoint of a matrix leads to an interesting and often useful construction of the inverse of a nonsingular matrix.

Theorem 1.9.32 For any square matrix A,

$$A \cdot \text{adj } A = \text{adj } A \cdot A = (\det A) I$$

In particular, when A is nonsingular,

$$A^{-1} = \frac{1}{\det A} \text{adj } A$$

PROOF We will first show that

$$A \cdot \text{adj } A = (\det A) I \qquad (1.9.33)$$

is a direct consequence of equation (1.9.24) and Corollary 1.9.13. Set

$$A \cdot \text{adj } A = C = [c_{ij}]$$

---

[1] The adjoint of a matrix as defined here should not be confused with the adjoint of a linear transformation of an abstract vector space (see, for example, Halmos [6] or Taylor [9]).

[2] It will be convenient to define the adjoint of any 1×1 matrix to be the 1×1 matrix [1].

According to Definition 1.9.30 and the definition of multiplication of matrices [allowing for the fact that A_{pq} is in the (q, p) position of adj A], we have

$$c_{ij} = a_{i1} A_{j1} + a_{i2} A_{j2} + \cdots + a_{in} A_{jn} \qquad i, j = 1, 2, \ldots, n \qquad (1.9.34)$$

But, by equation (1.9.24), the right-hand side of equation (1.9.34) is equal to the determinant of the matrix formed from A by replacing $A_{(j)}$ by $A_{(i)}$. When $i = j$, this matrix is simply A and hence

$$c_{ii} = \det A \qquad i = 1, 2, \ldots, n$$

This is equation (1.9.24) again.   On the other hand, when $i \neq j$, the right-hand side of equation (1.9.34) is equal to the determinant of a matrix whose ith and jth rows are identical, i.e., they are both equal to $A_{(i)}$.   By Corollary 1.9.13,

$$c_{ij} = 0 \qquad i \neq j$$

completing the proof of equation (1.9.33).

A proof that adj $A \cdot A = (\det A) I$ would be a repetition of the above argument beginning with equation (1.9.31) instead of equation (1.9.24).

When A is nonsingular, it follows from Corollary 1.9.17 that $\det A \neq 0$ and consequently

$$A \cdot \frac{1}{\det A} \text{ adj } A = \frac{1}{\det A} \text{ adj } A \cdot A = I$$

Thus

$$A^{-1} = \frac{1}{\det A} \text{ adj } A \qquad \text{////}$$

Note that when A is singular, then $\det A = 0$ and hence

$$A \cdot \text{adj } A = \text{adj } A \cdot A = 0$$

EXAMPLE 1.9.35   Let

$$A = \begin{bmatrix} 1 & -2 & 0 \\ 2 & -1 & 5 \\ 1 & 0 & 4 \end{bmatrix}$$

Find (a) the minor of $-2 \, (= a_{12})$, (b) the cofactor of -2, (c) adj A, and (d) A^{-1}.

(a) We will use the notation of Definition 1.9.30.   The minor of $a_{12} = -2$ is

$$B_{12} = \begin{bmatrix} 2 & 5 \\ 1 & 4 \end{bmatrix}$$

and is obtained by deleting the first row and the second column of A.

(b) the cofactor of -2 is

$$A_{12} = (-1)^{1+2} \det\begin{bmatrix} 2 & 5 \\ 1 & 4 \end{bmatrix} = -3$$

(c) Continuing in this way, we find that

$$\text{adj } A = \begin{bmatrix} -4 & 8 & -10 \\ -3 & 4 & -5 \\ 1 & -2 & 3 \end{bmatrix}$$

(d) Since det $A = 2$, we have

$$A^{-1} = \frac{1}{2}\begin{bmatrix} -4 & 8 & -10 \\ -3 & 4 & -5 \\ 1 & -2 & 3 \end{bmatrix} = \begin{bmatrix} -2 & 4 & -5 \\ -\frac{3}{2} & 2 & -\frac{5}{2} \\ \frac{1}{2} & -1 & \frac{3}{2} \end{bmatrix} \qquad ////$$

We conclude this section with a brief introduction to a special type of matrix which will be used extensively in later sections of this chapter.

Definition 1.9.36 For $i = 1, 2, \ldots, k$, let A_i be a square matrix of order n_i. The square matrix

$$A = \begin{bmatrix} A_1 & 0 & 0 & \cdots & 0 \\ 0 & A_2 & 0 & \cdots & 0 \\ 0 & 0 & A_3 & \cdots & 0 \\ \multicolumn{5}{c}{\cdots\cdots\cdots\cdots\cdots} \\ 0 & 0 & 0 & \cdots & A_k \end{bmatrix} \qquad (1.9.37)$$

of order $n_1 + n_2 + \cdots + n_k$ is called the *direct sum* of the matrices $A_1, A_2, \ldots, A_k$. A more convenient notation for (1.9.37) is

$$A = A_1 \oplus A_2 \oplus \cdots \oplus A_k$$

We have already encountered the direct sum. In terms of Definition 1.9.36, Theorem 1.7.8 states that a square matrix of order n and rank r can be transformed by elementary operations into the direct sum of an identity matrix of order r and a zero matrix of order $n - r$. (Of course, Theorem 1.7.8 applies to nonsquare matrices as well.) The direct sum of matrices will play a prominent role in Secs. 1.10, 1.12, and 1.13.

For any positive integer n and any square matrix B it can easily be shown that

$$\det(I_n \oplus B) = \det B = \det(B \oplus I_n) \qquad (1.9.38)$$

The first-row expansion of $\det(I_n \oplus B)$ yields

$$\det(I_n \oplus B) = 1 \cdot \det(I_{n-1} \oplus B) = \det(I_{n-1} \oplus B)$$

After repeating this process n times, we have the first equation of (1.9.38). The proof of the second equation of (1.9.38) is similar. This is a simple special case of:

Theorem 1.9.39 Let

$$A = A_1 \oplus A_2 \oplus \cdots \oplus A_k$$

Then
$$\det A = (\det A_1)(\det A_2) \cdots (\det A_k)$$

PROOF The proof is by an induction on k. When $k = 1$, we have $A = A_1$ and the theorem is immediate. Now assume the theorem is true for direct sums of $k - 1$ matrices and write

$$A = \begin{bmatrix} A_1 & 0 & \cdots & 0 & 0 \\ 0 & A_2 & \cdots & 0 & 0 \\ \hdotsfor{5} \\ 0 & 0 & \cdots & A_{k-1} & 0 \\ 0 & 0 & \cdots & 0 & A_k \end{bmatrix}$$

$$= \begin{bmatrix} A_1 & 0 & \cdots & 0 & 0 \\ 0 & A_2 & \cdots & 0 & 0 \\ \hdotsfor{5} \\ 0 & 0 & \cdots & A_{k-1} & 0 \\ 0 & 0 & \cdots & 0 & I \end{bmatrix} \begin{bmatrix} I & 0 & \cdots & 0 & 0 \\ 0 & I & \cdots & 0 & 0 \\ \hdotsfor{5} \\ 0 & 0 & \cdots & I & 0 \\ 0 & 0 & \cdots & 0 & A_k \end{bmatrix}$$

Using (1.9.38) twice, we see that

$$\det A = [\det(A_1 + A_2 + \cdots + A_{k-1})](\det A_k)$$

and, applying the induction hypothesis,

$$\det A = [(\det A_1)(\det A_2) \cdots (\det A_{k-1})](\det A_k) \qquad ////$$

EXERCISES

1 Find the determinant of each of the following matrices:

(a) $[-3]$
(b) $\begin{bmatrix} 1 & 2 \\ 2 & -5 \end{bmatrix}$
(c) $\begin{bmatrix} 1 & 2 & -1 \\ -7 & 0 & 5 \\ -4 & 1 & 3 \end{bmatrix}$

$$(d) \begin{bmatrix} 1 & 2 & -1 \\ 2 & 1 & 3 \\ 3 & 3 & 2 \end{bmatrix} \qquad (e) \begin{bmatrix} 1 & 2 & -2 & 0 \\ 3 & -3 & 5 & -1 \\ 1 & 3 & 5 & 12 \\ -3 & 0 & 1 & 4 \end{bmatrix}$$

(1) Directly from Definition 1.9.1.

(2) By transforming the matrix to the form (1.7.9) using elementary opera-tions.

(3) By using the first-row expansion.

(4) By using the last-column expansion.

2 Find the determinant of each of the following matrices:

$$(a) \begin{bmatrix} 1 & 2 & 0 & 0 \\ -1 & 2 & 0 & 0 \\ 0 & 0 & 1 & 2 \\ 0 & 0 & 2 & -5 \end{bmatrix} \qquad (b) \begin{bmatrix} 2 & 0 & 0 & 0 \\ 0 & 1 & 2 & -1 \\ 0 & 2 & 0 & 5 \\ 0 & -4 & 1 & 3 \end{bmatrix}$$

(1) Directly from Definition 1.9.1 and (2) By using Theorem 1.9.39.

3 For what values of x is

$$\begin{bmatrix} 1 & 2 & -1 \\ x & 1 & 3 \\ 3 & 3 & x \end{bmatrix}$$

nonsingular?

4 Find the inverse of the matrix in part (c) of Exercise 1:

(a) By using Theorem 1.9.32.

(b) By expressing the matrix as a product of elementary matrices and using Exercise 11 of Sec. 1.6.

(c) By treating the columns of the matrix as a basis for $\mathbb{F}^3$.

5 Find the adjoint of the matrix in part (d) of Exercise 1. Verify Theorem 1.9.32.

6 Let $A = [a_{ij}]$ be a square matrix of order n of the form

$$A = \begin{bmatrix} 0 & 0 & 0 & \cdots & 0 & a_{1n} \\ a_{21} & 0 & 0 & \cdots & 0 & 0 \\ 0 & a_{32} & 0 & \cdots & 0 & 0 \\ \cdots & \cdots & \cdots & \cdots & \cdots & \cdots \\ 0 & 0 & 0 & \cdots & a_{n,n-1} & 0 \end{bmatrix}.$$

Find det A (a) by an induction on n and (b) by showing that A is related to a permutation matrix.

7 Prove Lemma 1.9.15.

8 Let A be a square matrix of order n and let $\alpha \in \mathbb{F}$. Find (a) det αA and (b) adj αA.

9 Let A be a square matrix of order n. Show that the rank of adj A is:
  (a) n when A has rank n.
  (b) 1 when A has rank $n - 1$.
  (c) 0 when A has rank less than $n - 1$.

10 Show that

$$\text{adj } A^T = (\text{adj } A)^T$$

for every square matrix A.

11 Show that

$$\text{adj } A^{-1} = (\text{adj } A)^{-1}$$

for every nonsingular matrix A.

12 Show that

$$\det A = \det A^T$$

for every square matrix A by:
  (a) Using Lemma 1.9.6.
  (b) Using a first-row expansion of A, a first-column expansion of A^T, and an induction on the order of A.
  (c) Reducing A to the form (1.7.9) using elementary operations.

13 A matrix is *symmetric* if $A^T = A$ and *skew-symmetric* if $A^T = -A$. Prove that if $1 + 1 \neq 0$, then every skew-symmetric matrix with entries from $\mathbb{F}$ which is of odd order is singular. (See Exercise 20 of Sec. 1.7.)

14 Show that the adjoint of a symmetric matrix is symmetric. Show that the adjoint of a skew-symmetric matrix is skew-symmetric if its order is even and symmetric if its order is odd.

15 Show that for every square matrix A, the matrices $A + A^T$, AA^T, and $A^T A$ are symmetric and $A - A^T$ is skew-symmetric.

16 A square matrix $A = [a_{ij}]$ is *lower triangular* if $a_{ij} = 0$ whenever $i < j$. Prove that the determinant of a lower-triangular matrix is equal to the product of its diagonal entries.

17 Let A, B, C and D be $m \times m, m \times n, n \times m$, and $n \times n$ matrices respectively. Show that

$$\det \begin{bmatrix} A & 0 \\ C & D \end{bmatrix} = \det \begin{bmatrix} A & B \\ 0 & D \end{bmatrix} = (\det A)(\det D)$$

Hint: Show that

$$\begin{bmatrix} A & 0 \\ C & D \end{bmatrix} = \begin{bmatrix} A & 0 \\ 0 & I \end{bmatrix} \begin{bmatrix} I & 0 \\ C & I \end{bmatrix} \begin{bmatrix} I & 0 \\ 0 & D \end{bmatrix}$$

18   Let A be a square matrix of order n, $n \geq 2$. Show that

$$\text{adj(adj } A) = (\det A)^{n-2}A$$

[*Hint:* Consider three cases: (*a*) A is nonsingular (use Theorem 1.9.32 and Exercises 8 and 11); (*b*) A is singular, $n \geq 3$ (use Exercise 9); (*c*) $n = 2$.]

19   *Cramer's rule* (1750).   Let

$$
\begin{aligned}
a_{11}x_1 + a_{12}x_2 + \cdots + a_{1n}x_n &= b_1 \\
a_{21}x_1 + a_{22}x_2 + \cdots + a_{2n}x_n &= b_2 \\
&\cdots\cdots\cdots\cdots\cdots\cdots\cdots\cdots \\
a_{n1}x_1 + a_{n2}x_2 + \cdots + a_{nn}x_n &= b_n
\end{aligned}
$$

be a system of n simultaneous linear equations in n unknowns for which the matrix of coefficients

$$A = [a_{ij}] \qquad i,j = 1, 2, \ldots, n$$

is nonsingular.   Show that the unique solution of this system of equations is given by

$$x_k = \frac{\det B_k}{\det A}$$

where B_k is the $n \times n$ matrix formed from A by replacing the kth column by the column vector

$$\begin{bmatrix} b_1 \\ b_2 \\ \vdots \\ b_n \end{bmatrix}$$

(*Hint:* Express the system of equations in the matrix form $AX = B$ and multiply both sides of the equation by adj A.)

20   Let A be an $m \times n$ matrix of rank r.   Show that for each value of k, $k = 1$, $2, \ldots, r$, A has a nonsingular submatrix of order k.   (*Hint:* Show that if every $k \times k$ submatrix of A is singular, then, for all $l > k$, every $l \times l$ submatrix of A is singular.   Also, see Exercises 28 to 30 of Sec. 1.7.)

21   Let $A = [a_{ij}]$ be a square matrix of order n.   Treat $\det A$ as a quadratic polynomial in the entries of the first two rows of A.   What is the coefficient of $a_{11}a_{22}$?   What is the coefficient of $a_{12}a_{21}$?   In this way, find row and column expansions of $\det A$, expressing $\det A$ as a sum of $\frac{1}{2}n(n-1)$ terms, where each term consists of a product of a determinant of a 2×2 matrix and a determinant of an $(n-2) \times (n-2)$ matrix.   (This is a special case of Exercise 23 below.)

22 Show that for any $a_1, a_2, b_1, b_2, c_1, c_2, d_1, d_2 \in \mathbb{F}$

$$\det\begin{bmatrix} a_1 & c_1 \\ a_2 & c_2 \end{bmatrix}\det\begin{bmatrix} b_1 & d_1 \\ b_2 & d_2 \end{bmatrix} - \det\begin{bmatrix} b_1 & c_1 \\ b_2 & c_2 \end{bmatrix}\det\begin{bmatrix} a_1 & d_1 \\ a_2 & d_2 \end{bmatrix}$$

$$= \det\begin{bmatrix} a_1 & b_1 \\ a_2 & b_2 \end{bmatrix}\det\begin{bmatrix} c_1 & d_1 \\ c_2 & d_2 \end{bmatrix}$$

[*Hint:* Use Exercises 17 and 21 and the matrix

$$\begin{bmatrix} a_1 & 0 & c_1 & d_1 \\ a_2 & 0 & c_2 & d_2 \\ 0 & b_1 & c_1 & d_1 \\ 0 & b_2 & c_2 & d_2 \end{bmatrix}$$

23 *The Laplace expansion of a determinant* (1772).   Let A be an $n \times n$ matrix and let m be a fixed integer, $1 \leq m \leq n - 1$.   Partition A as

$$A = \begin{bmatrix} B \\ C \end{bmatrix}$$

where B has m rows and C has $n - m$ rows.   Let Θ be a subset of $\mathbb{N} = \{1, 2, \ldots, n\}$ consisting of m integers, $\Theta = \{\theta_1, \theta_2, \ldots, \theta_m\}$, $\theta_1 < \theta_2 < \cdots < \theta_m$. Further let $\overline{\Theta}$ be the set of $n - m$ integers of $\mathbb{N}$ which do not belong to Θ, $\overline{\Theta} = \{\bar{\theta}_1, \bar{\theta}_2, \ldots, \bar{\theta}_{n-m}\}$, $\bar{\theta}_1 < \bar{\theta}_2 < \cdots < \bar{\theta}_{n-m}$.   Define B_Θ to be the square matrix of order m formed from B by deleting all columns $B^{(j)}$ for $j \notin \Theta$.   Similarly, define $C_{\overline{\Theta}}$ to be the square matrix of order $n - m$ formed from C by deleting all columns $C^{(j)}$ for $j \notin \overline{\Theta}$.   Finally, let σ_Θ be the permutation

$$\sigma_\Theta = \begin{pmatrix} 1 & 2 & \cdots & m & m+1 & \cdots & n \\ \theta_1 & \theta_2 & \cdots & \theta_m & \bar{\theta}_1 & \cdots & \bar{\theta}_{n-m} \end{pmatrix}$$

Prove that

$$\det A = \sum_\Theta \text{sgn } \sigma_\Theta \, (\det B_\Theta)(\det C_{\overline{\Theta}})$$

where the sum is taken over the $n!/[m!\,(n-m)!]$ subsets Θ of $\mathbb{N}$.   What does this theorem become when $m = 1$?

24 *The Cauchy-Binet theorem*[1] (1812)   Let A and B be $m \times n$ and $n \times m$ matrices respectively, $m \leq n$, and let $C = AB$.   For each subset Θ of $\mathbb{N} = \{1, 2, \ldots, n\}$ consisting of m elements of $\mathbb{N}$, define A_Θ to be the square matrix of order m formed from A by deleting all columns $A^{(j)}$ for $j \notin \Theta$.

[1] This theorem was discovered and published independently by both Cauchy and Binet in 1812.

Similarly, define B_Θ to be the square matrix of order m formed from B by deleting all rows $B_{(j)}$ for $j \notin \Theta$.   Prove that

$$\det C = \sum_\Theta (\det A_\Theta)(\det B_\Theta)$$

where the sum is taken over the $n!/[m!(n-m)!]$ subsets Θ of $\mathbb{N}$.   (*Hint:* Set

$$D = \begin{bmatrix} 0 & -A \\ B & I \end{bmatrix} = \begin{bmatrix} AB & -A \\ 0 & I \end{bmatrix}\begin{bmatrix} I & 0 \\ B & I \end{bmatrix}$$

Apply Exercise 20 of Sec. 1.8 and Exercise 23 above.)

1.10   LINEAR TRANSFORMATIONS AND INVARIANT SPACES

In Sec. 1.1 we defined a matrix to be a rectangular array of numbers.   There is, however, another approach to the concept of a matrix which is of prime importance throughout mathematics.   A matrix is thought of not so much in terms of what it is but rather in terms of what it does.

For a (fixed) $m \times n$ matrix A and a (variable) vector $X \in \mathbb{F}^n$, a vector

$$Y = AX \qquad (1.10.1)$$

belonging to $\mathbb{F}^m$ is determined.   We can consider equation (1.10.1) as a mapping

$$\mathscr{A} : X \to Y \quad (= AX) \qquad (1.10.2)$$

of the vector X (belonging to $\mathbb{F}^n$) to the vector Y (belonging to $\mathbb{F}^m$).   The effect on X of the mapping $\mathscr{A}$ is obtained by multiplying X on the left by the matrix A.

Similarly, if $\mathbb{V}$ is a subspace of $\mathbb{F}^n$, then

$$\mathbb{W} = A\mathbb{V} = \{AX \mid X \in \mathbb{V}\}$$

is a subspace of $\mathbb{F}^m$ (see Exercise 13 of Sec. 1.3).   The mapping $\mathscr{A}$ of (1.10.2) induces a mapping

$$\mathscr{A} : \mathbb{V} \to \mathbb{W} \quad (= A\mathbb{V})$$

of the subspace $\mathbb{V}$ of $\mathbb{F}^n$ to the subspace $\mathbb{W}$ of $\mathbb{F}^m$.   As before, $\mathscr{A}$ is carried out by multiplying the elements of $\mathbb{V}$ on the left by the matrix A.

Not all mappings of $\mathbb{F}^n$ into $\mathbb{F}^m$ can be carried out by means of a matrix multiplication.   For example, it is certainly necessary (though not sufficient) that $\mathscr{A}$ map the zero vector of $\mathbb{F}^n$ into the zero vector of $\mathbb{F}^m$.   We will see that the mappings of one vector space into another which can be carried out by a matrix multiplication are characterized in Definition 1.10.3.

Definition 1.10.3[1]   A mapping $\mathscr{A}$ of a vector space $\mathbb{V}$ into a vector space $\mathbb{W}$ which satisfies the two conditions

$1\quad \mathscr{A}(X + Y) = \mathscr{A}(X) + \mathscr{A}(Y) \qquad X, Y \in \mathbb{V}$

$2\quad \mathscr{A}(\alpha X) = \mathscr{A}(X\alpha) = \alpha\mathscr{A}(X) \qquad \alpha \in \mathbb{F}, X \in \mathbb{V}$

$\qquad\qquad\qquad\qquad\qquad\qquad\qquad\qquad\qquad (1.10.4)$

is called a *linear transformation* of $\mathbb{V}$ into $\mathbb{W}$.

EXAMPLE 1.10.5   It is not difficult to verify that the mapping

$$\mathscr{A}: \begin{bmatrix} x_1 \\ x_2 \\ \vdots \\ x_n \end{bmatrix} \rightarrow \begin{bmatrix} x_n \\ x_{n-1} \\ \vdots \\ x_1 \end{bmatrix}$$

is a linear transformation of $\mathbb{F}^n$ into $\mathbb{F}^n$.   ////

EXAMPLE 1.10.6   The mapping

$$\mathscr{D}: \begin{bmatrix} x_1 \\ x_2 \\ x_3 \\ \vdots \\ x_{n-1} \\ x_n \end{bmatrix} \rightarrow \begin{bmatrix} x_2 \\ 2x_3 \\ 3x_4 \\ \vdots \\ (n-1)x_n \end{bmatrix}$$

is a linear transformation of $\mathbb{F}^n$ into $\mathbb{F}^{n-1}$. $\mathscr{D}$ is called the *derivative*.   Why?

                                                                                ////

For every $m \times n$ matrix A, the mapping $\mathscr{A}$ defined by (1.10.2) provides an example of a linear transformation of $\mathbb{F}^n$ into $\mathbb{F}^m$.   Moreover, two $m \times n$ matrices yield the same linear transformation of $\mathbb{F}^n$ into $\mathbb{F}^m$ if and only if they are equal. Conversely, we will prove (in Theorem 1.10.9) that every linear transformation of $\mathbb{F}^n$ into $\mathbb{F}^m$ is of this type; i.e., for each linear transformation

$$\mathscr{A} : X \rightarrow Y$$

of $\mathbb{F}^n$ into $\mathbb{F}^m$ there exists exactly one $m \times n$ matrix A for which

$$Y = AX$$

---

[1] All the material of this chapter can be formulated in terms of linear transformations rather than in terms of matrices.   The theorems concerning matrices then follow for the corresponding theorems about linear transformations.   Indeed, this is very often done; see, for example, Halmos [6].

for all $X \in \mathbb{F}^n$.   This will establish a one-to-one correspondence between the set of all linear transformations of $\mathbb{F}^n$ into $\mathbb{F}^m$ and the set $\mathbb{F}_n{}^m$ of all $m \times n$ matrices with entries in $\mathbb{F}$.   Because of this correspondence it is seldom necessary to study matrices and linear transformations separately.   Information about one can usually be translated into information about the other.   We have chosen matrices rather than linear transformations as a framework for this chapter primarily because matrices are more suitable for the subjects to be discussed in succeeding chapters.

Because of Equations (1.10.4), a linear transformation of one vector space into another is completely determined by its effect on a basis.   More specifically:

Lemma 1.10.7   Let $\mathbb{V}$ and $\mathbb{W}$ be vector spaces, let $\mathbb{X} = \{X_1, X_2, \ldots, X_k\}$ be a basis for $\mathbb{V}$ and let $Y_1, Y_2, \ldots, Y_k$ be any k (distinct or not) elements of $\mathbb{W}$.   Then there is exactly one linear transformation $\mathscr{A}$ of $\mathbb{V}$ into $\mathbb{W}$ for which

$$\mathscr{A} : X_i \rightarrow Y_i \qquad i = 1, 2, \ldots, k \qquad (1.10.8)$$

PROOF   Let us first show that there is at least one linear transformation $\mathscr{A}$ of $\mathbb{V}$ into $\mathbb{W}$ for which (1.10.8) holds.   Since $\mathbb{X}$ is a basis for $\mathbb{V}$, for each $X \in \mathbb{V}$ there exist unique numbers $\alpha_1, \alpha_2, \ldots, \alpha_k$ belonging to $\mathbb{F}$ such that

$$X = \alpha_1 X_1 + \alpha_2 X_2 + \cdots + \alpha_k X_k$$

We now define the mapping $\mathscr{A}$ of $\mathbb{V}$ into $\mathbb{W}$ by

$$\mathscr{A} : X \rightarrow \alpha_1 Y_1 + \alpha_2 Y_2 + \cdots + \alpha_k Y_k$$

Clearly this implies

$$\mathscr{A} : X_i \rightarrow Y_i \qquad i = 1, 2, \ldots, k$$

It only remains to verify that the mapping $\mathscr{A}$ defined in this way is a linear transformation, i.e., that $\mathscr{A}$ satisfies equations (1.10.4).   This is left as an exercise.   (Note that if $\alpha_1, \alpha_2, \ldots, \alpha_k$ were not uniquely defined by X, then this definition of $\mathscr{A}$ might be ambiguous.)

Conversely, in order to show that $\mathscr{A}$ is unique, let us suppose that $\mathscr{A}_1$ and $\mathscr{A}_2$ are linear transformations of $\mathbb{V}$ into $\mathbb{W}$ satisfying (1.10.8). Let $X \in \mathbb{V}$,

$$X = \alpha_1 X_1 + \alpha_2 X_2 + \cdots + \alpha_k X_k$$

Then, by Definition 1.10.3,

$$\begin{aligned}
\mathscr{A}_1(X) &= \mathscr{A}_1(\alpha_1 X_1 + \alpha_2 X_2 + \cdots + \alpha_k X_k) \\
&= \alpha_1 \mathscr{A}_1(X_1) + \alpha_2 \mathscr{A}_1(X_2) + \cdots + \alpha_k \mathscr{A}_1(X_k) \\
&= \alpha_1 Y_1 + \alpha_2 Y_2 + \cdots + \alpha_k Y_k
\end{aligned}$$

In the same way,

$$
\begin{aligned}
\mathscr{A}_2(X) &= \mathscr{A}_2(\alpha_1 X_1 + \alpha_2 X_2 + \cdots + \alpha_k X_k) \\
&= \alpha_1 \mathscr{A}_2(X_1) + \alpha_2 \mathscr{A}_2(X_2) + \cdots + \alpha_k \mathscr{A}_2(X_k) \\
&= \alpha_1 Y_1 + \alpha_2 Y_2 + \cdots + \alpha_k Y_k
\end{aligned}
$$

Consequently $\mathscr{A}_1(X) = \mathscr{A}_2(X)$ for every $X \in \mathbb{V}$. Thus $\mathscr{A}_1$ and $\mathscr{A}_2$ are the same mapping.   ////

Theorem 1.10.9   Let $\mathscr{A}$ be a linear transformation of $\mathbb{F}^n$ into $\mathbb{F}^m$. Then there exists exactly one $m \times n$ matrix A such that

$$
\mathscr{A} : X \to AX   \quad (1.10.10)
$$

for all $X \in \mathbb{F}^n$.

PROOF   We can prove that there exists at least one $m \times n$ matrix A satisfying (1.10.10) simply by constructing it.   According to Lemma 1.10.7, it is sufficient to show that (1.10.10) holds for the elements of some basis for $\mathbb{F}^n$.   We will select the basis consisting of the n unit column vectors, $I^{(1)}, I^{(2)}, \ldots, I^{(n)}$.   Denote the image of $I^{(i)}$ under the mapping $\mathscr{A}$ by Y_i, that is,

$$
\mathscr{A} : I^{(i)} \to Y_i   \quad i = 1, 2, \ldots, n
$$

Clearly $Y_i \in \mathbb{F}^m$.   Define A to be the $m \times n$ matrix whose ith column is Y_i.   Then

$$
AI^{(i)} = A^{(i)} = Y_i   \quad i = 1, 2, \ldots, n
$$

and hence $\mathscr{A} : I^{(i)} \to Y_i \ \ (= AI^{(i)})   \quad i = 1, 2, \ldots, n$

Thus A satisfies (1.10.10).

In order to show that A is unique, let us suppose that A_1 and A_2 are $m \times n$ matrices which satisfy (1.i0.10).   Then

$$
A_1 X = A_2 X
$$

for all $X \in \mathbb{F}^n$, and hence $\mathbb{N}(A_1 - A_2) = \mathbb{F}^n$.   It follows from Theorem 1.6.1 that the rank of $A_1 - A_2$ is zero and therefore

$$
A_1 - A_2 = 0   \quad ////
$$

Because of Theorem 1.10.9 we can say that A is the *matrix of the linear transformation* $\mathscr{A}$ and that $\mathscr{A}$ is the *linear transformation of the matrix* A.

There is a completely analogous development of linear transformations of $\mathbb{F}_n$ into $\mathbb{F}_m$. In this case there is a one-to-one correspondence between linear transformations $\mathscr{A}$ and $n \times m$ matrices A such that

$$\mathscr{A} : X \to Y \quad (= XA)$$

for all $X \in \mathbb{F}_n$. The details are left for the exercises.

The most important linear transformations are those which map a vector space into itself, and we will study these in more detail. When a linear transformation maps $\mathbb{F}^n$ into itself, its matrix is square and has order n. We will be able to obtain a great deal of information about a square matrix by studying the subspaces which are mapped into themselves by the linear transformation corresponding to the matrix.

Definition 1.10.11 Let A be a square matrix of order n. A subspace $\mathbb{V}$ of $\mathbb{F}^n$ is *A-invariant* (or when there is no doubt about the choice of A, simply *invariant*) under the linear transformation

$$\mathscr{A} : X \to AX$$

of $\mathbb{F}^n$ into itself if $AX \in \mathbb{V}$ for all $X \in \mathbb{V}$.

For every matrix A of order n, both $\mathbb{F}^n$ and 0 are A-invariant subspaces of $\mathbb{F}^n$. Another example of an A-invariant subspace of $\mathbb{F}^n$ is $\mathbb{N}(A)$. It is a simple matter to show that the sum and the intersection of any number of A-invariant subspaces of $\mathbb{F}^n$ are again A-invariant.

The most important invariant subspaces of $\mathbb{F}^n$ are those of dimension 1, and we will consider these in this section before turning our attention in Sec. 1.12 to the more complicated theory of invariant subspaces in general. If X is a nonzero vector of $\mathbb{F}^n$ and the (one-dimensional) vector space which has X as a basis is denoted by $\mathbb{V}$, then $\mathbb{V}$ is A-invariant if and only if there exists a scalar λ in $\mathbb{F}$ such that

$$AX = \lambda X \quad (1.10.12)$$

Note that although X is not permitted to be the zero vector, there is no restriction against λ being zero. In particular, then, every nonzero vector contained in $\mathbb{N}(A)$ is a basis for a one-dimensional A-invariant subspace of $\mathbb{F}^n$ [with $\lambda = 0$ in equation (1.10.12)].

Definition 1.10.13 Let A be a square matrix of order n. If X is a nonzero vector of $\mathbb{F}^n$ and λ is a number such that equation (1.10.12) is satisfied, then λ is called a *characteristic root* of A and X is called a *characteristic vector* of A corresponding to the characteristic root λ.

Clearly equation (1.10.12) is equivalent to

$$(A - \lambda I)X = 0$$

Thus X is a characteristic vector of A corresponding to the characteristic root λ if and only if X is a nonzero element of $\mathbb{N}(A - \lambda I)$. However, applying Theorem 1.6.1 and Corollary 1.9.17, we see that $\mathbb{N}(A - \lambda I) \neq 0$ if and only if $A - \lambda I$ is singular or, equivalently, if and only if

$$\det(A - \lambda I) = 0$$

Thus the characteristic roots of A are precisely the solutions of the equation

$$\det(A - xI) = \det \begin{bmatrix} a_{11} - x & a_{12} & \cdots & a_{1n} \\ a_{21} & a_{22} - x & \cdots & a_{2n} \\ \cdots\cdots\cdots\cdots\cdots\cdots\cdots\cdots \\ a_{n1} & a_{n2} & \cdots & a_{nn} - x \end{bmatrix} = 0 \quad (1.10.14)$$

Evaluating this determinant, we find that $\det(A - xI)$ is a polynomial of degree n (where n is the order of A). The coefficient of x^n is $(-1)^n$, and the constant term of $\det(A - xI)$, obtained by setting $x = 0$ in equation (1.10.14), is $\det A$.

Definition 1.10.15 Let A be a square matrix. Then the polynomial $\det(A - xI)$ is called the *characteristic polynomial* of A. The equation

$$\det(A - xI) = 0$$

is called the *characteristic equation* of A.

EXAMPLE 1.10.16 Let

$$A = \begin{bmatrix} 2 & 1 \\ 3 & 4 \end{bmatrix}$$

The characteristic polynomial of A is

$$\det \begin{bmatrix} 2 - x & 1 \\ 3 & 4 - x \end{bmatrix} = x^2 - 6x + 5 = (x - 1)(x - 5)$$

the characteristic equation of A is

$$x^2 - 6x + 5 = 0$$

and the characteristic roots of A are

$$\lambda_1 = 1 \qquad \lambda_2 = 5$$

The characteristic vectors of A can be found by examining $A - I$ and $A - 5I$,

$$A - I = \begin{bmatrix} 1 & 1 \\ 3 & 3 \end{bmatrix} \qquad A - 5I = \begin{bmatrix} -3 & 1 \\ 3 & -1 \end{bmatrix}$$

Clearly

$$X_1 = \begin{bmatrix} 1 \\ -1 \end{bmatrix} \in \mathbb{N}(A - I) \qquad X_2 = \begin{bmatrix} 1 \\ 3 \end{bmatrix} \in \mathbb{N}(A - 5I) \qquad (1.10.17)$$

Moreover X_1 is a basis for $\mathbb{N}(A - I)$, and X_2 is a basis for $\mathbb{N}(A - 5I)$. Thus X_1 is the unique (up to multiplication by a nonzero scalar) characteristic vector of A corresponding to the characteristic root $\lambda_1 = 1$ and X_2 is the unique (up to multiplication by a nonzero scalar) characteristic vector of A corresponding to the characteristic root $\lambda_2 = 5$. The two equations of (1.10.17) can also be written in the form of equation (1.10.12)

$$A \begin{bmatrix} 1 \\ -1 \end{bmatrix} = 1 \begin{bmatrix} 1 \\ -1 \end{bmatrix} \qquad A \begin{bmatrix} 1 \\ 3 \end{bmatrix} = 5 \begin{bmatrix} 1 \\ 3 \end{bmatrix}$$

The only A-invariant subspaces of $\mathbb{F}^2$ are 0, $\mathbb{F}^2$, $\mathbb{V}_1$, and $\mathbb{V}_2$, where $\mathbb{V}_1$ and $\mathbb{V}_2$ are the one-dimensional vector spaces spanned by X_1 and X_2, respectively.

////

Let us pause here to point out that we can define A-invariant subspaces of $\mathbb{F}_n$ in an analogous way and that one-dimensional A-invariant subspaces of $\mathbb{F}_n$ correspond to nonzero vectors X such that $XA = \lambda X$. It is evident that we are then led to define *row characteristic roots, row characteristic vectors,* the *row characteristic polynomial,* and the *row characteristic equation.* However, it follows from equation (1.10.14) and from Definition 1.10.15 that the (column) characteristic roots, polynomial, and equation are equal to their row counterparts. *Warning:* The corresponding characteristic vectors are quite different.

The roots of the characteristic equation of A are the characteristic roots of A. Clearly 0 is a characteristic root of A if and only if the constant term of its characteristic polynomial (which, as we pointed out above, is equal to det A) is 0. We can now add another useful necessary and sufficient condition that a matrix be nonsingular to the seven we already have (see Theorem 1.6.4 and Corollary 1.9.17).

Corollary 1.10.18 Let A be a square matrix. Then A is nonsingular if and only if 0 is not a characteristic root of A.

We have shown that to each characteristic root of A there corresponds at least one characteristic vector. It is clear from equation (1.10.12) that characteristic vectors corresponding to distinct characteristic roots are distinct. We can, in fact, prove that they are linearly independent.

Lemma 1.10.19 If $X_1, X_2, \ldots, X_k$ are characteristic vectors of A corresponding to the distinct characteristic roots $\lambda_1, \lambda_2, \ldots, \lambda_k$, then $X_1, X_2, \ldots, X_k$ are linearly independent.

PROOF   Let $\mathbb{V}$ be the vector space spanned by $X_1, X_2, \ldots, X_k$. We may assume that $\lambda_1, \lambda_2, \ldots, \lambda_k$ are ordered in such a way that $X_1, X_2, \ldots, X_l$ form a basis for $\mathbb{V}$. In order to show that $X_1, X_2, \ldots, X_k$ are linearly independent we will prove that $l = k$. Hence let us suppose that $l < k$. Then there exist numbers $\beta_1, \beta_2, \ldots, \beta_l$ such that

$$X_{l+1} = \beta_1 X_1 + \beta_2 X_2 + \cdots + \beta_l X_l \quad (1.10.20)$$

Multiply both sides of equation (1.10.20) on the left by A. Then

$$AX_{l+1} = \beta_1 AX_1 + \beta_2 AX_2 + \cdots + \beta_l AX_l$$

from which it follows that

$$\lambda_{l+1} X_{l+1} = \beta_1 \lambda_1 X_1 + \beta_2 \lambda_2 X_2 + \cdots + \beta_l \lambda_l X_l \quad (1.10.21)$$

On the other hand, when we multiply both sides of equation (1.10.20) by λ_{l+1}, we have

$$\lambda_{l+1} X_{l+1} = \beta_1 \lambda_{l+1} X_1 + \beta_2 \lambda_{l+1} X_2 + \cdots + \beta_l \lambda_{l+1} X_l \quad (1.10.22)$$

Subtracting equation (1.10.22) from equation (1.10.21) yields

$$0 = \beta_1(\lambda_1 - \lambda_{l+1})X_1 + \beta_2(\lambda_2 - \lambda_{l+1})X_2 + \cdots + \beta_l(\lambda_l - \lambda_{l+1})X_l$$

Since $X_1, X_2, \ldots, X_l$ are linearly independent, it follows that

$$\beta_1(\lambda_1 - \lambda_{l+1}) = \beta_2(\lambda_2 - \lambda_{l+1}) = \cdots = \beta_l(\lambda_l - \lambda_{l+1}) = 0$$

But $\lambda_1, \lambda_2, \ldots, \lambda_l, \lambda_{l+1}$ are distinct, and consequently

$$\beta_1 = \beta_2 = \cdots = \beta_l = 0$$

We have now shown that $X_{l+1} = 0$. This is a contradiction since X_{l+1} is a characteristic vector. Since the assumption that $l < k$ leads to a contradiction, we conclude that $l = k$; that is, $X_1, X_2, \ldots, X_k$ are linearly independent.   ////

The simplest situation that can arise is for the characteristic polynomial of A to be expressible as a product of n distinct linear factors. Then A has n distinct characteristic roots, and we can prove:

Theorem 1.10.23 If A is a square matrix of order n having n distinct characteristic roots, $\lambda_1, \lambda_2, \ldots, \lambda_n$, then there is a nonsingular matrix P such that

$$P^{-1}AP = \begin{bmatrix} \lambda_1 & 0 & \cdots & 0 \\ 0 & \lambda_2 & \cdots & 0 \\ & & \cdots & \\ 0 & 0 & \cdots & \lambda_n \end{bmatrix} \qquad (1.10.24)$$

PROOF   Let $X_1, X_2, \ldots, X_k$ be characteristic vectors of A corresponding to the characteristic roots $\lambda_1, \lambda_2, \ldots, \lambda_k$ respectively and let P be the $n \times n$ matrix whose columns are $X_1, X_2, \ldots, X_k$, that is

$$P^{(i)} = X_i \qquad i = 1, 2, \ldots, n$$

Then $\qquad AP^{(i)} = AX_i = \lambda_i X_i = \lambda_i P^{(i)} \qquad i = 1, 2, \ldots, n \quad (1.10.25)$

We can combine the n equations of (1.10.25) into the single matrix equation

$$AP = P \begin{bmatrix} \lambda_1 & 0 & \cdots & 0 \\ 0 & \lambda_2 & \cdots & 0 \\ & & \cdots & \\ 0 & 0 & \cdots & \lambda_n \end{bmatrix} \qquad (1.10.26)$$

Lemma 1.10.19 implies that P is nonsingular, and the theorem follows by multiplying equation (1.10.26) on the left by P^{-1}.   ////

The hypothesis of Theorem 1.10.23 may be relaxed somewhat. If A has n linearly independent characteristic vectors, corresponding to distinct characteristic roots or not, then the same construction yields a nonsingular matrix P for which $P^{-1}AP$ is a diagonal matrix whose distinct diagonal entries[1] are the characteristic roots of A. Conversely, if $P^{-1}AP$ is a diagonal matrix for some nonsingular matrix P, then the columns of P are a set of linearly independent characteristic vectors of A and the distinct diagonal entries of $P^{-1}AP$ are characteristic roots of A.

The corresponding theorem for row characteristic vectors states that a necessary and sufficient condition that there exist a nonsingular matrix Q such

---

[1] At this point of our discussion it is not at all clear what happens either when the characteristic polynomial of A has multiple roots or when the diagonal entries of $P^{-1}AP$ are not distinct. We will consider this problem shortly.

that QAQ^{-1} is a diagonal matrix is that A have n linearly independent row characteristic vectors. In this case, the rows of Q are the row characteristic vectors of A, and the distinct diagonal entries of QAQ^{-1} are characteristic roots of A. Setting $Q = P^{-1}$, we see that A has n linearly independent row characteristic vectors if and only if it has n linearly independent (column) characteristic vectors.

There is another way of looking at the row characteristic vectors of a matrix. We can identify them with the (column) characteristic vectors of A^T; that is, since

$$XA = \lambda X \quad \text{implies} \quad A^T X^T = \lambda X^T$$

(see Exercise 20 of Sec. 1.7) we see that X is a row characteristic vector of A if and only if X^T is a (column) characteristic vector of A^T. The characteristic polynomial, equation, and roots of A^T are identical to the characteristic polynomial, equation, and roots of A. However, the characteristic vectors may be quite different.

EXAMPLE 1.10.27   In Example 1.10.16 we showed that the matrix

$$A = \begin{bmatrix} 2 & 1 \\ 3 & 4 \end{bmatrix}$$

has characteristic vectors

$$\begin{bmatrix} 1 \\ -1 \end{bmatrix} \quad \text{and} \quad \begin{bmatrix} 1 \\ 3 \end{bmatrix}$$

corresponding to the characteristic roots

$$\lambda = 1 \quad \text{and} \quad \lambda = 5$$

respectively.   As in Theorem 1.10.23, we form

$$P = \begin{bmatrix} 1 & 1 \\ -1 & 3 \end{bmatrix}$$

Then
$$AP = \begin{bmatrix} 1 & 5 \\ -1 & 15 \end{bmatrix} = \begin{bmatrix} 1 & 1 \\ -1 & 3 \end{bmatrix}\begin{bmatrix} 1 & 0 \\ 0 & 5 \end{bmatrix} = P\begin{bmatrix} 1 & 0 \\ 0 & 5 \end{bmatrix}$$

Moreover P is nonsingular, and

$$P^{-1} = \frac{1}{4}\begin{bmatrix} 3 & -1 \\ 1 & 1 \end{bmatrix}$$

Thus
$$P^{-1}AP = \frac{1}{4}\begin{bmatrix} 3 & -1 \\ 1 & 1 \end{bmatrix}\begin{bmatrix} 2 & 1 \\ 3 & 4 \end{bmatrix}\begin{bmatrix} 1 & 1 \\ -1 & 3 \end{bmatrix} = \begin{bmatrix} 1 & 0 \\ 0 & 5 \end{bmatrix}$$

as expected.

In the same way we see that

$$[3 \quad -1] \quad \text{and} \quad [1 \quad 1]$$

are the row characteristic vectors of A corresponding to the characteristic roots

$$\lambda_1 = 1 \quad \text{and} \quad \lambda_2 = 5$$

respectively. Set

$$Q = \begin{bmatrix} 3 & -1 \\ 1 & 1 \end{bmatrix}$$

Then Q is nonsingular, and

$$Q^{-1} = \frac{1}{4} \begin{bmatrix} 1 & 1 \\ -1 & 3 \end{bmatrix}$$

Thus we again have

$$QAQ^{-1} = \frac{1}{4} \begin{bmatrix} 3 & -1 \\ 1 & 1 \end{bmatrix} \begin{bmatrix} 2 & 1 \\ 3 & 4 \end{bmatrix} \begin{bmatrix} 1 & 1 \\ -1 & 3 \end{bmatrix} = \begin{bmatrix} 1 & 0 \\ 0 & 5 \end{bmatrix} \qquad ////$$

It may be difficult at this time to see the significance of equation (1.10.24). Equation (1.10.24) is reminiscent of equation (1.7.13), reducing a given matrix to a simple form by means of left and right multiplications. However, in Theorem 1.10.23 we consider only square matrices, and the type of reduction that is permitted is severely restricted.

Definition 1.10.28 Let A and B be square matrices of the same order. If there exists a nonsingular matrix P such that

$$B = P^{-1}AP$$

we say that B is *similar* to A.[1]

---

[1] The similarity of matrices has many applications. For example, it is useful in solving a system of linear homogeneous first-order differential equations
$$X' = AX \qquad \text{(I)}$$
where A is a square matrix of real numbers, X is a column vector of unknown functions in the variable t, and X' is the vector of the first derivatives of these functions with respect to t. If we introduce a column vector Y of new unknowns related to the original unknowns by
$$Y = PX$$
where P is a nonsingular matrix of real numbers, then the system of differential equations (I) becomes
$$Y' = PAP^{-1}Y$$
For a more geometric interpretation of similarity of matrices, see Halmos [6, Secs. 46 and 47].

It follows from Exercises 10 and 11 of Sec. 1.6 that similarity of matrices is an equivalence relation. Clearly similar matrices are also equivalent (see Exercise 12 of Sec. 1.7), and consequently they have the same rank. In Corollary 1.10.29 we present a number of other properties shared by similar matrices.

Corollary 1.10.29 Similar matrices have the same characteristic polynomial and consequently the same characteristic equation, the same characteristic roots, and the same determinant.

PROOF Let $B = P^{-1}AP$. Then

$$\det(B - xI) = \det[P^{-1}AP - P^{-1}(xI)P]$$
$$= \det[P^{-1}(A - xI)P]$$
$$= (\det P^{-1})[\det(A - xI)](\det P)$$
$$= \det(A - xI) \qquad\qquad ////$$

It is not true, however, that similar matrices have the same characteristic vectors (see Exercise 20). Also the converse of Corollary 1.10.29 (namely, that matrices having the same characteristic polynomial are similar) is false. Essentially, Secs. 1.12 and 1.13 are devoted to determining necessary and sufficient conditions for two matrices to be similar and to selecting, from all the matrices similar to a given matrix, the one which has the "simplest" form. In the case where A is of order n and has n linearly independent characteristic vectors we have already solved the problem, for then A is similar to a diagonal matrix.

When a matrix does not have n linearly independent characteristic vectors (either because the characteristic polynomial has multiple roots or because it is not expressible as a product of linear factors), the simple form of equation (1.10.24) is no longer obtainable. In order to prepare for a discussion of this more complicated situation, let us interpret Lemma 1.10.19 and Theorem 1.10.23 in the language of invariant subspaces. In Lemma 1.10.19, let us denote by V_i the one-dimensional subspace of $\mathbb{F}^n$ having the characteristic vector X_i of A as a basis. When $k = n$, Lemma 1.10.19 states that

$$\mathbb{F}^n = V_1 \oplus V_2 \oplus \cdots \oplus V_n \qquad (1.10.30)$$

and it follows in Theorem 1.10.23 that A is similar to the direct sum of n 1×1 matrices,

$$P^{-1}AP = [\lambda_1] \oplus [\lambda_2] \oplus \cdots \oplus [\lambda_n]$$

In general, whenever $\mathbb{F}^n$ can be expressed as a direct sum of A-invariant subspaces [as in (1.10.30), though the subspaces need not be one-dimensional], the same basic argument as that used in Theorem 1.10.23 yields the same basic

result, namely, that A is similar to the direct sum of square matrices, one for
each of the A-invariant subspaces $\mathbb{V}_i$, and that the square matrix corresponding
to the subspace $\mathbb{V}_i$ has order equal to the dimension of $\mathbb{V}_i$. These results are
contained in:

Theorem 1.10.31   The fundamental theorem of direct sums $\mathbb{F}^n$ can be
expressed as

$$\mathbb{F}^n = \mathbb{V}_1 \oplus \mathbb{V}_2 \oplus \cdots \oplus \mathbb{V}_k$$

where $\mathbb{V}_i$ is an A-invariant subspace of $\mathbb{F}^n$ of dimension n_i if and only if
there exists a nonsingular matrix P such that

$$P^{-1}AP = B_1 \oplus B_2 \oplus \cdots \oplus B_k \quad (1.10.32)$$

where B_i is a square matrix of order n_i.

PROOF   First, let us suppose that

$$\mathbb{F}^n = \mathbb{V}_1 \oplus \mathbb{V}_2 \oplus \cdots \oplus \mathbb{V}_k \quad (1.10.33)$$

We wish to construct a nonsingular matrix P such that $P^{-1}AP$ is given by
(1.10.32). For each i, $i = 1, 2, \ldots, k$, let

$$\mathbb{X}_i = \{X_{i1}, X_{i2}, \ldots, X_{in_i}\}$$

be a basis for $\mathbb{V}_i$. Since $\mathbb{F}^n$ is a direct sum of $\mathbb{V}_1, \mathbb{V}_2, \ldots, \mathbb{V}_k$, it follows
from Definition 1.5.13 that

$$\mathbb{X} = \mathbb{X}_1 \cup \mathbb{X}_2 \cup \cdots \cup \mathbb{X}_k = \{X_{11}, X_{12}, \ldots, X_{1n_1}, X_{21}, \ldots, X_{kn_k}\}$$

is a basis for $\mathbb{F}^n$. Let P be the matrix whose columns are the n vectors of
$\mathbb{X}$, in the order $X_{11}, X_{12}, \ldots, X_{1n_1}, X_{21}, \ldots, X_{kn_k}$. (Note that this im-
poses an order on the bases $\mathbb{X}_1, \mathbb{X}_2, \ldots, \mathbb{X}_k$ but not on the vectors within
each basis.) Then P is a square matrix of order n and, in fact, is nonsingu-
lar. The columns of AP are $AX_{11}, AX_{12}, \ldots, AX_{1n_1}, AX_{21}, \ldots, AX_{kn_k}$.
Since each of the vector spaces $\mathbb{V}_1, \mathbb{V}_2, \ldots, \mathbb{V}_k$ is A-invariant, the first n_1
columns of AP are elements of $\mathbb{V}_1$, the next n_2 columns of AP are elements
of $\mathbb{V}_2$, etc. Hence each of the first n_1 columns of AP can be expressed as
a linear combination of the elements of $\mathbb{X}_1$, each of the next n_2 columns
of AP can be expressed as a linear combination of the elements of $\mathbb{X}_2$, and
so on. Therefore, for each j, $j = 1, 2, \ldots, n_1$, there exist numbers

$$\alpha_{1j}, \alpha_{2j}, \ldots, \alpha_{n_1 j}$$

such that

$$AX_{1j} = \alpha_{1j}X_{11} + \alpha_{2j}X_{12} + \cdots + \alpha_{n_1 j}X_{1n_1}$$
$$= \alpha_{1j}X_{11} + \alpha_{2j}X_{12} + \cdots + \alpha_{n_1 j}X_{1n_1} + 0X_{12}$$
$$+ \cdots + 0X_{kn_k} \quad (1.10.34)$$

Similarly, for each $j, j = 1, 2, \ldots, n_2$, there exist numbers

$$\beta_{1j}, \beta_{2j}, \ldots, \beta_{n_2 j}$$

such that

$$AX_{2j} = \beta_{1j}X_{21} + \beta_{2j}X_{22} + \cdots + \beta_{n_2 j}X_{2n_2}$$
$$= 0X_{11} + 0X_{12} + \cdots + 0X_{1n_1} + \beta_{1j}X_{21} + \cdots$$
$$+ \beta_{n_2 j}X_{2n_2} + \cdots + 0X_{kn_k} \quad (1.10.35)$$

Continuing in this way for each of the sets $X_3, X_4, \ldots, X_k$ and combining the resulting n equations into a single matrix equation, we have

$$AP = P\begin{bmatrix} B_1 & 0 & \cdots & 0 \\ 0 & B_2 & \cdots & 0 \\ \multicolumn{4}{c}{\dotfill} \\ 0 & 0 & \cdots & B_k \end{bmatrix} \quad (1.10.36)$$

where each B_i is a square matrix of order n_i. In particular, the n_1 equations of (1.10.34) and the n_2 equations of (1.10.35) yield the $n_1 \times n_1$ and $n_2 \times n_2$ matrices

$$B_1 = \begin{bmatrix} \alpha_{11} & \alpha_{12} & \cdots & \alpha_{1n_1} \\ \alpha_{21} & \alpha_{22} & \cdots & \alpha_{2n_1} \\ \multicolumn{4}{c}{\dotfill} \\ \alpha_{n_1 1} & \alpha_{n_1 2} & \cdots & \alpha_{n_1 n_1} \end{bmatrix} \qquad B_2 = \begin{bmatrix} \beta_{11} & \beta_{12} & \cdots & \beta_{1n_2} \\ \beta_{21} & \beta_{22} & \cdots & \beta_{2n_2} \\ \multicolumn{4}{c}{\dotfill} \\ \beta_{n_2 1} & \beta_{n_2 2} & \cdots & \beta_{n_2 n_2} \end{bmatrix}$$

respectively. Since P is nonsingular, we may multiply equation (1.10.36) on the left by P^{-1}, completing the proof.

Conversely, suppose P is a nonsingular matrix for which

$$P^{-1}AP = B_1 \oplus B_2 \oplus \cdots \oplus B_k$$

where B_i is a square matrix of order n_i. By expressing this equation in the form (1.10.36) we see that the first n_1 columns of P form a basis for an A-invariant subspace V_1 of F^n, the next n_2 columns of P form a basis for an A-invariant subspace V_2 of F^n, etc., and finally,

$$F^n = V_1 \oplus V_2 \oplus \cdots \oplus V_k \qquad \text{////}$$

EXAMPLE 1.10.37   Let

$$A = \begin{bmatrix} 3 & 1 & -2 \\ -1 & -1 & 4 \\ 0 & -1 & 3 \end{bmatrix}$$

The characteristic polynomial of A is

$$\det \begin{bmatrix} 3-x & 1 & -2 \\ -1 & -1-x & 4 \\ 0 & -1 & 3-x \end{bmatrix} = -x^3 + 5x^2 - 8x + 4 = -(x-2)^2(x-1)$$

and hence the characteristic roots of A are

$$\lambda_1 = \lambda_2 = 2 \qquad \lambda_3 = 1$$

The characteristic vectors of A can be found by examining the matrices

$$A - 2I = \begin{bmatrix} 1 & 1 & -2 \\ -1 & -3 & 4 \\ 0 & -1 & 1 \end{bmatrix} \qquad A - I = \begin{bmatrix} 2 & 1 & -2 \\ -1 & -2 & 4 \\ 0 & -1 & 2 \end{bmatrix}$$

Clearly $\left\{ \begin{bmatrix} 1 \\ 1 \\ 1 \end{bmatrix} \right\}$ is a basis for $\mathbb{N}(A - 2I)$, and hence $\begin{bmatrix} 1 \\ 1 \\ 1 \end{bmatrix}$ is the only characteristic vector of A (up to multiplication by a scalar) corresponding to the characteristic root 2.   In the same way we see that $\left\{ \begin{bmatrix} 0 \\ 2 \\ 1 \end{bmatrix} \right\}$ is a basis for $\mathbb{N}(A - I)$ and $\begin{bmatrix} 0 \\ 2 \\ 1 \end{bmatrix}$ is the only characteristic vector of A (up to multiplication by a scalar) corresponding to the characteristic root 1.   Moreover, A has no other characteristic roots (or characteristic vectors).   Thus $\mathbb{F}^3$ cannot be expressed as a direct sum of three one-dimensional A-invariant subspaces, and A is not similar to a diagonal matrix. However, if we let $\mathbb{V}_1$ and $\mathbb{V}_2$ be the vector spaces which have the sets

$$\mathbb{X}_1 = \left\{ X_{11} = \begin{bmatrix} 1 \\ 1 \\ 1 \end{bmatrix}, X_{12} = \begin{bmatrix} 2 \\ -1 \\ 0 \end{bmatrix} \right\} \qquad \mathbb{X}_2 = \left\{ \begin{bmatrix} 0 \\ 2 \\ 1 \end{bmatrix} \right\}$$

as bases, then $\mathbb{V}_1$ and $\mathbb{V}_2$ are A-invariant (since

$$AX_{12} = \begin{bmatrix} 5 \\ -1 \\ 1 \end{bmatrix} = 2X_{12} + X_{11} \in \mathbb{V}_1) \qquad \text{and} \qquad \mathbb{F}^3 = \mathbb{V}_1 \oplus \mathbb{V}_2$$

We now set

$$P = \begin{bmatrix} 1 & 2 & 0 \\ 1 & -1 & 2 \\ 1 & 0 & 1 \end{bmatrix}$$

Then

$$AP = \begin{bmatrix} 2 & 5 & 0 \\ 2 & -1 & 2 \\ 2 & 1 & 1 \end{bmatrix} = \begin{bmatrix} 1 & 2 & 0 \\ 1 & -1 & 2 \\ 1 & 0 & 1 \end{bmatrix}\begin{bmatrix} 2 & 1 & 0 \\ 0 & 2 & 0 \\ 0 & 0 & 1 \end{bmatrix} = P\begin{bmatrix} 2 & 1 & 0 \\ 0 & 2 & 0 \\ 0 & 0 & 1 \end{bmatrix}$$

and hence

$$P^{-1}AP = \begin{bmatrix} 2 & 1 & 0 \\ 0 & 2 & 0 \\ 0 & 0 & 1 \end{bmatrix} = B_1 \oplus B_2$$

where

$$B_1 = \begin{bmatrix} 2 & 1 \\ 0 & 2 \end{bmatrix} \qquad B_2 = [1] \qquad ////$$

Definition 1.10.38 Let λ be a characteristic root of A. The multiplicity of λ as a root of the characteristic polynomial of A is called the *algebraic multiplicity* of λ, and $v(A - \lambda I)$ is called the *geometric multiplicity* of λ.

In Example 1.10.37, 2 is a characteristic root of algebraic multiplicity 2, since $\lambda_1 = \lambda_2 = 2$, and geometric multiplicity 1, since $\begin{bmatrix} 1 \\ 1 \\ 1 \end{bmatrix}$ is the only characteristic vector of A corresponding to the characteristic root 2.

Let A and B be similar matrices, $B = P^{-1}AP$. Then

$$B - \lambda I = P^{-1}(A - \lambda I)P$$

for every $\lambda \in \mathbb{F}$, and it follows from Theorem 1.6.7 that the geometric multiplicities of the characteristic roots of A and B are equal. Moreover, since similar matrices have the same characteristic polynomial, the algebraic multiplicities of their characteristic roots are equal. In Exercise 24 we outline a proof that the algebraic multiplicity of a characteristic root of a matrix is never less than its geometric multiplicity. In Secs. 1.12 and 1.13 we will investigate further the properties of the algebraic and geometric multiplicities of the characteristic roots of a matrix.

As we have seen, similar matrices have many properties in common, and much information about a matrix A can be obtained by finding a matrix B, similar to A, which is the direct sum of smaller matrices $B_1, B_2, \ldots, B_k$. According to Theorem 1.10.31, this will be accomplished when we have expressed $\mathbb{F}^n$ as a direct sum of A-invariant subspaces $\mathbb{V}_1, \mathbb{V}_2, \ldots, \mathbb{V}_k$. Theorem 1.10.31 raises two problems:

1 How to express $\mathbb{F}^n$ as a direct sum of A-invariant subspaces $\mathbb{V}_1$, $\mathbb{V}_2$, $\ldots$, $\mathbb{V}_k$.

2 How to make the form of the matrices B_1, B_2, $\ldots$, B_k as simple as possible.

The first question is answered in Sec. 1.12.   The second question requires us to decide which of the possible forms for B_1, B_2, $\ldots$, B_k we want.   Two somewhat different "simple" forms are given in Sec. 1.13.

EXERCISES

1 Find the matrix of the linear transformation of Example 1.10.5.
2 Find the matrix of the linear transformation of Example 1.10.6.
3 Let $\mathscr{B}$ be the linear transformation of $\mathbb{F}^3$ into itself for which

$$\mathscr{B}: \quad \begin{aligned} \begin{bmatrix} 1 \\ 2 \\ 0 \end{bmatrix} &\rightarrow \begin{bmatrix} -1 \\ 3 \\ 2 \end{bmatrix} \\[6pt] \begin{bmatrix} 0 \\ 2 \\ 1 \end{bmatrix} &\rightarrow \begin{bmatrix} 3 \\ 4 \\ -2 \end{bmatrix} \\[6pt] \begin{bmatrix} -1 \\ 1 \\ 1 \end{bmatrix} &\rightarrow \begin{bmatrix} 3 \\ 1 \\ -3 \end{bmatrix} \end{aligned}$$

What vector is $\begin{bmatrix} 1 \\ 1 \\ 1 \end{bmatrix}$ mapped into by $\mathscr{B}$?   What vector is mapped into $\begin{bmatrix} 1 \\ 1 \\ 1 \end{bmatrix}$ by $\mathscr{B}$?   Find the matrix B of $\mathscr{B}$.   Show that the subspace of $\mathbb{F}^3$ which has

$$\left\{ \begin{bmatrix} -1 \\ 2 \\ 1 \end{bmatrix}, \begin{bmatrix} 0 \\ 1 \\ 0 \end{bmatrix} \right\}$$

as a basis is B-invariant.
4 Show that there is a nonzero vector X such that

$$\mathscr{A}(X) = \mathscr{B}(X)$$

where $\mathscr{A}$ and $\mathscr{B}$ are the linear transformations of Example 1.10.5 (with $n=3$) and Exercise 3 respectively.
5 Find all nonsingular matrices P such that

$$P^{-1} \begin{bmatrix} 1 & 1 \\ 0 & 2 \end{bmatrix} P = \begin{bmatrix} 1 & 0 \\ 0 & 2 \end{bmatrix}$$

Find all nonsingular matrices Q such that

$$Q^{-1} \begin{bmatrix} 1 & 1 \\ 0 & 2 \end{bmatrix} Q = \begin{bmatrix} 1 & 0 \\ 1 & 2 \end{bmatrix}$$

6   Show that

$$\begin{bmatrix} 1 & y \\ 0 & 1 \end{bmatrix} \text{ and } \begin{bmatrix} 1 & z \\ 0 & 1 \end{bmatrix}$$

are similar to each other for all nonzero numbers y and z.   What happens when y or z is zero?

7   Show that

$$\begin{bmatrix} 1 & 0 \\ w & 1 \end{bmatrix}$$

is similar to the matrices of Exercise 6 for all nonzero numbers w.

8   Find the characteristic roots and all the characteristic vectors of

$$A = \begin{bmatrix} 3 & 4 & 5 \\ 0 & 1 & -1 \\ -1 & -2 & -1 \end{bmatrix}$$

Find a matrix P such that $P^{-1}AP$ is a diagonal matrix.   Find the row characteristic vectors of A and use them to construct a nonsingular matrix Q such that QAQ^{-1} is a diagonal matrix.

9   Repeat Exercise 8 using the matrix

$$A = \begin{bmatrix} 2 & 3 & 0 \\ -4 & -5 & 0 \\ 3 & 3 & -1 \end{bmatrix}$$

10   Find the characteristic roots and all the corresponding row and column characteristic vectors of

$$A = \begin{bmatrix} 3 & 2 & -1 \\ 0 & -1 & 0 \\ 1 & -4 & 1 \end{bmatrix}$$

Is A similar to a diagonal matrix?

11   Let $\mathcal{A}$ be a linear transformation of the vector space $\mathbb{V}$ into the vector space $\mathbb{W}$.   We define the *image* of $\mathcal{A}$, denoted by im $\mathcal{A}$, as follows:

$$Y \in \text{im } \mathcal{A} \text{ if there exists } X \in \mathbb{V} \text{ such that } \mathcal{A}: X \to Y$$

Prove that im $\mathcal{A}$ is a subspace of $\mathbb{W}$.

12   Let $\mathscr{A}$ be a linear transformation of the vector space $\mathbb{V}$ into the vector space
     $\mathbb{W}$.   We define the *kernel* of $\mathscr{A}$, denoted by ker $\mathscr{A}$, as follows:

$$\ker \mathscr{A} = \{X \mid X \in \mathbb{V} \text{ and } \mathscr{A} : X \to 0 \in \mathbb{W}\}$$

Prove that ker $\mathscr{A}$ is a subspace of $\mathbb{V}$.

13   Let $\mathscr{A}$ be a linear transformation of $\mathbb{V}$ into $\mathbb{W}$.   Prove that

$$\dim \mathbb{V} = \dim(\ker \mathscr{A}) + \dim(\operatorname{im} \mathscr{A})$$

Interpret this result when $\mathbb{V} = \mathbb{F}^n$, $\mathbb{W} = \mathbb{F}^m$, and A is the matrix of $\mathscr{A}$.

14   Show that if B is a matrix which commutes with A, then $\mathbb{N}(B)$ is A-invariant.

15   Show that $\mathbb{C}(A)$ is A-invariant for every square matrix A.

16   Prove that the sum and intersection of A-invariant vector spaces are A-invariant.

17   Prove that the characteristic polynomial of a direct sum of matrices is equal to the product of the characteristic polynomials of the individual matrices.

18   Prove that if X is a characteristic vector of A corresponding to the characteristic root λ, then, for every positive integer k, X is a characteristic vector of A^k corresponding to the characteristic root λ^k.

19   Show that if X is a characteristic vector of A corresponding to the characteristic root λ, then, for all $\alpha, \beta \in \mathbb{F}$, X is a characteristic vector of $\alpha A + \beta I$ corresponding to the characteristic root $\alpha \lambda + \beta$.

20   Let $B = P^{-1}AP$.   Show that X is a characteristic vector of A corresponding to the characteristic root λ if and only if $P^{-1}X$ is a characteristic vector of B corresponding to the characteristic root λ.

21   Let A and B be square matrices of the same order and let A be nonsingular. Prove that AB is similar to BA.   Give an example to show that if A and B are singular, then AB need not be similar to BA.

22   Let A and B be $m \times n$ and $n \times m$ matrices respectively.   Prove that if X is a characteristic vector of AB corresponding to the characteristic root $\lambda \neq 0$, then BX is a characteristic vector of BA corresponding to the characteristic root λ.   Why is it necessary to specify that $\lambda \neq 0$?

23   Let A be a square matrix of order n and let V be an A-invariant subspace of $\mathbb{F}^n$ of dimension m.   Show that A is similar to a matrix of the form

$$\begin{bmatrix} B_{11} & B_{12} \\ 0 & B_{22} \end{bmatrix}$$

where B_{11}, B_{12}, and B_{22} are $m \times m$, $m \times (n-m)$, and $(n-m) \times (n-m)$ matrices respectively.

24 Prove that the algebraic multiplicity of a characteristic root of a matrix is greater than or equal to its geometric multiplicity. (*Hint:* Let $X_1, X_2, \ldots,$ X_l be linearly independent characteristic vectors of the matrix A corresponding to the characteristic root λ. Construct a nonsingular matrix P whose first l columns are $X_1, X_2, \ldots, X_l$ and consider the characteristic polynomial of $P^{-1}AP$. Apply Exercise 17 of Sec. 1.9.)

1.11   POLYNOMIALS

Let

$$f(x) = \alpha_m x^m + \alpha_{m-1} x^{m-1} + \cdots + \alpha_1 x + \alpha_0$$

be a polynomial with coefficients from $\mathbb{F}$. For a square matrix A, we define the *matrix polynomial $f(A)$* by

$$f(A) = \alpha_m A^m + \alpha_{m-1} A^{m-1} + \cdots + \alpha_1 A + \alpha_0 I$$

Then $f(A)$ is again a square matrix whose order is the same as the order of A. In the next section we will show that for a given square matrix A, we can find the A-invariant subspaces of $\mathbb{F}^n$ by considering polynomials in A with coefficients from $\mathbb{F}$. For this reason we first need to develop a number of theorems dealing with polynomials in one variable.

Theorem 1.11.1   The division algorithm   Let $f(x)$ and $g(x)$ be polynomials, $g(x) \neq 0$. Then there exist polynomials $q(x)$ and $r(x)$ such that

$$f(x) = q(x)g(x) + r(x)$$

and the degree of $r(x)$ is less than the degree of $g(x)$,[1] that is

$$\deg r(x) < \deg g(x)$$

PROOF   When $f(x) = 0$, then $q(x) = r(x) = 0$ satisfies the theorem. Thus we may henceforth assume that $f(x) \neq 0$. Let

$$f(x) = \alpha_m x^m + \alpha_{m-1}x^{m-1} + \cdots + \alpha_0 \qquad m \geq 0 \qquad \alpha_m \neq 0$$
$$g(x) = \beta_l x^l + \beta_{l-1}x^{l-1} + \cdots + \beta_0 \qquad l \geq 0 \qquad \beta_l \neq 0$$

The proof is by an induction on m, the degree of $f(x)$. It will be convenient for us to think of $g(x)$ as being fixed throughout the proof. First, suppose $m = 0$. Then $f(x) = \alpha_0$, and two distinct cases arise. If

---

[1] We define the degree of the zero polynomial to be $-\infty$.

$l = 0$, so that $g(x) = \beta_0 \neq 0$, then we set $q(x) = \alpha_0/\beta_0$, $r(x) = 0$. If $l \geq 1$, then $q(x) = 0$, $r(x) = f(x)$ satisfies the theorem. Thus we have established the theorem for $m = 0$.

Now suppose that for a fixed nonzero polynomial $g(x)$ the theorem has been proved for all polynomials $f(x)$ of degree less than k. Let $f(x)$ be a polynomial of degree k. If $k < l$, we again set $q(x) = 0$, $r(x) = f(x)$. On the other hand, if $k \geq l$, then we define the polynomial $\tilde{f}(x)$ by

$$\tilde{f}(x) = f(x) - \frac{\alpha_k}{\beta_l} x^{k-l} g(x) \qquad (1.11.2)$$

The degree of $\tilde{f}(x)$ is definitely less than k and hence, by the induction hypothesis, there exist polynomials $\tilde{q}(x)$ and $\tilde{r}(x)$ such that

$$\tilde{f}(x) = \tilde{q}(x)g(x) + \tilde{r}(x) \qquad (1.11.3)$$

and $$\deg \tilde{r}(x) < \deg g(x)$$

Substituting equation (1.11.3) into equation (1.11.2), we have

$$f(x) = \tilde{f}(x) + \frac{\alpha_k}{\beta_l} x^{k-l} g(x)$$

$$= \left[\tilde{q}(x) + \frac{\alpha_k}{\beta_l} x^{k-l} \right] g(x) + \tilde{r}(x)$$

The required polynomials, $q(x)$ and $r(x)$, are obtained by setting

$$q(x) = \tilde{q}(x) + \frac{\alpha_k}{\beta_l} x^{k-l}, \qquad r(x) = \tilde{r}(x) \qquad ////$$

EXAMPLE 1.11.4 Let

$$f(x) = x^5 - 4x^4 + 2x^3 + 5x^2 - 4x - 2 \qquad g(x) = -x^4 + 3x^3 - 4x$$

If we set

$$q(x) = -x + 1 \qquad r(x) = -x^3 + x^2 - 2$$

then $$f(x) = q(x)g(x) + r(x)$$

and $$\deg r(x) = 3 < 4 = \deg g(x) \qquad ////$$

The division algorithm is usually thought of as a division process, $q(x)$ being the quotient and $r(x)$ being the remainder after dividing $f(x)$ by $g(x)$.

When $r(x) = 0$, we have

$$f(x) = q(x)g(x)$$

and we say that $g(x)$ is a *divisor* of $f(x)$ and that $f(x)$ is a *multiple* of $g(x)$.

Definition 1.11.5 The polynomial $d(x)$ is a *greatest common divisor* of the polynomials $f_1(x), f_2(x), \ldots, f_k(x)$ provided:

1 $d(x)$ is a divisor of each of the polynomials $f_1(x), f_2(x), \ldots, f_k(x)$.
2 If $h(x)$ is a divisor of each of the polynomials $f_1(x), f_2(x), \ldots, f_k(x)$, then $h(x)$ is a divisor of $d(x)$.

Definition 1.11.5 does not tell us how many, if any, greatest common divisors a set of polynomials has. In fact, if $d_1(x)$ is a greatest common divisor of a set of polynomials and α is a nonzero element of $\mathbb{F}$, then

$$d_2(x) = \alpha d_1(x)$$

is also a greatest common divisor of these polynomials. Conversely, suppose that $d_1(x)$ and $d_2(x)$ are both greatest common divisors of the same set of polynomials. Since $d_2(x)$ is a divisor of each of the polynomials and $d_1(x)$ is a greatest common divisor, it follows that $d_1(x)$ divides $d_2(x)$. Interchanging the roles of $d_1(x)$ and $d_2(x)$ shows that $d_2(x)$ divides $d_1(x)$. Thus $d_1(x)$ and $d_2(x)$ divide each other, and consequently they differ by at most a multiplication by a constant; i.e., there exists $\alpha \in \mathbb{F}$ such that

$$d_2(x) = \alpha d_1(x) \qquad \alpha \neq 0 \qquad (1.11.6)$$

Any two polynomials $d_1(x)$ and $d_2(x)$ which satisfy equation (1.11.6) are called *associates*. Associateness of polynomials is an equivalence relation, and every nonzero polynomial has exactly one associate whose leading coefficient is 1. A polynomial whose leading coefficient is 1 is said to be *monic*. If we also define the zero polynomial to be monic, then every polynomial has a unique monic associate and we have proved:

Corollary 1.11.7 If a set of polynomials has any greatest common divisors, then they have a unique monic greatest common divisor.

The monic greatest common divisor of a set of polynomials is referred to as *the greatest common divisor*. We express the statement that $d(x)$ is the greatest common divisor of $f_1(x), f_2(x), \ldots, f_k(x)$ by writing

$$d(x) = (f_1(x), f_2(x), \ldots, f_k(x))$$

Note that we have not yet established the existence of any greatest common divisors. This will be done in Theorem 1.11.8, but first we will consider the case where some of the polynomials are the zero polynomial. Since

$$h(x) \cdot 0 = 0$$

for every polynomial $h(x)$, it follows that every polynomial $h(x)$, including $h(x) = 0$, is a divisor of 0. Moreover, 0 is the only polynomial which has every polynomial as a divisor. Thus, if

$$f_1(x) = f_2(x) = \cdots = f_k(x) = 0$$

then 0 is their only greatest common divisor. Conversely, the only polynomial which 0 divides is the zero polynomial. Thus if 0 is a greatest common divisor of $f_1(x), f_2(x), \ldots, f_k(x)$, then

$$f_1(x) = f_2(x) = \cdots = f_k(x) = 0$$

Finally, if $f_1(x), f_2(x), \ldots, f_l(x)$ are nonzero polynomials and $f_{l+1}(x) = f_{l+2}(x) = \cdots = f_k(x) = 0$, then the set of greatest common divisors of $f_1(x), f_2(x), \ldots, f_l(x)$ is exactly the same as the set of greatest common divisors of $f_1(x), f_2(x), \ldots, f_k(x)$.

Theorem 1.11.8   The euclidean algorithm[1]   Every pair of polynomials has a greatest common divisor.

PROOF   When $f(x) = g(x) = 0$, then 0 is their greatest common divisor. When $f(x) = 0$ and $g(x) \neq 0$, then the set of greatest common divisors of $f(x)$ and $g(x)$ is identical to the set of greatest common divisors of $g(x)$ alone, which is the set of associates of $g(x)$.

Finally, let $f(x)$ and $g(x)$ be nonzero polynomials. By Theorem 1.11.1 there exist polynomials $q(x)$ and $r(x)$ such that

$$f(x) = q(x)g(x) + r(x)$$

and $$\deg r(x) < \deg g(x)$$

Moreover, the set of greatest common divisors, if any, of $f(x)$ and $g(x)$ is exactly the same as the set of greatest common divisors of $g(x)$ and $r(x)$. If $r(x) = 0$, then $g(x)$ is a greatest common divisor of $g(x)$ and $r(x)$, and hence of $f(x)$ and $g(x)$. If $r(x) \neq 0$, we apply Theorem 1.11.1 again. There exist polynomials $q_1(x)$ and $r_1(x)$ such that

$$g(x) = q_1(x)r(x) + r_1(x)$$

and $$\deg r_1(x) < \deg r(x) < \deg g(x)$$

---

[1] This algorithm was discovered by Euclid, who used it to obtain the greatest common divisor of two integers. The theory of the greatest common divisor and the least common multiple are virtually identical for polynomials and for integers.

As before, if $r_1(x) = 0$, then $r(x)$ is a greatest common divisor of $r(x)$ and $r_1(x)$ and hence of $g(x)$ and $r(x)$ and, in the same way, of $f(x)$ and $g(x)$. If $r_1(x) \neq 0$, then we continue this process

$$r(x) = q_2(x)r_1(x) + r_2(x)$$
$$\cdots \cdots \cdots \cdots \cdots \cdots \cdots \cdots$$
$$r_{l-3}(x) = q_{l-1}(x)r_{l-2}(x) + r_{l-1}(x) \qquad (1.11.9)$$
$$r_{l-2}(x) = q_l(x)r_{l-1}(x) + r_l(x)$$

and $\qquad \deg r_l(x) < \deg r_{l-1}(x) < \cdots < \deg r(x) < \deg g(x)$

Since the degree of the remainder is decreasing at each step, it follows that for some value of l we have $r_l(x) = 0$. Then $r_{l-1}(x)$ is a greatest common divisor of $f(x)$ and $g(x)$. ////

Corollary 1.11.10   For each pair of polynomials the greatest common divisor exists and is unique.

The proof of Theorem 1.11.8 is constructive; i.e., it provides us with a method for actually constructing a greatest common divisor of two polynomials. Later we will see that it is not the only method.

EXAMPLE 1.11.11   Let us use the euclidean algorithm to find the greatest common divisor of the pair of polynomials

$$f(x) = x^5 - 4x^4 + 2x^3 + 5x^2 - 4x - 2 \qquad g(x) = -x^4 + 3x^3 - 4x$$

of Example 1.11.4.   According to Example 1.11.4,

$$x^5 - 4x^4 + 2x^3 + 5x^2 - 4x - 2 = (-x + 1)(-x^4 + 3x^3 - 4x) + (-x^3 + x^2 - 2)$$

and, continuing this process, we have

$$-x^4 + 3x^3 - 4x = (x - 2)(-x^3 + x^2 - 2) + (2x^2 - 2x - 4)$$
$$-x^3 + x^2 - 2 = (-\tfrac{1}{2}x)(2x^2 - 2x - 4) + (-2x - 2)$$
$$2x^2 - 2x - 4 = (-x + 2)(-2x - 2)$$

Thus $-2x - 2$ is a greatest common divisor of $x^5 - 4x^4 + 2x^3 + 5x^2 - 4x - 2$ and $-x^4 + 3x^3 - 4x$.   The monic associate of $-2x - 2$ is $x + 1$, and hence

$$x + 1 = (x^5 - 4x^4 + 2x^3 + 5x^2 - 4x - 2, -x^4 + 3x^3 - 4x) \qquad ////$$

There is an important consequence of this construction.   Working backward through equations (1.11.9), we can express the greatest common divisor of any two polynomials as a linear combination of them with polynomial coefficients; that is:

Theorem 1.11.12   If

$$d(x) = (f(x), g(x))$$

then there exist polynomials $s(x)$ and $t(x)$ such that

$$d(x) = s(x)f(x) + t(x)g(x) \quad (1.11.13)$$

PROOF   In equation (1.11.9) [with $r_l(x) = 0$], $d(x)$ is the monic associate of $r_{l-1}(x)$. Therefore it is sufficient to prove that there exist polynomials $\tilde{s}(x)$ and $\tilde{t}(x)$ such that

$$r_{l-1}(x) = \tilde{s}(x)f(x) + \tilde{t}(x)g(x)$$

The proof consists of successively eliminating $r_{l-2}(x)$, $r_{l-3}(x)$, ..., $r_1(x)$ and $r(x)$ from equations (1.11.9).   Initially we have

$$r_{l-1}(x) = r_{l-3}(x) - q_{l-1}(x)r_{l-2}(x)$$

and

$$r_{l-2}(x) = r_{l-4}(x) - q_{l-2}(x)r_{l-3}(x)$$

Substituting for $r_{l-2}(x)$ yields

$$r_{l-1}(x) = r_{l-3}(x) - q_{l-1}(x)[r_{l-4}(x) - q_{l-2}(x)r_{l-3}(x)]$$
$$= -q_{l-1}(x)r_{l-4}(x) + [q_{l-1}(x)q_{l-2}(x) + 1]r_{l-3}(x)$$

and thus $r_{l-2}(x)$ has been eliminated.   At the next stage we eliminate $r_{l-3}(x)$, and so on, until finally $r(x)$ is eliminated, leaving $r_{l-1}(x)$ expressed in terms of $f(x)$ and $g(x)$.   ////

EXAMPLE 1.11.14   Let us apply Theorem 1.11.12 to the polynomials of Examples 1.11.4 and 1.11.11.   We have already shown that the greatest common divisor is $x + 1$.

$$-2x - 2 = (-x^3 + x^2 - 2) - (-\tfrac{1}{2}x)(2x^2 - 2x - 4)$$
$$= (-x^3 + x^2 - 2)$$
$$- (-\tfrac{1}{2}x)[(-x^4 + 3x^3 - 4x) - (x - 2)(-x^3 + x^2 - 2)]$$
$$= -\tfrac{1}{2}(x^2 - 2x - 2)(-x^3 + x^2 - 2) + (\tfrac{1}{2}x)(-x^4 + 3x^3 - 4x)$$
$$= -\tfrac{1}{2}(x^2 - 2x - 2)[(x^5 - 4x^4 + 2x^3 + 5x^2 - 4x - 2)$$
$$- (-x + 1)(-x^4 + 3x^3 - 4x)] + (\tfrac{1}{2}x)(-x^4 + 3x^3 - 4x)$$
$$= -\tfrac{1}{2}(x^2 - 2x - 2)(x^5 - 4x^4 + 2x^3 + 5x^2 - 4x - 2)$$
$$+ \tfrac{1}{2}(-x^3 + 3x^2 + x - 2)(-x^4 + 3x^3 - 4x)$$

Thus we have equation (1.11.13), where

$$s(x) = \tfrac{1}{4}(x^2 - 2x - 2) \qquad t(x) = \tfrac{1}{4}(x^3 - 3x^2 - x + 2) \qquad ////$$

Thus far we have largely restricted our attention to a pair of polynomials. Now, for any finite set of polynomials we can prove that there exists a greatest common divisor and that it has an expression similar to the form of equation (1.11.13). To do this, we show that in the construction of a greatest common divisor of more than two polynomials we may successively replace two polynomials by their greatest common divisor. It will be helpful to prove the result first for a set of three polynomials.

Lemma 1.11.15   Let $f(x)$, $g(x)$, and $j(x)$ be polynomials and set

$$\tilde{d}(x) = (f(x), g(x)) \qquad d(x) = (\tilde{d}(x), j(x))$$

Then
$$d(x) = (f(x), g(x), j(x))$$

PROOF   By Corollary 1.11.10, $d(x)$ and $\tilde{d}(x)$ exist and are unique. Since $d(x)$ is a divisor of $\tilde{d}(x)$ and $\tilde{d}(x)$ is a divisor of $f(x)$ and $g(x)$, it follows that $d(x)$ is a divisor of $f(x)$ and $g(x)$.   Since $d(x)$ also divides $j(x)$, $d(x)$ satisfies the first part of Definition 1.11.5.

To establish the second part of the definition we must show that any polynomial $h(x)$ which is a divisor of $f(x)$, $g(x)$, and $j(x)$ is also a divisor of $d(x)$.   Since $h(x)$ divides $f(x)$ and $g(x)$, it follows that $h(x)$ divides $\tilde{d}(x)$. Also, since $h(x)$ divides $\tilde{d}(x)$ and $j(x)$, it divides $d(x)$.   Thus $d(x)$ is a monic greatest common divisor of $f(x)$, $g(x)$, and $j(x)$, and therefore

$$d(x) = (f(x), g(x), j(x)) \qquad ////$$

By repeated applications of Lemma 1.11.15 we have:

Theorem 1.11.16   Let $f_1(x), f_2(x), \ldots, f_k(x)$ be polynomials and set

$$d_1(x) = (f_1(x), f_2(x))$$
$$d_2(x) = (d_1(x), f_3(x))$$
$$\cdots\cdots\cdots\cdots\cdots\cdots\cdots$$
$$d_{k-1}(x) = (d_{k-2}(x), f_k(x))$$

Then
$$d_{k-1} = (f_1(x), f_2(x), \ldots, f_k(x))$$

We can obtain an expression, similar to equation (1.11.13), for the greatest common divisor of a set of polynomials by means of Theorem 1.11.16 and repeated use of Theorem 1.11.12.

Theorem 1.11.17   If

$$d(x) = (f_1(x), f_2(x), \ldots, f_k(x))$$

then there exist polynomials $s_1(x)$, $s_2(x)$, ..., $s_k(x)$ such that

$$d(x) = s_1(x)f_1(x) + s_2(x)f_2(x) + \cdots + s_k(x)f_k(x) \quad (1.11.18)$$

PROOF Define $d_1(x)$, $d_2(x)$, ..., $d_{k-1}(x)$ as in Theorem 1.11.16. Then $d(x) = d_{k-1}(x)$. By Theorem 1.11.12, there exist polynomials $s(x)$ and $t(x)$ such that

$$d_1(x) = s(x)f_1(x) + t(x)f_2(x)$$

Similarly there exist polynomials $u(x)$ and $v(x)$ such that

$$d_2(x) = u(x)d_1(x) + v(x)f_3(x)$$

Substituting the first of these equations into the second yields

$$d_2(x) = u(x)s(x)f_1(x) + u(x)t(x)f_2(x) + v(x)f_3(x) \quad (1.11.19)$$

As before, there exist polynomials $w(x)$ and $y(x)$ such that

$$d_3(x) = w(x)d_2(x) + y(x)f_4(x) \quad (1.11.20)$$

Substituting (1.11.19) into (1.11.20) yields an expression for $d_3(x)$ in terms of $f_1(x)$, $f_2(x)$, $f_3(x)$, and $f_4(x)$. Continuing this process, we eventually arrive at an expression for $d(x)$ in the form of equation (1.11.18).   ////

EXAMPLE 1.11.21 Find the greatest common divisor of

$$f(x) = x^5 - 4x^4 + 2x^3 + 5x^2 - 4x - 2$$
$$g(x) = -x^4 + 3x^3 - 4x \qquad h(x) = 2x^2 + 6x + 5$$

and express it in the form of equation (1.11.18).
     Set

$$d_1(x) = (f(x), g(x))$$

In Example 1.11.11 we showed that

$$d_1(x) = x + 1$$

By Theorem 1.11.17,

$$d(x) = (x + 1, 2x^2 + 6x + 5)$$

is the greatest common divisor of $f(x)$, $g(x)$, and $h(x)$. The euclidean algorithm can be used to show that

$$d(x) = 1$$

and, applying the method of Theorem 1.11.12, we have

$$1 = -\tfrac{1}{2}(x^3 - 6x - 4)(x^5 - 4x^4 + 2x^3 + 5x^2 - 4x - 2)$$
$$- \tfrac{1}{2}(x^4 - x^3 - 7x^2 + 4)(-x^4 + 3x^3 - 4x) + 1(2x^2 + 6x + 5) \qquad ////$$

Although the euclidean algorithm provides a constructive method for obtaining greatest common divisors, this is not the method commonly used when greatest common divisors are to be computed. Instead, each polynomial is expressed as a product of its divisors, and the greatest common divisor is immediately apparent.

Definition 1.11.22 The polynomials $f_1(x), f_2(x), \ldots, f_k(x)$ are *relatively prime* if

$$(f_1(x), f_2(x), \ldots, f_k(x)) = 1$$

and are *relatively prime in pairs* if

$$(f_i(x), f_j(x)) = 1 \qquad i, j = 1, 2, \ldots, k \qquad i \neq j$$

Lemma 1.11.23 If $f(x)$ is relatively prime to each of the polynomials $f_1(x), f_2(x), \ldots, f_k(x)$, then $f(x)$ is relatively prime to the product

$$f_1(x)f_2(x) \cdots f_k(x)$$

PROOF According to Theorem 1.11.12, for each $i, i = 1, 2, \ldots, k$, there exist polynomials $s_i(x)$ and $t_i(x)$ such that

$$t_i(x)f_i(x) = 1 - s_i(x)f(x)$$

Multiplying these k equations together yields

$$t_1(x)t_2(x) \cdots t_k(x)f_1(x)f_2(x) \cdots f_k(x) = \prod_{i=1}^{k} (1 - s_i(x)f(x))$$

$$= 1 - u(x)f(x)$$

where $u(x)$ is a function of $f(x), s_1(x), s_2(x), \ldots, s_k(x)$. We can write this equation as

$$1 = u(x)f(x) + t_1(x)t_2(x) \cdots t_k(x)f_1(x)f_2(x) \cdots f_k(x)$$

Now suppose that $h(x)$ divides both $f(x)$ and the product $f_1(x)f_2(x) \cdots f_k(x)$. Then $h(x)$ also divides 1 and hence is a nonzero number. Thus

$$(f(x), f_1(x)f_2(x) \cdots f_k(x)) = 1 \qquad ////$$

Definition 1.11.24   A polynomial is *irreducible* if it is not a constant and if its only divisors are its associates and the nonzero elements of $\mathbb{F}$.

The irreducibility of a polynomial may depend on the choice of $\mathbb{F}$. Although all polynomials of degree 1 are irreducible no matter what field $\mathbb{F}$ is, $x^2 + 1$ is irreducible when $\mathbb{F}$ is the real field but not when $\mathbb{F}$ is the complex field.
If a polynomial is irreducible, then all its associates are irreducible. Moreover, if $f(x)$ is a monic irreducible polynomial, then

$$(f(x), g(x)) = \begin{cases} f(x) & \text{when } f(x) \text{ divides } g(x) \\ 1 & \text{when } f(x) \text{ does not divide } g(x) \end{cases}$$

Lemma 1.11.25   If $p(x)$ is an irreducible polynomial which divides the product $f_1(x)f_2(x) \cdots f_k(x)$, then $p(x)$ divides at least one of the polynomials $f_1(x), f_2(x), \ldots, f_k(x)$.

PROOF   By Lemma 1.11.23, $p(x)$ is not relatively prime to at least one of the polynomials $f_i(x)$, that is,

$$d(x) = (p(x), f_i(x)) \neq 1$$

for some i.   Since $p(x)$ is irreducible, $d(x)$ is the monic associate of $p(x)$ and $p(x)$ divides $f_i(x)$.   ////

Theorem 1.11.26   The unique factorization theorem   Every polynomial of positive degree can be expressed in exactly one way (up to associates) as a product of irreducible polynomials.

PROOF   The proof that every polynomial of positive degree has at least one expression as a product of irreducible polynomials is by an induction on the degree.   Let $f(x)$ be a polynomial of degree n, $n \geq 1$.   If $n = 1$, then $f(x)$ is itself irreducible.   Now suppose that every polynomial of degree less than n can be expressed as a product of irreducible polynomials and consider $f(x)$.   If $f(x)$ is irreducible, then the required expression is simply $f(x)$ itself and we are finished.   On the other hand, if $f(x)$ is not irreducible, then, by Definition 1.11.24, $f(x)$ has a divisor, say $g(x)$, of degree k, where $1 \leq k \leq n - 1$.   Hence there exists a polynomial $h(x)$ such that

$$f(x) = g(x)h(x) \qquad 1 \leq \deg g(x), \deg h(x) \leq n - 1$$

By the induction hypothesis, both $g(x)$ and $h(x)$ can be expressed as a product of irreducible polynomials.   Consequently $f(x)$ is also expressible as a product of irreducible polynomials.   It remains only to show that such an expression is unique up to associates. Suppose that $f(x)$ can be expressed in two ways as a product of irreducible polynomials, say

$$p_1(x)p_2(x) \cdots p_r(x) = f(x) = q_1(x)q_2(x) \cdots q_s(x) \quad (1.11.27)$$

By Lemma 1.11.25, $p_1(x)$ is a divisor of $q_i(x)$ for some i, $1 \leq i \leq s$. We may assume that the terms of (1.11.27) are arranged so that $i = 1$; that is, $p_1(x)$ is a divisor of $q_1(x)$. But $q_1(x)$ is irreducible, and hence its only divisors are the nonzero numbers [and $p_1(x)$ is not one of these] and its associates. Thus there is a nonzero number α such that

$$p_1(x) = \alpha q_1(x)$$

It then follows from equation (1.11.27) that

$$\alpha p_2(x) \cdots p_r(x) = q_2(x) \cdots q_s(x)$$

Continuing this process, we see that $r = s$ and the polynomials $p_1(x)$, $p_2(x), \ldots, p_r(x)$ and $q_1(x), q_2(x), \ldots, q_s(x)$ are equal in some order, up to associates.   ////

By requiring that each of the irreducible polynomials be monic we eliminate the ambiguity caused by associates. Thus:

Theorem 1.11.28 Every monic polynomial of positive degree can be expressed in exactly one way as a product of irreducible monic polynomials of positive degree.

The unique factorization theorem provides an alternate method for finding the greatest common divisor of a set of polynomials. First we express each nonzero polynomial of the set as a product of monic irreducible polynomials and a nonzero number. We use these expressions to find the highest power of each irreducible polynomial which divides all the polynomials of the set. The product of these is the greatest common divisor.

EXAMPLE 1.11.29   Let

$$f(x) = x^5 - 4x^3 - 2x^2 + 3x + 2 \qquad g(x) = -x^4 + 3x^3 - 4x$$
$$h(x) = 2x^2 - 8x + 6$$

Find (a) $(f(x), g(x))$ and (b) $(f(x), g(x), h(x))$ using the method described in the last paragraph.

Since

$$f(x) = \quad 1 \cdot (x + 1)^3(x - 1)(x - 2)$$
$$g(x) = -1 \cdot (x + 1) \qquad\qquad (x - 2)^2 x$$
$$h(x) = \quad 2 \cdot \qquad\qquad (x - 1) \qquad\qquad (x - 3)$$

we immediately have

$$(f(x), g(x)) = (x + 1)(x - 2) = x^2 - x - 2$$
$$(f(x), g(x), h(x)) = 1 \qquad\qquad ////$$

One drawback of this method is that it does not provide a means for obtaining the expression (1.11.18) for the greatest common divisor. Another drawback is that not all polynomials factor as nicely as those in Example 1.11.29. It would not be an easy task to find the greatest common divisor of the polynomials in Example 1.11.4 in this manner.

There is a theory for the multiples of a set of polynomials which in many ways is analogous to the one we have just developed for divisors. Let us begin with a definition.

Definition 1.11.30   The polynomial $n(x)$ is a *least common multiple* of the polynomials $f_1(x), f_2(x), \ldots, f_k(x)$ provided:

1 $n(x)$ is a multiple of each of the polynomials $f_1(x), f_2(x), \ldots, f_k(x)$.
2 If the polynomial $h(x)$ is a multiple of $f_1(x), f_2(x), \ldots, f_k(x)$, then $h(x)$ is a multiple of $n(x)$.

If $n_1(x)$ and $n_2(x)$ are least common multiples of the same set of polynomials, then they are multiples of each other and hence are associates. Thus:

Corollary 1.11.31   If a set of polynomials has any least common multiples, then it has a unique monic least common multiple.

The monic least common multiple of a set of polynomials is called *the least common multiple*. We express the fact that $n(x)$ is the least common multiple of $f_1(x), f_2(x), \ldots, f_k(x)$ by writing

$$n(x) = [f_1(x), f_2(x), \ldots, f_k(x)]$$

Theorem 1.11.32   For every finite set of polynomials the least common multiple exists and is unique.

PROOF   Since the only multiple of 0 is 0, it follows that 0 is the unique least common multiple of any set of polynomials which contains 0.

Now, assume that $f_1(x), f_2(x), \ldots, f_k(x)$ are nonzero polynomials. By the unique factorization theorem there exist irreducible monic polynomials $p_1(x), p_2(x), \ldots, p_l(x)$ and nonzero numbers $\alpha_1, \alpha_2, \ldots, \alpha_l$ such that

$$f_1(x) = \alpha_1 p_1(x)^{e_{11}} p_2(x)^{e_{12}} \cdots p_l(x)^{e_{1l}}$$
$$f_2(x) = \alpha_2 p_1(x)^{e_{21}} p_2(x)^{e_{22}} \cdots p_l(x)^{e_{2l}}$$
$$\cdots\cdots\cdots\cdots\cdots\cdots\cdots\cdots\cdots$$
$$f_k(x) = \alpha_k p_1(x)^{e_{k1}} p_2(x)^{e_{k2}} \cdots p_l(x)^{e_{kl}}$$

where $e_{11}, e_{12}, \ldots, e_{kl}$ are nonnegative (possibly zero) integers. For each $j, j = 1, 2, \ldots, l$, we set

$$r_j = \max_{i=1,2,\cdots,k} e_{ij}$$

Then
$$n(x) = p_1(x)^{r_1} p_2(x)^{r_2} \cdots p_l(x)^{r_l}$$

is the least common multiple of $f_1(x), f_2(x), \ldots, f_k(x)$.   ////

EXAMPLE 1.11.33   Find (a) $[f(x), g(x)]$ and (b) $[f(x), g(x), h(x)]$ where $f(x)$, $g(x)$, and $h(x)$ are the polynomials of Example 1.11.29.

In Example 1.11.29 we expressed $f(x)$, $g(x)$, and $h(x)$ as products of a nonzero number and irreducible monic polynomials. Applying the method of construction of the least common multiple which we used to prove Theorem 1.11.32, we have

$$[f(x), g(x)] = (x + 1)^3 (x - 1)(x - 2)^2 x$$
$$[f(x), g(x), h(x)] = (x + 1)^3 (x - 1)(x - 2)^2 x(x - 3)$$ ////

In the next two sections we will apply the results of this section to matrix polynomials. For a square matrix A of order n, the matrices $I, A, A^2, \ldots,$ A^{n^2-1}, A^{n^2} are $n^2 + 1$ elements of the n^2-dimensional vector space $\mathbb{F}_n{}^n$ and hence are linearly dependent. Thus, there exist scalars $\alpha_{n^2}, \alpha_{n^2-1}, \ldots, \alpha_2, \alpha_1, \alpha_0$, not all zero, such that

$$\alpha_{n^2} A^{n^2} + \alpha_{n^2-1} A^{n^2-1} + \cdots + \alpha_2 A^2 + \alpha_1 A + \alpha_0 I = 0$$

In other words,

$$f(A) = 0$$

where $f(x)$ is the polynomial

$$f(x) = \alpha_{n^2} x^{n^2} + \alpha_{n^2-1} x^{n^2-1} + \cdots + \alpha_2 x^2 + \alpha_1 x + \alpha_0$$

In such a case we say that A *satisfies* the polynomial $f(x)$. Thus we have proved that a matrix of order n satisfies at least one nonzero polynomial of degree no greater than n^2. (See Exercise 15 of Sec. 1.5.) In the next section we will improve on this considerably.

If A and B are matrices which commute, then any power of A commutes with any power of B. More generally, when A and B commute, we have

$$f(A)g(B) = g(B)f(A)$$

for any polynomials $f(x)$ and $g(x)$. Since A always commutes with itself, it follows that

$$f(A)g(A) = g(A)f(A)$$

Likewise, if

$$f(x) = g(x)h(x)$$

then

$$f(A) = g(A)h(A)$$

EXERCISES

1 Find

(*a*) $(x^5 - x, 2x^5 - x^4 + 2x^3 - x^2)$

(*b*) $(x^5 - x, x^5 - 2x^4 + x^3 - 2x^2)$

(*c*) $(x^5 - x, x^4 - 4x^2 + 4)$

(1) by means of the euclidean algorithm and (2) by expressing each polynomial as a product or irreducible polynomials.

2 Express each of the greatest common divisors in Exercise 1 in the form of equation (1.11.13).

3 Find:

(*a*) $(x^5 - x, 2x^5 - x^4 + 2x^3 - x^2, x^5 - 2x^4 + x^3 - 2x^2)$

(*b*) $(x^3 + x, x^3 - x, 4x^2 - 1)$

(1) by means of the euclidean algorithm and Lemma 1.11.15 and (2) by expressing each polynomial as a product of irreducible polynomials. Express each greatest common divisor in the form of equation (1.11.18).

4 Let $f(x)$, $g(x)$, and $h(x)$ be polynomials. Prove that:

(*a*) $(f(x) + g(x), g(x)) = (f(x), g(x))$.

(*b*) $(h(x)f(x), h(x)g(x)) = h(x)(f(x), g(x))$.

(c) $(f(x) + g(x), f(x) - g(x)) = (f(x), g(x))$ provided $1 + 1 \neq 0$.

(d) $[f(x), g(x), h(x)] = [[f(x), g(x)], h(x)]$.

(e) $[f(x), g(x)] = \dfrac{f(x)g(x)}{(f(x), g(x))}$ provided $(f(x), g(x)) \neq 0$.

(f) $\dfrac{[f(x), g(x), h(x)]}{(f(x), g(x), h(x))} = \dfrac{[f(x), g(x)][f(x), h(x)][g(x), h(x)]}{f(x)g(x)h(x)}$

provided $f(x)g(x)h(x) \neq 0$.

5 Find the least common multiple of each of the sets of polynomials in Exercises 1 and 3
(1) by expressing each polynomial as a product of irreducible polynomials and (2) by using parts (e) and (f) of Exercise 4.

6 Prove that the polynomials $q(x)$ and $r(x)$ of Theorem 1.11.1 are unique.

7 Let $f(x)$ and $g(x)$ be polynomials where the degree of $g(x)$ is at least 1. Show that there exist polynomials $r_0(x), r_1(x), r_2(x), \ldots, r_k(x)$, each of degree less than the degree of $g(x)$, such that

$$f(x) = r_0(x) + r_1(x)g(x) + r_2(x)[g(x)]^2 + \cdots + r_k(x)[g(x)]^k \quad (1.11.34)$$

[*Hint:* Apply the euclidean algorithm to $f(x)$ and $g(x)$, obtaining $q(x)$ and $r(x)$. Next, apply the euclidean algorithm to $q(x)$ and $g(x)$.]

8 Express $f(x)$ in the form of equation (1.11.34) when

(a) $f(x) = x^7 + 2x^3 - x^2 + x + 3 \qquad g(x) = x^2 - 1$.

(b) $f(x) = 2x^4 - 3x^3 - 2x^2 + 2x - 7 \qquad g(x) = x - 2$.

9 Prove that the polynomials $r_0(x), r_1(x), r_2(x), \ldots, r_k(x)$ of Exercise 7 are unique. [What corresponds to equation (1.11.34) when the euclidean algorithm is applied to the rational integers rather than to polynomials; in particular, when $g(x)$ is replaced by 10?]

10 Find a polynomial of degree 2 satisfied by the matrix:

(a) $\begin{bmatrix} 2 & 1 \\ 3 & 4 \end{bmatrix}$
(b) $\begin{bmatrix} 2 & 1 \\ 0 & 2 \end{bmatrix}$

(c) $\begin{bmatrix} 3 & -3 \\ 3 & -3 \end{bmatrix}$
(d) $\begin{bmatrix} 2 & -2 & 1 \\ 3 & -3 & 1 \\ 0 & 0 & -1 \end{bmatrix}$

11 Find a polynomial of degree 3 satisfied by the matrix:

(a) $\begin{bmatrix} 2 & -4 & 4 \\ 1 & -2 & 3 \\ 0 & 0 & 1 \end{bmatrix}$
(b) $\begin{bmatrix} 3 & 1 & -2 \\ -1 & -1 & 4 \\ 0 & -1 & 3 \end{bmatrix}$

and show that the matrix does not satisfy any polynomial of lesser degree.

12   Let $f_1(x), f_2(x), \ldots, f_k(x)$ be polynomials satisfied by the matrix A.   Prove that A also satisfies $d(x) = (f_1(x), f_2(x), \ldots, f_k(x))$.

13   Let A be a square matrix of order n, let $Z \in \mathbb{F}^n$, and let $\mathbb{U}$ be an A-invariant subspace of $\mathbb{F}^n$.   Show that if $f(x)$ and $g(x)$ are polynomials for which

$$f(A)Z \in \mathbb{U} \qquad g(A)Z \in \mathbb{U}$$

and
$$d(x) = (f(x), g(x))$$

then
$$d(A)Z \in \mathbb{U}$$

14   Let A be a square matrix of order n and let $Z \in \mathbb{F}^n$, $Z \neq 0$.   Let $p(x)$ and $f(x)$ be polynomials for which

$$p(A)Z = f(A)Z = 0$$

Show that if $p(x)$ is irreducible, then $p(x)$ divides $f(x)$.

15   Let A be a square matrix of order n and let $Z \in \mathbb{F}^n$.   Let $p(x)$ and $f(x)$ be polynomials and let $p(x)$ be irreducible.   Show that if e is a positive integer for which

$$p(A)^e Z = f(A)Z = 0 \qquad p(A)^{e-1}Z \neq 0$$

then $p(x)^e$ divides $f(x)$.

16   Let A be a square matrix of order n and let $Z \in \mathbb{F}^n$.   Let $\mathbb{U}$ be an A-invariant subspace of $\mathbb{F}^n$ and let $p(x)$ and $f(x)$ be polynomials, $p(x)$ irreducible.   Show that if e is a positive integer for which

$$p(A)^e Z \in \mathbb{U} \qquad f(A)Z \in \mathbb{U} \qquad p(A)^{e-1}Z \notin \mathbb{U}$$

then $p(x)^e$ divides $f(x)$.

17   Let $f(x)$ and $g(x)$ be relatively prime polynomials and let A be a square matrix of order n.   Prove that

$$r(f(A)) + r(g(A)) \geq n$$

(*Hint:* Use Theorem 1.11.12).

18   Prove that for a given set of polynomials $f_1(x), f_2(x), \ldots, f_k(x)$ the polynomial $h(x)$ can be expressed in the form

$$h(x) = s_1(x)f_1(x) + s_2(x)f_2(x) + \cdots + s_k(x)f_k(x)$$

if and only if $d(x) = (f_1(x), f_2(x), \ldots, f_k(x))$ divides $h(x)$.

1.12   THE DECOMPOSITION THEOREMS

In the fundamental theorem of direct sums (Theorem 1.10.31) we proved that a decomposition of $\mathbb{F}^n$ into a direct sum of k A-invariant subspaces leads to a matrix B, similar to A, which is the direct sum of k submatrices. Moreover, we indicated how the matrices of the direct sum correspond to these invariant subspaces and showed that the order of a submatrix is equal to the dimension of the invariant subspace corresponding to it. However, the problem of actually decomposing $\mathbb{F}^n$ into a direct sum of invariant subspaces was not attacked. The main purpose of this section is to show how such a decomposition can be obtained. Our main tool will be the information about polynomials developed in the last section.

It was pointed out in Sec. 1.10 that when A is a square matrix, $\mathbb{N}(A)$ is an A-invariant vector space. Moreover, if B is any matrix which commutes with A, then $\mathbb{N}(B)$ is also A-invariant. (See Exercise 14 of Sec. 1.10.) In particular, when B can be expressed as a polynomial in A, say $B = f(A)$, then B commutes with A and $\mathbb{N}(B) = \mathbb{N}[f(A)]$ is A-invariant. In Theorem 1.12.6 we will show that $\mathbb{F}^n$ can be expressed as a direct sum of just such invariant subspaces.

Let A be a fixed $n \times n$ matrix. At the end of the last section we proved that A satisfies some nonzero polynomial of degree no greater than n^2. We also indicated that this result could be improved. Let h be the smallest integer for which there exists a nonzero polynomial of degree h satisfied by A. The last theorem of this section states that $h \leq n$. First, we need to establish that this polynomial is unique (up to associates). Hence let us suppose that

$$m(x) = a_h x^h + a_{h-1} x^{h-1} + \cdots + a_0 \qquad a_h \neq 0$$
$$\tilde{m}(x) = \tilde{a}_h x^h + \tilde{a}_{h-1} x^{h-1} + \cdots + \tilde{a}_0 \qquad \tilde{a}_h \neq 0$$

are both polynomials of degree h satisfied by A. Set

$$q(x) = \tilde{a}_h m(x) - a_h \tilde{m}(x)$$

Clearly

$$q(A) = 0$$

and the degree of $q(x)$ is less than h. By the definition of h, we see that $q(x) = 0$. Thus

$$\tilde{m}(x) = \frac{\tilde{a}_h}{a_h} m(x)$$

that is, $m(x)$ and $\tilde{m}(x)$ are associates. It follows that there is exactly one monic polynomial of degree h satisfied by A. We call this polynomial the *minimal polynomial* of A and denote it by $m(x)$.

Although we have shown that the minimal polynomial of a square matrix exists, we have not given any hints of how to find it (other than the obvious method, which involves determining which subsets of the collection of matrices $I, A, A^2, \ldots, A^{n^2}$ are linearly dependent).   In fact, this is a rather nasty problem.   However, at the end of this section we will obtain another characterization of the minimal polynomial, and in Sec. 1.13 we will derive a number of important and useful relations between the minimal polynomial and the characteristic polynomial of a matrix.   In particular, we will show that the minimal polynomial of a matrix divides the characteristic polynomial.   This fact will cut down considerably the work required to find the minimal polynomial of a given matrix.

Lemma 1.12.1   Let A be a square matrix whose minimal polynomial is $m(x)$ and let $f(x)$ be any polynomial.   Then $f(A) = 0$ if and only if $m(x)$ divides $f(x)$.

PROOF   First, suppose that $m(x)$ divides $f(x)$; that is, there is a polynomial $q(x)$ such that

$$f(x) = m(x)q(x)$$

Substituting A for x, we have

$$f(A) = m(A)q(A) = 0$$

Conversely, suppose that $f(x)$ is a polynomial which is satisfied by A. By the division algorithm, there exist polynomials $q(x)$ and $r(x)$ such that

$$f(x) = q(x)m(x) + r(x) \qquad \deg r(x) < \deg m(x)$$

Then $$0 = f(A) = q(A)m(A) + r(A) = r(A)$$

Since the degree of $r(x)$ is less than the degree of $m(x)$ and $m(x)$ is the minimal polynomial of A, it follows that $r(x) = 0$.   Thus $f(x) = q(x)m(x)$, and $m(x)$ is a divisor of $f(x)$.   ////

We will use the minimal polynomial of A to obtain a decomposition of $\mathbb{F}^n$ into the direct sum of A-invariant subspaces.   First, let us factor the minimal polynomial into its irreducible factors.   According to Theorem 1.11.28, $m(x)$ can be expressed as

$$m(x) = p_1(x)^{e_1}, p_2(x)^{e_2} \cdots p_k(x)^{e_k} \qquad (1.12.2)$$

where $p_1(x), p_2(x), \ldots, p_k(x)$ are distinct irreducible monic polynomials and $e_1, e_2, \ldots, e_k$ are positive integers.   Clearly $p_1(x)^{e_1}$, $p_2(x)^{e_2}, \ldots, p_k(x)^{e_k}$ are relatively prime in pairs, i.e.,

$$(p_i(x)^{e_i}, p_j(x)^{e_j}) = 1 \qquad i \neq j$$

Moreover, if we define the polynomials $q_i(x)$ by[1]

$$q_i(x) = \frac{m(x)}{p_i(x)^{e_i}} = \prod_{j \neq i} p_j(x)^{e_j} \qquad i = 1, 2, \ldots, k \qquad (1.12.3)$$

then

$$(p_i(x), q_i(x)) = 1 \qquad i = 1, 2, \ldots, k$$

and

$$(q_1(x), q_2(x), \ldots, q_k(x)) = 1$$

By Theorem 1.11.17, there exist polynomials $s_1(x), s_2(x), \ldots, s_k(x)$ such that

$$1 = s_1(x)q_1(x) + s_2(x)q_2(x) + \cdots + s_k(x)q_k(x) \qquad (1.12.4)$$

Observe also that for each i there exist polynomials $u_i(x)$ and $v_i(x)$ such that

$$1 = u_i(x)p_i(x)^{e_i} + v_i(x)q_i(x) \qquad i = 1, 2, \ldots, k \qquad (1.12.5)$$

Now that all the polynomials we will need have been defined, we can begin to apply some of the machinery built up here and in the last section.

Theorem 1.12.6   The primary decomposition theorem   Let A be a square matrix of order n whose minimal polynomial is given by equation (1.12.2).   Then

$$\mathbb{F}^n = \mathbb{N}[p_1(A)^{e_1}] \oplus \mathbb{N}[p_2(A)^{e_2}] \oplus \cdots \oplus \mathbb{N}[p_k(A)^{e_k}] \qquad (1.12.7)$$

PROOF   Let $q_i(x)$ and $s_i(x)$, $i = 1, 2, \ldots, k$, be defined by equations (1.12.3) and (1.12.4) respectively.   Substituting A for x in equation (1.12.4) yields

$$I = s_1(A)q_1(A) + s_2(A)q_2(A) + \cdots + s_k(A)q_k(A) \qquad (1.12.8)$$

Let X be an arbitrary element of $\mathbb{F}^n$.   Multiplying equation (1.12.8) on the right by X, we have

$$X = s_1(A)q_1(A)X + s_2(A)q_2(A)X + \cdots + s_k(A)q_k(A)X \qquad (1.12.9)$$

If we set

$$Y_i = s_i(A)q_i(A)X \qquad i = 1, 2, \ldots, k$$

---

[1] When $k = 1$, we set $q_1(x) = 1$ and the remainder of the discussion proceeds without difficulty.   However, in this case, the primary decomposition theorem (Theorem 1.12.6) does not yield a meaningful decomposition of $\mathbb{F}^n$.

then equation (1.12.9) becomes

$$X = Y_1 + Y_2 + \cdots + Y_k \quad (1.12.10)$$

Moreover, $\qquad\qquad\qquad\qquad\qquad Y_i \in \mathbb{N}[p_i(A)^{e_i}]$

since $\qquad\quad p_i(A)^{e_i} Y_i = p_i(A)^{e_i} s_i(A) q_i(A) X = m(A) s_i(A) X = 0$

Because X is an arbitrary element of $\mathbb{F}^n$, it follows from equation (1.12.10) that

$$\mathbb{F}^n = \mathbb{N}[p_1(A)^{e_1}] + \mathbb{N}[p_2(A)^{e_2}] + \cdots + \mathbb{N}[p_k(A)^{e_k}] \quad (1.12.11)$$

According to Theorem 1.5.14, in order to show that the sum on the right hand side of equation (1.12.11) is actually a direct sum we must prove that the expression for X of equation (1.12.10), namely,

$$X = Y_1 + Y_2 + \cdots + Y_k \qquad Y_i \in \mathbb{N}[p_i(A)^{e_i}]$$

is unique. Suppose we also have

$$X = Z_1 + Z_2 + \cdots + Z_k \qquad Z_i \in \mathbb{N}[p_i(A)^{e_i}]$$

Then $\qquad 0 = (Y_1 - Z_1) + (Y_2 - Z_2) + \cdots + (Y_k - Z_k) \quad (1.12.12)$

Now, let i be fixed, $1 \leq i \leq k$. Because of the way in which $q_i(x)$ was defined, $p_j(x)^{e_j}$ divides $q_i(x)$ for all $j \neq i$. Therefore

$$q_i(A)(Y_j - Z_j) = 0 \qquad j \neq i$$

and consequently if we multiply both sides of equation (1.12.12) on the left by $q_i(A)$, we have

$$0 = q_i(A)(Y_i - Z_i)$$

Moreover, equation (1.12.5) states that there exist polynomials $u_i(x)$ and $v_i(x)$ such that

$$I = u_i(A) p_i(A)^{e_i} + v_i(A) q_i(A) \quad (1.12.13)$$

Multiplying equation (1.12.13) on the right by $Y_i - Z_i$, we see that

$$Y_i - Z_i = u_i(A) p_i(A)^{e_i}(Y_i - Z_i) + v_i(A) q_i(A)(Y_i - Z_i) = 0$$

Thus $\qquad\qquad\qquad\qquad\qquad\qquad Y_i = Z_i$

We have now proved that for each $X \in \mathbb{F}^n$ the expression for X which is given by equations (1.12.9) and (1.12.10) is unique, and consequently

$$\mathbb{F}^n = \mathbb{N}[p_1(A)^{e_1}] \oplus \mathbb{N}[p_2(A)^{e_2}] \oplus \cdots \oplus \mathbb{N}[p_k(A)^{e_k}] \qquad ////$$

EXAMPLE 1.12.14 Let

$$A = \begin{bmatrix} 2 & 1 \\ 3 & 4 \end{bmatrix}$$

In Example 1.10.16 we showed that the characteristic polynomial of A is

$$x^2 - 6x + 5 = (x - 1)(x - 5)$$

It is easily verified that the minimal polynomial of A is also $x^2 - 6x + 5$.[1]
[See part (a) of Exercise 10 in Sec. 1.11.] Thus

$$p_1(x)^{e_1} = x - 1 \qquad p_2(x)^{e_2} = x - 5$$

We have already shown that

$\left\{ \begin{bmatrix} 1 \\ -1 \end{bmatrix} \right\}$ is a basis for $\mathbb{N}(A - I)$ $\left\{ \begin{bmatrix} 1 \\ 3 \end{bmatrix} \right\}$ is a basis for $\mathbb{N}(A - 5I)$

and that
$$\mathbb{F}^2 = \mathbb{N}(A - I) \oplus \mathbb{N}(A - 5I) \qquad \qquad ////$$

EXAMPLE 1.12.15 Let

$$A = \begin{bmatrix} 3 & 1 & -2 \\ -1 & -1 & 4 \\ 0 & -1 & 3 \end{bmatrix}$$

In Example 1.10.37 we showed that the characteristic polynomial of A is

$$-(x^3 - 5x^2 + 8x - 4) = -(x - 2)^2(x - 1)$$

It is easily verified that the minimal polynomial of A is $x^3 - 5x^2 + 8x - 4$.[1]
[See part (b) Exercise 11 in Sec. 1.11.] Thus

$$p_1(x)^{e_1} = (x - 2)^2 \qquad p_2(x)^{e_2} = x - 1$$

and we have already shown that

$\left\{ \begin{bmatrix} 0 \\ 2 \\ 1 \end{bmatrix} \right\}$ is a basis for $\mathbb{N}(A - I)$

It remains to find a basis for $\mathbb{N}[(A - 2I)^2]$ and to verify the primary decomposition theorem.   First we calculate

$$(A - 2I)^2 = \begin{bmatrix} 0 & 0 & 0 \\ 2 & 4 & -6 \\ 1 & 2 & -3 \end{bmatrix}$$

---

[1] It is not entirely a coincidence that the minimal polynomial and the characteristic polynomial of A are equal.   The underlying reasons for this will be made clear in Sec. 1.13.

It is not difficult to show that

$$\left\{ \begin{bmatrix} 1 \\ 1 \\ 1 \end{bmatrix}, \begin{bmatrix} 2 \\ -1 \\ 0 \end{bmatrix} \right\} \text{ is a basis for } \mathbb{N}[(A - 2I)^2]$$

Note that $\mathbb{N}[(A - 2I)^2]$ is the vector space which we denoted by $\mathbb{V}_1$ in Example 1.10.37.   Thus

$$\mathbb{F}^3 = \mathbb{N}[(A - 2I)^2] \oplus \mathbb{N}(A - I) \qquad \qquad ////$$

EXAMPLE 1.12.16   Let

$$A = \begin{bmatrix} 2 & -2 & 1 \\ 3 & -3 & 1 \\ 0 & 0 & -1 \end{bmatrix}$$

The minimal polynomial of A is

$$m(x) = x^2 + x = x(x + 1)$$

[See part (d) of Exercise 10 in Sec. 1.11.]   Thus

$$p_1(x)^{e_1} = x \qquad p_2(x)^{e_2} = x + 1$$

Clearly

$$\left\{ \begin{bmatrix} 1 \\ 1 \\ 0 \end{bmatrix} \right\} \text{ is a basis for } \mathbb{N}(A)$$

Since

$$A + I = \begin{bmatrix} 3 & -2 & 1 \\ 3 & -2 & 1 \\ 0 & 0 & 0 \end{bmatrix}$$

it follows that

$$\left\{ \begin{bmatrix} 1 \\ 2 \\ 1 \end{bmatrix}, \begin{bmatrix} 0 \\ 1 \\ 2 \end{bmatrix} \right\} \text{ is a basis for } \mathbb{N}(A + I)$$

Again, $$\mathbb{F}^3 = \mathbb{N}(A) \oplus \mathbb{N}(A + I) \qquad \qquad ////$$

EXAMPLE 1.12.17 Let

$$A = \begin{bmatrix} 3 & 4 & 5 \\ 0 & 1 & -1 \\ -1 & -2 & -1 \end{bmatrix}$$

The minimal polynomial of A is

$$m(x) = x^3 - 3x^2 + 2x = x(x - 1)(x - 2)$$

and therefore $p_1(x)^{e_1} = x$ $p_2(x)^{e_2} = x - 1$ $p_3(x)^{e_3} = x - 2$

Since

$$A - I = \begin{bmatrix} 2 & 4 & 5 \\ 0 & 0 & -1 \\ -1 & -2 & -2 \end{bmatrix} \qquad A - 2I = \begin{bmatrix} 1 & 4 & 5 \\ 0 & -1 & -1 \\ -1 & -2 & -3 \end{bmatrix}$$

we can verify that

$$\left\{ \begin{bmatrix} 3 \\ -1 \\ -1 \end{bmatrix} \right\} \qquad \left\{ \begin{bmatrix} 2 \\ -1 \\ 0 \end{bmatrix} \right\} \qquad \left\{ \begin{bmatrix} 1 \\ 1 \\ -1 \end{bmatrix} \right\}$$

are bases for $\mathbb{N}(A)$, $\mathbb{N}(A - I)$, and $\mathbb{N}(A - 2I)$ respectively. As before,

$$\mathbb{F}^3 = \mathbb{N}(A) \oplus \mathbb{N}(A - I) \oplus \mathbb{N}(A - 2I) \qquad\qquad ////$$

It can be proved that none of the vector spaces $\mathbb{N}[p_i(A)^{e_i}]$ on the right-hand side of equation (1.12.7) is a zero space. (See Exercise 8.) Thus, unless $k = 1$, equation (1.12.7) splits $\mathbb{F}^n$ nontrivially into a direct sum of invariant subspaces; that is, $\mathbb{F}^n$ is now expressed as a direct sum of invariant subspaces none of which is either 0 or $\mathbb{F}^n$ itself. On the other hand, when $k = 1$, we have

$$m(x) = p_1(x)^{e_1}$$

and the primary decomposition theorem states that

$$\mathbb{F}^n = \mathbb{N}[p_1(A)^{e_1}] = \mathbb{N}(0) = \mathbb{F}^n$$

which, while certainly true, adds no new information, In the secondary decomposition theorem we will determine when an invariant subspace of $\mathbb{F}^n$ of the form $\mathbb{N}[p(A)^e]$ can itself be expressed as the direct sum of invariant subspaces. Then the decomposition of $\mathbb{F}^n$ will be complete.

Recall that our original reason for splitting $\mathbb{F}^n$ into a direct sum of invariant subspaces was that such a decomposition of $\mathbb{F}^n$ enables us to obtain a matrix B, similar to A, having the form of a direct sum. Thus combining Theorems 1.10.31 and 1.12.6 yields:

Corollary 1.12.18   Let A be a square matrix whose minimal polynomial is given by (1.12.2).   For each i, $i = 1, 2, \ldots, k$, let

$$\mathbb{X}_i = \{X_{i1}, X_{i2}, \ldots, X_{in_i}\}$$

be a basis for $\mathbb{N}[p_i(A)^{e_i}]$ and let P be the $n \times n$ matrix whose columns are $X_{11}, X_{12}, \ldots, X_{1n_1}, X_{21}, \ldots, X_{kn_k}$ in that order.   Then

$$P^{-1}AP = B = \begin{bmatrix} B_1 & 0 & \cdots & 0 \\ 0 & B_2 & \cdots & 0 \\ \multicolumn{4}{c}{\dotfill} \\ 0 & 0 & \cdots & B_k \end{bmatrix}$$

where B_i is a square matrix of order n_i.

Examples 1.10.16 and 1.10.37 provide illustrations of Corollary 1.12.18. In Example 1.10.37 there is no indication of how either the two-dimensional A-invariant subspace $\mathbb{V}_1$ or its basis $\mathbb{X}_1$ is obtained.   It is easily verified that $\mathbb{V}_1 = \mathbb{N}[(A - 2I)^2]$.   The form of $P^{-1}AP$ then follows from the primary decomposition theorem and Corollary 1.12.18.   In Example 1.12.33 we will see that when some basis other than $\mathbb{X}_1$ is chosen for $\mathbb{N}[(A - 2I)^2]$, the form of the resulting matrix B_1 is changed.

In Corollary 1.12.18 no attempt was made to obtain any particular matrices for $B_1, B_2, \ldots, B_k$.   Nor, in fact, was an attempt made to determine whether any of the matrices $B_1, B_2, \ldots, B_k$ could itself be expressed as a direct sum.   If it is possible to split one of the subspaces, say $\mathbb{N}[p_i(A)^{e_i}]$, appearing on the right-hand side of equation (1.12.7) into a direct sum of A-invariant subspaces, then the corresponding matrix B_i can be made to have the form of a direct sum.   In Theorem 1.12.27 we will determine when this can be done and, at the same time, exhibit a special basis for a vector space which will lead (in Sec. 1.13) to particularly simple matrices $B_1, B_2, \ldots, B_k$.   We begin with:

Definition 1.12.19   Let $\mathbb{U}$ be a nonzero, A-invariant, q-dimensional subspace of $\mathbb{F}^n$.   Then $\mathbb{U}$ is a *cyclic space* (with respect to A) if there exists $Y \in \mathbb{U}$ such that the q vectors

$$Y, AY, A^2 Y, \ldots, A^{q-1} Y \quad (1.12.20)$$

form a basis for $\mathbb{U}$.   The vector Y is called a *generator* of $\mathbb{U}$, and the basis (1.12.20) is called a *cyclic basis* for $\mathbb{U}$.

The generator of a cyclic space is not unique (see Exercise 16).   Every nonzero vector of $\mathbb{F}^n$ is a generator of some cyclic space.   For, let Y be any nonzero vector, let q be the largest integer for which the vectors $Y, AY, A^2 Y$,

..., $A^{q-1}Y$ are linearly independent, and let $\cup$ be the vector space which has $Y, AY, A^2Y, \ldots, A^{q-1}Y$ as a basis. In order to prove that $\cup$ is a cyclic space with generator Y, it remains only to show that $\cup$ is A-invariant. It is sufficient to show that

$$A(A^jY) = A^{j+1}Y \in \cup \qquad j = 0, 1, 2, \ldots, q-1$$

For $j = 0, 1, 2, \ldots, q-2$ there is no difficulty since $A^{j+1}Y$ is one of the elements of the basis that we used to define $\cup$. Finally, we need to prove that

$$A(A^{q-1}Y) = A^qY \in \cup$$

Since $Y, AY, A^2Y, \ldots, A^{q-1}Y, A^qY$ are linearly dependent, there exist scalars, $\beta_0, \beta_1, \beta_2, \ldots, \beta_{q-1}, \beta_q$, not all zero, such that

$$\beta_0 Y + \beta_1 AY + \beta_2 A^2Y + \cdots + \beta_{q-1}A^{q-1}Y + \beta_q A^qY = 0 \qquad (1.12.21)$$

However, $Y, AY, A^2Y, \ldots, A^{q-1}Y$ are linearly independent, and it follows that in equation (1.12.21) we must have $\beta_q \neq 0$. Thus

$$A^qY = -\frac{\beta_0}{\beta_q}Y - \frac{\beta_1}{\beta_q}AY - \frac{\beta_2}{\beta_q}A^2Y - \cdots - \frac{\beta_{q-1}}{\beta_q}A^{q-1}Y \in \cup \qquad (1.12.22)$$

proving that $\cup$ is A-invariant.

According to equation (1.12.22), if we set

$$f(x) = x^q + \frac{\beta_{q-1}}{\beta_q}x^{q-1} + \cdots + \frac{\beta_2}{\beta_q}x^2 + \frac{\beta_1}{\beta_q}x + \frac{\beta_0}{\beta_q}$$

then

$$f(A)Y = 0 \qquad (1.12.23)$$

The (unique) monic nonzero polynomial of least degree, $n(x)$, for which

$$n(A)Y = 0$$

is called the *minimal polynomial* of Y (with respect to A). We have just proved that there is a monic polynomial of degree q satisfying equation (1.12.23). Moreover, since $Y, AY, A^2Y, \ldots, A^{q-1}Y$ are linearly dependent, no nonzero polynomial of degree less than q satisfies equation (1.12.23). Thus we have proved:

Lemma 1.12.24 The degree of the minimal polynomial of a vector is equal to the dimension of the cyclic space it generates.

The minimal polynomial of a vector behaves very much like the minimal polynomial of a matrix.

Lemma 1.12.25   If $n(x)$ is the minimal polynomial of Y (with respect to A) and $f(x)$ is any polynomial for which

$$f(A)Y = 0$$

then $n(x)$ divides $f(x)$.   In particular, $n(x)$ divides $m(x)$, the minimal polynomial of A.

PROOF   The proof is left as an exercise.   ////

In the next theorem we will apply Lemma 1.12.25 to the case where $p(x)$ is an irreducible polynomial of degree l and $Y \in \mathbb{N}[p(A)^e]$, $Y \neq 0$.   Then the minimal polynomial of Y has the form

$$n(x) = p(x)^d$$

for some integer d, $1 \leq d \leq l$, and the dimension of the cyclic space generated by Y is dl.

Theorem 1.12.26   Let $p(x)$ be an irreducible monic polynomial of degree l and let $p(x)^e$, $e \geq 1$, be the largest power of $p(x)$ which divides the minimal polynomial, $m(x)$, of A.[1]   Then there exists an el-dimensional cyclic subspace $\mathbb{U}$ of $\mathbb{N}[p(A)^e]$.

PROOF   Define the polynomials $q(x)$ and $\tilde{q}(x)$ by

$$q(x) = \frac{m(x)}{p(x)^e} \qquad \tilde{q}(x) = \frac{m(x)}{p(x)} = p(x)^{e-1}q(x)$$

Then $q(x)$ and $\tilde{q}(x)$ are polynomials such that

$$q(A) \neq 0 \qquad \tilde{q}(A) \neq 0$$

and $$p(A)^e q(A) = p(A)\tilde{q}(A) = 0$$

Since $\tilde{q}(A) \neq 0$, there exists a vector Z such that $\tilde{q}(A)Z \neq 0$.   Set $Y = q(A)Z$.   We wish to show that the cyclic space $\mathbb{U}$ generated by Y is a subspace of $\mathbb{N}[p(A)^e]$ and has dimension el.   Because

$$p(A)^e Y = p(A)^e q(A)Z = m(A)Z = 0$$

we have $Y \in \mathbb{N}[p(A)^e]$.   Moreover, since

$$p(A)^{e-1} Y = p(A)^{e-1}q(A)Z = \tilde{q}(A)Z \neq 0$$

---

[1] In other words, $p(x)$ is one of the polynomials $p_1(x), p_2(x), \ldots, p_k(x)$ of equation (1.12.2), and e is the corresponding exponent $e_1, e_2, \ldots, e_k$.

the minimal polynomial of Y is $p(x)^e$.   By Lemma 1.12.24, the cyclic space $\mathbb{U}$ generated by Y has dimension el.   ////

According to Theorem 1.12.26

$$\dim \mathbb{U} = el \leq \nu[p(A)^e] = \dim \mathbb{N}[p(A)^e]$$

If $el = \nu[p(A)^e]$, then $\mathbb{N}[p(A)^e]$ is an el-dimensional cyclic space.   It is not difficult to prove that a cyclic space is *irreducible*, i.e., it cannot be split into a direct sum of invariant subspaces (see Exercise 17).   Hence, if $el = \nu[p(A)^e]$, then the decomposition of $\mathbb{N}[p(A)^e]$ has been carried as far as possible.   On the other hand, we will see that if $el < \nu[p(A)^e]$, then $\mathbb{N}[p(A)^e]$ is expressible as a direct sum of (A-invariant) cyclic subspaces.   This result is contained in:

Theorem 1.12.27   The secondary decomposition theorem   Let $p(x)$ be an irreducible monic polynomial of degree l and let $p(x)^e$, $e \geq 1$, be the largest power of $p(x)$ which divides $m(x)$.   Then there exist positive integers $r_1 = e \geq r_2 \geq \cdots \geq r_h \geq 1$ and cyclic subspaces $\mathbb{U}_1 = \mathbb{U}, \mathbb{U}_2, \ldots,$ $\mathbb{U}_h$ of $\mathbb{N}[p(A)^e]$ of dimensions $r_1 l = el, r_2 l, \ldots, r_h l$ respectively such that

$$\mathbb{N}[p(A)^e] = \mathbb{U}_1 \oplus \mathbb{U}_2 \oplus \cdots \oplus \mathbb{U}_h$$

PROOF   Theorem 1.12.26 states that there exists a vector $Y_1 = Y$ belonging to $\mathbb{N}[p(A)^e]$ which generates a cyclic subspace $\mathbb{U}_1 = \mathbb{U}$ of $\mathbb{N}[p(A)^e]$ of dimension $r_1 l = el$.   If $\mathbb{N}[p(A)^e] = \mathbb{U}_1$, then we are finished.

Now suppose that $\mathbb{U}_1 \subsetneq \mathbb{N}[p(A)^e]$.   We will construct a cyclic space $\mathbb{U}_2$ with generator Y_2 such that $\mathbb{U}_1 \oplus \mathbb{U}_2 \subseteq \mathbb{N}[p(A)^e]$.   Let r_2 be the largest integer for which there exists a vector Z belonging to $\mathbb{N}[p(A)^e]$ such that

$$p(A)^{r_2}Z \in \mathbb{U}_1 \qquad p(A)^{r_2-1}Z \notin \mathbb{U}_1$$

Clearly $r_1 = e \geq r_2 \geq 1$.   If $p(A)^{r_2}Z = 0$, we set $Y_2 = Z$.   If $p(A)^{r_2}Z \neq 0$, we need to show that there exists a vector Y_2 such that

$$p(A)^{r_2}Y_2 = 0 \in \mathbb{U}_1 \qquad p(A)^{r_2-1}Y_2 \notin \mathbb{U}_1 \qquad (1.12.28)$$

Since $p(A)^{r_2}Z \in \mathbb{U}_1$ and $Y_1, AY_1, A^2Y_1, \ldots, A^{r_1 l-1}Y_1$ form a basis for $\mathbb{U}_1$, there exist scalars $\alpha_{r_1 l-1}, \ldots, \alpha_2, \alpha_1, \alpha_0$ for which

$$p(A)^{r_2}Z = \alpha_{r_1 l-1}A^{r_1 l-1}Y_1 + \cdots + \alpha_2 A^2 Y_1 + \alpha_1 AY_1 + \alpha_0 Y_1 \qquad (1.12.29)$$

Define the polynomial $f(x)$ by

$$f(x) = \alpha_{r_1 l-1}x^{r_1 l-1} + \cdots + \alpha_2 x^2 + \alpha_1 x + \alpha_0$$

Equation (1.12.29) then becomes

$$p(A)^{r_2}Z = f(A)Y_1 \quad (1.12.30)$$

Multiplying both sides of this equation by $p(A)^{r_1 - r_2}$, we have

$$p(A)^{r_1 - r_2}f(A)Y_1 = p(A)^{r_1 - r_2}p(A)^{r_2}Z = p(A)^{r_1}Z = p(A)^{e}Z = 0$$

Since the minimal polynomial of Y_1 is $p(x)^{r_1}$, it follows from Lemma 1.12.25 that $p(x)^{r_1}$ divides $p(x)^{r_1 - r_2}f(x)$ and consequently $f(x)$ is divisible by $p(x)^{r_2}$, say

$$f(x) = p(x)^{r_2}g(x)$$

Substituting this expression for $f(x)$ into equation (1.12.30), we have

$$p(A)^{r_2}Z = f(A)Y_1 = p(A)^{r_2}g(A)Y_1$$

If we now set

$$Y_2 = Z - g(A)Y_1$$

then Y_2 satisfies both parts of (1.12.28):

$$p(A)^{r_2}Y_2 = p(A)^{r_2}Z - p(A)^{r_2}g(A)Y_1 = 0 \in U_1$$
$$p(A)^{r_2 - 1}Y_2 = p(A)^{r_2 - 1}Z - p(A)^{r_2 - 1}g(A)Y_1 \notin U_1 \quad (1.12.31)$$

since $p(A)^{r_2 - 1}g(A)Y_1 \in U_1$ and $p(A)^{r_2 - 1}Z \notin U_1$.

Denote the cyclic space generated by Y_2 by U_2. We wish to prove that U_2 is a subspace of $N[p(A)^e]$ of dimension $r_2 l$ and that $U_1 + U_2$ is a direct sum. Combining the two parts of (1.12.31), we see that $p(x)^{r_2}$ is the minimal polynomial of Y_2, and applying Lemma 1.12.24 shows that the dimension of U_2 is $r_2 l$. Moreover, it follows from the first equation of (1.12.31) that

$$Y_2 \in N[p(A)^{r_2}] \subseteq N[p(A)^e]$$

and hence $U_2 \subseteq N[p(A)^e]$. Consequently

$$U_1 + U_2 \subseteq N[p(A)^e]$$

and it remains only to prove that this sum is actually a direct sum. Suppose there exists a nonzero vector W contained in $U_1 \cap U_2$. Because $W \in U_2$ and

$$\{Y_2, AY_2, A^2Y_2, \ldots, A^{r_2 l - 1}Y_2\}$$

is a basis for U_2, there exists a nonzero polynomial, say $g(x)$, of degree less that $r_2 l$ such that

$$W = g(A)Y_2$$

If we set

$$d(x) = (g(x), p(x)^{r_2})$$

then

$$d(x) = p(x)^m$$

for some integer m, $0 \leq m \leq r_2$. Moreover, the degree of $g(x)$ is less than the degree of $p(x)^{r_2}$, and we conclude that

$$0 \leq m < r_2$$

By Theorem 1.11.12, there exist polynomials $s(x)$ and $t(x)$ such that

$$d(x) = p(x)^m = s(x)g(x) + t(x)p(x)^{r_2} \quad (1.12.32)$$

Substituting A for x in equation (1.12.32) and multiplying on the right by Y_2 yields

$$p(A)^m Y_2 = s(A)g(A)Y_2 + t(A)p(A)^{r_2}Y_2 = s(A)W \in \mathbb{U}_1$$

since $W \in \mathbb{U}_1$. Then

$$p(A)^{r_2-1}Y_2 = p(A)^{r_2-m-1}p(A)^m Y_2 = p(A)^{r_2-m-1}s(A)W \in \mathbb{U}_1$$

which contradicts the second part of (1.12.31). Thus we have proved that $\mathbb{U}_1 \cap \mathbb{U}_2 = 0$ and consequently that $\mathbb{U}_1 + \mathbb{U}_2$ is a direct sum (see Exercise 11 of Sec. 1.5).

If $\mathbb{U}_1 \oplus \mathbb{U}_2 = \mathbb{N}[p(A)^e]$, we are finished. Otherwise we repeat the entire process. Let r_3 be the largest integer for which there exists a vector Z belonging to $\mathbb{N}[p(A)^e]$ such that

$$p(A)^{r_3}Z \in \mathbb{U}_1 \oplus \mathbb{U}_2 \qquad p(A)^{r_3-1}Z \notin \mathbb{U}_1 \oplus \mathbb{U}_2$$

As before, we can demonstrate that Z may be replaced by a vector Y_3 such that

$$p(A)^{r_3}Y_3 = 0 \qquad p(A)^{r_3-1}Y_3 \notin \mathbb{U}_1 \oplus \mathbb{U}_2$$

We then define $\mathbb{U}_3$ to be the cyclic space generated by Y_3 and show that $\mathbb{U}_3$ is a subspace of $\mathbb{N}[p(A)^e]$ of dimension $r_3 l$ and that $(\mathbb{U}_1 \oplus \mathbb{U}_2) + \mathbb{U}_3$ is a direct sum. Proceeding in this way, we eventually express $\mathbb{N}[p(A)^e]$ as $\mathbb{U}_1 \oplus \mathbb{U}_2 \oplus \cdots \oplus \mathbb{U}_h$, where each $\mathbb{U}_j$ is a cyclic subspace of $\mathbb{N}[p(A)^e]$ of dimension $r_j l$ with generator Y_j.   ////

It certainly cannot be inferred from what we have done that for a given matrix A the cyclic subspaces $\mathbb{U}_1, \mathbb{U}_2, \ldots, \mathbb{U}_h$ of the secondary decomposition theorem are unique. In fact, they are not. Furthermore, it does not even follow that the numbers h, $r_2, r_3, \ldots, r_h$ are uniquely determined. On the other hand, $r_1 = e$ is prescribed by the fact that $p(x)^e$ is the highest power of $p(x)$ which divides $m(x)$. We will consider this problem again in Theorem 1.12.42.

EXAMPLE 1.12.33   Let

$$A = \begin{bmatrix} 3 & 1 & -2 \\ -1 & -1 & 4 \\ 0 & -1 & 3 \end{bmatrix}$$

(See Examples 1.10.37 and 1.12.15.)   We have previously shown that

$$m(x) = (x - 2)^2(x - 1)$$

and that

$$\left\{ \begin{bmatrix} 1 \\ 1 \\ 1 \end{bmatrix}, \begin{bmatrix} 2 \\ -1 \\ 0 \end{bmatrix} \right\}$$

is a basis for $\mathbb{N}[(A - 2I)^2]$.   However it is not a cyclic basis, though $\mathbb{N}[(A - 2I)^2]$ is a cyclic space.   If we set

$$Y_1 = \begin{bmatrix} 2 \\ -1 \\ 0 \end{bmatrix}$$

then

$$AY_1 = \begin{bmatrix} 5 \\ -1 \\ 1 \end{bmatrix} \quad A^2Y_1 = \begin{bmatrix} 12 \\ 0 \\ 4 \end{bmatrix} = 4AY_1 - 4Y_1$$

and

$$\left\{ Y_1 = \begin{bmatrix} 2 \\ -1 \\ 0 \end{bmatrix}, AY_1 = \begin{bmatrix} 5 \\ -1 \\ 1 \end{bmatrix} \right\}$$

is a cyclic basis for $\mathbb{N}[(A - 2I)^2]$.

Clearly

$$\left\{ \begin{bmatrix} 0 \\ 2 \\ 1 \end{bmatrix} \right\}$$

is a cyclic basis for $\mathbb{N}(A - I)$.   ////

EXAMPLE 1.12.34   Let

$$A = \begin{bmatrix} 2 & -2 & 1 \\ 3 & -3 & 1 \\ 0 & 0 & -1 \end{bmatrix}$$

(See Example 1.12.16.)   We have previously shown that

$$m(x) = x(x + 1)$$

and that

$$\dim \mathbb{N}(A) = 1 \quad \dim \mathbb{N}(A + I) = 2$$

Then
$$\left\{ \begin{bmatrix} 1 \\ 1 \\ 1 \\ 0 \end{bmatrix} \right\}$$

is a cyclic basis for $\mathbb{N}(A)$.   Because the highest power of $x + 1$ which divides $m(x)$ is $x + 1$, we expect the cyclic subspaces of $\mathbb{N}(A + I)$ to be one-dimensional. Indeed, $\mathbb{N}(A + I)$ is not a cyclic space but can be expressed as the direct sum of two one-dimensional cyclic subspaces.   If Y_1 and Y_2 are any two linearly independent vectors of $\mathbb{N}(A + I)$, and if we denote the cyclic spaces generated by Y_1 and Y_2 by $\mathbb{U}_1$ and $\mathbb{U}_2$ respectively, then $\mathbb{U}_1$ and $\mathbb{U}_2$ each has dimension 1 and

$$\mathbb{N}(A + I) = \mathbb{U}_1 \oplus \mathbb{U}_2 \qquad ////$$

EXAMPLE 1.12.35   Let

$$A = \begin{bmatrix} -1 & 2 & 2 & 8 \\ 1 & 1 & 2 & 0 \\ 1 & -1 & 0 & -4 \\ -1 & 1 & 1 & 5 \end{bmatrix}$$

It can be verified that the minimal polynomial of A is

$$m(x) = x^3 - 4x^2 + 5x - 2 = (x - 1)^2(x - 2)$$

We compute

$$(A - I)^2 = \begin{bmatrix} 0 & 2 & 6 & 8 \\ 0 & 0 & 0 & 0 \\ 0 & -1 & -3 & -4 \\ 0 & 1 & 3 & 4 \end{bmatrix} \qquad A - 2I = \begin{bmatrix} -3 & 2 & 2 & 8 \\ 1 & -1 & 2 & 0 \\ 1 & -1 & -2 & -4 \\ -1 & 1 & 1 & 3 \end{bmatrix}$$

and we see that

$$\left\{ \begin{bmatrix} 1 \\ 0 \\ 0 \\ 0 \end{bmatrix}, \begin{bmatrix} 0 \\ 3 \\ -1 \\ 0 \end{bmatrix}, \begin{bmatrix} 0 \\ 4 \\ 0 \\ -1 \end{bmatrix} \right\} \qquad \text{and} \qquad \left\{ \begin{bmatrix} -2 \\ 0 \\ 1 \\ -1 \end{bmatrix} \right\}$$

are bases for $\mathbb{N}[(A - I)^2]$ and $\mathbb{N}(A - 2I)$ respectively.   The primary decomposition theorem yields

$$\mathbb{F}^4 = \mathbb{N}[(A - I)^2] \oplus \mathbb{N}(A - 2I)$$

Let us apply the secondary decomposition theorem to $\mathbb{N}[(A - I)^2]$ with $p(x) = x - 1$, $l = 1$, and $r_1 = e = 2$.   The largest cyclic subspace of $\mathbb{N}[(A - I)^2]$ has dimension 2. Since the dimension of $\mathbb{N}[(A - I)^2]$ is 3, we expect to be able to express it as

$$\mathbb{N}[(A - I)^2] = \mathbb{U}_1 \oplus \mathbb{U}_2 \quad (1.12.36)$$

where $\mathbb{U}_1$ and $\mathbb{U}_2$ are cyclic spaces of dimension 2 and 1, respectively.

Any vector Y_1 belonging to $\mathbb{N}[(A - I)^2]$ for which Y_1 and AY_1 are linearly independent may be used as a generator for $\mathbb{U}_1$ Specifically, if we set

$$Y_1 = \begin{bmatrix} 1 \\ 0 \\ 0 \\ 0 \end{bmatrix}$$

then $\qquad AY_1 = \begin{bmatrix} -1 \\ 1 \\ 1 \\ -1 \end{bmatrix} \qquad A^2Y_1 = \begin{bmatrix} -3 \\ 2 \\ 2 \\ -2 \end{bmatrix} = 2AY_1 - Y_1$

and the vector space which has

$$\left\{ \begin{bmatrix} 1 \\ 0 \\ 0 \\ 0 \end{bmatrix}, \begin{bmatrix} -1 \\ 1 \\ 1 \\ -1 \end{bmatrix} \right\}$$

as a basis can be taken for $\mathbb{U}_1$. In order to find a basis for a cyclic space $\mathbb{U}_2$ which satisfies equation (1.12.36), it suffices to find a vector Y_2 such that

$$Y_2 \notin \mathbb{U}_1 \qquad (A - I)Y_2 = 0 \in \mathbb{U}_1$$

Let us try

$$Z = \begin{bmatrix} 0 \\ 3 \\ -1 \\ 0 \end{bmatrix}$$

It is not difficult to show that $Z \notin \mathbb{U}_1$. However,

$$(A - I)Z = \begin{bmatrix} -2 & 2 & 2 & 8 \\ 1 & 0 & 2 & 0 \\ 1 & -1 & -1 & -4 \\ -1 & 1 & 1 & 4 \end{bmatrix} \begin{bmatrix} 0 \\ -3 \\ 1 \\ 0 \end{bmatrix} = \begin{bmatrix} 4 \\ -2 \\ -2 \\ 2 \end{bmatrix} \neq 0$$

At this point we can construct Y_2 as we did in the proof of Theorem 1.12.27.

A simple computation shows that

$$(A - I)Z = \begin{bmatrix} 4 \\ -2 \\ -2 \\ 2 \end{bmatrix} = -2AY_1 + 2Y_1 = -2(A - I)Y_1$$

If we set

$$Y_2 = Z + 2Y_1 = \begin{bmatrix} 2 \\ 3 \\ -1 \\ 0 \end{bmatrix}$$

then we have

$$Y_2 \notin U_1 \qquad (A - I)Y_2 = 0$$

and the cyclic space generated by Y_2, which we call U_2, satisfies equation (1.12.36).

The dimension of $N(A - 2I)$ is 1, and hence it is already a cyclic space. Any nonzero element of $N(A - 2I)$ is a generator.   ////

When we combine the primary and secondary decomposition theorems, each of the subspaces $N[p_i(A)^{e_i}]$ of equation (1.12.7) is expressed as a direct sum of cyclic spaces and thus F^n itself is expressed as a direct sum of cyclic spaces. Since a cyclic space is irreducible (see Exercise 17), this decomposition can go no further. If a basis is selected for each of the cyclic spaces and a matrix P is formed from them, as in the fundamental theorem of direct sums, then $P^{-1}AP$ is a direct sum of matrices, one corresponding to each of the cyclic spaces. In the next section we will discuss in greater detail the exact form of $P^{-1}AP$ when the columns of P are the vectors in the cyclic bases of the cyclic subspaces.

It was pointed out earlier in discussing the secondary decomposition theorem that there is no reason to suppose that the decomposition of $N[p(A)^e]$ into the direct sum of cyclic subspaces $U_1, U_2, \ldots, U_h$ is unique. Example 1.12.35 provides a simple illustration of this phenomenon. We chose, rather arbitrarily, the vector

$$Y_1 = \begin{bmatrix} 1 \\ 0 \\ 0 \\ 0 \end{bmatrix}$$

as the generator of the two-dimensional cyclic space U_1. The only requirements placed on Y_1 were that Y_1 belong to $N[(A - I)^2]$ and that Y_1 and $A Y_1$
be linearly independent.   We leave it as an exercise to verify that

$$\tilde{Y}_1 = Y_1 + A Y_1 = \begin{bmatrix} 1 \\ 0 \\ 0 \\ 0 \end{bmatrix} + \begin{bmatrix} -1 \\ 1 \\ 1 \\ -1 \end{bmatrix} = \begin{bmatrix} 0 \\ 1 \\ 1 \\ -1 \end{bmatrix}$$

can be used to generate the same cyclic space U_1 and that

$$\hat{Y}_1 = Y_1 + Y_2 = \begin{bmatrix} 1 \\ 0 \\ 0 \\ 0 \end{bmatrix} + \begin{bmatrix} 2 \\ 3 \\ -1 \\ 0 \end{bmatrix} = \begin{bmatrix} 3 \\ 3 \\ -1 \\ 0 \end{bmatrix}$$

can be used to generate a different two-dimensional cyclic subspace of
$N[(A - I)^2]$.   It is interesting that we can still choose Y_2 as the generator of U_2.

   Since neither the vectors $Y_1, Y_2, \ldots, Y_h$ nor the cyclic spaces $U_1, U_2, \ldots,$
U_h which they generate are unique in the secondary decomposition theorem, it is
not unreasonable to expect the values of $h, r_2, \ldots, r_h$ to depend on the choice of
$U_1, U_2, \ldots, U_h$.   In order to show that this is in fact not the case we introduce
a second type of basis for a cyclic space.

Definition 1.12.37   Let $p(x)$ be an irreducible monic polynomial of degree l
and let U be a dl-dimensional cyclic subspace of $N[p(A)^e]$ with generator Y.
Then the dl vectors

$Y,$	$AY,$	$A^2 Y,$	$\ldots,$	$A^{l-1}Y$
$p(A)Y,$	$p(A)AY,$	$p(A)A^2 Y,$	$\ldots,$	$p(A)A^{l-1}Y$
$\cdots$	$\cdots$	$\cdots$		$\cdots$
$p(A)^{d-1}Y,$	$p(A)^{d-1}AY,$	$p(A)^{d-1}A^2 Y,$	$\ldots,$	$p(A)^{d-1}A^{l-1}Y$

$$(1.12.38)$$

form a basis for U called the *canonical basis*.

   When $d = 1$, the cyclic basis and the canonical basis generated by Y are
identical.   For $d > 1$ it is necessary to verify that the dl vectors of (1.12.38) do
actually form a basis for U.   Clearly these vectors belong to U, and since the
dimension of U is dl, it is sufficient to show that they span U.   For any $X \in U$
there exists a polynomial of degree less than dl, say $f(x)$, such that

$$X = f(A)Y$$

By repeated use of the division algorithm we can show that there exist polynomials $r_0(x), r_1(x), r_2(x), \ldots, r_{d-1}(x)$, each of degree less than l, such that

$$f(x) = r_0(x) + p(x)r_1(x) + p(x)^2 r_2(x) + \cdots + p(x)^{d-1} r_{d-1}(x)$$

(See Exercise 7 of Sec. 1.11.)   Then

$$X = f(A)Y = r_0(A)Y + p(A)r_1(A)Y + p(A)^2 r_2(A)Y + \cdots + p(A)^{d-1} r_{d-1}(A)Y$$

and hence X can be expressed as a linear combination of the vectors of (1.12.38).

One advantage the canonical basis has over the cyclic basis for a dl-dimensional space $\mathbb{U}$ is that for each value of k, $k = 0, 1, 2, \ldots, d$, we can exhibit a basis for $p(A)^k \mathbb{U}$.   The $(d - k)l$ vectors

$$
\begin{array}{ccccc}
p(A)^k Y, & p(A)^k A Y, & p(A)^k A^2 Y, & \ldots, & p(A)^k A^{l-1} Y \\
p(A)^{k+1} Y, & p(A)^{k+1} A Y, & p(A)^{k+1} A^2 Y, & \ldots, & p(A)^{k+1} A^{l-1} Y \\
\multicolumn{5}{c}{\cdots\cdots\cdots\cdots\cdots\cdots\cdots\cdots\cdots\cdots} \\
p(A)^{d-1} Y, & p(A)^{d-1} A Y, & p(A)^{d-1} A^2 Y, & \ldots, & p(A)^{d-1} A^{l-1} Y
\end{array}
$$

$$(1.12.39)$$

are linearly independent elements of $p(A)^k \mathbb{U}$ since they constitute a subset of the canonical basis for $\mathbb{U}$.   In order to show that they form a basis for $p(A)^k \mathbb{U}$ it remains only to show that they span $p(A)^k \mathbb{U}$.   If $Z \in p(A)^k \mathbb{U}$, then Z can be expressed as

$$Z = p(A)^k X$$

for some $X \in \mathbb{U}$.   As before, we can express X as

$$X = f(A)Y = r_0(A)Y + p(A)r_1(A)Y + p(A)^2 r_2(A)Y + \cdots + p(A)^{d-1} r_{d-1}(A)Y$$

and hence

$$Z = p(A)^k X = p(A)^k r_0(A)Y + p(A)^{k+1} r_1(A)Y + $$
$$p(A)^{k+2} r_2(A)Y + \cdots + p(A)^{k+d-1} r_{d-1}(A)Y \quad (1.12.40)$$

By Lemma 1.12.24, the minimal polynomial of Y is $p(x)^d$, and consequently equation (1.12.40) becomes

$$Z = p(A)^k r_0(A)Y + p(A)^{k+1} r_1(A)Y + p(A)^{k+2} r_2(A)Y + \cdots + p(A)^{d-1} r_{d-k-1}(A)Y$$

$$(1.12.41)$$

Because each of the polynomials $r_0(x), r_1(x), \ldots, r_{d-1}(x)$ is of degree less than l, equation (1.12.41) expresses Z as a linear combination of the vectors of (1.12.39). Thus the vectors of (1.12.39) form a basis for $p(A)^k \mathbb{U}$, and we have $p(A)^k \mathbb{U}$ as a $(d - k)$ l-dimensional cyclic space with generator $p(A)^k Y$.

Making use of the canonical basis for a cyclic space, we can now prove:

Theorem 1.12.42 The numbers $h, r_2, r_3, \ldots, r_h$ of the secondary decomposition theorem are uniquely determined by A.

PROOF The method just used to determine the dimension of each of the vector spaces $p(A)^k \mathbb{U}$, where $\mathbb{U}$ is a cyclic subspace of $\mathbb{N}[p(A)^e]$, can also be used to find the dimension of the vector spaces $p(A)^k\{\mathbb{N}[p(A)^e]\}$, $k = 0, 1, 2, \ldots, e$. Since

$$\mathbb{N}[p(A)^e] = \mathbb{U}_1 \oplus \mathbb{U}_2 \oplus \cdots \oplus \mathbb{U}_h$$

and $\mathbb{U}_1, \mathbb{U}_2, \ldots, \mathbb{U}_h$ are A-invariant, it follows that for each k we have

$$p(A)^k\{\mathbb{N}[p(A)^e]\} = p(A)^k\mathbb{U}_1 \oplus p(A)^k\mathbb{U}_2 \oplus \cdots \oplus p(A)^k\mathbb{U}_h$$

and hence

$$\dim(p(A)^k\{\mathbb{N}[p(A)^e]\}) = \dim[p(A)^k\mathbb{U}_1] + \dim[p(A)^k\mathbb{U}_2] + \cdots$$
$$+ \dim[p(A)^k\mathbb{U}_h] \quad (1.12.43)$$

In the discussion above we showed that for a cyclic space $\mathbb{U}$ of dimension dl,

$$\dim[p(A)^k\mathbb{U}] = \begin{cases} (d - k)l & k = 0, 1, 2, \ldots, d \\ 0 & k \geq d \end{cases} \quad (1.12.44)$$

Let us define $t_e, t_{e-1}, \ldots, t_1$ by

$$e = r_1 \geq r_2 \geq \cdots \geq r_{t_k} \geq k > r_{t_k+1} \geq \cdots \geq r_h \geq 1 \quad (1.12.45)$$

and set $t_{e+1} = 0$; i.e., t_k is the largest integer for which $r_{t_k} \geq k$. Note that $t_1 = h$. If we set

$$s_k = t_k - t_{k+1}$$

then among the cyclic spaces $\mathbb{U}_1, \mathbb{U}_2, \ldots, \mathbb{U}_h$ there are exactly s_k whose dimension is kl. Substituting equation (1.12.44) into equation (1.12.43) yields

$$\dim(p(A)^k\{\mathbb{N}[p(A)^e]\}) = [(r_1 - k) + (r_2 - k) + \cdots + (r_{t_k} - k)]l \quad (1.12.46)$$

Clearly A uniquely determines the dimension of the $e + 1$ vector spaces

$$p(A)^k\{\mathbb{N}[p(A)^e]\} \quad k = 0, 1, 2, \ldots, e$$

and it remains to prove that $h, r_2, r_3, \ldots, r_h$ are uniquely determined by the numbers on the right-hand side of the various equations of (1.12.46). Setting $k = e$ in (1.12.46) produces no information. However, setting $k = e - 1$, we see that

$$\dim(p(A)^{e-1}\{\mathbb{N}[p(A)^e]\}) = t_e l = s_e l$$

which proves that s_e is completely determined by A. Similarly, setting $k = e - 2$ in (1.12.46), we have

$$\dim(p(A)^{e-2}\{\mathbb{N}[p(A)^e]\}) = (2s_e + s_{e-1})l$$

from which we can calculate s_{e-1}. Continuing in this way, we see that the dimensions of the cyclic spaces $\mathbb{U}_1, \mathbb{U}_2, \ldots, \mathbb{U}_h$ can be obtained from the equations of (1.12.46). It follows immediately that h is also uniquely determined by A. ////

EXAMPLE 1.12.47   Let

$$A = \begin{bmatrix} -1 & 2 & 2 & 8 \\ 1 & 1 & 2 & 0 \\ 1 & -1 & 0 & -4 \\ -1 & 1 & 1 & 5 \end{bmatrix}$$

We have seen that the minimal polynomial of A is

$$m(x) = (x - 1)^2(x - 2)$$

(see Example 1.12.35) and that

$$\left(\begin{bmatrix} 1 \\ 0 \\ 0 \\ 0 \end{bmatrix}, \begin{bmatrix} 0 \\ 3 \\ -1 \\ 0 \end{bmatrix}, \begin{bmatrix} 0 \\ 4 \\ 0 \\ -1 \end{bmatrix} \right)$$

is a basis for $\mathbb{N}[(A - I)^2]$. Multiplying each of these vectors by $A - I$, we see that

$$\left(\begin{bmatrix} -2 \\ 1 \\ 1 \\ -1 \end{bmatrix} \right)$$

is a basis for $(A - I)\{\mathbb{N}[(A - I)^2]\}$. Thus, with $p(x) = x - 1$, $e = 2$, and $l = 1$ and setting $k = 1, 0$ we have

$$k = 1: \quad \dim[(A - I)\{\mathbb{N}[(A - I)^2]\}] = 1 = s_2$$
$$k = 0: \quad \dim\{\mathbb{N}[(A - I)^2]\} = 3 = 2s_2 + s_1$$

verifying that $\mathbb{N}[(A - I)^2]$ can be expressed as the direct sum of a two-dimensional cyclic space and a one-dimensional cyclic space. ////

Since the numbers $h, r_1 = e, r_2, \ldots, r_h$ are uniquely determined by A, we can make the following definition.

Definition 1.12.48 Let $p(x)$ be an irreducible monic polynomial which divides the minimal polynomial of A. The polynomials $p(x)^{r_1}, p(x)^{r_2}, \ldots, p(x)^{r_h}$ of the secondary decomposition theorem are called the $p(x)$-*elementary divisors of A*. The totality of $p(x)$-elementary divisors of A for all irreducible monic polynomials $p(x)$ which divide $m(x)$, that is, the polynomials $p_1(x), p_2(x), \ldots, p_k(x)$ of equation (1.12.2), is called the *set of elementary divisors of A*.

The $p(x)$-elementary divisor $p(x)^{r_j}$ arises from the $r_j l$-dimensional cyclic subspace $\mathbb{U}_j$ of $\mathbb{N}[p(A)^e]$ of the secondary decomposition theorem.[1] The elementary divisors of the matrix of Example 1.12.47 are $(x-1)^2$, $x-1$, and $x-2$.

Corollary 1.12.49 Similar matrices have the same elementary divisors.

PROOF Let $B = P^{-1}AP$. Then A and B have the same minimal polynomial. Moreover, if

$$\mathbb{N}[p(A)^e] = \mathbb{U}_1 \oplus \mathbb{U}_2 \oplus \cdots \oplus \mathbb{U}_h$$

where $\mathbb{U}_i$ is a cyclic space with generator Y_i, then

$$\mathbb{N}[p(B)^e] = P^{-1}\{\mathbb{N}[p(A)^e]\} = P^{-1}\mathbb{U}_1 \oplus P^{-1}\mathbb{U}_2 \oplus \cdots \oplus P^{-1}\mathbb{U}_h$$

and $P^{-1}\mathbb{U}_i$ is a cyclic space with generator $P^{-1}Y_i$. (See Exercise 16 of Sec. 1.6.) The cyclic spaces $\mathbb{U}_i$ and $P^{-1}\mathbb{U}_i$ have the same dimension and therefore give rise to the same elementary divisor.   ////

Corollary 1.12.50 Let A be a square matrix. Then A and A^T have the same elementary divisors.

PROOF A and A^T have the same minimal polynomial. It follows from the proof of Theorem 1.12.42 that in order to prove that A and A^T have the same elementary divisors it is sufficient to establish

$$\dim(p(A)^k\{\mathbb{N}[p(A)^e]\}) = \dim(p(A^T)^k\{\mathbb{N}[p(A^T)^e]\}) \qquad k = 1, 2, \ldots, e$$

Clearly

$$\mathbb{N}[p(A)^k] \cap \mathbb{N}[p(A)^e] = \mathbb{N}[p(A)^k]$$

so that $$\dim\{\mathbb{N}[p(A)^k] \cap \mathbb{N}[p(A)^e]\} = v[p(A)^k] \quad (1.12.51)$$

[1] For an entirely different approach to the study of elementary divisors see Albert [1].

Making use of Exercise 19 of Sec. 1.5 (with $\mathbb{V} = \mathbb{N}[p(A)^e]$) and equation (1.12.51), we have

$$\dim(p(A)^k\{\mathbb{N}[p(A)^e]\}) = v[p(A)^e] - v[p(A)^k]$$
$$= v[p(A^T)^e] - v[p(A^T)^k] = \dim(p(A^T)^k\{\mathbb{N}[p(A^T)^e]\})$$

as required.   ////

The minimal polynomial of A is equal to the least common multiple of the elementary divisors of A. It follows from the secondary decomposition theorem that

$$v[p(A)^e] = \dim(\mathbb{N}[p(A)^e]) = \dim \mathbb{U}_1 + \dim \mathbb{U}_2 + \cdots + \dim \mathbb{U}_h$$
$$= \deg p(x)^{r_1} + \deg p(x)^{r_2} + \cdots + \deg p(x)^{r_h}$$
$$= \deg[p(x)^{r_1}p(x)^{r_2} \cdots p(x)^{r_h}] = (r_1 + r_2 + \cdots + r_h)l$$

The degree of the product of the $p(x)$-elementary divisors of A is equal to $v[p(A)^e]$, and, combining this result with equation (1.12.7), we see that the degree of the product of all the elementary divisors of A is equal to the order of A. Since the degree of the least common multiple of a set of polynomials is at most the degree of the product of these polynomials, we can now conclude:

Theorem 1.12.52   The degree of the minimal polynomial of a matrix of order n is at most n.

The product of all the elementary divisors of A is a monic polynomial whose degree is n. In the next section we will show that, up to a possible change in sign, the product of all the elementary divisors of A is equal to the characteristic polynomial of A.

EXERCISES

1 Find the minimal polynomial and the elementary divisors of each of the following matrices:

(a) $\begin{bmatrix} 1 & 0 \\ 0 & 1 \end{bmatrix}$ (b) $\begin{bmatrix} 1 & 0 \\ 0 & 2 \end{bmatrix}$ (c) $\begin{bmatrix} 1 & 1 \\ 0 & 1 \end{bmatrix}$ (d) $\begin{bmatrix} 1 & 0 \\ 1 & 1 \end{bmatrix}$

(e) $\begin{bmatrix} 1 & 1 \\ 0 & 2 \end{bmatrix}$ (f) $\begin{bmatrix} 1 & 1 \\ 1 & 1 \end{bmatrix}$ (g) $\begin{bmatrix} 1 & -1 \\ -1 & 1 \end{bmatrix}$ (h) $\begin{bmatrix} 1 & 2 \\ 4 & 3 \end{bmatrix}$

In each case express $\mathbb{F}^2$ as a direct sum of irreducible cyclic spaces.

2 For each of the following matrices

(a) $\begin{bmatrix} -1 & 1 \\ -2 & 1 \end{bmatrix}$ (b) $\begin{bmatrix} 1 & -2 \\ 2 & 1 \end{bmatrix}$

show that $\mathbb{F}^2$ can be expressed as a direct sum of two one-dimensional cyclic spaces when $\mathbb{F}$ is the complex field but cannot be so expressed when $\mathbb{F}$ is the real field.   Show that $\mathbb{F}^2$ is a cyclic space with generator

$$Y = \begin{bmatrix} 1 \\ 0 \end{bmatrix}$$

when $\mathbb{F}$ is either the real or complex field. (Note that when $\mathbb{F}$ is the complex field, $\mathbb{F}^2$ is a two-dimensional cyclic space even though it can be expressed as the direct sum of two one-dimensional cyclic spaces; see Exercises 15 and 17.)

3 Find the minimal polynomial and the elementary divisors of the matrix

$$A = \begin{bmatrix} -3 & 1 & 3 \\ 3 & -1 & -3 \\ -5 & 1 & 5 \end{bmatrix}$$

Show that A is similar to a diagonal matrix and use Corollary 1.12.49 to find the minimal polynomial of A.   Express $\mathbb{F}^3$ as a direct sum of three one-dimensional cyclic spaces.

4 Find the minimal polynomial and the elementary divisors of the matrix

$$A = \begin{bmatrix} 0 & 0 & 1 \\ 1 & -1 & 1 \\ -1 & 0 & -2 \end{bmatrix}$$

Show that A is similar to the direct sum of a 1×1 matrix and a 2×2 matrix but is not similar to a diagonal matrix.   Express $\mathbb{F}^3$ as a direct sum of cyclic spaces.   (*Hint:* Use Theorem 1.13.16 to find the minimal polynomial of A.)

5 Let

$$A = \begin{bmatrix} 2 & -2 & 1 \\ 1 & 5 & -1 \\ 2 & 3 & 2 \end{bmatrix}$$

Show that the minimal polynomial of A is $(x - 3)^3$.   Show that $\mathbb{F}^3$ is a cyclic space with respect to A and that

$$Y = \begin{bmatrix} 1 \\ 0 \\ 0 \end{bmatrix}$$

is a generator. Show also that

$$\tilde{Y} = \begin{bmatrix} 0 \\ 0 \\ 1 \end{bmatrix}$$

does not generate $\mathbb{F}^3$. Find the canonical and cyclic bases for $\mathbb{F}^3$ using Y and obtain $P^{-1}AP$ where:

(a) P is the 3×3 matrix whose columns are the vectors in the canonical basis.

(b) P is the 3×3 matrix whose columns are the vectors in the cyclic basis.

6 Let

$$A = \begin{bmatrix} 1 & 1 & 0 & 0 \\ -2 & 0 & 1 & 0 \\ 2 & 0 & 0 & 1 \\ -2 & -1 & -1 & -1 \end{bmatrix}$$

Show that the minimal polynomial of A is $(x^2 + 1)^2$. Show that if $\mathbb{F}$ is the real field then $\mathbb{F}^4$ is a cyclic space. Find a generator for $\mathbb{F}^4$. Find a nonzero vector of $\mathbb{F}^4$ which does *not* generate $\mathbb{F}^4$. Using the generator previously found, obtain the cyclic and canonical bases for $\mathbb{F}^4$, Apply the method of (1.12.39) to construct a canonical basis for $(A^2 + I)\mathbb{F}^4$. Find $P^{-1}AP$ when:

(a) P is the 4×4 matrix whose columns are the vectors of the canonical basis of $\mathbb{F}^4$.

(b) P is the 4×4 matrix whose columns are the vectors of the cyclic basis of $\mathbb{F}^4$.

Show that if $\mathbb{F}$ is the complex field then $\mathbb{F}^4$ can be expressed as the direct sum of two two-dimensional cyclic subspaces.

7 Let A be a diagonal matrix and let the distinct diagonal entries of A be $\lambda_1, \lambda_2, \ldots, \lambda_k$. Prove that

$$m(x) = (x - \lambda_1)(x - \lambda_2) \cdots (x - \lambda_k)$$

8 Let $f(x)$ be a factor of the minimal polynomial of A of degree at least 1. Show that $f(A)$ is singular.

9 Let $p(x)$ be an irreducible polynomial such that $p(A)$ is singular. Show that $p(x)$ is a factor of the minimal polynomial of A.

10 Prove that a matrix and its transpose have the same minimal polynomial.

11 Prove that similar matrices have the same minimal polynomial.

12   Prove that the minimal polynomial of a direct sum of matrices is equal to the least common multiple of the minimal polynomials of the individual matrices. Prove that the elementary divisors of a direct sum of matrices is equal to the union of the elementary divisors of the individual matrices.

13   Let

$$d(x) = (f_1(x), f_2(x), \ldots, f_k(x))$$

Prove that

$$\mathbb{N}[d(A)] = \mathbb{N}[f_1(A)] \cap \mathbb{N}[f_2(A)] \cap \cdots \cap \mathbb{N}[f_k(A)]$$

14   Let $p(x)$ be an irreducible polynomial and let $p(x)^e$, $e \geq 1$, be the highest power of $p(x)$ which divides the minimal polynomial of A. Prove that

$$\mathbb{N}[p(A)^f] = \mathbb{N}[p(A)^{f+1}]$$

if and only if $f \geq e$.

15   Let $p_1(x)$ and $p_2(x)$ be distinct irreducible monic polynomials of degree l_1 and l_2 respectively. Let $\mathbb{U}_1$ and $\mathbb{U}_2$ be cyclic subspaces of $\mathbb{N}[p_1(A)^{e_1}]$ and $\mathbb{N}[p_2(A)^{e_2}]$ of dimension $d_1 l_1$ and $d_2 l_2$ with generators Y_1 and Y_2 respectively. Prove that $Y_1 + Y_2$ generates a cyclic subspace of $\mathbb{F}^n$ of dimension $d_1 l_1 + d_2 l_2$.

16   Let $\mathbb{U}$ be a cyclic subspace of $\mathbb{N}[p(A)^e]$ with generator Y and let $q(x)$ be a polynomial which is relatively prime to $p(x)$. Show that $Z = q(A)Y$ is also a generator of $\mathbb{U}$.

17   Let $p(x)$ be an irreducible polynomial, let $\mathbb{U}$ be a cyclic subspace of $\mathbb{N}[p(A)^e]$ with generator Y, and let $\mathbb{V}$ be a nonzero A-invariant subspace of $\mathbb{U}$. Prove that $\mathbb{V}$ has the form

$$\mathbb{V} = p(A)^k \mathbb{U}$$

for some k, $0 \leq k < e$, and that $\mathbb{V}$ is a cyclic space and is generated by $W = p(A)^k Y$. Show how it now follows that $\mathbb{U}$ is irreducible. [*Hint:* Let k be defined to be the largest integer for which there exists a vector $Z \in \mathbb{V}$ such that $p(A)^k Z = 0$ and $p(A)^{k-1} Z \neq 0$.] Is this result still true when $p(x)$ is not assumed to be irreducible? (See Exercise 15.)

18   Let A be a square matrix of order n. Prove that $\mathbb{F}^n$ is a cyclic space with respect to A if and only if the minimal polynomial of A has degree n.

19   Let A be a square matrix of order n. Prove that there exists a vector in $\mathbb{F}^n$ whose minimal polynomial is equal to the minimal polynomial of A.

1.13   THE CANONICAL FORMS

At the end of Sec. 1.10 we raised two problems.   An answer to the first, namely, how to decompose $\mathbb{F}^n$ into a direct sum of A-invariant subspaces, was provided by the two decomposition theorems of Sec. 1.12.   We have now expressed $\mathbb{F}^n$ as a direct sum of cyclic spaces corresponding to the elementary divisors of A. Since the elementary divisors are powers of irreducible polynomials, it follows from Exercise 17 of Sec. 1.12 that the cyclic spaces which correspond to them are irreducible.   Also in Sec. 1.12 we introduced two special types of bases for cyclic subspaces of $\mathbb{N}[p(A)^e]$.   In this section we will see that each of these bases provides an answer to the second problem raised at the end of Sec. 1.10, namely, how to make the forms of the matrices $B_1, B_2, \ldots, B_k$ of equation (1.10.32) as simple as possible.

Let us assume that for a given matrix A, each of the subspaces $\mathbb{N}[p_i(A)^{e_i}]$ of the primary decomposition theorem has been expressed as in the secondary decomposition theorem

$$\mathbb{N}[p_i(A)^{e_i}] = \mathbb{U}_{i1} \oplus \mathbb{U}_{i2} \oplus \cdots \oplus \mathbb{U}_{ih_i}$$

If for each of the subspaces $\mathbb{U}_{ij}, j = 1, 2, \ldots, h_i$, of $\mathbb{N}[p_i(A)^{e_i}]$ we select some basis, say $\mathbb{Y}_{ij}$, then the union of these bases

$$\mathbb{X}_i = \mathbb{Y}_{i1} \cup \mathbb{Y}_{i2} \cup \cdots \cup \mathbb{Y}_{ih_i} \qquad (1.13.1)$$

is a basis for $\mathbb{N}[p_i(A)^{e_i}]$.   Similarly, if for each $\mathbb{N}[p_i(A)^{e_i}], i = 1, 2, \ldots, k$, we select the basis $\mathbb{X}_i$ of (1.13.1), then their union

$$\mathbb{X} = \mathbb{X}_1 \cup \mathbb{X}_2 \cup \cdots \cup \mathbb{X}_k$$

is a basis for $\mathbb{F}^n$.   Finally, if we form the matrix P from the vectors of $\mathbb{X}$, then P is nonsingular and $P^{-1}AP$ is a direct sum of k matrices $B_1, B_2, \ldots, B_k$, one for each of the subspaces $\mathbb{N}[p_i(A)^{e_i}]$.   Moreover, each of the B_i's is itself the direct sum of h_i matrices, say

$$B_i = C_{i1} \oplus C_{i2} \oplus \cdots \oplus C_{ih_i}$$

one for each of the cyclic subspaces of $\mathbb{N}[p_i(A)^{e_i}]$.   Thus $P^{-1}AP$ has the form

$$\begin{aligned} P^{-1}AP &= B_1 \oplus B_2 \oplus \cdots \oplus B_k \\ &= \{C_{11} \oplus C_{12} \oplus \cdots \oplus C_{1h_1}\} \oplus \{C_{21} \oplus C_{22} \oplus \cdots \oplus C_{2h_2}\} \\ &\qquad \oplus \cdots \oplus \{C_{k1} \oplus C_{k2} \oplus \cdots \oplus C_{kh_k}\} \qquad (1.13.2) \end{aligned}$$

where the matrices $C_{i1}, C_{i2}, \ldots, C_{ih_i}$ correspond to the $p_i(x)$-elementary divisors of A.   We will investigate the form of each matrix C_{ij} of equation (1.13.2) when $\mathbb{Y}_{ij}$ is the cyclic basis for $\mathbb{U}_{ij}$ and when $\mathbb{Y}_{ij}$ is the canonical basis for $\mathbb{U}_{ij}$.

We begin with a *dl*-dimensional cyclic subspace $\mathbb{U}$ of $\mathbb{N}[p(A)^e]$ with generator Y. The matrix C which corresponds to $\mathbb{U}$ [and to the elementary divisor $p(x)^d$] will have the simplest form when we use the cyclic basis

$$\mathbb{Y} = \{Y, AY, A^2Y, \ldots, A^{dl-1}Y\} \qquad (1.13.3)$$

for $\mathbb{U}$. Let us set

$$p(x)^d = x^{dl} + \alpha_{dl-1}x^{dl-1} + \alpha_{dl-2}x^{dl-2} + \cdots + \alpha_1 x + \alpha_0 \qquad (1.13.4)$$

Multiplying each vector belonging to the basis $\mathbb{Y}$ of (1.13.3) on the left by A and expressing the resulting vector (which, of course, is contained in $\mathbb{U}$) as a linear combination of the vectors of $\mathbb{Y}$, we have

$$
\begin{aligned}
A \cdot Y &= 0Y &&+ 1AY &&+ 0A^2Y &&+ \cdots + 0A^{dl-1}Y \\
A \cdot AY &= 0Y &&+ 0AY &&+ 1A^2Y &&+ \cdots + 0A^{dl-1}Y
\end{aligned}
$$

$$\cdots\cdots\cdots\cdots\cdots\cdots\cdots\cdots\cdots\cdots\cdots\cdots\cdots\cdots\cdots\cdots\cdots\cdots \qquad (1.13.5)$$

$$
\begin{aligned}
A \cdot A^{dl-2}Y &= 0Y &&+ 0AY &&+ 0A^2Y &&+ \cdots + 1A^{dl-1}Y \\
A \cdot A^{dl-1}Y &= -\alpha_0 Y - \alpha_1 AY - \alpha_2 A^2 Y - \cdots - \alpha_{dl-1}A^{dl-1}Y
\end{aligned}
$$

The entries of the matrix C which corresponds to $\mathbb{U}$ are the coefficients in the equations of (1.13.5).   Thus

$$C = \begin{bmatrix} 0 & 0 & 0 & \cdots & 0 & -\alpha_0 \\ 1 & 0 & 0 & \cdots & 0 & -\alpha_1 \\ 0 & 1 & 0 & \cdots & 0 & -\alpha_2 \\ \multicolumn{6}{c}{\cdots\cdots\cdots\cdots\cdots\cdots\cdots\cdots} \\ 0 & 0 & 0 & \cdots & 1 & -\alpha_{dl-1} \end{bmatrix} \qquad (1.13.6)$$

Each of the matrices C_{ij} of equation (1.13.2) has the form (1.13.6).

Definition 1.13.7   The matrix C of (1.13.6) is called the *companion matrix* of the monic polynomial

$$x^{dl} + \alpha_{dl-1}x^{dl-1} + \alpha_{dl-2}x^{dl-2} + \cdots + \alpha_1 x + \alpha_0$$

A matrix consisting of the direct sum of the companion matrices of the elementary divisors of A, in any order, is called the *rational canonical form* of A.

The order of the companion matrix of a polynomial is equal to the degree of the polynomial.   Equation (1.13.6) and Definition 1.13.7 can be used to construct the companion matrix of any monic polynomial and hence the rational canonical form of any matrix.   According to the discussion leading up to Definition 1.13.7, we have proved:

Theorem 1.13.8   A matrix is similar to its rational canonical form.

EXAMPLE 1.13.9   Let

$$A = \begin{bmatrix} 2 & 1 \\ 3 & 4 \end{bmatrix}$$

We have already shown (in Examples 1.10.16, 1.10.27, and 1.12.14) that the elementary divisors of A are $x - 1$ and $x - 5$, corresponding to the cyclic subspaces

$$\mathbb{U}_{11} = \mathbb{N}(A - I) \qquad \mathbb{U}_{21} = \mathbb{N}(A - 5I)$$

with generators

$$Y_{11} = \begin{bmatrix} 1 \\ -1 \end{bmatrix} \qquad Y_{21} = \begin{bmatrix} 1 \\ 3 \end{bmatrix}$$

respectively.   The companion matrices of $x - 1$ and $x - 5$ are the 1×1 matrices [1] and [5] respectively.   We then set

$$P = \begin{bmatrix} 1 & 1 \\ -1 & 3 \end{bmatrix}$$

and obtain

$$P^{-1}AP = \begin{bmatrix} 1 & 0 \\ 0 & 5 \end{bmatrix} \qquad \qquad ////$$

EXAMPLE 1.13.10   Let

$$A = \begin{bmatrix} 3 & 1 & -2 \\ -1 & -1 & 4 \\ 0 & -1 & 3 \end{bmatrix}$$

In Examples 1.10.37, 1.12.15, and 1.12.33 we showed that the elementary divisors of A are

$$(x - 2)^2 = x^2 - 4x + 4 \qquad x - 1$$

corresponding to the cyclic subspaces

$$\mathbb{U}_{11} = \mathbb{N}[(A - 2I)^2] \qquad \mathbb{U}_{21} = \mathbb{N}(A - I)$$

with generators

$$Y_{11} = \begin{bmatrix} 2 \\ -1 \\ 0 \end{bmatrix} \qquad Y_{21} = \begin{bmatrix} 0 \\ 2 \\ 1 \end{bmatrix}$$

respectively.   The cyclic basis generated by Y_{11} is

$$Y_1 = \left\{ \begin{bmatrix} 2 \\ -1 \\ 0 \end{bmatrix}, \begin{bmatrix} 5 \\ -1 \\ 1 \end{bmatrix} \right\}$$

The companion matrices of $(x - 2)^2$ and $x - 1$ are

$$\begin{bmatrix} 0 & -4 \\ 1 & 4 \end{bmatrix} \qquad [1]$$

and it is not difficult to verify that if we set

$$P = \begin{bmatrix} 2 & 5 & 0 \\ -1 & -1 & 2 \\ 0 & 1 & 1 \end{bmatrix}$$

then
$$P^{-1}AP = \begin{bmatrix} 0 & -4 & 0 \\ 1 & 4 & 0 \\ 0 & 0 & 1 \end{bmatrix} = \begin{bmatrix} 0 & -4 \\ 1 & 4 \end{bmatrix} \oplus [1] \qquad ////$$

In Corollary 1.12.49 we proved that similar matrices have the same elementary divisors.   Conversely, according to Theorem 1.13.8, two matrices which have the same elementary divisors are each similar to the same rational canonical form and hence are similar to each other.   Combining these results, we have a characterization of similar matrices; that is:

Theorem 1.13.11   Two matrices are similar if and only if they have the same elementary divisors.

Corollary 1.13.12 now follows from Corollary 1.12.50:

Corollary 1.13.12   Let A be a square matrix.   Then A is similar to A^T.

Surprisingly, the computational problem of transforming A into A^T by means of a similarity is generally no easier than transforming A into any other matrix similar to it.

We can use Theorem 1.13.8 to obtain a great deal of information about any square matrix.   Since the rational canonical form is a direct sum of matrices, we can investigate each of these individually.   In particular, because of the simple form of the companion matrix we can easily compute its minimal polynomial and its characteristic polynomial.

Theorem 1.13.13   If C is the companion matrix of the qth degree monic polynomial $g(x)$, then the minimal polynomial of C is $g(x)$ and the characteristic polynomial of C is $(-1)^q g(x)$.

PROOF   If we set

$$g(x) = x^q + \alpha_{q-1}x^{q-1} + \alpha_{q-2}x^{q-2} + \cdots + \alpha_1 x + \alpha_0$$

then
$$C = \begin{bmatrix} 0 & 0 & 0 & \cdots & -\alpha_0 \\ 1 & 0 & 0 & \cdots & -\alpha_1 \\ 0 & 1 & 0 & \cdots & -\alpha_2 \\ \cdots\cdots\cdots\cdots\cdots\cdots\cdots \\ 0 & 0 & 0 & \cdots & -\alpha_{q-1} \end{bmatrix}$$

$\mathbb{F}^q$ is a cyclic space with respect to C, and the q unit column vectors form a cyclic basis with generator $I^{(1)}$ since

$$CI^{(1)} = I^{(2)}, \qquad C^2 I^{(1)} = CI^{(2)} = I^{(3)}, \qquad \ldots, \qquad C^{q-1}I^{(1)} = I^{(q)}$$

$$C^q I^{(1)} = CI^{(q)} = \begin{bmatrix} -\alpha_0 \\ -\alpha_1 \\ -\alpha_2 \\ \vdots \\ -\alpha_{q-1} \end{bmatrix} = (-\alpha_0 I - \alpha_1 C - \alpha_2 C^2 - \cdots - \alpha_{q-1}C^{q-1})I^{(1)}$$

Consequently the minimal polynomial of $I^{(1)}$ with respect to C is $g(x)$.   By Lemma 1.12.25, $g(x)$ divides $m(x)$, the minimal polynomial of C.   On the other hand, it follows from Theorem 1.12.52 that the degree of $m(x)$ is at most q.   Thus we have shown that $m(x) = g(x)$.

We use an induction on q to prove that the characteristic polynomial of C is $(-1)^q g(x)$.   First, let us set $q = 1$, say

$$g(x) = x + \alpha_0$$

Then
$$C = [-\alpha_0] \qquad C - xI = [-\alpha_0 - x]$$

and the result is immediate.   Now assume that the result has been established for all polynomials of degree $q - 1$ and that

$$g(x) = x^q + \alpha_{q-1}x^{q-1} + \cdots + \alpha_2 x^2 + \alpha_1 x + \alpha_0$$

When C is the companion matrix of $g(x)$, we have

$$C - xI = \begin{bmatrix} -x & 0 & 0 & \cdots & -\alpha_0 \\ 1 & -x & 0 & \cdots & -\alpha_1 \\ 0 & 1 & -x & \cdots & -\alpha_2 \\ \cdots\cdots\cdots\cdots\cdots\cdots\cdots \\ 0 & 0 & 0 & \cdots & -x - \alpha_{q-1} \end{bmatrix}$$

and, expanding the determinant of $C - xI$ along the first row, we see that

$$\det(C - xI) = (-1)^2(-x) \det \begin{bmatrix} -x & 0 & \cdots & -\alpha_1 \\ 1 & -x & \cdots & -\alpha_2 \\ \cdots\cdots\cdots\cdots\cdots\cdots\cdots\cdots \\ 0 & 0 & \cdots & -x-\alpha_{q-1} \end{bmatrix}$$

$$+ 0 + \cdots + 0 + (-1)^{q+1}(-\alpha_0) \det \begin{bmatrix} 1 & -x & 0 & \cdots & 0 \\ 0 & 1 & -x & \cdots & 0 \\ \cdots\cdots\cdots\cdots\cdots\cdots\cdots\cdots \\ 0 & 0 & 0 & \cdots & 1 \end{bmatrix} \qquad (1.13.14)$$

The first determinant on the right-hand side of equation (1.13.14) is the determinant of a matrix of the form $\tilde{C} - xI$, where $\tilde{C}$ is the companion matrix of

$$\tilde{g}(x) = x^{q-1} + \alpha_{q-1}x^{q-2} + \cdots + \alpha_2 x + \alpha_1$$

By the induction hypothesis,

$$\det(C - xI) = -x[(-1)^{q-1}\tilde{g}(x)] + (-1)^q \alpha_0 = (-1)^q g(x) \qquad ////$$

Corollary 1.13.15   If $p(x)$ is an irreducible monic polynomial, then the companion matrix of $p(x)^e$ has the single elementary divisor $p(x)^e$.

PROOF   We have just shown that $I^{(1)}$ generates a cyclic space (with respect to C) whose dimension is equal to the degree of $p(x)^e$.   ////

In Sec. 1.12 we proved that the minimal polynomial of a matrix is the least common multiple of its elementary divisors. Theorem 1.13.13 permits us to express the characteristic polynomial of a matrix in terms of its elementary divisors.

Theorem 1.13.16   The characteristic polynomial of a matrix of order n is equal to the product of its elementary divisors multiplied by $(-1)^n$.

PROOF   Since similar matrices have the same characteristic polynomial and the same elementary divisors, it is sufficient to find the characteristic polynomial of a matrix A which is in the rational canonical form. According to Exercise 17 of Sec. 1.10, the characteristic polynomial of A is equal to the product of the characteristic polynomials of the companion matrices of the elementary divisors of A. However, by Theorem 1.13.13 and Corollary 1.13.15, the characteristic polynomial of the companion matrix of an elementary divisor of A is equal to the elementary divisor multiplied by $(-1)^q$, where q is the degree of the elementary divisor.   ////

Since the minimal polynomial of a matrix is equal to the least common multiple of the elementary divisors, it divides the characteristic polynomial and we have proved the following celebrated theorem.

Theorem 1.13.17   The Hamilton-Cayley theorem[1]   A matrix satisfies its characteristic polynomial.

In this section we also wish to determine the form of the matrix C of (1.13.6) when the canonical basis for a cyclic space is used instead of the cyclic basis. Although the computations involved are somewhat more complicated in this case, the resulting matrix is considerably simpler and consequently more useful. Let

$$p(x) = x^l + \gamma_{l-1}x^{l-1} + \gamma_{l-2}x^{l-2} + \cdots + \gamma_1 x + \gamma_0$$

be an irreducible polynomial and let

$$\mathbb{Y} = \{Y, AY, A^2Y, \ldots, A^{l-1}Y, p(A)Y, p(A)AY, \ldots, p(A)^{d-1}A^{l-1}Y\} \quad (1.13.18)$$

be the canonical basis for the dl-dimensional cyclic subspace $\mathbb{U}$ of $\mathbb{N}[p(A)^e]$ generated by $\mathbb{Y}$. [Note that $p(x)^d$ is the polynomial of (1.13.4).] Multiplying each vector belonging to the basis $\mathbb{Y}$ of (1.13.18) by A and expressing the resulting vector (which, of course, belongs to $\mathbb{U}$) as a linear combination of the vectors of $\mathbb{Y}$, we have

$$A \cdot Y = \; 0Y + \; 1AY + \; 0A^2Y + \cdots + \; 0A^{l-1}Y + 0p(A)Y \\ + 0p(A)AY + \qquad \cdots \qquad\quad + 0p(A)^{d-1}A^{l-1}Y$$

$$A \cdot AY = \; 0Y + \; 0AY + \; 1A^2Y + \cdots + \; 0A^{l-1}Y + 0p(A)Y \\ + 0p(A)AY + \qquad \cdots \qquad\quad + 0p(A)^{d-1}A^{l-1}Y$$

. . . .

$$A \cdot A^{l-1}Y = -\gamma_0 Y - \gamma_1 AY - \gamma_2 A^2Y - \cdots - \gamma_{l-1}A^{l-1}Y + 1p(A)Y \\ + 0p(A)AY + \qquad \cdots \qquad\quad + 0p(A)^{d-1}A^{l-1}Y$$

$$A \cdot p(A)Y = \; 0Y + \; 0AY + \; 0A^2Y + \cdots + \; 0A^{l-1}Y + 0p(A)Y \\ + 1p(A)AY + \qquad \cdots \qquad\quad + 0p(A)^{d-1}A^{l-1}Y$$

$$(1.13.1$$

. . . .

$$A \cdot p(A)^{d-1}A^{l-2}Y = \; 0Y + \; 0AY + \; 0A^2Y + \cdots + \; 0A^{l-1}Y + 0p(A)Y \\ + 0p(A)AY + \qquad \cdots \qquad\quad + 1p(A)^{d-1}A^{l-1}Y$$

$$A \cdot p(A)^{d-1}A^{l-1}Y = \; 0Y + \; 0AY + \; 0A^2Y + \cdots + \; 0A^{l-1}Y + 0p(A)Y \\ + 0p(A)AY + \qquad \cdots \\ - \gamma_0 p(A)^{d-1}Y - \gamma_1 p(A)^{d-1}AY - \cdots - \gamma_{l-1}p(A)^{d-1}A^{l-1}Y$$

----

[1] This theorem was first proved for quaternions by Hamilton in 1853. In 1858 Cayley announced the theorem for all square matrices though he presented proofs only for 2×2 and 3×3 matrices. Since then many different proofs have been given.

From the coefficients in equations (1.13.19) we obtain

$$
C = \left[
\begin{array}{c}
\begin{array}{cccccc}
0 & 0 & 0 & \cdots & 0 & -\gamma_0 \\
1 & 0 & 0 & \cdots & 0 & -\gamma_1 \\
0 & 1 & 0 & \cdots & 0 & -\gamma_2 \\
\multicolumn{6}{c}{\cdots\cdots\cdots\cdots\cdots\cdots\cdots} \\
0 & 0 & 0 & \cdots & 1 & -\gamma_{l-1}
\end{array} \\
\qquad 1 \quad
\begin{array}{cccccc}
0 & 0 & 0 & \cdots & 0 & -\gamma_0 \\
1 & 0 & 0 & \cdots & 0 & -\gamma_2 \\
0 & 1 & 0 & \cdots & 0 & -\gamma_2 \\
\multicolumn{6}{c}{\cdots\cdots\cdots\cdots\cdots\cdots} \\
0 & 0 & 0 & \cdots & 1 & -\gamma_{l-1}
\end{array} \\
\qquad\qquad 1 \\
\qquad\qquad \cdots \\
\qquad\qquad 1 \quad
\begin{array}{cccccc}
0 & 0 & 0 & \cdots & 0 & -\gamma_0 \\
1 & 0 & 0 & \cdots & 0 & -\gamma_1 \\
0 & 1 & 0 & \cdots & 0 & -\gamma_2 \\
\multicolumn{6}{c}{\cdots\cdots\cdots\cdots\cdots\cdots} \\
0 & 0 & 0 & \cdots & 1 & -\gamma_{l-1}
\end{array}
\end{array}
\right]
\tag{1.13.20}
$$

that is, C is a $dl \times dl$ matrix and consists of d copies of the $l \times l$ companion matrix of $p(x)$ in the diagonal block positions, 1s in each of the positions immediately below the last column of the companion matrix (except, of course, for the last of the companion matrices) and 0s elsewhere. Thus C contains 1s in all the $dl - 1$ positions just below the diagonal positions. The matrix C of (1.13.20) is called the *hypercompanion matrix* of the polynomial $p(x)^d$. A matrix consisting of the direct sum of the hypercompanion matrices of the elementary divisors of A (in any order) is called the *classical canonical form of A*. Thus we also have proved.

Theorem 1.13.21 Every matrix is similar to its classical canonical form.

EXAMPLE 1.13.22 Let

$$
A = \begin{bmatrix} 3 & 1 & -2 \\ -1 & -1 & 4 \\ 0 & -1 & 3 \end{bmatrix}
$$

(See Examples 1.10.37, 1.12.15, 1.12.33, and 1.13.10). We have previously shown that

$$m(x) = (x - 2)^2(x - 1)$$

and that

$$\left\{ Y_1 = \begin{bmatrix} 2 \\ -1 \\ 0 \end{bmatrix}, AY_1 = \begin{bmatrix} 5 \\ -1 \\ 1 \end{bmatrix} \right\} \quad \left(\left\{ \begin{bmatrix} 0 \\ 2 \\ 1 \end{bmatrix} \right\} \right)$$

are cyclic bases for $\mathbb{N}[(A - 2I)^2]$ and $\mathbb{N}(A - I)$ respectively. The canonical basis for $\mathbb{N}[(A - 2I)^2]$ generated by Y_1 is given by

$$\left\{ Y_1 = \begin{bmatrix} 2 \\ -1 \\ 0 \end{bmatrix}, (A - 2I)Y_1 = \begin{bmatrix} 1 \\ 1 \\ 1 \end{bmatrix} \right\}$$

To obtain the classical canonical form of A we set

$$P = \begin{bmatrix} 2 & 1 & 0 \\ -1 & 1 & 2 \\ 0 & 1 & 1 \end{bmatrix}$$

and then

$$P^{-1}AP = \begin{bmatrix} 2 & 0 & 0 \\ 1 & 2 & 0 \\ 0 & 0 & 1 \end{bmatrix} = \begin{bmatrix} 2 & 0 \\ 1 & 2 \end{bmatrix} \oplus [1] \qquad ////$$

One special advantage of the classical canonical form occurs when the degree of $p(x)$ is 1. The companion matrix of $x - \lambda$ is the 1×1 matrix $[\lambda]$, and therefore the hypercompanion matrix of $(x - \lambda)^q$ is the $q \times q$ matrix

$$\begin{bmatrix} \lambda & 0 & 0 & \cdots & 0 & 0 \\ 1 & \lambda & 0 & \cdots & 0 & 0 \\ 0 & 1 & \lambda & \cdots & 0 & 0 \\ & & & \cdots & & \\ 0 & 0 & 0 & \cdots & 1 & \lambda \end{bmatrix} \qquad (1.13.23)$$

When each of the elementary divisors of A is of the form $(x - \lambda)^q$, the classical canonical form of A is the direct sum of matrices of the form (1.13.23) and is called the *Jordan canonical form of A*.

EXERCISES

1 Find the rational canonical form and the classical canonical form of each of the matrices of Exercise 1 of Sec. 1.12. In each case find a matrix P such that $P^{-1}AP$ is the rational canonical form of A. Find a matrix P such that $P^{-1}AP$ is the classical canonical form of A.

2   Repeat Exercise 1 using the matrices of Exercises 3 to 6 of Sec. 1.12.

3   The elementary divisors of A are x, x, x^2, $x^2 - 7x - 2$, and $x - 1$.   What is the minimal polynomial of A?   What is the characteristic polynomial of A?   What is the rational canonical form of A?   What is the classical canonical form of A?

4   Repeat Exercise 3 when the elementary divisors of A are $(x - 2)^2$, $(x - 2)^2$, and $(x - 2)^3$.

5   Prove that a matrix is similar to a diagonal matrix if and only if each of its elementary divisors is of the first degree.

6   Prove that the minimal polynomial of a matrix is equal to the characteristic polynomial (up to a possible change in sign) if and only if the elementary divisors are relatively prime in pairs.

7   Let $\mathbb{Y}$ of (1.13.3) be a cyclic basis for a cyclic space $\mathbb{U}$.   Find the matrix C of (1.13.6) when the columns of P are the vectors of $\mathbb{Y}$, but in the reversed order.   Find C when the columns of P are the vectors of a canonical basis, but in reversed order.

8   Let C be the companion matrix of a monic polynomial of degree n.   Show that $\mathbb{F}^n$ is a cyclic space with respect to C with generator $I^{(1)}$ and is a cyclic space with respect to C^T with generator $I_{(n)}$.

9   Let $X \in \mathbb{F}_n$ and $Y \in \mathbb{F}^n$ such that $XY \neq 0$.   Set $B = YX$.   Prove that the elementary divisors of B are $x - XY$, x, x, ..., x.   Also prove that the characteristic vectors of B are (a) the nonzero elements of $\mathbb{N}(X)$, corresponding to the characteristic root 0 and (b) Y, corresponding to the characteristic root XY.

10   Let

$$A = \begin{bmatrix} 0 & 1 \\ 2 & 1 \end{bmatrix}$$

Show that

$$A^n = \frac{2^n - (-1)^n}{3} A + \frac{2^n + 2(-1)^n}{3} I \qquad n = 0, 1, 2, \ldots \quad (1.13.24)$$

[*Hint:* Show that for each n, $n \geq 2$, there exist scalars α_n and β_n and a polynomial $q_n(x)$ such that

$$x^n = q_n(x)(x^2 - x - 2) + \alpha_n x + \beta_n$$

Where did $x^2 - x - 2$ come from?   It then follows that

$$A^n = \alpha_n A + \beta_n I \qquad n = 0, 1, 2, \ldots$$

Show that α_n and β_n can be evaluated by comparing the characteristic roots of A^n and those of $\alpha_n A + \beta_n I$.   See Exercises 18 and 19 of Sec. 1.10].

Note that equation (1.13.24) can easily be verified by means of an induction on n. However, an induction can be used only when the coefficients are known.

11 Let $a_1 = 4$, $a_2 = 2$ and let $a_3, a_4, a_5, \ldots$ be defined recursively by

$$a_{n+2} = a_{n+1} + 2a_n \qquad n = 1, 2, \ldots$$

Prove that

$$a_n = \frac{2^{n-1} - (-1)^{n-1}}{3} 2 + \frac{2^{n-1} + 2(-1)^{n-1}}{3} 4 \qquad n = 1, 2, \ldots$$

[*Hint:* Set

$$X_n = \begin{bmatrix} a_n \\ a_{n+1} \end{bmatrix} \qquad n = 1, 2, \ldots$$

Show that

$$X_n = A^{n-1} X_1$$

where A is the matrix of Exercise 10.]

12 The Fibonacci numbers

$$1, 1, 2, 3, 5, 8, 13, 21, 34, \ldots$$

are defined recursively by

$$a_1 = a_2 = 1 \qquad a_{n+2} = a_{n+1} + a_n \qquad n = 1, 2, \ldots$$

Show that

$$a_n = \frac{[(1 + \sqrt{5})/2]^n - [(1 - \sqrt{5})/2]^n}{\sqrt{5}} \qquad n = 1, 2, \ldots$$

(*a*) by the method of Exercise 11 and (*b*) by an induction on n.

REFERENCES

The standard one-semester course in linear algebra is designed to contain, in addition to all the material of this chapter, a discussion of inner-product spaces; see Halmos [6, chap. 3] or Perlis [8, chap. 9], among many others. This basic course prepares the reader to explore several quite different advanced topics in linear algebra. We will name just a few. Greub [5] presents a detailed and highly abstract account of multilinear algebra. Gantmacher's treatise [2] considers a large number of classical topics, among which are the Perron-Frobenius theorem and the application of matrices to systems of linear differential equations. Graybill [3] presents a careful discussion of the generalized inverse of a matrix and summarizes in textbook form many important research papers of recent years. Taylor [9] develops the theory of vector spaces of arbitrary dimension, pointing out both the similarities and the differences with the finite-dimensional case studied here.

1 Albert, A. A.: "Introduction to Algebraic Theories," University of Chicago Press, Chicago, 1941.
2 Gantmacher, F. R.: "The Theory of Matrices," vols. I and II, trans. from the Russian by K. A. Hirsch, Chelsea Publishing Company, New York, 1959.
3 Graybill, F. A.: "Introduction to Matrices with Applications in Statistics, "Wadsworth Publishing Company, Belmont, Calif. 1969.
4 Greub, W. H.: "Linear Algebra," 3d ed., Springer-Verlag New York Inc., New York, 1967.
5 Greub, W. H.: "Multilinear Algebra," Springer-Verlag New York Inc., New York, 1967.
6 Halmos, P. R.: "Finite-dimensional vector spaces," 2d ed., D. Van Nostrand Company, Inc., Princeton, N.J., 1958.
7 Hoffman, K., and R. Kunze: "Linear Algebra," 2d ed., Prentice-Hall, Inc., Englewood Cliffs, N.J., 1971.
8 Perlis, S.: "Theory of Matrices," Addison-Wesley Press, Inc., Cambridge, Mass. 1952.
9 Taylor, A. E.: "Introduction to Functional Analysis," John Wiley & Sons, Inc., New York, 1958.

2

THE THEORY OF GAMES

2.1 EXAMPLES OF GAMES

This chapter is devoted to an introduction to the theory of games; in particular, to the theory of two-person games, in which there are only a finite number of possible ways in which the game can be played and in which the amount won by one player is equal to the amount lost by the other player. We refer to such games as *two-person finite zero-sum games*.[1] In this section we will present an intuitive discussion of some simple two-person finite zero-sum games, including some very specific and not uncommon children's games. Then we will discuss some other competitive situations which can be treated within the same general format as games. We will speak about the gambling philosophy of our players, their decision-making processes, and, in the case of competitions which are not games and which do not have payoffs in terms of money, their assignment of values.

[1] Although earlier attempts had been made to analyze decision-making and competitive situations mathematically, it is fair to say that the theory of games dates from 1928, when von Neumann [6] gave the first complete proof of the fundamental theorem (our Theorem 2.5.6). The subject took a giant leap forward in 1944 with the publication of the book "Theory of Games and Economic Behavior" by von Neumann and Morgenstern and has continued to grow rapidly since then.

After this rather informal discussion we will begin the subject afresh, this time giving a careful mathematical treatment.   We will then prove rigorously some of the conclusions arrived at earlier in the less formal treatment of this section.

Our attention throughout will be restricted to a game G (or G' or $\tilde{G}$) between two players, known simply as player I and player II (or as player I' and player II').   A game will consist of a succession of "moves" made according to rules set down before the game begins.   Every move belongs to one or the other of two distinct types:

1 A choice among a finite number of alternatives made by one or the other of the players, with or without information of the previous moves that have been made but with full knowledge of the rules of the game, e.g., the moves in chess, deciding how many cards to exchange in poker, deciding whether or not to bet in poker, or deciding what to bid or play in games such as bridge.

2 A chance occurrence which neither player can control, e.g., the numbers that appear as a result of throwing dice, the cards received in an exchange in poker, the cards dealt in any card game.

A game consisting only of chance moves is quite uninteresting, at least from the mathematical point of view, since neither player is ever called upon to make a decision, and the theory of games is essentially a device for making decisions. All the games we consider will contain decision-type moves, i.e., type *1* above. After all the moves have been made, the game is over and a payment is made from one player to the other.   In accordance with some prearranged set of rules, the amount of the payoff and the direction in which the payoff is to go are determined by the set of moves that has been made.

Our players will maintain an exceptionally conservative philosophy, taking absolutely no unnecessary risks even when there is the possibility of large winnings.   They seek not the largest possible windfall but the largest possible *guarantee*.

Let us begin by describing the rules for a few very simple games and the means of *solving* them.   A solution of a game will consist of a *best*, or *optimal*, way for each player to play and of the expectation each player will have, using his *optimal strategy*.

EXAMPLE 2.1.1 **Matching pennies (simple form)** Each player is to place one penny on a table.   He has the option of putting his penny down with the head side showing or with the tail side showing.   Neither player is to know what the other player has done until after his own penny is down. If both pennies show heads, or if both pennies show tails, player I takes both pennies from the table, i.e., player I wins one penny. On the other hand, if one penny shows heads and the other shows tails, player II wins one penny.

In this game, each player makes one move, and no chance moves occur. Each player must make a decision between two alternatives, namely, (1) play heads, H, and (2) play tails, T. We can describe this game in terms of player I's prospects by means of the array

II	**H**	**T**
H	$+1$	-1
T	-1	$+1$

There are two ways of arriving at the outlook toward the game that we assume each player adopts. We might call them the "susceptibility to espionage" method and the "need for a guarantee" method. It is clear that should either player be able, in some way, to determine the decision his opponent has made before he himself has played, then he would have no difficulty in winning. Thus if a player is fearful that his espionage system is not as good as his opponent's (or that his opponent is a mind reader), he must look for a way of playing which will minimize the effect of his opponent's advantage. In such a case (or for any other reason) each player is anxious to determine how to play in such a way that his expectation is maximized no matter what his opponent does. Clearly a player with no espionage system can do nothing to guarantee that he will fare any better than losing every time the game is played, i.e., every time the pennies are matched. However, we will show that there is a means whereby he can guarantee that his *expectation* is to win half of the games played, no matter how much better his opponent's espionage system is. The method is very simple. He simply throws the coin into the air and permits it to come to rest with either the head or the tail showing, purely by chance (of course, taking care that his opponent does not see which it is). Even with an opponent who is a mind reader, he can expect to win half of the games.[1] Thus, after a series of games, he expects neither to win nor to lose. We say that his expectation is 0. Another way of expressing this is to say that, from player I's point of view, the value of the game is 0. Because of the symmetry of this game, all the above remarks apply equally well to player II.   ////

---

[1] Because we have not properly developed the basic notions of probability theory, we must here tacitly assume that the two statements "the probability that player I will win any one game is $\frac{1}{2}$" and "player I expects to win half of the games played" mean the same thing or at least imply each other.

EXAMPLE 2.1.2   **Matching pennies (complex form)**   The rules of play for
the complex form of matching pennies are the same as the rules for the simple
form.   The difference lies in the payoff arrangements.   Let us suppose that the
game is described (again in terms of player I's prospects) by means of the array

I \ II	H	T
H	+1	−2
T	−3	+6

Most of the comments made concerning the simple form of matching
pennies apply equally well to the complex form.   However, the solution is not
the same.   Suppose player I adopts the following procedure to decide which
alternative (H or T) to choose.   He puts three white balls and one black ball in
a bag.   He then pulls out a ball at random.   If it is white, he plays heads.   If
it is black, he plays tails.

We say that player I has combined the *pure strategies*, H and T, to form
the (*mixed*) *strategy*

Play H $\frac{3}{4}$ of the time   Play T $\frac{1}{4}$ of the time

Suppose that player II learns of this.   How can he use this information?   If
he constantly plays heads, he can expect to win one game out of every four.
However, when he wins, he wins 3, and when he loses, he loses 1.   Thus, constantly
playing heads, he expects to neither win nor lose but to come out even.   On
the other hand, if he constantly plays tails, he can expect to win three games out
of every four.   However, when he wins, he wins only 2, and when he loses, he
loses 6.   As before, by constantly playing tails, his expectation is 0.   Thus, as
long as player I continues this procedure for making his move, as far as player II
is concerned, it is immaterial whether he plays heads or tails—or, for that matter,
any arrangement of heads or tails.   Player II's espionage gives him no advantage
whatsoever.

We have found a method whereby player I can guarantee that his expectation is 0, irrespective of player II's machinations.   What is the best that player
II can guarantee for himself, irrespective of player I?   Certainly he cannot
guarantee that he will expect to win.   But does he have a strategy whereby he can
guarantee that he will not lose?   That is, does there exist a strategy for player
II such that his expectation is 0 no matter what player I does?   The answer is
yes.   If he plays heads two-thirds of the time and tails one-third of the time
(but in a random fashion, using colored balls), his expectation is 0.

There is still more to be said about this game.   Suppose player I plays
heads with probability p (for some p, $0 \leqq p \leqq 1$) and tails with probability

$1 - p$ and player II plays heads with probability q (for some q, $0 \leq q \leq 1$) and tails with probability $1 - q$. Then the probability that both players will show heads is pq, and when this happens, player I wins 1. The probability that player I will play heads and that player II will play tails is $p(1 - q)$, and when this happens, player I loses 2. Similarly, the probability that player I will play tails and player II will play heads is $(1 - p)q$, and then player I loses 3. Finally, the probability that both players will play tails is $(1 - p)(1 - q)$, and then player I wins 6. Thus player I can expect to win

$$(1)pq + (-2)p(1 - q) + (-3)(1 - p)q + (6)(1 - p)(1 - q) =$$
$$12pq - 8p - 9q + 6 = 12(p - \tfrac{3}{4})(q - \tfrac{2}{3}) \qquad (2.1.3)$$

We express this by saying that player I's expectation when he uses the strategy

$$P = [p \quad 1 - p]$$

and player II uses the strategy

$$Q = \begin{bmatrix} q \\ 1 - q \end{bmatrix}$$

is given by (2.1.3). It is clear that (2.1.3) can be expressed as the matrix product

$$P \begin{bmatrix} 1 & -2 \\ -3 & 6 \end{bmatrix} Q$$

More generally, if G is a game with matrix A, and players I and II use strategies P and Q, respectively, then *player I's expectation with respect to P and Q is* denoted by $E(P, Q)$ and is given by

$$E(P, Q) = PAQ$$

In the game we have been discussing, suppose player II decides to use the strategy

$$Q = \begin{bmatrix} q \\ 1 - q \end{bmatrix}$$

for some q, $0 \leq q \leq 1$. It is not difficult to see [from the last expression of (2.1.3)] that unless $q = \tfrac{2}{3}$, player I can improve his expectations. In fact, if $q > \tfrac{2}{3}$, player I would do well to always play heads, i.e., set $p = 1$; for then he would expect to win

$$q(+1) + (1 - q)(-2) = -2 + 3q > 0$$

each game. If $q < \tfrac{2}{3}$, player I would play tails each game, i.e., set $p = 0$; for then his expectation is

$$q(-3) + (1 - q)(6) = 6 - 9q > 0$$

On the other hand, if player II sets $q = \frac{2}{3}$, then player I's expectation is always 0 (no matter what strategy he chooses).   Thus the mixed strategy for player II,

$$Q = \begin{bmatrix} \frac{2}{3} \\ \frac{1}{3} \end{bmatrix}$$

(obtained by setting $q = \frac{2}{3}$) is his unique optimal strategy.   Similarly we can show that

$$P = [\tfrac{3}{4} \quad \tfrac{1}{4}]$$

is the unique optimal strategy for player I.   ////

Let us discuss another, still more complex variant of the game of matching pennies, in which player II has three possible pure strategies.   In the discussion of this game, we will present another very simple method (the *graphical method*) for solving the games thus far introduced.   Because for each player we must be able to draw several accurate graphs in a space of as many dimensions as he has pure strategies, this method has extreme limitations.   Later we will show that it is often possible to eliminate some of the obviously unsound strategies each player might (but will not) make and thereby arrive at a "new" game which can be solved by graphical means.

EXAMPLE 2.1.4   **Matching pennies (more complex form)**   In this game player I will have the same two alternatives to decide between as in the first two games, namely, to play heads or to play tails.   However, player II will have three alternatives: to play heads, to play tails, or, to show no coin at all (which we denote by N).   As before, let us suppose that the game is described (in terms of the outcome for player I) using the array

I \ II	H	T	N
H	1	2	4
T	3	1	0

Player I must now decide how often he should play heads and how often he should play tails.   From our discussion of the previous games it is clear that once each player has decided how often each of the alternatives open to him should be played, it is always desirable for him to make his moves in a random way (as, for example, by using a bag containing various colored balls).   Let us suppose that player I plays heads with probability p $(0 \leq p \leq 1)$.   Then he must play tails with probability $1 - p$. If player II plays heads, then player I's expectation for each game is $(p)(1) + (1 - p)(3) = 3 - 2p$. If player II plays tails, then player I's expectation for each game is $(p)(2) + (1 - p)(1) = 1 + p$. Finally, if player II shows no coin, then player I's expectation is $(p)(4) + (1 - p)(0) = 4p$. For each of these three situations, let us plot the expectation of player I, using the strategy $P = [p \quad 1 - p]$.

Player I is now able to determine for each value of p the best expectations that he is able to guarantee. This is found by taking, for each value of p, the minimum of the three numbers, $3 - 2p$, $1 + p$, and $4p$. The function

$$m(p) = \min\{3 - 2p, 1 + p, 4p\}$$

is defined for each value of p, $0 \leq p \leq 1$, and is described by the line segments OB, BC, CD in Figs. 2.1.5 and 2.1.6. The best guaranteed expectation for player I is given by

$$\max_{0 \leq p \leq 1} m(p)$$

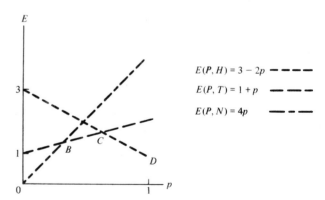

$E(P, H) = 3 - 2p$ — — — —

$E(P, T) = 1 + p$ — — — —

$E(P, N) = 4p$ — — —

FIGURE 2.1.5

From Figs. 2.1.5 and 2.1.6 it is easily seen that this best guaranteed expectation occurs at the point C, which is the intersection of the lines

$$E = 3 - 2p, \qquad E = 1 + p$$

Solving these two equations simultaneously yields $p = \frac{2}{3}$, $E = \frac{5}{3}$. If player I uses the mixed strategy $P = [\frac{2}{3} \quad \frac{1}{3}]$, his expectation is at least $\frac{5}{3}$, irrespective of players II's strategy. Unlike the previous examples, however, it is clear from

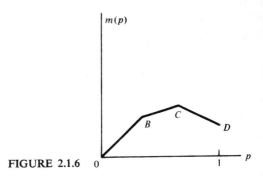

FIGURE 2.1.6

Fig. 2.1.5 that if player II uses the alternative N at all, then player I's expectation becomes greater than $\frac{5}{3}$.

Player II may go through the same process.   He must determine with what probability he should play heads, tails, or no coin.   Let us denote these probabilities by q_1, q_2, q_3, respectively.   Then

$$0 \leq q_i \leq 1 \qquad i = 1, 2, 3 \qquad q_1 + q_2 + q_3 = 1 \qquad (2.1.7)$$

If player I plays heads, then his expectation is

$$q_1 + 2q_2 + 4q_3 = 4 - 3q_1 - 2q_2$$

while if player I plays tails, then his expectation is

$$3q_1 + q_2$$

Let us plot player I's expectation, in each case, assuming that player II uses the strategy

$$Q = \begin{bmatrix} q_1 \\ q_2 \\ q_3 \end{bmatrix}$$

For any choice of q_1, q_2, the least expectation for player I that player II is able to guarantee is

$$n(q_1, q_2) = \max\{4 - 3q_1 - 2q_2, 3q_1 + q_2\}$$

Clearly $n(q_1, q_2)$ is defined for all q_1, q_2 satisfying equations (2.1.7) and is described by the planar segments $RSUT$ and TUV of Figs. 2.1.8 and 2.1.9.   Player II wishes to keep player I's expectation as small as possible, and the smallest he can make player I's expectation, solely by means of his own strategy, is

$$\min_{\substack{0 \leq q_1, q_2 \leq 1 \\ q_1 + q_2 \leq 1}} n(q_1, q_2)$$

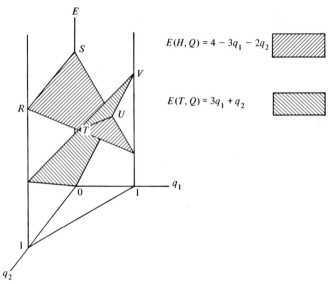

$$E(H, Q) = 4 - 3q_1 - 2q_2$$

$$E(T, Q) = 3q_1 + q_2$$

FIGURE 2.1.8

According to Figs. 2.1.8 and 2.1.9, the least guaranteed expectation for player I occurs at T, which is the intersection of the three planes

$$E = 4 - 3q_1 - 2q_2 \qquad E = 3q_1 + q_2 \qquad q_1 + q_2 = 1$$

Solving these equations simultaneously yields

$$q_1 = \tfrac{1}{3} \qquad q_2 = \tfrac{2}{3} \qquad q_3 = 0 \qquad E = \tfrac{5}{3}$$

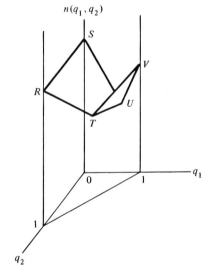

FIGURE 2.1.9

If player II uses the strategy

$$Q = \begin{bmatrix} \frac{1}{3} \\ \frac{2}{3} \\ 0 \end{bmatrix}$$

then player II's expectation is $\frac{5}{3}$, no matter how player I plays.   ////

Again we have a situation where this is a number v (in this case, $\frac{5}{3}$) such that:

1   Player I has an optimal strategy (in this case $P = \begin{bmatrix} \frac{2}{3} & \frac{1}{3} \end{bmatrix}$) whereby he is able to guarantee an expectation of winning at least v, irrespective of the play of player II.

2   Player II has an optimal strategy $\left(\text{in this case } Q = \begin{bmatrix} \frac{1}{3} \\ \frac{2}{3} \\ 0 \end{bmatrix} \right)$ whereby he is able to guarantee for himself an expectation of losing no more than v, irrespective of the play of player I.

In such a case, we call v the *value* of the game.   The main theorem of this chapter (the von Neumann minimax theorem of Sec. 2.5) states that every game has a value and that both players have optimal strategies.

Note that in Example 2.1.4 player II's (unique) optimal strategy calls upon him never to play N, even though his best possible expectation (namely, 0) occurs when he plays N and player I plays T.   Since player II is never going to make use of the pure strategy N, the game would not be changed if this strategy were removed.   Thus the game we have just finished discussing is essentially the same game as the one with the array

I \ II	H	T
H	1	2
T	3	1

Unfortunately, it does not become clear that player II never plays N until after the game is solved.   In the next game, there will be some alternatives which are so obviously bad that they can be eliminated before we attempt to solve the game.

EXAMPLE 2.1.10 Two-fingered morra   There are two players.   At some
signal, each player puts out either one finger or two fingers and at the same time
calls out one of the numbers 2, 3, 4.   If the number which one player calls
out is equal to the total number of fingers put out and the other player calls out
some other number, the first player wins.   If both call out incorrect numbers, or
if both players call out the correct number, the game is a tie.

Each player has a choice of six alternatives (or pure strategies):

Strategy	Fingers put out	Numbers called
S_1	1	2
S_2	1	3
S_3	1	4
S_4	2	2
S_5	2	3
S_6	2	4

Two of these alternatives have absolutely no chance of winning, namely,
S_3 (for player II cannot put out three fingers) and S_4 (since player II must put
out some fingers).   As far as we are concerned, the game would be unchanged
if these strategies were removed.   In the same way, if each player were permitted
to call any number he wishes (still being restricted to showing one or two fingers),
it is clear that no (sensible) player would use any combination of fingers and
numbers other than S_1, S_2, S_5, and S_6.

We can represent this game in the matrix form (from player I's point of
view) by

I \ II	S_1	S_2	S_5	S_6
S_1	0	1	-1	0
S_2	-1	0	0	1
S_5	1	0	0	-1
S_6	0	-1	1	0

Because of the size of the matrix involved, it would be, at best, extremely cumbersome to try to find a solution to this game by the methods we have previously
introduced.   However, solutions will be discussed in the exercises.   ////

Each of the games we have examined thus far has consisted solely of one
move made by each player.   At the time of each move, the players had no

information available to them (other than, of course, the rules of the game, including the payoff arrangements). Let us now examine a simple game in which (1) the first move is made by chance, (2) the players move in sequence (rather than simultaneously), and (3) one of the players has available some usable information at the time his move is to be made.

EXAMPLE 2.1.11 **Bluffing** Bluffing is a two-person card game, modeled vaguely after poker, played with a deck of two cards, called high (H) and low (L). Both cards have identical backs and are placed face down on a table. As the first move in the game, player I takes one of the cards. He inspects it (to determine which card he has drawn), making certain that player II does not have an opportunity to see which card has been chosen.

At this point, player I has two alternatives: (1) he may *fold*, in which case he pays one penny to player II and the game is over, or (2) he may *bet* two pennies. Now player II has two alternatives: (1) he may fold, in which case he pays one penny to player I and the game is over, or (2) he may *call*. In the event that player II calls, player I exposes the card he has chosen. If the card is H, player II pays player I two pennies. If the card is L, player I pays player II two pennies. In either case the game is over.

The first move in bluffing is made by chance. Player I receives either H or L. We must assume that these two possibilities are equally likely. We can change this assumption by altering the deck of cards. If the deck consists of two cards marked H and one card marked L, we would assume that the probability that player I's card would be H would be $\frac{2}{3}$ and the probability that it would be L would be $\frac{1}{3}$.

The second move of the game is made by player I. If player I folds, player II has no moves to make. On the other hand, if player I bets, player II has a move to make, either to fold or to bet. In addition, at the time that he is called upon to make his move, player I has some information at his disposal. He knows which card he has drawn.

Suppose player I bets every time he has H and folds every time he has L. He can expect to bet half of the time and fold half of the time. Should player II learn of this, he will never call, since he can never win by calling. In this event, in half of the games, player I will win one penny, and, in the other half of the games, he will lose one penny. Thus his expectation is zero. On the other hand, suppose player I bets no matter what he has drawn. If player II learns of this, he will always call, for then he will lose two pennies half of the time (when player I has H) and he will win two pennies half of the time (when player I has L). Again his expectation (and also player I's expectation) is zero.

Once we have shown that bluffing is a matrix game, it will be immediately clear that there are mixed strategies player I can use which will make his expectation positive, no matter what player II does. We have said that player I has two sensible pure strategies. These are

$$S_1 = \begin{cases} \text{if the card drawn is H, then bet} \\ \text{if the card drawn is L, then bet} \end{cases}$$

$$S_2 = \begin{cases} \text{if the card drawn is H, then bet} \\ \text{if the card drawn is L, then fold} \end{cases}$$

Player I has two other pure strategies. Neither of them is sensible since each would require him to fold when he has drawn H, and so we will ignore them. If we did include them in the array we are going to construct, it would be easy to show that player I would never use them.

We can describe player II's pure strategies more simply. They are

$$T_1 = \text{call (when the option arises)}$$
$$T_2 = \text{fold (when the option arises)}$$

Summarizing the discussion above, player I's expectation is zero either (1) when player I plays S_1 and player II plays T_1 or (2) when player I plays S_2 and player II plays T_2. There are two other possibilities, namely, (3) player I plays S_1, and player II plays T_2, in which case it is clear that player I wins 1, and (4) player I plays S_2, and player II plays T_1. In this last situation, player I will win 2 half of the time (when he draws H), and he will lose 1 half of the time (when he draws L). Thus his expectation is $\frac{1}{2}(2) + \frac{1}{2}(-1) = \frac{1}{2}$. With all of this information, we see that player I's expectation is given by the array

I \ II	T_1	T_2
S_1	0	1
S_2	$\frac{1}{2}$	0

Once we have put this game in matrix form, it is clear that the solution is easily obtained by using the graphical methods introduced in Example 2.1.4.   ////

A strategy for a player is a set of rules which tells him which of several alternatives to choose whenever he has such a decision to make. In the games we have studied thus far each player is called upon to make a decision, i.e., a move, exactly once. In the game of Example 2.1.12, player I may have to make two separate decisions, and his strategies must tell him how to make each one.

EXAMPLE 2.1.12   Three-card Poker   Each player is dealt one card from a deck consisting of the three cards, high (H), medium (M), and low (L). The cards are *ordered* from *strongest* to *weakest* by the ordering $H > M > L$. Player I has two choices:

1   He may *fold*, in which case he pays one penny to player II and the game is over.
2   He may *bet.*

If player I bets, player II has three choices:

1   He may fold, in which case player II pays one penny to player I and the game is over.
2   He may *call.*
3   He may *raise.*

If player II calls, both players expose the card they have been dealt and the player with the stronger card wins two pennies from the other player. The game is then over. On the other hand, if player II raises, player I has two choices:

1   He may fold, in which case he pays player II two pennies and the game is over.
2   He may call.

Should player I now call, the cards are exposed and the player with the stronger card wins three pennies from the other player.

If player I folds initially, player II needs no strategy since he has no decisions to make. On the other hand, if player I bets, then player II must decide between the three alternatives: (1) fold (F), (2) call (C), and (3) raise (R). The information that player II has available to help him make this decision is the identity of the card he has been dealt. A pure strategy for player II consists of a plan for all the eventualities he might be called upon to face. In this case there are three such eventualities:

1 Player I has bet, and he (player II) holds H
2 Player I has bet, and he holds M
3 Player I has bet, and he holds L

and for each of these eventualities, player II must assign either F, C, or R. Thus player II has a total of 27 pure strategies. However, some of these strategies are not sensible, and so we will ignore them. If player II has been dealt H, he can gain nothing by folding or by calling. In the same way, if he has been dealt L, he can gain nothing by calling. By folding, he keeps his losings down and by raising, i.e., bluffing, he retains the possibility of winning. Of the 27 pure strategies available to player II, only 6 are sensible. These are tabulated.

PLAYER II

Strategy	Card drawn	Move (if needed)
T_1	H	R
	M	R
	L	R
T_2	H	R
	M	R
	L	F
T_3	H	R
	M	C
	L	R
T_4	H	R
	M	C
	L	F
T_5	H	R
	M	F
	L	R
T_6	H	R
	M	F
	L	F

Player I may have two decisions to make. At the time of these decisions he is aware of which card he has been dealt, and he can make use of this information. When player I bets and player II raises, player I is called upon to make a second decision, namely, whether to call or to fold. Thus player I also has 27 pure strategies available. Clearly, if he has been dealt H, he should never fold (either on the first or on the second move). Also, if he has been dealt L, he should never call. Thus, of the 27 pure strategies available, only 6 are sensible. These are tabulated.

PLAYER I

Strategy	Card drawn	First move	Second move (if needed)
S_1	H	B	C
	M	B	C
	L	B	F
S_2	H	B	C
	M	B	F
	L	B	F
S_3	H	B	C
	M	B	C
	L	F	
S_4	H	B	C
	M	F	
	L	F	
S_5	H	B	C
	M	B	F
	L	F	
S_6	H	B	C
	M	F	
	L	B	F

We are now faced with the prospect of filling in the entries of the 6×6 matrix $A = [a_{ij}]$, where a_{ij} is player I's expectation when he plays S_i and player II plays T_j. There are six possible ways in which the cards can be dealt, namely

	Player I	Player II
1	H	L
2	L	H
3	H	M
4	M	H
5	M	L
6	L	M

We assume that all these possibilities are equally likely. Then

$$a_{11} = \tfrac{1}{6}(3 - 2 + 3 - 3 + 3 - 2) = \tfrac{1}{3},$$
$$a_{21} = \tfrac{1}{6}(1 - 2 + 3 - 3 + 1 - 2) = -\tfrac{1}{3},$$
$$a_{31} = \tfrac{1}{6}(3 - 2 + 2 - 3 + 3 - 2) = \tfrac{1}{6},$$

. . . .

$$a_{12} = \tfrac{1}{6}(3 - 2 + 3 - 2 - 2 - 2) = -\tfrac{1}{3},$$
$$a_{22} = \tfrac{1}{6}(1 - 2 + 3 - 2 + 1 - 2) = -\tfrac{1}{6},$$
$$a_{32} = \tfrac{1}{6}(3 - 2 + 2 - 2 - 2 - 2) = -\tfrac{1}{2},$$

. . . .   ////

There are several examples of competitive situations which are not games but which, with the assignment of reasonable numerical values to the possible outcomes, can be analyzed by the methods of this chapter. A simple example of such a nongame competition is:

EXAMPLE 2.1.13 **Duelling** Two players stand facing each other, $2n$ paces apart, where n is a positive integer. Each has a gun containing a single bullet. At a signal from the referee each player may fire at the other if he so wishes. If either player is hit or if both players fire and miss, the "game" is over. Otherwise both players advance one pace, and the procedure is repeated. After taking n paces forward, the players have reached each other and the game is over. It will be convenient to say that a player has taken zero paces forward when he is at his initial position. Also, we will assume that the probability that either player will be able to hit the other increases as the two players approach each other.

At this point we must decide how much information should be made available to each player. A duel in which each player knows when his opponent has shot at him and missed is called a *noisy duel*. A duel in which neither player knows he has been shot at unless he has been hit (as if both guns were equipped with silencers) is called a *silent duel*.

In a noisy duel each player has $n + 1$ sensible pure strategies. They are

$$S_i = \begin{cases} \text{shoot after the } i\text{th pace forward if the other player} \\ \text{has not previously shot and missed} \\ \text{shoot after the } n\text{th pace forward if the other player} \\ \text{has previously shot and missed} \end{cases} \quad i = 0, 1, 2, \dots, n$$

If the other player has shot and not missed, then no decision need be made.

In a silent duel each player has $n + 1$ pure strategies, all of which appear to be sensible. They are

$$S_i = \text{shoot after the } i\text{th pace forward} \quad i = 0, 1, 2, \dots, n$$

In order to apply the methods of game theory to duelling it is only necessary to assign values to the possible outcomes. It is reasonable to assign the value of $+1$ when player I survives and player II does not and the value -1 when player II survives and player I does not. It is also reasonable to assign the value 0 when both players survive, especially if the duel must then be repeated. Finally, if neither player survives, we must assign a value of 0. If each player feels that his death should be valued at -1 (in terms of his expectation) irrespective of the survival of his opponent, we no longer have a zero-sum game, for if both players are killed, both lose and no one wins. ////

A game in which the rules of play and the payoff arrangements are the same for both players, e.g., two-fingered morra, is called a *symmetric game*.[1] In a symmetric game the set of pure strategies and the set of all (mixed) strategies are the same for both players. An optimal strategy for one player is an optimal strategy for the other player. Moreover, if a symmetric game has a value, it must be zero since it is clear that neither player can find a strategy which will give him a positive expectation against every one of his opponent's strategies.

If, in a symmetric game, S_1 and S_2 are strategies for player I (and hence also for player II), then $E(S_1, S_2) = -E(S_2, S_1)$; for, clearly, the first player's expectation when he plays S_1 and player II plays S_2 is the same as player II's expectation when he plays S_1 and player I plays S_2. But, in any instance, player I's expectation is the negative of player II's expectation. Thus, the matrix of a symmetric game is skew-symmetric, that is, $A^T = -A$.

Any game can be changed into a symmetric game in a number of different ways (see Exercise 16). Let us examine here a particular method of *symmetrization* which requires the addition of a single chance move. In general, symmetrization is itself not a game but a method for transforming any game into a symmetric game.

The symmetrization G' of the game G is a game between the two players, player I' and player II', and is played as follows. A coin is tossed. If it shows heads, then player I' in G' becomes player I in G and player II' becomes player II in G; if the coin shows tails, then player I' becomes player II in G and player II' becomes player I in G. If the pure strategies in G are $S_1, S_2, \ldots, S_m$ for player I and $T_1, T_2, \ldots, T_n$ for player II, then the pure strategies for both of the players in G' are

$$S'_{ij} = \begin{cases} \text{when playing as player I, play } S_i & i = 1, 2, \ldots, m \\ \text{when playing as player II, play } T_j & j = 1, 2, \ldots, n. \end{cases}$$

Thus, in G', each player has mn pure strategies, and the matrix A' of G' is a skew-symmetric $mn \times mn$ matrix.

EXAMPLE 2.1.14   The symmetrization of bluffing

In the symmetrization of bluffing, players I' and II' each have four pure strategies:

$$S'_{11} = \begin{cases} \text{when playing as player I, play } S_1 \ (=\text{always bet}) \\ \text{when playing as player II, play } T_1 \ (=\text{call}) \end{cases}$$

$$S'_{12} = \begin{cases} \text{when playing as player I, play } S_1 \ (=\text{always bet}) \\ \text{when playing as player II, play } T_2 \ (=\text{fold}) \end{cases}$$

---

[1] Later we shall define a symmetric game in a somewhat different manner.

$$S'_{21} = \begin{cases} \text{when playing as player I, play } S_2 \text{ (= bet on H only)} \\ \text{when playing as player II, play } T_1 \text{ (= call)} \end{cases}$$

$$S'_{22} = \begin{cases} \text{when playing as player I, play } S_2 \text{ (= bet on H only)} \\ \text{when playing as player II, play } T_2 \text{ (= fold)} \end{cases}$$

Let us fill in some of the entries of the matrix A'. We can find the entry in the $(1, 4)$ position of A' by supposing that player I$'$ selects the strategy S'_{11} and player II$'$ selects the strategy S'_{22}. There are now two possibilities, and the probability of each is $\frac{1}{2}$. When the tossed coin shows heads, player I$'$ (who has become player I and has played S_1, that is, always bet) can expect to win 1 from player II$'$ (who has become player II and has played T_2, that is, fold). When the coin that is tossed shows tails, player I$'$ (who has become player II and has played T_1) will lose $\frac{1}{2}$ to player II$'$ (who has become player I and has played S_2). Thus

$$a'_{14} = E(S'_{11}, S'_{22}) = \tfrac{1}{2}(1 - \tfrac{1}{2}) = \tfrac{1}{4}$$

We can show very quickly that the entry in the $(4, 1)$ position of A' is $-\frac{1}{4}$. For if player I$'$ selects the strategy S'_{22} and player II$'$ selects the strategy S'_{11}, the expectations are the same as in the discussion above except that the roles of the players have been reversed. Thus

$$a'_{41} = E(S'_{22}, S'_{11}) = \tfrac{1}{2}(\tfrac{1}{2} - 1) = -\tfrac{1}{4} = -E(S'_{11}, S'_{22}) = -a'_{14}$$

The entries along the diagonal of A' are all zero; for if players I$'$ and II$'$ both decide to use the strategy $S'_{ij} (1 \le i, j \le 2)$, then half of the time (when the coin shows heads) player I$'$ can expect to win a_{ij} and half the time (when it shows tails) player II$'$ can expect to win a_{ij}. Thus

$$E(S'_{ij}, S'_{ij}) = \tfrac{1}{2}(a_{ij} - a_{ij}) = 0$$

We leave the task of finding the remaining entries of A' as an exercise. ////

In Sec. 2.8 we shall discuss linear-programming problems and show that they can be transformed into games.

EXERCISES

1 Solve the following games:

(a) $A = \begin{bmatrix} 1 & -1 \\ -1 & 3 \end{bmatrix}$ (b) $A = \begin{bmatrix} 1 & 1 \\ 3 & -2 \end{bmatrix}$

(c) $A = \begin{bmatrix} 0 & 3 \\ -2 & 2 \end{bmatrix}$ (d) $A = \begin{bmatrix} 2 & -2 \\ -1 & 3 \end{bmatrix}$

(1) by the method of Example 2.1.2 and (2) by the graphical method.

2 Solve the following games by the methods of this section

(a)

II I	T_1	T_2	T_3
S_1	1	-3	4
S_2	4	10	-6

(b)

II I	T_1	T_2	T_3
S_1	5	-1	-3
S_2	-3	-1	5

(c)

II I	T_1	T_2	T_3
S_1	6	0	-1
S_2	-4	2	4
S_3	-9	-5	-6

3 Show that in Example 2.1.10 playing S_1, S_2, S_5, S_6 with equal probability is an optimal strategy for each player. Are there any other optimal strategies?

4 Change the rules in Example 2.1.10 so that the winner wins an amount equal to the total number of fingers put out. Find the matrix for this game. Show that the solution given in Exercise 3 is no longer a solution. Show that $\bar{P} = [0 \quad 3/5 \quad 2/5 \quad 0]$ is now an optimal strategy for both players. Are there any other optimal strategies?

5 Solve the game of Example 2.1.11.

6 Write out the matrix in Example 2.1.11 showing all of player I's pure strategies (the nonsensible ones as well as the sensible ones). Show that the solutions are the same, with and without taking the nonsensible pure strategies into account.

7 Set up and solve the game of bluffing when the deck used consists of two H's and three L's.

8 Show that, in Example 2.1.12, S_3 is at least as good a strategy as S_4 for player I under all circumstances. (It follows then that player I need never use S_4.)

9 Let G be a noisy duel with the players beginning $2n$ paces apart in which the probability that a player will hit his opponent when firing after his ith pace forward is i/n, $i = 0, 1, 2, \ldots, n$. Show that:
(a) For n odd, $n = 2k - 1$, S_k is an optimal strategy for each player.
(b) For n even, $n = 2k$, S_k and S_{k+1} are both optimal strategies for each player.

10 Let G be a silent duel with the players beginning $2n$ paces apart in which the probability that a player will hit his opponent when firing after taking his ith pace forward is i/n, $i = 0, 1, 2, \ldots, n$. Show that
(a) For $n = 4$, S_2 is an optimal strategy.

(b) For $n = 5$, no pure strategy is optimal, but $\bar{P} = [0 \quad 0 \quad \frac{5}{11} \quad \frac{5}{11} \quad 0 \quad \frac{1}{11}]$ is an optimal strategy for both players.

11 Solve the silent duel in which the players begin six paces apart and the probability of each player's hitting the other is given in the table.

Pace	Player I	Player II
0	1/4	1/5
1	2/4	2/5
2	3/4	4/5
3	4/4	5/5

12 Find the pure strategies and the matrix for the symmetrization of matching pennies (Example 2.1.1). Show that playing each pure strategy with equal probability is an optimal strategy for both players.

13 Find the pure strategies and the matrix for the symmetrization of bluffing. Ignore the nonsensible strategies.

14 Show what happens to the matrix of a game when player I becomes player II and player II becomes player I.

15 *Scissors, paper, and stone* This is an ancient Italian game. At a signal each player puts out either (1) two fingers (scissors), (2) five fingers (paper), or (3) fist (stone). If both players put out the same thing, the game is a draw. Otherwise:

(a) Two fingers beats five fingers (scissors cut paper).

(b) Five fingers beats fist (paper covers stone).

(c) Fist beats two fingers (stone breaks scissors).

If the winner always wins $+1$, set up the matrix of the game and show that playing each strategy with equal probability is an optimal strategy for both players.

16 Let G be a game and form the game G' between the players I$'$ and II$'$ as follows. Players I$'$ and II$'$ can:

(a) Choose a strategy of player I in G.

(b) Choose a strategy of player II in G.

(c) Choose neither.

If players I$'$ and II$'$ both choose a player I strategy or both choose a player II strategy or both choose neither, the game is a tie. If one player chooses a player I strategy and the other chooses a player II strategy, they play that way and the payoff is determined by the payoff in G. If either player selects a player I strategy and the other player chooses no strategy, the player who chose the player I strategy loses 1. If either player selects a player II strategy and the other player chooses no strategy, the player who chose the player II strategy wins 1. Set up the matrix for this game.

2.2 MAXMIN AND MINMAX

It was pointed out at the beginning of Sec. 2.1 that the discussion in that section
is very informal and that no attempt would be made there to justify any of the
methods introduced.   Since we will begin a more formal development of game
theory from the very beginning in the next section, where many of the same
problems will arise, we devote this section to justifying some of the practices of
Sec. 2.1.   We will consider here some nonalgebraic fundamentals of game
theory. These come primarily from topology and analysis. A thorough
discussion of this material, including definitions and proofs for all the theorems
stated, would lead us too far astray.   In addition, for many readers such a
discussion would present little that is new.   For these reasons, only those facts
which are needed later are presented here, and some of these are given without
proof.

Definition 2.2.1   Let $X, X_1, X_2, \ldots, X_k$ be elements of the vector space $\mathbb{V}$.[1]
Then X is a *convex combination* of $X_1, X_2, \ldots, X_k$ if there exist real numbers
$\alpha_1, \alpha_2, \ldots, \alpha_k$ such that

$$\alpha_1 + \alpha_2 + \cdots + \alpha_k = 1 \qquad \begin{array}{c} 0 \leq \alpha_j \leq 1 \\ j = 1, 2, \ldots, k \end{array} \qquad (2.2.2)$$

for which

$$X = \alpha_1 X_1 + \alpha_2 X_2 + \cdots + \alpha_k X_k$$

A subset $\mathbb{S}$ of the vector space $\mathbb{V}$ is *convex* if for any k vectors $X_1, X_2, \ldots,$
X_k belonging to $\mathbb{S}$ and any k scalars $\alpha_1, \alpha_2, \ldots, \alpha_k$ satisfying (2.2.2) the vector

$$\alpha_1 X_1 + \alpha_2 X_2 + \cdots + \alpha_k X_k$$

also belongs to $\mathbb{S}$; that is, $\mathbb{S}$ is convex if it contains all the convex combinations
of its elements.

The set of all vectors P,

$$P = [p_1, p_2, \ldots, p_m]$$

of $\mathbb{R}_m$ for which

$$p_1 + p_2 + \cdots + p_m = 1 \qquad \begin{array}{c} 0 \leq p_i \leq 1 \\ i = 1, 2, \ldots, m \end{array} \qquad (2.2.3)$$

---

[1] Throughout this chapter we will consider only vector spaces and matrices whose
entries are from the field of real numbers $\mathbb{R}$.

is called the *strategy space for player* I (in the context of this chapter[1]) and is denoted by $\mathbb{P}_m$. Clearly $\mathbb{P}_m$ consists of all convex combinations of the n unit row vectors and hence is convex. In the same way the set of all vectors Q

$$Q = \begin{bmatrix} q_1 \\ q_2 \\ \vdots \\ q_n \end{bmatrix}$$

of $\mathbb{R}^n$ for which

$$q_1 + q_2 + \cdots + q_n = 1 \qquad \begin{matrix} 0 \leq q_j \leq 1 \\ j = 1, 2, \ldots, n \end{matrix}$$

is a convex set called the *strategy space for player* II and is denoted by $\mathbb{Q}^n$.

Note that neither $\mathbb{P}_m$ nor $\mathbb{Q}^n$ is a vector space. However, since all the unit vectors of $\mathbb{R}_m$ belong to $\mathbb{P}_m$ and all of the unit vectors of $\mathbb{R}^n$ belong to $\mathbb{Q}^n$, the only vector space which contains $\mathbb{P}_m$ is $\mathbb{R}_m$ and the only vector space which contains $\mathbb{Q}^n$ is $\mathbb{R}^n$. Equations (2.2.2) describe the entries of an element of $\mathbb{P}_m$. Thus, for $A \in \mathbb{R}_n{}^m$, $Z \in \mathbb{R}_n$, Z is a convex combination of the rows of A if and only if there exists a vector P belonging to $\mathbb{P}_m$ such that $Z = PA$. (Compare with Theorem 1.4.4.)

The vector spaces $\mathbb{R}_n$ and $\mathbb{R}^n$ have an important geometric interpretation, as n-dimensional euclidean space.[2] The vector $X = [x_1, x_2, \ldots, x_n]$ is interpreted as the point whose coordinates are $x_1, x_2, \ldots, x_n$. For $X, Y \in \mathbb{R}_n$,

$$X = [x_1, x_2, \ldots, x_n] \qquad Y = [y_1, y_2, \ldots, y_n]$$

we define the *distance* between X and Y by

$$d(X, Y) = [(x_1 - y_1)^2 + (x_2 - y_2)^2 + \cdots + (x_n - y_n)^2]^{1/2} \qquad (2.2.4)$$

Then $d(X, Y)$ defines a metric (or metric topology) on $\mathbb{R}_n$. Under the topology on $\mathbb{R}_n$ induced by the metric of (2.2.4), $\mathbb{P}_n$ and $\mathbb{Q}^n$ are compact. Thus, *every continuous real-valued function on $\mathbb{P}_n$ is bounded and assumes both a maximum and a minimum value.* Similarly, *every continuous real-valued function on $\mathbb{Q}^n$ is bounded and assumes both a maximum and minimum value.*

---

[1] This set of vectors will appear again in Chap. 3 but with a different name.

[2] When studying the geometry of these spaces no distinction need be made between the vector space of n-dimensional row vectors $\mathbb{R}_n$ and the vector space of n-dimensional column vectors $\mathbb{R}_n$ or, in fact, between any of the vector spaces $R_q{}^p$ provided only that $pq = n$.

Lemma 2.2.5 Let $Z = [z_i] \in \mathbb{R}^m$. Then[1]

1 $\displaystyle \min_{P \in \mathbb{P}_m} PZ = \min_{P \in \{I_{(1)}, I_{(2)}, ..., I_{(m)}\}} PZ = \min_{i=1, 2, ..., m} z_i$

2 $\displaystyle \max_{P \in \mathbb{P}_m} PZ = \max_{P \in \{I_{(1)}, I_{(2)}, ..., I_{(m)}\}} PZ = \max_{i=1, 2, ..., m} z_i$

PROOF It is sufficient to present the proof of part 1 since the proof of part 2 is essentially identical. Suppose that the rth entry of Z is the smallest, i.e.,

$$z_r = \min_{i=1, 2, ..., m} z_i$$

Let $P = [p_i]$ be any element of $\mathbb{P}_m$. Then

$$PZ = p_1 z_1 + p_2 z_2 + \cdots + p_m z_m$$
$$\geqq p_1 z_r + p_2 z_r + \cdots + p_m z_r$$
$$= (p_1 + p_2 + \cdots + p_m) z_r = z_r$$

Hence $PZ \geqq z_r$

for all $P \in \mathbb{P}_m$. On the other hand,

$$I_{(r)} Z = z_r$$

where $I_{(r)}$ is the rth unit vector of $\mathbb{R}_m$. Since $I_{(r)} \in \mathbb{P}_m$, it follows that

$$\min_{P \in \mathbb{P}_m} PZ = \min_{P \in \{I_{(1)}, I_{(2)}, ..., I_{(m)}\}} PZ = \min_{i=1, 2, ..., m} z_i = z_j \qquad ////$$

Lemma 2.2.5 was tacitly assumed throughout Sec. 2.1. In Example 2.1.4, to find an optimal strategy for player II, we considered all his possible strategies

$$Q = \begin{bmatrix} q_1 \\ q_2 \\ q_3 \end{bmatrix}$$

Then we computed player I's expectation when he plays each of his two pure strategies

$$E(H, Q) = I_{(1)} \begin{bmatrix} 1 & 2 & 4 \\ 3 & 1 & 0 \end{bmatrix} Q = q_1 + 2q_2 + 4q_3 = 4 - 3q_1 - 2q_2$$

$$E(T, Q) = I_{(2)} \begin{bmatrix} 1 & 2 & 4 \\ 3 & 1 & 0 \end{bmatrix} Q = 3q_1 + q_2$$

---

[1] It is also possible to prove that $\min_{P \in \mathbb{P}_m} PZ$ and $\max_{P \in \mathbb{P}_m} PZ$ exist by showing that $f(P) = PZ$ is a continuous real-valued function of P and then applying the results of the last paragraph.

We did not investigate what would happen should player I use a mixed strategy, say $P = [p \quad 1-p]$ against the strategy Q. In this case his expectation would be

$$E(P, Q) = P\begin{bmatrix} 1 & 2 & 4 \\ 3 & 1 & 0 \end{bmatrix} Q$$

Part 2 of Lemma 2.2.5 (with $Z = \begin{bmatrix} 1 & 2 & 4 \\ 3 & 1 & 0 \end{bmatrix} Q$) justifies this procedure. For each (fixed) strategy Q for player II, the maximum expectation for player I using all his possible strategies is the same as his maximum expectation using only his pure strategies. Thus, in Example 2.1.4, player II was able to compute the expectation against the two pure strategies of player I and then obtain his optimal strategy directly from Fig. 2.1.6.

There is one further difficulty concerning the graphical methods used to solve Example 2.1.4. We stated, after examining Fig. 2.1.9, that $n(q_1, q_2)$ assumes a minimum value (which it achieves at the point T). However, no attempt was made to show that the minimum value actually exists. To this end, we prove:

Theorem 2.2.6 Let A be an $m \times n$ matrix. There exist $\tilde{P} \in \mathbb{P}_m$ and $\tilde{Q} \in \mathbb{Q}^n$ such that

$$\min_{Q \in \mathbb{Q}^n} \left(\max_{P \in \mathbb{P}_m} PAQ \right) = \max_{P \in \mathbb{P}_m} PA\tilde{Q} = \tilde{P}A\tilde{Q}$$

PROOF   For each i, $i = 1, 2, \ldots, m$, the function $E(I_{(i)}, Q)$ defined by

$$E(I_{(i)}, Q) = I_{(i)}AQ$$

is a continuous (in fact, linear) real-valued function whose domain is $\mathbb{Q}^n$. It follows that the function $n(Q)$ defined by

$$n(Q) = \max_{P \in \{I_{(1)}, I_{(2)}, \ldots, I_{(m)}\}} PAQ$$

being the maximum of only finitely many continuous functions, is also a real-valued continuous function whose domain is $\mathbb{Q}^n$. Moreover, by part 2 of Lemma 2.2.5

$$n(Q) = \max_{P \in \mathbb{P}_m} PAQ$$

Since $\mathbb{Q}^n$ is compact, $n(Q)$ assumes a minimum value at some point, say $\tilde{Q}$, of $\mathbb{Q}^n$. Thus

$$\min_{Q \in \mathbb{Q}^n} \left(\max_{P \in \mathbb{P}_m} PAQ \right) = \min_{Q \in \mathbb{Q}^n} n(Q) = n(\tilde{Q}) = \max_{P \in \mathbb{P}_m} PA\tilde{Q}$$

Similarly, with $\tilde{Q}$ held fixed,

$$m(P) = PA\tilde{Q}$$

defines a continuous real-valued function whose domain is the compact set $\mathbb{P}_m$. There exists a vector $\tilde{P} \in \mathbb{P}_m$ such that

$$m(\tilde{P}) = \max_{P \in \mathbb{P}_m} PA\tilde{Q} = \tilde{P}A\tilde{Q} \qquad ////$$

Theorem 2.2.6 describes the process used to locate the point T in Fig. 2.1.9 and the optimal strategy

$$\bar{Q} = \begin{bmatrix} \frac{1}{3} \\ \frac{2}{3} \\ 0 \end{bmatrix}$$

for player II. The corresponding process used to find the optimal strategies for player I depends on:

Theorem 2.2.7 Let A be an $m \times n$ matrix. There exist $\tilde{\tilde{P}} \in \mathbb{P}_m$ and $\tilde{\tilde{Q}} \in \mathbb{Q}^n$ such that

$$\max_{P \in \mathbb{P}_m} \left(\min_{Q \in \mathbb{Q}^n} PAQ \right) = \min_{Q \in \mathbb{Q}^n} \tilde{\tilde{P}}AQ = \tilde{\tilde{P}}A\tilde{\tilde{Q}}$$

For any choice of $Q \in \mathbb{Q}^n$, player I's best expectation is given by

$$n(Q) = \max_{P \in \mathbb{P}_m} PAQ$$

Thus, player II has it in his power, by selecting Q properly, to limit player I's expectation to winning no more than

$$\min_{Q \in \mathbb{Q}^n} n(Q) = \min_{Q \in \mathbb{Q}^n} \left(\max_{P \in \mathbb{P}_m} PAQ \right)$$

and, moreover, no better, i.e., lower, limitation to player I's expectation exists. Hence, if $\tilde{P} \in \mathbb{P}_m$, $\tilde{Q} \in \mathbb{Q}^n$ satisfy Theorem 2.2.6, it follows that $\tilde{Q}$ is an optimal strategy for player II in the sense that it limits player I's expectation as effectively as possible. Similarly, for any choice of $P \in \mathbb{P}_m$, player II's best expectation is to lose no more than

$$m(P) = \min_{Q \in \mathbb{Q}^n} PAQ$$

Thus, by selecting P properly, player I can obtain an expectation for himself of

$$\max_{P \in \mathbb{P}_m} m(P) = \max_{P \in \mathbb{P}_m} \left(\min_{Q \in \mathbb{Q}^n} PAQ \right)$$

and this is, in fact, the best expectation he is able to guarantee for himself. As before, if $\tilde{P} \in \mathbb{P}_m$, $\tilde{Q} \in \mathbb{Q}^n$ satisfy Theorem 2.2.7, then $\tilde{P}$ is an optimal strategy for player I. (Note that $\tilde{P}$ and $\tilde{Q}$ may or may not be optimal strategies for players I and II respectively.)

On the basis of the discussion above, it is reasonable to expect:

Theorem 2.2.8 For every $m \times n$ matrix A,

$$\min_{Q \in \mathbb{Q}^n} \left(\max_{P \in \mathbb{P}_m} PAQ \right) \geq \max_{P \in \mathbb{P}_m} \left(\min_{Q \in \mathbb{Q}^n} PAQ \right)$$

Theorem 2.2.8 is a special case of the more general Theorem 2.2.9 which makes no use of the special structure of $\mathbb{P}_m$, $\mathbb{Q}^n$, or A.

Theorem 2.2.9 Let $\mathbb{S}$, $\mathbb{T}$ be nonempty sets and let $F(s, t)$ be a real-valued function defined for all $s \in \mathbb{S}$, $t \in \mathbb{T}$. If $\max_{s \in \mathbb{S}} F(s, t_1)$ and $\min_{t \in \mathbb{T}} F(s_1, t)$ exist for all $s_1 \in \mathbb{S}$, $t_1 \in \mathbb{T}$, and if $\min_{t \in \mathbb{T}} (\max_{s \in \mathbb{S}} F(s, t))$ and $\max_{s \in \mathbb{S}} (\min_{t \in \mathbb{T}} F(s, t))$ both exist, then

$$\min_{t \in \mathbb{T}} \left(\max_{s \in \mathbb{S}} F(s, t) \right) \geq \max_{s \in \mathbb{S}} \left(\min_{t \in \mathbb{T}} F(s, t) \right)$$

PROOF. For each (fixed) $t_1 \in \mathbb{T}$, $F(s, t_1)$ and $\min_{t \in \mathbb{T}} F(s, t)$ are both real-valued functions whose domain is $\mathbb{S}$, and for each $s \in \mathbb{S}$ we have

$$F(s, t_1) \geq \min_{t \in \mathbb{T}} F(s, t)$$

Then, for every $t_1 \in \mathbb{T}$,

$$\max_{s \in \mathbb{S}} F(s, t_1) \geq \max_{s \in \mathbb{S}} \left(\min_{t \in \mathbb{T}} F(s, t) \right) \qquad (2.2.10)$$

The right-hand side of (2.2.10) is a fixed real number, while the left-hand side depends on the choice of t_1. Since (2.2.10) is satisfied for every $t_1 \in \mathbb{T}$, it follows that the minimum value which the left-hand side assumes also satisfies this inequality, i.e.,

$$\min_{t_1 \in \mathbb{T}} \left(\max_{s \in \mathbb{S}} F(s, t_1) \right) \geq \max_{s \in \mathbb{S}} \left(\min_{t \in \mathbb{T}} F(s, t) \right)$$

Replacing t_1 by t throughout the left-hand side of this inequality completes the proof. ////

In each of the examples solved in Sec. 2.1 we found that

$$\min_{Q \in \mathbb{Q}^n} \left(\max_{P \in \mathbb{P}_m} PAQ \right) = \max_{P \in \mathbb{P}_m} \left(\min_{Q \in \mathbb{Q}^n} PAQ \right) \qquad (2.2.11)$$

It was this which permitted us to say in each case that the game had a value and that the optimal strategies for each player that we found could not be improved upon (in the sense that no better guaranteed expectation exists).   On the other hand, it is clear that if the choices of P and Q are restricted (e.g., to being unit vectors), then we need not have equality in (2.2.11).   (See Exercises 1 and 2.)

The statement that equation (2.2.11) holds for every matrix A is called the *minimax theorem* and will be proved in Sec. 2.5.   It is the fundamental theorem for finite zero-sum two-person games and establishes that every game has a well-defined solution.

EXERCISES

1   In the games of Examples 2.1.1, 2.1.2, 2.1.4, and 2.1.10, find

$$\max_{i=1, 2, \ldots, m} \left(\min_{j=1, 2, \ldots, n} I_{(i)}AI^{(j)} \right)$$

and

$$\min_{j=1, 2, \ldots, n} \left(\max_{i=1, 2, \ldots, m} I_{(i)}AI^{(j)} \right)$$

2   Find

$$\max_{i=1, 2, 3, 4} \left(\min_{j=1, 2, 3, 4, 5} I_{(i)}AI^{(j)} \right)$$

and

$$\min_{j=1,2,3,4,5} \left(\max_{i=1,2,3,4} I_{(i)}AI^{(j)} \right)$$

when

(a) $A = \begin{bmatrix} 1 & 2 & -1 & 0 & 0 \\ 1 & 1 & -1 & -2 & 0 \\ 2 & 4 & -8 & -6 & 1 \\ 1 & 7 & 3 & 2 & 5 \end{bmatrix}$ (b) $A = \begin{bmatrix} -3 & 0 & 1 & 7 & 1 \\ 6 & -2 & 2 & 1 & -1 \\ -2 & 0 & -3 & 0 & 0 \\ 8 & -1 & 2 & 4 & -6 \end{bmatrix}$

3   Let $F(x, y) = x^2 - 4x - 4xy + 2y^2$.   Show that $\max_{x \in \mathbb{R}}(\min_{y \in \mathbb{R}} F(x, y))$ exists and evaluate it.   [*Hint:* Find $n(x) = \min_{y \in \mathbb{R}} F(x, y)$ by holding x fixed.] Show that $\min_{y \in \mathbb{R}}(\max_{x \in \mathbb{R}} x^2 - 4x - 4xy + 2y^2)$ does not exist.

4   Which of the following sets in the plane are convex?
(a) $\{(x, y) \mid y = x^2\}$ (b) $\{(x, y) \mid y \geq x^2\}$
(c) $\{(x, y) \mid y \leq x^2)\}$ (d) $\{(x, y) \mid y > x^2\}$
(e) $\{(x, y) \mid x^2 + y^2 = 1\}$ (f) $\{(x, y) \mid x^2 + y^2 \geq 1\}$
(g) $\{(x, y) \mid x^2 + y^2 \leq 1\}$ (h) $\{(x, y) \mid x^2 + y^2 > 1\}$

5   Show that $\mathbb{P}_m$ and $\mathbb{Q}^n$ are convex.

6   Show that $\mathbb{S}$ is convex if $\mathbb{S}$ contains all convex combinations of pairs of elements of $\mathbb{S}$.

7 Let $\mathbb{S}$ be the collection of all convex combinations of the vectors $X_1, X_2,$ $\ldots, X_k$. Show that $\mathbb{S}$ is convex.

8 Prove that the intersection of convex sets is convex.

9 Let $Z \in \mathbb{R}^m$ and define $\mathbb{O}(Z)$ to be

$$\mathbb{O}(Z) = \{X \mid XZ \geq 0, \quad X \in \mathbb{R}_m\}$$

Show that $\mathbb{O}(Z)$ is convex. Show that $\mathbb{O}(Z)$ contains m linearly independent vectors but $\mathbb{O}(Z) \neq \mathbb{R}_m$ unless $Z = 0$.

10 A point X of a convex set $\mathbb{S}$ is an *extreme point* if it cannot be expressed as

$$X = \tfrac{1}{2}Y + \tfrac{1}{2}Z$$

for some $Y, Z \in \mathbb{S}$, $Y \neq Z$. Show that the unit vectors are the extreme points of $\mathbb{P}_m$ and $\mathbb{Q}^n$. Find the extreme points of those sets in Exercise 4 which are convex. Show that a nonzero vector space has no extreme points.

11 Let X, Y, Z be elements of the convex set $\mathbb{S}$, $Y \neq Z$. Show that if X can be expressed as

$$X = \lambda Y + (1 - \lambda)Z$$

for some real number λ, $0 < \lambda < 1$, then X is not an extreme point of $\mathbb{S}$. (This leads to an alternate definition of extreme point.)

12 Let $\mathbb{S}$ be a convex subset of $\mathbb{R}^n$ and let $X = [x_i]$ be a point of $\mathbb{S}$ whose distance from 0 is maximal; that is, X belongs to $\mathbb{S}$ and

$$\sum_{i=1}^{n} x_i^2 \geq \sum_{i=1}^{n} p_i^2$$

for every $P = [P_i]$ belonging to $\mathbb{S}$. Prove that X is an extreme point of $\mathbb{S}$. *Hint:* Suppose that $X = \tfrac{1}{2}Y + \tfrac{1}{2}Z$, where $Y \neq Z$ and $Y, Z \in \mathbb{S}$. Let $Y = [y_i]$, $Z = [z_i]$ and set $W = Y - Z$, $W = [w_i]$. Show that

$$\sum_{i=1}^{n} y_i^2 + \sum_{i=1}^{n} z_i^2 = \sum_{i=1}^{n} w_i^2 + 2\sum_{i=1}^{n} y_i z_i > 2\sum_{i=1}^{n} y_i z_i$$

13 Let $\mathbb{S}$ be a convex subset of $\mathbb{R}^n$ and let A be an $m \times n$ matrix. Show that $A\mathbb{S}$ is convex.

2.3 A MATHEMATICAL INTRODUCTION TO THE THEORY OF GAMES

In the previous sections we began our study of the theory of games with a discussion relying heavily on specific examples and intuitive arguments. To put the subject on a firmer foundation and also to develop more sophisticated

methods for obtaining information about games we now develop the subject
again, this time in a more formal way, but preserving as much of the language
and notation of the previous sections as possible.   In order to produce a self-
contained and logical development some repetition of the material of the first
two sections is necessary.

We will be concerned with a game G which can be represented by an
$m \times n$ matrix $A = [a_{ij}]$.   Henceforth we will not be concerned with specific
games, as in Sec. 2.1.   The rules of a game determine its matrix.   However,
once the matrix is determined, the rules of the game no longer interest us.   Thus
we will consider any two games which yield the same matrix as being the same
game.

Definition 2.3.1 A *strategy* for player I is an element of $\mathbb{P}_m$. A *strategy*
for player II is an element of $\mathbb{Q}^n$.   The unit vectors of $\mathbb{P}_m$ are the *pure strategies*
for player I, and the unit vectors of $\mathbb{Q}^n$ are the *pure strategies* for player II.

Let P, Q be strategies for players I and II respectively in a game G whose
matrix is A.   Then the *expected value* of G for player I with respect to P and Q
is defined to be PAQ and is denoted by $E(P, Q)$.

Definition 2.3.2 Let G be a game with matrix A.   If there exist strategies
$\bar{P} \in \mathbb{P}_m$ and $\bar{Q} \in \mathbb{Q}^n$ and a real number v such that

$$E(\bar{P}, Q) \geqq v \geqq E(P, \bar{Q}) \qquad (2.3.3)$$

for all $P \in \mathbb{P}_m$, $Q \in \mathbb{Q}^n$, then the triple $(\bar{P}, \bar{Q}, v)$ is a *solution* of G. We say that
v is a *value* of G and that $\bar{P}$, $\bar{Q}$ are *optimal strategies* (with respect to v) for
players I and II, respectively.

Suppose $(\bar{P}, \bar{Q}, v)$ is a solution of some game.   Since $\bar{Q} \in \mathbb{Q}^n$, we may set
$Q = \bar{Q}$ in the first inequality of (2.3.3).   Thus

$$E(\bar{P}, \bar{Q}) \geqq v$$

Similarly, setting $P = \bar{P}$ in the second inequality of (2.3.3) yields

$$v \geqq E(\bar{P}, \bar{Q})$$

Hence $$E(\bar{P}, \bar{Q}) = v \qquad (2.3.4)$$

According to Definition 2.3.2, a game may have no values, one value, or
more than one value.   Also optimal strategies are defined with respect to a
particular value of the game.   This ambiguity is removed by:

Lemma 2.3.5   Let $(\bar{P}_1, \bar{Q}_1, v_1)$ and $(\bar{P}_2, \bar{Q}_2, v_2)$ be solutions of the same game. Then

$$v_1 = v_2 \quad (=v)$$

and $(\bar{P}_1, \bar{Q}_2, v)$ and $(\bar{P}_2, \bar{Q}_1, v)$ are also solutions.

PROOF   Since $(\bar{P}_1, \bar{Q}_1, v_1)$ and $(\bar{P}_2, \bar{Q}_2, v_2)$ are both solutions of the game,

$$E(\bar{P}_1, Q) \geqq v_1 \geqq E(P, \bar{Q}_1) \qquad (2.3.6)$$
$$E(\bar{P}_2, Q) \geqq v_2 \geqq E(P, \bar{Q}_2) \qquad (2.3.7)$$

for all $P \in \mathbb{P}_m$, $Q \in \mathbb{Q}^n$. Substituting $Q = \bar{Q}_2$ in the first part of (2.3.6) and $P = \bar{P}_1$ in the second part of (2.3.7) yields

$$v_2 \geqq E(\bar{P}_1, \bar{Q}_2) \geqq v_1$$

Substituting $P = \bar{P}_2$ in the second part of (2.3.6) and $Q = \bar{Q}_1$ in the first part of (2.3.7) yields

$$v_1 \geqq E(\bar{P}_2, \bar{Q}_1) \geqq v_2$$

Hence $\qquad\qquad\qquad\qquad\qquad\qquad\qquad v_1 = v_2$

The remainder of the lemma is immediate, for clearly

$$E(\bar{P}_1, Q) \geqq v_1 = v_2 \geqq E(P, \bar{Q}_2)$$
$$E(\bar{P}_2, Q) \geqq v_2 = v_1 \geqq E(P, \bar{Q}_1)$$

for all $P \in \mathbb{P}_m$, $Q \in \mathbb{Q}^n$.   ////

We have proved that a game has at most one value (which we will denote by v or v_G or v_A).   It is not until Sec. 2.5 that we will be able to prove that every game has a value and that each player has at least one optimal strategy.   Actually finding a solution of a game is another matter entirely.   Some elementary methods for obtaining solutions were developed in Sec. 2.1, and more will be added in Sec. 2.6.[1]

In Sec. 2.2 we discussed the best expectations that each of the players is able to guarantee and showed that these expectations can be expressed in terms of

$$\min_{Q \in \mathbb{Q}^n} \left(\max_{P \in \mathbb{P}_m} E(P, Q) \right) \quad \text{and} \quad \max_{P \in \mathbb{P}_m} \left(\min_{Q \in \mathbb{Q}^n} E(P, Q) \right)$$

---

[1] There are still other schemes for finding solutions of a given game.   Undoubtedly the most important of these is the simplex method.   Here a game is first transformed into a linear-programming problem.   (See Owen [7] and Sec. 2.8.)   The simplex method is particularly adaptable for use with electronic digital computers.

The existence of a solution for a game G is equivalent to these best expectations being equal; that is:

Theorem 2.3.8 The game G has a solution $(\bar{P}, \bar{Q}, v)$ if and only if

$$\max_{P \in \mathbb{P}_m} \left(\min_{Q \in \mathbb{Q}^n} E(P, Q) \right) = \min_{Q \in \mathbb{Q}^n} \left(\max_{P \in \mathbb{P}_m} E(P, Q) \right)$$

PROOF   First, suppose

$$\max_{P \in \mathbb{P}_m} \left(\min_{Q \in \mathbb{Q}^n} E(P, Q) \right) = v = \min_{Q \in \mathbb{Q}^n} \left(\max_{P \in \mathbb{P}_m} E(P, Q) \right)$$

By Theorem 2.2.6 there exists $\bar{Q} \in \mathbb{Q}^n$ such that

$$v = \min_{Q \in \mathbb{Q}^n} \left(\max_{P \in \mathbb{P}_m} E(P, Q) \right) = \max_{P \in \mathbb{P}_m} E(P, \bar{Q})$$

Similarly, by Theorem 2.2.7 there exists $\bar{P} \in \mathbb{P}_n$ such that

$$\max_{P \in \mathbb{P}_n} \left(\min_{Q \in \mathbb{Q}^n} E(P,Q) \right) = \min_{Q \in \mathbb{Q}^n} E(\bar{P}, Q) = v$$

Hence, for all $Q \in \mathbb{Q}^n$, $P \in \mathbb{P}_m$, we have

$$E(\bar{P}, Q) \geq \min_{Q \in \mathbb{Q}^n} E(\bar{P}, Q) = v = \max_{P \in \mathbb{P}_m} E(P, \bar{Q}) \geq E(P, \bar{Q})$$

and it follows from Definition 2.3.2 that $(\bar{P}, \bar{Q}, v)$ is a solution of G.
    Conversely, let $(\bar{P}, \bar{Q}, v)$ be a solution of G. We wish to show that

$$\max_{P \in \mathbb{P}_m} \left(\min_{Q \in \mathbb{Q}^n} E(P, Q) \right) = v = \min_{Q \in \mathbb{Q}^n} \left(\max_{P \in \mathbb{P}_m} E(P, Q) \right)$$

In view of Theorem 2.2.8 it is only necessary to show that

$$\max_{P \in \mathbb{P}_m} \left(\min_{Q \in \mathbb{Q}^n} E(P, Q) \right) \geq \min_{Q \in \mathbb{Q}^n} \left(\max_{P \in \mathbb{P}_m} E(P, Q) \right)$$

Since $(\bar{P}, \bar{Q}, v)$ is a solution of G, we have

$$E(\bar{P}, Q) \geq v \geq E(P, \bar{Q})$$

for all $P \in \mathbb{P}_m$, $Q \in \mathbb{Q}^n$. From the first of these inequalities it follows that

$$\min_{Q \in \mathbb{Q}^n} E(\bar{P}, Q) \geq v$$

and from the second of these inequalities, it follows that

$$v \geq \max_{P \in \mathbb{P}_m} E(P, \bar{Q})$$

But then

$$\max_{P \in \mathbb{P}_m} \left(\min_{Q \in \mathbb{Q}^n} E(P, Q) \right) \geq \min_{Q \in \mathbb{Q}^n} E(\bar{P}, Q) \geq v \geq$$

$$\max_{P \in \mathbb{P}_m} E(P, \bar{Q}) \geq \min_{Q \in \mathbb{Q}^n} \left(\max_{P \in \mathbb{P}_m} E(P, Q) \right)$$

as required.   ////

EXERCISES

1 Verify that

$$\left([\tfrac{2}{3} \ \ \tfrac{1}{3}], \begin{bmatrix} \tfrac{1}{3} \\ \tfrac{2}{3} \\ 0 \end{bmatrix}, \tfrac{5}{3} \right)$$

is a solution for the game of Example 2.1.4 using Definition 2.3.2.

2 Prove that the set of optimal strategies for each player is convex.

3 Let A be the matrix of a game whose value is v. Prove that a necessary and sufficient condition that $\bar{P} \in \mathbb{P}_m$ be an optimal strategy for player I is that every entry of $\bar{P}A$ is greater than or equal to v. State and prove the corresponding condition for $\bar{Q} \in \mathbb{Q}^n$ to be an optimal strategy for player II.

4 Show that $(I_{(r)}, I^{(s)}, a_{rs})$ is a solution of a game with matrix A if and only if

$$a_{rj} \geq a_{rs} \geq a_{is} \qquad \begin{aligned} i &= 1, 2, \ldots, m \\ j &= 1, 2, \ldots, n \end{aligned} \qquad (2.3.9)$$

An entry a_{rs} of A which satisfies (2.3.9) is called a *saddle point* of A.

5 Show that A has a saddle point if and only if

$$\max_{i = 1, 2, \ldots, m} \left(\min_{j = 1, 2, \ldots, n} E(I_{(i)}, I^{(j)}) \right) = \min_{j = 1, 2, \ldots, n} \left(\max_{i = 1, 2, \ldots, m} E(I_{(i)}, I^{(j)}) \right)$$

(See Exercises 1 and 2 of Sec. 2.2.)   *Hint:* Show

$$\max_{i = 1, 2, \ldots, m} \left(\min_{j = 1, 2, \ldots, n} a_{ij} \right) = \min_{j = 1, 2, \ldots, n} a_{rj} = a_{rs}$$

6 Solve the following games by finding all the saddle points

(a) $A = \begin{bmatrix} 4 & 2 & 1 \\ -9 & 7 & -1 \end{bmatrix}$ (b) $A = \begin{bmatrix} -7 & 15 & 0 & 8 \\ 12 & -6 & -1 & -8 \\ 3 & 4 & 3 & 4 \\ 5 & -1 & -3 & 10 \end{bmatrix}$

$$(c)\ A = \begin{bmatrix} 1 & -2 & -1 & 1 \\ -4 & -1 & -1 & 3 \\ 1 & -1 & -3 & -1 \\ 3 & -1 & -1 & 2 \end{bmatrix} \quad (d)\ A = \begin{bmatrix} -5 & -2 & 2 & 2 & 0 \\ 2 & -5 & -4 & 0 & -7 \\ 2 & 3 & 4 & 2 & 5 \\ -1 & 4 & 2 & -4 & 2 \end{bmatrix}$$

7 Examine the games of Examples 2.1.1, 2.1.2, and 2.1.10 for saddle points.

8 Let a_{rs} be a saddle point of the matrix A for which

$$a_{rj} > a_{rs} > a_{is} \qquad \begin{matrix} i \neq r \\ j \neq s \end{matrix}$$

Show that $(I_{(r)}, I^{(s)}, a_{rs})$ is the only solution of the game. (*Hint:* Use Exercise 3.)

9 Let a_{rs} and a_{tn} be saddle points of A. Show that a_{rn} and a_{ts} are saddle points of A and that $a_{rs} = a_{tn} = a_{rn} = a_{ts}$.

10 Let the matrix A' be formed from the matrix A by deleting one of the columns of A. Show that $v_{A'} \geqq v_A$.

11 Let E be the $m \times n$ matrix all of whose entries are 1. Let G be a game whose matrix is the $m \times n$ matrix A and let G' be the game whose matrix is $A + \alpha E$. Show that $(\bar{P}, \bar{Q}, v)$ is a solution of G if and only if $(\bar{P}, \bar{Q}, v + \alpha)$ is a solution of G'.

12 Let $(\bar{P}_1, \bar{Q}_1, v_1)$ and $(\bar{P}_2, \bar{Q}_2, v_2)$ be solutions of games whose matrices are A_1 and A_2 respectively. Set

$$A = \begin{bmatrix} A_1 & 0 \\ 0 & A_2 \end{bmatrix}$$

Show that
(a) If $v_1 > 0$, $v_2 > 0$, then

$$\left(\alpha_1 [\bar{P}_1 \quad 0] + \alpha_2 [0 \quad \bar{P}_2],\ \alpha_1 \begin{bmatrix} \bar{Q}_1 \\ 0 \end{bmatrix} + \alpha_2 \begin{bmatrix} 0 \\ \bar{Q}_2 \end{bmatrix},\ \frac{v_1 v_2}{v_1 + v_2} \right)$$

where $\alpha_1 = \dfrac{v_2}{v_1 + v_2} \qquad \alpha_2 = \dfrac{v_1}{v_1 + v_2}$

is a solution of A.

(b) If $v_1 > 0$, $v_2 < 0$, then $\left([\bar{P}_1 \quad 0],\ \begin{bmatrix} 0 \\ \bar{Q}_2 \end{bmatrix},\ 0 \right)$ is a solution of A.

13 Show that the following game has a saddle point. Player I is dealt a card from a deck consisting of the two cards V and X. After examining the card, player I calls out either

"I have V"   or   "I have X"

Player II then calls out either

"I believe you"   or   "I do not believe you"

(a) If player I has told the truth and player II believes him, then player II wins 5 when player I holds V and 10 when player I holds X.

(b) If player I has told the truth and player II does not believe him, then player I wins 5 when he holds V and 10 when he holds X.

(c) On the other hand, if player I has lied and player II believes him, then player I wins 10 if he holds V and 20 if he holds X.

(d) Finally, if player I lied and player II catches the lie, then player II wins 10 if Player I holds V and 20 if player I holds X.

14 Suppose the payoff in the game of Exercise 13 is made to depend upon the card player I calls (rather than the card he holds). Does the resulting game have a saddle point?

2.4 LINEAR INEQUALITIES

In Sec. 2.5 we will prove that every (finite zero-sum two-person) game has a solution.[1] The main tool we will need for the proof presented here is a theorem concerning the existence of nonnegative solutions of a system of simultaneous linear inequalities. In Chap. 1 we saw that the behavior of a system of simultaneous linear equations is closely connected with the theory of vector spaces (see, for example, Exercises 23 and 24 of Sec. 1.3, Exercises 14 and 15 of Sec. 1.4, and Exercise 9 of Sec. 1.5). In the same way, problems dealing with the existence of nonnegative solutions of either simultaneous equations or simultaneous inequalities (or a mixture of both) lead to the theory of cones.

First we need some definitions.

Definition 2.4.1 A matrix A is said to be:

1 Positive if every entry of A is positive. We express this by writing $A > 0$.

2 Semipositive if every entry of A is nonnegative and $A \neq 0$. We express this by writing $A \geq 0$.

3 Nonnegative if every entry of A is nonnegative. In this case we write $A \geqq 0$.

---

[1] The first proof of this theorem was given by von Neumann [6] in 1928. Since then several other proofs have appeared by different authors. Most of these, though they require a knowledge either of topology (in particular, of fixed-point theorems) or of the geometry of convex sets, have the virtue of being intuitively clear—probably clearer than the proof presented here. (See Karlin [4] and McKinsey [5].) The main advantage of the proof we have chosen is that it requires few preliminaries, all of which can be presented in algebraic guise. For another proof using matrix methods see Dresher [1].

For two $m \times n$ matrices A and B we write

$$A > B \quad \text{if} \quad A - B > 0$$
$$A \geq B \quad \text{if} \quad A - B \geq 0$$
$$A \gneq B \quad \text{if} \quad A - B \gneq 0$$

All the vectors belonging to $\mathbb{P}_m$ and $\mathbb{Q}^n$ are semipositive. Further, it is not difficult to see that a vector of $\mathbb{R}_m$ (or $\mathbb{R}^n$) is semipositive if and only if it can be expressed as a positive multiple of some vector which belongs to $\mathbb{P}_m$ (or $\mathbb{Q}^n$).

Definition 2.4.2 Let $\mathbb{K}$ be a nonempty set of $m \times n$ matrices. $\mathbb{K}$ is called a *cone* if

1 $X + Y$ belongs to $\mathbb{K}$ whenever X and Y belong to $\mathbb{K}$.
2 αX belongs to $\mathbb{K}$ whenever X belongs to $\mathbb{K}$ and $\alpha \geq 0$.

The first part of Definition 2.4.2 is identical to the first part of Definition 1.3.1. The second parts of the two definitions differ only in that a cone is required to be closed under multiplication by a nonnegative scalar. Consequently every vector space is a cone. However, the set of all $m \times n$ matrices which have nonnegative entries is a cone which is not a vector space.

Definition 2.4.3 Let $A_1, A_2, \ldots, A_k, B$ be elements of a vector space $\mathbb{V}$. We say that B is a nonnegative linear combination of $A_1, A_2, \ldots, A_k$ if there exist nonnegative numbers $\alpha_1, \alpha_2, \ldots, \alpha_k$ such that

$$B = \alpha_1 A_1 + \alpha_2 A_2 + \cdots + \alpha_k A_k$$

Corollary 2.4.4. Let $\mathbb{V}$ be a vector space and let $A_1, A_2, \ldots, A_k$ be elements of $\mathbb{V}$. The set $\mathbb{K}$ of all nonnegative linear combinations of $A_1, A_2, \ldots, A_k$ is a cone.

PROOF The proof is left as an exercise.   ////

We say that the cone $\mathbb{K}$ is spanned by $A_1, A_2, \ldots, A_k$.

Definition 2.4.5 Let A be an $m \times n$ matrix. The cone spanned by the columns of A is called the *column cone* of A and is denoted by $\mathbb{K}(A)$. The cone spanned by the rows of A is called the *row cone* of A and is denoted by $\mathbb{K}_R(A)$.

Corollary 2.4.6 Let A be an $m \times n$ matrix and let $Z \in \mathbb{R}^m$. Then Z belongs to $\mathbb{K}(A)$ if and only if there exists $X \in \mathbb{R}^n$ such that

$$Z = AX \qquad X \geq 0$$

PROOF The proof of the corollary, as well as the statement and proof of the corresponding result for $\mathbb{K}_R(A)$, is left as an exercise.   ////

Corollary 2.4.6 is the analog of Theorem 1.4.4(b).

Definition 2.4.7 Let $\mathbb{S}$ be any subset of $\mathbb{R}_n{}^m$. The set of vectors $Y \in \mathbb{R}_m$ such that

$$YX \geq 0$$

for all $X \in \mathbb{S}$ is a cone called the *nonnegative row complement* of $\mathbb{S}$ (or simply the *complement* of $\mathbb{S}$) and is denoted by $\mathbb{O}(\mathbb{S})$.

When $\mathbb{S}$ consists of a single matrix A, we write $\mathbb{O}(A)$ instead of $\mathbb{O}(\mathbb{S})$. For $A = [a_i] \in \mathbb{R}^m$ it is clear that $A^T \subseteqq \mathbb{O}(A)$ since

$$A^T A = a_1{}^2 + a_2{}^2 + \cdots + a_m{}^2 \geq 0$$

Lemma 2.4.8 Let A be an $m \times n$ matrix and let Z be an element of $\mathbb{R}^m$. If

$$\mathbb{O}(A) \subseteq \mathbb{O}(Z)$$

then $Z \in \mathbb{C}(A)$; that is, the equation

$$Z = AX$$

has a solution.

PROOF It is sufficient to show that the row null space of A is contained in the row null space of Z, for then the lemma follows from Theorem 1.7.16. Let Y be a vector in the row null space of A. Clearly, the row null space of A is a vector space contained in $\mathbb{O}(A)$. Hence $-Y$ also belongs to $\mathbb{O}(A)$, and we have

$$YZ \geqq 0 \qquad (-Y)Z \geqq 0$$

Thus $YZ = 0$, and consequently Y belongs to the row null space of Z.   ////

Lemma 2.4.8 can be strengthened considerably. We will in fact show that if $\mathbb{O}(A) \subseteq \mathbb{O}(Z)$, then $Z \in \mathbb{K}(A)$; that is, the system

$$Z = AX \qquad X \geq 0$$

has a solution. This result is a part of Theorem 2.4.10, which states that $\mathbb{K}(Z) \subseteq \mathbb{K}(A)$ if and only if $\mathbb{O}(A) \subseteq \mathbb{O}(Z)$. Theorem 2.4.10 is the analog for cones of Theorem 1.7.16. First, however, we need a lemma.

Lemma 2.4.9 Let S, T be elements of $\mathbb{R}_m$, $\mathbb{R}^m$ respectively such that $ST \neq 0$ and let $B = TS$. Then the characteristic roots of B are 0, with geometric and algebraic multiplicity $m - 1$, and ST, with geometric and algebraic multiplicity 1. The characteristic vectors of B corresponding to the characteristic root 0 are the nonzero elements of $\mathbb{N}(S)$, and the unique (up to nonzero multiples) characteristic vector of B corresponding to ST is T.[1]

PROOF   For $X \in \mathbb{N}(S)$, we have

$$BX = TSX = T0 = 0$$

Thus the nonzero elements of the $(m - 1)$-dimensional subspace $\mathbb{N}(S)$ of $\mathbb{R}^m$ are all characteristic vectors of B corresponding to the characteristic root 0. Consequently, 0 is a characteristic root of A of geometric multiplicity $m - 1$. (See Exercise 24 of Sec. 1.10). Also,

$$[B - (ST)I]T = (TS)T - (ST)T = (TS)T - T(ST) = 0$$

since ST is of order 1 and hence commutes with T. Thus T is a characteristic vector of B corresponding to the characteristic root ST. Since $ST \neq 0$, we have now determined n characteristic roots for B, counting each root according to its geometric multiplicity. Thus, in each case, the algebraic multiplicity is equal to the geometric multiplicity, and B has no other characteristic roots. Consequently, the geometric multiplicity of ST is 1. Hence T is the unique (up to nonzero multiples) characteristic vector of B corresponding to the characteristic root ST.   ////

Now we are able to prove:

Theorem 2.4.10 (Farkas, 1902) Let A be an $m \times n$ matrix and let $Z \in \mathbb{R}^m$. Then

$$\mathbb{K}(Z) \subseteq \mathbb{K}(A)$$

if and only if

$$\mathbb{O}(A) \subseteq \mathbb{O}(Z)$$

---

[1] Note that we are purposely confusing the 1×1 matrix ST with the entry of this matrix.

PROOF   First suppose that $\mathbb{K}(Z) \subseteq \mathbb{K}(A)$; that is, suppose there exists a vector $X \in \mathbb{R}^n$ such that

$$Z = AX \qquad X \geq 0$$

For $Y \in \mathbb{O}(A)$ we have

$$YA \geq 0$$

and hence

$$YZ = YAX \geq 0$$

Thus $Y \in \mathbb{O}(Z)$.

Conversely, suppose $\mathbb{O}(A) \subseteq \mathbb{O}(Z)$. By Lemma 2.4.8, Z belongs to $\mathbb{C}(A)$, and hence there exists $X \in \mathbb{R}^n$ such that $Z = AX$. It remains to show that a nonnegative vector X exists. The proof is by an induction on n. For $n = 1$, $X = [x]$, $x \in \mathbb{R}$, and Z is a scalar multiple of A. Since $A^T \in \mathbb{O}(A)$, it follows that $A^T \in \mathbb{O}(Z)$ and we have

$$A^T Z = A^T A X = (a_1{}^2 + a_2{}^2 + \cdots + a_m{}^2)x \geq 0$$

If $A = 0$, then x is completely arbitrary and may be chosen to be nonnegative. If $A \neq 0$, then x is uniquely defined and is nonnegative.

Now assume that the theorem has been proved for all $m \times (n-1)$ matrices and let A be an $m \times n$ matrix. Denote by $\tilde{A}$ the $m \times (n-1)$ matrix formed from A by deleting the last column of A. Clearly $\mathbb{O}(A) \subseteq \mathbb{O}(\tilde{A})$. Here two cases arise. If $\mathbb{O}(\tilde{A}) \subseteq \mathbb{O}(Z)$, then, by the induction hypothesis, there is a nonnegative vector $\tilde{X}$ belonging to $\mathbb{R}^{n-1}$ such that $Z = \tilde{A}\tilde{X}$. Set

$$X = \begin{bmatrix} \tilde{X} \\ 0 \end{bmatrix}$$

Then X is nonnegative and

$$AX = [\tilde{A} \quad A^{(n)}] \begin{bmatrix} \tilde{X} \\ 0 \end{bmatrix} = \tilde{A}\tilde{X} = Z$$

proving the theorem.

Finally, let us assume $\mathbb{O}(\tilde{A}) \not\subseteq \mathbb{O}(Z)$; that is, there exists a vector $\tilde{Y} \in \mathbb{O}(\tilde{A})$ such that $\tilde{Y}Z < 0$. Then the induction hypothesis cannot be applied to $\tilde{A}$ and Z. Since $\tilde{Y} \notin \mathbb{O}(Z)$, it follows that $\tilde{Y} \notin \mathbb{O}(A)$; that is,

$$\tilde{Y}A = \tilde{Y}[\tilde{A} \quad A^{(n)}] = [\tilde{Y}\tilde{A} \quad \tilde{Y}A^{(n)}] \not\geq 0$$

But $\tilde{Y} \in \mathbb{O}(\tilde{A})$, and hence $\tilde{Y}\tilde{A} \geq 0$. Thus, $\tilde{Y}A^{(n)} < 0$.

Set

$$B = A^{(n)}\tilde{Y}$$
$$C = A^{(n)}\tilde{Y} - (\tilde{Y}A^{(n)})I$$

B and C are square matrices of order m. By Lemma 2.4.9, $A^{(n)}$ is a characteristic vector of B corresponding to the characteristic root $\tilde{Y}A^{(n)}$. Hence

$$CA^{(n)} = [B - (\tilde{Y}A^{(n)})I]A^{(n)} = 0 \qquad (2.4.11)$$

Set $\bar{A} = C\tilde{A}$ and $\bar{Z} = CZ$. We wish to show that the induction hypothesis can be applied to $\bar{A}$ and $\bar{Z}$; that is, we wish to show that $\mathbb{O}(\bar{A}) \subseteq \mathbb{O}(\bar{Z})$. Let $Y_0 \in \mathbb{O}(\bar{A})$ and set $\bar{Y} = Y_0 C$. Then, by equation (2.4.11),

$$\begin{aligned}\bar{Y}A = \bar{Y}[\tilde{A} \quad A^{(n)}] &= [\bar{Y}\tilde{A} \quad \bar{Y}A^{(n)}]\\ &= [Y_0 C\tilde{A} \quad Y_0 CA^{(n)}] \qquad (2.4.12)\\ &= [Y_0 \bar{A} \quad 0] \geq 0\end{aligned}$$

Equation (2.4.12) states that $\bar{Y} \in \mathbb{O}(A)$. By hypothesis $\mathbb{O}(A) \subseteq \mathbb{O}(Z)$, and hence

$$\bar{Y}Z = Y_0 CZ = Y_0 \bar{Z} \geq 0$$

Thus $Y_0 \in \mathbb{O}(\bar{Z})$. We may now apply the induction hypothesis to $\bar{A}$ and $\bar{Z}$. There exists a nonnegative vector $\bar{X} \in \mathbb{R}^{n-1}$ such that

$$\bar{Z} = \bar{A}\bar{X}$$

Then $CZ = C\tilde{A}\bar{X}$ which can be written

$$0 = C(Z - \tilde{A}\bar{X}) = [B - (\tilde{Y}A^{(n)})I](Z - \tilde{A}\bar{X})$$

Thus $Z - \tilde{A}\bar{X}$ is zero or is a characteristic vector of B corresponding to the characteristic root $\tilde{Y}A^{(n)}$. But by Lemma 2.4.9 $A^{(n)}$ is the only characteristic vector of B corresponding to the root $\tilde{Y}A^{(n)}$ (up to nonzero multiples). Hence

$$Z - \tilde{A}\bar{X} = \alpha A^{(n)} \qquad (2.4.13)$$

for some real number α. Equation (2.4.13) can be written

$$Z = [\tilde{A} \quad A^{(n)}]\begin{bmatrix}\bar{X}\\ \alpha\end{bmatrix} = A\begin{bmatrix}\bar{X}\\ \alpha\end{bmatrix}$$

and in order to complete the proof it remains only to show that $\alpha \geq 0$. Multiply equation (2.4.13) on the left by $\tilde{Y}$. Then

$$\tilde{Y}Z - \tilde{Y}\tilde{A}\bar{X} = \tilde{Y}\alpha A^{(n)} \qquad (2.4.14)$$

However, $\tilde{Y}$ was specifically defined as a vector satisfying the two conditions

$$\tilde{Y}Z < 0 \qquad \tilde{Y}\tilde{A} \geq 0$$

Since we also have $\overline{X} \geqq 0$, the left-hand side of equation (2.4.14) is negative. But we proved earlier that $\tilde{Y}A^{(n)} < 0$, and hence it follows that $\alpha > 0$.

$////$

Theorem 2.4.15 Let A be an $m \times n$ matrix and let $Z \in \mathbb{R}^m$. Then the system

$$Z = AX \qquad X \geqq 0$$

has a solution if and only if

$$\mathbb{O}(A) \subseteq \mathbb{O}(Z)$$

By means of a simple device we can also use Farkas' theorem to study the existence of nonnegative solutions of simultaneous linear inequalities. Let A be an $m \times n$ matrix and let $Z \in \mathbb{R}^m$. We wish to determine when the system of simultaneous linear inequalities

$$Z \geqq AX \qquad X \geqq 0 \qquad (2.4.16)$$

has a solution. We can change (2.4.16) into a system of m simultaneous linear equations in $m + n$ unknowns requiring a nonnegative solution by introducing the *slack vector W*. Set

$$W = Z - AX$$

Clearly (2.4.16) is equivalent to finding a vector $\begin{bmatrix} X \\ W \end{bmatrix}$ belonging to $\mathbb{R}^{m+n}$ such that

$$[A \quad I] \begin{bmatrix} X \\ W \end{bmatrix} = Z \qquad \begin{bmatrix} X \\ W \end{bmatrix} \geqq 0 \qquad (2.4.17)$$

Conditions for the existence of such a vector are given by Theorem 2.4.15 We have:

Corollary 2.4.18 Let A be an $m \times n$ matrix and let $Z \in \mathbb{R}^m$. A necessary and sufficient condition that there exist $X \in \mathbb{R}^n$ satisfying

$$Z \geqq AX \qquad X \geqq 0$$

is that

$$YZ \geqq 0$$

for all nonnegative vectors Y belonging to $\mathbb{O}(A)$; that is, every nonnegative element of $\mathbb{O}(A)$ belongs to $\mathbb{O}(Z)$.

PROOF   According to Theorem 2.4.15, the system (2.4.17) has a solution if and only if $\mathbb{O}([A \quad I] \subseteq \mathbb{O}(Z)$. However, Y is contained in $\mathbb{O}([A \quad I])$ if and only if

$$YA \geqq 0 \qquad Y \geqq 0 \qquad ////$$

In the particular situation where we use Corollary 2.4.18 we will be concerned only with nonzero vectors X which satisfy the system of homogeneous inequalities

$$0 \geqq AX \qquad X \geqq 0 \qquad (2.4.19)$$

A nonnegative vector X is a zero vector if and only if the sum of its entries is zero. Thus in order to avoid the trivial solution, $X = 0$, for (2.4.19) we add the condition $E_{(n)} X > 0$ [where $E_{(n)}$ is the n-dimensional row vector all of whose entries are 1]. Clearly if X satisfies (2.4.19), then αX also satisfies (2.4.19) for every nonnegative number α. Hence we may replace the condition $E_{(n)} X > 0$ by the more easily handled condition

$$E_{(n)} X \geqq 1$$

In order to make this condition compatible with the form of Corollary 2.4.18 we write it as

$$-1 \geqq (-E_{(n)})X$$

We have now shown that (2.4.19) has a semipositive (i.e., nonzero nonnegative) solution if and only if the system of inequalities

$$\begin{bmatrix} 0 \\ -1 \end{bmatrix} \geqq \begin{bmatrix} A \\ -E_{(n)} \end{bmatrix} X \qquad X \geqq 0 \qquad (2.4.20)$$

has a solution.

Theorem 2.4.21   Let A be an $m \times n$ matrix. Exactly one of the systems

$$0 \geqq AX \qquad X \geq 0 \qquad (2.4.22)$$

and

$$YA > 0 \qquad Y \geq 0 \qquad (2.4.23)$$

has a solution.

PROOF   Let us assume that both (2.4.22) and (2.4.23) have solutions. From the pair of inequalities

$$0 \geqq AX \qquad Y \geq 0$$

it follows that

$$0 \geqq YAX$$

On the other hand, from the inequalities

$$YA > 0 \qquad X \geq 0$$

it follows that

$$YAX > 0$$

since some entry of X is positive and every entry of YA is positive. Thus we have

$$0 \geq YAX > 0$$

which is a contradiction. Therefore (2.4.22) and (2.4.23) cannot both have solutions.

It remains to show that if (2.4.22) does not have a solution, then (2.4.23) does have a solution. According to the discussion preceding the theorem, the existence of a solution of (2.4.22) is equivalent to the existence of a solution of (2.4.20). Thus let us suppose (2.4.20) does not have a solution. By Corollary 2.4.18 [applied to (2.4.20)] there exists a non-negative vector $Y_0 \in \mathbb{R}_{m+1}$ such that

$$Y_0 \begin{bmatrix} 0 \\ -1 \end{bmatrix} \not\geq 0 \qquad Y_0 \in \mathbb{O}\left(\begin{bmatrix} A \\ -E_{(n)} \end{bmatrix} \right) \qquad (2.4.24)$$

Let

$$Y_0 = [Y, y_{m+1}]$$

where $Y \in \mathbb{R}_m$, $y_{m+1} \in \mathbb{R}$. Then the two parts of (2.4.24) become

$$-y_{m+1} \not\geq 0 \qquad YA - y_{m+1}E_{(n)} \geq 0$$

which in turn can be written

$$y_{m+1} > 0 \qquad YA \geq y_{m+1}E_{(n)}$$

Hence Y satisfies (2.4.23). ////

Corollary 2.4.25 If A is a skew-symmetric matrix, then

$$0 \geq AX \qquad X \geq 0 \qquad (2.4.26)$$

has a solution.

PROOF Suppose no solution of (2.4.26) exists. By Theorem 2.4.21, the system of inequalities

$$YA > 0 \qquad Y \geq 0$$

has a solution. Taking the transpose of the first of these inequalities yields

$$A^T Y^T > 0$$

Since $A^T = -A$, we now have

$$AY^T < 0 \qquad Y^T \geq 0$$

Thus Y^T satisfies (2.4.26).   ////

If we interpret the matrix A of Corollary 2.4.25 as the matrix of a symmetric game, the corollary states that there is an optimal strategy X (up to multiplication by some positive number) for player II.

EXERCISES

1 Which of the following subsets of $\mathbb{R}^3$ are cones?

(a) $\left\{ \begin{bmatrix} x \\ y \\ z \end{bmatrix} \middle| z \geq x^2 + y^2 \right\}$

(b) $\left\{ \begin{bmatrix} x \\ y \\ z \end{bmatrix} \middle| z^2 \geq x^2 + y^2, z \geq 0 \right\}$

(c) $\left\{ \begin{bmatrix} x \\ y \\ z \end{bmatrix} \middle| z = x^2 + y^2 \right\}$

(d) $\left\{ \begin{bmatrix} x \\ y \\ z \end{bmatrix} \middle| z^2 > x^2 + y^2, z \geq 0 \right\}$

(e) $\left\{ \begin{bmatrix} x \\ y \\ z \end{bmatrix} \middle| z^2 > x^2 + y^2 \right\}$

(f) $\left\{ \begin{bmatrix} x \\ y \\ z \end{bmatrix} \middle| z^2 \geq x^2 + y^2, z \leq 0 \right\}$

(g) $\left\{ \begin{bmatrix} x \\ y \\ z \end{bmatrix} \middle| x \geq 0, y > 0 \right\}$

(h) $\left\{ \begin{bmatrix} x \\ y \\ z \end{bmatrix} \middle| x \geq 0, y \geq 0, z = 0 \right\}$

2 Let

$$A = \begin{bmatrix} 1 & -1 \\ 2 & 1 \end{bmatrix} \qquad B = \begin{bmatrix} 0 & -1 \\ 1 & 4 \end{bmatrix}$$

Show that $\mathbb{K}(B) \subseteq \mathbb{K}(A)$:
(a) By expressing each column of B as a nonnegative linear combination of the columns of A.
(b) By sketching $\mathbb{K}(A)$ and $\mathbb{K}(B)$.
Show that $\mathbb{K}(B) \neq \mathbb{K}(A)$.   Find all 2×2 matrices C such that $\mathbb{K}(A) = \mathbb{K}(C)$.

3 Let

$$A = \begin{bmatrix} 1 & -1 \\ 2 & 1 \end{bmatrix} \qquad B = \begin{bmatrix} 0 & -1 \\ 1 & 4 \end{bmatrix}$$

Show that $\mathbb{O}(A) \subseteq \mathbb{O}(B)$:
(a) By using Exercise 2 and Farkas' theorem.
(b) By sketching $\mathbb{O}(A)$ and $\mathbb{O}(B)$.

4 Let

$$A = \begin{bmatrix} 2 & 3 \\ -1 & 1 \end{bmatrix} \qquad Z = \begin{bmatrix} 1 \\ -2 \end{bmatrix}$$

Show that there does not exist a nonnegative vector X such that $Z \geq AX$:
(a) By working directly with the inequalities.
(b) By showing that the condition of Corollary 2.4.18 is contradicted.
(c) By sketching Z and $\mathbb{K}(A)$.
Find a (necessarily not nonnegative) vector X such that $Z \geq AX$. Find
the set of all vectors Z for which there does exist a nonnegative vector X
such that $Z \geq AX$.

5 Let

$$A = \begin{bmatrix} 2 & 7 \\ 1 & -1 \end{bmatrix} \qquad Z = \begin{bmatrix} 1 \\ 3 \end{bmatrix}$$

Show that there does not exist a nonnegative vector X such that $Z = AX$:
(a) By working directly with the simultaneous equations.
(b) By using Farkas' theorem.
(c) By sketching Z and $\mathbb{K}(A)$.
Show that there exists a semipositive vector X such that $Z \geq AX$.

6 Let

$$A = \begin{bmatrix} -3 & 1 & -5 \\ 2 & -1 & 3 \end{bmatrix}$$

Show that

$$Z \geq AX \qquad X \geq 0$$

has a solution for every $Z \in \mathbb{R}^2$. Find $Z \in \mathbb{R}^2$ such that

$$Z = AX \qquad X \geq 0$$

does not have a solution.

7 Prove Corollary 2.4.4.
8 Prove Corollary 2.4.6.
9 Prove that $\mathbb{O}(A)$ is a cone for every matrix A.

10 Prove that the intersection of two cones is a cone.

11 Prove that $\mathbb{O}(A)$ is a vector space if and only if $\mathbb{C}(A) = \mathbb{K}(A)$.

12 Let $A \in \mathbb{R}^n$, $A \neq 0$. Show that every element of $\mathbb{O}(A)$ can be expressed in the form

$$\alpha A^T + X$$

where $\alpha \geq 0$ and X belongs to the row null space of A.

13 Let A be an $m \times n$ matrix. Show that

$$AX \geq Z \qquad X \geq 0$$

has a solution if and only if

$$YZ \geq 0$$

for all $Y \in \mathbb{O}(A)$ such that $Y \leq 0$.

14 Let A be an $m \times n$ matrix. Prove that exactly one of the systems

$$0 = AX \qquad X \geq 0 \qquad \text{and} \qquad YA > 0$$

has a solution.

2.5   THE FUNDAMENTAL THEOREM

In Sec. 2.1 we defined a game to be symmetric if the rules of play and the payoff are the same for each player. Clearly such a game has a skew-symmetric matrix; that is, $A^T = -A$. Two-fingered morra (both as described in Example 2.1.10 and in the variant of Exercise 4) and the duels of Example 2.1.13 (provided the accuracy of the duelists are equal, as in Exercises 9 and 10 of Sec. 2.1) are examples of symmetric games. However, it is not difficult to devise a game in which the rules of play for the two players are completely different but which has the same matrix as one of the symmetric games we have discussed. Clearly the theory of such a game would be identical to the theory of the symmetric game with the same matrix. For this reason we now make the following definition.

Definition 2.5.1 A game is *symmetric* if its matrix is skew-symmetric.

In discussing symmetric games in the past, we have said that certain strategies were optimal for both players. The matrix of a symmetric game is square, and hence the strategies for each player have the same number of entries. Thus P is a strategy vector for player I (that is, $P \in \mathbb{P}_n$) if and only if P^T is a strategy vector for player II (that is, $P^T \in \mathbb{Q}^n$). At this point it is inconvenient that the strategies for player I and for player II have been written as row and column vectors respectively. In discussing symmetric games, we will identify the two strategies $P \in \mathbb{P}_n$ and $P^T \in \mathbb{Q}^n$ for players I and II and speak of them as being the same strategy.

Lemma 2.5.2 The value of a symmetric game (if it exists) is zero. Moreover, a strategy is an optimal strategy for player I if and only if it is an optimal strategy for player II, that is, P is optimal for player I if and only if P^T is optimal for player II.

PROOF Let G be a symmetric game with (skew-symmetric) matrix A and let $(\bar{P}, \bar{Q}, v)$ be a solution of G. Then

$$E(\bar{P}, Q) \geq v \geq E(P, \bar{Q}) \qquad (2.5.3)$$

for all $P \in \mathbb{P}_n$, $Q \in \mathbb{Q}^n$. Since A is skew-symmetric, it follows [from part (c) of Exercise 20 in Sec. 1.7] that

$$E(\bar{P}, Q) = \bar{P}AQ = (\bar{P}AQ)^T = Q^T A^T \bar{P}^T = -Q^T A \bar{P}^T = -E(Q^T, \bar{P}^T)$$
$$E(P, \bar{Q}) = PA\bar{Q} = (PA\bar{Q})^T = \bar{Q}^T A^T P^T = -\bar{Q}^T A P^T = -E(\bar{Q}^T, P^T)$$

Substituting these equations into (2.5.3) yields

$$E(\bar{Q}^T, P^T) \geq -v \geq E(Q^T, \bar{P}^T)$$

for all $Q^T \in \mathbb{P}_n$, $P^T \in \mathbb{Q}^n$. Thus $(\bar{Q}^T, \bar{P}^T, -v)$ is a solution of G. Since the value of a game is unique, we have

$$v = -v$$

and hence $v = 0$. Moreover $\bar{Q}^T$ and $\bar{P}^T$ are optimal strategies for players I and II respectively. ////

Corollary 2.5.4 Let G be a symmetric game with matrix A and let $Q \in \mathbb{Q}^n$. Then Q is an optimal strategy (for either player) in G if and only if

$$0 \geq AQ$$

Theorem 2.5.5 Every symmetric game has a solution.

PROOF By Corollary 2.4.25 the system of inequalities

$$0 \geq AX \qquad X \geq 0$$

has a solution. Since X is semipositive, some positive multiple of X is a strategy vector, say

$$Q = \alpha X$$

Then $\qquad\qquad\qquad\qquad\qquad 0 \geq AQ \qquad ////$

We now have all the machinery needed to prove the fundamental theorem of game theory.

Theorem 2.5.6 Every game has a solution.

PROOF Let A be an $m \times n$ matrix. Since $(\bar{P}, \bar{Q}, v)$ is a solution of the game with matrix A if and only if $(\bar{P}, \bar{Q}, v + \alpha)$ is a solution of the game with matrix $A + \alpha E$ (see Exercise 11 of Sec. 2.3), it is sufficient to prove the theorem for games with positive matrices. Hence we may assume that $A > 0$.

Define B to be the matrix

$$B = \begin{bmatrix} 0 & A & -E^{(m)} \\ -A^T & 0 & E^{(n)} \\ E_{(m)} & -E_{(n)} & 0 \end{bmatrix}$$

where $E_{(k)}$ and $E^{(k)}$ denote the k-dimensional row and column vectors respectively, all of whose entries are 1. B is a skew-symmetric matrix of order $m + n + 1$. Let G' be the game whose matrix is B. G' is symmetric, and, according to Theorem 2.5.5, G' has a solution. Let X be an optimal strategy for the first player in G' and partition X as

$$X = [U \quad V \quad w]$$

where $U \in \mathbb{R}_m$, $V \in \mathbb{R}_n$, $w \in \mathbb{R}$. Since the value of G' is zero and X is an optimal strategy for the first player, we have

$$XB \geq 0$$
$$wE_{(m)} \geq VA^T \geq 0 \qquad (2.5.7)$$

Thus
$$UA \geq wE_{(n)} \geq 0 \qquad (2.5.8)$$
$$VE^{(n)} \geq UE^{(m)} \geq 0 \qquad (2.5.9)$$

We wish to show that U, V, and w are all nonzero. Suppose

$$V = 0$$

Then $VE^{(n)} = 0$, and it follows from (2.5.9) that $UE^{(m)} = 0$. Since U is nonnegative, we have

$$U = 0$$

Then (2.5.8) yields

$$w = 0$$

and consequently $X = 0$. This is a contradiction. Hence

$$V \geq 0$$

Clearly

$$VA^T > 0$$

since A is positive and it now follows from (2.5.7) that

$$w > 0$$

Finally, from (2.5.8) we obtain

$$U \geq 0$$

Let us rewrite (2.5.7) as

$$wE^{(m)} \geq AV^T \qquad (2.5.10)$$

Multiply (2.5.8) and (2.5.10) on the right by V^T and on the left by U respectively. Then

$$wUE^{(m)} \geq UAV^T \geq wE_{(n)}V^T$$

Combining this with (2.5.9), we have

$$UE^{(m)} = VE^{(n)} > 0$$

Denote this common value by λ and set

$$\bar{P} = \frac{1}{\lambda} U \qquad \bar{Q} = \frac{1}{\lambda} V^T$$

Clearly $\bar{P} \in \mathbb{P}_m$, $\bar{Q} \in \mathbb{Q}^n$. Moreover,

$$\bar{P}A = \frac{1}{\lambda} UA \geq \frac{w}{\lambda} E_{(n)}$$

$$A\bar{Q} = \frac{1}{\lambda} AV^T \leq \frac{w}{\lambda} E^{(m)}$$

Thus $(\bar{P}, \bar{Q}, w/\lambda)$ is a solution of the game with matrix A. (See Exercise 3 of Sec. 2.3.) ////

Although Theorem 2.5.6 guarantees the existence of a solution of a game, it does not provide us with a means for finding it. In the next two sections we will turn our attention to the problem of actually obtaining solutions or, at least, to the task of getting useful information about solutions.

EXERCISES

1 Let $(\bar{P}, \bar{Q}, v)$, $v \geq 0$, be a solution of the game with matrix A. Show that

$$X = \left[\frac{1}{2+v} \bar{P} \quad \frac{1}{2+v} \bar{Q} \quad \frac{v}{2+v} \right]$$

is an optimal strategy in the game with matrix

$$B = \begin{bmatrix} 0 & A & -E^{(m)} \\ -A^T & 0 & E^{(n)} \\ E_{(m)} & -E_{(n)} & 0 \end{bmatrix}$$

2 Solve the game whose matrix is

$$\begin{bmatrix} 0 & 0 & 4 & 2 & -1 \\ 0 & 0 & 3 & 5 & -1 \\ -4 & -3 & 0 & 0 & 1 \\ -2 & -5 & 0 & 0 & 1 \\ 1 & 1 & -1 & -1 & 0 \end{bmatrix}$$

Exercises 3 to 7 describe the successive parts of a proof of Theorem 2.5.6 using Theorem 2.5.5 and the symmetrization described in Example 2.1.14. Let G be a game with m × n matrix A and let G' be the symmetrization of G described in Sec. 2.1.

3 Show that if S_i and T_j are optimal pure strategies for players I and II respectively in G, then

$$S'_{ij} = \begin{cases} \text{when playing as player I, play } S_i \\ \text{when playing as player II, play } T_j \end{cases}$$

is an optimal strategy in G'.

4 Let S'_{ij} of Exercise 3 be an optimal pure strategy in G'. Show that S_i, T_j are optimal strategies for players I and II respectively in G.

5 Let

$$P = [p_i] \in \mathbb{P}_m \qquad Q = [q_j] \in \mathbb{Q}^n$$

and let X be the strategy in G' in which S'_{ij} is played with probability $p_i q_j$. We say that X is the *composition* of P and Q and write

$$X = P \times Q$$

Verify that X is a strategy vector in G'. Show that for $P_1, P_2 \in \mathbb{P}_m$, Q_1, $Q_2 \in \mathbb{Q}^n$, we have

$$E(P_1 \times Q_1, P_2 \times Q_2) = \tfrac{1}{2}[E(P_1, Q_2) - E(P_2, Q_1)]$$

6 Let X be the strategy in G' in which S'_{ij} is played with probability x_{ij}. Define $P = [p_i]$ and $Q = [q_j]$ by

$$p_i = \sum_{j=1}^{n} x_{ij} \qquad i = 1, 2, \ldots, m$$

$$q_j = \sum_{i=1}^{m} x_{ij} \qquad j = 1, 2, \ldots, n$$

We say that X has been *decomposed* into P and Q. Show that

$$E(X, Y) = E(P \times Q, Y)$$

for every strategy Y for player II'. (Note that if X decomposes into P and Q, it does not follow that $X = P \times Q$.)

7 Prove:

(a) If $(\bar{P}, \bar{Q}, v)$ is a solution of G, then $\bar{P} \times \bar{Q}$ is an optimal strategy in G'.

(b) If $\bar{X}$ is an optimal strategy in G' and $\bar{X}$ decomposes into $\bar{P}$ and $\bar{Q}$, then $(\bar{P}, \bar{Q}, E(\bar{P}, \bar{Q}))$ is a solution of G.

(c) Every game has a solution.

8 Show that the matrix

$$B = \begin{bmatrix} 0 & -3 & 2 & -1 \\ 3 & 0 & 0 & -3 \\ -2 & 0 & 0 & 2 \\ 1 & 3 & -2 & 0 \end{bmatrix}$$

is the symmetrization of a game with a 2×2 matrix. Use this fact to find a solution of the game whose matrix is B. Show that $[\frac{1}{5}\ \frac{1}{5}\ \frac{2}{5}\ \frac{1}{5}]$ is a solution which is not of the form $\bar{P} \times \bar{Q}$.

9 Show that the variant of two-fingered morra in Exercise 4 of Sec. 2.1 is not the symmetrization of any game having a 2×2 matrix (with the order of the strategies in G' given by $S'_{11}, S'_{12}, S'_{21}, S'_{22}$):

(a) By working with the entries of the matrix directly.

(b) By decomposing the solution given in Exercise 4 of Sec. 2.1 into strategies P and Q and then finding $P \times Q$.

10 Is the variant of two-fingered morra of Example 2.1.10 the symmetrization of a game having a 2×2 matrix?

11 Show that if G' is the symmetrization of the game with matrix A, then G' is the symmetrization of the game with matrix $A + \beta E$ for every β.

2.6 RELATIONS AMONG GAMES

Now that the existence of solutions of a game has been established, we turn our attention to the problem of obtaining information about the sets of optimal strategies for each player. It is often possible to find solutions for a given game G by first considering some simpler game G' which has the same value as G and in which the optimal strategies for each player yield optimal strategies for the corresponding player in G. The game G' is formed from G by eliminating those pure strategies in G which the players will never use. The dimensions of the resulting matrix A' (of G') are then smaller than the dimensions of the matrix A (of G).   In several of the examples of Sec. 2.1, e.g., bluffing, three-card poker, duels, we decided (without any real justification) to ignore certain pure strategies which seemed obviously bad, and in this way we reduced the dimensions of the matrix.   One way of finding pure strategies which are never used is given by:

Lemma 2.6.1   Let $\bar{P}$ be an optimal strategy for player I in a game with matrix A and value v.   If, for some l, the lth entry of $\bar{P}A$ is greater than v, that is, if

$$\bar{P}A^{(l)} > v$$

then the lth entry of every optimal strategy for player II is zero.

   Similarly, if $\bar{Q}$ is an optimal strategy for player II, and if for some k the kth entry of $A\bar{Q}$ is less than v, that is, if

$$A_{(k)}\bar{Q} < v$$

then the kth entry of every optimal strategy for player I is zero.

PROOF   It is sufficient to prove the first half of the lemma.   Set

$$\bar{P}A = [w_1 \quad w_2 \quad \cdots \quad w_n]$$

and let $\bar{Q} = [q_j]$ be an optimal strategy for player II.   Since $(\bar{P}, \bar{Q}, v)$ is a solution of the game, we have

$$w_j = \bar{P}A^{(j)} = \bar{P}AI^{(j)} = E(\bar{P}, I^{(j)}) \geq v \qquad j = 1, 2, \ldots, n$$
$$w_l = \bar{P}A^{(l)} > v$$

and $$v = E(\bar{P}, \bar{Q}) = \bar{P}A\bar{Q} = \sum_{j=1}^{n} w_j q_j$$

Set

$$e_j = w_j - v \qquad j = 1, 2, \ldots, n$$

Then $\sum_{j=1}^{n} e_j q_j = \sum_{j=1}^{n} (w_j - v)q_j = \sum_{j=1}^{n} w_j q_j - v\sum_{j=1}^{n} q_j = v - v = 0$ (2.6.2)

Since $e_j \geq 0$ and $q_j \geq 0$, it follows from equation (2.6.2) that

$$e_1 q_1 = e_2 q_2 = \cdots = e_n q_n = 0$$

Clearly, if $e_l > 0$, then $q_l = 0$.   ////

Definition 2.6.3   A pure strategy is said to be *essential* if it is played with positive probability in some optimal strategy.   A pure strategy which is never played, i.e., played with probability zero in every optimal strategy, is said to be *inessential*.

The pure strategy $I^{(l)}$ for player II is essential when player II has at least one optimal strategy whose lth component is not zero. Lemma 2.6.1 states that if $\bar{P}A^{(l)} > v$ for some optimal strategy $\bar{P}$ for player I, then $I^{(l)}$ is inessential. We wish to prove the converse of Lemma 2.6.1. First it is necessary to transform our game into one which has the same sets of optimal strategies and which, in addition, has the value zero. If the value of the game with matrix A is v, the value of the game with matrix $A - vE$ is zero. Moreover the two games have the same sets of optimal strategies (see Exercise 11 of Sec. 2.3).

Theorem 2.6.4 If $\bar{P}A^{(l)} = v$ for every optimal strategy $\bar{P}$ for player I, then $I^{(l)}$ is essential for player II. Similarly, if $A_{(k)}\bar{Q} = v$ for every optimal strategy $\bar{Q}$ for player II, then $I_{(k)}$ is essential for player I.

PROOF It will be sufficient to prove the first assertion of the theorem. Suppose $\bar{P}A^{(l)} = v$ for every optimal strategy $\bar{P}$ According to the discussion above, we may assume that $v = 0$.

Let us assume that $I^{(l)}$ is not an essential strategy for player II. This assumption is equivalent to stating that there does not exist a vector $X \in \mathbb{R}^n$ such that

$$0 \geqq AX \qquad I_{(l)}X \geqq 1 \qquad X \geq 0 \qquad (2.6.5)$$

because some positive multiple of a vector X satisfying (2.6.5) would be an optimal strategy for player II with a nonzero lth entry. The inequalities of (2.6.5) can be expressed as

$$\begin{bmatrix} 0 \\ -1 \end{bmatrix} \geqq \begin{bmatrix} A \\ -I_{(l)} \end{bmatrix} X \qquad X \geq 0 \qquad (2.6.6)$$

Since $I^{(l)}$ is assumed to be inessential, we require that inequalities (2.6.6) do not have a solution. By Corollary 2.4.18, there exists a nonnegative vector $Y \in \mathbb{R}_{m+1}$ such that

$$Y \in \mathbb{O}\left(\begin{bmatrix} A \\ -I_{(l)} \end{bmatrix}\right) \qquad Y\begin{bmatrix} 0 \\ -1 \end{bmatrix} < 0 \qquad (2.6.7)$$

Let $Y = [\tilde{Y} \quad y_{m+1}]$, where $\tilde{Y} \in \mathbb{R}_m$, $y_{m+1} \in \mathbb{R}$. Then (2.6.7) becomes

$$\tilde{Y}A - y_{m+1}I_{(l)} \geqq 0 \qquad -y_{m+1} < 0$$

Hence $\qquad\qquad\qquad \tilde{Y}A \geqq y_{m+1}I_{(l)} \qquad y_{m+1} > 0 \qquad (2.6.8)$

Some positive multiple of $\tilde{Y}$ is a strategy vector, say $\bar{P} = \alpha\tilde{Y}$. By (2.6.8)

$$\bar{P}A \geqq \alpha y_{m+1}I_{(l)} \geqq 0$$

and hence $\bar{P}$ is an optimal strategy for player I. Moreover,

$$\bar{P}A^{(l)} \geq \alpha y_{m+1} I_{(l)} I^{(l)} = \alpha y_{m+1} > 0 = v$$

This contradicts the hypothesis of the theorem. Consequently the assumption that $I^{(l)}$ is not essential is false.   ////

For convenience, we will say that the row $A_{(k)}$ (or the column $A^{(l)}$) of A is *essential* or *inessential* as the pure strategy $I_{(k)}$ (or $I^{(l)}$) is essential or inessential.

Definition 2.6.9 Let A be an $m \times n$ matrix. The matrix which results from deleting all the inessential rows and columns of A is called the *essential submatrix* of A and denoted by A^*. We will denote the dimensions of A^* by m^* and n^*. The game G^* whose matrix is A^* is called the *essential subgame* of G.

Let P be a strategy for player I in G. If the kth entry of P is deleted for all values of k for which $I_{(k)}$ is inessential, then the resulting vector, which we denote by P^*, is an m^*-dimensional strategy vector since all the deleted entries are zeros. Similarly, for a strategy Q for player II, the vector formed by deleting the lth entry whenever $I^{(l)}$ is inessential is an n^*-dimensional strategy vector and is denoted by Q^*. Clearly:

Theorem 2.6.10 If $(\bar{P}, \bar{Q}, v)$ is a solution of G, then $(\bar{P}^*, \bar{Q}^*, v)$ is a solution of G^*.

PROOF From $\bar{P}A \geq vE_{(n)}$ it follows that $\bar{P}^*A^* = vE_{(n^*)}$. Similarly, from $A\bar{Q} \leq vE^{(m)}$ it follows that $\bar{A}^*\bar{Q}^* = vE^{(m^*)}$.   ////

Theorem 2.6.10, in addition to reducing the size of the matrix under consideration, also replaces a system of linear inequalities by a system of linear equations. There still remains the difficult task of finding the essential submatrix of a given matrix.

EXAMPLE 2.6.11   Find the essential submatrix of

$$A = \begin{bmatrix} 6 & 8 & 9 & 10 \\ 8 & 4 & 3 & 0 \end{bmatrix}$$

We can use the graphical method to find all of player I's optimal strategies. **Let**

$$P = [p \quad 1-p] \qquad 0 \leq p \leq 1$$

A simple computation shows that

$$E(P, I^{(1)}) = 8 - 2p$$
$$E(P, I^{(2)}) = 4 + 4p$$
$$E(P, I^{(3)}) = 3 + 6p$$
$$E(P, I^{(4)}) = 10p$$

Sketching these equations (in Fig. 2.6.12), we see that player I has the unique optimal strategy

$$\bar{P} = [\tfrac{2}{3} \quad \tfrac{1}{3}] \qquad v = \tfrac{20}{3}$$

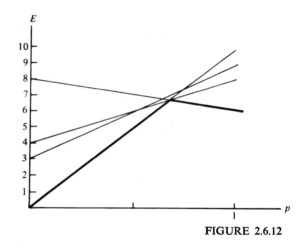

FIGURE 2.6.12

Both rows of A are essential. It follows from the equation

$$\bar{P}A = [\tfrac{20}{3} \quad \tfrac{20}{3} \quad \tfrac{21}{3} \quad \tfrac{20}{3}]$$

that the third column of A is inessential. Moreover, by the first part of Theorem 2.6.4 the other three columns of A are essential. Thus

$$A^* = \begin{bmatrix} 6 & 8 & 10 \\ 8 & 4 & 0 \end{bmatrix}$$

To obtain the optimal strategies for player II (or, rather, for player II*) we solve the equation

$$A^*\bar{Q}^* = \begin{bmatrix} \tfrac{20}{3} \\ \tfrac{20}{3} \end{bmatrix}$$

This yields

$$\bar{Q}^* = \begin{bmatrix} q_1 \\ \frac{5}{3} - 2q_1 \\ q_1 - \frac{2}{3} \end{bmatrix} \quad \frac{2}{3} \leq q_1 \leq \frac{5}{6}$$

and a computation shows that every vector

$$\bar{Q} = \begin{bmatrix} q_1 \\ \frac{5}{3} - 2q_1 \\ 0 \\ q_1 - \frac{2}{3} \end{bmatrix} \quad \frac{2}{3} \leq q_1 \leq \frac{5}{6}$$

is an optimal strategy for player II.   ////

Unfortunately, the converse of Theorem 2.6.10 is not generally true (see Example 2.6.21). It is possible that A^* will have optimal strategies which do not arise from optimal strategies of A. However, we will show in Theorem 2.6.16 that every optimal strategy in G^* can be expressed as a linear combination of optimal strategies in G (after deleting zeros corresponding to inessential strategies). First we need:

Lemma 2.6.13 There exists an optimal strategy $\bar{P}_0$ for player I such that

1 $\bar{P}_0^* > 0$ and
2 $\bar{P}_0 A^{(l)} > v$

whenever $I^{(l)}$ is inessential.
There exists an optimal strategy $\bar{Q}_0$ for player II such that

1 $\bar{Q}_0^* > 0$ and
2 $A_{(k)} \bar{Q}_0 < v$

whenever $I_{(k)}$ is inessential.

PROOF   It is sufficient to prove the first part of the lemma. By Definition 2.6.3, for each of the m^* essential strategies $I_{(i)}$ for player I, there exists an optimal strategy $\bar{P}_i$ whose ith entry is positive. Also, by Theorem 2.6.4, for each of the $n - n^*$ inessential strategies $I^{(j)}$ for player II, there exists an optimal strategy $\bar{P}_j$ for player I such that $\bar{P}_j A^{(j)} > v$. We will define $\bar{P}_0$ as a convex combination of these optimal strategies with positive coefficients, say

$$\bar{P}_0 = \frac{1}{2m^*} \sum_{i=1}^{m^*} \bar{P}_i + \frac{1}{2(n - n^*)} \sum_{j=1}^{n - n^*} \bar{P}_j$$

Clearly $\bar{P}_0$ is an optimal strategy for player I, $\bar{P}_0^* > 0$, and $\bar{P}_0 A^{(l)} > v$ for each l for which $I^{(l)}$ is inessential.   ////

We call the strategies $\bar{P}_0$ and $\bar{Q}_0$ satisfying the conditions of Lemma 2.6.13 *central optimal strategies* for players I and II respectively.   Because the positive vectors $\bar{P}_0^*$ and $\bar{Q}_0^*$ are optimal strategies in G^*, it follows that G^* has no inessential strategies and therefore the essential submatrix of A^* is A^*.   Hence

$$\bar{P}A^* = vE_{(n^*)} \qquad A^*\bar{Q} = vE^{(m^*)} \qquad (2.6.14)$$

for all optimal strategies $\bar{P}$, $\bar{Q}$ of G^*, whether or not they arose from optimal strategies of A.

After all the solutions of the essential subgame of a given game have been found [by solving the equations of (2.6.14) for $\bar{P}$ and $\bar{Q}$], it is still necessary to determine the solutions of the original game.   In order to do this, we let $\mathbb{U}$ be the subspace of $\mathbb{R}^{n^*}$ spanned by the optimal strategies for player II* in G^* and $\mathbb{V}$ be the subspace of $\mathbb{R}^n$ spanned by the optimal strategies for player II in G.   For each inessential strategy $I^{(l)}$ of G the lth entry of every element of $\mathbb{V}$ is zero. Thus we may set

$$\mathbb{V}^* = \{Y^* \mid Y \in \mathbb{V}\}$$

Since optimal strategies of G yield optimal strategies of G^*, we have

$$\mathbb{V}^* \subseteq \mathbb{U}$$

Moreover, we can prove

Theorem 2.6.15 $\qquad\qquad\qquad\qquad \mathbb{V}^* = \mathbb{U}$

PROOF   We may assume that $v = 0$.   Then every optimal strategy $\bar{Q}$ for player II* in G^* satisfies

$$A^*\bar{Q} = 0$$

that is, $\bar{Q} \in \mathbb{N}(A^*)$.   Hence we have

$$\mathbb{V}^* \subseteq \mathbb{U} \subseteq \mathbb{N}(A^*) \subseteq \mathbb{R}^{n^*}$$

In order to complete the proof it is sufficient to show that $\mathbb{N}(A^*) \subseteq \mathbb{V}^*$. Let $\mathbb{W}$ be the subspace of $\mathbb{R}^n$ spanned by the essential strategies for player II in G.   Clearly

$$\mathbb{V} \subseteq \mathbb{W}$$

and

$$\mathbb{W}^* = \mathbb{R}^{n^*}$$

Thus, for $X \in \mathbb{N}(A^*)$, there exists $Z \in \mathbb{W}$ such that

$$X = Z^*$$

It remains only to show that Z belongs to $\mathbb{V}$ (from which it follows that Z^* belongs to $\mathbb{V}^*$). Let $\bar{Q}_0$ be a central optimal strategy for player II and set

$$Y = Z + \alpha \bar{Q}_0 \quad (2.6.16)$$

where α is some (as yet undetermined) positive number. Clearly $Y \in \mathbb{W}$. Also, for sufficiently large values of α,

$$Y^* > 0 \quad (2.6.17)$$

since $\bar{Q}_0$ has a positive entry in each of player II's essential positions. When $I_{(k)}$ is an essential strategy for player I, we have

$$A_{(k)}Z = (A_{(k)})^*Z^* = (A_{(k)})^*X = 0$$
$$A_{(k)}\bar{Q}_0 = v = 0$$

since $(A_{(k)})^*$ is one of the rows of A^* (though not necessarily the kth row). Consequently

$$A_{(k)}Y = 0 \quad (2.6.18)$$

When $I_{(k)}$ is an inessential strategy for player I,

$$A_{(k)}\bar{Q}_0 < v = 0$$

and hence, for sufficiently large values of α, we have

$$A_{(k)}Y < 0 \quad (2.6.19)$$

We now fix α to be some number large enough so that (2.6.17) and all the inequalities of (2.6.19) are satisfied. Then we can combine (2.6.17), (2.6.18), and (2.6.19) into

$$AY \leq 0 = vE^{(m)} \qquad Y \geq 0$$

and some positive multiple of Y is a (central) optimal strategy for player II, say

$$\beta Y = \bar{Q}_1$$

Substituting for Y in equation (2.6.16) and solving for Z yields

$$Z = \frac{1}{\beta}\bar{Q}_1 - \alpha \bar{Q}_0 \in \mathbb{V}$$

Consequently $X = Z^* \in \mathbb{V}^*$ ////

In Theorem 2.6.15 we proved that when $v = 0$, we have

$$\mathbb{V}^* = \mathbb{U} = \mathbb{N}(A^*)$$

Similarly, if $\mathbb{Z}$ is the subspace of $\mathbb{R}_m$ spanned by the optimal strategies for player I, then $\mathbb{Z}^*$ is the row null space of A^*. We can now conclude that

Theorem 2.6.20   For every game

$$n^* - \dim \mathbb{V} = m^* - \dim \mathbb{Z}$$

Moreover, when $v = 0$,

$$n^* - \dim \mathbb{V} = r(A^*) = m^* - \dim \mathbb{Z}$$

PROOF   The first part of the theorem follows from the second. Hence, assume $v = 0$. Since only zero entries are deleted in passing from $\mathbb{V}$ to $\mathbb{V}^*$ it follows that

$$\dim \mathbb{V} = \dim \mathbb{V}^* = \dim \mathbb{N}(A^*) = v(A^*)$$

But by Theorem 1.6.1

$$r(A^*) = n^* - v(A^*) = n^* - \dim \mathbb{V}$$

Similarly, $$r(A^*) = m^* - \dim \mathbb{Z} \qquad ////$$

Note that Theorem 2.6.20 does not tell us when we have found all the optimal strategies of a game. Even when m^*, n^*, $\dim \mathbb{V}$, and $\dim \mathbb{Z}$ are known, Theorem 2.6.20 indicates only the number of linearly independent optimal strategies we can expect for each player. In the next section a method is given for finding all optimal strategies.

EXAMPLE 2.6.21   Find all solutions of the game with matrix

$$A = \begin{bmatrix} 1 & 1 \\ 4 & -3 \end{bmatrix}$$

Using the graphical method, we can show that the value of the game is 1, player I has the unique optimal strategy

$$\bar{P} = [1 \quad 0]$$

and every vector

$$\bar{Q} = \begin{bmatrix} q \\ 1 - q \end{bmatrix} \qquad 0 \leq q \leq \tfrac{4}{7}$$

is an optimal strategy for player II.   The essential submatrix of A is

$$A^* = [1 \quad 1]$$

Hence, $$m = n = 2 \qquad m^* = 1 \qquad n^* = 2$$

Note that every strategy for player II* is optimal in G^* but not every optimal strategy for player II* is an optimal strategy for player II.   On the other hand (using the notation of Theorems 2.6.15 and 2.6.20),

$$\mathbb{U} = \mathbb{V} = \mathbb{V}^* = \mathbb{R}^2 \qquad \mathbb{Z}^* = \mathbb{R}$$

$\mathbb{Z}$ is the one-dimensional subspace of $\mathbb{R}_2$ which has $\bar{P} = [1 \quad 0]$ as a basis.   Thus

$$n^* - \dim \mathbb{V} = m^* - \dim \mathbb{Z} = 0$$

Clearly every optimal strategy for player II can be expressed as a convex combination of the two optimal strategies

$$\bar{Q}_1 = \begin{bmatrix} \frac{4}{7} \\ \frac{3}{7} \end{bmatrix} \qquad \bar{Q}_2 = \begin{bmatrix} 0 \\ 1 \end{bmatrix} \qquad ////$$

EXERCISES

1 Find the essential submatrices of the matrices in Examples 2.1.1, 2.1.2, and 2.1.4.

2 Examine Example 2.1.10.   Show that the strategies which were eliminated (S_3 and S_4) are actually inessential.   Are any of the other strategies inessential?

3 Change the rules in Example 2.1.10 so that the winner wins an amount equal to the total number of fingers put out.   (See Exercise 4 of Sec. 2.1.) Find the essential strategies.

4 Find all solutions of

(a) $\begin{bmatrix} 2 & 4 \\ 2 & -1 \end{bmatrix}$
(b) $\begin{bmatrix} 2 & -1 & -4 \\ -3 & -1 & 5 \end{bmatrix}$

Find all solutions of the essential subgame and verify Theorems 2.6.15 and 2.6.20.

5 Let

$$A = \begin{bmatrix} 0 & 1 & 2 & 3 & 4 \\ 1 & 0 & 1 & 2 & 3 \\ 2 & 1 & 0 & 1 & 2 \\ 3 & 2 & 1 & 0 & 1 \\ 4 & 3 & 2 & 1 & 0 \end{bmatrix}$$

Given that $[\frac{1}{2} \ \ 0 \ \ 0 \ \ 0 \ \ \frac{1}{2}]$ is an optimal strategy for player I, find the essential submatrix and a maximal set of linearly independent optimal strategies for each player.

6   Prove:

(a) The essential submatrix of a symmetric game is skew-symmetric.

(b) $(A + vE)^* = A^* + vE$.

7   Let A be a matrix in which every column is essential.   Show that if Q is a strategy vector for which

$$AQ = wE^{(m)}$$

then $v_A = w$ and Q is an optimal strategy for player II.   Show that this result does not hold if one drops the hypothesis that every column of A is essential.

For X, $Y \in \mathbb{R}^m$ (or $\mathbb{R}_n$) we say that X dominates Y if $X \geqq Y$ and X strictly dominates Y if $X > Y$.

8   Show that if $A_{(k)}$ strictly dominates $A_{(i)}$, then $I_{(i)}$ is an inessential strategy for player I.   Formulate and prove the corresponding theorem for player II. (*Warning:* See Exercise 13.)

9   Show that if $A_{(i)}$ is strictly dominated by a convex combination of the other rows of A, then $I_{(i)}$ is inessential.

10   Show that if the ith row of A strictly dominates a convex combination of the other rows of A, then at least one of these rows is inessential.

11   Let A be a matrix in which $A_{(k)}$ dominates $A_{(i)}$, Let $\tilde{A}$ be the matrix formed from A by deleting $A_{(i)}$, and let $\tilde{G}$ be the game with matrix $\tilde{A}$.   Prove:

(a) $v_G = v_{\tilde{G}}$.

(b) The optimal strategies for player II in G are identical with the optimal strategies for player $\tilde{\text{II}}$ in $\tilde{G}$.

(c) Every optimal strategy for player $\tilde{\text{I}}$ in $\tilde{G}$ can be transformed into an optimal strategy for player I in G by inserting a zero entry at the ith position.

Formulate and prove the corresponding theorem when $A^{(l)}$ dominates $A^{(j)}$.

12   Find a solution of the games

$$(a) \quad \begin{bmatrix} 0 & 2 & -1 \\ 1 & 2 & 1 \\ -1 & 4 & 0 \end{bmatrix} \qquad (b) \quad \begin{bmatrix} -1 & -1 & -2 \\ -2 & -1 & -1 \\ -4 & 0 & 1 \\ 13 & 0 & -3 \end{bmatrix}$$

$$(c) \begin{bmatrix} -1 & 0 & -3 & -1 \\ 1 & -2 & 3 & -2 \\ -1 & 0 & 1 & -4 \\ -2 & 2 & 0 & 1 \end{bmatrix}$$

using the concept of dominance.

13 Construct a 2×2 matrix in which $A_{(2)}$ dominates $A_{(1)}$ and $I_{(1)}$ is essential.

2.7   EXTREME OPTIMAL STRATEGIES

In Sec. 2.6 we showed that if $(\bar{P}, \bar{Q}, v)$ is any solution of the game whose matrix is A, then

$$\bar{P}^*A^* = vE_{(n^*)} \qquad A^*\bar{Q}^* = vE^{(m^*)} \qquad (2.7.1)$$

When the essential submatrix A^* of A is nonsingular (of order n^*), equations (2.7.1) can be solved uniquely for $\bar{P}^*$ and $\bar{Q}^*$,

$$\bar{P}^* = vE_{(n^*)}(A^*)^{-1} \qquad \bar{Q}^* = v(A^*)^{-1}E^{(n^*)} \qquad (2.7.2)$$

It then follows from equations (2.7.2) that G^* (and hence G) has only one solution.   Further, it follows from equations (2.7.2) that

$$v \neq 0$$

and that $\qquad\qquad 1 = \bar{P}^*E^{(n^*)} = vE_{(n^*)}(A^*)^{-1}E^{(n^*)}$

Hence we have explicit formulas for v, $\bar{P}^*$, $\bar{Q}^*$, namely,

$$v = \frac{1}{E_{(n^*)}(A^*)^{-1}E^{(n^*)}}$$

$$\bar{P}^* = \frac{E_{(n^*)}(A^*)^{-1}}{E_{(n^*)}(A^*)^{-1}E^{(n^*)}} \qquad \bar{Q}^* = \frac{(A^*)^{-1}E^{(n^*)}}{E_{(n^*)}(A^*)^{-1}E^{(n^*)}} \qquad (2.7.3)$$

When A^* is singular, equations (2.7.1) cannot be solved explicitly for $\bar{P}^*$ and $\bar{Q}^*$ and the results in the paragraph above are no longer valid.   We will, however, prove that when $v \neq 0$, relations such as (2.7.3) exist for some (special) optimal strategies.   Moreover the remaining optimal strategies can easily be obtained from these.

Definition 2.7.4   An optimal strategy, which is an extreme point of the set of optimal strategies, i.e., which cannot be written as a convex combination of other optimal strategies, is an *extreme optimal strategy*.

When a player has a unique optimal strategy, it is clearly an extreme optimal strategy. It can be shown (see Theorem 2.7.23) that extreme optimal strategies always exist.

Theorem 2.7.5 Let $(\bar{P}, \bar{Q}, v)$, $v \neq 0$, be a solution of the game with $m \times n$ matrix A. Further, let $\bar{P}$ and $\bar{Q}$ be extreme optimal strategies. Then there exists a nonsingular submatrix $A^{\cdot}$ (having order $n^{\cdot}$) of A formed by deleting certain of the rows and columns of A such that

$$v = \frac{1}{E_{(n^{\cdot})}(A^{\cdot})^{-1}E^{(n^{\cdot})}}$$

and $\bar{P}$ and $\bar{Q}$ are given (up to zero entries in the deleted rows and columns) by

$$\bar{P}^{\cdot} = \frac{E_{(n^{\cdot})}(A^{\cdot})^{-1}}{E_{(n^{\cdot})}(A^{\cdot})^{-1}E^{(n^{\cdot})}} \qquad \bar{Q}^{\cdot} = \frac{(A^{\cdot})^{-1}E^{(n^{\cdot})}}{E_{(n^{\cdot})}(A^{\cdot})^{-1}E^{(n^{\cdot})}}$$

PROOF   It will simplify the proof greatly to break it up into a series of lemmas. First let us reorder the rows and columns of A (and hence of $\bar{P}$ and $\bar{Q}$) so that

$$\bar{P} = [p_1 \quad p_2 \quad \cdots \quad p_r \quad 0 \quad 0 \quad \cdots \quad 0] \qquad \begin{matrix} p_i > 0 \\ i = 1, 2, \ldots, r \end{matrix}$$

$$\bar{Q} = \begin{bmatrix} q_1 \\ q_2 \\ \vdots \\ q_s \\ 0 \\ 0 \\ \vdots \\ 0 \end{bmatrix} \qquad \begin{matrix} q_j > 0 \\ j = 1, 2, \ldots, s \end{matrix}$$

Set

$$\bar{P}A = Y = [y_i] \geq vE_{(n)} \qquad A\bar{Q} = Z = [z_j] \leq vE^{(m)}$$

Since the first r pure strategies for player I (after the reordering) are all essential, it follows that

$$z_1 = z_2 = \cdots = z_r = v \qquad z_{r+1} \leq v, z_{r+2} \leq v, \ldots, z_m \leq v$$

Similarly

$$y_1 = y_2 = \cdots = y_s = v \qquad y_{s+1} \geq v, y_{s+2} \geq v, \ldots, y_n \geq v$$

We now reorder the last $n - r$ rows of A (leaving the first r rows of A fixed) so that

$$z_1 = z_2 = \cdots = z_r = z_{r+1} = \cdots = z_{r+\bar{r}} = v$$

and $\qquad z_{r+\bar{r}+1} < v, \qquad z_{r+\bar{r}+2} < v, \ldots, z_m < v$

Similarly we reorder the last $n - s$ columns of A (leaving the first s columns of A fixed) so that

$$y_1 = y_2 = \cdots = y_s = y_{s+1} = \cdots = y_{s+\bar{s}} = v$$

and $\qquad y_{s+\bar{s}+1} > v, \qquad y_{s+\bar{s}+2} > v, \qquad \ldots, \qquad y_n > v$

Note that the last $m - r - \bar{r}$ rows and $n - s - \bar{s}$ columns of A are all inessential.

We now partition $\bar{P}$, $\bar{Q}$, Y, Z, and A as

$$A = \begin{bmatrix} A_{11} & A_{12} & A_{13} \\ A_{21} & A_{22} & A_{23} \\ A_{31} & A_{32} & A_{33} \end{bmatrix} \begin{matrix} r \\ \bar{r} \\ m-r-\bar{r} \end{matrix} \qquad \bar{Q} = \begin{bmatrix} \bar{Q}_1 \\ 0 \\ 0 \end{bmatrix} \qquad Z = \begin{bmatrix} Z_1 \\ Z_2 \\ Z_3 \end{bmatrix} \begin{matrix} r \\ \bar{r} \\ m-r-\bar{r} \end{matrix}$$

with column labels $s \quad \bar{s} \quad n-s-\bar{s}$ over A, and 1 over $\bar{Q}$ and Z.

$$\bar{P} = \begin{bmatrix} \bar{P}_1 & 0 & 0 \end{bmatrix} 1$$
$$Y = \begin{bmatrix} Y_1 & Y_2 & Y_3 \end{bmatrix} 1$$

where A_{11} is an $r \times s$ matrix, A_{12} is an $r \times \bar{s}$ matrix, A_{21} is an $\bar{r} \times s$ matrix, $\ldots$, $\bar{Q}_1$ and Z_1 are $r \times 1$ matrices, $\ldots$, and Y_3 is a $1 \times (n - s - \bar{s})$ matrix.[1] Set

$$\tilde{A} = \begin{bmatrix} A_{11} & A_{12} \\ A_{21} & A_{22} \end{bmatrix}$$

and note that

$$[\bar{P}_1 \quad 0]\tilde{A} = [Y_1 \quad Y_2] = vE_{(s+\bar{s})} \qquad (2.7.6)$$

$$\tilde{A}\begin{bmatrix} \bar{Q}_1 \\ 0 \end{bmatrix} = \begin{bmatrix} Z_1 \\ Z_2 \end{bmatrix} = vE^{(r+\bar{r})} \qquad (2.7.7)$$

$$Y_3 > vE_{(n-s-\bar{s})} \qquad Z_3 < vE^{(m-r-\bar{r})}$$

If $\tilde{A}$ is nonsingular, then it will serve for $A^{\cdot}$ and equations (2.7.6) and (2.7.7) can be solved for

$$\bar{P}^{\cdot} = [\bar{P}_1 \quad 0] \qquad \bar{Q}^{\cdot} = \begin{bmatrix} \bar{Q}_1 \\ 0 \end{bmatrix}$$

[1] Some of the numbers $\bar{r}$, $\bar{s}$, $m - r - \bar{r}$, $n - s - \bar{s}$ may be zero, in which case the corresponding matrices do not actually appear.

respectively. On the other hand, if $\tilde{A}$ is singular, we will delete some of its rows and columns until the resulting matrix is nonsingular and then apply the same process. In doing this it will be important not to delete any of the first r rows or s columns of $\tilde{A}$, that is, rows or columns through A_{11}. To show that there is a nonsingular submatrix of $\tilde{A}$ which contains A_{11} we first prove:

Lemma 2.7.8 The rows of the $r \times (s + \tilde{s})$ submatrix

$$B = [A_{11} \quad A_{12}]$$

of $\tilde{A}$ are linearly independent.

PROOF Let $X = [x_i]$ be any element of $\mathbb{R}_r$ which satisfies

$$XB = 0 \qquad (2.7.9)$$

We wish to show that $X = 0$. Multiply equation (2.7.9) on the right by the $(s + \tilde{s})$-dimensional column vector

$$\begin{bmatrix} \bar{Q}_1 \\ 0 \end{bmatrix}$$

Combining equations (2.7.9) and (2.7.7) yields

$$0 = XB \begin{bmatrix} \bar{Q}_1 \\ 0 \end{bmatrix} = vXE^{(r)}$$

Since $v \neq 0$, it follows that

$$XE^{(r)} = 0$$

i.e., the sum of the entries of X is 0. Let W be the m-dimensional row vector

$$W = [X \quad 0 \quad 0]$$

(with the same partition as above). For every $\varepsilon > 0$ the sum of the entries of

$$\bar{P} \pm \varepsilon W = [\bar{P}_1 \pm \varepsilon X \quad 0 \quad 0]$$

is 1. Moreover, since $\bar{P}_1 > 0$, both $\bar{P} + \varepsilon W$ and $\bar{P} - \varepsilon W$ are strategy vectors when ε is sufficiently small. In particular, when

$$0 < \varepsilon \leq \alpha = \min_{i = 1, 2, \dots, r} \frac{p_i}{1 + |x_i|}$$

we have $p_i \pm \varepsilon x_i \geqq 0$. (Note that $\alpha > 0$ since $\bar{P}_1 > 0$.) In fact a positive ε can be chosen so that $\bar{P} + \varepsilon W$ and $\bar{P} - \varepsilon W$ are both optimal strategies for player I. For, by equation (2.7.9)

$$(\bar{P} \pm \varepsilon W)A = \bar{P}A \pm \varepsilon WA = [Y_1 \quad Y_2 \quad Y_3 \pm \varepsilon X A_{13}]$$

According to equation (2.7.6),

$$[Y_1 \quad Y_2] = v E_{(s+\bar{s})}$$

Also, each entry of Y_3 is strictly greater than v, and thus

$$Y_3 \pm \varepsilon X A_{13} \geqq v E_{(n-s-\bar{s})}$$

when ε is sufficiently small (but positive). Clearly

$$\bar{P} = \tfrac{1}{2}(\bar{P} - \varepsilon W) + \tfrac{1}{2}(P + \varepsilon W) \qquad (2.7.10)$$

However by the hypotheses of Theorem 2.7.5, $\bar{P}$ is an extreme optimal strategy for player I. This contradicts equation (2.7.10) unless

$$\bar{P} = \bar{P} + \varepsilon W = \bar{P} - \varepsilon W$$

Thus $W = 0$. Consequently $X = 0$, and the rows of B are linearly independent, completing the proof of the lemma.   ////

The same argument can be used to prove that the columns of the submatrix

$$C = \begin{bmatrix} A_{11} \\ A_{21} \end{bmatrix}$$

of $\tilde{A}$ are linearly independent.

Recall that we wish to delete some of the last $\bar{r}$ rows and $\bar{s}$ columns of $\tilde{A}$ so that the resulting matrix is nonsingular. Of course, in the resulting matrix the first s columns and the first r rows must still be linearly independent. In particular, we wish to delete as many columns of $\tilde{A}$ as possible subject to the condition that the first r rows remain linearly independent. That is, form

$$\tilde{B} = [A_{11} \quad A'_{12}]$$

from B by deleting columns of A_{12} so that:

1 The rows of $\tilde{B}$ are linearly independent.
2 Either A'_{12} does not appear, or when any of the columns of A'_{12} are deleted from $\tilde{B}$, the rows of the resulting matrix are linearly dependent.

Similarly form

$$\tilde{C} = \begin{bmatrix} A_{11} \\ A'_{21} \end{bmatrix}$$

from C by deleting rows of A_{21} so that:

1 The columns of $\tilde{C}$ are linearly independent.
2 Either A'_{21} does not appear, or when any of the rows of A'_{21} are deleted from $\tilde{C}$, the columns of the resulting matrix are linearly dependent.

Finally, set

$$A^{\cdot} = \begin{array}{c} \\ \begin{bmatrix} A_{11} & A'_{12} \\ A'_{21} & A'_{22} \end{bmatrix} \end{array} \begin{array}{c} r \\ r' \end{array}$$

where A'_{22} is formed from A_{22} by deleting the same rows and columns as were deleted to form A'_{21} and A'_{12} respectively.

It remains now to prove:

Lemma 2.7.11 $A^{\cdot}$ is nonsingular.

PROOF It is sufficient to show that the rows of $A^{\cdot}$ are linearly independent. If $r' = 0$, that is, A'_{21} and A'_{22} do not actually appear, then $A^{\cdot} = \tilde{B}$ and the result is immediate.

Now let us assume that $r' > 0$ and let

$$Z = [z_i] = \begin{array}{cc} r & r' \\ [Z_1 & Z_2] \end{array} \; _1 \qquad (2.7.12)$$

satisfy

$$ZA^{\cdot} = 0 \qquad (2.7.13)$$

Let k be a (fixed) integer, $1 \leq k \leq r'$, and let A''_{12} be formed by deleting the kth row of A'_{21}. By the construction above there exists a nonzero vector Y such that

$$\begin{bmatrix} A_{11} \\ A''_{21} \end{bmatrix} Y = 0 \qquad W = [w_j] = \begin{bmatrix} A_{11} \\ A'_{21} \end{bmatrix} Y = \tilde{C} Y \neq 0 \qquad (2.7.14)$$

Hence $w_{r+k} \neq 0$, and every other entry of W is zero. Combining equations (2.7.13) and (2.7.14) yields

$$0 = (ZA^{\cdot}) \begin{bmatrix} Y \\ 0 \end{bmatrix} = Z \left(A^{\cdot} \begin{bmatrix} Y \\ 0 \end{bmatrix} \right) = ZW = z_{r+k} w_{r+k}$$

Thus we have $z_{r+k} = 0$. Therefore the matrix Z_2 of (2.7.12) is a zero matrix, and equation (2.7.13) becomes

$$0 = [Z_1 \quad 0]A^{.} = Z_1\tilde{B}$$

Since the rows of $\tilde{B}$ are linearly independent, it follows that $Z_1 = 0$. Hence $Z = 0$, and the rows of $A^{.}$ are linearly independent.   ////

The proof of Theorem 2.7.5 is now immediate. $A^{.}$ has been obtained by deleting some of the rows and columns of A. We now form $\bar{P}^{.}$ by deleting the entries of $\bar{P}$ corresponding to those rows of A which were deleted to form $A^{.}$. Since none of the rows of A through A_{11} have been removed, it follows that $\bar{P}^{.}$ contains all the nonzero entries of $\bar{P}$. Similarly $\bar{Q}^{.}$ is formed by deleting the entries of $\bar{Q}$ corresponding to those columns of A which have been deleted, and $\bar{Q}^{.}$ contains all the nonzero entries of $\bar{Q}$. Clearly

$$\bar{P}^{.}A^{.} = vE_{(n.)} \qquad A^{.}\bar{Q}^{.} = vE^{(n.)}$$

Since $A^{.}$ is nonsingular, we have

$$v = \frac{1}{E_{(n.)}(A^{.})^{-1}E^{(n.)}}$$

$$\bar{P}^{.} = \frac{E_{(n.)}(A^{.})^{-1}}{E_{(n.)}(A^{.})^{-1}E^{(n.)}} \qquad \bar{Q}^{.} = \frac{(A^{.})^{-1}E^{(n.)}}{E_{(n.)}(A^{.})^{-1}E^{(n.)}} \qquad (2.7.15)$$

    ////

EXAMPLE 2.7.16 Let

$$A = \begin{bmatrix} 6 & 8 & 9 & 10 \\ 8 & 4 & 3 & 0 \end{bmatrix}$$

Clearly A contains six 2×2 submatrices and eight 1×1 submatrices. We can consider each of them in turn as a candidate for $A^{.}$. (It is clear, even if we did not have the results of Example 2.6.11, that $v > 0$.) First let us assume

$$A^{.} = \begin{bmatrix} 6 & 8 \\ 8 & 4 \end{bmatrix}$$

Then

$$(A^{.})^{-1} = \begin{bmatrix} -\frac{2}{20} & \frac{4}{20} \\ \frac{4}{20} & -\frac{3}{20} \end{bmatrix}$$

and equations (2.7.15) yield

$$v = \frac{1}{\frac{3}{20}} = \frac{20}{3}$$

$$\bar{P}^{.} = \frac{20}{3}[\frac{2}{20} \quad \frac{1}{20}] = [\frac{2}{3} \quad \frac{1}{3}] \qquad \bar{Q}^{.} = \frac{20}{3}\begin{bmatrix} \frac{2}{20} \\ \frac{1}{20} \end{bmatrix} = \begin{bmatrix} \frac{2}{3} \\ \frac{1}{3} \end{bmatrix}$$

Thus, we have actually obtained a solution. Note that

$$\bar{P} = [\tfrac{2}{3} \ \tfrac{1}{3}] \qquad \bar{Q} = \begin{bmatrix} \tfrac{2}{3} \\ \tfrac{1}{3} \\ 0 \\ 0 \end{bmatrix}$$

are extreme optimal strategies for players I and II, respectively.

Next, let us assume

$$A = \begin{bmatrix} 6 & 10 \\ 8 & 0 \end{bmatrix}$$

Then

$$(A')^{-1} = \begin{bmatrix} 0 & \tfrac{5}{40} \\ \tfrac{4}{40} & -\tfrac{3}{40} \end{bmatrix}$$

and equations (2.7.15) yield

$$v = \frac{1}{\tfrac{6}{40}} = \frac{20}{3}$$

$$\bar{P}' = \tfrac{20}{3}[\tfrac{4}{40} \ \tfrac{2}{40}] = [\tfrac{2}{3} \ \tfrac{1}{3}] \qquad \bar{Q}' = \frac{20}{3} \begin{bmatrix} \tfrac{5}{40} \\ \tfrac{1}{40} \end{bmatrix} = \begin{bmatrix} \tfrac{5}{6} \\ \tfrac{1}{6} \end{bmatrix}$$

Again we have found extreme optimal strategies

$$\bar{P} = [\tfrac{2}{3} \ \tfrac{1}{3}] \qquad \bar{Q} = \begin{bmatrix} \tfrac{5}{6} \\ 0 \\ 0 \\ \tfrac{1}{6} \end{bmatrix}$$

for players I and II, respectively. Note that the two choices for A' yield the same $\bar{P}$.

Now, assume

$$A' = \begin{bmatrix} 9 & 10 \\ 3 & 0 \end{bmatrix}$$

Then

$$(A')^{-1} = \begin{bmatrix} 0 & \tfrac{10}{30} \\ \tfrac{3}{30} & -\tfrac{9}{30} \end{bmatrix}$$

and equations (2.7.15) yield

$$v = \frac{1}{\tfrac{4}{30}} = \frac{15}{2}$$

$$\bar{P}' = \tfrac{15}{2}[\tfrac{3}{30} \ \tfrac{1}{30}] = [\tfrac{3}{4} \ \tfrac{1}{4}] \qquad \bar{Q}' = \frac{15}{2} \begin{bmatrix} \tfrac{10}{30} \\ -\tfrac{6}{30} \end{bmatrix} = \begin{bmatrix} \tfrac{5}{2} \\ -\tfrac{3}{2} \end{bmatrix}$$

which is clearly not a solution. We leave it as an exercise to test the remaining 11 square submatrices of A. ////

In order to apply Theorem 2.7.5 we must be certain that $v \neq 0$, so that $A^{\cdot}$ is nonsingular and $(A^{\cdot})^{-1}$ exists. However, in Theorem 1.9.32 an expression is given for the inverse of a nonsingular matrix in terms of its adjoint,

$$A^{-1} = \frac{1}{\det A} \operatorname{adj} A$$

Substituting this expression into equations (2.7.15) yields

$$v = \frac{\det A^{\cdot}}{E_{(n\cdot)}(\operatorname{adj} A^{\cdot})E^{(n\cdot)}}$$

$$\bar{P}^{\cdot} = \frac{E_{(n\cdot)}(\operatorname{adj} A^{\cdot})}{E_{(n\cdot)}(\operatorname{adj} A^{\cdot})E^{(n\cdot)}} \qquad \bar{Q}^{\cdot} = \frac{(\operatorname{adj} A^{\cdot})E^{(n\cdot)}}{E_{(n\cdot)}(\operatorname{adj} A^{\cdot})E^{(n\cdot)}} \qquad (2.7.17)$$

The restriction that $v \neq 0$ will, of course, carry over from Theorem 2.7.5 to equations (2.7.17). We can prove, in fact, that this restriction is now unnecessary. First we need the following matrix identities.

Lemma 2.7.18 Let A be a square matrix of order n. Then

$$E_{(n)} \operatorname{adj}(A + xE) = E_{(n)} \operatorname{adj} A \quad (2.7.19a)$$
$$[\operatorname{adj}(A + xE)]E^{(n)} = (\operatorname{adj} A)E^{(n)} \quad (2.7.19b)$$
$$E_{(n)}[\operatorname{adj}(A + xE)]E^{(n)} = E_{(n)}(\operatorname{adj} A)E^{(n)} \quad (2.7.19c)$$

for every number x.

PROOF It is sufficient to prove (2.7.19a). Set

$$C = A + xE$$

and denote the entry in the (i, j) position of adj C and adj A by C_{ji} and A_{ji} respectively. The ith entries of the n-dimensional row vectors $E_{(n)} \operatorname{adj} A$ and $E_{(n)} \operatorname{adj} C$ are

$$\sum_{p=1}^{n} A_{ip} \qquad \text{and} \qquad \sum_{p=1}^{n} C_{ip}$$

respectively.
    Set

$$B = \begin{bmatrix} a_{11} + x & a_{12} + x & \cdots & a_{1n} + x \\ \cdots\cdots\cdots\cdots\cdots\cdots\cdots\cdots\cdots\cdots\cdots\cdots \\ a_{i-1,1} + x & a_{i-1,2} + x & \cdots & a_{i-1,n} + x \\ 1 & 1 & \cdots & 1 \\ a_{i+1,1} + x & a_{i+1,2} + x & \cdots & a_{i+1,n} + x \\ \cdots\cdots\cdots\cdots\cdots\cdots\cdots\cdots\cdots\cdots\cdots\cdots \\ a_{n1} + x & a_{n2} + x & \cdots & a_{nn} + x \end{bmatrix}$$

Since B is obtained from C by replacing the entries in the ith row by 1s, the ith row expansion for det B yields

$$\det B = \sum_{p=1}^{n} C_{ip}$$

However, det B is independent of x; for when the ith row multiplied by x is subtracted from every other row, the determinant is unchanged.   When $x = 0$, this expansion becomes

$$\det B = \sum_{p=1}^{n} A_{ip}$$

and hence
$$\sum_{p=1}^{n} A_{ip} = \sum_{p=1}^{n} C_{ip} \qquad i = 1, 2, \ldots, n \qquad ////$$

Corollary 2.7.20

$$\det(A + xE) = \det A + xE_{(n)}(\text{adj } A)E^{(n)}$$

PROOF   Multiplying $(2.7.19a)$ on the right by $A + xE$ and applying Theorem 1.9.32 yields

$$E_{(n)}[(\det(A + xE))I] = E_{(n)}[(\det A)I] + xE_{(n)}(\text{adj } A)E$$

Since $E = E^{(n)}E_{(n)}$, we have

$$[\det(A + xE)]E_{(n)} = (\det A)E_{(n)} + xE_{(n)}(\text{adj } A)(E^{(n)}E_{(n)})$$
$$= [(\det A + xE_{(n)}(\text{adj } A)E^{(n)}]E_{(n)} \qquad ////$$

We can now establish equations (2.7.17) without any restrictions on v.

Theorem 2.7.21   Let $(\bar{P}, \bar{Q}, v)$ be a solution of the game with matrix A and let $\bar{P}$ and $\bar{Q}$ be extreme optimal strategies for players I and II respectively.   Then there exists a submatrix $A^{\cdot}$ (having order $n^{\cdot}$) of A such that

$$v = \frac{\det A^{\cdot}}{E_{(n^{\cdot})}(\text{adj } A^{\cdot})E^{(n^{\cdot})}}$$

and $\bar{P}$ and $\bar{Q}$ are given (up to zero entries) by

$$\bar{P}^{\cdot} = \frac{E_{(n^{\cdot})}(\text{adj } A^{\cdot})}{E_{(n^{\cdot})}(\text{adj } A^{\cdot})E^{(n^{\cdot})}} \qquad \bar{Q}^{\cdot} = \frac{(\text{adj } A^{\cdot})E^{(n^{\cdot})}}{E_{(n^{\cdot})}(\text{adj } A^{\cdot})E^{(n^{\cdot})}}$$

PROOF   If $v \neq 0$, the theorem follows from Theorem 2.7.5 [as in (2.7.17)].
Now assume $v = 0$.   For any real number x, $(\bar{P}, \bar{Q}, x)$ is a solution of the game with matrix $A + xE$, and $\bar{P}$ and $\bar{Q}$ are extreme optimal

strategies.   Let $x \neq 0$.   According to (2.7.17), there exists a submatrix $A^{\cdot} + xE$ of $A + xE$ such that

$$x = \frac{\det(A^{\cdot} + xE)}{E_{(n \cdot)}[\text{adj}(A^{\cdot} + xE)]E^{(n \cdot)}}$$

$$\bar{P}^{\cdot} = \frac{E_{(n \cdot)}\,\text{adj}(A^{\cdot} + xE)}{E_{(n \cdot)}[\text{adj}(A^{\cdot} + xE)]E^{(n \cdot)}} \qquad \bar{Q}^{\cdot} = \frac{[\text{adj}(A^{\cdot} + xE)]E^{(n \cdot)}}{E_{(n \cdot)}[\text{adj}(A^{\cdot} + xE)]E^{(n \cdot)}} \qquad (2.7.22)$$

Using Lemma 2.7.18 and Corollary 2.7.20, equations (2.7.22) become

$$0 = \frac{\det A^{\cdot}}{E_{(n \cdot)}(\text{adj } A^{\cdot})E^{(n \cdot)}}$$

$$\bar{P}^{\cdot} = \frac{E_{(n \cdot)}(\text{adj } A^{\cdot})}{E_{(n \cdot)}(\text{adj } A^{\cdot})E^{(n \cdot)}} \qquad \bar{Q}^{\cdot} = \frac{(\text{adj } A^{\cdot})E^{(n \cdot)}}{E_{(n \cdot)}(\text{adj } A^{\cdot})E^{(n \cdot)}} \qquad ////$$

Theorem 2.7.21 says nothing about the existence of extreme optimal strategies (except that there can be only a finite number of them).   It can, however, be shown that:

Theorem 2.7.23   Every optimal strategy can be expressed as a convex combination of extreme optimal strategies.[1]

Theorem 2.7.23 follows from the fact that for each player the set of optimal strategies is convex and compact.   The proof that in $\mathbb{R}^n$ every element of a convex compact set can be expressed as a convex combination of extreme points requires a knowledge of the geometry of convex sets that is beyond the scope of this book.   Theorems 2.7.21 and 2.7.23, taken together, provide a (theoretical, at least) method for obtaining all solutions of a game.

EXERCISES

1   Find all extreme optimal strategies of each of the games:

(a) $\begin{bmatrix} 2 & 4 \\ 2 & -1 \end{bmatrix}$
(b) $\begin{bmatrix} 2 & -1 & -4 \\ -3 & -1 & 5 \end{bmatrix}$

(c) $\begin{bmatrix} 3 & -2 & -3 \\ -2 & 4 & 5 \end{bmatrix}$
(d) $\begin{bmatrix} 2 & 2 & 2 \\ 4 & 5 & -6 \end{bmatrix}$

by means of Theorem 2.7.21.   (See Exercise 4 of Sec. 2.6.)
2   Find all solutions of the game of Exercise 5 of Sec. 2.6.

[1] For a proof of this theorem see Karlin [4, vol. I, p. 401].

3 Solve:

(a) $\begin{bmatrix} 4 & -2 & -1 \\ 3 & -1 & 1 \\ -3 & 1 & 1 \end{bmatrix}$ (b) $\begin{bmatrix} 0 & 1 & -1 \\ -1 & -1 & 0 \\ 1 & -1 & 1 \end{bmatrix}$

by means of Theorem 2.7.21.

4 Let G be a game in which each player has a positive extreme optimal strategy. Show that the matrix of the game is square and that these are the only optimal strategies for each player.

5 Find all the extreme optimal strategies in the game of Exercise 2 of Sec. 2.5.

6 Let $\bar{P}$ and $\bar{Q}$ be optimal strategies satisfying equations (2.7.15) for some matrix A'. Show that $\bar{P}$ and $\bar{Q}$ are extreme optimal strategies.

2.8 LINEAR PROGRAMMING

Linear programming, which we study in this section, and the theory of games both involve finding nonnegative solutions for a system of simultaneous linear inequalities. Although they were originally treated as completely separate subjects, we can use the machinery of game theory to develop the theory of linear programming. However, unlike a game, a linear-programming problem need not have a solution.

EXAMPLE 2.8.1 **Maximum-profit problem** A farmer has 400 acres of land which is suitable for the growing of alfalfa, corn, and wheat. He also has 640 man-hours of labor per week at his disposal. Suppose that each acre devoted to alfalfa requires 1 man-hour of labor per week and will yield a profit of $40, each acre devoted to corn requires 2 man-hours of labor per week and will yield a profit of $60, and each acre of wheat requires 4 man-hours of labor per week and will yield a profit of $80. How many acres each of alfalfa, corn, and wheat should the farmer plant in order to maximize his profit?

Let a, c, and w be the number of acres to be planted with alfalfa, corn, and wheat respectively. Then the limitations under which the farmer must operate are described by the system of inequalities

$$a \geq 0 \qquad c \geq 0 \qquad w \geq 0$$

Land: $\qquad a + c + w \leq 400 \qquad (2.8.2)$

Labor: $\qquad a + 2c + 4w \leq 640$

A set of numbers a, c, w which satisfy the system of inequalities (2.8.2) will yield a profit of

$$P = 40a + 60c + 80w$$

Thus the farmer wishes to find values of a, c, w satisfying inequalities (2.8.2) which maximize $40a + 60c + 80w$.

Because there are only three unknowns in this problem, we can use graphical means to find all points (a, c, w) satisfying the inequalities (2.8.2).

The solid with vertices 0, A, B, C, D, E in Fig. 2.8.3 consists of all points (a, c, w) satisfying inequalities (2.8.2). We call these points the *feasible points* of the problem. It remains to determine at which feasible points the value of $P = 40a + 60c + 80w$ is a maximum. This can be done by intersecting the set of planes

$$40a + 60c + 80w = k$$

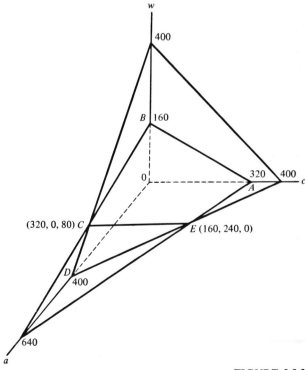

FIGURE 2.8.3

for various values of k with the set of feasible points. It is easily seen from Fig. 2.8.4 that the maximum value of P occurs at E and that the farmer is able to make a profit of \$20,800 by planting 160 acres of alfalfa and 240 acres of corn (and no wheat). $E = (160, 240, 0)$ is called an *optimal point* or *solution* of the maximum (linear-programming) problem.

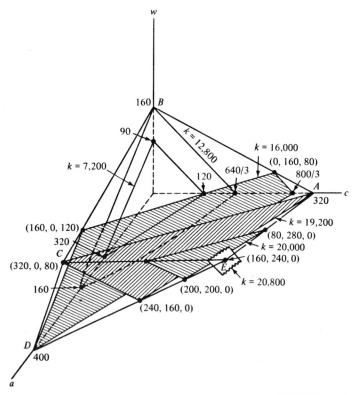

FIGURE 2.8.4

In order to apply the methods we have developed earlier in this chapter, we must restate Example 2.8.1 in the language of matrices. Thus we wish to find vectors

$$X = \begin{bmatrix} a \\ c \\ w \end{bmatrix}$$

belonging to $\mathbb{R}^3$ satisfying

$$\begin{bmatrix} 1 & 1 & 1 \\ 1 & 2 & 4 \end{bmatrix} X \leq \begin{bmatrix} 400 \\ 640 \end{bmatrix} \qquad X \geq 0 \qquad (2.8.5)$$

which maximize the expression

$$P = [40 \quad 60 \quad 80] X \qquad ////$$

EXAMPLE 2.8.6 **Minimum-cost problem** The same farmer is able to raise some livestock.   Each animal will eat alfalfa and hay, and from these he must obtain at least 40 units of minerals, 60 units of protein, and 80 units of vitamins every week.   Suppose that alfalfa and hay each have a mineral content of 1 unit per hundredweight.   Suppose also that the protein contents per hundredweight of alfalfa and hay are 1 and 2 units respectively and that the vitamin contents per hundredweight of alfalfa and hay are 1 and 4 units respectively.   In addition, the price of alfalfa is \$4.00 per hundredweight, and the price of hay is \$6.40 per hundredweight.   What diet will provide the animals with adequate amounts of minerals, protein, and vitamins at the least cost?

Let a and h be the number of hundredweights of alfalfa and hay that the animals are to be fed.   Then the system of inequalities

$$a \geq 0 \qquad h \geq 0$$

Minerals: $\qquad\qquad\qquad\qquad\qquad a + h \geq 40$

Protein: $\qquad\qquad\qquad\qquad\quad a + 2h \geq 60 \qquad (2.8.7)$

$$a + 4h \geq 80$$

describes the dietary requirements of the animals.   Thus the farmer wishes to find the values of a and h which satisfy inequalities (2.8.7) and which minimize the cost

$$C = 400a + 640h$$

Again we can solve the problem graphically.   First we find all the feasible points of the problem (the shaded area of Fig. 2.8.8).

The set of feasible points for this problem is an unbounded portion of the plane.   As before, we intersect the set of lines

$$400a + 640h = k$$

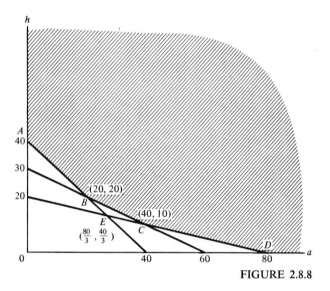

FIGURE 2.8.8

for various values of k, with the set of feasible points.   From Fig. 2.8.9 it is clear that the minimum value of the cost occurs at $B = (20, 20)$.   Thus the farmer is able to give his animals a satisfactory diet at a cost of $208.00 per week by using 20 hundredweight of alfalfa and 20 hundredweight of hay.

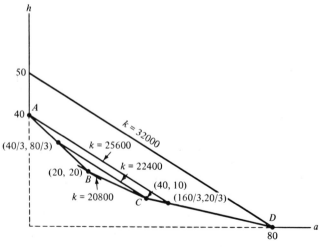

FIGURE 2.8.9

In matrix terminology, Example 2.8.6 requires us to find vectors

$$Y = [a \quad h]$$

belonging to $\mathbb{R}_2$ satisfying

$$Y\begin{bmatrix} 1 & 1 & 1 \\ 1 & 2 & 4 \end{bmatrix} \geq [40 \quad 60 \quad 80] \qquad Y \geq 0 \qquad (2.8.10)$$

which minimizes

$$C = Y\begin{bmatrix} 400 \\ 640 \end{bmatrix} \qquad \text{////}$$

Let us now begin again and put the study of linear-programming problems on a more solid foundation.

Definition 2.8.11 Let A, Z_1, Z_2 be $m \times n$, $m \times 1$, $1 \times n$ matrices respectively. The two problems

PROBLEM 1   MAXIMIZATION PROBLEM   Find a vector $X \in \mathbb{R}^n$ satisfying

$$AX \leq Z_1 \qquad X \geq 0 \qquad (2.8.12)$$

which maximizes $Z_2 X$.

PROBLEM 2:   MINIMIZATION PROBLEM   Find a vector $Y \in \mathbb{R}_m$ satisfying

$$YA \geq Z_2 \qquad Y \geq 0 \qquad (2.8.13)$$

which minimizes YZ_1.

form a *dual pair of linear-programming problems.*

We denote this pair of problems by (A, Z_1, Z_2). Thus Examples 2.8.1 and 2.8.6 together constitute a dual pair of linear-programming problems (A, Z_1, Z_2) in which

$$A = \begin{bmatrix} 1 & 1 & 1 \\ 1 & 2 & 4 \end{bmatrix} \qquad Z_1 = \begin{bmatrix} 400 \\ 640 \end{bmatrix} \qquad Z_2 = [40 \quad 60 \quad 80]$$

Definition 2.8.14 A vector $X \in \mathbb{R}^n$ which satisfies the system of inequalities (2.8.12) is called a *feasible vector* of the maximization problem of the dual pair (A, Z_1, Z_2). A feasible vector X which maximizes the quantity $Z_2 X$ is called an *optimal vector* or *solution* of the maximization problem, and $Z_2 X$ is called the *value* of the problem.

A vector $Y \in \mathbb{R}_m$ which satisfies the system of inequalities (2.8.13) is called a *feasible vector* of the minimization problem of the dual pair (A, Z_1, Z_2). A feasible vector which minimizes the quantity YZ_1 is called an *optimal vector* or *solution* of the minimization problem, and YZ_1 is called the *value* of the problem.

It is possible that the inequalities of a linear-programming problem are inconsistent, i.e., that no feasible vectors exist. In this case, of course, the problem has no solution. We call such a problem an *inconsistent problem*. Another way in which a linear-programming problem might fail to have a solution is for $Z_2 X$ to be unbounded from above or for YZ_1 to be unbounded from below. Such a problem is said to be *unbounded*. In the exercises we shall give examples to show that both these situations can actually occur.

Maximization problems and minimization problems are essentially the same thing, for any problem can be expressed either way. The minimization problem

PROBLEM 2′   Find a vector $X^T \in \mathbb{R}_n$ satisfying

$$X^T(-A^T) \geqq -Z_1{}^T \qquad X^T \geqq 0$$

which minimizes $X^T(-Z_2{}^T)$.

is identical with the maximization problem (Problem 1) of Definition 2.8.11.

Similarly, Problem 2 of Definition 2.8.11 can be expressed as a maximization problem.

PROBLEM 1′   Find a vector $Y^T \in \mathbb{R}^m$ satisfying

$$(-A^T)Y^T \leqq -Z_2{}^T \qquad Y^T \geqq 0$$

which maximizes $(-Z_1{}^T)Y^T$.

Clearly Problems 1′ and 2′ also form a dual pair of linear-programming problems $(-A^T, -Z_2{}^T, -Z_1{}^T)$. Thus we need not make any distinction between maximization and minimization problems in a dual pair.

Lemma 2.8.15   If X, Y are feasible vectors of the dual pair of linear programming problems (A, Z_1, Z_2), then

$$Z_2 X \leqq YZ_1$$

PROOF   Since X and Y are feasible vectors of the two problems of (A, Z_1, Z_2), we have

$$AX \leq Z_1 \quad X \geq 0 \quad YA \geq Z_2 \quad Y \geq 0$$

and hence $$Z_2 X \leq YAX \leq YZ_1 \qquad ////$$

Corollary 2.8.16   If both problems of a dual pair have feasible vectors, then neither problem is unbounded.

According to Lemma 2.8.15, when both problems of a dual pair have solutions, then the value of the maximization problem is less than or equal to the value of the minimization problem.   We shall show, in fact, in Theorem 2.8.19 that when both problems have solutions, their values are equal.   (Note Examples 2.8.1 and 2.8.6).   Also, it follows from Corollary 2.8.16 that the dual problem of an unbounded problem can have no feasible vectors and hence is inconsistent. Thus, neither problem will have a solution.   Conversely we have:

Theorem 2.8.17   If a linear-programming problem is inconsistent, then its dual does not have a solution.

PROOF   It is sufficient to treat the case in which the maximization problem is inconsistent.   If the dual problem is also inconsistent, the theorem is immediate.   Hence, we suppose that the maximization problem of (A, Z_1, Z_2) is inconsistent and that Y is a feasible vector of the minimization problem.   Then

$$Y \geq 0 \quad YA \geq Z_2$$

and there does not exist a vector $X \in \mathbb{R}^n$ such that

$$X \geq 0 \quad AX \leq Z_1$$

By Corollary 2.4.18 there exists a vector $\tilde{Y} \in \mathbb{R}_m$ such that

$$\tilde{Y} \geq 0 \quad \tilde{Y}Z_1 < 0 \quad \tilde{Y}A \geq 0$$

Clearly $Y + \alpha\tilde{Y}$ is a feasible vector of the minimization problem for every nonnegative real number α.   Since $\tilde{Y}Z_1 < 0$, it follows that $(Y + \alpha\tilde{Y})Z_1$ is unbounded from below.   Hence the minimization problem is unbounded and so does not have a solution.   ////

In the exercises we will give examples to show that it is possible for the dual of an inconsistent problem to be either inconsistent or unbounded.

Thus far we have not given any indication when optimal vectors exist. To this end we first prove Lemma 2.8.18, which gives an efficient method for recognizing optimal vectors.

Lemma 2.8.18 If $\overline{X}$, $\overline{Y}$ are feasible vectors of the dual pair of linear-programming problems (A, Z_1, Z_2) for which

$$\overline{Y}Z_1 \leqq Z_2\overline{X} \qquad (2.8.19)$$

then $\overline{X}$, $\overline{Y}$ are optimal vectors and $\overline{Y}Z_1 = Z_2\overline{X}$.

PROOF For any feasible vectors X, Y of the dual problems (A, Z_1, Z_2) we have, from Lemma 2.8.15,

$$Z_2 X \leqq \overline{Y}Z_1 \leqq Z_2\overline{X}$$
$$\overline{Y}Z_1 \leqq Z_2\overline{X} \leqq YZ_1$$

Since $\overline{X}$, $\overline{Y}$ are themselves feasible vectors, it follows that

$$Z_2\overline{X} = \max\{Z_2 X \mid X \text{ is a feasible vector of Problem 1}\}$$
$$\overline{Y}Z_1 = \min\{YZ_1 \mid Y \text{ is a feasible vector of Problem 2}\}$$

Thus $\overline{X}$ and $\overline{Y}$ are optimal vectors.

It is an immediate consequence of equation (2.8.19) and Lemma 2.8.15 that $Z_2\overline{X} = \overline{Y}Z_1$.   ////

Note that Lemma 2.8.18 describes a situation in which the values of the problems in a dual pair are equal. However, Lemma 2.8.18 does not give any indication that such a situation can ever occur, much less that it is the only way in which optimal vectors can exist. That this is, in fact, the case is contained in:

Theorem 2.8.20 If both linear-programming problems of a dual pair have feasible vectors, then both have optimal vectors and the values of the two problems are equal.

PROOF Let (A, Z_1, Z_2) be a dual pair of problems both of which have feasible vectors. We wish to show that there exist vectors $X \in \mathbb{R}^n$, $Y \in \mathbb{R}_m$ such that

$$AX \leqq Z_1 \qquad X \geqq 0 \qquad YA \geqq Z_2 \qquad Y \geqq 0$$
$$Z_2 X - YZ_1 \geqq 0 \qquad (2.8.21)$$

for, according to Lemma 2.8.18, a pair of vectors satisfying all these inequalities would be a pair of optimal vectors. It is more convenient to express the inequalities of (2.8.21) in the form

$$AX \leqq Z_1 \qquad X \geqq 0 \qquad (-A^T)Y^T \leqq -Z_2{}^T \qquad Y^T \geqq 0$$
$$-Z_2 X + Z_1{}^T Y^T \leqq 0$$

for these inequalities can then be combined into the pair of inequalities

$$
\begin{bmatrix} 0 & A \\ -A^T & 0 \\ Z_1^{\ T} & -Z_2 \end{bmatrix} \begin{bmatrix} Y^T \\ X \end{bmatrix} \leqq \begin{bmatrix} Z_1 \\ -Z_2^{\ T} \\ 0 \end{bmatrix} \qquad \begin{bmatrix} Y^T \\ X \end{bmatrix} \geqq 0 \qquad (2.8.22)
$$

Suppose the inequalities (2.8.22) do not have a solution. It follows from Corollary 2.4.18 that there exists a nonnegative vector $W \in \mathbb{R}_{m+n+1}$ such that

$$
W \begin{bmatrix} 0 & A \\ -A^T & 0 \\ Z_1^{\ T} & -Z_2 \end{bmatrix} \geqq 0 \qquad W \begin{bmatrix} Z_1 \\ -Z_2^{\ T} \\ 0 \end{bmatrix} < 0 \qquad (2.8.23)
$$

Partition W as

$$
W = [W_1 \quad W_2 \quad w_3] \qquad W_1 \in \mathbb{R}_m \qquad W_2 \in \mathbb{R}_n \qquad w_3 \in \mathbb{R}
$$

Then inequalities (2.8.23) become

$$
-W_2 A^T + w_3 Z_1^{\ T} \geqq 0 \qquad W_1 A - w_3 Z_2 \geqq 0 \\
W_1 Z_1 - W_2 Z_2^{\ T} < 0 \qquad\qquad (2.8.24)
$$

Now, two cases arise.

CASE 1　First let us suppose that $w_3 \neq 0$. We can rewrite (2.8.24) as

$$
A \left(\frac{1}{w_3} W_2^{\ T} \right) \leqq Z_1 \qquad \left(\frac{1}{w_3} W_1 \right) A \geqq Z_2
$$

$$
\left(\frac{1}{w_3} W_1 \right) Z_1 - Z_2 \left(\frac{1}{w_3} W_2^{\ T} \right) < 0
$$

Consequently　　　　　$\overline{X} = \frac{1}{w_3} W_2^{\ T} \qquad \overline{Y} = \frac{1}{w_3} W_1$

are feasible vectors for the dual pair of problems, and

$$
\overline{Y} Z_1 - Z_2 \overline{X} < 0
$$

which contradicts Lemma 2.8.15.

CASE 2　Next, let us suppose that $w_3 = 0$. Let X, Y be any pair of feasible vectors. The first two inequalities of (2.8.24) yield

$$
W_1 Z_1 \geqq W_1 A X \geqq 0 \cdot X = 0 = Y \cdot 0 \geqq Y A W_2^{\ T} \geqq Z_2 W_2^{\ T} = W_2 Z_2^{\ T}
$$

which contradicts the last inequality of (2.8.24).

Thus, for either supposition for w_3, the original supposition that no solutions of inequalities (2.8.22) exist has led to a contradiction. By Lemma 2.8.18, both problems have solutions, and their values are equal. ////

Corollary 2.8.25 Either both linear-programming problems of a dual pair have optimal vectors or neither has optimal vectors.

PROOF According to Theorem 2.8.20, either both linear-programming problems of a dual pair have optimal vectors or at least one of the problems is inconsistent. By Theorem 2.8.17, if one of the problems is inconsistent, then its dual does not have optimal vectors. ////

We conclude this chapter with a theorem which shows the close connection between linear-programming problems and the theory of games. It is possible to associate a symmetric game with a dual pair of linear-programming problems in such a way that all the optimal vectors (if any) of the linear-programming problems are obtainable from the optimal strategies of the game. We associate the game whose matrix is the $(m + n + 1) \times (m + n + 1)$ skew-symmetric matrix

$$B = \begin{bmatrix} 0 & A & -Z_1 \\ -A^T & 0 & Z_2^T \\ Z_1^T & -Z_2 & 0 \end{bmatrix}$$

with the dual pair (A, Z_1, Z_2).

Theorem 2.8.26 If X, Y are optimal vectors for the dual pair of linear-programming problems (A, Z_1, Z_2), then some positive multiple of

$$U = [Y \quad X^T \quad 1]$$

is an optimal strategy for each player in the game whose matrix is B. Conversely, if

$$P = [W_1 \quad W_2 \quad w_3] \quad W_1 \in \mathbb{R}_m \quad W_2 \in \mathbb{R}_n \quad w_3 \in \mathbb{R}$$

is an optimal strategy in the game whose matrix is B, and if $w_3 \neq 0$, then

$$X = \frac{1}{w_3} W_2^T \quad Y = \frac{1}{w_3} W_1$$

are optimal vectors of the dual pair of linear-programming problems (A, Z_1, Z_2).

Finally, the dual pair (A, Z_1, Z_2) have no optimal vectors if and only if $w_3 = 0$ for every optimal strategy of B.

PROOF   Let X, Y be optimal vectors for the dual pair (A, Z_1, Z_2).   By Theorem 2.8.20 we have

$$AX \leq Z_1 \qquad YA \geq Z_2 \qquad Z_2 X = Y Z_1$$

and hence
$$UB = [Y \quad X^T \quad 1] \begin{bmatrix} 0 & A & -Z_1 \\ -A^T & 0 & Z_2^T \\ Z_1^T & -Z_2 & 0 \end{bmatrix} \geq 0$$

Since U is semipositive and the value of the game with matrix B is zero, it follows that some positive multiple of U is an optimal strategy of B. It only remains to divide U by the sum of its entries (which is positive) in order to obtain a strategy vector.

Conversely, if $P = [W_1 \quad W_2 \quad w_3]$ is an optimal strategy of B, then $PB \geq 0$ and hence

$$- W_2 A^T + w_3 Z_1^T \geq 0 \qquad W_1 A - w_3 Z_2 \geq 0 \qquad - W_1 Z_1 + W_2 Z_2^T \geq 0$$
$$(2.8.27)$$

As before, if $w_3 \neq 0$, then inequalities (2.8.27) can be written as

$$A\left(\frac{1}{w_3} W_2^T\right) \leq Z_1 \qquad \left(\frac{1}{w_3} W_1\right) A \geq Z_2 \qquad \left(\frac{1}{w_3} W_1\right) Z_1 \leq \left(\frac{1}{w_3} W_2\right) Z_2^T$$

and, by Lemma 2.8.18,

$$\bar{X} = \left(\frac{1}{w_3} W_2^T\right) \qquad \bar{Y} = \left(\frac{1}{w_3} W_1\right)$$

are optimal vectors of (A, Z_1, Z_2).

It is now clear that $w_3 = 0$ for every optimal strategy P of B if and only if the dual pair of linear-programming problems do not have any optimal vectors, since every pair of optimal vectors leads to an optimal strategy of B in which $w_3 \neq 0$, and conversely.   ////

EXERCISES

1   Prove that the system of inequalities

$$x + 2y + z - w \leq -1$$
$$-x + y - 2z + 5w \leq 4$$

has no nonnegative solutions (*a*) directly from the inequalities, and (*b*) by constructing an inconsistent linear-programming problem with these inequalities. (*Hint:* Try to maximize $x + y + z + w$. Show that the dual problem is unbounded.)

2 Show that the pair of linear-programming problems

$$\left(\begin{bmatrix} 2 & -5 \\ -7 & 10 \end{bmatrix}, \begin{bmatrix} \frac{1}{2} \\ -\frac{1}{4} \end{bmatrix}, [2 \quad -1] \right)$$

does not have any solutions (a) by the graphical methods of this section, and (b) by showing that the last pure strategy of the associated game is inessential. Determine whether each problem is inconsistent or unbounded.

3 Solve:

(a) $\left(\begin{bmatrix} 2 & 1 & -1 \\ 1 & -1 & -1 \end{bmatrix}, \begin{bmatrix} -2 \\ 1 \end{bmatrix}, [4 \quad 1 \quad 1] \right)$.

(b) $\left(\begin{bmatrix} 1 & 3 \\ 2 & 1 \end{bmatrix}, \begin{bmatrix} 2 \\ 3 \end{bmatrix}, [2 \quad 5] \right)$.

(c) $\left(\begin{bmatrix} 1 & 4 & -2 \\ 1 & 1 & 1 \\ 1 & -2 & 0 \end{bmatrix}, \begin{bmatrix} 4 \\ 4 \\ 1 \end{bmatrix}, [3 \quad 4 \quad 2] \right)$.

4 A farmer has 7 acres of land and $2,400 capital. He can raise either livestock or vegetables. An acre of land is sufficient for either 10 animals or 10 tons of vegetables. Raising livestock requires an investment of $20 per animal, while raising vegetables requires a capital outlay of $60 per ton. In addition, the farmer must raise at least enough vegetables to feed his livestock, and each ton of vegetables feeds six animals. How can the farmer maximize his profit if he realizes a profit of $45 for each animal he sells and $30 for each ton of vegetables he sells (after feeding his livestock)? Can the farmer make more money if he is able to buy vegetables to feed his livestock at $30 per ton?

5 Show that the set of feasible vectors of a linear-programming problem is convex.

6 Show that the set of optimal vectors of a linear-programming problem is convex.

7 Show that if A, Z_1, and Z_2 are positive matrices, both problems of (A, Z_1, Z_2) have solutions.

8 Show that (A, Z_1, Z_2) does not have any solutions if and only if there exist vectors $W_1 \in \mathbb{R}_m$ and $W_2 \in \mathbb{R}^n$ such that

$$W_1 A \geq 0 \qquad W_1 \geq 0 \qquad A W_2 \leq 0 \qquad W_2 \geq 0$$
$$Z_2 W_2 > W_1 Z_1$$

9 An optimal vector is *extreme* if it cannot be expressed as a convex combination of other optimal vectors. Let X and Y be optimal vectors of (A, Z_1, Z_2) and let

$$P = \frac{1}{\alpha} [Y \quad X^T \quad 1]$$

where α is chosen so that P is a strategy vector.   Show that P is an extreme optimal strategy of the associated game if and only if both X and Y are extreme optimal vectors.

REFERENCES

Although the theory of games is a relatively new branch of mathematics, there is already an extensive literature on the subject dealing not only with the finite zero-sum two-person games in this chapter but also with several more general situations.   The fundamental theorem remains valid for a large and important class of infinite zero-sum two-person games, i.e., games in which there are infinitely many pure strategies. Many of the results of this chapter can be extended to these games.   A number of interesting infinite games are described by Karlin [4] and Dresher [1].   The latter discusses in great detail duels in which each player can fire at any time.   In addition to studying infinite games, Owen [7] treats non-zero-sum games, e.g., a duel in which both players lose if both are killed, and n-person games.   In a finite non-zero-sum game there is a payoff matrix for each player.   The players are often not fighting each other, and it is possible that both can benefit by cooperating with each other.   The possibility of players forming coalitions is even more important in n-person games.

We have chosen to study linear-programming problems in terms of games. Hillier and Lieberman [3] show that linear programming can be treated independently and that a game can be solved by transforming it into a linear-programming problem. In addition they develop the fundamentals of nonlinear programming.   The theory of linear programming is also discussed by Gass [2], who also presents both the theory and the computational aspects of the simplex method.

1 Dresher, M.: "Games of Strategy," Prentice-Hall, Inc., Englewood Cliffs, N.J., 1961.
2 Gass, S. I.: "Linear Programming," 2d ed., McGraw-Hill Book Company, New York, 1964.
3 Hillier, F. S., and G. J. Lieberman: "Introduction to Operations Research," Holden-Day, Inc., San Francisco, 1967.
4 Karlin, S.: "Mathematical Methods and Theory in Games, Programming and Economics,". vols. I and II, Addison-Wesley Publishing Company, Inc., Reading, Mass., 1959
5 McKinsey, J. C. C.: "Introduction to the Theory of Games," McGraw-Hill Book Company, New York, 1952.
6 Neumann, J. von: Zur Theorie der Gesellschaftsspiele, *Math. Ann.*, *100*: 295–320 (1928).
7 Owen, G.: "Game Theory," W. B. Saunders Company, Philadelphia, 1968.

3

FINITE MARKOV CHAINS

3.1 EXAMPLES OF FINITE MARKOV CHAINS

A stochastic process is concerned with events that change in a random way with time. A typical stochastic process might predict the motion of an object which is constrained to be at any one time in exactly one of a number of possible states. It would then be a scheme for determining the probability of the object's being in a specific state at a specific time. This probability would generally depend on a large number of factors, e.g., (1) the state, (2) the time, (3) some or all of the previous states the object has been in, (4) the states other objects are in or have been in. For example, the object might be the reader, and the states might be the countries of the world. Clearly aspects of all four factors mentioned above will influence the probability that the reader will be in a specific state at a specific time. If the probability that an object in one state will move to another state depends only on the two states involved (and not on earlier states, or time, or other factors), then the stochastic process is called a *Markov process*.[1] If the number of states the object can be in is finite (in which case we denote them by $S_1, S_2, \ldots, S_n$), the Markov process is called a *finite Markov chain*.

[1] Named for the Russian mathematician A. A. Markov (1856–1922), who introduced it in 1907.

The probability that an object in state S_j at a given time will be in state S_i at the end of a specified time interval is called the *transition probability* and will be denoted by a_{ij}. Thus the stochastic process is a Markov chain if a_{ij} is a constant (depending only on i and j). Returning to the example above, a_{ij} might describe the probability that the reader, if he is in state S_j (say, Japan) on the first day of any month, will be in state S_i (say Mexico) on the first day of the next month. For this example to be a Markov chain the probability of being in Mexico a month after being in Japan must not depend on anything other than the two countries involved.

The $n \times n$ matrix $A = [a_{ij}]$ which has the transition probability a_{ij} in the (i, j) position is called the *transition matrix* of the Markov chain. For a Markov chain with n states, A is a real square semipositive matrix of order n. Since, for each j, a Markov chain moves an object from state S_j to some state, it follows that

$$a_{1j} + a_{2j} + \cdots + a_{nj} = 1$$

i.e., the sum of the entries in any column of A is 1.

To define a Markov chain completely it is necessary to specify two things:

1   The transition matrix
2   For each j, the probability $p_j^{(0)}$ that the object is initially in the state S_j

The vector

$$P^{(0)} = \begin{bmatrix} p_1^{(0)} \\ p_2^{(0)} \\ \vdots \\ p_n^{(0)} \end{bmatrix}$$

is called the *initial* or *0-step probability vector*. Clearly

$$p_j^{(0)} \geq 0 \quad j = 1, 2, \ldots, n$$
$$p_1^{(0)} + p_2^{(0)} + \cdots + p_n^{(0)} = 1 \tag{3.1.1}$$

More generally, for $k = 0, 1, 2, \ldots$, we define the *k-step probability vector* to be

$$P^{(k)} = \begin{bmatrix} p_1^{(k)} \\ p_2^{(k)} \\ \vdots \\ p_n^{(k)} \end{bmatrix}$$

where $p_j^{(k)}$ is the probability that the object is in state S_j after the Markov chain process has been applied k times, beginning with the initial probability vector $P^{(0)}$. Any column vector whose entries satisfy conditions (3.1.1) is called a *probability vector.*[1] Thus every column of a transition matrix is a probability vector.   The collection of all n-dimensional probability vectors will be denoted by $\mathbb{P}^n$.

Let $A = [a_{ij}]$ be the transition matrix of a Markov chain process and let $P^{(0)} = [p_j^{(0)}]$ be the initial probability vector.   We wish to find the probability $p_i^{(1)}$ that the object will be in state S_i after a single application of the Markov chain process.   Let (S_i, S_j) denote the event "the object is initially in the state S_j and is in the state S_i after the Markov chain process has acted once."   Thus, in order for the event (S_i, S_j) to occur the object must initially be in state S_j (and this happens with probability $p_j^{(0)}$) and must then move to state S_i (and this happens with probability a_{ij}).   Hence, the probability of the event (S_i, S_j) is $a_{ij}p_j^{(0)}$.   The probability that the object is in the state S_i after a single application of the Markov chain process is the sum of the probabilities of the events (S_i, S_j) for all j, that is,

$$p_i^{(1)} = \sum_{j=1}^{n} a_{ij}p_j^{(0)}$$

Thus:

Theorem 3.1.2   If A is the transition matrix of a Markov chain with initial probability vector $P^{(0)}$, then

$$P^{(1)} = AP^{(0)}$$

Corollary 3.1.3   For any nonnegative integers k, l

$$P^{(k+l)} = A^k P^{(l)} \qquad (3.1.4)$$

EXAMPLE 3.1.5   The men in a certain community are classified as either smokers or nonsmokers.   It is found that among the sons of smokers, $\frac{3}{5}$ are smokers and $\frac{2}{5}$ are nonsmokers, while among the sons of nonsmokers, $\frac{4}{5}$ are nonsmokers and $\frac{1}{5}$ are smokers.   If $\frac{2}{3}$ of the current population smokes, what portion of each of the next two generations will smoke?

A research organization wishes to study a sample population of this community.   Is it possible to construct a sample such that the proportion of smokers in the next generation will be the same as the proportion of smokers in the current generation?

---

[1] In Chap. 2 we called such a vector a strategy, but the term probability vector is more appropriate here.

Note that the probability that a man will be a smoker depends only on the state, i.e., smoker or nonsmoker, of his father. We will classify the men of the present generation as either:

1 S, smokers
2 N, nonsmokers

and the men of the next generation as:

1 SS, smoking sons of smokers
2 SN, smoking sons of nonsmokers
3 NS, nonsmoking sons of smokers
4 NN, nonsmoking sons of nonsmokers

Since $\frac{2}{3}$ of the present generation smokes, it follows that $\frac{2}{3}$ of the next generation will be in either the SS or NS categories.[1] But, $\frac{3}{5}$ of these will be smokers (SS), and $\frac{2}{5}$ will be nonsmokers (NS). Hence the proportion of SS and NS in the next generation will be $\frac{2}{3} \cdot \frac{3}{5} = \frac{6}{15}$ and $\frac{2}{3} \cdot \frac{2}{5} = \frac{4}{15}$, respectively. In the same way, $\frac{1}{3}$ of the present generation are nonsmokers, and hence $\frac{1}{3}$ of the next generation will be in the SN and NN categories. Since $\frac{1}{5}$ of these will be smokers (SN) and $\frac{4}{5}$ will be nonsmokers (NN), the proportion of SN and NN in the next generation will be $\frac{1}{3} \cdot \frac{1}{5} = \frac{1}{15}$ and $\frac{1}{3} \cdot \frac{4}{5} = \frac{4}{15}$, respectively. Thus, in the next generation the smokers (SS and SN combined) will constitute $\frac{6}{15} + \frac{1}{15} = \frac{7}{15}$ of the population while the nonsmokers (NS and NN combined) will constitute $\frac{4}{15} + \frac{4}{15} = \frac{8}{15}$ of the population. We can find the proportion of smokers and nonsmokers in all succeeding generations in the same manner.

Let us denote the proportion of smokers and nonsmokers in the kth generation by $S^{(k)}$ and $N^{(k)}$ respectively.[2] Using the method of the last paragraph, it is clear that

$$S^{(k+1)} = \tfrac{3}{5}S^{(k)} + \tfrac{1}{5}N^{(k)}$$
$$N^{(k+1)} = \tfrac{2}{5}S^{(k)} + \tfrac{4}{5}N^{(k)}$$

Thus we have a Markov chain in which the transition matrix is

$$A = \begin{bmatrix} \frac{3}{5} & \frac{1}{5} \\ \frac{2}{5} & \frac{4}{5} \end{bmatrix}$$

and the initial probability vector is

$$P^{(0)} = \begin{bmatrix} S^{(0)} \\ N^{(0)} \end{bmatrix} = \begin{bmatrix} \frac{2}{3} \\ \frac{1}{3} \end{bmatrix}$$

[1] We are assuming that smoking and nonsmoking fathers have the same number of sons. In probabilistic terms, we are assuming that smoking and fertility are independent attributes.

[2] It will be convenient for us to refer to the present generation as the zeroth generation.

The requirement that the proportion of smokers in the next generation be the same as the proportion of smokers in the current generation can be expressed as

$$P^{(1)} = P^{(0)}$$

Since $P^{(1)} = AP^{(0)}$, we are looking for a solution $P^{(0)}$ of the equation

$$AP^{(0)} = P^{(0)} \qquad (3.1.6)$$

i.e., we are looking for a probability vector which is a characteristic vector of A corresponding to the characteristic root $+1$. A simple calculation will show that the characteristic roots of A are $+1$ and $\frac{2}{3}$ and that the characteristic vector corresponding to $+1$ is

$$\begin{bmatrix} \frac{1}{3} \\ \frac{2}{3} \end{bmatrix}$$

Thus if any sample of the community contains $\frac{1}{3}$ smokers and $\frac{2}{3}$ nonsmokers, the next generation (and, in fact, every succeeding generation) will also contain $\frac{1}{3}$ smokers and $\frac{2}{3}$ nonsmokers. We call a probability vector satisfying equation (3.1.6) a *stationary vector* of the Markov chain. $\quad////$

When dealing with a number $N > 1$ of objects all of which are subject to the same Markov chain process, we may replace the initial probability vector by the *initial population vector*

$$Q^{(0)} = \begin{bmatrix} q_1^{(0)} \\ q_2^{(0)} \\ \vdots \\ q_n^{(0)} \end{bmatrix}$$

where $q_j^{(0)}$ is the number of objects initially in state S_j. Similarly we replace the k-step probability vector $P^{(k)}$ by the *k-step population vector*

$$Q^{(k)} = \begin{bmatrix} q_1^{(k)} \\ q_2^{(k)} \\ \vdots \\ q_n^{(k)} \end{bmatrix}$$

where $q_j^{(k)}$ is the number of objects in state S_j after the Markov chain process has been applied k times, beginning with the initial population vector $Q^{(0)}$. Clearly

$$q_j^{(k)} = N p_j^{(k)} \qquad \begin{array}{l} j = 1, 2, \ldots, n \\ k = 0, 1, 2, \ldots \end{array} \qquad (3.1.7)$$

and equation (3.1.4) becomes

$$Q^{(k+l)} = A^k Q^{(l)} \qquad k, l = 0, 1, 2, \ldots \qquad (3.1.8)$$

EXAMPLE 3.1.9   The population of Ruritania is 1,800.   There are three cities in Ruritania, S, T, and U, and every year the entire population of each city moves, half to each of the other two cities.   If the populations of S, T, and U are currently 200, 600, and 1,000 respectively, what will the population of each city be next year and the year following?   What are the long-range population expectations for each city?

Let $s^{(k)}$, $t^{(k)}$, $u^{(k)}$ be the population of S, T, U respectively after k years and set

$$Q^{(k)} = \begin{bmatrix} s^{(k)} \\ t^{(k)} \\ u^{(k)} \end{bmatrix} \qquad k = 0, 1, 2, \ldots$$

Clearly we have a Markov chain with three states.   The transition matrix and the initial population vector are

$$A = \begin{bmatrix} 0 & \frac{1}{2} & \frac{1}{2} \\ \frac{1}{2} & 0 & \frac{1}{2} \\ \frac{1}{2} & \frac{1}{2} & 0 \end{bmatrix} \qquad Q^{(0)} = \begin{bmatrix} 200 \\ 600 \\ 1{,}000 \end{bmatrix}$$

respectively.   It follows from Theorem 3.1.2 and equations (3.1.8) that

$$Q^{(1)} = A Q^{(0)} = \begin{bmatrix} 800 \\ 600 \\ 400 \end{bmatrix} \qquad Q^{(2)} = A Q^{(1)} = A^2 Q^{(0)} = \begin{bmatrix} 500 \\ 600 \\ 700 \end{bmatrix}$$

We can show by an induction[1] on k that

$$A^k = \begin{bmatrix} a_k & a_{k+1} & a_{k+1} \\ a_{k+1} & a_k & a_{k+1} \\ a_{k+1} & a_{k+1} & a_k \end{bmatrix} \qquad k = 0, 1, 2, \ldots \qquad (3.1.10)$$

where

$$a_k = \frac{1}{3}\left[1 + \frac{(-1)^k}{2^{k-1}} \right]$$

[The verification of equation (3.1.10) is left as an exercise.]   Substituting these values into equation (3.1.8), we have

$$Q^{(k)} = \begin{bmatrix} 600 + \dfrac{400(-1)^{k+1}}{2^k} \\[2ex] 600 \\[2ex] 600 - \dfrac{400(-1)^{k+1}}{2^k} \end{bmatrix} \qquad k = 0, 1, 2, \ldots$$

---

[1] For a more explicit method for obtaining equation (3.1.10) see Exercises 10 to 12 of Sec. 1.13.

Thus, the population of T remains at 600. The population of S and U both approach 600 as a limit, and each is alternately greater than and less than 600.　　　　////

EXAMPLE 3.1.11   Player I draws a card from a three-card deck consisting of two cards marked H and one card marked L. If player I draws L, he pays player II \$1; while if he draws H, player II pays him \$1. At the outset, player I has \$1, and player II has \$3. The game is over when one of the players has won all the other player's money. Find the probability of each player's emerging the winner. What is the probability that a tenth draw will be required, i.e., that after each of the first nine draws, both players have some money?

　　We will take player I to be the object and use the amount of money that he has to describe the various states. The most natural way to do this would be to define S_1 to be the state "player I has \$0," S_2 to be the state "player I has \$1," S_3 to be the state "player I has \$2," etc. This will lead to the transition matrix

$$\begin{bmatrix} 1 & \frac{1}{3} & 0 & 0 & 0 \\ 0 & 0 & \frac{1}{3} & 0 & 0 \\ 0 & \frac{2}{3} & 0 & \frac{1}{3} & 0 \\ 0 & 0 & \frac{2}{3} & 0 & 0 \\ 0 & 0 & 0 & \frac{2}{3} & 1 \end{bmatrix}$$

However, it will be much more convenient to order the states, still in terms of player I's holdings, \$1, \$2, \$3, \$0, \$4, for then the transition matrix becomes

$$A = \begin{bmatrix} 0 & \frac{1}{3} & 0 & 0 & 0 \\ \frac{2}{3} & 0 & \frac{1}{3} & 0 & 0 \\ 0 & \frac{2}{3} & 0 & 0 & 0 \\ \frac{1}{3} & 0 & 0 & 1 & 0 \\ 0 & 0 & \frac{2}{3} & 0 & 1 \end{bmatrix}$$

We now partition A into

$$A = \begin{bmatrix} B & 0 \\ C & I \end{bmatrix}$$

where
$$C = \begin{bmatrix} \frac{1}{3} & 0 & 0 \\ 0 & 0 & \frac{2}{3} \end{bmatrix} \qquad B = \begin{bmatrix} 0 & \frac{1}{3} & 0 \\ \frac{2}{3} & 0 & \frac{1}{3} \\ 0 & \frac{2}{3} & 0 \end{bmatrix}$$

It can easily be verified, using an induction on k, that

$$A^k = \begin{bmatrix} B^k & 0 \\ C(I + B + B^2 + \cdots + B^{k-1}) & I \end{bmatrix} \qquad k = 1, 2, \ldots \qquad (3.1.12)$$

By a direct computation we can show that

$$B^3 = \tfrac{4}{9}B$$

and hence

$$B + B^3 + B^5 + \cdots + B^{2l+1} = [1 + \tfrac{4}{9} + (\tfrac{4}{9})^2 + \cdots + (\tfrac{4}{9})^l]B$$

$$B^2 + B^4 + B^6 + \cdots + B^{2l} = [1 + \tfrac{4}{9} + (\tfrac{4}{9})^2 + \cdots + (\tfrac{4}{9})^l]B^2$$

$$l = 1, 2, \ldots$$

Since the infinite geometric series $1 + \tfrac{4}{9} + (\tfrac{4}{9})^2 + (\tfrac{4}{9})^3 + \cdots$ converges to $\tfrac{9}{5}$, it is reasonable to say that the infinite series of matrices $I + B + B^2 + B^3 + \cdots$ converges to[1]

$$I + \tfrac{9}{5}B + \tfrac{9}{5}B^2 = \begin{bmatrix} \tfrac{7}{5} & \tfrac{3}{5} & \tfrac{1}{5} \\ \tfrac{6}{5} & \tfrac{9}{5} & \tfrac{3}{5} \\ \tfrac{4}{5} & \tfrac{6}{5} & \tfrac{7}{5} \end{bmatrix}$$

and that

$$\lim_{k \to \infty} C(I + B + B^2 + \cdots + B^{k-1}) =$$

$$\begin{bmatrix} \tfrac{1}{3} & 0 & 0 \\ 0 & 0 & \tfrac{2}{3} \end{bmatrix} \begin{bmatrix} \tfrac{7}{5} & \tfrac{3}{5} & \tfrac{1}{5} \\ \tfrac{6}{5} & \tfrac{9}{5} & \tfrac{3}{5} \\ \tfrac{4}{5} & \tfrac{6}{5} & \tfrac{7}{5} \end{bmatrix} = \begin{bmatrix} \tfrac{7}{15} & \tfrac{3}{15} & \tfrac{1}{15} \\ \tfrac{8}{15} & \tfrac{12}{15} & \tfrac{14}{15} \end{bmatrix}$$

Moreover, from $B^3 = \tfrac{4}{9}B$ it follows that

$$B^k = \begin{cases} \left(\dfrac{2}{3}\right)^k \begin{bmatrix} 0 & \tfrac{1}{2} & 0 \\ 1 & 0 & \tfrac{1}{2} \\ 0 & 1 & 0 \end{bmatrix} & k \text{ odd} \\[3ex] \left(\dfrac{2}{3}\right)^k \begin{bmatrix} \tfrac{1}{2} & 0 & \tfrac{1}{4} \\ 0 & 1 & 0 \\ 1 & 0 & \tfrac{1}{2} \end{bmatrix} & k \text{ even} \end{cases} \qquad (3.1.13)$$

Thus,
$$\lim_{k \to \infty} B^k = 0$$

Combining these results, we have

$$\lim_{k \to \infty} A^k = \begin{bmatrix} 0 & 0 & 0 & 0 & 0 \\ 0 & 0 & 0 & 0 & 0 \\ 0 & 0 & 0 & 0 & 0 \\ \tfrac{7}{15} & \tfrac{3}{15} & \tfrac{1}{15} & 1 & 0 \\ \tfrac{8}{15} & \tfrac{12}{15} & \tfrac{14}{15} & 0 & 1 \end{bmatrix} \qquad (3.1.14)$$

---

[1] All the material introduced here concerning infinite series of matrices, limits of matrix functions, etc., will be reintroduced and treated more rigorously in Sec. 3.2.

Up to this point we have made no mention of the initial probability vector. Since, by hypothesis, player I is initially in state S_1 (he has \$1), the initial probability vector for the Markov chain is

$$P^{(0)} = \begin{bmatrix} 1 \\ 0 \\ 0 \\ 0 \\ 0 \end{bmatrix}$$

Multiplying $P^{(0)}$ by the matrix of equation (3.1.14) shows us that the probability of player I's eventually emerging the loser, i.e., being in state S_4, is $\frac{7}{15}$ and the probability of player I's eventually emerging the winner, i.e., being in state S_5, is $\frac{8}{15}$. It also shows that the probability of the game's continuing endlessly is zero.

In the same way, it follows from equations (3.1.12) and (3.1.13) (with $k = 9$) that there is no possibility of player I's being in states S_1 or S_3 after nine draws and that the probability of his being in state S_2 is $(\frac{2}{3})^9$. Thus the probability that a tenth draw will be required is $(\frac{2}{3})^9$, which is approximately 0.026.   ////

EXAMPLE 3.1.15   A small mouse and an enormous cheese are placed in the maze shown (though not necessarily in the same room). We make four assumptions:

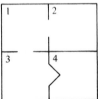

1 Rooms 1 and 3 have two exits, room 2 has one exit, and room 4 has no exits. The mouse cannot leave room 4 once he has entered it.
2 The mouse changes rooms exactly once every minute (provided he is not in room 4).
3 The mouse will choose each of the exits of the room he is in with equal probability.
4 The mouse is given a piece of cheese every time he enters the room which has the cheese.

How many pieces of cheese can the mouse be expected to get if he is put into room j and the cheese is put in room i, provided this process continues indefinitely?

We can express the location of the mouse at any time by means of a Markov chain process with the four states

$$S_j = \text{the mouse is in room } j \qquad j = 1, 2, 3, 4$$

The transition matrix is

$$A = [a_{ij}] = \begin{bmatrix} 0 & 1 & \frac{1}{2} & 0 \\ \frac{1}{2} & 0 & 0 & 0 \\ \frac{1}{2} & 0 & 0 & 0 \\ 0 & 0 & \frac{1}{2} & 1 \end{bmatrix}$$

and the initial probability vector is $I^{(j)}$. The mouse will get a piece of cheese at the 0-minute point, i.e., before he makes his first move, provided he is placed in the same room as the cheese, that is, $j = i$. The probability that the mouse will be in room i and hence get a piece of cheese at the 1-minute point, i.e., after his first move, is equal to the ith entry of

$$AI^{(j)}$$

Similarly, for $k = 1, 2, \ldots$, the probability that he will be in room i at the k-minute point is given by the ith entry of

$$A^k I^{(j)}$$

Since the mouse is given one piece of cheese every time he is fed, the total number of pieces of cheese he can expect is equal to the ith entry of

$$(I + A + A^2 + \cdots)I^{(j)}$$

Let us partition A into

$$A = \begin{bmatrix} B & 0 \\ C & I \end{bmatrix}$$

where
$$B = \begin{bmatrix} 0 & 1 & \frac{1}{2} \\ \frac{1}{2} & 0 & 0 \\ \frac{1}{2} & 0 & 0 \end{bmatrix} \qquad C = [0 \quad 0 \quad \tfrac{1}{2}]$$

As in (3.1.12) of Example 3.1.11,

$$A^k = \begin{bmatrix} B^k & 0 \\ C(I + B + B^2 + \cdots + B^{k-1}) & I \end{bmatrix} \qquad k = 1, 2, \ldots$$

and consequently

$$I + A + A^2 + \cdots + A^k =$$
$$\begin{bmatrix} I + B + B^2 + \cdots + B^k & 0 \\ C[kI + (k-1)B + (k-2)B^2 + \cdots + B^{k-1}] & (k+1)I \end{bmatrix}$$

A direct computation shows that

$$B^3 = \tfrac{3}{4}B$$

and hence

$$C[kI + (k-1)B + (k-2)B^2 + \cdots + B^{k-1}] = C(kI + \alpha B + \beta B^2)$$

where
$$\alpha = (k-1) + \tfrac{3}{4}(k-3) + (\tfrac{3}{4})^2(k-5) + \cdots$$
$$\beta = (k-2) + \tfrac{3}{4}(k-4) + (\tfrac{3}{4})^2(k-6) + \cdots$$

Moreover, since

$$1 + \tfrac{3}{4} + (\tfrac{3}{4})^2 + \cdots = 4$$

it follows that

$$I + B + B^2 + \cdots = I + 4B + 4B^2 = \begin{bmatrix} 4 & 4 & 2 \\ 2 & 3 & 1 \\ 2 & 2 & 2 \end{bmatrix}$$

Combining these results, we have

$$I + A + A^2 + \cdots = \begin{bmatrix} 4 & 4 & 2 & 0 \\ 2 & 3 & 1 & 0 \\ 2 & 2 & 2 & 0 \\ \infty & \infty & \infty & \infty \end{bmatrix} \qquad (3.1.16)$$

The mouse's expectation can be read directly from the matrix of (3.1.16). Thus, if the cheese is placed in room 4, the mouse will get infinitely many pieces of cheese irrespective of his initial position; if the cheese is placed in room 1 and the mouse is placed in either room 1 or room 2, he can expect four pieces of cheese, etc. ////

We will call a state which once it is entered can never be left, e.g., the states $0 and $4 in Example 3.1.11 and room 4 in Example 3.1.15, an *absorbing state*. The state S_q of a Markov chain process is absorbing if and only if the qth column of the transition matrix is $I^{(q)}$, that is

$$a_{qq} = 1 \qquad a_{pq} = 0 \qquad p \neq q$$

Let $a_{ij}^{(k)}$ denote the entry in the (i,j) position of A^k. Then

$$a_{ij}^{(k+1)} = \sum_{l=1}^{n} a_{il}^{(k)} a_{lj}^{(1)} = \sum_{m=1}^{n} a_{im}^{(1)} a_{mj}^{(k)} \qquad (3.1.17)$$

When S_j is an absorbing state, the first part of equation (3.1.17) becomes

$$a_{ij}^{(k+1)} = a_{ij}^{(k)} \qquad i = 1, 2, \ldots, n$$

and it follows that

$$a_{ij} = a_{ij}^{(1)} = a_{ij}^{(2)} = \cdots = 0 \qquad i \neq j$$
$$a_{jj} = a_{jj}^{(1)} = a_{jj}^{(2)} = \cdots = 1$$

Moreover, when S_i is an absorbing state, equation (3.1.17) yields

$$a_{ij}^{(k+1)} = \sum_{m=1}^{n} a_{im}^{(1)} a_{mj}^{(k)} = a_{ij}^{(k)} + \sum_{m \neq i} a_{im}^{(1)} a_{mj}^{(k)} \geqq a_{ij}^{(k)} \qquad j = 1, 2, \ldots, n$$

That is, for each value of j, $a_{ij}^{(k)}$ is a nondecreasing function of k (which has, of course, an upper bound of 1). Thus when S_i is an absorbing state, we know that $\lim_{k \to \infty} a_{ij}^{(k)}$ exists. We will be able to make use of this in:

EXAMPLE 3.1.18   A mouse is placed in room j of the maze. Let us assume that:

1 The mouse cannot leave room 5 or room 6 once he has entered it.
2 The mouse changes rooms exactly once every minute (provided he is not in room 5 or 6).
3 The mouse will choose each exit of the room he is in with equal probability.

Find the probability (which we denote by f_{ij}) that the mouse will ever enter room i.

Let us first consider the problem when room i corresponds to an absorbing state S_i. Note that the absorbing states of this Markov chain process are S_5 (room 5) and S_6 (room 6). Since the mouse may never leave either of these rooms, the probability that he will have entered one of them at or before the kth minute is the same as the probability that he will be in that room at the kth minute. Hence, for an absorbing state S_i, the probability that the mouse will have entered S_i (being originally in S_j) at or before the kth minute is given by

$$a_{ij}^{(k)}$$

and the probability that he will ever enter S_i is

$$f_{ij} = \lim_{k \to \infty} a_{ij}^{(k)} \qquad (3.1.19)$$

The transition matrix of this Markov chain process is

$$A = \begin{bmatrix} 0 & 0 & 0 & \frac{1}{2} & 0 & 0 \\ 0 & 0 & \frac{1}{2} & 0 & 0 & 0 \\ 0 & \frac{1}{2} & 0 & 0 & 0 & 0 \\ 1 & 0 & 0 & 0 & 0 & 0 \\ 0 & \frac{1}{2} & 0 & \frac{1}{2} & 1 & 0 \\ 0 & 0 & \frac{1}{2} & 0 & 0 & 1 \end{bmatrix}$$

We partition A as

$$A = \begin{bmatrix} B & 0 \\ C & I \end{bmatrix}$$

where

$$B = \begin{bmatrix} 0 & 0 & 0 & \frac{1}{2} \\ 0 & 0 & \frac{1}{2} & 0 \\ 0 & \frac{1}{2} & 0 & 0 \\ 1 & 0 & 0 & 0 \end{bmatrix} \qquad C = \begin{bmatrix} 0 & \frac{1}{2} & 0 & \frac{1}{2} \\ 0 & 0 & \frac{1}{2} & 0 \end{bmatrix}$$

As in (3.1.12) of Example 3.1.11,

$$A^k = \begin{bmatrix} B^k & 0 \\ C(I + B + B^2 + \cdots + B^{k-1}) & I \end{bmatrix}$$

Since

$$B^4 = \tfrac{3}{4}B^2 - \tfrac{1}{8}I$$

for each k, $k = 4, 5, 6, \ldots$ there exist scalars α_k, β_k, γ_k, δ_k such that

$$B^k = \alpha_k B^3 + \beta_k B^2 + \gamma_k B + \delta_k I \qquad (3.1.20)$$

The characteristic roots of B are $\pm\frac{1}{2}$, $\pm\sqrt{2}/2$, and it follows from equation (3.1.20) that

$$(\tfrac{1}{2})^k = \alpha_k(\tfrac{1}{2})^3 + \beta_k(\tfrac{1}{2})^2 + \gamma_k(\tfrac{1}{2}) + \delta_k$$

$$(-\tfrac{1}{2})^k = \alpha_k(-\tfrac{1}{2})^3 + \beta_k(-\tfrac{1}{2})^2 + \gamma_k(-\tfrac{1}{2}) + \delta_k$$

$$\left(\frac{\sqrt{2}}{2}\right)^k = \alpha_k\left(\frac{\sqrt{2}}{2}\right)^3 + \beta_k\left(\frac{\sqrt{2}}{2}\right)^2 + \gamma_k\left(\frac{\sqrt{2}}{2}\right) + \delta_k$$

$$\left(-\frac{\sqrt{2}}{2}\right)^k = \alpha_k\left(-\frac{\sqrt{2}}{2}\right)^3 + \beta_k\left(-\frac{\sqrt{2}}{2}\right)^2 + \gamma_k\left(-\frac{\sqrt{2}}{2}\right) + \delta_k$$

Solving these equations yields

$$
\alpha_k = \begin{cases} 0 & \text{when } k \text{ is even} \\ \dfrac{2^l - 1}{2^{2(l-1)}} & \text{when } k \text{ is odd, } k = 2l + 1 \end{cases}
$$

$$
\beta_k = \begin{cases} \dfrac{2^l - 1}{2^{2(l-1)}} & \text{when } k \text{ is even, } k = 2l \\ 0 & \text{when } k \text{ is odd} \end{cases}
$$

$$
\gamma_k = \begin{cases} 0 & \text{when } k \text{ is even} \\ \dfrac{2 - 2^l}{2^{2l}} & \text{when } k \text{ is odd, } k = 2l + 1 \end{cases}
$$

$$
\delta_k = \begin{cases} \dfrac{2 - 2^l}{2^{2l}} & \text{when } k \text{ is even, } k = 2l \\ 0 & \text{when } k \text{ is odd} \end{cases}
$$

Consequently

$$
\lim_{k \to \infty} B^k = 0
$$

$$
\lim_{k \to \infty} (I + B + B^2 + \cdots + B^{k-1})
$$

$$
= I + B + B^2 + B^3 + I\left(\sum_{l=2}^{\infty} \frac{2 - 2^l}{2^{2l}} \right) + B\left(\sum_{l=2}^{\infty} \frac{2 - 2^l}{2^{2l}} \right)
$$

$$
+ B^2\left(\sum_{l=2}^{\infty} \frac{2^l - 1}{2^{2(l-1)}} \right) + B^3\left(\sum_{l=2}^{\infty} \frac{2^l - 1}{2^{2(l-1)}} \right)
$$

$$
= \tfrac{2}{3}I + \tfrac{2}{3}B + \tfrac{8}{3}B^2 + \tfrac{8}{3}B^3 = \begin{bmatrix} 2 & 0 & 0 & 1 \\ 0 & \tfrac{4}{3} & \tfrac{2}{3} & 0 \\ 0 & \tfrac{2}{3} & \tfrac{4}{3} & 0 \\ 2 & 0 & 0 & 2 \end{bmatrix}
$$

and

$$
\lim_{k \to \infty} C(I + B + B^2 + B^3 + \cdots + B^{k-1}) = \begin{bmatrix} 1 & \tfrac{2}{3} & \tfrac{1}{3} & 1 \\ 0 & \tfrac{1}{3} & \tfrac{2}{3} & 0 \end{bmatrix}
$$

Thus we have

$$
f_{51} = 1 \qquad f_{52} = \tfrac{2}{3} \qquad f_{53} = \tfrac{1}{3} \qquad f_{54} = 1
$$

$$
f_{61} = 0 \qquad f_{62} = \tfrac{1}{3} \qquad f_{63} = \tfrac{2}{3} \qquad f_{64} = 0
$$

i.e., if the mouse is placed in room 2, the probability that he will ever enter (and hence forever remain in) room 5 is $\frac{2}{3}$, while the probability that he will ever enter room 6 is $\frac{1}{3}$, etc.

Now let us consider a case in which the state corresponding to room i is not absorbing. In particular we determine f_{1j}, for $j = 2, 3, \ldots, 6$. First, construct another maze, identical to the given one except that the mouse can never leave room 1 after he has entered it. If the mouse is not originally in

room 1, the probability that he will enter room 1 for the first time at the kth minute is the same for both mazes (though the mouse's subsequent movements are completely different). Consequently the probability that the mouse will ever enter room 1 is the same in both mazes. The transition matrix for the Markov chain process in which S_1 is absorbing is

$$A' = \begin{bmatrix} 1 & 0 & 0 & \frac{1}{2} & 0 & 0 \\ 0 & 0 & \frac{1}{2} & 0 & 0 & 0 \\ 0 & \frac{1}{2} & 0 & 0 & 0 & 0 \\ 0 & 0 & 0 & 0 & 0 & 0 \\ 0 & \frac{1}{2} & 0 & \frac{1}{2} & 1 & 0 \\ 0 & 0 & \frac{1}{2} & 0 & 0 & 1 \end{bmatrix}$$

which we partition as

$$A' = \begin{bmatrix} 1 & D' & 0 \\ 0 & B' & 0 \\ 0 & C' & I \end{bmatrix}$$

where

$$D' = [0 \quad 0 \quad \tfrac{1}{2}] \qquad B' = \begin{bmatrix} 0 & \frac{1}{2} & 0 \\ \frac{1}{2} & 0 & 0 \\ 0 & 0 & 0 \end{bmatrix} \qquad C' = \begin{bmatrix} \frac{1}{2} & 0 & \frac{1}{2} \\ 0 & \frac{1}{2} & 0 \end{bmatrix}$$

Clearly

$$A'^k = \begin{bmatrix} 1 & D'(I + B' + B'^2 + \cdots + B'^{k-1}) & 0 \\ 0 & B'^k & 0 \\ 0 & C'(I + B' + B'^2 + \cdots + B'^{k-1}) & I \end{bmatrix}$$

Since

$$B'^3 = \tfrac{1}{4}B'$$

and

$$I + B' + B'^2 + \cdots = I + \tfrac{4}{3}B' + \tfrac{4}{3}B'^2 = \begin{bmatrix} \frac{4}{3} & \frac{2}{3} & 0 \\ \frac{2}{3} & \frac{4}{3} & 0 \\ 0 & 0 & 1 \end{bmatrix}$$

we find that

$$D'(I + B' + B'^2 + \cdots) = [0 \quad 0 \quad \tfrac{1}{2}]$$

$$C'(I + B' + B'^2 + \cdots) = \begin{bmatrix} \tfrac{2}{3} & \tfrac{1}{3} & \tfrac{1}{2} \\ \tfrac{1}{3} & \tfrac{2}{3} & 0 \end{bmatrix}$$

Thus
$$\lim_{k \to \infty} A'^k = \begin{bmatrix} 1 & 0 & 0 & \tfrac{1}{2} & 0 & 0 \\ 0 & 0 & 0 & 0 & 0 & 0 \\ 0 & 0 & 0 & 0 & 0 & 0 \\ 0 & 0 & 0 & 0 & 0 & 0 \\ 0 & \tfrac{2}{3} & \tfrac{1}{3} & \tfrac{1}{2} & 1 & 0 \\ 0 & \tfrac{1}{3} & \tfrac{2}{3} & 0 & 0 & 1 \end{bmatrix}$$

and we see that

$$f_{12} = f_{13} = 0 \qquad f_{14} = \tfrac{1}{2} \qquad f_{15} = f_{16} = 0$$

Finally, let us find the probability that the mouse if originally placed in room 1 will be in room 1 at any later time. (We will denote this probability by f_{11}.) At the first minute, the mouse must leave room 1. In this example his only choice is to enter room 4. We have already shown that after entering room 4 the probability that he will ever subsequently be in room 1 is $\tfrac{1}{2}$. Hence

$$f_{11} = \tfrac{1}{2}$$

We leave it to the reader to find f_{ij}, $i \neq 1, 5, 6$.   ////

EXAMPLE 3.1.21   A mouse is placed in room j ($j \neq 3$) of the maze. What is

the probability that the mouse will ever enter room 3? If the mouse dies as soon as he enters room 3, what is his average life expectancy?

We have just seen that the first problem can be solved by the methods of Example 3.1.18. Here we present an alternate way of solving the first problem which at the same time provides a solution for the second problem. Let us denote the probability that the mouse will enter room 3 for the first time at the kth minute by

$$f_{3j}^{(k)}, \qquad k = 1, 2, \ldots$$

Then
$$f_{3j}^{(1)} + f_{3j}^{(2)} + \cdots + f_{3j}^{(k)}$$

is the probability that he will enter room 3 at least once during the first k minutes, and the probability that he will ever enter room 3 (which we have previously denoted by f_{3j}) is given by the infinite series

$$f_{3j} = f_{3j}{}^{(1)} + f_{3j}{}^{(2)} + f_{3j}{}^{(3)} + \cdots \qquad (3.1.22)$$

In order to evaluate $f_{3j}{}^{(k)}$ we transform the state corresponding to room 3

into an absorbing state.   The transition matrix of the new maze is

$$A' = \begin{bmatrix} 0 & \frac{1}{2} & 0 \\ 1 & 0 & 0 \\ 0 & \frac{1}{2} & 1 \end{bmatrix}$$

which we partition as

$$A' = \begin{bmatrix} B' & 0 \\ C' & 1 \end{bmatrix}$$

where

$$B' = \begin{bmatrix} 0 & \frac{1}{2} \\ 1 & 0 \end{bmatrix} \qquad C' = [0 \quad \tfrac{1}{2}]$$

$f_{3j}{}^{(k)}$ is also the probability that the mouse will enter room 3′ of the second maze at the kth minute.   Because he can never leave room 3′ after he has entered it, the probability that the mouse will be in room 3′ at the kth minute is the sum of the probabilities of entering room 3′ for the first time in each of the first k minutes, i.e.,

$$a'_{3j}{}^{(k)} = f_{3j}{}^{(1)} + f_{3j}{}^{(2)} + \cdots + f_{3j}{}^{(k)} \qquad (3.1.23)$$

[Compare equation (3.1.23) with equation (3.1.19).]   It follows from equation (3.1.23) that

$$a'_{3j}{}^{(k)} - a'_{3j}{}^{(k-1)} = f_{3j}{}^{(k)} \qquad k = 2, 3, \ldots$$

and consequently $f_{3j}{}^{(k)}$ is equal to the entry in the $(3, j)$ position of

$$(A')^k - (A')^{k-1}$$

Moreover, by equation (3.1.12),

$$(A')^k - (A')^{k-1} = \begin{bmatrix} (B')^k - (B')^{k-1} & 0 \\ C'(B')^{k-1} & 0 \end{bmatrix}$$

and $f_{3j}{}^{(k)}$ is the jth entry of the two-dimensional row vector $C'(B')^{k-1}$.   A simple computation shows that

$$(B')^2 = \tfrac{1}{2}I$$

and consequently
$$(B')^{k-1} = \begin{cases} (\tfrac{1}{2})^{(k-1)/2} I & \text{when } k \text{ is odd} \\ (\tfrac{1}{2})^{(k-2)/2} B' & \text{when } k \text{ is even} \end{cases}$$

Thus
$$C'(B')^{k-1} = \begin{cases} [0 \quad (\tfrac{1}{2})^{(k+1)/2}] & \text{when } k \text{ is odd} \\ [(\tfrac{1}{2})^{k/2} \quad 0] & \text{when } k \text{ is even} \end{cases}$$

and
$$f_{31}{}^{(k)} = \begin{cases} 0 & \text{when } k \text{ is odd} \\ (\tfrac{1}{2})^{k/2} & \text{when } k \text{ is even} \end{cases}$$

$$f_{32}{}^{(k)} = \begin{cases} (\tfrac{1}{2})^{(k+1)/2} & \text{when } k \text{ is odd} \\ 0 & \text{when } k \text{ is even} \end{cases}$$
(3.1.24)

Substituting equations (3.1.24) into equation (3.1.22) yields

$$f_{31} = f_{32} = \tfrac{1}{2} + \tfrac{1}{4} + \tfrac{1}{8} + \cdots = 1 \qquad (3.1.25)$$

Equation (3.1.25) states that the mouse is certain to die. The probability that he will live exactly k minutes (provided he is initially in room j) is

$$f_{3j}{}^{(k)}$$

The mouse's average life expectancy is given by

$$E_j = \sum_{k=1}^{\infty} k f_{3j}{}^{(k)} \qquad (3.1.26)$$

For $j = 1$, substituting the first equation of (3.1.24) into equation (3.1.26) and setting $k = 2l$ yields[1]

$$E_1 = \sum_{l=1}^{\infty} 2l(\tfrac{1}{2})^l = 4 \text{ minutes}$$

For $j = 2$, substituting the second equation of (3.1.24) into equation (3.1.26) and setting $k = 2l - 1$ yields

$$E_2 = \sum_{l=1}^{\infty} (2l - 1)(\tfrac{1}{2})^l = 3 \text{ minutes} \qquad ////$$

EXERCISES

1 Show that in Example 3.1.9 the same ultimate population will occur (namely, 600 people in each city) no matter where the 1,800 citizens of Ruritania are initially located.

---

[1] It can be shown by an induction on p that

$$1x + 2x^2 + 3x^3 + \cdots + px^p = \frac{px^{p+2} - (p+1)x^{p+1} + x}{(1-x)^2}$$

and hence, for $|x| < 1$,

$$1x + 2x^2 + 3x^3 + \cdots = \frac{x}{(1-x)^2}$$

2 Verify equation (3.1.10).

3 Show that the stationary vectors in the Markov chain of Example 3.1.11 are exactly those vectors in the set of all five-dimensional probability vectors whose first three entries are zero.

4 Repeat Example 3.1.11 with a deck consisting of two cards marked H and two cards marked L.

5 Find the probability that a mouse initially in room 2 of the maze of Example 3.1.15 will ever enter room 3. Find the probability that a mouse initially in room 3 will ever enter room 2.

6 In example 3.1.15 the maze is replaced by the maze shown. Find the number of pieces of cheese the mouse can be expected to get if he is placed in room *j* and the cheese is in room *i*.

7 Show that a mouse initially in room 2 of the maze of Exercise 6 is certain to enter either room 1 or room 4. What is the probability that he will enter room 4? What is the probability that he will enter room 1?

8 If the mouse of Exercise 7 dies when he enters either room 1 or room 4, what is his life expectancy? (*Hint*: Show that the maze can be replaced by one with three rooms, one of which has two one-way doors leading into it.)

9 Repeat Example 3.1.18 with the maze replaced by the one sketched.

10 There are two coins, of which one when tossed shows heads and tails with equal probability and the other shows heads two-thirds of the time. If a coin is tossed and shows tails, the other coin is used for the next toss. If a coin is tossed and shows heads, the same coin is used for the next toss. The first coin to be tossed is picked at random. What is the probability that the biased coin will be used for the third toss? What is the probability that the third toss will be heads? (*Hint:* There are four states.)

11 The following problem describes a common model used to explain diffusion in gases. We begin with two baskets, each of which contains five balls. The balls in one basket are all white, and those in the other basket are all black.   Each second one ball is selected from each basket and moved to the other basket.   Show that this is a Markov chain process and find the transition matrix and the initial vector.   What is the probability that both baskets will have at least one white ball after 3 seconds?   After 5 seconds?

12 In a certain community it is found that half the sons of college-educated fathers finish college, a quarter finish high school but not college, and a quarter do not finish high school.   Among sons of fathers who finish high school but not college one-third finish college, one-third finish high school but not college, and one-third do not finish high school.   Among sons of fathers who did not finish high school one-sixth finish college, one-half finish high school but not college, and one-third do not finish high school.   Find a transition matrix which expresses the relation between one generation and the next.   What proportion of each type of man will yield a stationary population?

13 Prove that the product of two transition matrices of the same order is a transition matrix.

14 Prove that the product of a transition matrix and a probability vector is a probability vector.

15 Let A be a square matrix of order n.   Prove that if AX is a probability vector for every probability vector X, then A is a transition matrix.

16 Prove that the set of stationary vectors of a Markov chain is convex.

17 Let A be the transition matrix of a Markov chain and let $\tilde{A}$ be the transition matrix of the same Markov chain after the order of the states has been changed .   Show that A and $\tilde{A}$ are similar.

18 A set $\mathbb{S}$ of states is *closed* if it is impossible to go from a state in $\mathbb{S}$ to a state not in $\mathbb{S}$.   (Thus a state is absorbing if the set consisting of that state is closed.)   Describe the transition matrix of a Markov chain with n states in which the set $\mathbb{S} = \{S_1, S_2, \ldots, S_k\}$ is closed.   Describe the transition matrix if the set $\mathbb{S}' = \{S_{k+1}, S_{k+2}, \ldots, S_n\}$ is also closed.

3.2   INFINITE PROCESSES

In Sec. 3.1 we introduced several procedures involving limits of matrix expressions in order to solve problems that arose in Examples 3.1.11, 3.1.15, and 3.1.18. No attempt was made to justify what was done or even to define explicitly the terms introduced.   This will be rectified now.

It is possible to impose a meaningful *distance*, or *metric*, function on the set $\mathbb{C}_n^m$ of all $m \times n$ matrices with complex entries.   It is the existence of this distance function which makes it possible for us to define limits of certain matrix expressions.

Definition 3.2.1 Let $A = [a_{ij}] \in \mathbb{C}_n{}^m$. We define the *norm* of A, written $\|A\|$, by

$$\|A\| = \max_{\substack{i=1,2,\cdots,m \\ j=1,2,\cdots,n}} \{|a_{ij}|\}$$

When $m = n = 1$, the norm of A is simply the absolute value of the single entry of A.

Lemma 3.2.2 Let A and B be $m \times n$ matrices, C be an $n \times p$ matrix and α be a complex number. Then:

1 $\|A\|$ is a nonnegative real number and $\|A\| = 0$ if and only if $A = 0$.
2 $\|\alpha A\| = |\alpha|\,\|A\|$.
3 $\|A + B\| \leq \|A\| + \|B\|$ (the triangle law for matrices).
4 $\|AC\| \leq n\|A\|\,\|C\|$.

PROOF   Parts 1 and 2 are immediate.   The remainder of the lemma follows from the triangle law for complex numbers.

(3)   Let $A = [a_{ij}]$, $B = [b_{ij}]$.   Then

$$\|A + B\| = \max_{\substack{i=1,2,\cdots,m \\ j=1,2,\cdots,n}} \{|a_{ij} + b_{ij}|\}$$

$$\leq \max_{\substack{i=1,2,\cdots,m \\ j=1,2,\cdots,n}} \{|a_{ij}| + |b_{ij}|\}$$

$$\leq \max_{\substack{i=1,2,\cdots,m \\ j=1,2,\cdots,n}} |a_{ij}| + \max_{\substack{i=1,2,\cdots,m \\ j=1,2,\cdots,n}} |b_{ij}|$$

$$= \|A\| + \|B\|$$

(4)   Let $C = [c_{jk}]$ and set

$$AC = D = [d_{ik}] \quad \begin{array}{l} i = 1, 2, \ldots, m \\ j = 1, 2, \ldots, n \\ k = 1, 2, \ldots, p \end{array}$$

Then

$$\|AC\| = \|D\| = \max_{\substack{i=1,2,\cdots,m \\ k=1,2,\cdots,p}} \{|d_{ik}|\} = \max_{\substack{i=1,2,\cdots,m \\ k=1,2,\cdots,p}} \left\{ \left| \sum_{j=1}^{n} a_{ij} c_{jk} \right| \right\}$$

$$\leq \max_{\substack{i=1,2,\cdots,m \\ k=1,2,\cdots,p}} \left\{ \sum_{j=1}^{n} |a_{ij} c_{jk}| \right\} = \max_{\substack{i=1,2,\cdots,m \\ k=1,2,\cdots,p}} \left\{ \sum_{j=1}^{n} |a_{ij}||c_{jk}| \right\}$$

$$\leq \max_{\substack{i=1,2,\cdots,m \\ k=1,2,\cdots,p}} \left\{ \sum_{j=1}^{n} \|A\|\|C\| \right\} = n\|A\| \cdot \|C\| \qquad ////$$

The norm of a matrix generally behaves very much like the absolute value and will play the same role in matrix situations as the absolute value does when treating real- or complex-valued functions.

Definition 3.2.3   Let $L, A_0, A_1, A_2, \ldots$ be $m \times n$ matrices.   We say that the sequence of matrices $A_0, A_1, A_2 \ldots$, *converges* to L and write

$$\lim_{k \to \infty} A_k = L$$

if for every $\varepsilon > 0$ there exists a positive integer K (depending on ε) such that

$$\|A_k - L\| < \varepsilon$$

for every integer $k > K$.

Most frequently we shall define A_k to be A^k, where A is a square matrix. The main result of this section (Theorem 3.2.9) gives a set of necessary and sufficient conditions for

$$\lim_{k \to \infty} A^k$$

to exist.   First we must show that many of the familiar limit properties of functions of a real or complex variable can be proved for matrix functions as well, usually in the same way.

Lemma 3.2.4   Let B be an $m \times n$ matrix and let $A_0, A_1, A_2, \ldots$ be a sequence of $n \times p$ matrices.   If $\lim_{k \to \infty} A_k$ exists, then $\lim_{k \to \infty} BA_k$ exists and

$$\lim_{k \to \infty} BA_k = B\left(\lim_{k \to \infty} A_k \right)$$

PROOF   If $B = 0$, the result is clear.   Now suppose that $B \neq 0$.   Then $\|B\| \neq 0$.   Set

$$\lim_{k \to \infty} A_k = L$$

and let $\varepsilon > 0$.   There exists a positive integer K such that

$$\|A_k - L\| < \frac{\varepsilon}{n\|B\|}$$

whenever $k > K$.   It follows from part 4 of Lemma 3.2.2 that

$$\|BA_k - BL\| = \|B(A_k - L)\| \leq n\|B\| \cdot \|A_k - L\| < \varepsilon$$

whenever $k > K$.   Thus

$$\lim_{k \to \infty} BA_k = BL = B\left(\lim_{k \to \infty} A_k \right) \qquad \text{////}$$

An important consequence of Lemma 3.2.4 is:

Corollary 3.2.5 If $\lim_{k \to \infty} A^k$ exists and P is nonsingular, then $\lim_{k \to \infty} (P^{-1}AP)^k$ exists and

$$P^{-1}\left(\lim_{k \to \infty} A^k\right)P = \lim_{k \to \infty} (P^{-1}AP)^k$$

PROOF Clearly

$$(P^{-1}AP)^k = P^{-1}A^kP \qquad k = 1, 2, \ldots$$

By two applications of Lemma 3.2.4, one on the left and the other on the right, we have

$$P^{-1}\left(\lim_{k \to \infty} A^k\right)P = \lim_{k \to \infty} (P^{-1}A^kP) = \lim_{k \to \infty} (P^{-1}AP)^k \qquad ////$$

Note that Lemma 3.2.4 and Corollary 3.2.5 each state two things: (1) the limits exist, and (2) they are equal. Lemma 3.2.4 was tacitly assumed while solving Example 3.1.11 of Sec. 3.1. The probability of each player's ultimately emerging the winner in Example 3.1.11 is given by the entries of

$$\lim_{k \to \infty} P^{(k)} = \lim_{k \to \infty} (A^k P^{(0)})$$

However, we actually computed

$$\left(\lim_{k \to \infty} A^k\right)P^{(0)}$$

Lemma 3.2.4 shows that these two expressions (when they exist) are equal.

Corollary 3.2.5 was also tacitly assumed in Example 3.1.11 of Sec. 3.1 when we changed the order of (i.e., permuted) the states. The resulting transition matrix A is related to the original transition matrix $\tilde{A}$ by

$$\tilde{A} = P^{-1}AP$$

where P is the permutation matrix corresponding to the permutation of the states.

Lemma 3.2.6 For $i = 1, 2, \ldots, l$, let A_{ii} be a square matrix of order n_i and let $A = A_{11} \oplus A_{22} \oplus \cdots \oplus A_{ll}$. Then $\lim_{k \to \infty} A^k$ exists if and only if $\lim_{k \to \infty} A_{ii}^{\ k}$ exists for each i. In this event

$$\lim_{k \to \infty} A^k = \lim_{k \to \infty} A_{11}^{\ k} \oplus \lim_{k \to \infty} A_{22}^{\ k} \oplus \cdots \oplus \lim_{k \to \infty} A_{ll}^{\ k}$$

PROOF   The proof is left as an exercise.   ////

Let A be an arbitrary square matrix. We wish to determine whether or not $\lim_{k\to\infty} A^k$ exists. According to Corollary 3.2.5, A may be replaced by any matrix similar to it; i.e., instead of examining $\lim_{k\to\infty} A^k$, we may examine $\lim_{k\to\infty}(P^{-1}AP)^k$ for any nonsingular matrix P. In Chapter 1 we discussed various simple forms $P^{-1}AP$ can have. According to Theorem 1.13.21, we can construct a nonsingular matrix P such that $P^{-1}AP$ is the direct sum of the hyper-companion matrices of the elementary divisors of A. By virtue of Lemma 3.2.6, it is sufficient to determine under what conditions $\lim_{k\to\infty} C^k$ exists when C has the form of (1.13.23).

It is convenient to consider the field from which the entries of A are taken to be the complex field $\mathbb{C}$ (rather than the real field $\mathbb{R}$). The fundamental theorem of algebra[1] states that *in the set of polynomials with complex coefficients, the irreducible polynomials are exactly the polynomials of degree 1.* The companion matrix of

$$p(x) = x - \lambda$$

is the 1×1 matrix $[\lambda]$. Hence, A is similar to the direct sum of matrices of the form

$$C = \begin{bmatrix} \lambda & 0 & 0 & \cdots & 0 & 0 \\ 1 & \lambda & 0 & \cdots & 0 & 0 \\ 0 & 1 & \lambda & \cdots & 0 & 0 \\ & & \cdots\cdots\cdots & & \\ 0 & 0 & 0 & \cdots & 1 & \lambda \end{bmatrix} \quad (1.13.23)$$

and it is sufficient to determine under what conditions $\lim_{k\to\infty} C^k$ exists when C is a matrix of the form (1.13.23). Set

$$N = \begin{bmatrix} 0 & 0 & 0 & \cdots & 0 & 0 \\ 1 & 0 & 0 & \cdots & 0 & 0 \\ 0 & 1 & 0 & \cdots & 0 & 0 \\ & & \cdots\cdots\cdots & & \\ 0 & 0 & 0 & \cdots & 1 & 0 \end{bmatrix}$$

Clearly

$$C = \lambda I + N$$

[1] For a proof of this result see, for example, E. Hille, "Analytic Function Theory," vol. I, p. 207, Ginn and Co., Boston, 1959.

Moreover, if the order of C (and hence N) is r then $N^r = 0$. Hence, for $k \geq 0$ we have

$$C^k = \lambda^k I + \frac{k}{1}\lambda^{k-1}N + \frac{k(k-1)}{1\cdot 2}\lambda^{k-2}N^2 + \cdots + \frac{k(k-1)\cdots(k-r+2)}{1\cdot 2\cdots(r-1)}\lambda^{k-r+1}N^{r-1} =$$

$$\begin{bmatrix}
\lambda^k & 0 & 0 & \cdots & 0 & 0 & 0 \\
k\lambda^{k-1} & \lambda^k & 0 & \cdots & 0 & 0 & 0 \\
\frac{k(k-1)}{2}\lambda^{k-2} & k\lambda^{k-1} & \lambda^k & \cdots & 0 & 0 & 0 \\
 & \frac{k(k-1)}{2}\lambda^{k-2} & k\lambda^{k-1} & \cdots & & & \\
 & & \frac{k(k-1)}{2}\lambda^{k-2} & \cdots & & & \\
 & & & & & & \\
\frac{k(k-1)\cdots(k-r+2)}{1\cdot 2\cdots(r-1)}\lambda^{k-r+1} & & & \cdots & \frac{k(k-1)}{2}\lambda^{k-2} & k\lambda^{k-1} & \lambda^k
\end{bmatrix}$$

$$(3.2.7)$$

Lemma 3.2.8 Let C be a square matrix of the form of (1.13.23). Then $\lim_{k\to\infty} C^k$ exists if and only if

$$|\lambda| < 1$$

in which case $\lim_{k\to\infty} C^k = 0$, or

$$\lambda = +1 \quad \text{and} \quad r = 1$$

in which case $\lim_{k\to\infty} C^k = [1]$.

PROOF   Since $\lim_{k\to\infty}\lambda^k$ does not exist unless $|\lambda| < 1$ or $\lambda = +1$, it follows from equation (3.2.7) that $\lim_{k\to\infty} C^k$ does not exist unless $|\lambda| < 1$ or $\lambda = +1$.

Conversely, if $|\lambda| < 1$, it can be shown, using L'Hospital's rule, that

$$\lim_{k\to\infty}\lambda^k = \lim_{k\to\infty} k\lambda^{k-1} = \lim_{k\to\infty}\frac{k(k-1)}{2}\lambda^{k-1} = \cdots$$

$$= \lim_{k\to\infty}\frac{k(k-1)\cdots(k-r+2)}{1\cdot 2\cdots(r-1)}\lambda^{k-r+1} = 0$$

Hence $\lim_{k\to\infty} C^k = 0$.

Finally, let us suppose $\lambda = +1$. If $r \geq 2$, then $\lim_{k\to\infty} C^k$ does not exist since $\lim_{k\to\infty} k\lambda^{k-1} = \lim_{k\to\infty} k$ does not exist. On the other hand, if $r = 1$, then $C = [1]$ and $\lim_{k\to\infty} C^k = [1]$.   ////

Since the minimal polynomial of the $r \times r$ matrix of (1.13.23) is $(x - \lambda)^r$, it follows that, in the event that $\lambda = +1$, $\lim_{k \to \infty} C^k$ exists if and only if $(x - 1)^2$ does not divide the minimal polynomial of C.

Now, let A be an arbitrary $n \times n$ matrix with entries from $\mathbb{C}$. Each elementary divisor of A is of the form $(x - \lambda)^r$ and gives rise (by Theorem 1.13.21) to an $r \times r$ matrix of the form of (1.13.23), having $(x - \lambda)^r$ as its minimal polynomial. By Lemma 3.2.8 and the discussion preceding it, $\lim_{k \to \infty} A^k$ exists if and only if each elementary divisor of A is of the form

$$(x - \lambda)^r \qquad \text{where} \quad |\lambda| < 1$$

or is of the form

$$x - 1$$

However, the minimal polynomial of A is the least common multiple of the elementary divisors of A. Thus we have proved the main theorem of this section.

Theorem 3.2.9 Let A be a square matrix with entries from $\mathbb{C}$. Then $\lim_{k \to \infty} A^k$ exists if and only if the following conditions are satisfied:

1 The absolute value of every characteristic root of A is at most 1.
2 The only characteristic root of A of absolute value 1 (if any) is $+1$.
3 $(x - 1)^2$ does not divide the minimal polynomial of A.

In Examples 3.1.11, 3.1.15, and 3.1.18 we were also faced with an infinite series of matrices. Just as when dealing with infinite series of numbers, we define convergence of an infinite series by means of partial sums. Thus, if

$$\alpha_0 I + \alpha_1 A + \alpha_2 A^2 + \cdots$$

is an infinite series in the square matrix A, we first define the *partial sums*

$$A_0 = \alpha_0 I$$
$$A_1 = \alpha_0 I + \alpha_1 A$$
$$A_2 = \alpha_0 I + \alpha_1 A + \alpha_2 A^2$$
$$\cdots \cdots \cdots \cdots \cdots \cdots \cdots$$

and then say that the infinite series $a_0 I + a_1 A + a_2 A^2 + \cdots$ *converges* to B if the sequence $A_0, A_1, A_2, \ldots$ converges to B.

Theorem 3.2.10 For a square matrix A, the following conditions are equivalent:

1 $\lim_{k \to \infty} A^k = 0$.
2 The absolute value of every characteristic root of A is less than 1.
3 $I + A + A^2 + \cdots$ converges.

Moreover, when one (and hence all) of these conditions are satisfied, we also have

4 $I - A$ is nonsingular.
5 $I + A + A^2 + \cdots$ converges to $(I - A)^{-1}$.

PROOF The equivalence of conditions 1 and 2 is contained in Lemma 3.2.8. Now, let A be a matrix which satisfies conditions 1 and 2. Clearly λ is a characteristic root of A if and only if $1 - \lambda$ is a characteristic root of $I - A$. Hence 0 is not a characteristic root of $I - A$ and it follows that $I - A$ is nonsingular. Then conditions 3 and 5 are immediate consequences of the identity

$$(I - A)^{-1} - (I + A + A^2 + \cdots + A^{k-1}) = (I - A)^{-1} A^k$$

Finally, let A be a matrix which satisfies condition 3. We may assume that A has the form of the direct sum of hypercompanion matrices. It is sufficient to consider the infinite series

$$I + C + C^2 + \cdots$$

where C has the form of (1.13.23). The diagonal entries of $I + C + C^2 + \cdots$ are

$$1 + \lambda + \lambda^2 + \cdots$$

which converges only when $|\lambda| < 1$. Hence A satisfies condition 2. ////

EXERCISES

1 Find $\lim_{k \to \infty} A^k$ where A is

(a) $\begin{bmatrix} 1 & -\frac{1}{2} & \frac{1}{2} \\ 0 & \frac{3}{2} & -1 \\ 0 & 1 & -\frac{1}{2} \end{bmatrix}$
(b) $\begin{bmatrix} -\frac{1}{2} & \frac{3}{2} & -\frac{3}{2} \\ 1 & 0 & -\frac{1}{2} \\ 1 & -1 & \frac{1}{2} \end{bmatrix}$

(a) $\begin{bmatrix} 2 & -4 & -8 \\ -3 & 12 & 21 \\ 2 & -8 & -14 \end{bmatrix}$
(d) $\begin{bmatrix} -\frac{1}{2} & 1 & -3 \\ 1 & -\frac{1}{2} & 3 \\ \frac{1}{2} & -\frac{1}{2} & 2 \end{bmatrix}$

2 Let

$$A = \begin{bmatrix} 0 & \frac{1}{2} & 0 \\ \frac{1}{2} & 0 & \frac{3}{2} \\ \frac{1}{2} & -\frac{1}{2} & \frac{3}{2} \end{bmatrix}$$

Find $\lim_{k \to \infty} A^k$ by (a) using Corollary 3.2.5 and (b) proving and using the recurrence relation

$$A^{k+2} = \left(2 - \frac{1}{2^k}\right)A^2 - \left(1 - \frac{1}{2^k}\right)A \qquad k = 1, 2, \dots$$

3 Let

$$A = \begin{bmatrix} 0 & -1 & 0 \\ 0 & \frac{1}{2} & 0 \\ 2 & 0 & 0 \end{bmatrix}$$

Find $I + A + A^2 + \cdots$ (a) by a direct calculation and (b) by means of Theorem 3.2.10.

4 Let

$$A = \begin{bmatrix} 1 & 0 & 1 \\ \frac{1}{2} & 0 & \frac{1}{2} \\ 0 & -\frac{1}{2} & 0 \end{bmatrix}$$

Show that $\lim_{k \to \infty} A^k = 0$ by (a) Theorem 3.2.10 and (b) proving and using the recurrence relation

$$A^{k+2} = \frac{k+1}{2^k} A^2 - \frac{k}{2^{k+1}} A \qquad k = 1, 2, 3, \dots$$

Find $I + A + A^2 + \cdots$ by (a) using Theorem 3.2.10 and (b) means of the above recurrence relation. (*Hint:* $\sum_{k=1}^{\infty} \frac{k}{2^k} = 2$.)

5 Prove Lemma 3.2.6.

6 Verify equation (3.2.7).

7 Show that if $\lim_{k \to \infty} A^k$ exists, then:

(a) $A\left(\lim_{k \to \infty} A^k\right) = \left(\lim_{k \to \infty} A^k\right)A = \lim_{k \to \infty} A^k$.

(b) $\left(\lim_{k \to \infty} A^k\right)^2 = \lim_{k \to \infty} A^k$.

(c) $\left(\lim_{k \to \infty} A^k\right)^T = \lim_{k \to \infty} (A^T)^k$.

(d) $\lim_{k \to \infty} (det\ A)^k = det\left(\lim_{k \to \infty} A^k\right)$.

8 Show that if λ is a characteristic root of the square matrix A of order n, then

$$|\lambda| \leq n\|A\|$$

9 Let A and B be square matrices of the same order such that $AB = BA$. Show that if $\lim_{k \to \infty} A^k$ and $\lim_{k \to \infty} B^k$ both exist, then

$$\lim_{k \to \infty} (AB)^k = \left(\lim_{k \to \infty} A^k \right) \left(\lim_{k \to \infty} B^k \right)$$

10 Find a pair of 2×2 matrices, A and B, such that $\lim_{k \to \infty} A^k$ and $\lim_{k \to \infty} B^k$ both exist but $\lim_{k \to \infty}(AB)^k$ does not exist.

11 Prove that $\alpha_0 I + \alpha_1 C + \alpha_2 C^2 + \cdots$ converges to B if and only if

$$\alpha_0 I + \alpha_1 P^{-1}CP + \alpha_2(P^{-1}CP)^2 + \cdots$$

converges to $P^{-1}BP$.

12 Show that the infinite series

$$e^A = I + \frac{1}{1!} A + \frac{1}{2!} A^2 + \frac{1}{3!} A^3 + \cdots$$

converges for every square matrix A. Find e^A when

(a) $A = I$ (b) $A = \begin{bmatrix} 2 & 0 \\ 1 & 2 \end{bmatrix}$ (c) $A = \begin{bmatrix} 3 & 1 \\ -1 & 1 \end{bmatrix}$

What are the characteristic roots of e^A?

13 Let A be the matrix of (1.13.23). Prove that

$$A^k = \begin{bmatrix} \frac{1}{0!} x^k|_{x=\lambda} & 0 & 0 & \cdots & 0 \\ \frac{1}{1!} Dx^k|_{x=\lambda} & \frac{1}{0!} x^k|_{x=\lambda} & 0 & \cdots & 0 \\ \frac{1}{2!} D^2x^k|_{x=\lambda} & \frac{1}{1!} Dx^k|_{x=\lambda} & \frac{1}{0!} x^k|_{x=\lambda} & \cdots & 0 \\ \cdots & \cdots & \cdots & \cdots & \cdots \\ \frac{1}{r!} D^r x^k|_{x=\lambda} & \frac{1}{(r-1)!} D^{r-1}x^k|_{x=\lambda} & \frac{1}{(r-2)!} D^{r-2}x^k|_{x=\lambda} & \cdots & \frac{1}{0!} x^k|_{x=\lambda} \end{bmatrix}$$

where $D^s x^k|_{x=\lambda}$ is the sth derivative of x^k evaluated at $x = \lambda$ and $0! = 1$.

14 Let A be a square matrix for which $L = \lim_{k \to \infty} A^k$ exists. Let Y, AY, $A^2 Y$, ..., $A^{r-1} Y$ be a cyclic basis for an r-dimensional A-invariant cyclic subspace $\mathbb{U}$ of $\mathbb{N}[(A - \lambda I)^e]$. Show that if $|\lambda| < 1$, then Y, AY, $A^2 Y$, ..., $A^{r-1} Y$ are characteristic vectors of L corresponding to the characteristic root 0. State and prove a corresponding result for $\lambda = 1$.

15 Prove that det A is a continuous function of A with respect to $\| \quad \|$; that is, show that for any $\varepsilon > 0$ there exists $\delta > 0$ such that

$$|\det B - \det A| < \varepsilon$$

whenever $\|B - A\| < \delta$.

3.3 A MATHEMATICAL INTRODUCTION TO THE THEORY OF FINITE MARKOV CHAINS

In Sec. 3.1 we began our study of the theory of finite Markov chains with a heuristic discussion and then presented a number of numerical examples. Throughout we drew extensively from the elementary theory of probability. It is beyond the scope of this book to present rigorous proofs (or even rigorous definitions) when dealing with the basics of probability theory. Thus, we now develop the subject of finite Markov chains once again, from the beginning, obtaining all our results as theorems about matrices and then interpreting them in the language of probability theory. The entries of these matrices will be taken either from the real field $\mathbb{R}$ or the complex field $\mathbb{C}$.

Definition 3.3.1 A vector $P = [p_j]$ belonging to $\mathbb{R}^n$ is a *probability vector* if

$$P \geq 0 \qquad \sum_{j=1}^{n} p_j = 1$$

A matrix $A = [a_{ij}]$ belonging to $\mathbb{R}_n{}^n$ is a *transition matrix* if each of its columns is a probability vector.

It is equivalent to define a transition matrix to be a nonnegative real matrix of order n whose transpose has

$$E^{(n)} = \begin{bmatrix} 1 \\ 1 \\ \vdots \\ 1 \end{bmatrix}$$

as a characteristic vector corresponding to the characteristic root $+1$.

Definition 3.3.2 A *Markov chain* is a pair $(A, P^{(0)})$, where A is a transition matrix and $P^{(0)}$ is a probability vector, called the *initial vector*.

We will interpret a_{ij} as the probability that an object in state S_j at a given time will be in state S_i at the "next" time. Also we will interpret the ith entry of $P^{(0)}$ as the probability of the object's initially being in the state S_i, that is, before the process begins. Further, we define[1] the probability vector $P^{(k)}$ by

$$P^{(k)} = A^k P^{(0)} \qquad k = 0, 1, 2, \ldots \qquad (3.3.3)$$

and say that the ith entry of $P^{(k)}$ is the probability of the object's being in the state S_i at time k.

Definition 3.3.4 A probability vector P is a *stationary vector* of the transition matrix A if $AP = P$, that is, if P is a characteristic vector of A corresponding to the characteristic root $+1$.

Clearly, if $P^{(0)}$ is a stationary vector of A, then

$$P^{(0)} = P^{(1)} = P^{(2)} = \cdots$$

The main theorem of this section (Theorem 3.3.26) states that every transition matrix has at least one stationary vector. When $\lim_{k \to \infty} A^k$ exists, the theorem follows immediately from part (a) of Exercise 7 in Sec. 3.2 and each column of $\lim_{k \to \infty} A^k$ is a stationary vector of A. In Theorem 3.3.18 we will show that $\lim_{k \to \infty} A^k$ exists for a large class of transition matrices, and in Theorem 3.3.22 we show how the results of Theorem 3.3.18 can be used to prove the more general Theorem 3.3.25.

Before beginning the study of transition matrices we present a few lemmas which will be needed later. These lemmas provide us with upper bounds for the absolute value of the characteristic roots of a matrix in terms of the entries of the matrix. Let $A = [a_{ij}]$ be a square matrix of order n. We set

$$\rho_i(A) = \sum_{j=1}^{n} |a_{ij}| \qquad i = 1, 2, \ldots, n$$

$$\gamma_j(A) = \sum_{i=1}^{n} |a_{ij}| \qquad j = 1, 2, \ldots, n$$

$$\rho(A) = \max_{i=1, 2, \cdots, n} \{\rho_i(A)\} \qquad \gamma(A) = \max_{j=1, 2, \cdots, n} \{\gamma_j(A)\}$$

and call $\rho(A)$ and $\gamma(A)$ the *row sum* and the *column sum* of A respectively.

Lemma 3.3.5 If λ is a characteristic root of A, then

$$|\lambda| \leq \min \{\rho(A), \gamma(A)\}$$

---

[1] There are many texts in which the development of finite Markov chains is based on probability theory and which make virtually no use of matrix methods (see, for example, Chung [1] or Parzen [8]). In such a treatment the basic tool is the Chapman-Kolmogorov theorem, which is equivalent to equation (3.3.3). A proof of this theorem would lead us too far afield into probability theory, and so we choose to "define" our way around this difficulty.

PROOF   Let $X = [x_i]$ be a characteristic vector of A corresponding to λ. Then X satisfies the matrix equation

$$AX = \lambda X$$

which can be written as a system of n simultaneous linear equations

$$\sum_{j=1}^{n} a_{ij} x_j = \lambda x_i \qquad i = 1, 2, \ldots, n \qquad (3.3.6)$$

Set

$$M = \max_{j=1,2,\ldots,n} |x_j|$$

Suppose this maximum takes place at $j = k$, that is,

$$M = |x_k| \geq |x_j| \qquad j = 1, 2, \ldots, n$$

From the equation of (3.3.6) in which $i = k$ we have

$$|\lambda| M = |\lambda| |x_k| = |\lambda x_k| = \left| \sum_{j=1}^{n} a_{kj} x_j \right| \leq \sum_{j=1}^{n} |a_{kj} x_j|$$

$$= \sum_{j=1}^{n} |a_{kj}| |x_j| \leq \sum_{j=1}^{n} |a_{kj}| M = \rho_k(A) M \leq \rho(A) M \qquad (3.3.7)$$

However, $M > 0$ since X, being a characteristic vector of A, is not the zero vector.   Thus

$$|\lambda| \leq \rho(A)$$

Now let us repeat this process using A^T and a characteristic vector of A^T corresponding to the characteristic root λ.   (Recall that A and A^T have the same characteristic roots.)   Then

$$|\lambda| \leq \rho(A^T) = \gamma(A)$$

and consequently

$$|\lambda| \leq \min \{\rho(A), \gamma(A)\} \qquad ////$$

Since every column sum of a transition matrix is $+1$, we have:

Corollary 3.3.8   If λ is a characteristic root of a transition matrix, then

$$|\lambda| \leq +1$$

Every transition matrix A does in fact have $+1$ as a characteristic root since, by definition, $+1$ is a characteristic root of A^T and $E^{(n)}$ is a corresponding characteristic vector.

It will be important for us to know when there is equality throughout (3.3.7). An inequality can occur at three places. In order for the first inequality to be an equality, i.e., in order that

$$\left| \sum_{j=1}^{n} a_{kj} x_j \right| = \sum_{j=1}^{n} |a_{kj} x_j|,$$

it is both necessary and sufficient that all the terms

$$a_{kj} x_j \qquad j = 1, 2, \ldots, n$$

be nonnegative real multiples of some fixed complex number, i.e., that there exist a complex number α and nonnegative real numbers $b_1, b_2, \ldots, b_n$ such that

$$a_{kj} x_j = b_j \alpha \qquad j = 1, 2, \ldots, n \qquad (3.3.9)$$

For the second inequality to be an equality it is necessary and sufficient to have, for each j, either

$$a_{kj} = 0 \qquad \text{or} \qquad |x_j| = M \qquad (3.3.10)$$

Finally, the third inequality is an equality if and only if

$$\rho_k(A) = \rho(A)$$

We will apply these results in proving:

Lemma 3.3.11   Let A be a positive square matrix of order n and let λ be a characteristic root of A for which

$$|\lambda| = \rho(A)$$

Then

1   $\lambda = \rho(A)$.
2   The absolute value of every other characteristic root of A is less than $\rho(A)$.
3   $E^{(n)}$ is the only characteristic vector of A corresponding to λ.

PROOF   Since $|\lambda| = \rho(A)$, the three inequalities of (3.3.7) are all equalities. Moreover, since every entry of A is positive, it follows from (3.3.10) that

$$|x_j| = M \qquad j = 1, 2, \ldots, n$$

Finally, according to equations (3.3.9), there exist nonnegative real numbers $b_1, b_2, \ldots, b_n$ and a number α such that

$$a_{kj} x_j = b_j \alpha \qquad j = 1, 2, \ldots, n$$

Then
$$x_j = \frac{b_j}{a_{kj}}\alpha \qquad j = 1, 2, \ldots, n$$

and hence $M = |x_j| = \left|\dfrac{b_j}{a_{kj}}\right| |\alpha| = \dfrac{b_j}{a_{kj}} |\alpha| \qquad j = 1, 2, \ldots, n$

since b_j and a_{kj} are real, b_j is nonnegative and a_{kj} is positive.   Consequently

$$x_j = \frac{b_j}{a_{kj}}\alpha = \frac{M}{|\alpha|}\alpha \qquad j = 1, 2, \ldots, n$$

that is
$$x_1 = x_2 = \cdots = x_n = \frac{M}{|\alpha|}\alpha$$

However, a nonzero multiple of a characteristic vector corresponding to the characteristic root λ is again a characteristic vector corresponding to λ. Thus we may set

$$x_1 = x_2 = \cdots = x_n = 1$$

and hence $X = E^{(n)}$.

We have now proved that every characteristic root of A whose absolute value is $\rho(A)$ has $E^{(n)}$ as a characteristic vector, and it follows that there is only one such characteristic root.   Combining this with Lemma 3.3.5, we may conclude that the absolute value of every other characteristic root of A is less than $\rho(A)$.

It remains only to show that $\lambda = \rho(A)$.   Since

$$AE^{(n)} = \lambda E^{(n)} \qquad (3.3.12)$$

and the left-hand side of equation (3.3.12) is a positive element of $\mathbb{R}^n$, λ is a positive number [whose absolute value is $\rho(A)$].   Thus $\lambda = \rho(A)$.   ////

Corollary 3.3.13   Let A be a positive transition matrix.   The only characteristic root of A of absolute value $+1$ is $+1$.   Moreover, the only characteristic vector of A^T corresponding to the root $+1$ is $E^{(n)}$.

In order to be able to conclude from Theorem 3.2.9 that $\lim_{k \to \infty} A^k$ exists when A is a positive transition matrix, it remains only to show that $(x - 1)^2$ does not divide the minimal polynomial of A.   Note that the first two conditions in Theorem 3.2.9 are contained in Corollaries 3.3.8 and 3.3.13.   According to Theorem 1.12.26, if $(x - 1)^e$ is the largest power of $x - 1$ which divides the minimal polynomial of A (or, equivalently, the minimal polynomial of A^T), then there exists a vector Y such that

$$(A^T - I)^{e-1}Y \neq 0 \qquad (A^T - I)^e Y = 0$$

Thus $(A^T - I)^{e-1}Y$ is a characteristic vector of A^T corresponding to the characteristic root $+1$. By Corollary 3.3.13, $(A^T - I)^{e-1}Y$ is a nonzero multiple of $E^{(n)}$. Hence we have shown that if $e \geq 2$, there exists a vector $X = (A^T - I)^{e-2}Y$ such that $(A^T - I)X = E^{(n)}$. Consequently, in order to show that $e = 1$, it is sufficient to prove:

Lemma 3.3.14 Let A be a transition matrix. There does not exist a vector X such that $(A^T - I)X = E^{(n)}$.

PROOF Suppose there does exist a vector X such that

$$A^T X = X + E^{(n)}$$

Multiplying this equation on the left by $E_{(n)}$, we have

$$E_{(n)}A^T X = E_{(n)}X + E_{(n)}E^{(n)} = E_{(n)}X + n$$

We can show by an induction on k that

$$E_{(n)}(A^T)^k X = E_{(n)}X + kn \qquad k = 1, 2, \ldots \qquad (3.3.15)$$

For, if we suppose that equation (3.3.15) has been established for k, then

$$E_{(n)}(A^T)^{k+1}X = E_{(n)}(A^T)^k[A^T X] = E_{(n)}(A^T)^k[X + E^{(n)}]$$

$$= E_{(n)}(A^T)^k X + E_{(n)}(A^T)^k E^{(n)}$$

$$= (E_{(n)}X + kn) + n = E_{(n)}X + (k+1)n$$

since $(A^T)^k E^{(n)} = E^{(n)}$.

Rewrite equation (3.3.15) as

$$kn = E_{(n)}(A^T)^k X - E_{(n)}X \qquad k = 1, 2, \ldots$$

By Lemma 3.2.2 we have

$$kn = \|E_{(n)}(A^T)^k X - E_{(n)}X\| \leq \|E_{(n)}(A^T)^k X\| + \|E_{(n)}X\|$$

$$\leq n^2\|E_{(n)}\| \|(A^T)^k\| \|X\| + n\|E_{(n)}\| \|X\|$$

$$\leq (n^2 + n)\|X\|$$

But n and $\|X\|$ are fixed while k is permitted to become arbitrarily large. Clearly this is impossible, and we have arrived at a contradiction. Hence, no such vector X exists. ////

Theorem 3.3.16 If A is a positive transition matrix, then $\lim_{k \to \infty} A^k$ exists. Moreover, $\lim_{k \to \infty} A^k > 0$.

PROOF   According to the discussion above, $\lim_{k \to \infty} A^k$ exists.   Since $A^k \geq 0$ for $k = 1, 2, \ldots$, it follows that $\lim_{k \to \infty} A^k \geq 0$.   Moreover, by Lemma 3.2.4, with $B = E_{(n)}$, $\lim_{k \to \infty} A^k$ is a transition matrix. Applying part (*a*) of Exercise 7 in Sec. 3.2, we have $\lim_{k \to \infty} A^k > 0$.   ////

Theorem 3.3.16 can be strengthened considerably.   Instead of A being positive, we will show that it is sufficient for A^h to be positive for some positive integer h.

Definition 3.3.17   A transition matrix is *regular* if some power of it is positive. A Markov chain is *regular* if its transition matrix is regular.

The transition matrix of Example 3.1.5 is positive and consequently regular.   The transition matrix of Example 3.1.9 is not positive but is regular since its square is positive.   The Markov chains of the remaining examples of Sec. 3.1 are nonregular.

Theorem 3.3.18   If A is a regular transition matrix, then:

1   the only characteristic root of A of absolute value $+1$ is $+1$.
2   the only characteristic vector of A^T corresponding to the root $+1$ is $E^{(n)}$
3 $\lim_{k \to \infty} A^k$ exists and is positive.
4   The columns of the matrix $L = \lim_{k \to \infty} A^k$ are identical, and each is the unique stationary vector of A.

PROOF   Let h be a positive integer for which

$$H = A^h > 0$$

H is also a transition matrix, and we may apply Corollary 3.3.13 to H. Let X be a characteristic vector of A^T corresponding to the characteristic root λ and let $|\lambda| = +1$.   Then X is also a characteristic vector of H^T corresponding to the characteristic root λ^h, and $|\lambda^h| = |\lambda|^h = +1$.   By Corollary 3.3.13

$$X = E^{(n)}$$

However, $E^{(n)}$ is a characteristic vector of A corresponding to the characteristic root $+1$.   Thus we have proved that $+1$ is the only characteristic root of A of absolute value $+1$ and that the only characteristic vector of A^T corresponding to the root $+1$ is $E^{(n)}$

If λ is a characteristic root of A and $\lambda \neq +1$, then $|\lambda| \neq +1$ and, by Corollary 3.3.8,

$$|\lambda| < 1$$

Furthermore, according to Lemma 3.3.14, $(x - 1)^2$ does not divide the minimal polynomial of A. It now follows from Theorem 3.2.9 that $\lim_{k \to \infty} A^k$ exists. Let

$$L = \lim_{k \to \infty} A^k$$

By Lemma 3.2.4 and part (a) of Exercise 7 in sec 3.2, L is a positive transition matrix.

In order to determine the rank of L it is necessary to examine the rational canonical form of A^T. Since $(x - 1)^2$ does not divide the minimal polynomial of A^T, every $(x - 1)$-elementary divisor of A^T gives rise to a characteristic vector corresponding to $+1$. But we have just shown that $E^{(n)}$ is the only such characteristic vector, and thus $x - 1$ appears only once among the elementary divisors of A^T. The other elementary divisors of A^T, and hence of A, are of the form $(x - \lambda)^e$, where $|\lambda| < 1$. By Theorem 1.13.8 there exists a nonsingular matrix P such that

$$A = P \begin{bmatrix} 1 & 0 \\ 0 & \tilde{A} \end{bmatrix} P^{-1}$$

where $\tilde{A}$ has order $n - 1$ and the absolute value of each of its characteristic roots is less than 1. By Lemma 3.2.8

$$\lim_{k \to \infty} \tilde{A}^k = 0$$

and hence

$$L = \lim_{k \to \infty} A^k = \lim_{k \to \infty} P \begin{bmatrix} 1 & 0 \\ 0 & \tilde{A} \end{bmatrix}^k P^{-1} = P \left(\lim_{k \to \infty} \begin{bmatrix} 1 & 0 \\ 0 & \tilde{A}^k \end{bmatrix} \right) P^{-1}$$

$$= P \begin{bmatrix} 1 & 0 \\ 0 & 0 \end{bmatrix} P^{-1} \tag{3.3.19}$$

The rank of L is 1, and each column of L is a scalar multiple of the first column. But the sum of the entries in each column of L is $+1$. Thus the columns of L are identical. Also,

$$AL = A \lim_{k \to \infty} A^k = \lim_{k \to \infty} A^{k+1} = L$$

and consequently each column of L is a stationary vector of A.

Since the rank of $A^T - I$ (and therefore of $A - I$) is $n - 1$, the stationary vector of A is unique.   ////

If S_j is a fixed state of a Markov chain with transition matrix A, then

$$\mathbb{U}_1 = \{S_i \mid a_{ij} \neq 0\}$$

is the set of states to which it is *possible* to go (i.e., the probability is positive) from S_j in exactly one step. Clearly a transition matrix is positive if and only if it is possible to go from any state to any state in one step. Similarly,

$$\mathbb{U}_2 = \{S_i \mid a_{ik} a_{kj} \neq 0 \text{ for some } k\}$$
$$= \{S_i \mid a_{ij}^{(2)} \neq 0\}$$

is the set of states to which it is possible to go from S_j in exactly two steps. More generally

$$\mathbb{U} = \{S_i \mid a_{ik_1} a_{k_1 k_2} \cdots a_{k_{l-1} j} \neq 0 \text{ for some } k_1, k_2, \ldots, k_{l-1}\}$$
$$= \{S_i \mid a_{ij}^{(l)} \neq 0\}$$

is the set of states to which it is possible to go from S_j in exactly l steps. A transition matrix is regular if and only if there is a positive number h such that it is possible to go from any state to any state in exactly h moves.

The main theorem of this section (Theorem 3.3.30) states that every Markov chain has a stationary vector. The existence of a unique stationary vector for a regular Markov chain is guaranteed by Theorem 3.3.18. In order to prove Theorem 3.3.30 we must first be able to transform an arbitrary transition matrix into one with a simple form. This is accomplished by a process which is equivalent to reordering the states of the Markov chain. (Actually the same results can be obtained without this reordering, but the notation would be terribly cumbersome and make even the simplest proof seem opaque.) We have seen in Example 3.1.11 that when the states are reordered, both the rows and the columns of the transition matrix must undergo the same reordering as was performed on the states. A reordering is, of course, a permutation, and the operation or performing the same permutation simultaneously on the rows and columns of a matrix is a similarity transformation using the corresponding permutation matrix. Similarity transformations leave unchanged the properties of transition matrices that we are primarily interested in, namely, the characteristic roots, the existence of characteristic vectors (though not the characteristic vectors themselves), the existence of the limit of powers, etc.

292 MATRIX THEORY AND FINITE MATHEMATICS

Definition 3.3.20 An $n \times n$ matrix is *reducible* if it can be brought into the form

$$\begin{bmatrix} C & 0 \\ B & D \end{bmatrix} \qquad (3.3.21)$$

where B, C, and D are $(n - l) \times l$, $l \times l$ and $(n - l) \times (n - l)$ matrices, respectively, by applying the same permutation to its rows and columns. A matrix which is not reducible is *irreducible*.

A Markov chain is *reducible* or *irreducible* as its transition matrix is reducible or irreducible.

If A is reducible, then every power of A is also reducible, for if

$$P^{-1}AP = \begin{bmatrix} C & 0 \\ B & D \end{bmatrix}$$

then $P^{-1}A^kP$ has the form

$$P^{-1}A^kP = \begin{bmatrix} C^k & 0 \\ B_k & D^k \end{bmatrix} \qquad k = 1, 2, \ldots$$

Consequently, a regular transition matrix is irreducible.

Since the transition matrices of Examples 3.1.5 and 3.1.9 are regular, they are also irreducible. On the other hand, Example 3.1.21 furnishes an example of an irreducible Markov chain which is not regular. The Markov chain of the remaining examples of Sec. 3.1 are reducible.

Note that if the matrix of (3.3.21) is a transition matrix, D is also a transition matrix, acting on the states $S_{l+1}, S_{l+2}, \ldots, S_n$. Once an object is in any of the states $S_{l+1}, S_{l+2}, \ldots, S_n$, it can *never* (i.e., with probability 0) afterwards appear in any of the states $S_1, S_2, \ldots, S_l$. We call such a set of states a *closed* set, i.e., a set of states with the property that whenever an object enters the set, it can never leave. Clearly the intersection and union of closed sets of states are closed.

A reducible Markov chain contains a proper closed set of states. Conversely, if there is a proper closed set of states in a Markov chain, we may reorder the states in such a way that the resulting transition matrix has the form of (3.3.20). Hence, *a transition matrix is irreducible if and only if the only closed set of states is the set of all states.*[1] A minimal closed set of states, i.e., a closed set of states which does not properly contain another closed set of states, is called an *ergodic set*. Since the intersection of two closed sets of states is again

---

[1] A more formal proof of a somewhat stronger version of this description of irreducible transition matrices is contained in Theorem 3.3.22.

a closed set (or is empty), it follows that two distinct ergodic sets of a Markov chain are disjoint. A state is called an *ergodic state* if it belongs to some ergodic set. A state which is not ergodic is called *transient*. Thus, a Markov chain is irreducible if and only if it contains one ergodic set consisting of all the states.

Since an object can never leave an ergodic set of states, it can never go from an ergodic state to a transient state or from one ergodic set to another ergodic set. It is possible for all the states of a Markov chain to be ergodic, e.g., when the transition matrix is irreducible.[1] However it is not possible for all the states of a Markov chain to be transient.

Theorem 3.3.22   Let A be a transition matrix of order n and let $B = \frac{1}{2}(I + A)$. Then A is irreducible if and only if B is regular. In this case, $B^n > 0$.

PROOF   Clearly B is a transition matrix. If A is reducible, then B is also reducible and consequently not regular. Thus, if B is regular, then A is irreducible.

Conversely, let A be irreducible. We wish to show that $B^n > 0$. Let j be fixed and define

$$\mathbb{U}_1 = \{S_i \mid b_{ij} \neq 0\}$$
$$\mathbb{U}_2 = \{S_i \mid b_{ik_1} b_{k_1 j} \neq 0 \text{ for some } k_1\}$$
$$\mathbb{U}_3 = \{S_i \mid b_{ik_1} b_{k_1 k_2} b_{k_2 j} \neq 0 \text{ for some } k_1, k_2\}$$
$$\cdots \cdots \cdots \cdots \cdots \cdots \cdots \cdots \cdots \cdots \cdots \cdots \cdots$$

We wish to show that

$$S_j \in \mathbb{U}_1 \subsetneqq \mathbb{U}_2 \subsetneqq \cdots \subsetneqq \mathbb{U}_h = \{S_1, S_2, \ldots, S_n\}$$

for some h; that is, we wish to show that

1 $\mathbb{U}_q \subseteq \mathbb{U}_{q+1}$ for all q.
2 $\mathbb{U}_q = \mathbb{U}_{q+1}$ if and only if $\mathbb{U}_q$ consists of all the states.

Let $S_i \in \mathbb{U}_q$. Then

$$b_{ik_1} b_{k_1 k_2} \cdots b_{k_{q-1} j} \neq 0 \qquad \text{for some } k_1, k_2, \ldots, k_{q-1}$$

Hence

$$b_{ik_1} b_{k_1 k_2} \cdots b_{k_{q-1} j} b_{jj} \neq 0$$

since $b_{jj} = \frac{1}{2}(1 + a_{jj}) > 0$. Thus $S_i \in \mathbb{U}_{q+1}$ and

$$\mathbb{U}_q \subseteq \mathbb{U}_{q+1}$$

----

[1] In a Markov chain with infinitely many states it is possible to have closed sets of states but no minimal closed sets of states. Then every state is transient (see Feller [5]).

Now let us assume that $U_h = U_{h+1}$ for some h. We will show that U_h is closed. For, suppose $S_r \in U_h$, $S_s \notin U_h$. There exists a nonzero product of h terms,

$$b_{rk_1} b_{k_1 k_2} \cdots b_{k_{h-1} j} \neq 0$$

But

$$b_{sr} b_{rk_1} b_{k_1 k_2} \cdots b_{k_{h-1} j} = 0$$

since $S_s \notin U_h = U_{h+1}$. Consequently

$$0 = b_{sr} = \tfrac{1}{2} a_{sr}$$

and an object can never go from a state S_r belonging to U_h to a state S_s not belonging to U_h. Thus U_h is closed. However since A is irreducible, no proper subset of states is closed and it follows that U_h contains all the states.

When U_q does not consist of all the states, then U_{q+1} contains at least one more state than U_q. Thus $h \leq n$, and we have

$$B^n > 0 \qquad ////$$

A Markov chain with transition matrix B can be obtained from the Markov chain whose transition matrix is A in the following way. At the beginning of each step of the new Markov chain some experiment (say, tossing a coin) which has two equally likely outcomes (say, H and T) is performed. If the outcome of the experiment is H, the step consists of the objects remaining in the state it is in. On the other hand, if the outcome of the experiment is T, then, for that step, the object behaves as though the transition matrix were A. Note that the sets of states of the two Markov chains are the same.

In Theorem 3.3.22 and the discussion preceding it we were interested in deciding whether it was possible to go from one state to another. We were not concerned with the numerical probability of this event, only with whether or not it was zero. When this is the case, we can describe the Markov chain by means of a diagram.

EXAMPLE 3.3.23 The movements that are possible for the mouse of Example 3.1.18 to make in one step are described by the arrows of Fig. 3.3.24. These can be summarized in the table

	S_1	S_2	S_3	S_4	S_5	S_6
First step	S_4	S_3	S_2	S_1	S_5	S_6
		S_5	S_6	S_5		

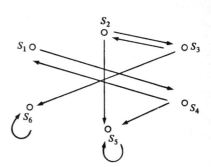

FIGURE 3.3.24

We can follow the arrows of Fig. 3.3.24 to find the additional states that can be reached from each state when more steps are taken.

	S_1	S_2	S_3	S_4	X	X
Second step	S_5	S_6	S_5			
Third step	X	X	X	X	X	X

Thus, no more states can be reached in three (or more) steps than can be reached in two.   Clearly $\{S_1, S_4, S_5\}, \{S_2, S_3, S_5, S_6\}, \{S_5\}$, and $\{S_6\}$ are closed sets of states.   Consequently $\{S_5\}, \{S_6\}$ are the ergodic sets and S_1, S_2, S_3, S_4 are transient states   ////

Theorem 3.3.22 implies that in an irreducible Markov chain it is possible to go from any state to any state in at most $n + 1$ moves.   For if

$$B^n = [\tfrac{1}{2}(I + A)]^n = \alpha_0 I + \alpha_1 A + \alpha_2 A^2 + \cdots + \alpha_n A^n > 0$$

where

$$\alpha_i = \frac{n!}{2^n i!(n - i)!} > 0$$

then

$$\alpha_0 A + \alpha_1 A^2 + \alpha_2 A^3 + \cdots + \alpha_n A^{n+1} > 0$$

and for any pair of states S_i and S_j there exists an integer r, $1 \leq r \leq n + 1$, such that the (i, j) entry of A^r is positive.   Hence it is possible to go from S_j to S_i in exactly r steps.

Theorem 3.3.18 can be extended (in somewhat weakened form) to irreducible transition matrices by means of Theorem 3.3.22.

Theorem 3.3.25   If A is an irreducible transition matrix, then:

1   The only characteristic vector of A^T corresponding to the root $+1$ is $E^{(n)}$.

2   A has a unique stationary vector.

3   The stationary vector of A is positive.

In fact, if $B = \frac{1}{2}(I + A)$, then $\lim_{k \to \infty} B^k$ exists and is positive and each of its columns is the stationary vector of A.

PROOF   Note that X is a characteristic vector of A (or A^T) corresponding to the root λ if and only if X is a characteristic vector of B (or B^T) corresponding to $\frac{1}{2}(1 + \lambda)$.   But[1]

$$\frac{1}{2}(1 + \lambda) = 1 \qquad \text{if and only if} \qquad \lambda = 1$$

It now follows from Theorem 3.3.18 that $E^{(n)}$ is the only characteristic vector of A^T corresponding to the characteristic root $+1$.

Since B is a regular transition matrix, Theorem 3.3.18 states that $\lim_{k \to \infty} B^k$ exists and is positive and each of its columns is the unique stationary vector of B (and hence of A).   ////

In Sec. 3.5 we will need:

Corollary 3.3.26   If X is the stationary vector of an irreducible transition matrix A and L is the square matrix each of whose columns is X, then there is a nonsingular matrix P such that

$$A = P\begin{bmatrix} 1 & 0 \\ 0 & \tilde{A} \end{bmatrix}P^{-1} \qquad L = P\begin{bmatrix} 1 & 0 \\ 0 & 0 \end{bmatrix}P^{-1}$$

PROOF   Let $B = \frac{1}{2}(I + A)$.   In the proof of Theorem 3.3.18 we showed that

$$B = P\begin{bmatrix} 1 & 0 \\ 0 & \tilde{B} \end{bmatrix}P^{-1} \qquad L = P\begin{bmatrix} 1 & 0 \\ 0 & 0 \end{bmatrix}P^{-1}$$

for some nonsingular matrix P.   Set $\tilde{B} = \frac{1}{2}(I + \tilde{A})$.   ////

Let $S_1, S_2, \ldots, S_r$ be the ergodic sets of the transition matrix A and let S_i consist of n_i states.   Then A has $n_1 + n_2 + \cdots + n_r$ ergodic states and $l = n - n_1 - n_2 - \cdots - n_r$ transient states.   Since the sets $S_1, S_2, \ldots, S_r$ are disjoint, we can permute the states so that A has the form

$$A = \begin{bmatrix} C & 0 & 0 & \cdots & 0 \\ B_1 & D_1 & 0 & \cdots & 0 \\ B_2 & 0 & D_2 & \cdots & 0 \\ \cdots\cdots\cdots\cdots\cdots\cdots\cdots \\ B_r & 0 & 0 & \cdots & D_r \end{bmatrix} \qquad (3.3.27)$$

---

[1] It does not follow from part 1 of Theorem 3.3.18 that the only characteristic root of an irreducible transition matrix of absolute value 1 is 1.   For if λ is a characteristic root of A such that

$$|\lambda| = 1 \qquad \lambda \neq 1$$

then

$$|\tfrac{1}{2}(1 + \lambda)| < 1$$

where D_i is an irreducible $n_i \times n_i$ transition matrix, $i = 1, 2, \ldots, r$, C is a square matrix of order l, and the states corresponding to C are transient. We call (3.3.27) the *standard form* for a transition matrix. Recall that in Example 3.1.11 we found it convenient to put the transition matrix into standard form in order to facilitate the computations.

For any state S_j, the set of states consisting of S_j and all states which can be reached from S_j in a finite number of steps is a closed set. If S_j is a transient state, this set contains a proper closed subset, say $\mathbb{T}_j$, which does not contain S_j. Moreover, some subset of $\mathbb{T}_j$ is an ergodic set, and hence it is always possible to go from a transient state to some ergodic state in a finite number of steps. Let us interpret this result in terms of the matrix of (3.3.27).

Lemma 3.3.28   There exists a positive integer k such that $\gamma(C^k) < 1$.

PROOF   When the matrix of (3.3.27) is expressed as

$$A = \begin{bmatrix} C & 0 \\ B & D \end{bmatrix}$$

then $\qquad\qquad A^k = \begin{bmatrix} C^k & 0 \\ B_k & D^k \end{bmatrix} \qquad k = 1, 2, \ldots$

According to the discussion above, for each transient state S_j, there is a positive integer k_j for which at least one of the entries in the jth column of B_{k_j} is positive. Hence the sum of the entries in the jth column of C^{k_j} is less than 1. Moreover, the sum of the entries in the jth column of C^p is less than 1 for every integer $p \geq k_j$. If we set

$$k = \max_{j=1,2,\ldots,l} \{k_j\}$$

then $\qquad\qquad\qquad\qquad \gamma(C^k) < 1 \qquad\qquad$ ////

Corollary 3.3.29 $\lim_{k\to\infty} C^k = 0$, and consequently $1 + C + C^2 + \cdots$ converges to $(I - C)^{-1}$.

PROOF   According to Theorem 3.2.10, it is sufficient to show that the absolute value of every characteristic root of C is less than 1. Let λ be a characteristic root of C. Then λ^k is a characteristic root of C^k and, by Lemma 3.3.5, $|\lambda^k| < 1$. Consequently

$$|\lambda| < 1 \qquad\qquad$$ ////

Once an object has entered one of the states of the ergodic set $\mathbb{S}_i$, it may never leave the states of $\mathbb{S}_i$. The states of the Markov chain other than those of $\mathbb{S}_i$ can thereafter be ignored, and the object can be thought of as being in a Markov chain whose transition matrix is D_i.

Theorem 3.3.30 To each ergodic set of states $\mathbb{S}_i$ of a finite Markov chain there corresponds a stationary vector X_i having the property that the positions corresponding to the states of $\mathbb{S}_i$ contain positive entries and all other entries of X_i are zero.   Every stationary vector is a convex combination of these.

PROOF   Let A be a transition matrix in standard form and let Y_i be the (unique) stationary vector of the irreducible transition matrix D_i, $i = 1, 2, \ldots, r$.   Then the r n-dimensional vectors

$$X_1 = \begin{bmatrix} 0 \\ Y_1 \\ 0 \\ 0 \\ \vdots \\ 0 \end{bmatrix}, \quad X_2 = \begin{bmatrix} 0 \\ 0 \\ Y_2 \\ 0 \\ \vdots \\ 0 \end{bmatrix}, \quad X_3 = \begin{bmatrix} 0 \\ 0 \\ 0 \\ Y_3 \\ \vdots \\ 0 \end{bmatrix}, \quad \ldots, \quad X_r = \begin{bmatrix} 0 \\ 0 \\ 0 \\ 0 \\ \vdots \\ Y_r \end{bmatrix} \begin{matrix} l \\ n_1 \\ n_2 \\ n_3 \\ \\ n_r \end{matrix}$$

(3.3.31)

are linearly independent stationary vectors of A.

Conversely, suppose that

$$X = \begin{bmatrix} Z_0 \\ Z_1 \\ Z_2 \\ \vdots \\ Z_r \end{bmatrix}$$

is a stationary vector of the matrix A of (3.3.27).   A direct calculation shows that

$$CZ_0 = Z_0 \qquad (3.3.32)$$

However, by Lemma 3.3.5, $+1$ is not a characteristic root of C^k for the value of k found in Corollary 3.3.29.   Consequently $+1$ is not a characteristic root of C, and equation (3.3.32) yields

$$Z_0 = 0$$

It is now clear that for each i, $i = 1, 2, \ldots, r$,

$$D_i Z_i = Z_i$$

Hence Z_i is a nonnegative multiple of the unique stationary vector Y_i of D_i, say $Z_i = \alpha_i Y_i$, $\alpha_i \geq 0$.   Since the sum of the entries of each of the vectors X, Y_1, Y_2, $\ldots$, Y_r is $+1$, it follows that

$$\alpha_1 + \alpha_2 + \cdots + \alpha_r = 1$$

and X is a convex combination of the vectors of (3.3.31).   ////

EXERCISES

1 Find the ergodic sets and the transient states of the Markov chains in Examples 3.1.15, 3.1.18, and 3.1.21. Find all stationary vectors of these transition matrices by applying Theorem 3.3.30.

2 Find the ergodic sets, the transient states, and the stationary vectors of the Markov chains in Exercises 6 and 9 to 12 of Sec. 3.1.

3 Identify the ergodic sets and the transient states of the Markov chain corresponding to the following mazes:

A   B

Find the standard form of the transition matrix and the stationary vectors.

4 Find the ergodic sets and the transient states of the Markov chain whose transition matrix is:

$$(a) \begin{bmatrix} 0 & 0 & 0 & \frac{1}{2} & 0 \\ 1 & 0 & 0 & \frac{1}{2} & 0 \\ 0 & \frac{1}{2} & 1 & 0 & 0 \\ 0 & \frac{1}{4} & 0 & 0 & 1 \\ 0 & \frac{1}{4} & 0 & 0 & 0 \end{bmatrix} \quad (b) \begin{bmatrix} 0 & 0 & \frac{1}{2} & 0 & 0 \\ 0 & 0 & 0 & \frac{1}{3} & 1 \\ 1 & 0 & \frac{1}{2} & 0 & 0 \\ 0 & 1 & 0 & \frac{1}{3} & 0 \\ 0 & 0 & 0 & \frac{1}{3} & 0 \end{bmatrix} \quad (c) \begin{bmatrix} \frac{1}{2} & \frac{2}{3} & 0 & 1 & 0 \\ \frac{1}{2} & 0 & \frac{1}{2} & 0 & 0 \\ 0 & 0 & \frac{1}{2} & 0 & 1 \\ 0 & \frac{1}{6} & 0 & 0 & 0 \\ 0 & \frac{1}{6} & 0 & 0 & 0 \end{bmatrix}.$$

Find the standard form of each transition matrix and a reordering of the states which yields it.

5 Let A be a regular transition matrix and let X be its stationary vector. Show that

$$\lim_{k \to \infty} P^{(k)} = X$$

for every initial probability vector $P^{(0)}$.

6 Let S_i be a transient state of a Markov chain. Show that

$$\lim_{k \to \infty} p_i^{(k)} = 0$$

for every initial probability vector $P^{(0)}$.

7 Let A_1, A_2 be transition matrices of the same order neither of which has a row consisting only of 0s. Show that $A_1 A_2$ is regular if and only if $A_2 A_1$ is regular. Show that $A_1 A_2$ is irreducible if and only if $A_2 A_1$ is irreducible.

8 Refine the argument of Theorem 3.3.21 to show that $B^{n-1} > 0$.

9 Let A be a regular transition matrix and let

$$L = \lim_{k \to \infty} A^k$$

Show that:

(a) $(A - L)^k = A^k - L$ for $k = 1, 2, \dots$.

(b) $I + (A - L) + (A^2 - L) + \cdots$ converges to $[I - (A - L)]^{-1}$.

Set $Z = [I - (A - L)]^{-1}$ so that $Z = \lim_{n \to \infty}[(\sum_{k=0}^{n} A^k) - nL]$. Show that

(c) $E_{(n)}Z = E_{(n)}$.

(d) $ZL = LZ = L$.

(e) $I - Z = L - ZA$.

10   Let A be a transition matrix of order n and let $X = [x_i]$ be an n-dimensional row vector. Set $Y = XA = [y_j]$ and

$$M_0 = \max_{i=1,2,\dots,n} x_i \qquad m_0 = \min_{i=1,2,\dots,n} x_i$$

$$M_1 = \max_{j=1,2,\dots,n} y_j \qquad m_1 = \min_{j=1,2,\dots,n} y_j$$

$$a = \min_{i,j=1,2,\dots\,n} a_{ij}$$

Show that $M_1 \leq M_0$, $m_1 \geq m_0$ and

$$M_1 - m_1 \leq (1 - 2a)(M_0 - m_0)$$

Use this result to give another proof that if A is a positive transition matrix, then $\lim_{k \to \infty} A^k$ exists and is positive and all its columns are identical.

11   Complete the following proof that the geometric multiplicity of $+1$ as a characteristic root of a positive transition matrix is 1.

If X is a characteristic vector of A^T corresponding to the characteristic root of $+1$ which is linearly independent with $E^{(n)}$, then there exists a non-negative characteristic vector Y of A^T corresponding to the root $+1$ which has zero components. Contradiction.

Extend this method to prove the corresponding theorem for regular transition matrices.

12   Show that the closed sets of states in the Markov chains with matrices A and $B = \frac{1}{2}(I + A)$ are identical. (Hence the ergodic sets and transient states are identical.)

13   Let A be a transition matrix and define

$$B_m = \frac{1}{m}(I + A + A^2 + \cdots + A^{m-1})$$

Show that A is irreducible if and only if B is regular. Show, moreover, that if A is irreducible, then $\lim_{k \to \infty}(B_m)^k$ exists for $m \geq 2$ and

$$\lim_{k \to \infty}(B_m)^k = \lim_{k \to \infty}(B_q)^k = L \qquad \text{for } m, q \geq 2$$

14 Let A be a transition matrix. Prove that A is singular if and only if there exists distinct *probability vectors* X_1, X_2 such that

$$AX_1 = AX_2$$

[*Hint:* Show that a characteristic vector of A corresponding to a characteristic root other than $+1$ belongs to $\mathbb{N}(E_{(n)})$.]

15 Prove that an irreducible transition matrix A for which $\lim_{k \to \infty} A^k$ exists is regular. (*Hint:* Prove $\lim_{k \to \infty} A^k > 0$.)

16 Show that the first column of the matrix P of Theorem 3.3.18 is a scalar multiple of the stationary vector of A, say $P^{(1)} = \alpha X$. Show that the first row of P^{-1} is a scalar multiple of $E_{(n)}$. In fact, $[P^{-1}]_{(1)} = (1/\alpha)E_{(n)}$.

17 Show that every closed set of states in the Markov chain with transition matrix A is a closed set of states in the Markov chain with transition matrix A^2 but that the converse need not be true. Formulate this statement in terms of the transient states and the ergodic sets of each Markov chain.

18 Prove that if A is a nonsingular transition matrix and $A^{-1} \geqq 0$, then A^{-1} is a transition matrix.

19 Show that if A and A^{-1} are both transition matrices, then A is a permutation matrix.

20 Let A be an irreducible transition matrix of order n. Prove that every submatrix of $I - A$ of order $n - 1$ is nonsingular.

21 Let A be an irreducible nonregular transition matrix. Show that A has a characteristic root λ such that $|\lambda| = 1$, $\lambda \neq +1$.

22 Construct an irreducible nonregular transition matrix of order 2. Show that it is unique.

3.4 PERIODIC TRANSITION MATRICES

In this section we will prove that an irreducible nonregular transition matrix has (possibly after a simultaneous permutation of its rows and columns) a particularly simple form. First we will show (in Theorem 3.4.2) one way in which irreducible nonregular transition matrices can be found. Then (in Theorem 3.4.7) we will prove that all irreducible transition matrices are of this type.

We can construct an irreducible nonregular transition matrix as follows. Let $m_1, m_2, \ldots, m_l, l \geqq 2$, be positive integers and let $B_1, B_2, \ldots, B_l$ be $m_2 \times m_1$, $m_3 \times m_2, \ldots, m_l \times m_{l-1}, m_1 \times m_l$ matrices respectively satisfying the conditions

1 $B_i \geqq 0$, $i = 1, 2, \ldots, l$.
2 The sum of the entries in each column of each B_i is 1.

Set

$$A = \begin{bmatrix} 0 & 0 & 0 & \cdots & 0 & B_l \\ B_1 & 0 & 0 & \cdots & 0 & 0 \\ 0 & B_2 & 0 & \cdots & 0 & 0 \\ 0 & 0 & B_3 & \cdots & 0 & 0 \\ \multicolumn{6}{c}{\dotfill} \\ 0 & 0 & 0 & \cdots & B_{l-1} & 0 \end{bmatrix} \quad (3.4.1)$$

Then:

Theorem 3.4.2 The matrix A of (3.4.1) is a nonregular transition matrix. If λ is any lth root of unity[1], then λ is a characteristic root of A (and consequently of A^T) and the vector

$$Y = \begin{bmatrix} E^{(m_1)} \\ \lambda E^{(m_2)} \\ \vdots \\ \lambda^{l-1} E^{(m_l)} \end{bmatrix}$$

is a characteristic vector of A^T corresponding to λ.

Moreover

$$A_1 = B_l B_{l-1} \quad \cdots \quad B_2 B_1$$
$$A_2 = {}_1 B_l \quad \cdots \quad B_3 B_2$$
$$\dotfill$$
$$A_l = B_{l-1} B_{l-2} \quad \cdots \quad B_1 B_l$$

are transition matrices and A is irreducible if and only if $A_1, A_2, \ldots, A_l$ are irreducible.

PROOF A direct computation will verify that if λ is an lth root of unity, then Y is a characteristic vector of A^T corresponding to the characteristic root λ. Also, it is clear that $A_1, A_2, \ldots, A_l$ and A are square nonnegative matrices of orders $m_1, m_2, \ldots, m_l$ and $m_1 + m_2 + \cdots + m_l$, respectively. The sum of the entries in each column of each matrix is 1, and hence $A_1, A_2, \ldots, A_l$ and A are transition matrices.

It remains only to show that A is not regular and that A is irreducible if and only if $A_1, A_2, \ldots, A_l$ are irreducible. First let us compute $A^2, A^3, \ldots, A^l$ from (3.4.1). For each j, $j = 1, 2, \ldots, l$, A^j has exactly

---

[1] Recall that λ is an lth *root of unity* if $\lambda^l = 1$. The lth roots of unity are

$$\exp \frac{2\pi i p}{l} = \cos \frac{2\pi i p}{l} + i \sin \frac{2\pi i p}{l} \qquad p = 0, 1, 2, \ldots, l-1$$

l nonzero blocks, and these are located in the positions j places below the main diagonal and $l - j$ places above the main diagonal. For example,

$$A^2 = \begin{bmatrix} 0 & 0 & \cdots & 0 & B_1 B_{l-1} & 0 \\ 0 & 0 & \cdots & 0 & 0 & B_1 B_l \\ B_2 B_1 & 0 & \cdots & 0 & 0 & 0 \\ 0 & B_3 B_2 & \cdots & 0 & 0 & 0 \\ \cdots\cdots\cdots\cdots\cdots\cdots\cdots\cdots\cdots\cdots\cdots \\ 0 & 0 & \cdots & B_{l-1} B_{l-2} & 0 & 0 \end{bmatrix}$$

and

$$A^l = \begin{bmatrix} A_1 & 0 & \cdots & 0 \\ 0 & A_2 & \cdots & 0 \\ \cdots\cdots\cdots\cdots\cdots \\ 0 & 0 & \cdots & A_l \end{bmatrix}$$

Since A^{ql} is a diagonal block matrix for $q = 1, 2, \ldots$, it follows that A is not regular.

Finally, suppose that $A_1, A_2, \ldots, A_l$ are irreducible and set

$$m = \max_{i=1, 2, \ldots, l} \{m_i\}$$

According to Theorem 3.3.21,

$$(I + A_i)^m > 0 \qquad i = 1, 2, \ldots, l$$

Hence each of the l diagonal blocks of $(I + A^l)^m$ is positive. More generally, for each j, the l nonzero blocks of $(I + A^l)^m A^j$ are positive. Thus

$$(I + A^l)^m (I + A + A^2 + \cdots + A^{l-1}) > 0$$

If we set $B = \frac{1}{2}(I + A)$, then B is regular and hence A is irreducible.

Conversely, let A be irreducible. Then $(I + A)^n > 0$. By expanding $(I + A)^n$ we see that the ith diagonal block of $(I + A)^n$ is of the form

$$I + \alpha_1 A_i + \alpha_2 A_i^2 + \cdots + \alpha_k A_i^k$$

where $\alpha_1, \alpha_2, \ldots, \alpha_k$ are positive integers and k is the largest integer such that kl is less than or equal to n. Thus

$$(I + A_i)^k > 0$$

and consequently A_i is irreducible.   ////

Definition 3.4.3 A transition matrix is *periodic* if it can be brought into the form (3.4.1) by performing the same permutation of its rows and columns.

A Markov chain whose transition matrix is periodic is a *periodic Markov chain*. An object initially in one of the first m_1 states of a Markov chain whose matrix is given by (3.4.1) will be in one of the next m_2 states after the first time period. The object will be in one of the following m_3 states after the second time period, etc. After l time periods have elapsed, the object will be in one of the first m_1 states and the process is repeated. The largest integer l for which the set of states of a Markov chain with matrix A can be partitioned into l disjoint sets such that at each step the object passes from one set of states into the next (and after l steps has returned to the first set of states) is called the *period* of A. That is, A is a periodic transition matrix with period l if it can be brought into the $l \times l$ block form (3.4.1), but it cannot be brought into this form for any integer greater than l by performing the same permutation of its rows and columns.

Theorem 3.4.2 gives a specific example of an irreducible nonregular transition matrix. In Theorem 3.4.7 we will prove that every irreducible nonregular transition matrix can be brought to form (3.4.1) and hence is periodic. The proof of Theorem 3.4.7 is based upon a characteristic root which is not equal to 1 but whose absolute value is 1. In order to establish the existence of this characteristic root we need

Lemma 3.4.4 If A is an irreducible nonregular transition matrix, then $\lim_{k \to \infty} A^k$ does not exist.

PROOF Let $B = \frac{1}{2}(I + A)$. According to Theorem 3.3.25, $\lim_{k \to \infty} B^k$ exists and is positive and each of its columns is the unique stationary vector of B and likewise of A. Now, suppose that $\lim_{k \to \infty} A^k$ exists. Then each column of $\lim_{k \to \infty} A^k$ is a stationary vector of A, and hence

$$\lim_{k \to \infty} A^k = \lim_{k \to \infty} B^k > 0 \qquad (3.4.5)$$

Equation (3.4.5) implies that A is regular, which is a contradiction. ////

Theorem 3.2.9 gives a convenient set of necessary and sufficient conditions that $\lim_{k \to \infty} A^k$ exists for an arbitrary square matrix A with elements from the complex field $\mathbb{C}$. When A is an irreducible nonregular transition matrix, at least one of these conditions must be violated. According to Corollary 3.3.8 the first condition (namely, that the characteristic roots of A have absolute value at most 1) is satisfied for every transition matrix. By Theorem 3.3.25, $E^{(n)}$ is the only characteristic vector of A^T corresponding to the root $+1$. Applying Lemma 3.3.14, we see that the third condition [namely, that $(x - 1)^2$ does not divide the minimal polynomial of A] is also satisfied. Hence the second condition must be violated; that is:

Lemma 3.4.6 An irreducible nonregular transition matrix has a characteristic root λ such that

$$|\lambda| = 1 \qquad \lambda \neq 1$$

It is the existence of this characteristic root that yields form (3.4.1) for an irreducible nonregular transition matrix. The proof is similar to the proof of Lemma 3.3.11 and is applied to A^T since

$$|\lambda| = 1 = \gamma(A) = \rho(A^T)$$

Theorem 3.4.7 An irreducible nonregular transition matrix is periodic.

PROOF Let A be an irreducible nonregular transition matrix. We wish to show that A^T can be brought into a form which is the transpose of (3.4.1) by a simultaneous permutation of its rows and columns. Such a permutation induces the same permutation among the entries of each characteristic vector.

Let $A = [a_{ij}]$, $A^T = [c_{ij}]$ (that is, $c_{ij} = a_{ji}$) and let $Y = [y_j]$ be a characteristic vector of A^T, corresponding to the characteristic root λ of Lemma 3.4.7. Then

$$\sum_{j=1}^{n} c_{ij} y_j = \lambda y_i \qquad i = 1, 2, \ldots, n \qquad (3.4.8)$$

Set

$$M = \max_{j=1,2,\ldots,n} \{|y_j|\} > 0$$

We may assume (possibly after a permutation of the rows and columns of A^T and multiplying Y by a scalar of absolute value $+1$) that

$$M = y_1 = y_2 = \cdots = y_{m_1} \qquad y_j \neq M \quad \text{for} \quad j > m_1 \qquad (3.4.9)$$

Since $\lambda \neq 1$, it follows that $Y \neq E^{(n)}$ and $m_1 < n$. The first m_1 equations of (3.4.8) now become

$$\sum_{j=1}^{n} c_{ij} y_j = \lambda y_i = \lambda M \qquad i = 1, 2, \ldots, m_1 \qquad (3.4.10)$$

and, as in (3.3.7), we have

$$M = |\lambda| M = \left| \sum_{j=1}^{n} c_{ij} y_j \right| \leq \sum_{j=1}^{n} |c_{ij} y_j| = \sum_{j=1}^{n} c_{ij} |y_j|$$

$$\leq \left(\sum_{j=1}^{n} c_{ij} \right) M = \rho_i(A^T) M = M \qquad (3.4.11)$$

Thus there is equality throughout each of the m_1 inequalities of (3.4.11).
Since the first inequality of (3.4.11) is actually an equality, it follows
that:

> *1* For each i, $1 \leq i \leq m_1$, there exists a complex number α_i such that
> each of the terms $c_{ij} y_j$, $j = 1, 2, \ldots, n$, is a real, nonnegative multiple
> of α_i. By equation (3.4.10), we may take
>
> $$\alpha_1 = \alpha_2 = \cdots = \alpha_{m_1} = \lambda M$$
>
> and hence, for $1 \leq i \leq m_1$, each of the terms $c_{ij} y_j$ is a real non-
> negative multiple of λM.

Since the second inequality of (3.4.11) is actually an equality, it follows
that:

> *2* $|y_j| = M$ whenever $c_{ij} \neq 0$ for some i, $1 \leq i \leq m_1$.

From 1 and 2 we conclude that for $1 \leq i \leq m_1$,

$$c_{ij} = 0 \quad \text{whenever} \quad y_j \neq \lambda M$$

In particular, according to equation (3.4.8), we have

$$c_{ij} = 0 \quad i, j = 1, 2, \ldots, m_1$$

Let us now suppose that

$$y_{m_1+1} = y_{m_1+2} = \cdots = y_{m_1+m_2} = \lambda M$$
$$y_j \neq \lambda M \quad \text{for} \quad j > m_1 + m_2 \tag{3.4.12}$$

Then we also have

$$c_{ij} = 0 \quad \begin{array}{l} i = 1, 2, \ldots, m_1 \\ j > m_1 + m_2 \end{array}$$

and hence the first m_1 rows of A^T can be partitioned into

$$[0 \quad B_1 \quad 0]$$

where the two zero matrices have dimensions $m_1 \times m_1$ and $m_1 \times (n - m_1 - m_2)$, respectively, and B_1 is an $m_1 \times m_2$ real nonnegative matrix for which the sum of the entries in each row is 1. Moreover, the first $m_1 + m_2$ entries of Y have the form

$$\begin{bmatrix} ME^{(m_1)} \\ \lambda ME^{(m_2)} \end{bmatrix}$$

We can continue this process. When we substitute equation (3.4.12) into equation (3.4.8), the next m_2 equations of (3.4.8) (after the first m_1 equations) become

$$\sum_{j=1}^{n} c_{ij} y_j = \lambda y_i = \lambda^2 M \qquad i = m_1 + 1, m_1 + 2, \ldots, m_1 + m_2$$

As before, we conclude that

$$c_{ij} = 0 \qquad \text{whenever} \qquad y_j \neq \lambda^2 M$$

In particular, we know from equation (3.4.12) that

$$y_j = \lambda M \neq \lambda^2 M \qquad j = m_1 + 1, m_1 + 2, \ldots, m_1 + m_2$$

Thus $\quad c_{ij} = 0 \qquad i,j = m_1 + 1, m_1 + 2, \ldots, m_1 + m_2$

At this point two cases arise.

CASE 1 Suppose that $\lambda^2 \neq 1$. When $\lambda^2 \neq 1$, then

$$y_j \neq \lambda^2 M \qquad j = 1, 2, \ldots, m_1$$

and hence
$$c_{ij} = 0 \qquad \begin{array}{l} i = m_1 + 1, m_1 + 2, \ldots, m_1 + m_2 \\ j = 1, 2, \ldots, m_1 \end{array}$$

We may assume

$$y_{m_1 + m_2 + 1} = y_{m_1 + m_2 + 2} = \cdots = y_{m_1 + m_2 + m_3} = \lambda^2 M$$
$$y_j \neq \lambda^2 M \qquad j > m_1 + m_2 + m_3$$

the first $m_1 + m_2$ rows of A^T can be partitioned into

$$\begin{array}{ccccc} & m_1 & m_2 & m_3 & n - m_1 - m_2 - m_3 \\ m_1 & \begin{bmatrix} 0 & B_1 & 0 & 0 \\ m_2 & 0 & 0 & B_2 & 0 \end{bmatrix} \end{array}$$

and the first $m_1 + m_2 + m_3$ entries of Y have the form

$$\begin{bmatrix} ME^{(m_1)} \\ \lambda ME^{(m_2)} \\ \lambda^2 ME^{(m_3)} \end{bmatrix}$$

We examine the next block row of A^T in exactly the same way and are again faced with two cases, the first of which (namely, $\lambda^3 \neq 1$) can be treated exactly as before. Since the order of A^T is finite, we eventually reach a point at which the first $m_1 + m_2 + \cdots + m_{l-1}$ rows of A^T are given by

$$
\begin{array}{c}
\begin{array}{cccccccc}
m_1 & m_2 & m_3 & & m_{l-1} & m_l & n - m_1 - m_2 - \cdots - m_l
\end{array} \\
\begin{array}{c}
m_1 \\ m_2 \\ m_3 \\ \\ m_{l-1}
\end{array}
\left[
\begin{array}{ccccccc}
0 & B_1 & 0 & \cdots & 0 & 0 & 0 \\
0 & 0 & B_2 & \cdots & 0 & 0 & 0 \\
0 & 0 & 0 & \cdots & 0 & 0 & 0 \\
\multicolumn{7}{c}{\cdots\cdots\cdots\cdots\cdots\cdots\cdots\cdots\cdots\cdots\cdots} \\
0 & 0 & 0 & \cdots & 0 & B_{l-1} & 0
\end{array}
\right]
\end{array}
$$

and the first $m_1 + m_2 + \cdots + m_{l-1} + m_l$ entries of Y are given by

$$
\begin{bmatrix}
ME^{(m_1)} \\
\lambda ME^{(m_2)} \\
\lambda^2 ME^{(m_3)} \\
\vdots \\
\lambda^{l-1} ME^{(m_l)}
\end{bmatrix}
\tag{3.4.13}
$$

CASE 2 $\lambda^l = 1$. For each i, $m_1 + m_2 + \cdots + m_{l-1} \leq i \leq m_1 + m_2 + \cdots + m_{l-1} + m_l$, we have

$$
c_{ij} \neq 0 \qquad \text{whenever} \qquad y_j \neq \lambda^l M = M
$$

But we previously required [in equation (3.4.9)] that

$$
M = y_1 = y_2 = \cdots = y_{m_1} \qquad y_j \neq M \qquad \text{for} \quad j > m_1
$$

Thus the first $m_1 + m_2 + \cdots + m_l$ rows of A^T have the form

$$
\begin{array}{c}
\begin{array}{cccccc}
m_1 & m_2 & m_3 & & m_l & n - m_1 - \cdots - m_l
\end{array} \\
\begin{array}{c}
m_1 \\ m_2 \\ \\ m_{l-1} \\ m_l
\end{array}
\left[
\begin{array}{cccccc}
0 & B_1 & 0 & \cdots & 0 & 0 \\
0 & 0 & B_2 & \cdots & 0 & 0 \\
\multicolumn{6}{c}{\cdots\cdots\cdots\cdots\cdots\cdots\cdots\cdots\cdots\cdots} \\
0 & 0 & 0 & \cdots & B_{l-1} & 0 \\
B_l & 0 & 0 & \cdots & 0 & 0
\end{array}
\right]
\end{array}
\tag{3.4.14}
$$

It remains to show that

$$
n = m_1 + m_2 + \cdots + m_l
$$

[so that A^T has no further rows, and, in fact, the last column of the matrix of (3.4.14) does not actually appear]. If $n > m_1 + m_2 + \cdots + m_l$, we have proved that A^T has the form

$$
A^T = \begin{bmatrix} C_{11} & 0 \\ C_{21} & C_{22} \end{bmatrix}
$$

where C_{11} is the matrix of (3.4.14) and C_{22} is a square matrix of order $n - (m_1 + m_2 + \cdots + m_l)$. But, A (and consequently A^T) is irreducible, and hence C_{22} does not actually appear, i.e.,

$$n = m_1 + m_2 + \cdots + m_l$$

The matrix A^T of (3.4.14) clearly has the form we required, namely, the transpose of the matrix of (3.4.1). Hence A is periodic.   ////

Incidentally Theorem 3.4.7 also shows that a characteristic root λ of an irreducible nonregular (and hence periodic) transition matrix of absolute value 1 is a root of unity. If λ is an hth root of unity, the proof of Theorem 3.4.7 shows that the period l of A is at least h; that is,

$$h \le l$$

This is the only conclusion we can draw at this time. In Theorem 3.4.21 we will prove that h divides l and consequently λ is an lth root of unity.

Lemma 3.4.15 If λ, μ are characteristic roots of an irreducible periodic matrix A and

$$|\lambda| = 1$$

then μ/λ is also a characteristic root of A.

PROOF According to the proof of Theorem 3.4.7, there exists a characteristic vector

$$Y = [y_j]$$

of A^T corresponding to the characteristic root λ such that

$$|y_j| = 1 \qquad j = 1, 2, \ldots, n \qquad (3.4.16)$$

[Note that we may set $M = 1$ in (3.4.13).] Clearly

$$Y = DE^{(n)}$$

where D is the diagonal matrix

$$D = \begin{bmatrix} y_1 & 0 & \cdots & 0 \\ 0 & y_2 & \cdots & 0 \\ \multicolumn{4}{c}{\dotfill} \\ 0 & 0 & \cdots & y_n \end{bmatrix}$$

and the equation

$$A^T Y = \lambda Y$$

becomes

$$A^T D E^{(n)} = \lambda D E^{(n)}$$

which we may write as

$$\lambda^{-1} D^{-1} A^T D E^{(n)} = E^{(n)} \qquad (3.4.17)$$

Set

$$\lambda^{-1} D^{-1} A^T D = F \qquad (3.4.18)$$

so that (3.4.17) becomes

$$F E^{(n)} = E^{(n)} \qquad (3.4.19)$$

Since μ is a characteristic root of A, it is a characteristic root of A^T and hence of $D^{-1} A^T D$. Thus μ/λ is a characteristic root of F.

Let $|F|$ denote the matrix whose entries are the absolute values of the entries of F, that is,

$$|F| = [\,|f_{ij}|\,] \qquad i, j = 1, 2, \ldots, n$$

It follows from (3.4.16) and (3.4.18) that

$$|F| = A^T$$

Combining (3.4.19) and

$$A^T E^{(n)} = E^{(n)}$$

yields

$$1 = \sum_{j=1}^{n} f_{ij} = \sum_{j=1}^{n} a_{ij} = \sum_{j=1}^{n} |f_{ij}| \qquad i = 1, 2, \ldots, n$$

However, this is possible only when

$$f_{ij} = |f_{ij}| \qquad i, j = 1, 2, \ldots, n$$

that is,

$$F = |F| = A^T$$

Consequently μ/λ is a characteristic root of A. ////

If we set $\mu = 1$ in Lemma 3.4.15, it follows that λ^{-1} is also a characteristic root of A. In the same way, if we set

$$\mu = \lambda \qquad \lambda = \lambda^{-1}$$

we see that λ^2 is a characteristic root of A. Continuing this process, we see that $\lambda^3, \lambda^4, \ldots$ are characteristic roots of A. In order to show that $\lambda^{-2}, \lambda^{-3}, \ldots$ are characteristic roots of A we substitute 1 for μ and $\lambda^2, \lambda^3, \ldots$ for λ. Finally we have:

Corollary 3.4.20 If λ, μ are characteristic roots of an irreducible periodic transition matrix A and $|\lambda| = |\mu| = 1$, then

$$\lambda^p \mu^q$$

is a characteristic root of A for all integers p and q.

Now we can prove:

Theorem 3.4.21 Every characteristic root of absolute value 1 of an irreducible periodic transition matrix of period l is an lth root of unity.

PROOF Let v be a characteristic root of absolute value 1 of the irreducible periodic transition matrix A of period l and let $\tilde{l}$ be the smallest positive integer for which

$$v^{\tilde{l}} = 1$$

that is, v is a *primitive $\tilde{l}$th root* of unity. By the discussion above, all the $\tilde{l}$th roots of unity are characteristic roots of A. The $\tilde{l}$th roots of unity can be expressed as

$$\exp \frac{2\pi i \cdot q}{\tilde{l}} = \cos \frac{2\pi \cdot q}{\tilde{l}} + i \sin \frac{2\pi \cdot q}{\tilde{l}} \qquad q = 0, 1, 2, \ldots, \tilde{l} - 1.$$

Since the period of A is l, the lth roots of unity

$$\exp \frac{2\pi i \cdot p}{l} = \cos \frac{2\pi \cdot p}{l} + i \sin \frac{2\pi \cdot p}{l} \qquad p = 0, 1, 2, \ldots, l - 1$$

are also characteristic roots of A. Let d and h be the greatest common divisor and least common multiple of l and $\tilde{l}$, respectively,

$$d = (l, \tilde{l}) \qquad h = [l, \tilde{l}]$$

(See Sec. 1.11.) Recall that

$$h = \frac{l\tilde{l}}{d}$$

There exist integers p and q such that

$$d = p\tilde{l} + ql \qquad (3.4.22)$$

Set

$$\lambda = \exp \frac{2\pi i \cdot q}{\tilde{l}} \qquad \mu = \exp \frac{2\pi i \cdot p}{l}$$

where p and q satisfy (3.4.22). By Corollary 3.4.20,

$$\lambda\mu = \exp \frac{2\pi i \cdot d}{l\tilde{l}} = \exp \frac{2\pi i}{h}$$

is a characteristic root of A. Clearly $\lambda\mu$ is a primitive hth root of unity. However, since A has period l, it follows that

$$h \leqq l$$

Consequently $h = l$ and l divides l. Thus

$$\lambda^l = 1 \qquad ////$$

The converse of Theorem 3.4.21 is contained in Theorem 3.4.2. Hence:

Corollary 3.4.23 The characteristic roots of absolute value of 1 of an irreducible periodic transition matrix whose period is l are exactly the lth roots of unity.

In Theorem 3.4.2 we showed that A^l is the direct sum of l irreducible transition matrices $A_1, A_2, \ldots, A_l$. According to Corollary 3.4.23, none of the matrices $A_1, A_2, \ldots, A_l$ has a characteristic root of absolute value 1 other than 1. It now follows from Lemma 3.4.6 that:

Corollary 3.4.24 If the matrix A of (3.4.1) is an irreducible transition matrix of period l and $A_1, A_2, \ldots, A_l$ are defined as in Theorem 3.4.2, then $A_1, A_2, \ldots, A_l$ are regular.

EXERCISES

1 Show that the transition matrices of the Markov chains corresponding to the following mazes are periodic. Find the period and the transient states.

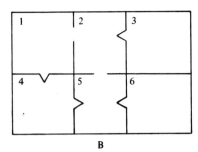

A   B

2 Show that the period of a nonsingular periodic transition matrix divides the order of the matrix.

3 Let A be a periodic transition matrix of period m. Show that if l divides m, then the states of the Markov chain can be numbered so that A has the

$l \times l$ block form (3.4.1).   (How has this difficulty been avoided when defining the period of a transition matrix?)

4   Illustrate Exercise 3 using the maze of part (a) of Exercise 1.

5   Let A be an irreducible periodic transition matrix of period l.   Show that A^{rl+1} is irreducible for $r = 0, 1, 2, \ldots$.

6   Show that if A is an irreducible periodic transition matrix of period l, then A^q is irreducible if and only if q is relatively prime to l.

7   Show that if A is a transition matrix and $|\lambda| = 1$, then $(x - \lambda)^2$ does not divide the minimal polynomial of A.

8   Let A be an irreducible periodic transition matrix in the form of (3.4.1). Let $Z = [Z_j]$ be the stationary vector of A partitioned to be compatible with A; that is, Z_j is an m_j-dimensional column vector.   Prove that

$$E_{(m_j)}Z_j = \frac{1}{l}$$

9   Let A be an irreducible period transition matrix of the form (3.4.1) and let $Z_1, Z_2, \ldots, Z_l$ be the stationary vectors of the matrices $A_1, A_2, \ldots, A_l$ of Theorem 3.4.2.   Set

$$Z = \frac{1}{l}[Z_j] \qquad j = 1, 2, \ldots, l$$

Show that Z is the stationary vector of A.

3.5   RECURRENCE PROBABILITIES

In Sec. 3.3 we showed that for any transition matrix A, any initial probability vector $P^{(0)}$, and any transient state S_q

$$\operatorname*{Lim}_{k \to \infty} p_q^{(k)} = 0$$

i.e., as the number of steps of a Markov chain increases, the probability that an object will be in a transient state approaches 0 and the probability that an object will be in some ergodic state approaches 1.   In this section we will study several aspects of the behavior of a Markov chain as the number of steps increases. Not surprisingly we will see that this behavior is quite different for transient and ergodic states.   For example, we will study f_{pq}, the probability that an object initially in state S_q will be in state S_p at least once if the process is permitted to continue indefinitely.   (See Example, 3.1.18; f_{pq} is called the *first-passage probability* from S_q to S_p.   In particular, when $p = q$, we call f_{pp} the *recurrence probability* of the state S_p.)   In Theorem 3.5.15 we will show an object initially in an ergodic state S_p is *certain* to be in S_p at some subsequent time, that is, $f_{pp} = 1$, while if S_p is transient, the recurrence of S_p is *uncertain*, that is, $f_{pp} < 1$.

Another example of the difference between the long-range behavior of transient and ergodic states is provided by n_{pq}, the number of times an object initially in S_q can be expected to be in S_p. (See Example 3.1.15.) We will show in Theorems 3.5.27 and 3.5.29 that if it is possible to go from S_q to S_p at all, then the object can be expected to be in S_p infinitely many times (in which case we write $n_{pq} = \infty$) if and only if S_p is an ergodic state. In particular, S_p is ergodic if and only if $n_{pp} = \infty$.

For a given transition matrix A, we know that the entry in the (p, q) position of A^s is the probability of being in the state S_p exactly s steps after being in the state S_q, no consideration being made of the intervening steps. By considering these intervening steps we are led to:

Definition 3.5.1   Let $A = [a_{ij}]$ be a transition matrix and let

$$A^s = [a_{ij}^{(s)}] \qquad s = 0, 1, 2, \ldots$$

We say that the probability[1] $f_{pq}^{(s)}$ that the *first passage* of an object from state S_q to state S_p will take place at the sth step is given recursively by

$$f_{pq}^{(1)} = a_{pq}$$

$$f_{pq}^{(s)} = a_{pq}^{(s)} - a_{pp}^{(s-1)} f_{pq}^{(1)} - a_{pp}^{(s-2)} f_{pq}^{(2)} - \cdots - a_{pp} f_{pq}^{(s-1)} \qquad (3.5.2)$$

$$= a_{pq}^{(s)} - \sum_{h=1}^{s-1} a_{pp}^{(s-h)} f_{pq}^{(h)} \qquad s = 2, 3, \ldots$$

For each h, $1 \leq h \leq s - 1$, $a_{pp}^{(s-h)} f_{pq}^{(h)}$ is the probability that the object entered S_p from S_q for the first time at the hth step and was again in S_p $s - h$ steps later. Definition 3.5.1 states that the probability that an object initially in S_q will enter S_p for the first time at the sth step can be found as follows: subtract from $a_{pq}^{(s)}$ (that is, the probability that an object initially in S_q will be in S_p at the sth step for the first time or not) all the probabilities that the object entered S_p at some earlier step.

---

[1] In order to justify calling $f_{pq}^{(s)}$ a probability it should be shown that

$$0 \leq f_{pq}^{(s)} \leq 1$$

In fact, by means of equation (3.5.4) and an induction on s it can be shown that

$$0 \leq f_{pq}^{(s)} \leq a_{pq}^{(s)} \leq 1$$

An alternate proof of this result follows as an immediate consequence of Exercise 8.

We shall need the identity:

Lemma 3.5.3   For $s = 1, 2, 3, \ldots$

$$f_{pq}^{(s+1)} = \sum_{k \neq p} f_{pk}^{(s)} a_{kq} \qquad (3.5.4)$$

PROOF   The proof is by an induction on s.   For $s = 1$ we have

$$\sum_{k \neq p} f_{pk}^{(1)} a_{kq} = \sum_{k \neq p} a_{pk} a_{kq} = \sum_{k=1}^{n} a_{pk} a_{kq} - a_{pp} a_{pq}$$

$$= a_{pq}^{(2)} - a_{pp} f_{pq}^{(1)} = f_{pq}^{(2)}$$

as required.   Now, suppose the lemma has been proved for all $h < s$. According to equation (3.5.2),

$$\sum_{k \neq p} f_{pk}^{(s)} a_{kq} = \sum_{k \neq p} \left(a_{pk}^{(s)} - \sum_{h=1}^{s-1} a_{pp}^{(s-h)} f_{pk}^{(h)} \right) a_{kq}$$

$$= \sum_{k \neq p} a_{pk}^{(s)} a_{kq} - \sum_{h=1}^{s-1} a_{pp}^{(s-h)} \left(\sum_{k \neq p} f_{pk}^{(h)} a_{kq} \right)$$

Using the induction hypothesis

$$\sum_{k \neq p} f_{pk}^{(s)} a_{kq} = \sum_{k \neq p} a_{pk}^{(s)} a_{kq} - \sum_{h=1}^{s-1} a_{pp}^{(s-h)} f_{pq}^{(h+1)}$$

and adding and subtracting $a_{pp}^{(s)} a_{pq}$, we have

$$\sum_{k \neq p} f_{pk}^{(s)} a_{kq} = \sum_{k=1}^{n} a_{pk}^{(s)} a_{kq} - \sum_{h=1}^{s} a_{pp}^{(s+1-h)} f_{pq}^{(h)}$$

$$= a_{pq}^{(s+1)} - \sum_{h=1}^{s} a_{pp}^{(s+1-h)} f_{pq}^{(h)} = f_{pq}^{(s+1)}$$

since $a_{pp}^{(s)} a_{pq} = a_{pp}^{(s)} f_{pq}^{(1)}$.   ////

Clearly $f_{pk}^{(s)} a_{kq}$ is the probability that the first passage from S_q to S_p will take place at the $(s + 1)$th step by means of going from S_q to S_k in one step and then entering S_p from S_k for the first time in exactly s steps.
It will be convenient to write equation (3.5.4) as

$$f_{pq}^{(s+1)} = \sum_{k=1}^{n} f_{pk}^{(s)} a_{kq} - f_{pp}^{(s)} a_{pq} \qquad (3.5.5)$$

For each s, let us define the $n \times n$ matrix $F^{(s)}$ by

$$F^{(s)} = [f_{pq}^{(s)}] \qquad p, q = 1, 2, \ldots, n$$

The term on the left side of equation (3.5.5) is the entry in the (p, q) position of $F^{(s+1)}$. Similarly, the first term on the right side of equation (3.5.5) is the entry in the (p, q) position of $F^{(s)}A$. For any $n \times n$ matrix $X = [x_{pq}]$ let us define dg X (called *diagonal X*) to be the $n \times n$ matrix having the same diagonal entries as X [that is, dg X has x_{ii} in the (i, i) position] and 0s elsewhere. Then the second term on the right side of equation (3.5.5) is the entry in the (p, q) position of $(\text{dg } F^{(s)})A$. Thus equation (3.5.5) is equivalent to

$$F^{(s+1)} = F^{(s)}A - (\text{dg } F^{(s)})A \qquad (3.5.6)$$

When S_q is an ergodic state and S_p is either a transient state or is an ergodic state which is not in the same ergodic set as S_q, then $a_{pq}^{(s)} = 0$ for all s. Consequently $f_{pq}^{(s)} = 0$ for all s, and, for A in standard form (3.3.24) with transient states $S_1, S_2, \ldots, S_l$ and ergodic sets $\mathbb{S}_1, \mathbb{S}_2, \ldots, \mathbb{S}_r$, $F^{(s)}$ has the same form

$$F^{(s)} = \begin{bmatrix} F_0^{(s)} & 0 & 0 & \cdots & 0 \\ G_1^{(s)} & E_1^{(s)} & 0 & \cdots & 0 \\ G_2^{(s)} & 0 & E_2^{(s)} & \cdots & 0 \\ \cdots & \cdots & \cdots & \cdots & \cdots \\ G_r^{(s)} & 0 & 0 & \cdots & E_r^{(s)} \end{bmatrix}$$

which we call the *standard form* of $F^{(s)}$.

Definition 3.5.7 We say that the *first-passage probability*[1] f_{pq} from the state S_q to the state S_p is given by

$$f_{pq} = \sum_{s=1}^{\infty} f_{pq}^{(s)} \qquad (3.5.8)$$

f_{pp} is called the *recurrence probability* of the state S_p and the $n \times n$ matrix

$$F = [f_{pq}] \qquad p, q = 1, 2, \ldots, n$$

is called the *recurrence matrix* or the *first-passage matrix* of the Markov chain.

Equation (3.5.8) can also be written in matrix form

$$F = \sum_{s=1}^{\infty} F^{(s)}$$

---

[1] It will follow from Corollary 3.5.24 that f_{pq} exists and $0 \leq f_{pq} \leq 1$.

When A is in standard form (3.3.24), F has the same form

$$F = \begin{bmatrix} F_0 & 0 & 0 & \cdots & 0 \\ G_1 & E_1 & 0 & \cdots & 0 \\ G_2 & 0 & E_2 & \cdots & 0 \\ \cdots\cdots\cdots\cdots\cdots\cdots \\ G_r & 0 & 0 & \cdots & E_r \end{bmatrix} \qquad (3.5.9)$$

which we call the *standard form* of F. In Theorem 3.5.15 we will show that the transient and ergodic states can be characterized by their recurrence probabilities. First we need:

Lemma 3.5.10 For any p and q,

$$f_{pq} = \sum_{k \neq p} f_{pk} a_{kq} + a_{pq} \qquad (3.5.11)$$

PROOF Equation (3.5.11) is a direct consequence of equation (3.5.4) since

$$\sum_{k \neq p} f_{pk} a_{kq} + a_{pq} = \sum_{k \neq p} \left(\sum_{s=1}^{\infty} f_{pq}{}^{(s)} \right) a_{kq} + a_{pq}$$

$$= \sum_{s=1}^{\infty} \left(\sum_{k \neq p} f_{pk}{}^{(s)} a_{kq} \right) + a_{pq}$$

$$= \sum_{s=1}^{\infty} f_{pq}{}^{(s+1)} + f_{pq}{}^{(1)} = f_{pq} \qquad ////$$

$f_{pk} a_{kq}$ is the probability of going from S_q to S_k in one step and then entering S_p (at least once) from S_k. Equation (3.5.11) can be written as

$$f_{pq} = \sum_{k=1}^{n} f_{pk} a_{kq} + a_{pq} - f_{pp} a_{pq}$$

which is equivalent to the matrix equation

$$F = FA + A - (\text{dg } F)A \qquad (3.5.12)$$

Equation (3.5.12) is a special case of a more general relation connecting A and F.

Theorem 3.5.13 For each t, $t = 1, 2, \ldots,$

$$F = FA^t + (A + A^2 + \cdots + A^t) - (\text{dg } F)(A + A^2 + \cdots + A^t) \qquad (3.5.14)$$

PROOF Equation (3.5.12) is the case of equation (3.5.14) in which $t = 1$. Equation (3.5.14) can be established by an induction on t and is left as an exercise. ////

As an immediate consequence of Theorem 3.5.13 we have:

Theorem 3.5.15   If S_p is an ergodic state, then $f_{pp} = 1$.   In fact, if S_q is any state in the same ergodic set as S_p, then $f_{pq} = f_{qp} = 1$.   (For any other state, S_q, $f_{qp} = 0$.)
  If S_p is a transient state, then $f_{pp} < 1$.

PROOF   First, let S_p be an ergodic state.   According to Theorem 3.3.26, there exists a stationary vector

$$X = [x_j]$$

of A for which $x_p \neq 0$.   Multiplying equation (3.5.12) on the right by X, we have

$$FX = FAX + AX - (\text{dg } F)AX$$
$$= FX + X - (\text{dg } F)X$$

and consequently
$$0 = x_p - f_{pp}x_p$$

Since $x_p \neq 0$, it follows that

$$f_{pp} = 1$$

Now suppose that S_q is a state in the same ergodic set as S_p.   Since it is possible to reach S_q from S_p, there exists a positive integer t such that $a_{qp}^{(t)} \neq 0$.   Examining the (p, p) position of the matrices of equation (3.5.14) yields

$$1 = f_{pp} = \sum_{k=1}^{n} f_{pk} a_{kp}^{(t)} + \sum_{m=1}^{t} a_{pp}^{(m)} - f_{pp}\left(\sum_{m=1}^{t} a_{pp}^{(m)}\right)$$

$$= \sum_{k=1}^{n} f_{pk} a_{kp}^{(t)} \leq \sum_{k=1}^{n} a_{kp}^{(t)} = 1 \qquad (3.5.16)$$

Since the first and last terms of (3.5.16) are equal, there is equality throughout and

$$f_{pk} a_{kp}^{(t)} = a_{kp}^{(t)} \qquad k = 1, 2, \ldots, n$$

But $a_{qp}^{(t)} \neq 0$, and consequently

$$f_{pq} = 1$$

Finally, let S_p be a transient state.   There is an (ergodic) state S_q and a positive integer t such that

$$a_{qp}^{(t)} \neq 0 \qquad f_{pq} = 0 \qquad (3.5.17)$$

As before, the (p, p) position of the matrices of equation (3.5.14) yields

$$f_{pp} = \sum_{k=1}^{n} f_{pk} a_{kp}^{(t)} + \sum_{m=1}^{t} a_{pp}^{(m)} - f_{pp}\left(\sum_{m=1}^{t} a_{pp}^{(m)}\right)$$

which can be written in the form

$$\left(1 + \sum_{m=1}^{t} a_{pp}^{(m)}\right)(f_{pp} - 1) = \sum_{k=1}^{n} f_{pk} a_{kp}^{(t)} - 1 \qquad (3.5.18)$$

It follows from (3.5.17) that

$$\sum_{k=1}^{n} f_{pk} a_{kp}^{(t)} - 1 = \sum_{k \neq q} f_{pk} a_{kp}^{(t)} - 1 \leq \sum_{k \neq q} a_{kp}^{(t)} - 1 = -a_{qp}^{(t)} < 0 \qquad (3.5.19)$$

Comparing (3.5.18) and (3.5.19), we have

$$f_{pp} < 1 \qquad \qquad ////$$

It is now clear that the choice of notation, $E_1, E_2, \ldots, E_r$ for the diagonal blocks of F in (3.5.9) corresponding to the ergodic sets is consistent with our usual notation since every entry of each of these matrices is 1. Thus far nothing has been said about f_{pq} when S_q is transient (other than the fact that recurrence to S_q is uncertain). It is not difficult to construct examples of Markov chains with passage from a transient state S_q to a state S_p (transient or ergodic) which is certain or uncertain. We can obtain a precise expression for f_{pq} by substituting the standard forms of A and F into equation (3.5.14). Then

$$F_0 = F_0 C^t + (C + C^2 + \cdots + C^t) - (\text{dg } F_0)(C + C^2 + \cdots + C^t)$$

Moreover, for $i = 1, 2, \ldots, r$ equation (3.5.12) yields

$$G_i = G_i C + E_i B_i + B_i - (\text{dg } E_i) B_i = G_i C + E_i B_i$$

According to Theorem 3.3.26, $\lim_{t \to 0} C^t = 0$ and the infinite series

$$C + C^2 + \cdots$$

converges to $(I - C)^{-1} - I$. Thus

$$F_0 = [I - (\text{dg } F_0)][(I - C)^{-1} - I]$$
$$G_i = E_i B_i (I - C)^{-1}$$

For reasons that will become clear shortly we define the $l \times l$ matrix N_0 by

$$N_0 = [n_{ij}] = (I - C)^{-1} = I + C + C^2 + \cdots$$

Corollary 3.5.20 (a) $F_0 = [I - (\text{dg } F_0)](N_0 - I)$,

(b) $G_i = E_i B_i N_0$, $i = 1, 2, \ldots, r$,

from which it follows that:

1 When S_p and S_q are transient,

$$f_{pq} = \frac{n_{pq} - \delta_{pq}}{n_{pp}} \quad \text{where} \quad \delta_{pq} = \begin{cases} 1 & \text{when } p = q \\ 0 & \text{when } p \neq q \end{cases}$$

2 When S_q is transient and S_p and S_s belong to the same ergodic set, then

$$f_{pq} = f_{sq}$$

PROOF The proof is left as an exercise. ////

The second conclusion of the corollary is intuitively clear. For, once an object has entered one state of an ergodic set, it is certain to enter all other states of that set. Thus we can speak of the probability $f_{\mathbb{S}_i, q}$ that an object initially in the transient state S_q will ever enter (and thereafter remain in) the ergodic set $\mathbb{S}_i$. $f_{\mathbb{S}_i, q}$ is the qth entry of the l-dimensional row vector

$$E(n_i) B_i N_0$$

At the end of Example 3.1.18 the special case of equation (3.5.11)

$$f_{11} = \sum_{k \neq 1} f_{1k} a_{k1} + a_{11}$$

was used to evaluate f_{11}. However, a different method was used to obtain f_{1k} when $k \neq 1$, namely, changing S_1 into an absorbing state. We now wish to show that this method and definition 3.5.7 agree. Let A be a transition matrix, S_p a fixed state, and $\tilde{A}$ the matrix formed from A by replacing the pth column of A by $I^{(p)}$, that is, making S_p an absorbing state. Set

$$\tilde{A}^k = [\tilde{a}_{ij}{}^{(k)}] \qquad i, j = 1, 2, \ldots, n$$

A direct computation shows that

$$\tilde{A}^{s+1} = \tilde{A}^s \tilde{A} = \begin{bmatrix} \tilde{a}_{11}{}^{(s)} & \tilde{a}_{12}{}^{(s)} & \cdots & \tilde{a}_{1n}{}^{(s)} \\ \vdots & & & \vdots \\ \vdots & & & \vdots \\ \vdots & & & \vdots \\ \tilde{a}_{n1}{}^{(s)} & \tilde{a}_{n2}{}^{(s)} & \cdots & \tilde{a}_{nn}{}^{(s)} \end{bmatrix} \begin{bmatrix} a_{11} & a_{12} & \cdots & 0 & \cdots & a_{1n} \\ \vdots & & & & & \vdots \\ a_{p1} & a_{p2} & \cdots & 1 & \cdots & a_{pn} \\ \vdots & & & & & \vdots \\ a_{n1} & a_{n2} & \cdots & 0 & \cdots & a_{nn} \end{bmatrix}$$

and hence

$$\tilde{a}_{ij}{}^{(s+1)} = \sum_{k=1}^{n} \tilde{a}_{ik}{}^{(s)} a_{kj} \qquad j \neq p$$

$$\tilde{a}_{ip}{}^{(s+1)} = \tilde{a}_{ip}{}^{(s)} = \begin{cases} 1 & \text{when } i = p \\ 0 & \text{when } i \neq p \end{cases} \qquad (3.5.21)$$

Lemma 3.5.22   Let A be a transition matrix and S_p a fixed state and let $\tilde{A}$ be the matrix defined above.   For $q \neq p$ and $s = 1, 2, \ldots$

$$\tilde{a}_{pq}^{(s)} - \tilde{a}_{pq}^{(s-1)} = f_{pq}^{(s)} \qquad (3.5.23)$$

PROOF   The proof is by an induction on s and is immediate for $s = 1$ since

$$\tilde{a}_{pq}^{(1)} - \tilde{a}_{pq}^{(0)} = a_{pq} - 0 = f_{pq}^{(1)}$$

Now suppose equation (3.5.23) holds for some value of s.   Substituting (3.5.21) into the left-hand side of equation (3.5.23) and applying Lemma 3.5.3 yields

$$\tilde{a}_{pq}^{(s+1)} - \tilde{a}_{pq}^{(s)} = \sum_{k=1}^{n} \tilde{a}_{pk}^{(s)} a_{kq} - \sum_{k=1}^{n} \tilde{a}_{pk}^{(s-1)} a_{kq}$$

$$= \left[\sum_{k=1}^{n} (\tilde{a}_{pk}^{(s)} - \tilde{a}_{pk}^{(s-1)}) \right] a_{kq}$$

$$= \left(\sum_{k \neq p} f_{pk}^{(s)} + \tilde{a}_{pp}^{(s)} - \tilde{a}_{pp}^{(s-1)} \right) a_{kq}$$

$$= \sum_{k \neq p} f_{pk}^{(s)} a_{kq} = f_{pq}^{(s+1)} \qquad\qquad ////$$

We can now justify the procedures used in Example 3.1.18 to find the first-passage probabilities.   As an immediate consequence of Lemma 3.5.22 we have:

Corollary 3.5.24   For $q \neq p$

$$f_{pq} = \lim_{m \to \infty} \tilde{a}_{pq}^{(m)}$$

PROOF   It follows from Lemma 3.5.22 that

$$\sum_{s=1}^{m} f_{pq}^{(s)} = \tilde{a}_{pq}^{(m)} - \tilde{a}_{pq}^{(0)} = \tilde{a}_{pq}^{(m)} \qquad m = 1, 2, \ldots \qquad ////$$

In Example 3.1.15 we had a Markov chain in which a mouse was given a piece of cheese every time he was in a certain room of the maze.   We were able to calculate the number of pieces of cheese the mouse could be expected to get (i.e., the number of times he could be expected to be in a particular state) if the process were permitted to run indefinitely.   We will show that an object will be in a transient state only a finite number of times no matter what its initial state is. On the other hand, we will show that if an object can reach an ergodic state at all, it will be found there infinitely often.

Definition 3.5.25 We say that an object initially in state S_q can be expected to be in state S_p

$$n_{pq} = \sum_{s=0}^{\infty} a_{pq}{}^{(s)} \qquad (3.5.26)$$

times. When the sum on the right-hand side of equation (3.5.26) is unbounded, we say that an object initially in S_q can be expected to be in S_p *infinitely often* and we write

$$n_{pq} = \infty$$

Equation (3.5.25) defined n_{pq} to be the entry in the (p, q) position of the infinite series of matrices

$$I + A + A^2 + \cdots$$

[and $n_{pq} = \infty$ when the sum of the entries in the (p, q) positions of the terms of this series is unbounded]. Clearly, if S_p is an absorbing state, then

$$a_{pp}{}^{(s)} = 1 \qquad s = 1, 2, \ldots$$

and consequently $n_{pp} = \infty$. On the other hand, when S_q is an ergodic state and S_p is either a transient state or an ergodic state not belonging to the same ergodic set as S_q, then

$$a_{pq}{}^{(s)} = 0 \qquad s = 1, 2, \ldots$$

and consequently $n_{pq} = 0$. When S_p and S_q are both transient states (and A is in standard form), then equation (3.5.26) defines n_{pq} to be the entry in the (p, q) position of

$$I + C + C^2 + \cdots$$

that is, n_{pq} is the entry in the (p, q) position of the matrix N_0 which we defined earlier. We now have:

Theorem 3.5.27 If S_p is a transient state, then n_{pq} is finite for every state S_q.

It remains to find n_{pq} when S_q is an ergodic state. For this we need:

Lemma 3.5.28 For any states S_p and S_q we have

$$n_{pq} - \delta_{pq} = f_{pq} n_{pp}$$

where

$$\delta_{pq} = \begin{cases} 1 & \text{when } p = q \\ 0 & \text{when } p \neq q \end{cases}$$

(If $n_{pp} = \infty$, we interpret $f_{pq} n_{pp}$ to be ∞ if $f_{pq} \neq 0$ and 0 if $f_{pq} = 0$.)

PROOF   (Note that the special case in which S_p and S_q are both transient has already been proved in Corollary 3.5.20.)   Let us write equation (3.5.2) as

$$a_{pq}^{(s)} = f_{pq}^{(s)} + a_{pp}^{(1)} f_{pq}^{(s-1)} + a_{pp}^{(2)} f_{pq}^{(s-2)} + \cdots + a_{pp}^{(s-1)} f_{pq}^{(1)}$$

$$s = 1, 2, \ldots$$

Summing these equations vertically yields

$$n_{pq} - \delta_{pq} = \sum_{s=1}^{\infty} a_{pq}^{(s)} = f_{pq} + a_{pp}^{(1)} \sum_{s=1}^{\infty} f_{pq}^{(s)} + a_{pp}^{(2)} \sum_{s=1}^{\infty} f_{pq}^{(s)} + \cdots$$

$$= f_{pq}(1 + a_{pp}^{(1)} + a_{pp}^{(2)} + \cdots)$$

$$= f_{pq} n_{pp} \qquad\qquad\qquad ////$$

Theorem 3.5.29   Let S_p be an ergodic state.   If S_q is a state in the same ergodic set as S_p, then

$$n_{pq} = \infty$$

In particular,

$$n_{pp} = \infty$$

PROOF   When $p = q$, Lemma 3.5.28 states that

$$n_{pp} - 1 = f_{pp} n_{pp}$$

Since S_p is ergodic, $f_{pp} = 1$ and hence

$$n_{pp} - 1 = n_{pp}$$

Clearly this equation has no finite solutions, and we conclude that

$$n_{pp} = \infty$$

If S_q is a state in the same ergodic set as S_p, then $f_{pq} = 1$ and it follows from Lemma 3.5.28 that $n_{pq} = \infty$.   ////

In Theorem 3.5.15 we showed that if S_p and S_q are states belonging to the same ergodic set, then it is certain that an object initially in S_q will enter S_p at some time.   However, we did not consider the number of steps that can be expected to elapse before the object enters S_p.   (See Example 3.1.21.)

Definition 3.5.30 For states S_p and S_q belonging to the same ergodic set we define the *mean-first-passage time* from S_q to S_p by

$$m_{pq} = \sum_{s=1}^{\infty} sf_{pq}^{(s)}$$

For an irreducible Markov chain, i.e., when all states belong to the same ergodic set, then,

$$M = [m_{pq}] = \sum_{s=1}^{\infty} sF^{(s)}$$

is called the *mean-first-passage matrix*.

The mean-first-passage time from one state to another is not defined unless the passage is certain. It will follow from Exercises 11 to 14 that $m_{pq} < \infty$.

Lemma 3.5.31 Let A be an irreducible transition matrix. Then

$$M - E = MA - (\text{dg } M)A \qquad (3.5.32)$$

PROOF Multiplying equation (3.5.6) by s yields

$$sF^{(s+1)} = sF^{(s)}A - [\text{dg}(sF^{(s)})]A$$

which we can write as

$$(s + 1)F^{(s+1)} - F^{(s+1)} = sF^{(s)}A - [\text{dg}(sF^{(s)})]A$$

Summing on s, we have

$$\sum_{s=1}^{\infty}(s + 1)F^{(s+1)} - \sum_{s=1}^{\infty} F^{(s+1)} = \left(\sum_{s=1}^{\infty} sF^{(s)}\right)A - \left(\text{dg}\sum_{s=1}^{\infty} sF^{(s)}\right)A$$

that is,
$$M - F = MA - (\text{dg } M)A \qquad (3.5.33)$$

Since all states belong to the same ergodic set,

$$F = E$$

proving the lemma. ////

From Lemma 3.5.31 we can obtain a simple expression for the diagonal elements of M, which will in turn be used to find the nondiagonal elements of M.

Theorem 3.5.34 Let $X = [x_j]$ be the (unique, positive) stationary vector of an irreducible Markov chain with mean-first-passage matrix M. Then

$$m_{jj} = \frac{1}{x_j} \qquad j = 1, 2, \ldots, n$$

PROOF   Multiplying equation (3.5.32) on the right by X,

$$MX - EX = MAX - (\text{dg } M)AX$$
$$MX - E^{(n)} = MX - (\text{dg } M)X$$
$$E^{(n)} = (\text{dg } M)X$$

Thus $1 = m_{jj}x_j \quad j = 1, 2, \ldots, n$ ////

If L is the $n \times n$ matrix each of whose columns is the stationary vector of A, then Theorem 3.5.33 states that

$$\text{dg } M = (\text{dg } L)^{-1} \qquad (3.5.35)$$

In order to find the nondiagonal elements of M we will need a lemma.

Lemma 3.5.36[1]   If A is an irreducible matrix, then $I - A + L$ is non-singular and

$$Z = (I - A + L)^{-1}$$

satisfies

$$Z(I - A) = I - L$$

PROOF   In Sec. 3.3 we proved that there exists a nonsingular matrix P such that

$$A = P \begin{bmatrix} 1 & 0 \\ 0 & \tilde{A} \end{bmatrix} P^{-1} \qquad L = P \begin{bmatrix} 1 & 0 \\ 0 & 0 \end{bmatrix} P^{-1}$$

where $\tilde{A}$ is an $(n - 1) \times (n - 1)$ matrix which does not have 1 as a characteristic root.   Then

$$I - A + L = P \begin{bmatrix} 1 & 0 \\ 0 & I - \tilde{A} \end{bmatrix} P^{-1}$$

which is nonsingular.   A direct computation shows that

$$(I - A + L)(I - L) = I - A + L - L + AL - L^2$$
$$= I - A$$

and the lemma follows.   ////

---

[1] An alternate proof of Lemma 3.5.36 for the case in which A is regular is described in Exercise 9 of Sec. 3.3.

In preparation for the next theorem let us write equation (3.5.32) as

$$M(I - A) = E - (\mathrm{dg}\,M)A \qquad (3.5.37)$$

Theorem 3.5.38

$$M = (\mathrm{dg}\,L)^{-1}[I - Z + (\mathrm{dg}\,Z)E] \qquad (3.5.39)$$

PROOF The proof consists of two parts. First we verify that the matrix M defined by equation (3.5.39) satisfies equations (3.5.35) and (3.5.37). Using Lemma 3.5.36, we have

$$
\begin{aligned}
M(I - A) &= (\mathrm{dg}\,L)^{-1}[I - Z + (\mathrm{dg}\,Z)E](I - A) \\
&= (\mathrm{dg}\,L)^{-1}[(I - A) - Z(I - A) + (\mathrm{dg}\,Z)E(I - A)] \\
&= (\mathrm{dg}\,L)^{-1}[(I - A) - (I - L)] = (\mathrm{dg}\,L)^{-1}(L - A) \\
&= E - (\mathrm{dg}\,M)A
\end{aligned}
$$

Also

$$
\begin{aligned}
\mathrm{dg}\,M &= (\mathrm{dg}\,L)^{-1}[\mathrm{dg}\,I - \mathrm{dg}\,Z + (\mathrm{dg}\,Z)(\mathrm{dg}\,E)] \\
&= (\mathrm{dg}\,L)^{-1}[I - \mathrm{dg}\,Z + (\mathrm{dg}\,Z)I] = (\mathrm{dg}\,L)^{-1}
\end{aligned}
$$

Second we show that equations (3.5.35) and (3.5.37) have only one solution. Suppose M and $\tilde{M}$ are both solutions of these equations. Then

$$M(I - A) = \tilde{M}(I - A)$$
$$\mathrm{dg}\,M = \mathrm{dg}\,\tilde{M} = (\mathrm{dg}\,L)^{-1}$$

and consequently

$$(M - \tilde{M})(I - A) = 0$$

By Corollary 3.3.13 each row of $M - \tilde{M}$ is a scalar multiple of $E_{(n)}$. However,

$$\mathrm{dg}(M - \tilde{M}) = \mathrm{dg}\,M - \mathrm{dg}\,\tilde{M} = 0$$

and consequently each row of $M - \tilde{M}$ is the zero vector, i.e.,

$$M = \tilde{M} \qquad\qquad ////$$

It is not difficult to show that an object initially in a transient state of a Markov chain is certain to enter some ergodic state. (See Exercise 7.) Let A be a transition matrix in standard form

$$
A = \begin{bmatrix}
C & 0 & 0 & \cdots & 0 \\
B_1 & D_1 & 0 & \cdots & 0 \\
B_2 & 0 & D_2 & \cdots & 0 \\
\multicolumn{5}{c}{\dotfill} \\
B_r & 0 & 0 & \cdots & D_r
\end{bmatrix}
$$

where C is an $l \times l$ matrix.   If we lump together all the ergodic states into a single (absorbing) state (which we will call S'_{l+1}) without disturbing the transient states, the resulting Markov chain has an $(l + 1) \times (l + 1)$ transition matrix

$$A' = \begin{bmatrix} C & 0 \\ B & 1 \end{bmatrix}$$

where B is a $1 \times l$ matrix. B is uniquely determined by the requirement that the sum of the entries of each column of A' be 1.   An object initially in a transient state S_p is certain to enter (and remain in) S'_{l+1}.   We define the *mean time before absorption* of an object initially in S_p by

$$m_p = \sum_{s=1}^{\infty} s f_{l+1, p}^{(s)'}$$

where $f_{l+1, p}^{(s)'}$ refers to the Markov chain with transition matrix A'.   The $1 \times l$ matrix

$$M'_{21} = [m_1 \quad m_2 \quad \cdots \quad m_l] \qquad (3.5.40)$$

is called the *mean-time-before-absorption matrix*.   We can express M'_{21} in terms of C.

Theorem 3.5.41

$$M'_{21} = E_{(l)}(I - C)^{-1} = E_{(l)}N_0$$

PROOF   Let M' be the mean-first-passage matrix of A' and partition M' as

$$M' = \begin{bmatrix} M'_{11} & M'_{12} \\ M'_{21} & m'_{22} \end{bmatrix}$$

where M'_{11} is an $l \times l$ matrix, M'_{21} is a $1 \times l$ matrix, etc.   According to Definition 3.5.30, the matrix M'_{21} defined by equation (3.5.40) is identical to the submatrix M'_{21} of M'.   In addition it is clear that

$$M'_{12} = 0 \qquad m'_{22} = 1$$

Applying equation (3.5.33) and recalling that an object in a transient state is certain to be absorbed in S'_{l+1}, we have

$$M'_{21} - E_{(l)} = (M'_{21}C + m'_{22} B) - m'_{22} B$$

Thus

$$M'_{21}(I - C) = E_{(l)}$$

and the theorem follows.   ////

EXERCISES

1 A man lives at one end of a street 40 yards long. There is a bar at the other end of the street, and streetlight poles are placed at either end of the street and at 10-yard intervals in between. When the man leaves the bar, he heads for home but is only able to get to the next pole before stopping. At each pole except the one in front of his house and the one in front of the bar he is equally likely to start off in either direction, and he keeps going until he reaches the next pole, whereupon the same thing happens. However, when he gets to the pole in front of his house, he goes in and goes to bed. When he finds himself at the pole in front of the bar, he goes in for a drink and starts home. If the man is initially in the bar and has had one drink, how many drinks can he be expected to have before getting home? Is he certain of getting home eventually?

2 A gambler plays the following game. A coin is flipped repeatedly, and a tally is taken of the number of heads and tails which have appeared. The game is over when there are either 3 more heads than tails or 3 more tails than heads. The gambler wins an amount equal to the number of times the coin is flipped. How much can he be expected to win?

3 A cat has fallen down a 3-foot-deep well in which there is a ladder with rungs 1 foot apart. The bottom rung is on the floor of the well, so that if the cat can reach the fourth rung of the ladder, he is out of the well. The cat tries to get out of the well by jumping up the ladder from rung to rung. However, the probability that he can successfully make each jump is $\frac{1}{2}$. If he misses any jump, he falls back to the bottom of the well and has to begin again. Is the cat certain to get out of the well? If so, how many jumps can he be expected to make, and how many times can he be expected to find himself at the bottom of the well before he manages to get out?

4 Let

$$A = \begin{bmatrix} 0 & \frac{1}{2} \\ 1 & \frac{1}{2} \end{bmatrix}$$

Express $F^{(s)}$ as a function of s. Find F by means of Definition 3.5.7 and verify Theorem 3.5.15. Find M by using Definition 3.5.30 and verify equation (3.5.32).

5 Prove Theorem 3.5.13.

6 Prove Corollary 3.5.20.

7 Prove that

$$\sum_{i=1}^{r} f_{\mathbb{S}_i, q} = 1 \qquad (3.5.42)$$

for every transient state S_q. Equation (3.5.42) states that an object initially in a transient state is certain to enter some ergodic set.

8 Prove the following relations:

(a) $f_{ij}^{(s)} = \sum\limits_{k_1, k_2, \cdots, k_{s-1} \neq i} a_{ik_1} a_{k_1 k_2} \cdots a_{k_{s-1} j}$.

(b) $F_0 = (\text{dg } N_0)^{-1}(N_0 - I)$.

9 For an irreducible transition matrix A

(a) prove that

$M = MA^k + kE - (\text{dg } M)(A + A^2 + A^3 + \cdots + A^k)$ for $k = 1, 2, \ldots$

(b) Use this result to prove that

$$\lim_{k \to \infty} \frac{A + A^2 + A^3 + \cdots + A^k}{k} = L$$

where L is the matrix each of whose columns is the stationary vector of A.

10 Let A be an irreducible transition matrix and let A_i denote the submatrix of A obtained by deleting the ith row and ith column of A. Let $M_{(i)}$ be the ith row of M and let $\tilde{M}_{(i)}$ be obtained from $M_{(i)}$ by deleting the entry m_{ii}. Show that

$$\tilde{M}_{(i)} = E_{(n-1)}(I - A_i)^{-1}$$

(*Hint:* See Exercise 20 of Sec. 3.3.)

11 For states S_p and S_q of an irreducible Markov chain define

$$m_{pq}^{(k)} = \sum_{s=1}^{k} s f_{pq}^{(s)} \qquad k = 1, 2, 3, \ldots$$

and set

$$M^{(k)} = [m_{pq}^{(k)}]$$

Show that

$$M^{(k+1)} = M^{(k)}A + \sum_{s=1}^{k+1} F^{(s)} - (\text{dg } M^{(k)})A \qquad k = 1, 2, 3, \ldots \qquad (3.5.43)$$

12 Use equation (3.5.43) to show that

$$(\text{dg } M^{(k)})X \leq FX = E$$

and consequently

$$m_{pp} \leq \frac{1}{x_p} \qquad p = 1, 2, \ldots, n$$

13 Show that

$$M^{(k+l)} = M^{(k)}A^l + \sum_{s=1}^{k+l} F^{(s)} + \sum_{s=1}^{k+l-1} F^{(s)}A + \cdots + \sum_{s=1}^{k+1} F^{(s)}A^{l-1}$$
$$- \sum_{r=0}^{l-1} (\text{dg } M^{(k+r)}) A^{l-r}$$

330 MATRIX THEORY AND FINITE MATHEMATICS

for all k and l and consequently

$$(\mathrm{dg}\, M)(A + A^2 + A^3 + \cdots + A^l) + M^{(k+l)} \geqq M^{(k)}A^l$$

14 For fixed p, q there exists l such that $a_{qp}^{(l)} \neq 0$. Show that

$$m_{pq} \leqq \frac{1}{x_p\, a_{qp}^{(l)}} (1 + a_{pp} + a_{pp}^{(2)} + \cdots + a_{pp}^{(l)})$$

REFERENCES

There are two quite different ways to approach the subject of finite Markov chains. We have chosen to emphasize the algebraic aspects of the subject in order to make use of the matrix theory developed in Chap. 1. Romanovsky [9] also restricts himself to matrix methods and studies numerous topics related to finite Markov chains in addition to those covered here. He discusses in detail the original papers of Markov and several other Russian mathematicians. The second approach to finite Markov chains treats them as a special type of stochastic process and relies on probability theory rather than matrices. This treatment is used by Chung [1] and Parzen [8]. Karlin [6] and Feller [5] present both approaches to the subject and discuss the relation between them.

It is not a great jump from the Markov chains studied here to those which have a countable number of states [7]. The finite sums of this chapter are then replaced by infinite sums, and many of the results of this chapter can be carried over. Chung [1] and Karlin [6] consider finite Markov chains as a special case of Markov chains with countably many states. In the more general Markov process the state space is continuous [3].

In the Markov chains we have considered, the set of times at which the object may move is discrete. Cox and Miller [3] and Doob [4] treat continuous-parameter Markov chains in which an object may move at any time. Parzen [8] and Feller [5] develop the birth-and-death process as a continuous-parameter Markov chain, and Saaty [10] applies these ideas to the behavior of a queue. Cox [2] studies semi-Markov chains, which consist of a Markov chain together with a probability function describing the amount of time it takes an object to go from one state to another in one step.

1 Chung, K. L.: "Markov Chains with Stationary Transition Probabilities," 2d ed., Springer-Verlag New York Inc., New York, 1967.

2 Cox, D. R. "Renewal Theory," John Wiley & Sons, Inc., New York, 1962.

3 Cox, D. R., and H. D. Miller: "The Theory of Stochastic Processes," John Wiley & Sons, Inc., New York, 1965.

4 Doob, J. L.: "Stochastic Processes," John Wiley & Sons, Inc., New York, 1953.

5 Feller, W.: "An Introduction to Probability Theory and Its Applications," 3d ed., vol. 1, John Wiley & Sons, Inc., New York, 1968.

6 Karlin, S. "A First Course in Stochastic Processes," Academic Press Inc., New York, 1966.

7 Kemeny, J. G., J. L. Snell, and A. W. Knapp: "Denumerable Markov Chains," D. Van Nostrand Company Inc., Princeton, N.J., 1966.

8 Parzen, E.: "Stochastic Processes," Holden-Day, Inc., San Francisco, 1962.

9 Romanovsky, V. I.: "Discrete Markov Chains," trans. from the Russian by E. Seneta, Wolters-Noordhoff Publishing, Groningen, 1960.

10 Saaty, T. L.: "Elements of Queueing Theory," McGraw-Hill Book Company, New York, 1967.

4

THE THEORY OF GRAPHS

4.1 SOME EXAMPLES

In this chapter we introduce the concept of a graph and prove most of the basic theorems of elementary graph theory.   We will investigate undirected graphs at some length and then show that an analogous theory can be developed for directed graphs.   We will see that there are quite a number of matrices that can be associated with a graph and although the language of graph theory is geometric and we will often find it convenient to employ a diagram, we will strive to obtain as many of the results as we can by matrix methods.   In many cases all the entries of these matrices are 0 or 1.   For such matrices we may consider 0 and 1 to be real numbers or we may consider them to be elements of the field $\mathbb{Z}_2$ consisting only of the two elements 0 and 1.   We will find that unless we are counting things, the latter point of view is usually more fruitful.

This section is devoted to some definitions and a number of examples illustrating various types of problems which can be analyzed and solved using graphs.   We will meet two basic types of problems.   In one we are primarily concerned with the structure of a graph and, perhaps, of its subgraphs.   The other type of problem is a counting problem and asks how many ways some process can be carried out.   Unfortunately, many of the most interesting and important applications of graphs require a great deal of " outside " information, e.g., electric networks, or molecular structure of compounds, and these are beyond the scope of this book.

EXAMPLE 4.1.1   The communication network connecting the points (or vertices) $v_1, v_2, \ldots, v_7$ is represented by the *directed graph* of Fig. 4.1.2.   Is it

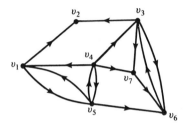

FIGURE 4.1.2

possible to reach v_1 from v_7?   If so, which path passes through the smallest number of points; or, equivalently, if each link of the network (or *arc* of the graph) is assumed to have length 1, what is the shortest path from v_1 to v_7?   Is it possible to go from v_7 to v_1?

Although these questions can be answered by searching through the graph and testing all possible paths, we are more concerned with finding a systematic way of arriving at an answer.   To this end we define the vertex-adjacency matrix of the graph.   For each pair of vertices v_i and v_j, let a_{ij} be the number (possibly zero) of arcs from v_i to v_j.   Then $A = [a_{ij}]$ is a square matrix called the *vertex-adjacency matrix*[1] of the graph.   The order of A is equal to the number of vertices of the graph.   Since we will insist that the two vertices at the ends of an arc are distinct, the diagonal elements of the vertex-adjacency matrix

[1] If we assign a probability to each arc of the graph in such a way that for each vertex the sum of the assigned probabilities of the arcs directed from that vertex is 1, then we have a Markov chain whose transition matrix has 0s in the same positions as A and which has these probabilities in the other positions. However, in this section we will be primarily interested in counting the number of possible arrangements and not in the probability that each will actually occur. Nevertheless, a number of important concepts from Chap. 3 will reappear in this chapter, though in slightly disguised form.

are 0.   For the graph of Fig. 4.1.2, the vertex-adjacancy matrix is the 7×7 matrix

$$A = \begin{bmatrix} 0 & 1 & 0 & 0 & 1 & 0 & 0 \\ 0 & 0 & 0 & 0 & 0 & 0 & 0 \\ 0 & 1 & 0 & 0 & 0 & 1 & 1 \\ 1 & 0 & 1 & 0 & 1 & 0 & 1 \\ 1 & 0 & 0 & 1 & 0 & 1 & 0 \\ 0 & 0 & 1 & 0 & 0 & 0 & 0 \\ 0 & 0 & 0 & 0 & 0 & 1 & 0 \end{bmatrix}$$

The matrix A completely describes the network of links connecting the seven points of Fig. 4.1.2.

Let

$$A^k = [a_{ij}^{(k)}] \qquad k = 0, 1, 2, \ldots$$

Then

$$a_{ij}^{(2)} = \sum_{l=1}^{7} a_{il} a_{lj}$$

is equal to the number of paths of length 2 from v_i to v_j, because

$$a_{il} a_{lj} = 1$$

if there are links from v_i to v_l and from v_l to v_j and

$$a_{il} a_{lj} = 0$$

otherwise.   Similarly, $a_{ij}^{(k)}$ is the number of paths of length k from v_i to v_j. A computation shows that

$$A^2 = \begin{bmatrix} 1 & 0 & 0 & 1 & 0 & 1 & 0 \\ 0 & 0 & 0 & 0 & 0 & 0 & 0 \\ 0 & 0 & 1 & 0 & 0 & 1 & 0 \\ 1 & 2 & 0 & 1 & 1 & 3 & 1 \\ 1 & 1 & 2 & 0 & 2 & 0 & 1 \\ 0 & 1 & 0 & 0 & 0 & 1 & 1 \\ 0 & 0 & 1 & 0 & 0 & 0 & 0 \end{bmatrix} \quad \text{and} \quad A^3 = \begin{bmatrix} 1 & 1 & 2 & 0 & 2 & 0 & 1 \\ 0 & 0 & 0 & 0 & 0 & 0 & 0 \\ 0 & 1 & 1 & 0 & 0 & 1 & 1 \\ 2 & 1 & 4 & 1 & 2 & 2 & 1 \\ 2 & 3 & 0 & 2 & 1 & 5 & 2 \\ 0 & 0 & 1 & 0 & 0 & 1 & 0 \\ 0 & 1 & 0 & 0 & 0 & 1 & 1 \end{bmatrix}$$

There are no paths of length 1 or 2 from v_1 to v_7 since

$$a_{17}^{(1)} = a_{17}^{(2)} = 0$$

However, there is exactly one path of length 3 from v_1 to v_7 since

$$a_{17}^{(3)} = 1$$

By examining the graph of Fig. 4.1.2 or, preferably, by determining how the 1 in the $(1, 7)$ position of A^3 was formed, we see that

$$v_1 \rightarrow v_5 \rightarrow v_4 \rightarrow v_7$$

is the unique path of length 3 from v_1 to v_7.

Now let us determine whether there is a path from v_7 to v_1. From v_7 it is possible to reach only v_6 in one step. According to the last row of A^2, only v_3 can be reached from v_7 along a path of length 2. The last row of A^3 tells us that v_2, v_6, and v_7 can be reached from v_7 in exactly three steps. Thus, in three or fewer steps we can reach v_2, v_3, v_6, and v_7 from v_7. By examining the last row of A^4 we see that the only points which can be reached from v_7 in four steps are v_3 and v_6. Since each of these points can be reached in fewer than four steps, no further points can be reached by taking further steps. Thus it is not possible to reach v_1 from v_7.

The same argument can be demonstrated more forcefully using the vertex-adjacency matrix. First we must permute the rows and columns of A. Although this would follow naturally from a renumbering of the vertices of the graph, such a step is not really necessary. If we impose the order v_1, v_4, v_5, v_2, v_3, v_6, v_7 on the vertices (rather than the natural order defined by the subscripts), the vertex-adjacency matrix becomes

$$
A = \left[
\begin{array}{ccc:cccc}
0 & 0 & 1 & 1 & 0 & 0 & 0 \\
1 & 0 & 1 & 0 & 1 & 0 & 1 \\
1 & 1 & 0 & 0 & 0 & 1 & 0 \\
\hdashline
0 & 0 & 0 & 0 & 0 & 0 & 0 \\
0 & 0 & 0 & 1 & 0 & 1 & 1 \\
0 & 0 & 0 & 0 & 1 & 0 & 0 \\
0 & 0 & 0 & 0 & 0 & 1 & 0
\end{array}
\right]
\begin{array}{l}
v_1 \\ v_4 \\ v_5 \\ \\ v_2 \\ v_3 \\ v_6 \\ v_7
\end{array}
$$

with column labels $v_1\ v_4\ v_5\ \ v_2\ v_3\ v_6\ v_7$.

If we express this matrix as

$$A = \begin{bmatrix} A_{11} & A_{12} \\ 0 & A_{22} \end{bmatrix}$$

it is clear that A^k, $k = 1, 2, \ldots$, has the same form

$$A^k = \begin{bmatrix} A_{11}{}^k & A_{12}{}^{(k)} \\ 0 & A_{22}{}^k \end{bmatrix}$$

(It is not necessary to compute the exact value of $A_{12}{}^{(k)}$.) Thus

$$a_{71}{}^{(k)} = 0$$

for all k, and consequently it is impossible to reach v_1 from v_7.

One apparent drawback of this approach is that it seems necessary to know the answer to the problem before we begin looking for it. However, the re-ordering of the rows and columns of A is merely a convenience, and the same results could have been obtained without it, though with somewhat more labor.

////

EXAMPLE 4.1.3   The following well-known puzzle can be solved by methods similar to those of Example 4.1.1.

Two missionaries and two cannibals are on a journey. They reach a river which they must cross. There is a rowboat available which can carry two people. All four men can row, but the missionaries feel safer if they are not outnumbered by the cannibals at any time (except when there are no mission-aries). Is it possible for the four men to cross the river in such a way that one missionary is never in the company of both cannibals? If so, how many trips across the river are necessary? Suppose it is also required that the two cannibals never be left alone. Can the crossing still be made, and if so, how many trips are necessary?

It will be convenient to think of the missionaries and cannibals as travelling from east to west. We will designate the situation "there are m missionaries and c cannibals and the boat on the east bank of the river" by the triple (m, c, e) and the situation "there are m missionaries and c cannibals and the boat on the west bank of the river" by (m, c, w). Clearly $m = 0, 1, 2$ and $c = 0, 1, 2$, and so there are 18 triples. However, $(1, 2, e)$, $(1, 2, w)$, $(1, 0, e)$, and $(1, 0, w)$ would require one missionary to be with two cannibals, and hence these triples are not permitted. The remaining 14 triples are allowed (at least, until the restrictions added at the end of the problem are considered). We can model this problem as a directed graph whose vertices are the 14 allowable triples and whose arcs describe the possible trips across the river. Set

$$
\begin{array}{ll}
v_1 = (2, 2, e) & v_8 = (2, 2, w) \\
v_2 = (2, 1, e) & v_9 = (2, 1, w) \\
v_3 = (1, 1, e) & v_{10} = (1, 1, w) \\
v_4 = (2, 0, e) & v_{11} = (2, 0, w) \\
v_5 = (0, 2, e) & v_{12} = (0, 2, w) \\
v_6 = (0, 1, e) & v_{13} = (0, 1, w) \\
v_7 = (0, 0, e) & v_{14} = (0, 0, w)
\end{array}
$$

We can now construct the 14×14 vertex adjacency matrix A and use the method of Example 4.1.1 to find the smallest value of k for which

$$
a_{18}{}^{(k)} \neq 0
$$

or show that $a_{18}{}^{(k)} = 0$ for all values of k. Note that $a_{18}{}^{(k)}$ can be nonzero only when k is odd. Because the boat crosses the river at each step, A has the form

$$A = \begin{bmatrix} 0 & A_{12} \\ A_{21} & 0 \end{bmatrix} \qquad (4.1.4)$$

where A_{12} and A_{21} are 7×7 matrices. Moreover, because the same conditions prevail on the east bank of the river as on the west bank, we have

$$A_{12} = A_{21} \qquad (4.1.5)$$

Let us set

$$A_{12} = A_{21} = \tilde{A} = [a_{ij}]$$

It is not difficult to see that for odd values of k

$$A^k = \begin{bmatrix} 0 & \tilde{A}^k \\ \tilde{A}^k & 0 \end{bmatrix}$$

and hence

$$a_{18}{}^{(k)} = \tilde{a}_{11}{}^{(k)}$$

Let us compute $\tilde{A}$. From v_1 it is possible in one step to reach:

1 v_{10} by having one missionary and one cannibal cross the river
2 v_{11} by having both missionaries cross the river
3 v_{12} by having both cannibals cross the river
4 v_{13} by having one cannibal cross the river

Continuing in this way we see that

$$\tilde{A} = \begin{bmatrix} 0 & 0 & 1 & 1 & 1 & 1 & 0 \\ 0 & 1 & 1 & 0 & 1 & 0 & 0 \\ 1 & 1 & 0 & 0 & 0 & 0 & 0 \\ 1 & 0 & 0 & 0 & 0 & 0 & 0 \\ 1 & 1 & 0 & 0 & 0 & 0 & 0 \\ 1 & 0 & 0 & 0 & 0 & 0 & 0 \\ 0 & 0 & 0 & 0 & 0 & 0 & 0 \end{bmatrix}$$

and consequently

$$\tilde{A}^3 = \begin{bmatrix} 0 & 2 & 6 & 4 & 6 & 4 & 0 \\ 2 & 5 & 5 & 2 & 5 & 2 & 0 \\ 6 & 5 & 1 & 0 & 1 & 0 & 0 \\ 4 & 2 & 0 & 0 & 0 & 0 & 0 \\ 6 & 5 & 1 & 0 & 1 & 0 & 0 \\ 4 & 2 & 0 & 0 & 0 & 0 & 0 \\ 0 & 0 & 0 & 0 & 0 & 0 & 0 \end{bmatrix} \quad \text{and} \quad \tilde{A}^5 = \begin{bmatrix} 4 & 18 & 34 & 20 & 34 & 20 & 0 \\ 18 & 29 & 29 & 14 & 29 & 14 & 0 \\ 34 & 29 & 9 & 2 & 9 & 2 & 0 \\ 20 & 14 & 2 & 0 & 2 & 0 & 0 \\ 34 & 29 & 9 & 2 & 9 & 2 & 0 \\ 20 & 14 & 2 & 0 & 2 & 0 & 0 \\ 0 & 0 & 0 & 0 & 0 & 0 & 0 \end{bmatrix}$$

Thus the minimum number of trips in which the missionaries and cannibals can cross the river is 5.   However, there are four different ways in which the crossing can be effected in five trips, namely,

$$v_1 \to v_{12} \to v_2 \to v_9 \to v_5 \to v_8$$
$$v_1 \to v_{12} \to v_2 \to v_9 \to v_3 \to v_8$$
$$v_1 \to v_{10} \to v_2 \to v_9 \to v_5 \to v_8$$
$$v_1 \to v_{10} \to v_2 \to v_9 \to v_3 \to v_8$$

Only the last of the four solutions above satisfies the restriction that the two cannibals are never to be alone.   Thus this problem can also be solved with a crossing requiring five trips.   In this case, the solution is unique.   If we had wished to solve only the restricted problem, our computations could have been considerably simplified; for then the situations corresponding to the four points v_4, v_5, v_{11}, and v_{12} would not have been permitted and the same methods could have been used with a 5×5 matrix.   The details are left as an exercise.   Incidentally, a little thought would have shown that the boat cannot be on a bank of the river with no people, permitting us to eliminate the points v_7 and v_{14}.

Throughout this problem every step that is carried out can be reversed; i.e., for any two vertices of the graph, say v_i and v_j, if v_j can be reached from v_i in one step, then v_i can be reached from v_j in one step.   Thus

$$a_{ij} = a_{ji}$$

for all i and j, which means that A is a symmetric matrix.   It now follows from equations (4.1.4) and (4.1.5) that $\tilde{A}$ is also symmetric.   This property leads to an interesting phenomenon in the graph.   Because no arc of the graph can be traversed in only one direction, it is not necessary to specify a direction at all. The 14-vertex graph of the problem without the restriction that the two cannibals

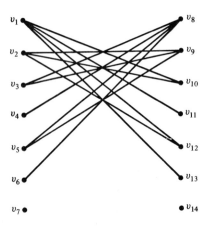

are never alone is illustrated.   (The symmetry about the vertical of this graph is
due to the symmetry of $\tilde{A}$.)   A graph in which no directions are assigned to the
arcs is called an *undirected graph*.   ////

EXAMPLE 4.1.6   The first serious study of graphs grew out of a puzzle solved
by Euler in 1736.   The river Pregel flows through the city of Königsberg (now
called Kaliningrad) in East Prussia.   In the river there are two islands, which
are connected to each other and to the banks of the river by seven bridges.

FIGURE 4.1.7

Is it possible to find a path which uses each of the bridges exactly once and
which ends at the same place it began?   (Such a path is called an *Euler line*.)
Euler proved that such a path does not exist for the Königsberg bridges, and, in
fact, he showed that even if the requirement that the path begin and end at the
same place is dropped, there still is no solution.

We may represent the four-land area of Fig. 4.1.7 as vertices of a graph
and the bridges as arcs of the graph.   Since we have not restricted the direction
in which a bridge may be crossed, the graph is undirected.

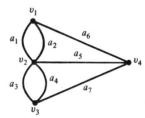

FIGURE 4.1.8

For a path in any graph, beginning at a vertex v_i and ending at a vertex v_j
(using each edge of the graph exactly once or not), the number of times each
vertex other than v_i and v_j is entered is equal to the number of times it is left.
If v_i is the same vertex as v_j, every vertex has the property that the number of
times it is entered is equal to the number of times it is left.   On the other hand,
if v_i and v_j are distinct, then v_i is left one more time than it is entered and v_j is

entered one more time than it is left. When each edge of the graph is used exactly once, the total of the number of times any vertex is entered and is left is equal to the number of arcs incident with that vertex. Thus we have proved that if a graph has a path which uses each arc exactly once, then at every vertex (except possibly for the vertices where the path begins and ends) there are an even number of arcs. Since there are an odd number of arcs at each of the four vertices of the graph of Fig. 4.1.8, no such path exists.

The number of arcs incident with a vertex of a graph is called the *degree* of the vertex. In Sec. 4.2 we will prove that a connected graph in which every vertex has even degree has an Euler line. Also, we will show that a connected graph in which all but two of the vertices have even degree has a path beginning at one of these vertices, ending at the other, and using each arc exactly once.

////

EXAMPLE 4.1.9   Figure 4.1.10 is a map of the oil-pipeline system in South Transylvania. Each pipe is capable of carrying 1 ton of oil per hour. Oil can be transferred from one pipe to another only where pipes meet at one of the pumping stations $v_1, v_2, \ldots, v_{10}$; that is, the pipes between v_1 and v_6 and between v_2 and v_8 do not intersect, and oil cannot be transferred directly from one to the other.

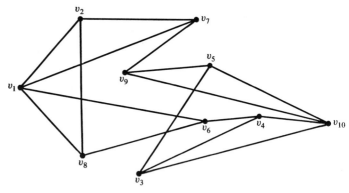

FIGURE 4.1.10

How much oil per hour can the South Transylvanians move from v_1 to v_{10}? What is the smallest number of pipes in Fig. 4.1.10 that the North Transylvanians can block which will prevent the South Transylvanians from shipping any oil from v_1 to v_{10}?

There are numerous ways in which 2 tons of oil per hour can be sent from v_1 to v_{10}. For example, 1 ton per hour can be piped using the route v_1, v_7, v_9, v_{10}, while another ton per hour can be sent using the route $v_1, v_8, v_6, v_4, v_{10}$.

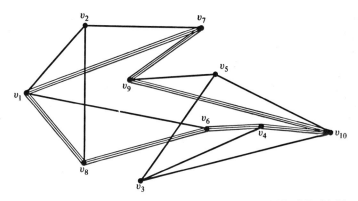

FIGURE 4.1.11

On the other hand, no oil can flow from v_1 to v_{10} if the pipe connecting v_7 and v_9 and the pipe connecting v_6 and v_4 are both blocked, since the remaining network of pipes is no longer connected. The graph of Fig. 4.1.12 consists of two disjoint (connected) components, one of which contains v_1 while the other contains v_{10}.

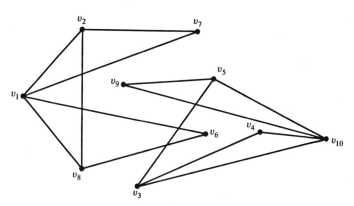

FIGURE 4.1.12

South Transylvania cannot ship more than 2 tons of oil per hour from v_1 to v_{10} since the North Transylvanians are able to stop the flow of oil completely by blocking only two pipes. On the other hand, if the South Transylvanians can ship 2 tons of oil per hour from v_1 to v_{10}, the entire flow of oil cannot be stopped by blocking only one pipe. More generally, it is clear that if in a graph there are k *arc-disjoint paths* (i.e., no arc is used in more than one path) from v_i to v_j and if by the removal of l arcs of the graph there are no paths from v_i to v_j then $k \leqq l$. Ford and Fulkerson [4] have shown that if k is the largest number of arc-disjoint paths that exist between v_i and v_j and if l is the smallest number of arcs whose removal from the graph results in there being no paths between v_i and v_j, then $k = l$. ////

EXAMPLE 4.1.13 The South Transylvanians are anxious to defend their oil-pipeline system (Fig. 4.1.10) against attacks by the North Transylvanians but do not wish to commit more of their small army than is absolutely necessary. If one army unit is needed to protect each pipeline segment, and if the South Transylvanians must maintain a system whereby oil can be sent from any city to any other city, how many army units are required?

If any set of pipes form a loop, for example, $v_3 v_4$, $v_4 v_{10}$, $v_{10} v_3$ or $v_1 v_2$, $v_2 v_8$, $v_8 v_6$, $v_6 v_1$, then any one of these pipes is superfluous and need not be defended. For example, if the pipe $v_3 v_4$ is blocked, oil can be shipped from v_3 to v_4 via $v_3 v_{10}$ and $v_{10} v_4$. Thus the system that the South Transylvanians will defend will not contain any loops. Figure 4.1.14 provides an example of a network of pipes connecting all cities in which no pipe may be left undefended. There are many others.

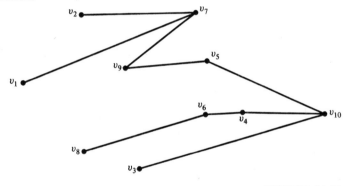

FIGURE 4.1.14

The graph of the pipeline system of Fig. 4.1.14 has nine arcs, and thus nine army units are required. In Sec. 4.4 we will prove that every set of arcs which connects all the vertices of the graph of Fig. 4.1.14 and which has no loops contains precisely nine arcs.[1] ////

The relationship between a graph and the various matrices associated with it is the main topic of this chapter. In Example 4.1.1 we defined the vertex-adjacency matrix A of a directed graph G. If G' is formed from G by reversing the direction of each arc, the vertex-adjacency matrix of G' is A^T, the transpose

---

[1] This problem can also be solved in the more realistic situation where the pipeline segments are not all of the same length; i.e., not all require the same number of army units for protection. See Ore [9].

of A. In Example 4.1.3 we defined the vertex-adjacency matrix of an undirected graph. The two definitions agree if we consider an undirected arc as being directed toward both its vertices. Clearly the vertex-adjacency matrix of an undirected graph is symmetric, i.e.,

$$A^T = A$$

In Example 4.1.6 we defined the degree d_i of the vertex v_i of an undirected graph to be the number of arcs incident with v_i. The *degree matrix* (to be denoted by D) is the diagonal matrix whose diagonal entries are the degree of the vertices. For a directed graph we define the *in degree* δ_i and the *out degree* Δ_i of v_i to be the number of arcs directed to and the number of arcs directed from v_i respectively. As in the undirected case, the *in-degree matrix* δ and the *out-degree matrix* Δ are the diagonal matrices whose diagonal entries are the degrees of the vertices. We leave it as an exercise to show that:

1 For an undirected graph d_i is equal to the entry in the (i, i) position of A^2.
2 For a directed graph δ_i is equal to the entry in the (i, i) position of $A^T A$ whereas Δ_i is equal to the entry in the (i, i) position of AA^T.

Another important matrix associated with a graph is the *vertex-arc incidence matrix* $B = [b_{ij}]$. For a directed graph we define B by

$$b_{ij} = \begin{cases} 1 & \text{if arc } a_j \text{ is directed toward vertex } v_i \\ -1 & \text{if arc } a_j \text{ is directed from vertex } v_i \\ 0 & \text{otherwise} \end{cases}$$

In each column of B there is one 1 and one -1, and the remaining entries are 0. The number of rows of B is equal to the number of vertices of the graph, and the number of columns of B is equal to the number of arcs. For an undirected graph we define B by

$$b_{ij} = \begin{cases} 1 & \text{if } v_i \text{ is one of the vertices of the arc } a_j \\ 0 & \text{otherwise} \end{cases}$$

In each column of B there are two 1s, and the remaining entries are 0.

EXAMPLE 4.1.15 For the graph of Fig. 4.1.8, the vertex-adjacancy matrix, the degree matrix, and the vertex-arc incidence matrix are

$$A = \begin{bmatrix} 0 & 2 & 0 & 1 \\ 2 & 0 & 2 & 1 \\ 0 & 2 & 0 & 1 \\ 1 & 1 & 1 & 0 \end{bmatrix} \quad D = \begin{bmatrix} 3 & 0 & 0 & 0 \\ 0 & 5 & 0 & 0 \\ 0 & 0 & 3 & 0 \\ 0 & 0 & 0 & 3 \end{bmatrix}$$

$$B = \begin{bmatrix} 1 & 1 & 0 & 0 & 0 & 1 & 0 \\ 1 & 1 & 1 & 1 & 1 & 0 & 0 \\ 0 & 0 & 1 & 1 & 0 & 0 & 1 \\ 0 & 0 & 0 & 0 & 1 & 1 & 1 \end{bmatrix} \qquad ////$$

Since the matrices defined above reflect different properties of the same graph, it is not surprising that the matrices themselves are related. A typical relation is described in:

Theorem 4.1.16 Let G be an undirected graph with m vertices and n arcs. Then

$$BB^T = A + D$$

PROOF For convenience set

$$BB^T = H = [h_{ij}]$$

Then
$$h_{ij} = \sum_{k=1}^{n} b_{ik} b_{jk} \qquad i, j = 1, 2, \ldots, m \qquad (4.1.17)$$

CASE 1 Let $i = j$. Clearly

$$b_{ik} b_{ik} = \begin{cases} 1 & \text{if } v_i \text{ is one of the vertices of } a_k \\ 0 & \text{otherwise} \end{cases}$$

It follows from equation (4.1.17) that h_{ij} is equal to the number of arcs which have v_i as one of its vertices, i.e.,

$$h_{ii} = d_{ii}$$

CASE 2 Let $i \neq j$. Then

$$b_{ik} b_{jk} = \begin{cases} 1 & \text{if } v_i \text{ and } v_j \text{ are the vertices of } a_k \\ 0 & \text{otherwise} \end{cases}$$

and consequently h_{ij} is equal to the number of arcs which have v_i and v_j as vertices. Thus

$$h_{ij} = a_{ij} \qquad ////$$

EXERCISES

1 Show that the minimum number of trips in which three missionaries and three cannibals using a boat carrying two people can cross a river in such a way that on each of the banks of the river the missionaries are never outnumbered (except when there are no missionaries) is 11. In how many different ways can they cross the river in 11 trips? Show also that if the three cannibals are never permitted to be together, except in the presence of the three missionaries, they cannot get across the river.

2 Use 5×5 matrices to show that there is exactly one way in which two missionaries and two cannibals can cross a river in five trips if the two cannibals are not permitted to be alone together.

3   A farmer has two bulls and three goats and wishes to bring them to town
    from his farm.   The road to town is downhill, and his truck can carry any
    two animals on each trip.   However, on the trips back to the farm his truck
    can carry only one animal.   Unless the farmer is present, there must be more
    goats than bulls or no goats at all.   Show that the farmer can get his live-
    stock to town but once there he cannot get them back to the farm.   (*Hint:*
    Define a matrix to describe the possible trips up the hill and another to
    describe the possible trips down the hill.)

4   Show that in Example 4.1.1 it would still be possible to reach v_7 from v_1 if
    the arc from v_4 to v_7 is removed from the graph.   What is the shortest path
    in this case, and how many such paths are there?

5   Show that the analog of Theorem 4.1.16 for directed graphs is

    $$BB^T = \delta + \Delta - A - A^T$$

6   For an undirected graph G with m vertices and n arcs, set $\tilde{a}_{ij}$ equal to the
    number (possibly zero but at most 2) vertices common to the arcs a_i and a_j.
    The $n \times n$ matrix $\tilde{A} = [a_{ij}]$ is called the *arc-adjacency matrix* of G.   Prove
    that:

    (*a*) $B^T B = \tilde{A}$ (*b*) $B\tilde{A} - AB = DB$

7   For a directed graph G we define $\tilde{a}_{ij}$ by

    $\tilde{a}_{ij} = e_{ij}$ times the number of vertices common to the arcs a_i and a_j,

    and when a_i and a_j have vertices in common, e_{ij} is defined by

    $$e_{ij} = \begin{cases} 1 & \text{if } a_i \text{ and } a_j \text{ are either both directed toward a common} \\ & \text{vertex or both directed from a common vertex} \\ -1 & \text{if one of } a_i \text{ and } a_j \text{ is directed toward a common vertex and} \\ & \text{the other is directed from the common vertex} \end{cases}$$

    Prove that $B^T B = \tilde{A}$.

8   Construct the arc-adjacency, vertex-adjacency, and incidence matrices of the
    graph of Fig. 4.1.8 and verify Theorem 4.1.16 and Exercise 6.

9   Transform the graph of Fig. 4.1.8 into a directed graph by orienting a_1, a_2,
    a_3, and a_4 toward v_2 and a_5, a_6, and a_7 toward v_4.   Construct the arc-
    adjacency, vertex-adjacency, and incidence matrices and verify Exercises 5
    and 7.

4.2   UNDIRECTED GRAPHS

A number of examples involving graphs were presented and analyzed in the last
section.   A diagram was part of the statement of the problem in several of these
examples, and the solution was found by treating the diagram as an abstract

graph. In Example 4.1.3 a diagram was not mentioned in the statement of the problem, but we were able to display all the pertinent information by means of a graph and this led to a solution. In order to put the entire subject on a firmer foundation and decrease our reliance on (though not our use of) diagrams we will give some precise definitions and prove some of the "obvious" properties of graphs. In this section we will consider only undirected graphs.

Definition 4.2.1 An $m \times n$ (undirected) *graph* $G = (\mathbb{V}, \mathbb{A})$ consists of a set $\mathbb{V}$ of m objects called *vertices*

$$\mathbb{V} = \{v_1, v_2, \ldots, v_m\}$$

and a set $\mathbb{A}$ of n *arcs*

$$\mathbb{A} = \{a_1, a_2, \ldots, a_n\}$$

where each arc is an unordered pair of distinct vertices. If an arc a consists of the vertices v_1 and v_2, we say that v_1 and v_2 are *incident* with a and that v_1 and v_2 are *adjacent*. Two arcs which have a vertex in common are *adjacent*. When two arcs have both vertices in common, they are *parallel*.

Each arc of a graph is incident with exactly two vertices. A vertex may be incident with any number of arcs, however, and the number of arcs incident with a vertex is called its *degree*.

In order to specify a graph it is sufficient to enumerate the vertices and the adjacency relations among them. It is an easy matter (perhaps too easy) to put all this information into the form of a diagram. As we have seen, this information can also be presented in terms of matrices, and, of course, this is the approach that we will emphasize.

EXAMPLE 4.2.2 The graph with vertices v_1, v_2, v_3, v_4, v_5 and arcs

$$a_1 = (v_1, v_3) \qquad a_2 = (v_1, v_4)$$
$$a_3 = (v_2, v_6) \qquad a_4 = (v_7, v_1)$$
$$a_5 = (v_1, v_3) \qquad a_6 = (v_7, v_4)$$

can be represented by Fig. 4.2.3.   ////

The graph of Fig. 4.2.3 is composed of three separate subgraphs:

1 G_1 with vertices v_1, v_4, v_7, v_3 and arcs a_1, a_2, a_4, a_5, a_6
2 G_2 with vertices v_2, v_6 and arc a_3
3 G_3 with vertex v_5 and no arcs

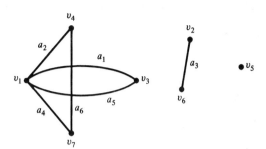

FIGURE 4.2.3

It is much easier to investigate the properties of each of these subgraphs separately than it would be to restrict our attention to the entire graph. In order to characterize these subgraphs we make the following definition.

Definition 4.2.4   A graph is *connected* if it is not possible to partition the set of vertices into two disjoint nonempty subsets in such a way that no arc of the graph is incident with vertices from both subsets.

The graph of Fig. 4.2.3 is clearly not connected. On the other hand, each of the subgraphs G_1, G_2, G_3 described above is connected. An important characterization of connected graphs follows from:

Definition 4.2.5   Let v_1 and v_2 be vertices of the graph G. An ordered set of l distinct arcs of G,

$$(v_1, v_{h_1}), (v_{h_1}, v_{h_2}), (v_{h_2}, v_{h_3}), \ldots, (v_{h_{l-1}}, v_2)$$

is called a *path of length l* between v_1 and v_2. If the vertices $v_{h_1}, v_{h_2}, \ldots, v_{h_{l-1}}$ are distinct from each other and also from v_1 and v_2, the path is said to be *simple*. A simple path between a vertex and itself, that is, $v_1 = v_2$, is called a *loop*.

In addition, we define the empty set of arcs to be a simple path between every vertex and itself. However, the empty set of arcs is not defined to be a loop. An arc is a simple path of length 1 between its vertices. In Fig. 4.2.3 there are two loops, namely,

$$\{a_1, a_5\} \quad \text{and} \quad \{a_2, a_6, a_4\}$$

Also, $\{a_1, a_2, a_6, a_4\}$ is a path between v_3 and v_1, but it is not simple.

If

$$\{(v_1, v_{h_1}), (v_{h_1}, v_{h_2}), (v_{h_2}, v_{h_3}), \ldots, (v_{h_{l-1}}, v_2)\}$$

is a path of length l between v_1 and v_2, then

$$\{(v_2, v_{h_{l-1}}), \ldots, (v_{h_3}, v_{h_2}), (v_{h_2}, v_{h_1}), (v_{h_1}, v_1)\}$$

is a path of length l between v_2 and v_1. It will be convenient not to consider these paths as being distinct (just as we did not consider $\{a_1, a_5\}$ and $\{a_5, a_1\}$ to be distinct loops). Henceforth we will say that a path between v_1 and v_2 is also a path between v_2 and v_1, and we will not specify the direction of the path. However, if another path between v_1 and v_2 can be obtained by rearranging the order of the arcs, this path will be considered to be a different path. For example, in Fig. 4.2.3,

$$\{a_1, a_2, a_6, a_4\} \qquad \text{and} \qquad \{a_4, a_6, a_2, a_1\}$$

are the same path between v_3 and v_1 (or v_1 and v_3) but

$$\{a_1, a_2, a_6, a_4\} \qquad \text{and} \qquad (a_1, a_4, a_6, a_2\}$$

are distinct paths between v_1 and v_3.

Theorem 4.2.6 Let v_1, v_2, v_3 be vertices of the graph G.

1 If there is a path between v_1 and v_2, then there is a simple path between v_1 and v_2.

2 If there is a path between v_1 and v_2 and there is a path between v_2 and v_3, then there is a path between v_1 and v_3.

PROOF The proofs of the two parts of the theorem are similar and illustrate a technique frequently used in graph theory.

(1) Let

$$\{a_1 = (v_1, v_{h_1}), a_2 = (v_{h_1}, v_{h_2}), \ldots, a_{l-1} = (v_{h_{l-2}}, v_{h_{l-1}}), a_l = (v_{h_{l-1}}, v_2)\}$$

be a path of length l between v_1 and v_2. Although the arcs of the path are distinct, some of the vertices may occur more than once. For convenience of notation, set

$$v_1 = v_{h_0} \qquad v_2 = v_{h_l}$$

If the l vertices $v_{h_0}, v_{h_1}, v_{h_2}, \ldots, v_{h_{l-1}}$ are distinct, the path is simple. If the l vertices are not distinct, let i be the smallest integer such that v_{h_i} occurs more than once. Let j be the largest integer such that

$$v_{h_i} = v_{h_j}$$

(Note that the special case $j = l$ is not excluded.)   Then

$$\{a_1, a_2, \ldots, a_i, a_{j+1}, \ldots, a_l\} \qquad (4.2.7)$$

is a path between v_1 and v_2 containing fewer than l arcs.   When $i = 0$ and $j \neq l$, (4.2.7) is interpreted to mean $\{a_{j+1}, \ldots, a_l\}$.   When $i \neq 0$ and $j = l$, (4.2.7) is interpreted to mean $\{a_1, a_2, \ldots, a_i\}$.   Finally, when $i = 0$ and $j = l$, (4.2.7) is interpreted to be the empty set of arcs.   If the path of (4.2.7) is simple we have finished.   If not, the process is repeated.   In a finite number of steps we reach a single path between v_1 and v_2.

   (2) Let

$$\{a_1 = (v_1, v_{h_1}), a_2 = (v_{h_1}, v_{h_2}), \ldots, a_k = (v_{h_{k-1}}, v_2)\}$$

and $\{a_{k+1} = (v_2, v_{h_{k+1}}), a_{k+2} = (v_{h_{k+1}}, v_{h_{k+2}}), \ldots, a_l = (v_{h_{l-1}}, v_3)\}$

be paths between v_1 and v_2 and between v_2 and v_3 respectively.   For convenience of notation set

$$v_1 = v_{h_0} \qquad v_2 = v_{h_k} \qquad v_3 = v_{h_l}$$

According to Definition 4.2.5, $a_1, a_2, \ldots, a_k$ are distinct.   Likewise, $a_{k+1}, a_{k+2}, \ldots, a_l$ are distinct.   If these l arcs are distinct, they form a path between v_1 and v_3.   If these arcs are not distinct, let i be the smallest integer such that the arc a_i occurs more than once.   Consequently a_i occurs twice, once in the path between v_1 and v_2 and once in the path between v_2 and v_3.   Let a_i appear in the form

$$a_i = (v_{h_p}, v_{h_{p+1}}) \qquad 0 \leq p \leq k - 1$$

in the path between v_1 and v_2 and as

$$a_i = (v_{h_q}, v_{h_{q+1}}) \qquad k \leq q \leq l - 1$$

in the path between v_2 and v_3.   Now we must consider two cases.

CASE 1   Suppose

$$v_{h_p} = v_{h_q}$$

and consequently

$$v_{h_{p+1}} = v_{h_{q+1}}$$

Then $\{a_1, a_2, \ldots, a_i, a_{j+1}, \ldots, a_l\}$ is the required path.

CASE 2   Suppose

$$v_{h_p} = v_{h_{q+1}}$$

and consequently

$$v_{h_{p+1}} = v_{h_q}$$

Then $\{a_1, a_2, \ldots, a_{i-1}, a_{j+1}, \ldots, a_l\}$ is the required path.

The interpretation of the cases $i = 0$ and $j = l$ follows the same pattern as in part 1 of the theorem.   ////

Let us define the relation $\sim$ on the vertices of a graph G by

$$v_i \sim v_j \Leftrightarrow \text{there is a path between } v_i \text{ and } v_j$$

Clearly $\sim$ is reflexive. Because we have defined the empty set of arcs to be a path between each vertex of a graph and itself, $\sim$ is symmetric. Also, part 2 of Theorem 4.2.6 states that $\sim$ is transitive. Thus $\sim$ is an equivalence relation and partitions the vertices of G into disjoint equivalence classes. We can extend this equivalence to the arcs of G since an arc is a path of length 1 between its two vertices and consequently both vertices of an arc belong to the same equivalence class. Let us now add each arc to the equivalence class of its vertices and denote the resulting sets (of vertices and arcs) by $G_1, G_2, \ldots, G_k$. Then:

1 Each G_i is a graph, $i = 1, 2, \ldots, k$.
2 For $i \neq j$, the graphs G_i and G_j are both arc- and vertex-distinct.
3 G is the union of $G_1, G_2, \ldots, G_k$.

We call $G_1, G_2, \ldots, G_k$ the *components* of G.

Theorem 4.2.8 A graph G is connected if and only if there is a path between any two vertices of G; that is, G consists of only one component.

PROOF Let v be a vertex of the connected graph G. Denote by $\mathbb{C}(v)$ the set of all vertices of G which are equivalent to v and by $\mathbb{D}(v)$ the remaining vertices. We wish to show that $\mathbb{D}(v)$ is empty. Let us suppose that $\mathbb{D}(v)$ is not empty. Since G is connected, there is at least one arc, say (v_s, v_t), of G which is incident with both a vertex belonging to $\mathbb{C}(v)$, say v_s, and a vertex belonging to $\mathbb{D}(v)$, say v_t. Since $v_s \in \mathbb{C}(v)$, there is a path between v and v_s. Moreover the single arc (v_s, v_t) is a path between v_s and v_t. It follows from part 2 of Theorem 4.2.6 that there is a path between v and v_t and hence $v_t \notin \mathbb{D}(v)$. This is a contradiction, and consequently $\mathbb{D}(v)$ is empty.

Conversely, let G be a graph such that between any two vertices there is a path. Let us suppose that G is not connected. Then the vertices of G can be partitioned into two disjoint nonempty sets, say $\mathbb{V}_1$ and $\mathbb{V}_2$, in such a way that no arc of G is incident with vertices from both $\mathbb{V}_1$ and $\mathbb{V}_2$. Let v_1 be any vertex belonging to $\mathbb{V}_1$ and v_2 be any vertex belonging to $\mathbb{V}_2$. By hypothesis there is a path

$$\{(v_1, v_{h_1}), (v_{h_1}, v_{h_2}), \ldots, (v_{h_{l-1}}, v_2)\}$$

between v_1 and v_2.   Set

$$v_1 = v_{h_0} \qquad v_2 = v_{h_l}$$

Let i be the smallest integer for which v_{h_i} belongs to $\mathbb{V}_2$.   Since $v_{h_l} \in \mathbb{V}_2$, such an integer exists.   Also, because $v_{h_0} \notin \mathbb{V}_2$, it follows that $1 \leqq i \leqq l$. We now have found an arc $(v_{h_{i-r}}, v_{h_i})$ which is incident with a vertex of $\mathbb{V}_1$ and with a vertex of $\mathbb{V}_2$.   This is a contradiction, and hence the supposition that G is not connected is false.   ////

Corollary 4.2.9   Each component of a graph is connected.   Consequently every graph is a union of disjoint connected subgraphs.

EXAMPLE 4.2.10   Applying Corollary 4.2.9 to the graph of Fig. 4.2.3 shows that there are three components:

$$G_1 = \{\mathbb{V}_1 = \{v_1, v_3, v_4, v_7\}, \mathbb{A}_1 = \{a_1, a_2, a_4, a_5\}\}$$
$$G_2 = \{\mathbb{V}_2 = \{v_2, v_6\}, \mathbb{A}_2 = \{a_3\}\}$$
$$G_3 = \{\mathbb{V}_3 = \{v_5\}, \mathbb{A}_3 = \varnothing\} \qquad\qquad ////$$

Henceforth, unless we specify otherwise, we will assume that all graphs under consideration are connected.   In Sec. 4.4 we will present an alternate way of describing connected graphs.

We conclude this brief introduction to undirected graphs with a proof of the theorem we used (without proof) in Example 4.1.6 to show that the Königsberg bridge problem has no solution.

Theorem 4.2.11   If the degree of every vertex of a graph G is even, then G has an Euler line, i.e., a path between a vertex and itself which contains all the arcs of G.

PROOF   Let v be any vertex of G.   Among all the paths between v and itself there is one which is of maximal length, say

$$P = \{a_1 = (v, v_{h_1}), a_2 = (v_{h_1}, v_{h_2}), \ldots, a_l = (v_{h_{l-1}}, v)\}$$

It will be convenient to set

$$v = v_{h_0} = v_{h_l}$$

We wish to prove that P is an Euler line.

Let us suppose that P is not an Euler line.   First we need to show that at least one vertex of P is incident with fewer arcs than its degree.   Let us suppose that this is false. Then both vertices of every arc of G which is not in P do not belong to P. Consequently G is not connected. This is a contradiction, and we have now shown that at least one vertex of P is incident with fewer arcs than its degree. Let us call this vertex v_{k_0} (so that $v_{k_0} = v_{h_i}$ for some i, $0 \le i \le l$). There is an arc $\bar{a}_1 = (v_{k_0}, v_{k_1})$ which does not belong to P. We now see that v_{k_1} is incident with an even number of arcs of P and with $\bar{a}_1$. (Note that we have not specified whether or not v_{k_1} belongs to P.) Since the degree of v_{k_1} is even, there is an arc other than $\bar{a}_1$ which is incident with v_{k_1} and which does not belong to P. Denote this arc by $\bar{a}_2 = (v_{k_1}, v_{k_2})$. Unless $v_{k_2} = v_{k_0}$, the same argument can be applied to v_{k_2} to prove that there is an arc $\bar{a}_3 = (v_{k_2}, v_{k_3})$ which does not belong to P. Continuing in this way, we eventually return to v_{k_0}, and then we have a path $\bar{P}$,

$$\bar{P} = \{\bar{a}_1 = (v_{k_0}, v_{k_1}), \bar{a}_2 = (v_{k_1}, v_{k_2}), \ldots, \bar{a}_m = (v_{k_{m-1}}, v_{k_0})\}$$

between v_{k_0} and itself consisting of arcs which do not belong to P. If we now form

$$P' = \{a_1, a_2, \ldots, a_i, \bar{a}_1, \bar{a}_2, \ldots, \bar{a}_m, a_{i+1}, \ldots, a_l\}$$

P' is a path between v and itself and, moreover, the length of P' is equal to the sum of the lengths of P and $\bar{P}$. This contradicts the maximality of the length of P. Hence P is an Euler line. ////

Corollary 4.2.12 If G has exactly two vertices whose degree is odd, then there is a path between them which contains all the arcs of G.

PROOF Let v_1 and v_2 be the two vertices of odd degree. Form the graph G' from G by adding the arc $a_0 = (v_1, v_2)$. Every vertex of G' has even degree, and hence G' has an Euler line P'. We may consider a_0 to be the last arc of P'. If a_0 is removed from P', the remaining path, P, satisfies the conditions of the corollary. ////

EXAMPLE 4.2.13 The graph in Fig. 4.2.14 contains an Euler line since

$$d_3 = 4 \qquad d_1 = d_2 = d_4 = d_5 = 2$$

Although there are two Euler lines between v_1 and itself

$$P_1 = \{a_1, a_4, a_5, a_6, a_2, a_3\} \qquad P_2 = \{a_1, a_6, a_5, a_4, a_2, a_3\}$$

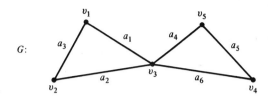

G:

FIGURE 4.2.14

there are four Euler lines between v_3 and itself

$$Q_1 = \{a_1, a_3, a_2, a_6, a_5, a_4\} \qquad Q_2 = \{a_2, a_3, a_1, a_6, a_5, a_4\}$$
$$Q_3 = \{a_1, a_3, a_2, a_4, a_5, a_6\} \qquad Q_4 = \{a_2, a_3, a_1, a_4, a_5, a_6\} \qquad ////$$

An Euler line describes the order of the arcs in a path between a vertex and itself. If

$$P = \{a_1 = (v_{h_1}, v_{h_2}), a_2, \ldots, a_{k-1}, a_k = (v_{l_1}, v_{l_2}), \ldots, a_n\}$$

is an Euler line between v_{h_1} and itself, then

$$P' = \{a_k = (v_{l_1}, v_{l_2}), \ldots, a_n, a_1 = (v_{h_1}, v_{h_2}), a_2, \ldots, a_{k-1}\}$$

is an Euler line between v_{l_1} and itself. We do not exclude the case $v_{h_1} = v_{l_1}$. We will say that two Euler lines of a graph belong to the same *circuit* if one of them can be obtained from the other by the process described above. Thus, a circuit in an $m \times n$ graph consists of n Euler lines. Each Euler line in a circuit is uniquely determined by specifying its first arc. In Example 4.2.13, P_1, Q_1, and Q_4 belong to the same circuit. Similarly, P_2, Q_2, and Q_3 belong to the same circuit. These are the only circuits of the graph. Find the other Euler lines in each circuit.

Lemma 4.2.15 Let G be a graph which has c circuits. If the degree of the vertex v_i is d_i, there are $\frac{1}{2}d_i c$ Euler lines between v_i and itself.

PROOF Let

$$C = \{a_1 = (v_{h_1}, v_{h_2}), a_2 = (v_{h_2}, v_{h_3}), \ldots, a_n = (v_{h_n}, v_{h_1})\} \qquad (4.2.16)$$

be a circuit of G. The number of Euler lines between a fixed vertex, say v_i, and itself is equal to the number of arcs in (4.2.16) whose first vertex is v_i. But this number is $\frac{1}{2}d_i c$.   ////

EXERCISES

1 Find the circuits of the component G_1 of the graph of Fig. 4.2.3. Find the Euler lines of each circuit.

2 Let v_1 and v_2 be vertices of the graph G. We define the *distance* between v_1 and v_2 by

$$d(v_1, v_2) = \begin{cases} \text{length of shortest path} & \text{if } v_1 \text{ and } v_2 \text{ belong to same} \\ \text{between } v_1 \text{ and } v_2 & \text{component of } G \\ \infty & \text{If } v_1 \text{ and } v_2 \text{ do not belong to} \\ & \text{same component of } G \end{cases}$$

Show that d satisfies the properties of a *metric*, i.e.,
(a) $d(v_1, v_2) \geqq 0$ for all $v_1, v_2 \in G$ and $d(v_1, v_2) = 0$ if and only if $v_1 = v_2$.
(b) $d(v_1, v_2) = d(v_2, v_1)$ for all v_1, v_2.
(c) $d(v_1, v_2) + d(v_2, v_3) \geqq d(v_1, v_3)$ for all v_1, v_2, v_3.
For the graph of Fig. 4.1.10, find $d(v_1, v_3)$ and $d(v_7, v_{10})$.

3 Let G be a graph which has an Euler line and let v_1 and v_2 be vertices of G. Show that if P is a path between v_1 and v_2, then there is a path P' between v_1 and v_2 which is arc-disjoint from P. [*Hint:* Consider the case $v_1 = v_2$ separately. For $v_1 \neq v_2$ there is an arc $a_1 = (v_1, v_{h_1})$ which does not belong to P.]

4 Let G be a connected graph. We define the *diameter* of G by

$$\text{diam } G = \max_{v_1, v_2} d(v_1, v_2)$$

Find the diameter of the graphs of Figs. 4.1.8, 4.1.10, and 4.1.14.

5 Let G be a connected graph with m vertices and n arcs. Show that diam $G \leqq m - 1$. Also show that if G has an Euler line, then diam $G \leqq n/2$.

6 Let A be the vertex-adjacency matrix of the connected graph G. Prove that diam G is the smallest positive integer d such that every entry of $(I + A)^d$ is positive.

7 Show that a graph contains exactly

$$\frac{1}{2} \sum_{i=1}^{m} d_i(d_i - 1)$$

paths of length 2, where m is the number of vertices and the d_i's are the degrees of the vertices.

8 A *complete graph* is defined to be a graph such that between every pair of distinct vertices there is exactly one arc. Show that a complete graph with m vertices has $\frac{1}{2}m(m - 1)$ arcs and $\frac{1}{2}m!$ simple paths of length $m - 1$. For which values of m does a complete graph have an Euler line?

9 Let G be a connected graph with m vertices. Prove that the following conditions are equivalent:

(*a*) G has no loops.

(*b*) G has $m - 1$ arcs.

(*c*) Between every pair of vertices of G there is exactly one path.

A graph satisfying one of these conditions (and hence all) is called a *tree*

4.3 THE VECTOR SPACE OF SETS

In Exercise 6 of Chap. 0 a method was outlined for constructing an infinite number of fields, one for each positive prime integer. The field $\mathbb{Z}_p$ corresponding to the prime p contains p elements. In this chapter we will make extensive use of the field $\mathbb{Z}_2$ consisting of the two elements 0 and 1. Addition and multiplication within $\mathbb{Z}_2$ are given by the following tables:

<table>
<tr><td colspan="3">ADDITION</td><td colspan="3">MULTIPLICATION</td><td></td></tr>
<tr><td></td><td>0</td><td>1</td><td></td><td>0</td><td>1</td><td></td></tr>
<tr><td>0</td><td>0</td><td>1</td><td>0</td><td>0</td><td>0</td><td rowspan="2">(4.3.1)</td></tr>
<tr><td>1</td><td>1</td><td>0</td><td>1</td><td>0</td><td>1</td></tr>
</table>

Let S be a set consisting of n elements

$$S = \{s_1, s_2, \ldots, s_n\}$$

and let $\mathbb{S}$ be the collection of all subsets of S. Among the elements of $\mathbb{S}$ are

1 The empty set $\varnothing$

2 n elements $\{s_1\}, \{s_2\}, \ldots, \{s_n\}$, each consisting of a single element of S

3 $\frac{1}{2}n(n-1)$ elements $\{s_i, s_j\}$ each consisting of a pair of distinct elements of S, etc.

In total, $\mathbb{S}$ contains 2^n elements. We will show that with the proper definitions of addition and scalar multiplication (the scalars coming from the field $\mathbb{Z}_2$) $\mathbb{S}$ has the properties of a vector space given in Sec. 1.3. We will also show that there is a natural way of representing each element of $\mathbb{S}$ as an n-dimensional (row or column) vector with entries from $\mathbb{Z}_2$. Once this representation has been made, $\mathbb{S}$ and the vector space of n-dimensional vectors are (mathematically) identical.

Let x be an element of $\mathbb{S}$. Then x is a subset of S. The set of elements of S which do not belong to x is a subset of S and hence belongs to $\mathbb{S}$. We call this set the *complement* of x and denote it by x'. Clearly

$$(x')' = x$$

Thus if y is the complement of x, then x is the complement of y. Note that $\varnothing$ and S are complements of each other.

We can define an operation on the elements of $\mathbb{S}$ which, as we will see, behaves very much like addition.

Definition 4.3.2 Let x and y be subsets of S. The set of all elements of S which belong to exactly one of the subsets x and y is a subset of S called the *symmetric difference* of x and y and is denoted by $x + y$.

All the following properties are easy to verify.

Corollary 4.3.3. For all $x, y \in \mathbb{S}$,

1 $x + y = y + x$.
2 $x + (y + z) = (x + y) + z$.
3 $x + \varnothing = \varnothing + x = x$.
4 $x + S = S + x = x'$.
5 $x + x' = x' + x = S$.
6 $x + x = \varnothing$.

Thus $\varnothing$ acts as an additive identity, and every element of $\mathbb{S}$ is its own additive inverse.

The equations

$$1 + 1 = 0 \quad \text{and} \quad x + x = \varnothing$$

in (4.3.1) and part 6 of Corollary 4.3.3 respectively suggest that addition in $\mathbb{Z}_2$ and the symmetric difference behave in a similar manner. Let us define a *scalar multiplication* between elements of $\mathbb{Z}_2$ and elements of $\mathbb{S}$ by

$$1 \cdot x = x \qquad 0 \cdot x = \varnothing$$

for all $x \in \mathbb{S}$. Then

Lemma 4.3.4 For all $x, y \in \mathbb{S}$ and $\alpha, \beta \in \mathbb{Z}_2$,

1 $(\alpha + \beta)x = \alpha x + \beta x$.
2 $\alpha(x + y) = \alpha x + \alpha y$.
3 $(\alpha \beta)x = \alpha(\beta x)$.

With these operations we call $\mathbb{S}$ *the vector space of subsets of S over the field $\mathbb{Z}_2$.*   All the properties of a vector space of $m \times n$ matrices as defined in Sec. 1.3 are shared by $\mathbb{S}$.   In fact, there is a natural way whereby we can identify $\mathbb{S}$ with $\mathbb{F}^n$ or $\mathbb{F}_n$, where $\mathbb{F}$ is $\mathbb{Z}_2$.   For $x \in \mathbb{S}$, let $X = [x_i]$ be the n-dimensional column vector defined by

$$x_i = \begin{cases} 1 & \text{if } s_i \in x \\ 0 & \text{if } s_i \notin x \end{cases} \qquad (4.3.5)$$

Equation (4.3.5) gives rise to a mapping

$$\Pi : x \in \mathbb{S} \to X \in \mathbb{F}^n$$

of the 2^n elements of $\mathbb{S}$ to the 2^n elements of $\mathbb{F}^n$.   This mapping is a one-to-one correspondence between the elements of $\mathbb{S}$ and $\mathbb{F}^n$.   Moreover the arithmetic operations on $\mathbb{S}$ and $\mathbb{F}^n$ are preserved, i.e.,

$$\Pi(x + y) = \Pi(x) + \Pi(y)$$
$$\Pi(\alpha x) = \alpha \Pi(x)$$

for all $x, y \in \mathbb{S}$ and $\alpha \in \mathbb{Z}_2$.   We call Π the *natural correspondence* between the elements of $\mathbb{S}$ and $\mathbb{F}^n$.   In the same way there is a natural correspondence between the elements of $\mathbb{S}$ and $\mathbb{F}_n$.   We will phrase our discussion in terms of the natural correspondence between $\mathbb{S}$ and $\mathbb{F}^n$.   Clearly the same results are valid for $\mathbb{S}$ and $\mathbb{F}_n$.

Under the natural correspondence the zero vector of $\mathbb{F}^n$ corresponds to the empty subset of S, and E, the vector all of whose entries are equal to 1, corresponds to S.

We can use the natural correspondence to transfer properties of sets from $\mathbb{S}$ to $\mathbb{F}^n$ and to transfer properties of vectors and vector spaces from $\mathbb{F}^n$ to $\mathbb{S}$. For example, let $X, Y \in \mathbb{F}^n$ correspond to $x, y \in \mathbb{S}$ respectively.   We say that X is *contained* in Y and write $X \subseteq Y$ if $x \subseteq y$, considering x and y as subsets of S. Similarly we define X and Y to be *disjoint* if x and y are disjoint, and we say that X is the *complement* of Y and write $X = Y'$ if $x = y'$.   Thus, for every vector X of $\mathbb{F}^n$, X and X' are disjoint and $X + X' = E$.   Every vector of $\mathbb{F}^n$ is contained in E.   Clearly X is contained in Y if whenever $x_i = 1$, then $y_i = 1$.   Similarly X and Y are disjoint if whenever $x_i = 1$, then $y_i = 0$.

Let $\mathbb{T}$ be a subspace of $\mathbb{S}$.   An element $x \in \mathbb{T}$ is *minimal* if $x \neq \varnothing$ and there does not exist $y \in \mathbb{S}$, $y \neq \varnothing$ such that $y \subseteq x$.   Thus the minimal elements of $\mathbb{S}$ are those subsets of S which consist of a single element.   Let $\mathbb{V}$ be the set of vectors of $\mathbb{F}^n$ which correspond to the elements of $\mathbb{T}$ under the natural correspondence.   Then $\mathbb{V}$ is a subspace of $\mathbb{F}^n$, and we say that X is *minimal* in $\mathbb{V}$ if the element of $\mathbb{T}$ corresponding to X is minimal in $\mathbb{T}$.

As an example of a property of vectors being carried across the natural correspondence from $\mathbb{F}^n$ to $\mathbb{S}$ we say that the elements $x, y, \ldots, z$ of $\mathbb{S}$ are *linearly independent* if the corresponding elements $\Pi(x), \Pi(y), \ldots, \Pi(z)$ are linearly independent vectors of $\mathbb{F}^n$.

Lemma 4.3.6 Let $\mathbb{V}$ be a subspace of $\mathbb{F}^n$ and let $X \in \mathbb{V}$. Then X can be expressed as a sum of disjoint minimal vectors of $\mathbb{V}$ which are contained in X.

PROOF If X is not itself minimal, then it contains a minimal vector, say Y. Set

$$Z = X + Y \qquad (4.3.7)$$

Since $Y \subseteq X$, it follows that $Z \subseteq X$ and that Y and Z are disjoint. Moreover, equation (4.3.7) is equivalent to

$$X = Y + Z \qquad (4.3.8)$$

If Z is a minimal vector, we have expressed X as a sum of disjoint minimal vectors of $\mathbb{V}$ contained in X. If Z is not minimal, then we use the same argument to express Z as

$$Z = U + V \qquad (4.3.9)$$

where U and V are vectors contained in Z (and hence are contained in X) and U is minimal. Substituting equation (4.3.9) into equation (4.3.8) yields

$$X = Y + U + V$$

If V is a minimal vector, we are finished. If V is not minimal, the process is repeated once again. Since the number of entries which are equal to 1 is decreasing at each step (that is, X has more 1s than Z and Z has more 1s than V), this process must eventually stop.   ////

Note that we do not claim that the minimal elements are either unique for a given vector or are the same for different vectors. In Example 4.3.11 we will show that neither of these properties need be true.

It is an immediate consequence of Lemma 4.3.6 that the minimal vectors of a vector space form a spanning set, and hence:

Theorem 4.3.10 Every subspace of $\mathbb{F}^n$ has a basis of minimal vectors.

EXAMPLE 4.3.11   Let $\mathbb{V}$ be the set of all vectors of $\mathbb{F}^4$ (where $\mathbb{F} = \mathbb{Z}_2$) which contain an even number of 1s.   Then $\mathbb{F}^4$ contains 16 elements, and it is not difficult to verify that $\mathbb{V}$ is a subspace of $\mathbb{F}^4$ and has 8 elements.   There are 6 vectors belonging to $\mathbb{V}$ which have exactly two 1s, and these are the minimal vectors of $\mathbb{V}$.   It can easily be shown that $\mathbb{V}$ does not have a basis of disjoint vectors.   However, one basis for $\mathbb{V}$ consisting of minimal vectors is

$$\left\{ \begin{bmatrix} 1 \\ 1 \\ 0 \\ 0 \end{bmatrix}, \begin{bmatrix} 0 \\ 0 \\ 1 \\ 1 \end{bmatrix}, \begin{bmatrix} 0 \\ 1 \\ 1 \\ 0 \end{bmatrix} \right\} \qquad ////$$

In the next section we will make extensive use of the natural correspondence. The set S will be chosen in several different ways according to our needs at the time.   Among the choices we will make are: (1) the set of vertices of a graph, (2) the set of arcs of a graph, (3) the set of loops of a graph.

EXERCISES

1   For the vector space $\mathbb{V}$ of Example 4.3.11, express the following vectors in terms of the given basis:

$$(a) \begin{bmatrix} 1 \\ 0 \\ 0 \\ 1 \end{bmatrix} \qquad (b) \begin{bmatrix} 1 \\ 0 \\ 1 \\ 0 \end{bmatrix} \qquad (c) \begin{bmatrix} 1 \\ 1 \\ 1 \\ 1 \end{bmatrix}$$

2   Let x_1, x_2, and x_3 be subsets of S.   Show that $x_1 + x_2 + x_3$ consists of all elements of S which belong to exactly one of x_1, x_2, and x_3 or which belong to all three.   Formulate and prove a corresponding result for the sum of any finite number of subsets of S.

3   Show that for every choice of n the set $\mathbb{E}$ of all vectors of $\mathbb{F}^n$ (where $\mathbb{F} = \mathbb{Z}_2$) containing an even number of 1s is a vector space.   What are the minimal vectors of $\mathbb{E}$?   Find a basis for $\mathbb{E}$ consisting of minimal vectors.   What is the dimension of $\mathbb{E}$, and how many vectors does $\mathbb{E}$ contain?

4   Show that for $n \geq 5$ the subspace of $\mathbb{F}^n$ spanned by all vectors containing four 1s is $\mathbb{E}$.

5   Show that for $n \geq 4$ the subspace of $\mathbb{F}^n$ spanned by all vectors containing three 1s is $\mathbb{F}^n$.

6   Prove that every k-dimensional subspace of $\mathbb{F}^n$ contains 2^k vectors.

4.4   MATRICES AND UNDIRECTED GRAPHS

Our prime purpose in this chapter is to illustrate how the theory of matrices can be applied to graph theory.   As we have already shown, a matrix provides a systematic and efficient means for displaying and manipulating the properties of a graph.   This can be a particularly useful method when we are interested in counting the number of times, if any, that a specific type of configuration appears in a given graph (see, for example, Examples 4.1.1 and 4.1.3).   However, there is a second and more basic way in which the theory of matrices can be used to obtain information about graphs.   We will be able to formulate many of the important properties of a graph (e.g., connectedness, loops, etc.) in terms of properties of one or more of the matrices associated with it.   We will then be able to prove theorems about graphs by demonstrating the corresponding theorems about their matrices.   Often the theorem concerning graphs is considerably more complex than the corresponding theorem about matrices, e.g., Theorem 4.4.14.   In Sec. 4.1 we defined the vertex-adjacency matrix A, the arc-adjacency matrix $\tilde{A}$, the vertex-arc incidence matrix B, and the degree matrix D of an undirected graph.   For a graph with m vertices and n arcs these matrices have dimensions $m \times m$, $n \times n$, $m \times n$, and $m \times m$ respectively.   It will be convenient to call such a graph *an $m \times n$ graph*.

In Sec. 4.1 the field from which the entries of the matrices A, $\tilde{A}$, B, and D were taken was not specified, though certainly for A, $\tilde{A}$, and D the field of real (or rational) numbers was implied.   However, since all the entries of B are either 0 or 1 and every field contains both a 0 and a 1, we are at liberty to choose whichever field we wish to contain the entries of B.   If fact, we will occasionally decide to change fields. B is a typical example of an incidence matrix.   Its entries, 1 and 0, denote the answer yes or no to a question of incidence.   The entries of A, $\tilde{A}$, and D answer the question: How many?.   Later in this section we will define several other incidence matrices associated with a graph.   For this reason we will call B the *incidence matrix* of its graph.

In order for Theorem 4.1.15 to have meaning, the entries of B must be taken from the same field as those of A and D, namely, the real field.   Although we will occasionally revert to the real field again, we will find that it is more profitable to consider the entries of an incidence matrix as coming from the field $\mathbb{Z}_2$, which we discussed at length in the last section.   Thus, unless we specify otherwise, we will henceforth assume that the field $\mathbb{F}$ from which the entries of B, as well as the other incidence matrices to be introduced later, are taken is $\mathbb{Z}_2$.

The incidence matrix of an $m \times n$ graph is an $m \times n$ matrix.   Since every column of B contains exactly two 1s, we have

$$E_{(m)} B = 0 \qquad (4.4.1)$$

where $E_{(m)}$ is the m-dimensional row vector all of whose entries are 1.   The rows of B are linearly dependent, and it follows that the rank of the incidence matrix of an $m \times n$ graph is at most $m - 1$.   In Theorem 4.4.4 we will prove that the rank of an $m \times n$ incidence matrix is uniquely determined by the number of components of the graph.

Let $G_1, G_2, \ldots, G_k$ be the components of the graph G and let G_i consist of m_i vertices and n_i arcs.   If the vertices and arcs of G are numbered so that

$$G_1 = \{\mathbb{V}_1 = \{v_1, v_2, \ldots, v_{m_1}\}, \mathbb{A}_1 = \{a_1, a_2, \ldots, a_{n_1}\}\}$$
$$G_2 = \{\mathbb{V}_2 = \{v_{m_1+1}, v_{m_1+2}, \ldots, v_{m_1+m_2}\}, \mathbb{A}_2 = \{a_{n_1+1}, a_{n_1+2}, \ldots, a_{n_1+n_2}\}\}$$
$$\cdots\cdots\cdots\cdots\cdots\cdots\cdots\cdots\cdots\cdots\cdots\cdots\cdots\cdots$$

then the incidence matrix of G has the form

$$B = \begin{bmatrix} B_1 & 0 & \cdots & 0 \\ 0 & B_2 & \cdots & 0 \\ \multicolumn{4}{c}{\cdots\cdots\cdots\cdots\cdots} \\ 0 & 0 & \cdots & B_k \end{bmatrix} \qquad (4.4.2)$$

where B_i is the $m_i \times n_i$ incidence matrix of G_i.[1]   Conversely, a graph whose incidence matrix can be written in the form of (4.4.2) can be expressed as the union of k disjoint subgraphs, one for each of the matrices $B_1, B_2, \ldots, B_k$.   Since a connected graph is one which consists of a single component, we can prove:

Lemma 4.4.3 If B is the incidence matrix of a graph G with m vertices, then G is connected if and only if the only row vectors X such that

$$XB = 0$$

are $X = 0$ and $X = E_{(m)}$.

PROOF   If G is not connected, it has more than one component and we may assume that B has the form (4.4.2).   Clearly

$$X = [E_{(m_1)} \quad 0 \quad \cdots \quad 0]$$

satisfies the equation

$$XB = 0$$

and $X \neq 0$, $X \neq E_{(m_i)}$.

[1] A vertex is isolated if there are no arcs incident with it.   We are then faced with the paradox that the incidence matrix of an isolated point has one row and no columns. This need cause no trouble, however.   If G contains at least one arc but the component G_i does not, then there is a row of B corresponding to G_i and every entry of that row is 0.

Conversely, let G be a connected graph and suppose that there exists a vector X such that

$$XB = 0 \qquad X \neq 0 \qquad X \neq E_{(m)}$$

Since each entry of X is either 0 or 1, we may assume that the vertices of G are numbered so that

$$X = [E_{(q)} \quad 0] \qquad 1 \leq q \leq m - 1$$

We now partition B as

$$B = \begin{bmatrix} B_1 \\ B_2 \end{bmatrix}$$

where B_1 and B_2 contain q rows and $m - q$ rows respectively. Then

$$E_{(q)} B_1 = 0$$

and consequently every column of B_1 contains either two 1s or no 1s. We may assume that the arcs of G are numbered so that each of the first p columns of B_1 contains two 1s and each of the remaining columns of B_1 contains no 1s. We can now partition B as

$$B = \begin{bmatrix} B_{11} & 0 \\ 0 & B_{22} \end{bmatrix}$$

where B_{11} and B_{22} are $q \times p$ and $(m - q) \times (n - p)$ matrices respectively. (If $p = 0$, then

$$B = \begin{bmatrix} 0 \\ B_{22} \end{bmatrix}$$

whereas if $p = n$, then

$$B = \begin{bmatrix} B_{11} \\ 0 \end{bmatrix}$$

Either way, G contains isolated vertices.) Since the graph is connected, we have reached a contradiction. Thus, no such vector X exists. ////

Lemma 4.4.3 states that a graph is connected if and only if the rows of its incidence matrix are a *minimal linearly dependent set*; i.e., the m rows of B are linearly dependent, but every subset consisting of fewer than m rows is linearly independent. Thus we have an algebraic characterization of a connected graph. The same idea can be used to identify the components of a nonconnected graph. Thus Lemma 4.4.3 is a special case of the following theorem.

Theorem 4.4.4   If B is the incidence matrix of an $m \times n$ graph with k components the rank of B is given by

$$r(B) = m - k$$

PROOF   We may assume that the vertices and arcs of the graph are numbered so that B has the form (4.4.2).   The rank of B can be obtained by means of (the row version of) Theorem 1.6.1, i.e.,

$$r(B) = m - v_R(B)$$

where $v_R(B)$ is the row nullity of B.   Recall that the row nullity of B is the dimension of $\mathbb{N}_R(B)$, the row null space of B,

$$\mathbb{N}_R(B) = \{X \mid XB = 0, X \in \mathbb{F}_m\}$$

Thus, in order to prove the theorem it is sufficient to show that

$$v_R(B) = k \qquad (4.4.5)$$

It follows from equation (4.4.1) that the k vectors

$$
\begin{aligned}
Y_1 &= [E_{(m_1)} \quad 0 \quad\ 0 \quad \cdots \quad 0 \ ] \\
Y_2 &= [\ 0 \quad\ E_{(m_2)} \quad 0 \quad \cdots \quad 0 \ ] \\
&\ \cdot\ \cdot\ \cdot\ \cdot\ \cdot\ \cdot\ \cdot\ \cdot\ \cdot\ \cdot\ \cdot\ \cdot\ \cdot\ \cdot \\
Y_k &= [\ 0 \quad\quad 0 \quad\ 0 \quad \cdots \quad E_{(m_k)}]
\end{aligned}
$$

belong to $\mathbb{N}_R(B)$.   Clearly $Y_1, Y_2, \ldots, Y_k$ are linearly independent.   In order to establish equation (4.4.5) it remains only to prove that $Y_1, Y_2, \ldots, Y_k$ span $\mathbb{N}_R(B)$.

Let $X \in \mathbb{F}_m$.   Partition X as

$$X = [X_1 \quad X_2 \quad \cdots \quad X_k]$$

where X_i is an m_i-dimensional row vector.   Clearly $XB = 0$ if and only if

$$X_i B_i = 0 \qquad i = 1, 2, \ldots, k$$

and, by Lemma 4.4.3, we have

$$X_i = 0 \qquad \text{or} \qquad X_i = E_{(m_i)}$$

Consequently X can be expressed as a linear combination of $Y_1, Y_2, \ldots, Y_k$.   ////

Theorem 4.4.4 illustrates an important advantage gained by choosing the field $\mathbb{Z}_2$ for $\mathbb{F}$.   In proving the theorem we incidentally showed that in the natural correspondence between the vector space of subsets of the set of vertices of an

$m \times n$ graph G and $\mathbb{F}_m$, the components of G correspond to the minimal elements of $\mathbb{N}_R(B)$.[1] In this natural correspondence the elements of the vector space of sets of components of G correspond to the elements of $\mathbb{N}_R(B)$. The various natural correspondences associated with a graph form one of the strongest tools in graph theory.

Because of Theorem 4.4.4 and the simple form of (4.4.2) we can once again limit our discussion to graphs which are connected. In all cases the extension to graphs with more than one component is a trivial matter. It will be convenient for us to look at the incidence matrices of a simple path and a loop before investigating the properties of incidence matrices generally.

EXAMPLE 4.4.6   For a pair of vertices v_1 and v_l, $v_1 \neq v_l$, a path P of length $l - 1$ between v_1 and v_l is a graph which consists of l vertices and $l - 1$ arcs. When these vertices and arcs can be numbered so that

$$P = \{a_1 = (v_1, v_2), a_2 = (v_2, v_3), \ldots, a_{l-1} = (v_{l-1}, v_l)\}$$

then the incidence matrix of P is the $l \times (l - 1)$ matrix

$$B_P = \begin{bmatrix} 1 & & & & \\ 1 & 1 & & & \\ & 1 & 1 & & \\ & & \ddots & \ddots & \\ & & & 1 & 1 \\ & & & & 1 \end{bmatrix}$$

A loop L of length l is a graph which consists of l vertices and l arcs. When the vertices and arcs of L are numbered so that

$$L = \{a_1 = (v_1, v_2), a_2 = (v_2, v_3), \ldots, a_{l-1} = (v_{l-1}, v_l), a_l = (v_l, v_1)\}$$

then the incidence matrix of L is the $l \times l$ matrix

$$B_L = \begin{bmatrix} 1 & & & & & 1 \\ 1 & 1 & & & & \\ & 1 & 1 & & & \\ & & \ddots & \ddots & & \\ & & & 1 & 1 & \\ & & & & 1 & 1 \end{bmatrix} \qquad (4.4.7)$$

[1] More accurately, the natural correspondence is between the minimal elements of $\mathbb{N}_R(B)$ and those sets of vertices of G which are the vertices of a component of G. We have tacitly identified a component with the set of vertices of that component.

THE THEORY OF GRAPHS 365

Clearly L can be formed from P by adding the arc $a_l = (v_l, v_1)$, and therefore B_L can be formed from B_P by adding the column corresponding to a_l. Since P and L are connected graphs with l vertices, it follows from Theorem 4.4.4 that

$$r(B_P) = r(B_L) = l - 1$$

(The rank can also be obtained directly from the matrix.) Since B_P has $l - 1$ columns, they are linearly independent. However, each row of B_L has exactly two 1s, and consequently

$$B_L E^{(l)} = 0$$

According to Theorem 1.6.1, the (column) nullity of B_L is 1, and hence $X = E^{(l)}$ is the only nonzero vector such that

$$B_L X = 0$$

Thus the columns of B_L form a minimal linearly dependent set. We will see that this property characterizes a loop. ////

Lemma 4.4.8 Let B be the incidence matrix of the graph G. A necessary and sufficient condition that a set L of arcs of G (and the vertices incident with them) form a loop is that the columns of B corresponding to the arcs of L form a minimal linearly dependent set.

PROOF In the discussion above we showed that a set of columns of B which correspond to the arcs of a loop form a minimal linearly dependent set. Conversely, let L be any subgraph of G which has the property that the columns of B corresponding to the arcs contained in L form a minimal linearly dependent set. We wish to prove that L is a loop. Let L contain l arcs and assume that the arcs of G are numbered so that the arcs of L are $a_1, a_2, \ldots, a_l$. Then we can partition B as

$$B = [B_1 \quad B_2]$$

where B_1 consists of the l columns corresponding to the arcs of L. Since the columns of B_1 form a minimal linearly dependent set, we have

$$B_1 E^{(l)} = 0 \qquad (4.4.9)$$

and $X = E^{(l)}$ is the only nonzero vector such that

$$B_1 X = 0$$

It follows from equation (4.4.9) that every row of B_1 contains an even number of 1s.

We wish to show that there is some numbering of the vertices and arcs of G (or L) for which B_1 has the form (4.4.7), possibly augmented by rows consisting entirely of 0s. We may assume that the vertices of G are numbered so that

$$a_1 = (v_1, v_2)$$

Then the first column of B_1 is

$$\begin{bmatrix} 1 \\ 1 \\ 0 \\ 0 \\ \vdots \\ 0 \end{bmatrix}$$

Since the second row of B_1 possesses an even number of 1s, there is another column of B_1 which contains a 1 in the second row. We may assume that the second column of B_1 has this property. However, every column of B_1 (in particular, the second column) has two 1s, and we must consider two cases. If the other 1 of the second column is in the first row, then the first two columns of B_1 are identical. Hence they form a minimal linearly dependent set, and their arcs, being parallel, form a loop of length 2. In this case we are finished. On the other hand, if the other 1 of the second column of B_1 is not in the first row, then we may assume that it is in the third row, i.e.,

$$a_2 = (v_2, v_3)$$

and the first two columns of B_1 are

$$\begin{bmatrix} 1 & 0 \\ 1 & 1 \\ 0 & 1 \\ 0 & 0 \\ \vdots & \vdots \\ 0 & 0 \end{bmatrix}$$

Suppose now that we have continued this process until we have

$$a_1 = (v_1, v_2), \qquad a_2 = (v_2, v_3), \qquad \dots, \qquad a_k = (v_k, v_{k+1}) \qquad (4.4.10)$$

and the first k columns of B_1 are

$$
\begin{array}{c}
 \\
v_1 \\
v_2 \\
v_3 \\
v_4 \\
\\
v_k \\
v_{k+1} \\
\\
\\
\end{array}
\begin{array}{cccccc}
a_1 & a_2 & a_3 & & a_k \\
\left[\begin{array}{ccccc}
1 & 0 & 0 & \cdots & 0 \\
1 & 1 & 0 & \cdots & 0 \\
0 & 1 & 1 & \cdots & 0 \\
0 & 0 & 1 & \cdots & 0 \\
\multicolumn{5}{c}{\cdots\cdots\cdots\cdots} \\
0 & 0 & 0 & \cdots & 1 \\
0 & 0 & 0 & \cdots & 1 \\
\multicolumn{5}{c}{\cdots\cdots\cdots\cdots} \\
0 & 0 & 0 & \cdots & 0
\end{array}\right]
\end{array}
$$

As before, the $(k + 1)$st row of B_1 contains at least two 1s, and we may assume that the $(k + 1)$st column of B_1 has a 1 in the $(k + 1)$st row. Now several cases arise.

CASE 1   If the second 1 of the $(k + 1)$st column is in the first row, we may partition the first $k + 1$ columns of B_1 into

$$
B_1 = \begin{bmatrix} B_{11} \\ 0 \end{bmatrix}
$$

where B_{11} is a $(k + 1) \times (k + 1)$ matrix of the form of (4.4.7). The first $k + 1$ columns of B_1 are linearly dependent and hence constitute all of B_1, that is, $k + 1 = l$.

In this case we also have

$$
a_{k+1} = (v_{k+1}, v_1)
$$

and a_{k+1} together with the arcs of (4.4.10) form a loop of length $k + 1 = l$. Note that the rows of B_1 which do not belong to B_{11} correspond to the vertices of G which are not incident with the arcs of the loop.

CASE 2   Suppose that the second 1 of the $(k + 1)$st column of B_1 is in the rth row, where $2 \le r \le k$. Then the rth, $(r + 1)$st, $\ldots$, kth, and $(k + 1)$st columns of B_1 are linearly dependent since their sum is 0. This contradicts the fact that the columns of B_1 form a minimal linearly dependent set. Thus case 2 can never occur.

CASE 3   If the second 1 of the $(k + 1)$st column of B_1 is not in any of the first $k + 1$ rows, we may assume that it is in the $(k + 2)$nd row and this process continues.   Since there are only a finite number of columns in B_1, we must eventually reach case 1.   ////

Theorem 4.4.11   Under the natural correspondence between the vector space of sets of arcs of G and $\mathbb{F}^n$ the loops of G correspond to the minimal vectors of the vector space

$$\mathbb{N}(B) = \{X \mid BX = 0,\ X \in \mathbb{F}^n\}$$

[Recall that $\mathbb{N}(B)$ is the (column) null space of B.]

Lemma 4.4.8 and Theorem 4.4.11 apply equally to connected and non-connected graphs.   Since the rank of an $m \times n$ graph with k components is $m - k$, it follows that

$$\dim \mathbb{N}(B) = v(B) = n - m + k$$

For a connected $m \times n$ graph we have

$$\dim \mathbb{N}(B) = v(B) = n - m + 1$$

Unlike the situation we encountered with the minimal vectors of $\mathbb{N}_R(B)$, the minimal vectors of $\mathbb{N}(B)$ need not be disjoint.

EXAMPLE 4.4.12   The incidence matrix of the 5×7 graph G

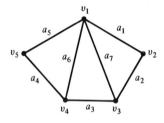

is the 5×7 matrix

$$B = \begin{array}{c} \begin{array}{ccccccc} a_1 & a_2 & a_3 & a_4 & a_5 & a_6 & a_7 \end{array} \\ \left[\begin{array}{ccccccc} 1 & 0 & 0 & 0 & 1 & 1 & 1 \\ 1 & 1 & 0 & 0 & 0 & 0 & 0 \\ 0 & 1 & 1 & 0 & 0 & 0 & 1 \\ 0 & 0 & 1 & 1 & 0 & 1 & 0 \\ 0 & 0 & 0 & 1 & 1 & 0 & 0 \end{array}\right] \begin{array}{c} v_1 \\ v_2 \\ v_3 \\ v_4 \\ v_5 \end{array} \end{array}$$

Since G is a connected graph, the rank of B is 4.   According to Theorem 1.6.1, the dimension of $\mathbb{N}(B)$ is 3.   It is not difficult to verify that the three vectors

$$
X_1 = \begin{bmatrix} 1 \\ 1 \\ 0 \\ 0 \\ 0 \\ 0 \\ 1 \end{bmatrix} \qquad
X_2 = \begin{bmatrix} 0 \\ 0 \\ 1 \\ 0 \\ 0 \\ 1 \\ 1 \end{bmatrix} \qquad
X_3 = \begin{bmatrix} 0 \\ 0 \\ 0 \\ 1 \\ 1 \\ 1 \\ 0 \end{bmatrix}
$$

corresponding to the three loops

$$L_1 = \{a_1, a_2, a_7\} \quad L_2 = \{a_3, a_6, a_7\} \quad L_3 = \{a_4, a_5, a_6\}$$

form a basis for $\mathbb{N}(B)$.   The remaining vectors belonging to $\mathbb{N}(B)$ are 0 and

$$
X_4 = X_1 + X_2 = \begin{bmatrix} 1 \\ 1 \\ 1 \\ 0 \\ 0 \\ 1 \\ 0 \end{bmatrix} \quad
X_5 = X_1 + X_3 = \begin{bmatrix} 1 \\ 1 \\ 0 \\ 1 \\ 1 \\ 1 \\ 1 \end{bmatrix} \quad
X_6 = X_2 + X_3 = \begin{bmatrix} 0 \\ 0 \\ 1 \\ 1 \\ 1 \\ 0 \\ 1 \end{bmatrix} \quad
X_7 = X_1 + X_2 + X_3 = \begin{bmatrix} 1 \\ 1 \\ 1 \\ 1 \\ 1 \\ 0 \\ 0 \end{bmatrix}
$$

All the nonzero vectors of $\mathbb{N}(B)$ are minimal except X_5. X_5 is not minimal since it contains both X_1 and X_3.   The loops corresponding to X_4, X_6, and X_7 are

$$L_4 = \{a_1, a_2, a_3, a_6\} \qquad L_6 = \{a_3, a_4, a_5, a_7\} \qquad L_7 = \{a_1, a_2, a_3, a_4, a_5\}$$

whereas X_5 corresponds to the set of arcs belonging to L_1 and L_3.   ////

The natural correspondence described in Theorem 4.4.11 between the minimal elements of $\mathbb{N}(B)$ and the loops of G can be extended to include all the elements of $\mathbb{N}(B)$.   For this we need:

Definition 4.4.13   A subgraph of a graph is called a *loop set* if its set of arcs can be expressed as the sum of the arcs of arc-disjoint loops.

Definition 4.4.13 is somewhat clumsily worded because we do not require the loops of a loop set to be vertex-disjoint.   In Example 4.4.12, the subgraph L_5 consisting of the set of arcs corresponding to the vector X_5, and the vertices incident with them, is a loop set.   In order to simplify our discussion we will say that

$$L_5 = L_1 + L_3$$

(ignoring the problem caused by the fact that L_1 and L_3 are not vertex-disjoint). Note that a loop is a loop set consisting of a single loop.

Theorem 4.4.14   Under the natural correspondence between the vector space of sets of arcs of G and $\mathbb{F}^n$ the loop sets of G correspond to the elements of $\mathbb{N}(B)$.

PROOF   Let L be a loop set of G.   Since a loop of G corresponds to a minimal vector of $\mathbb{N}(B)$, the vector corresponding to L is a sum of minimal vectors of $\mathbb{N}(B)$ and consequently belongs to $\mathbb{N}(B)$.

Conversely, let $X \in \mathbb{N}(B)$.   According to Lemma 4.3.6, X can be expressed as a sum of disjoint minimal vectors of $\mathbb{N}(B)$.   Therefore the graph whose set of arcs corresponds to X can be expressed as a sum of arc-disjoint loops and hence is a loop set.   ////

Since $\mathbb{N}(B)$ is closed under the operation of addition, it follows from Theorem 4.4.14 that the collection of loop sets of G has the same property.   Thus the sum of (possibly not arc-disjoint) loops of G can be expressed as the sum of arc-disjoint loops.   In Example 4.4.12, the sum $L_2 + L_7$ is not arc-disjoint, but it can be expressed as the arc-disjoint sum $L_1 + L_3$.

According to Theorem 4.4.14, the number of loop sets of a graph is equal to the number of vectors contained in $\mathbb{N}(B)$.   We have previously shown that for a connected $m \times n$ graph G the dimension of $\mathbb{N}(B)$ is $n - m + 1$ and $\mathbb{N}(B)$ contains 2^{n-m+1} vectors.   Therefore G contains 2^{n-m+1} loop sets, including the empty set.   It is generally not possible to specify the number of loops of G. The 5×7 graph of Example 4.4.12 has eight loop sets, of which six are loops. We leave it as an exercise to show that if the arc $a_7 = (v_1, v_3)$ of the graph of Example 4.4.12 is replaced by $a_7 = (v_2, v_5)$, the resulting 5×7 graph also has eight loop-sets but seven of them are loops.

It is an immediate consequence of Lemma 4.4.8 that a necessary and sufficient condition for a set of arcs of a graph not to contain a loop is that the columns of the incidence matrix corresponding to these arcs be linearly in-dependent.   This result allows us to characterize those subgraphs of a graph which do not contain loops.

Definition 4.4.15   A subgraph T of a connected graph G is called a *tree* if T is a maximal subgraph of G containing no loops; that is, T contains no loops, but every subgraph of G which contains T properly also contains a loop.

It does not follow from Definition 4.4.15 that all trees of a graph have the same number of arcs.   However, in Theorem 4.4.16 we shall show that this is, in fact, the case.   Lemma 4.4.8 states that a set of arcs of G form a tree if and only if the corresponding columns of B form a basis for $\mathbb{C}(B)$, the column space of B.   We can now apply some of the theorems concerning bases of a vector space which we proved in Chap. 1.   Then we have:

Theorem 4.4.16   For an $m \times n$ graph G:

1   Every tree of G has exactly $m - 1$ arcs.
2   Every subgraph of G which contains $m - 1$ arcs and has no loops is a tree.
3   If S is any set of arcs of G which contains no loops, then G has a tree which contains S.

PROOF   By Theorem 4.4.4,

$$r(B) = m - 1$$

Parts 1, 2, and 3 of the theorem are restatements of Corollary 1.5.3, Corollary 1.5.4 and Theorem 1.5.1 respectively.   ////

There are numerous trees in the graph of Example 4.4.12.   In Sec. 4.6 we shall calculate the exact number.   Every tree of the graph has four arcs; e.g.,

$$T_1 = \{a_1, a_2, a_3, a_4\} \qquad T_2 = \{a_1, a_7, a_5, a_6\}$$
$$T_3 = \{a_5, a_6, a_3, a_2\} \qquad T_4 = \{a_5, a_6, a_7, a_2\}$$

are trees.   Since the incidence matrix of a tree with m vertices has rank $m - 1$, a tree is connected.

Definition 4.4.17   Let T be a tree of the graph G.   An arc of G which does not belong to T is called a *chord* of G (with respect to T).

A graph with n arcs containing a tree with $m - 1$ arcs has $n - m + 1$ chords. Recall that

$$\dim \mathbb{N}(B) = v(B) = n - m + 1$$

In Theorem 4.4.21 we will show that the chords of a graph with respect to a given tree determine a basis for $\mathbb{N}(B)$.

Lemma 4.4.18   Each chord of a tree is contained in a unique loop which contains no other chords of the tree.

PROOF   Let T be a tree of G and let $c = (v_1, v_2)$ be a chord of G with respect to T.   Since T is connected, there is a path P between v_1 and v_2

consisting of arcs of T. Let L be the union of P and $\{c\}$. Then L is a loop consisting of c and some of the arcs of T. It is left as an exercise to show that L is the only loop with this property.   ////

The set of $n - m + 1$ loops of a graph corresponding to the $n - m + 1$ chords of a tree, as described by Lemma 4.4.18, is called a *fundamental set of loops*. Under the natural correspondence between sets of arcs of G and $\mathbb{F}^n$, the vectors corresponding to the loops of a fundamental set constitute a set of $n - m + 1$ vectors of $\mathbb{N}(B)$. We will show that they form a basis for $\mathbb{N}(B)$.

Definition 4.4.19 Let $L_1, L_2, \ldots, L_p$ be the loops of G (in some order). The $n \times p$ matrix $L = [l_{jk}]$, where

$$
l_{jk} = \begin{cases} 1 & \text{when arc } a_j \text{ belongs to loop } L_k \\ 0 & \text{when arc } a_j \text{ does not belong to loop } L_k \end{cases}
$$

is called the *loop matrix* of G.

The loop matrix of the graph of Example 4.4.12 is the 7×6 matrix

$$
L = \begin{array}{c} \begin{matrix} L_1 & L_2 & L_3 & L_4 & L_6 & L_7 \end{matrix} \\ \begin{bmatrix} 1 & 0 & 0 & 1 & 0 & 1 \\ 1 & 0 & 0 & 1 & 0 & 1 \\ 0 & 1 & 0 & 1 & 1 & 1 \\ 0 & 0 & 1 & 0 & 1 & 1 \\ 0 & 0 & 1 & 0 & 1 & 1 \\ 0 & 1 & 1 & 1 & 0 & 0 \\ 1 & 1 & 0 & 0 & 1 & 0 \end{bmatrix} \begin{matrix} a_1 \\ a_2 \\ a_3 \\ a_4 \\ a_5 \\ a_6 \\ a_7 \end{matrix} \end{array}
$$

There are several relations between the incidence matrix and the loop matrix of a graph.

Theorem 4.4.20

1 $BL = 0$.
2 $\mathbb{C}(L) = \mathbb{N}(B)$.
3 $r(L) = n - m + 1$.
4 $\mathbb{N}_R(L) = \mathbb{R}(B)$.

PROOF The kth column of L is the vector of $\mathbb{F}^n$ corresponding to the set of arcs of L_k under the natural correspondence. Applying Theorem 4.4.11, we see that the columns of L are the minimal elements of $\mathbb{N}(B)$. Thus

$BL = 0$.   Moreover, by Theorem 4.3.10, the columns of L span $\mathbb{N}(B)$, and consequently

$$\mathbb{C}(L) = \mathbb{N}(B)$$

Thus $$r(L) = v(B) = n - m + 1$$

Finally, it is clear from part 2 of the theorem that

$$\mathbb{R}(B) \subseteq \mathbb{N}_R(L)$$

Since $$r(B) = m - 1 = v_R(L)$$

it follows that $\mathbb{N}_R(L) = \mathbb{R}(B)$.   ////

In Definition 4.4.19 we did not specify any particular ordering of the arcs and loops of G.   Because of Lemma 4.4.18 there is an ordering which produces an interesting form for L.   Let T be a tree of G and let us number the arcs of G so that $a_1, a_2, \ldots, a_{n-m+1}$ are the chords of T.   Also, let us number the loops of G so that L_i is the unique loop contained in $T \cup \{a_i\}$, $i = 1, 2, \ldots, n - m + 1$. Then it is easily seen that L has the form

$$L = \begin{bmatrix} I & L_{12} \\ L_{21} & L_{22} \end{bmatrix}$$

where I is an identity matrix of order $n - m + 1$. $L_1, L_2, \ldots, L_{n-m+1}$ form a fundamental set of loops of T, and the vectors corresponding to them are the first $n - m + 1$ columns of B.   Clearly these columns are linearly independent. Since the rank of L is $n - m + 1$, we have proved:

Theorem 4.4.21   The vectors of a fundamental set of loops of a graph form a basis for $\mathbb{C}(L) = \mathbb{N}(B)$.

In Example 4.1.9 we were concerned with finding the smallest set of arcs of a connected graph such that the removal of these arcs leaves a graph which is not connected (and also has the property that two specified vertices are in different components).   Such a set of arcs is called a *cut*.   Let us give a formal definition.

Definition 4.4.22   Let G be a connected graph.   A set C of arcs of G is a *cut* if:

1 The removal of the arcs of C (but not the vertices of these arcs) from G leaves a graph which is not connected.
2 C is minimal; i.e., the removal of any proper subset of the arcs of C from G leaves a connected graph.

Note that a cut is defined only for a connected graph. Since a cut is defined to be a minimal set which disconnects a graph (rather than the smallest set with this property), different cuts of a graph may have different numbers of arcs. There are numerous cuts in the graph of Example 4.4.12; e.g.,

$$C_1 = \{a_2, a_5, a_6, a_7\} \qquad C_2 = \{a_3, a_5, a_6\}$$
$$C_3 = \{a_1, a_2\} \qquad\qquad C_4 = \{a_1, a_5, a_6, a_7\}$$

The cuts of a graph can be described in terms of its incidence matrix. Let C be a cut of the $m \times n$ graph G and let $\tilde{G}$ be the (not connected) graph which remains when the arcs of C are removed from G. It will be convenient to number the arcs of G so that $\tilde{G}$ contains $a_1, a_2, \ldots, a_r$ and C consists of a_{r+1}, $a_{r+2}, \ldots, a_n$. Then B can be partitioned as

$$B = [B_1 \quad B_2] \qquad (4.4.23)$$

where B_1 and B_2 are $m \times r$ and $m \times (n - r)$ matrices respectively. Moreover, by Definition 4.4.22, B_1 and B_2 satisfy the following two conditions:

1 $r(B_1) \leqq m - 2$ (since $\tilde{G}$ is not connected).
2 An $m \times (r + 1)$ matrix formed from B_1 and any column of   (4.4.24)
   B_2 has rank $m - 1$ (since the graph consisting of $\tilde{G}$ and any
   arc of C is connected).

Consequently $r(B_1) = m - 2$, and $\tilde{G}$ contains exactly two components.

Conversely, if B is an incidence matrix which can be partitioned as in (4.4.23) so that the two conditions of (4.4.24) are satisfied, then the set of arcs of G corresponding to the columns of B_2 form a cut.

Since the rank of B_1 is $m - 2$, there exists an m-dimensional row vector X, $X \neq 0$, $X \neq E_{(m)}$, such that

$$XB_1 = 0 \qquad (4.4.25)$$

Moreover the only other vector satisfying equation (4.4.25) is

$$\bar{X} = E_{(m)} - X$$

According to condition 2 of (4.4.24), if $\tilde{B}$ is an $m \times (r + 1)$ matrix formed from B_1 and any column of B_2, then

$$X\tilde{B} \neq 0$$

Thus, if we denote the columns of B_2 by $B^{(r+1)}, B^{(r+2)}, \ldots, B^{(n)}$, then

$$XB^{(r+1)} = XB^{(r+2)} = \cdots = XB^{(n)} = 1$$

that is,

$$XB_2 = E_{(n-r)} \qquad (4.4.26)$$

We can use this result to prove:

Theorem 4.4.27   For every loop L and every cut C of a graph the intersection of L and C contains an even number of arcs.

PROOF   We may assume that B has the form (4.4.23). We may further assume that $a_1, a_2, \ldots, a_s$ are the arcs of L which do not belong to C and that $a_{r+1}, a_{r+2}, \ldots, a_{r+t}$ are the arcs of L which do belong to C. Thus

$$L = \{a_1, a_2, \ldots, a_s, a_{r+1}, a_{r+2}, \ldots, a_{r+t}\}$$

and we wish to show that t is even. It follows from Lemma 4.4.8 that

$$0 = B^{(1)} + B^{(2)} + \cdots + B^{(s)} + B^{(r+1)} + B^{(r+2)} + \cdots + B^{(r+t)} \qquad (4.4.28)$$

Let X be a vector satisfying equations (4.4.25) and (4.4.26). Then

$$XB^{(1)} = XB^{(2)} = \cdots = XB^{(s)} = 0$$
$$XB^{(r+1)} = XB^{(r+2)} = \cdots = XB^{(r+t)} = 1$$

Multiplying equation (4.4.28) on the left by X yields

$$0 = XB^{(1)} + XB^{(2)} + \cdots + XB^{(s)} + XB^{(t+1)} + XB^{(r+2)} + \cdots + XB^{(r+t)}$$
$$= 0 + 0 + \cdots + 0 + 1 + 1 + \cdots + 1$$

Thus t is even.   ////

At this point we can introduce another important matrix associated with a graph.

Definition 4.4.29   Let $C_1, C_2, \ldots, C_q$ be the cuts of the connected $m \times n$ graph G. The $q \times n$ matrix $C = [c_{hj}]$, where

$$c_{hj} = \begin{cases} 1 & \text{when arc } a_j \text{ belongs to cut } C_h \\ 0 & \text{when arc } a_j \text{ does not belong to cut } C_h \end{cases}$$

is called the *cut matrix* of G.

The hth row of C is the vector of $\mathbb{F}_n$ which corresponds to the set of arcs of the cut C_i under the natural correspondence. Because of Theorem 4.4.27 we have:

Corollary 4.4.30 $CL = 0$.

As we will see, there are surprising similarities between the properties of the set of loops and the set of cuts of a graph. In Theorem 4.4.21 we showed that the chords of a tree determine a fundamental set of loops and that these loops form a basis for the vector space of loop sets. Similarly, let T be a tree of the connected graph G. Let a_i be an arc of T and denote by T_i the graph which remains when a_i is removed from T. Since T_i contains m vertices and $m - 2$ arcs, T_i is not connected. Moreover, because the addition of the single arc a_i to T_i yields a connected graph, T_i consists of exactly two components. Let us call these components $\tilde{T}_i$ and $\tilde{\tilde{T}}_i$. Since T_i is not connected, the set of arcs of G which do not belong to T_i contains a cut, say C_i. Moreover C_i contains a_i. The two components of the graph which remains when C_i is removed from G have the same sets of vertices as $\tilde{T}_i$ and $\tilde{\tilde{T}}_i$, though possibly they contain more arcs. We have now proved:

Lemma 4.4.31 Each arc of a tree of a connected graph is contained in a unique cut which contains no other arcs of the tree.

A set of $m - 1$ cuts corresponding to the $m - 1$ arcs of a tree, as described by Lemma 4.4.31, is called a *fundamental set of cuts*. If we number the arcs and cuts of a graph so that the arcs of a given tree are $a_1, a_2, \ldots, a_{m-1}$ and the corresponding cuts are $C_1, C_2, \ldots, C_{m-1}$, the cut matrix C has the form

$$C = \begin{bmatrix} I & C_{12} \\ C_{21} & C_{22} \end{bmatrix} \quad (4.4.32)$$

where I is an identity matrix of order $m - 1$. The first $m - 1$ rows of C are the vectors corresponding to the cuts of the fundamental set under the natural correspondence. We will show that they form a basis for $\mathbb{R}(C)$.

Lemma 4.4.33 $\mathbb{R}(C) = \mathbb{R}(B)$

PROOF Combining Corollary 4.4.30 and part 4 of Theorem 4.4.20, we have

$$\mathbb{R}(C) \subseteq \mathbb{N}_R(L) = \mathbb{R}(B) \quad (4.4.34)$$

In order to show that we have equality throughout equation (4.4.34) it is sufficient to show that

$$r(C) = r(B)$$

From equation (4.4.34) we have

$$r(C) \leq r(B)$$

However, when C is written in the form of (4.4.32), then the identity submatrix of order $m - 1$ is a submatrix of C and consequently

$$r(C) \geq m - 1 = r(B)$$

completing the proof. $/\!/\!/\!/$

Theorem 4.4.35   Under the natural correspondence between the vector space of sets of arcs of G and $\mathbb{F}_n$ the cuts of G correspond to the minimal vectors of $\mathbb{R}(B)$.

PROOF   It follows immediately from Corollary 4.4.30 that if X is the n-dimensional row vector corresponding to a cut of G, then

$$X \in \mathbb{R}(C) = \mathbb{N}_R(L) = \mathbb{R}(B)$$

It remains to show that X is a minimal element of $\mathbb{R}(B)$.

However, we will prove the converse first.   That is, we will show that the set of arcs of a minimal element of $\mathbb{R}(B)$ is a cut of G.   If Y is a minimal vector of $\mathbb{R}(B)$, then Y can be expressed as

$$Y = WB \qquad (4.4.36)$$

for some vector $W \in \mathbb{F}_m$.   Further, $W \neq 0, E$ since $Y \neq 0$.   Let us now assume that the arcs of G are numbered so that

$$Y = [0 \quad E] \qquad B = [B_1 \quad B_2] \qquad (4.4.37)$$

i.e., the columns forming B_2 are those which correspond to the arcs of Y whereas the columns forming B_1 are those which correspond to the arcs of G that do not belong to Y. B_1 is the incidence matrix of the graph, say G_1, which remains when the arcs of Y are removed from G.   Substituting equations (4.4.37) into equation (4.4.36), we have

$$0 = WB_1 \qquad E = WB_2$$

Since $W \in \mathbb{N}_R(B_1)$ and $W \neq 0, E$ it follows that

$$r(B_1) \leq m - 2$$

and consequently G_1 is not connected.   We have now shown that Y contains a cut.   We wish to show that, in fact, Y is a cut.   Let Z be any cut contained in Y.   By the first paragraph of the proof,

$$Z \in \mathbb{R}(C) = \mathbb{R}(B)$$

However, because Y is a minimal element of $R(B)$, we now have

$$Y = Z$$

and thus Y is a cut.

We now return to the first part of the proof.   It remains to show that a vector corresponding to a cut of G is a minimal element of $\mathbb{R}(B)$.   Let X be a cut.   If X is not minimal, then X contains a minimal element Y of $\mathbb{R}(B)$.   By the second part of the proof given above, Y is a cut of G.   But one cut can not properly contain another.   Thus $X = Y$ and X is a minimal element of $\mathbb{R}(B)$.   ////

The analogy between loops and cuts can be carried further. It follows from Theorem 4.4.35 that the vector corresponding to the sum of two cuts of G belongs to $\mathbb{R}(B)$ and, by Lemma 4.3.6, can be expressed as the disjoint union of minimal elements of $\mathbb{R}(B)$. In other words, the sum of two cuts of a graph can be expressed as the arc-disjoint union of cuts. For example, in the graph of Example 4.4.12,

$$C_1 + C_4 = \{a_1, a_2\} = C_3$$

Also, let us set

$$C_5 = \{a_1, a_4, a_6, a_7\} \qquad C_6 = \{a_4, a_5\}$$

then

$$C_1 + C_5 = \{a_1, a_2, a_4, a_5\} = C_3 + C_6$$

i.e., the nondisjoint sum of cuts $C_1 + C_5$ can be expressed as the disjoint sum of cuts $C_3 + C_6$. We define a *cut set* of a graph to be any set of arcs which can be expressed as a disjoint union of cuts. Then we have proved:

Theorem 4.4.38 Under the natural correspondence between the vector space of sets of arcs of G and $\mathbb{F}_n$, the cut sets of G correspond to the elements of $\mathbb{R}(B)$.

For a large class of graphs the similarity which we have observed in our study of the loops and the cuts of a graph can be illustrated more dramatically. At the same time we can present an interesting application of the dimensions of the vector spaces associated with a graph.

Definition 4.4.39 Let G and $\tilde{G}$ be connected graphs with the same number of arcs. We say that $\tilde{G}$ is a *dual* of G if the arcs of G and $\tilde{G}$ can be numbered so that

$$L = \{a_{h_1}, a_{h_2}, \ldots, a_{h_k}\}$$

is a loop of G if and only if

$$\tilde{C} = \{\tilde{a}_{h_1}, \tilde{a}_{h_2}, \ldots, \tilde{a}_{h_k}\}$$

is a cut of $\tilde{G}$.

We call $\tilde{G}$ a dual of G (rather than *the* dual of G) because a graph may have more than one dual (see Exercise 5). Although dual graphs have the same number of arcs, they generally do not have the same number of vertices. By combining Theorems 4.4.11 and 4.4.35 we can express the duality of graphs in terms of their incidence matrices. Specifically:

Theorem 4.4.40 Let G and $\tilde{G}$ be graphs with the same number of arcs. Then $\tilde{G}$ is a dual of G if and only if the arcs of G and $\tilde{G}$ can be numbered so that

$$\mathbb{N}(B) = \mathbb{R}(\tilde{B})^T$$

where B and $\tilde{B}$ are the incidence matrices of G and $\tilde{G}$ respectively and where $\mathbb{R}(\tilde{B})^T$ is the vector space consisting of the transposes of the vectors of $\mathbb{R}(\tilde{B})$; that is, $\mathbb{R}(\tilde{B})^T = \mathbb{C}(\tilde{B}^T)$.

EXAMPLE 4.4.41 It is not difficult to verify that a dual of the 6×9 graph

G:

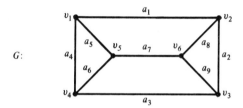

is the 5×9 graph

$\tilde{G}$:

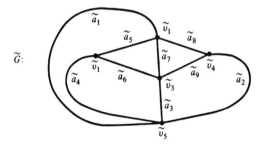

The relation of duality between G and $\tilde{G}$ can be seen more easily when the graphs are positioned as in Fig. 4.4.42.

Note that when the diagrams of G and $\tilde{G}$ are superimposed, as in Fig. 4.4.42, each arc of each graph crosses the corresponding arc of the other graph

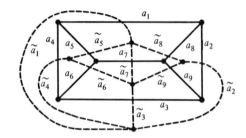

FIGURE 4.4.42

and intersects no other arc of either graph (except at the vertices). In other words, the arcs of each graph meet only at the vertices of the graph, and a_i and $\tilde{a}_i$ intersect for $i = 1, 2, \ldots, 9$, but a_i and $\tilde{a}_i$ do not intersect when $i \neq j$. Each loop L of G partitions the vertices of $\tilde{G}$ into two disjoint sets, namely, the set of vertices of $\tilde{G}$ which are "inside" the loop and the set of vertices of $\tilde{G}$ which are "outside" the loop. When all the arcs of $\tilde{G}$ which intersect L are removed from $\tilde{G}$, the resulting graph consists of two components, one inside L and the other outside L. The arcs that have been removed from $\tilde{G}$ are those which correspond to the arcs of L, and they form a cut of $\tilde{G}$. ////

The description we have just given of the properties of the two graphs of Example 4.4.41 is symmetric; that is, G bears the same relation to $\tilde{G}$ as $\tilde{G}$ does to G. We next show that the same thing can be said of all pairs of dual graphs.

Theorem 4.4.43 $\tilde{G}$ is a dual of G if and only if G is a dual of $\tilde{G}$.

PROOF According to Exercise 27 of Sec. 1.7,

$$\mathbb{N}(B) = \mathbb{C}(\tilde{B}^T) \quad \text{if and only if} \quad \mathbb{N}_R(\tilde{B}^T) = \mathbb{R}(B) \quad (4.4.44)$$

But the right-hand equation of (4.4.44) is equivalent to

$$\mathbb{N}(\tilde{B}) = \mathbb{C}(B^T)$$

The theorem now follows from Theorem 4.4.40. ////

The material to be presented from this point to the end of this section leans heavily on the geometric picture of a graph and is developed here in a more informal and less rigorous manner than we have previously used. A rigorous treatment of this subject requires large doses of algebraic topology and is beyond the scope of this book.

Definition 4.4.45 A graph G is *planar* if it can be drawn in the plane in such a way that the arcs of G intersect only at the vertices of G.

Clearly the graphs G and $\tilde{G}$ of Example 4.4.41 are planar.   It is not always apparent when a graph is planar.   The 4×6 graph

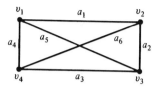

is planar since it can be drawn as

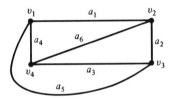

In Examples 4.4.50 and 4.4.51 we will construct two graphs which we will prove are not planar.   A planar graph often can be drawn in several different ways. For example, the graph G of Example 4.4.41 can also be drawn as

FIGURE 4.4.46

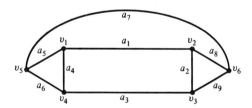

For reasons which will become clear shortly we will restrict our attention to connected planar graphs in which the degree of every vertex is at least 2. Since there are no vertices of degree 1, every arc of the graph belongs to at least one loop.   Such a graph divides the plane into a number of regions.   A *region*

can be defined as an area of the plane which is bounded by arcs of the graph and which does not contain any arcs in its interior. The regions of a graph depend on the way the graph is drawn. The regions of the graph G of Example 4.4.41 and the graph of Fig. 4.4.46 are not the same although the graphs are the same graph (in the sense that the incidence relations among the vertices and arcs are identical in the two cases). However, we will see that the number of regions of a planar graph does not depend on how the graph is drawn. The regions of the graph of Figure 4.4.46 are

R_1, bounded by the loop $\{a_4, a_5, a_6\}$
R_2, bounded by the loop $\{a_1, a_2, a_3, a_4\}$
R_3, bounded by the loop $\{a_2, a_8, a_9\}$
R_4, bounded by the loop $\{a_1, a_8, a_7, a_5\}$
R_5, outside the loop $\{a_3, a_9, a_7, a_6\}$

One of the regions of a planar graph will have infinite extent, for example, R_5 above, and is called the *infinite region*. All other regions are said to be *finite*. Every arc of a planar graph is a portion of the boundary of exactly two regions.

Theorem 4.4.47 The loops bounding the finite regions of a planar graph form a basis for the vector space of loop sets.

PROOF Let $R_1, R_2, \ldots, R_k$ be distinct finite regions of a planar graph and let $L_1, L_2, \ldots, L_k$ be the loops which enclose them. The area

$$R_1 \cup R_2 \cup \cdots \cup R_k$$

is bounded by the loop set

$$L_1 + L_2 + \cdots + L_k$$

Since $R_1 \cup R_2 \cup \cdots \cup R_k$ is neither the empty set nor the entire plane, it has a boundary consisting of a nonempty set of arcs, i.e.,

$$L_1 + L_2 + \cdots + L_k \neq 0$$

Thus $L_1, L_2, \ldots, L_k$ are linearly independent elements of $\mathbb{N}(B)$.

It remains only to show that every loop of the graph can be expressed as a sum of loops bounding finite regions. However, any loop of the graph encloses an area which is the union of a set of finite regions and therefore can be expressed as the sum of the loops of these regions.   ////

Since the dimension of the vector space of loop sets of an $m \times n$ graph is $n - m + 1$, we have established:

Theorem 4.4.48   Euler's formula.   A planar $m \times n$ graph contains $n - m + 2$ regions of which $n - m + 1$ are finite.

Thus, even though a planar graph may be represented by a diagram in several different ways, the number of regions is the same for all such representations. Euler's formula can be used to demonstrate that some graphs are not planar. First we need:

Lemma 4.4.49   If every loop of a planar $m \times n$ graph contains at least k arcs, then

$$n \geq \tfrac{1}{2}k(n - m + 2)$$

PROOF   We have shown that there are $n - m + 2$ regions. Each region is bounded by a loop containing at least k arcs. Thus the total number of arcs, counting each arc as many times as it belongs to loops which bound regions, is at least

$$k(n - m + 2)$$

However, each arc belongs to exactly two such loops. Thus the number of arcs is at least $\tfrac{1}{2}k(n - m + 2)$.   ////

EXAMPLE 4.4.50   Let us apply Lemma 4.4.49 to the complete graph with five vertices.

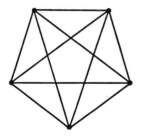

In this case

$$m = 5 \qquad n = 10 \qquad k = 3$$

and consequently $\tfrac{1}{2}k(n - m + 2) = \tfrac{1}{2} \cdot 3 \cdot 7 = 10\tfrac{1}{2} > 10 = n$

Thus the complete graph with five (or more) vertices is not planar.   ////

EXAMPLE 4.4.51   For the graph[1]

$$m = 6 \qquad n = 9 \qquad k = 4$$

This graph is not planar since

$$9 = n < \tfrac{1}{2}k(n - m + 2) = \tfrac{1}{2} \cdot 4 \cdot 5 = 10 \qquad \qquad ////$$

Clearly every graph which contains either the graph of Example 4.4.50 or the graph of Example 4.4.51 as a subgraph is not planar.   In 1930 the Polish mathematician Kazimierz Kuratowski [7] proved the converse, namely, that every nonplanar graph contains as a subgraph one of the two nonplanar graphs we have described.[2]   The proof of this theorem is difficult and will not be given here.   The graphs of Example 4.4.50 and 4.4.51 are called the *Kuratowski graphs*.

The dual pair of graphs G and $\tilde{G}$ of Example 4.4.41 are both planar.   The principle of superimposing a graph and its dual, as in Fig. 4.4.42, can be used to construct a dual of any planar graph.   Moreover, we will see that the dual graph obtained in this way is also planar.

Theorem 4.4.52   Every planar graph has a dual which is planar.

PROOF   Let G be a planar graph.   Form $\tilde{G}$ as follows.   Place a vertex of $\tilde{G}$ in each region (finite or infinite) of G.   For any two distinct regions of G, count the number of arcs common to the loops bounding them.   We will construct an equal number of arcs of $\tilde{G}$ between the vertices of $\tilde{G}$ which we have just placed inside these regions.   For each common arc between

---

[1] The nonplanarity of the graph of Example 4.4.51 is the basis for a well-known puzzle. "There are three houses on one side of the street and three utilities on the other side of the street.   Can lines be drawn from each utility to each house in such a way that no two lines cross?"

[2] If additional vertices are added along the arc of either of the nonplanar graphs, the resulting graph has a different set of incidence relations and is also nonplanar.   For Kuratowski's theorem all vertices of degree 2 are eliminated.   For a proof of the theorem see Berge [1].

two regions of G draw an arc of $\tilde{G}$ between the vertices of $\tilde{G}$ in these regions in such a way that it cuts the corresponding arc of G. (This will result in parallel arcs when two regions of G have more than one arc common to their bounding loops. What would happen in this construction if we permitted G to have arcs which do not belong to any loops?) It is easily seen that the resulting graph $\tilde{G}$ is a dual of G and is planar.

////

The converse of Theorem 4.4.52, namely, that any graph which has a dual is planar, was proved in 1932 by Hassler Whitney [11], an American mathematician. In addition he gave an alternate proof of Kuratowski's theorem by showing directly that a graph has a dual if and only if it does not contain either of the Kuratowski graphs as a subgraph. The proof of Whitney's theorem (in either version) is difficult and will not be given here.

EXERCISES

1 Find the cut matrix of the graph of Example 4.4.12. Using the incidence matrix and the loop matrix given in the text, verify Theorem 4.4.20, Corollary 4.4.30, and Lemma 4.4.33.

2 Find the incidence matrix, the loop matrix, and the cut matrix of the graphs G and $\tilde{G}$ of Example 4.4.41. Show that G and $\tilde{G}$ are duals of each other by verifying that their incidence matrices satisfy Theorem 4.4.40.

3 Use the construction of Theorem 4.4.52 to obtain a dual for the graph of Example 4.4.12.

4 Show that when the construction of the dual of a graph, as described in Theorem 4.4.52, is applied to the graph of Figure 4.4.46, the resulting graph is the graph $\tilde{G}$ of Example 4.4.41.

5 Let G be the graph of Example 4.4.12 and let G' be the graph

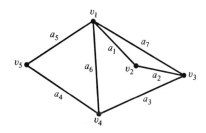

Show that G and G' are the same graph (in the sense that they have the same incidence relations). Use Theorem 4.4.52 to construct duals $\tilde{G}$ and $\tilde{G}'$ of G and G', respectively. Show that $\tilde{G}$ is a dual of G' and that $\tilde{G}'$ is a dual of G. Also, show that $\tilde{G}$ and $\tilde{G}'$ are not the same graph.

6 Construct a graph whose loop space has the vectors

$$
\begin{bmatrix} 1 \\ 1 \\ 1 \\ 1 \\ 1 \\ 1 \\ 0 \end{bmatrix}
\begin{bmatrix} 1 \\ 0 \\ 1 \\ 0 \\ 1 \\ 1 \\ 1 \end{bmatrix}
\begin{bmatrix} 1 \\ 0 \\ 1 \\ 1 \\ 0 \\ 0 \\ 1 \end{bmatrix}
$$

as a basis. Assuming that the graph has no isolated vertices, what are its dimensions? (*Hint:* In order to construct the graph, find the minimal vectors of the vector space spanned by the given vectors.)

7 Show that there does not exist a graph whose loop space has the vectors

$$
\begin{bmatrix} 1 \\ 1 \\ 1 \\ 1 \\ 0 \\ 0 \\ 0 \end{bmatrix}
\begin{bmatrix} 0 \\ 1 \\ 1 \\ 0 \\ 1 \\ 0 \\ 0 \end{bmatrix}
\begin{bmatrix} 0 \\ 0 \\ 1 \\ 1 \\ 0 \\ 1 \\ 0 \end{bmatrix}
\begin{bmatrix} 0 \\ 0 \\ 1 \\ 1 \\ 1 \\ 0 \\ 1 \end{bmatrix}
$$

as a basis. (*Hint:* Find the minimal vectors of the vector space spanned by the given vectors and try to construct the graph.)

8 Find a connected graph whose cut space has the vectors

$$
\begin{array}{ccccccc}
[1 & 1 & 1 & 1 & 0 & 1 & 0] \\
[1 & 0 & 0 & 1 & 1 & 1 & 0] \\
[1 & 0 & 1 & 1 & 1 & 0 & 1] \\
[0 & 1 & 1 & 1 & 1 & 1 & 1]
\end{array}
$$

as a basis. (*Hint:* Find the loop space of the graph.)

9 Prove that every arc of a graph belongs to at least one cut.

10 Prove that two arcs of a graph belong to exactly the same cut sets if and only if they are parallel.

11 Show that each arc of an $m \times n$ graph belongs to either no loop-sets or to 2^{n-m} loop sets. (*Hint:* Consider the graph which remains when the arc is removed. Find a necessary and sufficient condition that an arc belong to no loop sets.)

12 Show that each arc of an $m \times n$ graph belongs to exactly 2^{m-2} cut sets.

13 Show that a maximal subgraph with no loops of an $m \times n$ graph with k components has $n - m + k$ arcs and k components. (Such a subgraph is sometimes called a *forest*.)

14 Let G be an $m \times n$ graph and let $\bar{B}$ be the $(m - 1) \times n$ submatrix of B formed by deleting any one row of B. Show that there is a one-to-one correspondence between the trees of G and the nonsingular submatrices of $\bar{B}$ of order $m - 1$.

15 Prove that the subgraph H of G is a tree if and only if H contains all the vertices of G and between any two of these vertices there is exactly one path.

16 A vertex of a connected graph is called a *cut vertex* if when that vertex and all the arcs incident to it are removed from the graph, the remaining graph is not connected. Prove that a vertex is not a cut vertex if and only if the row of B corresponding to it is a minimal element of $\mathbb{R}(B)$.

17 Use Theorem 4.4.52 to give an alternate proof of Euler's formula. [*Hint:* If $\tilde{G}$ is the dual of G constructed using Theorem 4.4.52, how many vertices does $\tilde{G}$ have? What is $r(\tilde{B})$?]

18 Let $a = (v_1, v_2)$ be an arc of the connected graph G and let G have at least three vertices. Show that there is a cut in G such that v_1 and v_2 belong to the same component of the graph which results when the arcs of the cut are removed from G.

19 Show that every cut set of a complete graph is a cut.

20 Show that a connected graph with m vertices which has the property that every cut set is a cut is (or contains) the complete graph with m vertices.

4.5   DIRECTED GRAPHS

The purpose of this section is to do for directed graphs what we did for undirected graphs in the last section. There are, of course, many similarities between the two, but there are enough differences to warrant treating the two topics separately. One immediate difference is that we can no longer take the entries of our matrices from the field $\mathbb{F}_2$. Instead, throughout this section we will let $\mathbb{F}$ denote the field of real numbers unless we specify otherwise.

In a directed graph each arc is oriented, i.e., the arc is directed *from* one of its vertices *toward* the other. This orientation is most conveniently shown in a diagram for the graph by attaching an arrowhead to the arc. We understand then that the orientation of the arc is in the direction that the arrowhead is pointing. When we wish to show there is orientation both from the vertex v_1 to the vertex v_2 and from v_2 to v_1, we will use parallel arcs, each with a single direction (see Fig. 4.5.1).

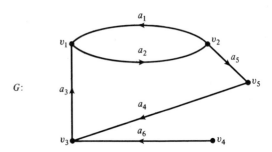

FIGURE 4.5.1

Let us give a precise definition of a directed graph.

Definition 4.5.2 An $m \times n$ *directed graph* $G = (\mathbb{V}, \mathbb{A})$ consists of a set $\mathbb{V}$ of m objects called *vertices*

$$\mathbb{V} = \{v_1, V_2, \ldots, v_m\}$$

and a set $\mathbb{A}$ of n *arcs*

$$\mathbb{A} = \{a_1, a_2, \ldots, a_n\}$$

where each arc is an ordered pair of distinct vertices. We write

$$a_j = (v_i, v_k)$$

to denote that the arc a_j is incident with the vertices v_i and v_k and is directed from v_i to v_k.

As with an undirected graph, the incidence relations between the vertices and arcs of a directed graph can be conveniently expressed in matrix form. In addition, the matrix contains the directional relations between the vertices and arcs.

Definition 4.5.3 Let G be a directed graph with vertices $v_1, v_2, \ldots, v_m$ and arcs $a_1, a_2, \ldots, a_n$. The $m \times n$ matrix

$$B = [b_{ij}] \qquad \begin{array}{l} i = 1, 2, \ldots, m \\ j = 1, 2, \ldots, n \end{array}$$

defined by

$$b_{ij} = \begin{cases} +1 & \text{if } a_j \text{ is directed toward } v_i \\ -1 & \text{if } a_j \text{ is directed from } v_i \\ 0 & \text{if } a_j \text{ is not incident with } v_i \end{cases}$$

is called the *(vertex-arc) incidence matrix* of G.[1]

---

[1] Since we will often be working with the real field and with $\mathbb{Z}_2$ simultaneously in this section, we will denote the real number one by $+1$ and the nonzero element of $\mathbb{Z}_2$ by 1.

Associated with any directed graph G there is an undirected graph (which we will denote by $|G|$) formed from G by ignoring the orientations of the arcs. $|G|$ is called the *underlying (undirected) graph* of G. Since $|G|$ is an undirected graph, the entries of its incidence matrix (denoted by $|B|$) are still elements of $\mathbb{Z}_2$. In fact, $|B|$ can be obtained from B by changing all -1s and $+1$s to 1s.

EXAMPLE 4.5.4   For the graph of G of Figure 4.5.1 we have

$$B = \begin{bmatrix} +1 & -1 & +1 & 0 & 0 & 0 \\ -1 & +1 & 0 & 0 & -1 & 0 \\ 0 & 0 & -1 & +1 & 0 & +1 \\ 0 & 0 & 0 & 0 & 0 & -1 \\ 0 & 0 & 0 & -1 & +1 & 0 \end{bmatrix}$$

The underlying graph of G is

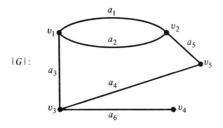

and

$$|B| = \begin{bmatrix} 1 & 1 & 1 & 0 & 0 & 0 \\ 1 & 1 & 0 & 0 & 1 & 0 \\ 0 & 0 & 1 & 1 & 0 & 1 \\ 0 & 0 & 0 & 0 & 0 & 1 \\ 0 & 0 & 0 & 1 & 1 & 0 \end{bmatrix}$$ ////

Many of the properties of a directed graph can be obtained from the incidence matrix. This includes properties which are, in reality, functions of the underlying undirected graph. The methods we will use for directed graphs often are precisely the same as those which we used in the last section. For example, we define a directed graph to be *connected* if its underlying graph is connected. More generally, the subgraph H of G is called a *component* of G if $|H|$ is a component of $|G|$. Then the analog of Lemma 4.4.3 is:

Lemma 4.5.5   If B is the incidence matrix of a directed graph G with m vertices, then G is connected if and only if the only row vectors X such that

$$XB = 0$$

are the scalar multiples of $E_{(m)} = [+1 \quad +1 \quad \cdots \quad +1]$.

PROOF   Since every column of B has exactly one $+1$, one -1, and $m - 2$ 0s, we have

$$E_{(m)}B = 0$$

If G is not connected, then the vertices and arcs of G can be ordered so that B has the form

$$B = \begin{bmatrix} B_1 & 0 & \cdots & 0 \\ 0 & B_2 & \cdots & 0 \\ \multicolumn{4}{c}{\dotfill} \\ 0 & 0 & \cdots & B_k \end{bmatrix}$$

where $B_1, B_2, \ldots, B_k$ are the incidence matrices of the components of G. If B_i is an $m_i \times n_i$ matrix for $i = 1, 2, \ldots, k$, then each of the k vectors

$$Y_1 = [E_{(m_1)} \quad 0 \quad 0 \quad \cdots \quad 0]$$
$$Y_2 = [0 \quad E_{(m_2)} \quad 0 \quad \cdots \quad 0]$$
$$\cdots\cdots\cdots\cdots\cdots\cdots\cdots\cdots$$
$$Y_k = [0 \quad 0 \quad 0 \quad \cdots \quad E_{(m_k)}]$$

satisfies

$$XB = 0$$

and is not a scalar multiple of $E_{(m)}$.

The proof of the converse requires one additional step.   Let G be a connected directed graph and suppose there exists a vector X such that

$$XB = 0 \qquad X \neq \alpha E_{(m)} \text{ for any real number } \alpha \qquad (4.5.6)$$

Set $X = [x_i]$ and note that since X satisfies (4.5.6), so does

$$Y = X - x_n E_{(m)}$$

Let the number of entries of Y which are equal to 0 be s.   Since the last entry of Y is 0, we have

$$s \geq 1$$

On the other hand, $Y \neq 0$, and hence

$$m - 1 \geq s$$

We may assume that the vertices of G are ordered so that

$$Y = [Z \quad 0]$$

where Z is an $(m - s)$-dimensional row vector none of whose entries is 0. Partition B as

$$B = \begin{bmatrix} B_1 \\ B_2 \end{bmatrix}$$

where B_1 and B_2 contain $m - s$ and s rows, respectively.   Then

$$ZB_1 = 0$$

Since none of the entries of Z is 0, each column of B_1 contains either two nonzero terms (and these must be $+1$ and -1) or contains no nonzero terms.   We may assume that the arcs of G are numbered so that the first p columns of B_1 have two nonzero terms and the remaining $n - p$ columns contain only 0's.   We can now partition B as

$$B = \begin{bmatrix} B_{11} & 0 \\ 0 & B_{22} \end{bmatrix}$$

where B_{11} and B_{22} are $(m - s) \times p$ and $s \times (n - p)$ matrices, respectively. Since G is a connected graph, we have reached a contradiction.   Consequently, no such vector X exists.   ////

The next theorem follows from Lemma 4.5.5 in the same way that Theorem 4.4.4 follows from Lemma 4.4.3.

Theorem 4.5.7 If B is the incidence matrix of an $m \times n$ directed graph with k components, then the rank of B is given by

$$r(B) = m - k$$

It is interesting to see that although we have changed the field from which the entries of our matrices are chosen, both the statement and the proof of Lemma 4.5.5 and Theorem 4.5.7 differ only slightly from their counterparts in the last section.   However, connectedness is actually a property of the underlying undirected graph rather than of the graph itself.   The field which is used will play a more prominent role when the direction of the arcs in a graph becomes a more significant factor.   Even then, however, many of the results of the last section can be extended to directed graphs with little or no change.

Definition 4.5.8 A set of l distinct arcs of a directed graph G,

$$\{(v_{h_0}, v_{h_1}), (v_{h_1}, v_{h_2}), \ldots, (v_{h_{l-1}}, v_{h_l})\}$$

is called a *directed path* of length l from v_{h_0} to v_{h_l}. If the vertices $v_{h_1}, v_{h_2}, \ldots, v_{h_{l-1}}$ are distinct from each other and from v_{h_0} and v_{h_l}, then the path is said to be *simple*. A simple path from a vertex to itself (that is, $v_{h_0} = v_{h_l}$) is called a *directed loop*. A subgraph H of G is a loop of *length* l if $|H|$ is a loop of length l of $|G|$.

If the directions of all the arcs of a directed loop are reversed, the resulting graph is again a directed loop. Every directed loop is a loop, and every loop differs from a directed loop only in the orientation of some of its arcs. In the graph of Example 4.5.4, there are three loops,

$$L_1 = \{a_1, a_2\} \qquad L_2 = \{a_3, a_2, a_5, a_4\} \qquad L_3 = \{a_1, a_5, a_4, a_3\}$$

L_1 and L_2 are directed loops. L_3 is not a directed loop.

EXAMPLE 4.5.9   When the vertices and arcs of a directed loop are numbered so that

$$L = \{a_1 = (v_1, v_2), a_2 = (v_2, v_3), \ldots, a_{l-1} = (v_{l-1}, v_l), a_l = (v_l, v_1)\}$$

then the incidence matrix of L is the $l \times l$ matrix

$$B_L = \begin{bmatrix} -1 & & & & & +1 \\ +1 & -1 & & & & \\ & +1 & -1 & & & \\ & & \cdots & \cdots & & \\ & & & +1 & -1 & \\ & & & & +1 & -1 \end{bmatrix} \qquad (4.5.10)$$

The incidence matrix of a loop has the same form as (4.5.10) except that within each column the $+1$ and the -1 may be interchanged, reflecting the change in the orientation of the corresponding arc. ////

For the incidence matrix B of a directed loop we have

$$BE = 0$$

and the only vectors X such that

$$BX = 0$$

are the scalar multiples of E.

On the other hand, suppose B is the incidence matrix of a loop L. We can form a directed loop $\tilde{L}$ from L by changing the direction of some of the arcs of L. (This may be done in two different ways.) If $F = [f_j]$ is the column vector defined by

$$
f_j = \begin{cases}
+1 & \text{when orientation of } a_j \text{ is the same in } L \\
   & \text{as in } \tilde{L} \\
-1 & \text{when orientation of } a_j \text{ in } L \text{ is opposite} \quad (4.5.11) \\
   & \text{to its orientation in } \tilde{L} \\
0 & \text{when } a_j \text{ is not an arc of } L
\end{cases}
$$

then
$$BF = 0$$

As before, the only vector X such that

$$BX = 0$$

are the scalar multiples of F.

Lemma 4.5.12   Let B be the incidence matrix of the directed graph G. A necessary and sufficient condition that a set L of arcs of G form a loop is that the columns of B corresponding to the arcs of L be a minimal linearly dependent set.

PROOF   The proof is similar to the proof of Lemma 4.4.8 and is left as an exercise.   ////

The natural correspondence between sets of arcs of G and n-dimensional vectors which was introduced in Sec. 4.3 and used extensively in Sec. 4.4 can be carried over to the case where the entries of our matrices come from the field $\mathbb{F}$ of real numbers.   More generally:

Definition 4.5.13   Let $S = \{s_1, s_2, \ldots, s_n\}$ be any set of n objects. An expression of the form

$$x = \alpha_1 s_1 + \alpha_2 s_2 + \cdots + \alpha_n s_n \qquad \alpha_1, \alpha_2, \ldots, \alpha_n \in \mathbb{F}$$

is called a *formal sum* of the elements of S with coefficients from $\mathbb{F}$.   Denote the set of all formal sums of elements of S be $\mathbb{S}$.   We define addition and scalar multiplication in $\mathbb{S}$ as follows.   Let

$$y = \beta_1 s_1 + \beta_2 s_2 + \cdots + \beta_n s_n$$

Set

$$x + y = (\alpha_1 + \beta_1)s_1 + (\alpha_2 + \beta_2)s_2 + \cdots + (\alpha_n + \beta_n)s_n$$

$$\alpha x = \alpha\alpha_1 s_1 + \alpha\alpha_2 s_2 + \cdots + \alpha\alpha_n s_n$$

With these operations we call $\mathbb{S}$ the *vector space of formal sums* of elements of S.

There is a natural way whereby we can identify $\mathbb{S}$ with $\mathbb{F}^n$ or $\mathbb{F}_n$. Let $x \in \mathbb{S}$,

$$x = \alpha_1 s_1 + \alpha_2 s_2 + \cdots + \alpha_n s_n$$

and let X be the n-dimensional column vector

$$X = \begin{bmatrix} \alpha_1 \\ \alpha_2 \\ \vdots \\ \alpha_n \end{bmatrix}$$

The mapping

$$\Pi : x \in \mathbb{S} \to X \in \mathbb{F}^n$$

is a one-to-one correspondence between the elements of $\mathbb{S}$ and $\mathbb{F}^n$. Moreover,

$$\Pi(x + y) = \Pi(x) + \Pi(y)$$

$$\Pi(\alpha x) = \alpha\Pi(x)$$

for all $x, y \in \mathbb{S}$ and $\alpha \in \mathbb{F}$. We call Π the *natural correspondence* between elements of $\mathbb{S}$ and $\mathbb{F}^n$. In the same way there is a natural correspondence between the elements of $\mathbb{S}$ and $\mathbb{F}_n$.

One more step is necessary before we can make use of the natural correspondence. Let L be a loop of a directed graph G and let $\tilde{L}$ be (either) one of the two directed loops that can be formed from L by changing the orientation of some of its arcs. We will identify L with the formal sum of arcs of G,

$$L = \varepsilon_1 a_1 + \varepsilon_2 a_2 + \cdots + \varepsilon_n a_n \qquad (4.5.14)$$

$$\text{where} \quad \varepsilon_j = \begin{cases} +1 & \text{when orientation of } a_j \text{ is the same in } L \\ & \text{as in } \tilde{L} \\ -1 & \text{when orientation of } a_j \text{ in } L \text{ is opposite} \\ & \text{to its orientation in } \tilde{L} \\ 0 & \text{when } a_j \text{ is not an arc of } L \end{cases}$$

These coefficients are precisely the ones described in (4.5.11). Thus both the

n-dimensional column vector $\mathbb{F}$ defined in (4.5.11) and the formal sum (4.5.14) correspond to the loop L, and these correspond to each other by means of the natural correspondence.   Later we will have reason to identify other subgraphs of a graph with formal sums of arcs.

In order to continue to preserve the language of Sec. 4.4 we will define a vector X of a subspace $\mathbb{V}$ of $\mathbb{F}^n$ or $\mathbb{F}_n$ to be *minimal* if the only vectors Y of $\mathbb{V}$ which have 0s in all the positions in which X has a zero are the scalar multiples of X.   Two minimal vectors of $\mathbb{V}$ which have 0s in the same positions are scalar multiples of each other.   For, suppose that $U = [u_j]$ and $V = [v_j]$ are two minimal vectors of $\mathbb{V}$ which have 0s in the same positions.   Let u_k be any non-zero entry of U and set

$$Z = u_k V - v_k U$$

Z has 0s wherever U and V have 0s and, moreover, in the kth component.   Since U and V are minimal vectors, it follows that $Z = 0$ and therefore

$$u_k V = v_k U$$

We will say that two minimal vectors are *distinct* if they are not scalar multiples of each other.

Combining Lemma 4.5.12 with the discussion above, we have:

Theorem 4.5.15   Under the natural correspondence between the vector space of formal sums of arcs of G and $\mathbb{F}^n$, the loops of G correspond to the distinct minimal vectors of $\mathbb{N}(B)$.   Each minimal element of $\mathbb{N}(B)$ is (a scalar multiple of) a vector containing only 0s, $+1$s, and -1s among its entries.   The directed loops of G are those which correspond to minimal elements of $\mathbb{N}(B)$ all of whose nonzero entries have the same sign.

EXAMPLE 4.5.16   The incidence matrix B of the graph of Fig. 4.5.1 is a 5×6 matrix whose rank is 4.   Hence $v(B) = 2$.   The distinct minimal vectors of $\mathbb{N}(B)$ are

$$L_1 = \begin{bmatrix} +1 \\ +1 \\ 0 \\ 0 \\ 0 \\ 0 \end{bmatrix} \quad L_2 = \begin{bmatrix} 0 \\ +1 \\ +1 \\ +1 \\ +1 \\ 0 \end{bmatrix} \quad L_3 = \begin{bmatrix} -1 \\ 0 \\ +1 \\ +1 \\ +1 \\ 0 \end{bmatrix}$$

corresponding to the formal sums

$$L_1 = a_1 + a_2$$
$$L_2 = a_2 + a_3 + a_4 + a_5$$
$$L_3 = -a_1 + a_3 + a_4 + a_5$$

which, in turn, correspond to the loops

$$L_1 = \{a_1, a_2\} \qquad L_2 = \{a_2, a_3, a_4, a_5\} \qquad L_3 = \{a_1, a_3, a_4, a_5\}$$

Note that

$$L_3 = L_2 - L_1 \qquad\qquad ////$$

We say that a vector is *unimodular* if all its entries are $+1$, -1, or 0. A vector space is *unimodular* if all its minimal vectors are scalar multiples of unimodular vectors. Clearly $\mathbb{F}^n$ and $\mathbb{F}_n$ are unimodular vector spaces. According to Theorem 4.5.15, $\mathbb{N}(B)$ is unimodular, and it follows from Lemma 4.5.5 and Theorem 4.5.7 that $\mathbb{N}_R(B)$ is unimodular.

The concept of a loop set can be extended to directed graphs. However, because the field of real numbers contains infinitely many elements, we will be able to form infinitely many loop sets from a set of loops.

Definition 4.5.17 A formal sum of arcs of a directed graph is a *loop set* if it can be expressed as a linear combination of loops.

Corollary 4.5.18 Under the natural correspondence between the vector space of formal sums of arcs of G and $\mathbb{F}^n$ the loop sets correspond to the elements of $\mathbb{N}(B)$.

Another important concept which can be carried over from undirected graphs to directed graphs is the loop matrix.

Definition 4.5.19 Let $L_1, L_2, \ldots, L_p$ be the loops of G. For each loop L_k, let $\tilde{L}_k$ be one of the two directed loops that can be formed from L_k by changing the direction of some of its arcs. The $n \times p$ matrix $L = [l_{jk}]$, where

$$l_{jk} = \begin{cases} +1 & \text{when orientation of } a_j \text{ is the same in } L_k \\ & \text{as in } \tilde{L}_k \\ -1 & \text{when orientation of } a_j \text{ in } L_k \text{ is opposite} \\ & \text{to its orientation in } \tilde{L}_k \\ 0 & \text{when } a_j \text{ is not an arc of } L_k \end{cases}$$

is called the *loop matrix* of G.

By comparing Definition 4.5.19 with (4.5.11), we see that the kth column of L is the vector corresponding to the loop L_k. The ambiguity in the definition of the vector corresponding to a loop is carried over to the loop matrix since either of the two directed loops associated with L_k may be chosen for $\tilde{L}_k$. When the choice of $\tilde{L}_k$ is changed, the sign of the entries of the kth column of L are changed. We will see that this causes no difficulty. The loop matrix of the graph of Fig. 4.5.1 is the 6×3 matrix

$$
L = \begin{matrix} & L_1 & L_2 & L_3 \\ \begin{bmatrix} +1 & 0 & -1 \\ +1 & +1 & 0 \\ 0 & +1 & +1 \\ 0 & +1 & +1 \\ 0 & +1 & +1 \\ 0 & 0 & 0 \end{bmatrix} & \begin{matrix} a_1 \\ a_2 \\ a_3 \\ a_4 \\ a_5 \\ a_6 \end{matrix} \end{matrix}
$$

At this point in Sec. 4.4 we defined a tree and showed how the chords of a tree determine a fundamental set of loops. We then proved that the loops of a fundamental set form a basis for $\mathbb{C}(L)$. We leave it as an exercise to show that the same treatment carries over to directed graphs. On the basis of these remarks we conclude that:

Theorem 4.5.20   For an $m \times n$ connected directed graph,

1 $BL = 0$.
2 $\mathbb{C}(L) = \mathbb{N}(B)$.
3 $r(L) = n - m + 1$.
4 $\mathbb{N}_R(L) = \mathbb{R}(B)$.

Just as we have defined loops and directed loops for a directed graph, we can define cuts and directed cuts. We will call a set C of arcs of a connected directed graph G a *cut* if $|C|$ is a cut of $|G|$. When the arcs of C are removed from G, the resulting graph, say G', consists of two components, say G_1 and G_2. Every arc of C has one vertex which belongs to G_1 and one vertex which belongs to G_2. We say that C is a *directed cut from G_1 to G_2* if every arc of C is directed from its vertex in G_1 to its vertex in G_2. If the orientation of every arc of a directed cut from G_1 to G_2 is reversed, then the resulting set of arcs form a directed cut from G_2 to G_1. Thus, every cut can be transformed into a directed cut by changing the direction of some of its arcs. In fact, this can be done in two different ways. We will see that this ambiguity causes no difficulty.

Let C be a cut and let $\tilde{C}$ be one of the two directed cuts associated with C. We will identify C with both the formal sum of arcs of G,

$$C = \delta_1 a_1 + \delta_2 a_2 + \cdots + \delta_n a_n$$

where
$$\delta_j = \begin{cases} +1 & \text{when orientation of } a_j \text{ is the same in } C \text{ as in } \tilde{C} \\ -1 & \text{when orientation of } a_j \text{ in } C \text{ is opposite to its orientation in } \tilde{C} \\ 0 & \text{when } a_j \text{ is not an arc of } C \end{cases}$$

and with the n-dimensional row vector

$$D = [\delta_1 \quad \delta_2 \quad \cdots \quad \delta_n]$$

As before, C and D correspond to each other under the natural correspondence between the vector space of formal sums of arcs of G and $\mathbb{F}_n$.

Definition 4.5.21 A formal sum of arcs of a connected directed graph is a *cut set* if it can be expressed as a linear combination of cuts.

The collection of cut sets of a graph is a subspace of $\mathbb{F}_n$ or, equivalently, of the vector space of formal sums of arcs. We will show that the cut sets of a directed graph play the same role as the cut sets of undirected graphs.

EXAMPLE 4.5.22 Let us denote the cuts of the graph G of Fig. 4.5.1 by

$$C_1 = \{a_1, a_2, a_3\} \qquad C_2 = \{a_3, a_4\}$$
$$C_3 = \{a_4, a_5\} \qquad C_4 = \{a_1, a_2, a_5\}$$
$$C_5 = \{a_1, a_2, a_4\} \qquad C_6 = \{a_3, a_5\}$$
$$C_7 = \{a_6\}$$

Let $\tilde{C}_1, \tilde{C}_2, \ldots, \tilde{C}_7$ be directed cuts corresponding to $C_1, C_2, \ldots, C_7$ respectively. We must specify an orientation for each of the directed cuts. In each case, let us select $\tilde{C}$ so that it is directed toward the component of G' which contains v_1. Then the cuts $C_1, C_2, \ldots, C_7$ correspond to the vectors

$$
\begin{aligned}
C_1 &= [+1 \quad -1 \quad +1 \quad 0 \quad 0 \quad 0] \\
C_2 &= [\ 0 \quad\ \ 0 \quad +1 \quad -1 \quad 0 \quad 0] \\
C_3 &= [\ 0 \quad\ \ 0 \quad\ \ 0 \quad +1 \quad -1 \quad 0] \\
C_4 &= [+1 \quad -1 \quad\ \ 0 \quad\ \ 0 \quad +1 \quad 0] \\
C_5 &= [+1 \quad -1 \quad\ \ 0 \quad +1 \quad 0 \quad 0] \\
C_6 &= [\ 0 \quad\ \ 0 \quad +1 \quad\ \ 0 \quad -1 \quad 0] \\
C_7 &= [\ 0 \quad\ \ 0 \quad\ \ 0 \quad\ \ 0 \quad 0 \quad +1]
\end{aligned}
$$

respectively. ////

Definition 4.5.23   Let $C_1, C_2, \ldots, C_q$ be the cuts of the connected directed graph G.   For each cut C_h let $\tilde{C}_h$ be either one of the two directed cuts that can be formed from C_h by changing the direction of some of its arcs.   The $q \times n$ matrix $C = [c_{hj}]$, where

$$c_{hj} = \begin{cases} +1 & \text{when orientation of } a_j \text{ is the same in } C_h \text{ as in } \tilde{C}_h \\ -1 & \text{when orientation of } a_j \text{ in } C_h \text{ is opposite to its orienta-} \\ & \text{tion in } \tilde{C}_h \\ 0 & \text{when } a_j \text{ is not an arc of } C_h \end{cases}$$

is called the *cut matrix* of G.

The 7×6 matrix

$$C = \begin{bmatrix} \begin{array}{cccccc} a_1 & a_2 & a_3 & a_4 & a_5 & a_6 \end{array} \\ \begin{bmatrix} +1 & -1 & +1 & 0 & 0 & 0 \\ 0 & 0 & +1 & -1 & 0 & 0 \\ 0 & 0 & 0 & +1 & -1 & 0 \\ +1 & -1 & 0 & 0 & +1 & 0 \\ +1 & -1 & 0 & +1 & 0 & 0 \\ 0 & 0 & +1 & 0 & -1 & 0 \\ 0 & 0 & 0 & 0 & 0 & +1 \end{bmatrix} \begin{array}{l} C_1 \\ C_2 \\ C_3 \\ C_4 \\ C_5 \\ C_6 \\ C_7 \end{array} \end{bmatrix}$$

is the cut matrix of the graph of Fig. 4.5.1.

Lemma 4.5.24   Let C be a cut of G and let the graph G' which results when the arcs of C are removed from G have the two components G_1 and G_2.   Let $\tilde{C}$ be the directed cut from G_2 to G_1 associated with C.   If $X = [x_i]$ is the m-dimensional row vector of the vertices of G_1, that is,

$$x_i = \begin{cases} +1 & \text{when } v_i \in G_1 \\ 0 & \text{when } v_i \notin G_1 \end{cases}$$

then the vector $C \in \mathbb{F}_n$ corresponding to the cut C (with respect to $\tilde{C}$) is given by

$$C = XB$$

where B is the incidence matrix of G.

PROOF   We may assume that the vertices and arcs of G are numbered so that

$$G_1 = \{v_1, v_2, \ldots, v_s; a_1, a_2, \ldots, a_t\}$$
$$G_2 = \{v_{s+1}, v_{s+2}, \ldots, v_m; a_{t+1}, a_{t+2}, \ldots, a_r\}$$
$$C = \{a_{r+1}, a_{r+2}, \ldots, a_n\}$$

Then B is partitioned as

$$B = [B_1 \quad B_2]$$

where B_1 is the $m \times r$ incidence matrix of G'. Moreover, since G_1 and G_2 are both arc- and vertex-disjoint,

$$B_1 = \begin{bmatrix} B_{11} & 0 \\ 0 & B_{22} \end{bmatrix}$$

where B_{11} and B_{22} are the $s \times t$ and $(m - s) \times (r - t)$ incidence matrices of G_1 and G_2 respectively. By definition, the vector of the vertices of G_1 is

$$X = [\overset{s}{E} \quad \overset{m-s}{0}]$$

According to Lemma 4.5.5, we have

$$XB_1 = 0$$

Since X is not a scalar multiple of E and the rank of any matrix formed with B_1 and a column of B_2 is $m - 1$, it follows that

$$XB^{(i)} \neq 0 \qquad i = r + 1, r + 2, \ldots, n$$

where $B^{(i)}$ is the column of B corresponding to the arc a_i of C. If a_i is directed toward G_1, then $B^{(i)}$ has a $+1$ in one of its first s rows, a -1 in one of its last $m - s$ rows, and 0s elsewhere. Consequently $XB^{(i)} = +1$. On the other hand, if a_i is directed away from G_1, then $B^{(i)}$ has a -1 in one of its first s rows, a $+1$ in one of its last $m - s$ rows, and 0s elsewhere. In this case $XB^{(i)} = -1$. We have now proved that the entries of XB are identical to the entries of the vector C corresponding to the cut C.
////

Theorem 4.5.25   For a connected directed graph:

1 $\mathbb{R}(C) \subseteq \mathbb{R}(B)$.
2 $CL = 0$.

PROOF   By Lemma 4.5.24, each row of C is contained in $\mathbb{R}(B)$. Hence $\mathbb{R}(C) \subseteq \mathbb{R}(B)$. The second part of the theorem follows immediately from part 4 of Theorem 4.5.20.   ////

Part 1 of Theorem 4.5.25 can be strengthened considerably. We can prove that $\mathbb{R}(C) = \mathbb{R}(B)$ in exactly the same way it was proved for undirected graphs. The arcs of any tree determine a fundamental set of cuts, consisting of $m - 1$ cuts. By numbering the arcs and cuts of the graph properly we can show that $r(C) \geq m - 1$. Combining this relation with Theorem 4.5.7 and the first part of Theorem 4.5.25, we have

$$\mathbb{R}(C) = \mathbb{R}(B) = \mathbb{N}_R(L)$$
$$\mathbb{N}(C) = \mathbb{N}(B) = \mathbb{C}(L)$$

We can complete the analogy between directed and undirected graphs with the following theorem.

Theorem 4.5.26   Under the natural correspondence between the vector space of sets of arcs of G and $\mathbb{F}_n$, the cuts of G correspond to the minimal vectors of $\mathbb{R}(B)$, and the cut sets of G correspond to the elements of $\mathbb{R}(B)$.

PROOF   The proof of part 1 of the theorem is similar to the proof of Theorem 4.4.35 and is left as an exercise.   Part 2 follows immediately from part 1.   ////

As a consequence of Theorem 4.5.26 we have:

Corollary 4.5.27 $\mathbb{R}(B)$ is unimodular.

In Sec. 4.4 we used Theorem 4.4.27 to establish the relation

$$CL = 0$$

We obtained the same result for directed graphs (in the second part of Theorem 4.5.25) by means of Lemma 4.5.24.   However, the analog of Theorem 4.4.27 for directed graphs has some interesting ramifications.   Let L be a loop and C be a cut of a connected directed graph G and let $\tilde{L}$ and $\tilde{C}$ be a directed loop and a directed cut respectively associated with L and C.   If $L = [l_j]$ and $C = [c_j]$ are the n-dimensional column and row vectors corresponding to L and C (with respect to $\tilde{L}$ and $\tilde{C}$), then according to the second part of Theorem 4.5.25,

$$\sum_{j=1}^{n} c_j l_j = CL = 0 \qquad (4.5.28)$$

Set

$$S = C \cap L = \{a_1, a_2, \ldots, a_s\}$$

Theorem 4.4.27 states that s is even.   For $a_j \in S$ the following situations are possible:

1   The orientation of a_j in L is the same as in $\tilde{L}$ (in which case $l_j = +1$).
2   The orientation of a_j in L is opposite to its orientation in $\tilde{L}$ (in which case $l_j = -1$).
1'   The orientation of a_j in C is the same as in $\tilde{C}$, (in which case $c_j = +1$).
2'   The orientation of a_j in C is opposite to its orientation in $\tilde{C}$ (in which case $c_j = -1$).

Then the term $c_j l_j$ in (4.5.28) is given by

$$c_j l_j = \begin{cases} +1 & \text{when either 1 and } 1' \text{ or 2 and } 2' \text{ occur} \\ -1 & \text{when either 1 and } 2' \text{ or 2 and } 1' \text{ occur} \\ 0 & \text{when } a_j \notin S \end{cases}$$

It follows from equation (2.5.28) that the number of arcs of S for which $c_j l_j = +1$ is equal to the number of arcs of S for which $c_j l_j = -1$, that is, $\frac{1}{2}s$. Thus we have proved:

Theorem 4.5.29 Let L be a loop and C be a cut of a connected directed graph and let $\tilde{L}$ and $\tilde{C}$ be a directed loop and a directed cut associated with L and C respectively. Set $S = C \cap L$. The number of arcs of S whose orientation is either (*a*) the same in both L and $\tilde{L}$ and in C and $\tilde{C}$ or (*b*) opposite in both L and $\tilde{L}$ and in C and $\tilde{C}$ is equal to the number of arcs of S for which the orientation is the same in one case and opposite in the other.

The most interesting situation occurs when L is a directed loop (so that we may set $L = \tilde{L}$). Then condition 1 above always occurs, and condition 2 never takes place. In this case we have:

Theorem 4.5.30 If L is a directed loop and C is a cut of the connected directed graph G, and if G_1 and G_2 are the components of the graph G' which results when the arcs of C are removed from G, then exactly half the arcs of $S = C \cap L$ are directed from G_1 to G_2 and exactly half are directed from G_2 to G_1.

PROOF Let $\tilde{C}$ be a directed cut, say from G_1 to G_2, associated with C. Since we have set $L = \tilde{L}$, case (*b*) of Theorem 4.5.29 cannot occur. Similarly, an arc of S for which the orientation is the same in one case and opposite in the other must have opposite orientations in C and $\tilde{C}$. Thus the number of arcs of S which have the same orientation in C and $\tilde{C}$ is the same as the number of arcs of S which have opposite orientation in C and $\tilde{C}$. ////

There is a simple geometric proof of Theorem 4.5.30. The components G_1 and G_2 can be thought of as being contained in the two regions K_1 and K_2 of Fig. 4.5.31. Each arc of C has one vertex in G_1 and one vertex in G_2. By following a path along the directed loop L (in the direction of the arrowheads) we alternately pass from G_1 to G_2 and from G_2 to G_1. The arcs of L which do not belong to C are indicated in G_1 and G_2 by wavy lines. The arcs of C which do not belong to L are not included in the diagram.

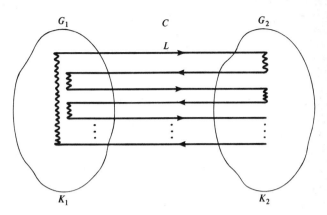

FIGURE 4.5.31

We have emphasized the close relation between directed and undirected graphs. The development of the theory of directed graphs in this section parallels the development of Sec. 4.3. However, we close this section with a theorem pertaining specifically to directed graphs. This theorem depends on one of the fundamental properties of the real field, namely, *a sum of squares of real numbers is zero only when the numbers themselves are all zero.*

Lemma 4.5.32   Let G be a connected directed $m \times n$ graph and let $\mathbb{S}$ be the vector space of formal sums of arcs of G. If $\mathbb{L}$ and $\mathbb{C}$ are the subspaces of $\mathbb{S}$ which consist of all loop sets and all cut sets respectively, then $\mathbb{L} \cap \mathbb{C} = 0$.

PROOF   Let $X \in \mathbb{L} \cap \mathbb{C}$. Since X is a loop set, X corresponds to an n-dimensional column vector belonging to $\mathbb{N}(B)$, say $Y = [y_j]$, under the natural correspondence between loop sets and elements of $\mathbb{N}(B)$. Similarly, since X is a cut set, X corresponds to an n-dimensional row vector belonging to $\mathbb{R}(B)$, say Z, under the natural correspondence between cut sets and elements of $\mathbb{R}(B)$. Thus we have

$$ZY = 0 \qquad (4.5.33)$$

However, because Y and Z correspond to the same formal sum of arcs under the two natural correspondences, we have

$$Z = Y^T$$

Then equation (4.5.33) becomes

$$ZY = Y^T Y = y_1{}^2 + y_2{}^2 + \cdots + y_n{}^2 = 0$$

and consequently $$y_1 = y_2 = \cdots = y_n = 0$$

Thus $X = 0$. ////

Theorem 4.5.34 Every formal sum of arcs of a connected directed graph G can be expressed in exactly one way as the sum of a loop set and a cut set.

PROOF   Let G be an $m \times n$ graph.   It follows from Corollary 4.5.18 and Theorem 4.5.26 that

$$\dim \mathbb{L} = n - m + 1 \qquad \dim \mathbb{C} = m - 1 \qquad (4.5.35)$$

Combining equations (4.5.35) with Lemma 4.5.32, we have

$$\mathbb{S} = \mathbb{L} \oplus \mathbb{C}$$

where $\mathbb{S}$ is the vector space of all formal sums of arcs of G. ////

EXERCISES

1 Denote by G the directed graph formed from the undirected graph of Example 4.4.12 by orienting each arc toward the vertex which has the larger subscript.   Form the incidence matrix, the loop matrix, and the cut matrix of G.   Verify Theorems 4.5.20 and 4.5.25.

2 Repeat Exercise 1 with the graph G of Example 4.4.41.

3 Express each of the arcs of the graph of Fig. 4.5.1 as a sum of loop sets and cut sets.

4 Show that Lemma 4.5.32 and Theorem 4.5.34 are not generally true for undirected graphs (and $\mathbb{F} = \mathbb{Z}_2$).   Specifically, show that Lemma 4.5.32 and Theorem 4.5.34 are false when G is a loop of even length.   Show, however, that the lemma and theorem remain true when G is a loop of odd length.

5 Give an example of two unimodular vector spaces whose sum is not unimodular.

6 Show that Theorem 4.5.29 can be obtained from Theorem 4.5.30 by changing the orientation of some of the arcs of the loop.   (This provides a proof of Theorem 4.5.29 based on the ideas suggested by Fig. 4.5.31.)

4.6 MATRICES AND TREES

We have seen in Secs. 4.4 and 4.5 that for a directed graph those properties which are not directly related to the orientations of the arcs, e.g., connectedness, undirected loops, and undirected cuts, can be studied equally well in terms of the matrices of the directed graph G or in terms of the matrices of the underlying undirected graph $|G|$. Recall that the entries of the matrices associated with G are from the real field whereas the entries of the matrices associated with $|G|$ are from $\mathbb{Z}_2$. In this section we will work with both fields simultaneously.

A subgraph T of a directed graph G is called a *tree* if $|T|$ is a tree of $|G|$. It follows from Lemmas 4.4.8 and 4.5.12 that a set T of $m-1$ arcs of an $m \times n$ graph form a tree:

1 If and only if the columns of B corresponding to the arcs of T are linearly independent.
2 If and only if the columns of $|B|$ corresponding to the arcs of $|T|$ are linearly independent.

We can conclude from the equivalence of these two conditions that a set of columns of an incidence matrix B are linearly independent if and only if the corresponding columns of $|B|$ are linearly independent. The same result can be obtained as a corollary to:

Theorem 4.6.1 The determinant of every square submatrix of an incidence matrix B is equal to 0, $+1$, or -1.

PROOF The proof is by an induction on the order of the submatrix and is immediate for square submatrices of order 1. Now suppose that the theorem is true for all $k \times k$ submatrices of B and let M be a square submatrix of B of order $k+1$. Each column of M contains at most two nonzero entries, at most one $+1$, and at most one -1. If every column of M contains either two nonzero entries or no nonzero entries, then M is singular since

$$EM = 0$$

and consequently det $M = 0$. On the other hand, suppose that some column of M, say the qth column, contains exactly one nonzero entry. Then, expressing det M by means of the qth column-expansion of the determinant [see equation (1.9.31)], we see that det M is equal (up to a possible change in sign) to the determinant of one of its submatrices, say N, of order k. By the induction hypothesis, det $N = 0$, $+1$, or -1, and therefore det $M = 0$, $+1$, or -1. ////

Corollary 4.6.2 Let M be a square submatrix of B and let $|M|$ be the corresponding submatrix of $|B|$. Then

1 det $|M| = 0$ if and only if det $M = 0$.
2 det $|M| = 1$ if and only if det $M = +1$ or -1.

PROOF   Let the order of M be k.   The determinant of M (or $|M|$) is a sum of $k!$ terms, each of which is $+1$, -1, or 0 (or is 1 or 0).   It follows from Theorem 4.6.1 that $\det M = 0$ if and only if an even number of the terms of the determinant are nonzero.   Similarly, $\det |M| = 0$ if and only if an even number of the $k!$ terms of the determinant are nonzero.   Finally, a term of $\det M$ is nonzero if and only if the corresponding term of $\det |M|$ is nonzero.   ////

Theorem 4.6.1 leads directly to a simple and beautiful formula for the number of trees contained in a connected graph.   First we need a lemma.

Lemma 4.6.3   Let B be the incidence matrix of an $m \times n$ connected graph G and let $\bar{B}$ be formed from B by deleting any one row of B.   (It does not matter which row of B is deleted.)   There is a one-to-one correspondence between the trees of G and the nonsingular submatrices of $\bar{B}$ of order $m - 1$.

PROOF   Since $\bar{B}$ has $m - 1$ rows, a submatrix, say M, of $\bar{B}$ of order $m - 1$ consists of $m - 1$ columns of $\bar{B}$.   By Corollary 4.6.2, M is nonsingular if and only if $|M|$ is nonsingular.   Hence, to complete the proof of the lemma it is sufficient to show that there is a one-to-one correspondence between the trees of G and those sets of $m - 1$ columns of $|\bar{B}|$ which are linearly independent.

According to part 2 of Theorem 4.4.16 and Lemma 4.4.8, a necessary and sufficient condition for a set of $m - 1$ arcs of G to form a tree is that the corresponding columns of $|B|$ be linearly independent.   We wish to prove that the same result holds true when $|B|$ is replaced by $|\bar{B}|$.   Specifically, we wish to show that if a set of columns of $|B|$ are linearly independent, then the corresponding columns of $|\bar{B}|$ are also linearly independent.   The converse of this statement is clear.

Let $|H|$ be an $m \times k$ submatrix of $|B|$ whose columns are linearly independent and let $|\bar{H}|$ be the corresponding submatrix of $|\bar{B}|$.   Also, let $H_{(0)}$ denote the row of H which is deleted from H to form $\bar{H}$.   According to equation (4.4.1),

$$E_{(m)}|H| = 0 \qquad (4.6.4)$$

Equation (4.6.4) can be expressed as

$$|H_{(0)}| + E_{(m-1)}|\bar{H}| = 0$$

and consequently

$$|H_{(0)}| = E_{(m-1)}|\bar{H}| \qquad (4.6.5)$$

Now, let us suppose that the columns of $|\bar{H}|$ are linearly dependent. Then there exists a vector X, $X \neq 0$, such that

$$|\bar{H}|\, X = 0$$

It follows from equation (4.6.5) that

$$|H_{(0)}|\, X = 0$$

and hence

$$|H|\, X = 0$$

This is a contradiction since the columns of $|H|$ are linearly independent.

$////$

Now we can prove:

Theorem 4.6.6   Let B be the incidence matrix of a connected graph G and let $\bar{B}$ be formed from B by deleting any one row of B.   Then the number of trees contained in G is equal to det $\bar{B}\bar{B}^T$ [1].

PROOF   The proof consists of evaluating det $\bar{B}\bar{B}^T$ by means of the Cauchy-Binet theorem (see Exercise 24 of Sec. 1.9).   For each subset Θ of $\mathbb{N} = \{1, 2, \ldots, n\}$ consisting of $m - 1$ elements of $\mathbb{N}$, define $\bar{B}_\Theta$ to be the square matrix of order $m - 1$ formed from $\bar{B}$ by deleting all columns $\bar{B}^{(j)}$ for $j \notin \Theta$.   Similarly, define $\bar{B}_\Theta{}^T$ to be the square matrix of order $m - 1$ formed from $\bar{B}^T$ by deleting all rows $\bar{B}_{(j)}^T$ for $j \notin \Theta$.   Clearly $\bar{B}_\Theta{}^T = (\bar{B}_\Theta)^T$.   Then

$$\det \bar{B}\bar{B}^T = \sum_\Theta (\det \bar{B}_\Theta)(\det \bar{B}_\Theta{}^T)$$
$$\sum_\Theta (\det \bar{B}_\Theta)[\det (\bar{B}_\Theta)^T]$$
$$\sum_\Theta (\det \bar{B}_\Theta)^2$$

where the sum is taken over all $n!/[(m - 1)!(n - m + 1)!]$ subsets of $\mathbb{N}$ which contain $m - 1$ elements.   However, combining Theorem 4.6.1 and Lemma 4.6.3, we have

$$\det \bar{B}_\Theta = \begin{cases} \pm 1 & \text{when arcs of } G \text{ corresponding to} \\ & \text{elements of } \Theta \text{ form a tree} \\ 0 & \text{otherwise} \end{cases}$$

Hence det BB^T is equal to a sum of 0s and $+1$s, and the number of $+1$s is the same as the number of trees of G. $////$

---

[1] This relation was known to Maxwell in 1892 and probably even to Kirchhoff in 1847.   However, the first formal proof was given by Brooks, Smith, Stone and Tutte [2] in 1940.

Theorem 4.6.6 remains true when G is not connected.   In this case

$$r(B) \leqq m - 2$$

and consequently

$$r(\bar{B}) \leqq m - 2$$

Thus $\bar{B}\bar{B}^T$ is singular, and det $\bar{B}\bar{B}^T = 0$.

The formula for the number of trees in a graph can be expressed in a somewhat different way.

Corollary 4.6.7   If B is the incidence matrix of a graph, then all the entries of adj BB^T are equal and each entry is equal to the number of trees in the graph.

PROOF   For convenience we will assume that in Theorem 4.6.6, $\bar{B}$ was formed from B by deleting the last row of B.   Then

$$B = \begin{bmatrix} \bar{B} \\ B_{(m)} \end{bmatrix} \quad \text{and} \quad BB^T = \begin{bmatrix} \bar{B}\bar{B}^T & \bar{B}B_{(m)}^T \\ B_{(m)}\bar{B}^T & B_{(m)}B_{(m)}^T \end{bmatrix}$$

Theorem 4.6.6 states that det $\bar{B}\bar{B}^T$ is equal to the number of trees contained in G.   However, det $\bar{B}\bar{B}^T$ is the cofactor of the entry in the last row and last column of BB^T and thus is equal to the entry in the last row and last column of adj BB^T.   In the same way, we are able to show that all the diagonal entries of adj BB^T are equal to the number of trees contained in G. It remains only to prove that all the entries of adj BB^T are equal.

According to part (a) of Exercise 25 in Sec. 1.7,

$$r(BB^T) \leqq r(B) \qquad (4.6.8)$$

We will consider two cases.   When G is not a connected graph, the rank of B is at most $m - 2$ and every submatrix of BB^T of order $m - 1$ is singular. Thus adj $BB^T = 0$, as required.   Now, let us suppose that G is connected. G contains some trees, and, according to the preceding paragraph, the diagonal entries of adj BB^T are positive.   By part (c) of Exercise 9 in Sec. 1.9,

$$r(BB^T) \geqq m - 1 \qquad (4.6.9)$$

Combining (4.6.8) and (4.6.9), we see that

$$r(BB^T) = m - 1 \qquad v_R(BB^T) = 1$$

But

$$E_{(m)}BB^T = 0$$

and consequently $E_{(m)}$ is a basis for $\mathbb{N}_R(BB^T)$. Applying Theorem 1.9.32, we have

$$[\text{adj } BB^T](BB^T) = (\det BB^T)I = 0$$

Every row of adj BB^T is contained in $\mathbb{N}_R(BB^T)$, and thus is a scalar multiple of $E_{(m)}$; that is all its entries are equal.   ////

EXAMPLE 4.6.10   For the graph of Example 4.4.12,

$$BB^T = \begin{bmatrix} 4 & -1 & -1 & -1 & -1 \\ -1 & 2 & -1 & 0 & 0 \\ -1 & -1 & 3 & -1 & 0 \\ -1 & 0 & -1 & 3 & -1 \\ -1 & 0 & 0 & -1 & 2 \end{bmatrix}$$

no matter what orientation is given to the arcs.   The entry in the first row and first column of adj BB^T is the cofactor of 4 in the matrix above.

$$\det BB^T = \det \begin{bmatrix} 2 & -1 & 0 & 0 \\ -1 & 3 & -1 & 0 \\ 0 & -1 & 3 & -1 \\ 0 & 0 & -1 & 2 \end{bmatrix} = 21$$

Verify that every entry of adj BB^T is 21.   List the 21 trees of G.   How many subgraphs does G contain consisting of four arcs? List the subgraphs of G containing four arcs which are not trees.   ////

The concept of a tree can be extended to a directed graph.   A subgraph T of a directed graph G is a *tree* if $|T|$ is a tree of $|G|$.   Moreover:

Definition 4.6.11   A tree T of a directed graph is a *directed tree, rooted at the vertex v*, if every vertex of T, other than v, has exactly one arc of T directed from it.

A directed tree has at least one arc directed to its root, and the (unique) path in the tree from any vertex to the root of the tree is a directed path, directed toward the root.

Let v be a vertex of the tree T of a directed graph. T can be transformed into a directed tree rooted at v by changing the orientations of some of the arcs of T.   Moreover, this can be done in exactly one way.

EXAMPLE 4.6.12   Let G be the directed graph

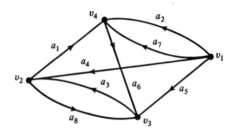

There are no directed trees rooted at v_1 since there are no arcs of G directed toward v_1.   On the other hand,

$$T_1 = (a_1, a_2, a_3) \qquad T_2 = (a_1, a_3, a_4)$$
$$T_3 = (a_1, a_3, a_5) \qquad T_4 = (a_1, a_3, a_7)$$

are the directed trees of G rooted at v_4.   Find the directed trees, if any, of G rooted at v_2 and at v_3.   ////

We will see that the number of trees of a directed graph rooted at a particular vertex can be expressed by means of determinants.   Recall that in Example 4.1.1 we defined the vertex-adjacency matrix $A = [a_{ij}]$ of a directed graph G by setting a_{ij} equal to the number of arcs of G directed from v_i to v_j.   Also, in the discussion preceding Example 4.1.15, we defined the out-degree matrix Δ of a directed graph.   For the graph of Example 4.6.12,

$$A = \begin{bmatrix} 0 & +1 & +1 & +2 \\ 0 & 0 & +1 & +1 \\ 0 & +1 & 0 & 0 \\ 0 & 0 & +1 & 0 \end{bmatrix} \qquad \Delta = \begin{bmatrix} +4 & 0 & 0 & 0 \\ 0 & +2 & 0 & 0 \\ 0 & 0 & +1 & 0 \\ 0 & 0 & 0 & +1 \end{bmatrix}$$

We set

$$M = \Delta - A$$

M is called the *tree matrix* of G.   The sum of the entries in each row of M is 0 and consequently

$$ME^{(m)} = 0$$

The tree matrix of the graph of Example 4.6.12. is

$$M = \begin{bmatrix} +4 & -1 & -1 & -2 \\ 0 & +2 & -1 & -1 \\ 0 & -1 & +1 & 0 \\ 0 & 0 & -1 & +1 \end{bmatrix}$$

In Theorem 4.6.14 we will show that the number of directed trees of a graph rooted at any vertex can be obtained from its tree matrix. First we will prove a special case of the general theorem which will be used later as a lemma.

Lemma 4.6.13   Let G be a directed $m \times n$ graph, let $\bar{G}$ be an $m \times (m-1)$ subgraph of G, and let $M = [m_{ij}]$ be the tree matrix of $\bar{G}$. For each vertex v_h of G, $1 \leq h \leq m$, the cofactor of m_{hh} in M is

$$\begin{cases} +1 & \text{if } \bar{G} \text{ is a directed tree rooted at } v_h \\ 0 & \text{otherwise} \end{cases}$$

PROOF   Partition M as

$$M = \begin{bmatrix} M_{11} & M_{12} \\ M_{21} & M_{22} \end{bmatrix}$$

where M_{11}, M_{12}, M_{21}, and $M_{22} = [m_{m\,m}]$ are $(m-1) \times (m-1)$, $(m-1) \times 1$, $1 \times (m-1)$, and 1×1 matrices, respectively. It is sufficient to prove that

$$\det M_{11} = \begin{cases} +1 & \text{if } \bar{G} \text{ is a directed tree rooted at } v_m \\ 0 & \text{otherwise} \end{cases}$$

If any vertex of G other than v_m does not have an arc of $\bar{G}$ directed from it, then the corresponding row of M (and of M_{11}) consists solely of 0s and we have $\det M_{11} = 0$, as required.

Henceforth we will assume that every vertex of G other than v_m does have an arc of $\bar{G}$ directed from it. Then $\bar{G}$ is a directed tree of G rooted at v_m if and only if $|\bar{G}|$ is a tree of $|G|$. All the diagonal entries of M_{11} are equal to $+1$. Moreover, each row of M_{11} contains one -1 unless the arc of $\bar{G}$ from the vertex corresponding to that row is directed toward v_m.

Let us suppose that $\bar{G}$ is not a directed tree of G rooted at v_m. Since $|\bar{G}|$ is not a tree of $|G|$, it follows that $|\bar{G}|$ (and therefore $\bar{G}$) is not connected. Let $\bar{G}_1$ be the component of $\bar{G}$ which does not contain v_m. M_{11} can be partitioned as

$$M_{11} = \begin{bmatrix} M_1 & 0 \\ 0 & M_2 \end{bmatrix}$$

where the rows and columns of M_1 correspond to the vertices of $\bar{G}_1$. Since

$$M_1 E = 0$$

M_1 is singular and consequently M_{11} is singular. Thus $\det M_{11} = 0$.

It remains to prove that if $\bar{G}$ is a directed tree of G rooted at v_m, then $\det M_{11} = +1$. Since all the diagonal entries of M_{11} are equal to $+1$, the term of $\det M_{11}$ corresponding to the identity permutation I, is $+1$. We will show that all other terms of $\det M_{11}$ are equal to 0. Suppose there exists a permutation σ, $\sigma \neq I$, for which the corresponding term of $\det M_{11}$, namely,

$$\text{sgn } \sigma \, m_{1\sigma(1)} \, m_{2\sigma(2)} \cdots m_{m-1 \, \sigma(m-1)}$$

is not 0.   Then there is an integer k_1 such that

$$\sigma(k_1) = k_2 \neq k_1$$

Hence $m_{k_1 k_2} = -1$, and $\bar{G}$ contains the arc $a_1 = (v_{k_1}, v_{k_2})$.
Let $\sigma(k_2) = k_3$.   Clearly $k_3 \neq k_2$, and two cases arise.

CASE 1   Suppose $k_3 = k_1$.   Then $m_{k_2 k_1} = -1$, and $\bar{G}$ contains the arc $a_2 = (v_{k_2}, v_{k_1})$.   However (a_1, a_2) is a loop.   Since $\bar{G}$ was assumed to be a tree, we have reached a contradiction.

CASE 2   Suppose $k_3 \neq k_1$.   Then $m_{k_2 k_3} = -1$, and $a_2 = (v_{k_2}, v_{k_3})$ is an arc of $\bar{G}$.   Let $\sigma(k_3) = k_4$.   Clearly $k_4 \neq k_2, k_3$.   If $k_4 = k_1$, we are again in case 1 and it follows that $\bar{G}$ contains a loop of length 3.   If $k_4 \neq k_1$, then this process is continued.   Eventually we must return to case 1 and reach the desired contradiction.   ////

Theorem 4.6.14   (Tutte, 1948)   Let M be the tree matrix of the directed $m \times n$ graph G.   For each h, $1 \leq h \leq m$, all the entries in the hth column of adj M are equal and are equal to the number of directed trees of G rooted at v_h.

PROOF   Since

$$ME^{(m)} = 0$$

M is singular and

$$M(\text{adj } M) = (\det M)I = 0$$

It follows that the entries in each column of adj M are equal.   For, if $r(M) \leq m - 2$, then adj $M = 0$, and if $r(M) = m - 1$, then $E^{(m)}$ forms a basis for $\mathbb{N}(M)$.
Let us partition M as

$$M = \begin{bmatrix} M_{11} & M_{12} \\ M_{21} & M_{22} \end{bmatrix}$$

where M_{11}, M_{12}, M_{21}, and M_{22} are $(m-1) \times (m-1)$, $(m-1) \times 1$, $1 \times (m-1)$, and 1×1 matrices, respectively. In order to prove the theorem it is sufficient to show that the number of directed trees of G rooted at v_m is equal to det M_{11}.

The number of directed trees of G rooted at v_m does not change when all arcs of the graph directed from v_m are removed. Moreover, when these arcs are removed, only the last row of M is affected. Thus we will henceforth assume that these arcs have been deleted, and M has the form

$$M = \begin{bmatrix} M_{11} & M_{12} \\ 0 & 0 \end{bmatrix}$$

Since there are Δ_i arcs of G directed from v_i [where Δ_i is the entry in the (i, i) position of Δ], there are $\Delta_1\Delta_2 \cdots \Delta_{m-1}$ $m \times (m-1)$ subgraphs of G each of which contains one arc directed from each vertex of G other than v_m. For convenience, set

$$\Delta_1\Delta_2 \cdots \Delta_{m-1} = d$$

Let us denote these graphs by $\bar{G}_1$, $\bar{G}_2$, ..., $\bar{G}_d$ and the corresponding tree matrices by $M^{(1)}$, $M^{(2)}$, ..., $M^{(d)}$. All the entries of the last row of each $M^{(q)}$, $q = 1, 2, \ldots, d$, are equal to 0. Partition $M^{(q)}$ as

$$M^{(q)} = \begin{bmatrix} M_{11}^{(q)} & M_{12}^{(q)} \\ 0 & 0 \end{bmatrix}$$

where $M_{11}^{(q)}$ and $M_{12}^{(q)}$ are $(m-1) \times (m-1)$ and $(m-1) \times 1$ matrices, respectively. In Lemma 4.6.13 we proved that

$$\det M_{11}^{(q)} = \begin{cases} +1 & \text{if } \bar{G}^{(q)} \text{ is a directed tree rooted at } v_m \\ 0 & \text{otherwise} \end{cases}$$

In order to complete the proof of the theorem it remains to show that

$$\det M_{11} = \sum_{q=1}^{d} \det M_{11}^{(q)} \qquad (4.6.15)$$

We will prove that equation (4.6.15) follows directly from the repeated use of Lemma 1.9.19. First let us apply Lemma 1.9.19 to the first row of M_{11}. The entry in the $(1, 1)$ position of M_{11} is Δ_1, and the remaining entries of the first row of M_{11} are nonpositive integers whose sum is at most $-\Delta_1$. The first row of M_{11} can be expressed (in exactly one way) as a sum of Δ_1 $(m-1)$-dimensional row vectors, say X_1, X_2, ..., X_{Δ_1}, each of which has a $+1$ in the first position and the remaining entries are either all 0s or consist of $m-2$ 0s and a single -1. There is a one-to-one correspondence between arcs of G directed from v_1 and the vectors $X_1, X_2, \ldots, X_{\Delta_1}$ such that:

1 $a_i = (v_1, v_m)$ corresponds to

$$X_i = [+1 \quad 0 \quad 0 \quad \cdots \quad 0]$$

2 $a_i = (v_1, v_k)$, $k \neq m$, corresponds to a vector X_i in which the first entry is $+1$, the kth entry is -1, and all other entries are 0.

Let

$$\tilde{M}^{(1)}, \tilde{M}^{(2)}, \ldots, \tilde{M}^{(\Delta_1)} \qquad (4.6.16)$$

be the square matrices of order $m - 1$ such that:

1 The first row of $\tilde{M}^{(q)}$ is X_q, $q = 1, 2, \ldots, \Delta_1$.
2 Each row of $\tilde{M}^{(q)}$, other than the first, is equal to the corresponding row of M_{11}.

According to Lemma 1.9.19,

$$\det M_{11} = \sum_{q=1}^{\Delta_1} \det \tilde{M}^{(q)}$$

Now let us apply the same procedure to the second row of each of the matrices of (4.6.16). Let $\tilde{M}^{(q)}$ be one of these matrices. For convenience we will set $q = 1$. From $\tilde{M}^{(1)}$ we construct Δ_2 square matrices of order $m - 1$ by expressing the second row of $\tilde{M}^{(1)}$ as a sum of Δ_2 vectors, each with a $+1$ in the second position, and either all the remaining entries are 0s or one of the remaining entries is a -1 and the rest are 0s. Denote these matrices by $\tilde{\tilde{M}}^{(1)}, \tilde{\tilde{M}}^{(2)}, \ldots, \tilde{\tilde{M}}^{(\Delta_2)}$. Then, by Lemma 1.9.19,

$$\det \tilde{M}^{(1)} = \sum_{s=1}^{\Delta_2} \det \tilde{\tilde{M}}^{(s)}$$

In this way we construct a total of $\Delta_1 \Delta_2$ matrices [Δ_2 matrices from each of the Δ_1 matrices of (4.6.16)], say

$$\tilde{\tilde{M}}^{(1)}, \tilde{\tilde{M}}^{(2)}, \ldots, \tilde{\tilde{M}}^{(\Delta_2)}, \ldots, \tilde{\tilde{M}}^{(\Delta_1 \Delta_2)} \qquad (4.6.17)$$

Again, by Lemma 1.9.19

$$\det M_{11} = \sum_{s=1}^{\Delta_1 \Delta_2} \det \tilde{\tilde{M}}^{(s)}$$

Moreover, there is a one-to-one correspondence between the matrices of (4.6.17) and all pairs of arcs of G, (a_i, a_j), such that a_i is directed from v_1 and a_j is directed from v_2.

Continuing this process, we express det M_{11} as the sum of the determinants of d square matrices of order $m - 1$. These are, in fact, the matrices $M_{11}^{(1)}$, $M_{11}^{(2)}$, ..., $M_{11}^{(d)}$ of equation (4.6.15).   ////

EXAMPLE 4.6.18   Let us illustrate the successive steps of the proof of Theorem 4.6.14 using the graph of Example 4.6.12.   When those arcs of G directed from v_4, (that is, a_6) are removed from G, the resulting graph is

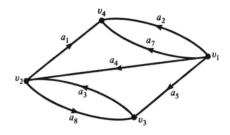

and the tree matrix is

$$\begin{bmatrix} +4 & -1 & -1 & -2 \\ 0 & +2 & -1 & -1 \\ 0 & -1 & +1 & 0 \\ 0 & 0 & 0 & 0 \end{bmatrix}$$

Then, maintaining the notation used in the proof of Theorem 4.6.14,

$$M_{11} = \begin{bmatrix} +4 & -1 & -1 \\ 0 & +2 & -1 \\ 0 & -1 & +1 \end{bmatrix}$$

and the matrices of (4.6.16) are

$$\tilde{M}_{11}^{(1)} = \begin{bmatrix} +1 & -1 & 0 \\ 0 & +2 & -1 \\ 0 & -1 & +1 \end{bmatrix} \qquad \tilde{M}_{11}^{(2)} = \begin{bmatrix} +1 & 0 & -1 \\ 0 & +2 & -1 \\ 0 & -1 & +1 \end{bmatrix}$$

$$\tilde{M}_{11}^{(3)} = \begin{bmatrix} +1 & 0 & 0 \\ 0 & +2 & -1 \\ 0 & -1 & +1 \end{bmatrix} = \tilde{M}_{11}^{(4)}$$

At the next stage we have the matrices of (4.6.17)

$$\tilde{\tilde{M}}_{11}^{(1)} = \begin{bmatrix} +1 & -1 & 0 \\ 0 & +1 & -1 \\ 0 & -1 & +1 \end{bmatrix} \quad \tilde{\tilde{M}}_{11}^{(2)} = \begin{bmatrix} +1 & -1 & 0 \\ 0 & +1 & 0 \\ 0 & -1 & +1 \end{bmatrix}$$

$$\tilde{\tilde{M}}_{11}^{(3)} = \begin{bmatrix} +1 & 0 & -1 \\ 0 & +1 & -1 \\ 0 & -1 & +1 \end{bmatrix} \quad \tilde{\tilde{M}}_{11}^{(4)} = \begin{bmatrix} +1 & 0 & -1 \\ 0 & +1 & 0 \\ 0 & -1 & +1 \end{bmatrix}$$

$$\tilde{\tilde{M}}_{11}^{(5)} = \begin{bmatrix} +1 & 0 & 0 \\ 0 & +1 & -1 \\ 0 & -1 & +1 \end{bmatrix} = M_{11}^{(7)} \quad M_{11}^{(6)} = \begin{bmatrix} +1 & 0 & 0 \\ 0 & +1 & 0 \\ 0 & -1 & +1 \end{bmatrix} = M_{11}^{(8)}$$

Since $\Delta_3 = +1$, another step is not needed. It is easily seen that

$$\det \tilde{\tilde{M}}_{11}^{(1)} = \det \tilde{\tilde{M}}_{11}^{(3)} = \det \tilde{\tilde{M}}_{11}^{(5)} = \det \tilde{\tilde{M}}_{11}^{(7)} = 0$$

and $\qquad \det \tilde{\tilde{M}}_{11}^{(2)} = \det \tilde{\tilde{M}}_{11}^{(4)} = \det \tilde{\tilde{M}}_{11}^{(6)} = \det \tilde{\tilde{M}}_{11}^{(8)} = +1$

G has four directed trees rooted at v_4. These are given in Example 4.6.12. However, we have also shown that G contains eight subgraphs which consist of one arc from each vertex of G other than v_4 (including the four trees). Find them (a) directly from the graph and (b) by using the matrices $\tilde{\tilde{M}}_{11}^{(q)}$, $q = 1, 2, \ldots, 8$. ////

There is an interesting relation between the directed trees and the Euler lines of a graph. In Sec. 4.2 we discussed undirected graphs which contain an Euler line. It is natural to extend the same idea to a directed graph. A directed path from a vertex to itself in a directed graph is called a *directed Euler line* if it contains all the arcs of the graph. For directed Euler lines the analog of Theorem 4.2.10 is:

Theorem 4.6.19 A connected directed graph contains a directed Euler line if and only if for each vertex v_k, the in degree δ_k is equal to the out degree Δ_k.

PROOF It is clear that a directed Euler line leaves each vertex of a graph as many times as it enters it. Thus, if a directed Euler line exists, the in degree and the out degree of every vertex are equal.

The proof that a connected directed graph in which

$$\delta_k = \Delta_k$$

for each vertex v_k has a directed Euler line is similar to the proof of Theorem 4.2.10 and is left as an exercise. ////

Corollary 4.6.20 In a connected directed graph for which:

1 $\Delta_1 - \delta_1 = +1$
2 $\Delta_2 - \delta_2 = -1$
3 $\Delta_i = \delta_i$ for $i \neq 1,2$

there is a directed path from v_1 to v_2 which contains all the arcs of the graph.

Let v_k be a vertex of a directed graph G and let P be a directed Euler line from v_k to itself. P induces an ordering on the arcs of G. We say that a_i *precedes* a_j with respect to P if it occurs before a_j in P. Similarly, for each vertex v_l of G, P induces an ordering on the Δ_l arcs of G directed from v_l; that is, $a_i = (v_l, v_{h_1})$ *precedes* $a_j = (v_l, v_{h_2})$ with respect to P if a_i occurs before a_j in P. For each vertex v of G, $v \neq v_k$, the arc by which P leaves v for the last time is called the *last exit* of v with respect to P. Note that v_k does not have a last exit with respect to P. The relation between directed trees and directed Euler lines of a graph is described in the next two lemmas.

Lemma 4.6.21 Let P be a directed Euler line from v_k to itself in the directed $m \times n$ graph G. The set of last exits with respect to P is a directed tree of G rooted at v_k.

PROOF Let the set of last exits of P be denoted by T. T contains $m - 1$ arcs, one directed from each vertex of G except v_k. In order to prove that T is a directed tree rooted at v_k it remains only to show that T does not contain a loop.

Let us suppose that L is a loop of G contained in T. Since every vertex in T has one arc directed from it, L is a directed loop. Furthermore, v_k does not belong to L because T does not contain an arc directed from v_k. Let us denote the arc of L which occurs last in P by a_i and set $a_i = (v_{h_1}, v_{h_2})$. Since L is a directed loop, it also contains an arc directed from v_{h_2}, say $a_j = (v_{h_2}, v_{h_3})$. According to the way in which T was formed, a_j is the last exit of v_{h_2} with respect to P. However, a_i occurs in P after a_j, and the arc in P immediately after a_i also leaves v_{h_2}. We have now reached a contradiction, and consequently we must conclude that T does not contain a loop.   ////

Lemma 4.6.21 describes a method for obtaining a directed tree rooted at v_k from a directed Euler line from v_k to itself. However, different Euler lines will yield the same directed tree if they have the same set of last exits.

Lemma 4.6.22   Let G be a connected directed $m \times n$ graph such that

$$\Delta_i = \delta_i \qquad i = 1, 2, \ldots, m$$

and let T be a directed tree of G rooted at v_k. The number of directed Euler lines from v_k to itself for which T is the set of last exits is

$$\Delta_k \prod_{i=1}^{m} [(\Delta_i - 1)!] \qquad (4.6.23)$$

PROOF   At each vertex of G we order the arcs directed from that vertex. The only requirement imposed on the ordering is that each of the arcs of T receive the highest number at the vertex from which it is directed. Thus, if a_j belongs to T and is directed from v_i, then a_j must be the Δ_ith arc at v_i. Since T does not contain an arc directed from v_k, the number of orderings which satisfy this requirement is given by (4.6.23). If P is a directed Euler line from v_k to itself which has T as its set of last exits, then the ordering of the arcs at each vertex induced by P satisfies the requirement above. We wish to show that, conversely, every ordering with this property determines a unique directed Euler line whose set of last exits is T.

We begin by selecting the first arc at v_k, say (v_k, v_{h_1}). Next we choose the first arc at v_{h_1}, say (v_{h_1}, v_{h_2}). We then take the first arc at v_{h_2} which has not been used. (If $v_{h_2} \neq v_k$, the desired arc is the first arc at v_{h_2}. If $v_{h_2} = v_k$, we take the second arc at v_k since the first one has already been used.) Continuing in this way, we choose the arc at each vertex which is first among the arcs that have not been previously used. Clearly the ordering of the arcs at each vertex uniquely determines the path from v_k which we construct. This path must end at v_k since, for every other vertex, whenever it is entered, there is an unused arc directed from it. Also, when the path ends, there remain no unused arcs directed from v_k, or else we would continue the process. Since there are no unused arcs directed from v_k when the process ends, there are also no unused arcs directed toward v_k. Finally, it is clear that if all the arcs of G are used, then the arcs of T is the set of last exits of the path. It now remains to show that all arcs of G are used.

Suppose that there is an arc of G which is never selected in the process described in the last paragraph. Let this arc be directed from v_{l_1}. We have already shown that $v_{l_1} \neq v_k$. Because of the method of selecting arcs at each vertex, the arc of T directed from v_{l_1}, say (v_{l_1}, v_{l_2}), is also never selected. Since (v_{l_1}, v_{l_2}) is directed toward v_{l_2}, it follows that there is an arc of G directed from v_{l_2} which is never used. As before, the arc of T directed from v_{l_2}, say (v_{l_2}, v_{l_3}), is never used. There is a directed path in

T from v_{l_1} to v_k, and hence this argument leads eventually to an arc directed toward v_k which is never selected. This is a contradiction. Therefore every arc of G is used in the construction described in the last paragraph, and the path we have formed in this way is a directed Euler line from v_k to itself. ////

If

$$P = \{a_1 = (v_{h_1}, v_{h_2}), a_2, \ldots, a_{k-1}, a_k = (v_{l_1}, v_{l_2}), \ldots, a_n\}$$

is a directed Euler line from v_{h_1} to itself, then

$$P' = \{a_k = (v_{l_1}, v_{l_2}), \ldots, a_n, a_1 = (v_{h_1}, v_{h_2}), a_2, \ldots, a_{k-1}\}$$

is a directed Euler line from v_{l_1} to itself. As in the case of undirected graphs, we say that two directed Euler lines of a graph belong to the same *circuit* if one of them can be obtained from the other in this way. A circuit in an $m \times n$ directed graph consists of n directed Euler lines, and each directed Euler line in a circuit is uniquely determined by specifying its first arc.

Theorem 4.6.24 (van Aardenne-Ehrenfest and de Brujin, 1950)
If G is a connected directed $m \times n$ graph in which

$$\Delta_i = \delta_i \qquad i = 1, 2, \ldots, m$$

then the number of directed trees of G rooted at any vertex is equal to the number of directed trees rooted at any other vertex. Denote this number by t. Moreover, the number of circuits in G is

$$t \prod_{i=1}^{m} [(\Delta_i - 1)!]$$

PROOF Let v_k be any vertex of G and let T be a directed tree rooted at v_k. By Lemma 4.6.22 there are

$$\Delta_k \prod_{i=1}^{m} [(\Delta_i - 1)!]$$

directed Euler lines from v_k to itself whose set of last exits is T. If G contains t_k directed trees rooted at v_k, then there are a total of

$$t_k \Delta_k \prod_{i=1}^{m} [(\Delta_i - 1)!]$$

directed Euler lines in G from v_k to itself. Since v_k has Δ_k arcs directed from it and each of these can be the first arc of a directed Euler line, it follows that each circuit contains Δ_k directed Euler lines from v_k to itself. Hence there are

$$t_k \prod_{i=1}^{m} [(\Delta_i - 1)!]$$

circuits in G. The number of circuits in G is independent of the vertex v_k and therefore

$$t_1 = t_2 = \cdots = t_m = t \qquad ////$$

See Exercise 8 for a different proof of Theorem 4.6.24.

EXERCISES

1 Let $G=$

G:

Find the tree matrix of G, the number of directed trees rooted at each vertex, the number of Euler lines from each vertex to itself, and the number of circuits. Find the number of trees in the underlying undirected graph of G.

2 Let m be an odd positive integer and let G be the complete graph with m vertices. For $1 \leq i < j \leq m$, define the orientation of the arc between v_i and v_j by

$$a = \begin{cases} (v_i, v_j) & \text{if } i - j \text{ is even} \\ (v_j, v_i) & \text{if } i - j \text{ is odd} \end{cases}$$

Prove that G has a directed Euler line.

3 Let $m = 2k + 1$ and let G be the complete graph with m vertices. For $1 \leq i < j \leq m$, define the orientation of the arc between v_i and v_j by

$$a = \begin{cases} (v_i, v_j) & \text{if } |i - j| \leq k \\ (v_j, v_i) & \text{if } |i - j| > k \end{cases}$$

Prove that G has a directed Euler line. Show that for $m = 5$ the graph described in Exercise 2 can be transformed into the graph described above by a renumbering of the vertices.

4   Show that a complete undirected graph with m vertices, $m \geq 2$, contains $m^{(m-2)}$ trees. [*Hint:* Let $f(n, a)$ denote the determinant of a square matrix of order n which has a's in the diagonal positions and -1s elsewhere. Prove that

$$f(n, a) = af(n - 1, a) - (n - 1)(a + 1)^{n-2}$$

and $f(n, n - 1) = 0$.

5   Show that if one edge is removed from a complete undirected graph with m vertices, $m \geq 2$, the resulting graph has $(m - 2)m^{(m-3)}$ trees. [*Hint:* Let $h(n, a)$ denote the determinant of a square matrix of order n which has $a - 1$ in the $(1, 1)$ position, a's in the remaining diagonal positions, and -1s elsewhere. Prove that

$$h(n, a) = (a - 1)f(n - 1, a) - (n - 1)(a + 1)^{n-2}$$

where $f(n, a)$ is defined in Exercise 4.]

6   Present an alternate proof of Exercise 5 by finding the number of trees in a complete undirected graph that contain a given arc.

7   Find the number of trees in a complete graph with m vertices which contain the arc $a = (v_1, v_2)$ but no other arc incident with v_1. Find the number of trees which contain both a and $a' = (v_1, v_3)$ but no other arcs incident with v_1. (*Hint:* Use Exercises 4 and 5.)

8   Let G be a connected graph. Prove that the number of directed trees rooted at a vertex is the same for all vertices if and only if

$$EM = 0$$

where M is the tree matrix of G, and that this equation is satisfied if and only if G contains a directed Euler line.

9   Prove that a directed complete graph with m vertices in which each arc is directed at the vertex with a lower subscript has $(m - 1)!$ directed trees rooted at v_1 and no directed trees rooted at any other vertex.

REFERENCES

Euler solved the Königsberg bridge problem in 1736, and since then there has been a vast amount written about graphs. However, until recently graph theory as an independent branch of mathematics and as a part of the mathematical curriculum has been largely ignored. Undoubtedly much of the impetus for the renewed interest in graphs which has taken place in recent years comes from the many varied applications to which graph theory has been put. In addition to the classical applications to the physical sciences, (e.g., electric network theory), graph theory now plays a prominent role in the social sciences, (e.g., communication and transportation problems), and in computer technology.   Moreover, it is only the advent of large-scale computers which has made many of the algorithms associated with solving graphical problems feasible.

Several texts present mathematical treatments of graph theory.   Harary [5] is primarily concerned with undirected graphs, whereas Harary, Norman, and Cartwright [6] is devoted to directed graphs.   Busacker and Saaty [3] treat both undirected and directed graphs and present a great variety of applications.   Seshu and Reed [10] develop the fundamentals of electrical analysis and network synthesis in terms of graphs.   To a large extent they follow Kirchhoff's development of Kirchhoff's laws. In Example 4.1.9 we discussed the maximal flow in a network and the minimum cut necessary to prevent any flow.   Ford and Fulkerson [4] combine graph theory with linear programming to solve these and related problems dealing with minimal cost and optimal assignment of flows.   Massey [8] applies the methods of algebraic topology to graphs and illustrates the relation between a graph and its fundamental group.

1   Berge, C.: "The Theory of Graphs and its Applications," trans. from the French by A. Doig, Methuen & Co., Ltd., London, 1962.

2   Brooks, R., C. Smith, A. Stone, and W. Tutte: The Dissection of Rectangles into Squares, *Duke Math. J.*, **7**: 312–340 (1940).

3   Busacker, R., and T. Saaty: "Finite Graphs and Networks: An Introduction with Applications," McGraw-Hill Book Company, New York, 1965.

4   Ford, L., and D. Fulkerson: "Flows in Networks," Princeton University Press, Princeton, N.J., 1962.

5   Harary, F.: "Graph Theory," Addison-Wesley Publishing Company, Inc., Reading, Mass., 1969.

6   Harary, F., R. Norman, and D. Cartwright: "Structural Models: An Introduction to the Theory of Directed Graphs," John Wiley & Sons, Inc., New York, 1965.

7   Kuratowski, K.: Sur le problème des courbes gauches en topologie, *Fund. Math.*, **15**: 271–283 (1930).

8   Massey, W. "Algebraic Topology: An Introduction," Harcourt, Brace & World, Inc., New York, 1967.

9   Ore, O., "Graphs and Their Uses," Random House, New York, 1963.

10 Seshu, S., and M. Reed: "Linear Graphs and Electrical Networks," Addison-Wesley Publishing Company, Inc., Reading, Mass., 1961.

11 Whitney, H.: Non-separable and Planar Graphs, *Trans. Am. Math. Soc.*, **34**: 339–362 (1932).

ANSWERS TO SELECTED EXERCISES

CHAPTER 0 FIELDS

0.1 Fields

1 (a), (b), (e)

CHAPTER 1 MATRICES

1.2 The Arithmetic of Matrices

1 (a) $\begin{bmatrix} x & y \\ 0 & x \end{bmatrix}$ x, y arbitrary (b) $\begin{bmatrix} x & y \\ 0 & x+y \end{bmatrix}$ x, y arbitrary

3 (a) 0, (b) 0, (c) 0, (d) 0, (e) $\dfrac{1}{12}\begin{bmatrix} 6 & -2 \\ 3 & 1 \end{bmatrix}$

4 (a) $\begin{bmatrix} 2x & 2y \\ -x & -y \end{bmatrix}$ x, y arbitrary

5 $x\begin{bmatrix} yz & y^2 \\ -z^2 & -yz \end{bmatrix}$ x, y, z arbitrary

1.3 Vector Spaces

1 $\begin{bmatrix} x \\ y \\ z \end{bmatrix} = \tfrac{1}{2}(3x + y - z)\begin{bmatrix} 1 \\ 1 \\ 0 \end{bmatrix} + \tfrac{1}{2}(-x + y + z)\begin{bmatrix} 0 \\ 1 \\ 1 \end{bmatrix} + \tfrac{1}{2}(-x - y + z)\begin{bmatrix} 1 \\ 2 \\ -1 \end{bmatrix}$

2 Since $2A_1 + 3A_2 - A_3 = 0$, $\mathbb{S}$ is not a basis for $\mathbb{V}$. Any two of these three vectors form a basis for $\mathbb{V}$.

3 $\left\{\begin{bmatrix} 0 & 0 \\ 1 & 1 \end{bmatrix}, \begin{bmatrix} 0 & 1 \\ -1 & 1 \end{bmatrix}, \begin{bmatrix} 1 & 1 \\ 1 & 0 \end{bmatrix}\right\}$ is a basis for $\mathbb{V} + \mathbb{W}$; $\left\{\begin{bmatrix} 1 & 2 \\ 0 & 1 \end{bmatrix}\right\}$ is a basis for $\mathbb{V} \cap \mathbb{W}$.

10 (a), (b), (d), (f)

1.4 The Row Space and the Column Space of a Matrix

1 Any two rows of A form a basis for $\mathbb{R}(A)$. Any two columns of A form a basis for $\mathbb{C}(A)$.

2 $A = \begin{bmatrix} -\frac{1}{2} & -2 & \frac{3}{2} \\ 0 & -1 & 2 \end{bmatrix}$ is the only matrix which satisfies $AB = C$.

3 Yes.

5 $\left\{ \begin{bmatrix} 4 \\ -1 \\ 5 \end{bmatrix} \right\}$ *6* $x \neq 9$ *7* $x = 1, y = \frac{1}{2}$

1.5 The Steinitz Replacement Theorem

1 (*a*) Either of the first two row of A forms a basis for $\mathbb{R}(A)$. Any column of A forms a basis for $\mathbb{C}(A)$. One basis for $\mathbb{N}(A)$ is

$$\left\{ \begin{bmatrix} 3 \\ -1 \\ 0 \end{bmatrix}, \begin{bmatrix} 2 \\ 0 \\ 1 \end{bmatrix} \right\}$$

2 (*a*) The row rank and the column rank are both 2, the nullity is 5.

3 $\left\{ \begin{bmatrix} 0 \\ 2 \\ -1 \\ 0 \end{bmatrix}, \begin{bmatrix} 0 \\ 1 \\ 1 \\ 0 \end{bmatrix}, \begin{bmatrix} 1 \\ 0 \\ 0 \\ 0 \end{bmatrix}, \begin{bmatrix} 0 \\ 0 \\ 0 \\ 1 \end{bmatrix} \right\}$

5 (*a*) $\left\{ \begin{bmatrix} 1 & 0 \\ 0 & 1 \end{bmatrix}, \begin{bmatrix} 0 & 1 \\ 4 & 0 \end{bmatrix} \right\}$

(*b*) Since $\{X \mid XA = BX\} = 0$, the empty set is the only basis.

6 (*a*) $\{[-1 \quad 2 \quad 0]\}$

(*b*) $\{[1 \quad 1 \quad 0], [0 \quad 1 \quad 0], [1 \quad 0 \quad 1]\}$. No.

(*c*) The empty set.   (*d*) $\left\{ \begin{bmatrix} 0 \\ 0 \\ 1 \end{bmatrix}, \begin{bmatrix} 2 \\ 1 \\ -2 \end{bmatrix} \right\}$. Yes.

7 $\left\{ \begin{bmatrix} 5 \\ 7 \\ 0 \end{bmatrix} \right\}$

1.6 Nonsingular Matrices

1 (*a*) $[x \quad y \quad z] = (x - y)[1 \quad 0 \quad 1] + (x - z)[1 \quad 1 \quad 0]$
$+ (-x + y - z)[1 \quad 1 \quad 1]$

2 $A^{-1} = \frac{1}{3}\begin{bmatrix} 0 & -3 \\ 1 & 1 \end{bmatrix}$

3 $A^{-1} = \begin{bmatrix} 1 & 1 & -1 \\ -1 & 0 & 1 \\ 0 & -1 & 1 \end{bmatrix}$

4 (a) The column rank is 2, the nullity is 5.

(b) The column rank is 2, the nullity is 0.

5 No.   **6** $x \neq 9$.

1.7 The Elementary Operations

1 (a) $\begin{bmatrix} 0 & 0 & 1 & 1 \\ 2 & -4 & 2 & 0 \\ -1 & 2 & 0 & 1 \end{bmatrix} \xrightarrow{C_{13}} \begin{bmatrix} 1 & 0 & 0 & 1 \\ 2 & -4 & 2 & 0 \\ 0 & 2 & -1 & 1 \end{bmatrix} \xrightarrow{R_{21}(-2)}$

$\begin{bmatrix} 1 & 0 & 0 & 1 \\ 0 & -4 & 2 & -2 \\ 0 & 2 & -1 & 1 \end{bmatrix} \xrightarrow{C_{41}(-1)} \begin{bmatrix} 1 & 0 & 0 & 0 \\ 0 & -4 & 2 & -2 \\ 0 & 2 & -1 & 1 \end{bmatrix} \xrightarrow{R_{32}(\frac{1}{2})}$

$\begin{bmatrix} 1 & 0 & 0 & 0 \\ 0 & -4 & 2 & -2 \\ 0 & 0 & 0 & 0 \end{bmatrix} \xrightarrow{C_2(-\frac{1}{4})} \begin{bmatrix} 1 & 0 & 0 & 0 \\ 0 & 1 & 2 & -2 \\ 0 & 0 & 0 & 0 \end{bmatrix} \xrightarrow{C_{32}(-2)}$

$\begin{bmatrix} 1 & 0 & 0 & 0 \\ 0 & 1 & 0 & -2 \\ 0 & 0 & 0 & 0 \end{bmatrix} \xrightarrow{C_{42}(2)} \begin{bmatrix} 1 & 0 & 0 & 0 \\ 0 & 1 & 0 & 0 \\ 0 & 0 & 0 & 0 \end{bmatrix}$

3 (a) The first two rows of A form a basis for $\mathbb{R}(A)$.

(b) The first two columns of A form a basis for $\mathbb{C}(A)$.

(c) $\left\{ \begin{bmatrix} 0 \\ 1 \\ 2 \\ 0 \\ 0 \end{bmatrix}, \begin{bmatrix} 1 \\ 0 \\ 0 \\ -1 \\ -1 \end{bmatrix}, \begin{bmatrix} 0 \\ 0 \\ 1 \\ -1 \\ 1 \end{bmatrix} \right\}$ is a basis for $\mathbb{N}(A)$.

5 No, because the last row of any such matrix would consist only of zeros.

6 $C_{rs} = R_{rs}$, $C_r(\lambda) = R_r(\lambda)$, $C_{rs}(\alpha) = R_{sr}(\alpha)$.

1.8 Permutations

1 (a) $\begin{pmatrix} 1 & 2 & 3 & 4 & 5 \\ 5 & 4 & 1 & 3 & 2 \end{pmatrix}$ (b) $\begin{pmatrix} 1 & 2 & 3 & 4 & 5 \\ 1 & 2 & 4 & 5 & 3 \end{pmatrix}$

(f) $\begin{pmatrix} 1 & 2 & 3 & 4 & 5 \\ 2 & 1 & 5 & 4 & 3 \end{pmatrix} = (1 \quad 2)(3 \quad 5)$

2 $(1 \quad 3)(1 \quad 7)(2 \quad 6)(2 \quad 4)(2 \quad 8)(2 \quad 9)$ is one expression for σ as a product of transpositions. σ is even.

3 If $\sigma(a) = b$, then let $\tilde{\sigma}(b) = a$.

4 (a) $\begin{pmatrix} 1 & 2 & 3 & 4 & 5 \\ 2 & 1 & 4 & 5 & 3 \end{pmatrix}$ (b) $\begin{pmatrix} 1 & 2 & 3 & 4 & 5 \\ 5 & 3 & 1 & 4 & 2 \end{pmatrix}$

5 $\sigma^{-1} = (2 \ 9)(2 \ 8)(2 \ 4)(2 \ 6)(1 \ 7)(1 \ 3)$

16 (a) $\begin{bmatrix} 0 & 1 & 0 & 0 & 0 \\ 1 & 0 & 0 & 0 & 0 \\ 0 & 0 & 0 & 1 & 0 \\ 0 & 0 & 0 & 0 & 1 \\ 0 & 0 & 1 & 0 & 0 \end{bmatrix}$ (b) $\begin{bmatrix} 0 & 0 & 0 & 0 & 1 \\ 0 & 0 & 1 & 0 & 0 \\ 1 & 0 & 0 & 0 & 0 \\ 0 & 0 & 0 & 1 & 0 \\ 0 & 1 & 0 & 0 & 0 \end{bmatrix}$ (c) $\begin{bmatrix} 0 & 0 & 1 & 0 & 0 \\ 0 & 0 & 0 & 0 & 1 \\ 0 & 0 & 0 & 1 & 0 \\ 0 & 1 & 0 & 0 & 0 \\ 1 & 0 & 0 & 0 & 0 \end{bmatrix}$

17 (a) $\begin{bmatrix} 0 & 1 & 0 & 0 & 0 \\ 1 & 0 & 0 & 0 & 0 \\ 0 & 0 & 0 & 0 & 1 \\ 0 & 0 & 1 & 0 & 0 \\ 0 & 0 & 0 & 1 & 0 \end{bmatrix}$ (b) $\begin{bmatrix} 0 & 0 & 1 & 0 & 0 \\ 0 & 0 & 0 & 0 & 1 \\ 0 & 1 & 0 & 0 & 0 \\ 0 & 0 & 0 & 1 & 0 \\ 1 & 0 & 0 & 0 & 0 \end{bmatrix}$ (c) $\begin{bmatrix} 0 & 0 & 0 & 0 & 1 \\ 0 & 0 & 0 & 1 & 0 \\ 1 & 0 & 0 & 0 & 0 \\ 0 & 0 & 1 & 0 & 0 \\ 0 & 1 & 0 & 0 & 0 \end{bmatrix}$

1.9 Determinants

1 (a) -3,   (b) -9,   (c) 4,   (e) 3

3 $2, -3$

4 $\dfrac{1}{4}\begin{bmatrix} -5 & -7 & 10 \\ 1 & -1 & 2 \\ -7 & -9 & 14 \end{bmatrix}$ 5 $\begin{bmatrix} -7 & -7 & 7 \\ 5 & 5 & -5 \\ 3 & 3 & -3 \end{bmatrix}$

6 $\det A = (-1)^{n+1} a_{1n} a_{21} a_{32} \cdots a_{n,\,n-1}$

8 (a) $\det \alpha A = \alpha^n \det A$ (b) $\operatorname{adj} \alpha A = \alpha^{n-1} \operatorname{adj} A$

1.10 Linear Transformations and Invariant Spaces

1 $\begin{bmatrix} & & & & 1 \\ & & & 1 & \\ & & 1 & & \\ & \cdots & & & \\ 1 & & & & \end{bmatrix}$ 2 $\begin{bmatrix} 0 & 1 & 0 & 0 & \cdots & 0 \\ 0 & 0 & 2 & 0 & \cdots & 0 \\ 0 & 0 & 0 & 3 & \cdots & 0 \\ \multicolumn{6}{c}{\dotfill} \\ 0 & 0 & 0 & 0 & \cdots & n-1 \end{bmatrix}$

3 $\mathscr{B}: \begin{bmatrix} 1 \\ 1 \\ 1 \end{bmatrix} \to \begin{bmatrix} 5 \\ 7 \\ -3 \end{bmatrix}$ $\mathscr{B}: \begin{bmatrix} 7 \\ -19 \\ -5 \end{bmatrix} \to \begin{bmatrix} 1 \\ 1 \\ 1 \end{bmatrix}$ $B = \begin{bmatrix} 1 & -1 & 5 \\ 3 & 0 & 4 \\ 0 & 1 & -4 \end{bmatrix}$

4 $\begin{bmatrix} 0 \\ 4 \\ 1 \end{bmatrix}$ 5 $P = \begin{bmatrix} x & y \\ 0 & y \end{bmatrix}$ $x, y \neq 0$ $Q = \begin{bmatrix} x & y \\ y & y \end{bmatrix}$ $x \neq y, y \neq 0$

8   $\lambda_1 = 0, \lambda_2 = 1, \lambda_3 = 2$

$$X_1 = \begin{bmatrix} -3 \\ 1 \\ 1 \end{bmatrix} \quad X_2 = \begin{bmatrix} 2 \\ -1 \\ 0 \end{bmatrix} \quad X_3 = \begin{bmatrix} 1 \\ 1 \\ -1 \end{bmatrix} \quad P = \begin{bmatrix} -3 & 2 & 1 \\ 1 & -1 & 1 \\ 1 & 0 & -1 \end{bmatrix}$$

$$Y_1 = [1 \quad 2 \quad 3] \qquad Y_2 = [1 \quad 1 \quad 2] \qquad Y_3 = [1 \quad 2 \quad 1]$$

$$Q = \begin{bmatrix} 1 & 2 & 3 \\ 1 & 1 & 2 \\ 1 & 2 & 1 \end{bmatrix}$$

9   $\lambda_1 = \lambda_2 = -1, \lambda_3 = -2$

$$X_1 = \begin{bmatrix} 1 \\ -1 \\ 0 \end{bmatrix} \quad X_2 = \begin{bmatrix} 0 \\ 0 \\ 1 \end{bmatrix} \quad X_3 = \begin{bmatrix} 3 \\ -4 \\ 3 \end{bmatrix} \quad P = \begin{bmatrix} 1 & 0 & 3 \\ -1 & 0 & -4 \\ 0 & 1 & 3 \end{bmatrix}$$

10   $\lambda_1 = -1, \lambda_2 = \lambda_3 = 2$ $X_1 = \begin{bmatrix} 0 \\ 1 \\ 2 \end{bmatrix} \quad X_2 = \begin{bmatrix} 1 \\ 0 \\ 1 \end{bmatrix} \quad$ No.

1.11   Polynomials

1   (a) $x^3 + x$,   (b) $x^3 + x$,   (c) 1
2   (a) $x^3 + x = (-\frac{2}{3}x - 1)(x^5 - x) + (\frac{1}{3}x - \frac{2}{3})(2x^5 - x^4 + 2x^3 - x^2)$
3   (a) $x^3 + x$,   (b) $1, 1 = 2x(x^3 + x) - 2x(x^3 - x) - (4x^2 - 1)$
5   (1a) $x^2(x - 1)(x + 1)(x^2 + 1)(2x - 1)$,
     (1b) $x^2(x - 1)(x + 1)(x^2 + 1)(x - 2)$,
     (1c) $(x^5 - x)(x^4 - 4x^2 + 4)$,
     (3a) $x^2(x - 1)(x + 1)(x^2 + 1)(2x - 1)(x - 2)$
8   (a) $x(x^2 - 1)^3 + 3x(x^2 - 1)^2 + (5x - 1)(x^2 - 1) + (4x + 2)$
     (b) $2(x - 2)^4 + 13(x - 2)^3 + 28(x - 2)^2 + 22(x - 2) - 3$
10   (a) $x^2 - 6x + 5$,   (b) $(x - 2)^2$,   (d) $x^2 + x$
11   (a) $x^3 - x^2$

1.12   The Decomposition Theorems

1   (a) $m(x) = x - 1$; the elementary divisors are $x - 1, x - 1$. Any two
     linearly independent vectors can be used to generate two one-dimen-
     sional cyclic spaces whose direct sum is $\mathbb{F}^2$.
   (b) $m(x) = (x - 1)(x - 2)$; the elementary divisors are $x - 1, x - 2$. $\mathbb{F}^2$ is
     the direct sum of the two one-dimensional cyclic spaces generated by
     $\begin{bmatrix} 1 \\ 0 \end{bmatrix}$ and $\begin{bmatrix} 0 \\ 1 \end{bmatrix}$.

(c) $m(x) = (x - 1)^2$; the elementary divisor is $(x - 1)^2$. $\mathbb{F}^2$ is an irreducible cyclic space. Any vector which is not a scalar multiple of $\begin{bmatrix} 0 \\ 1 \end{bmatrix}$ generates $\mathbb{F}^2$.

(f) $m(x) = x(x - 2)$; the elementary divisors are $x, x - 2$. $\mathbb{F}^2$ is the direct sum of the two one-dimensional cyclic spaces generated by $\begin{bmatrix} 1 \\ -1 \end{bmatrix}$ and $\begin{bmatrix} 1 \\ 1 \end{bmatrix}$.

2 (a) $m(x) = x^2 + 1$. When $\mathbb{F}$ is the complex field, the elementary divisors are $x - i$ and $x + i$. $\mathbb{F}^2$ is the direct sum of the two one-dimensional cyclic spaces generated by $\begin{bmatrix} 1 \\ 1 - i \end{bmatrix}$ and $\begin{bmatrix} 1 \\ 1 + i \end{bmatrix}$.

3 $m(x) = x(x - 2)(x + 1)$. The elementary divisors are $x, x - 2$ and $x + 1$. $\mathbb{F}^3$ is the direct sum of the three one-dimensional cyclic spaces generated by $\begin{bmatrix} 1 \\ 0 \\ 1 \end{bmatrix} \begin{bmatrix} 1 \\ -1 \\ 2 \end{bmatrix}$ and $\begin{bmatrix} 1 \\ -1 \\ 1 \end{bmatrix}$.

4 $m() = (x + 1)^2$. The elementary divisors are $(x + 1)^2$ and $x + 1$. $\mathbb{F}^3$ is the direct sum of the two-dimensional cyclic space generated by $\begin{bmatrix} 1 \\ 0 \\ 0 \end{bmatrix}$ and the one-dimensional cyclic space generated by $\begin{bmatrix} 1 \\ 0 \\ -1 \end{bmatrix}$.

5 $m(x) = (x - 3)^3$. The canonical basis for $\mathbb{F}^3$ with generator $Y = \begin{bmatrix} 1 \\ 0 \\ 0 \end{bmatrix}$ is

$$Y = \begin{bmatrix} 1 \\ 0 \\ 0 \end{bmatrix} \qquad (A - 3I)Y = \begin{bmatrix} -1 \\ 1 \\ 2 \end{bmatrix} \qquad (A - 3I)^2 Y = \begin{bmatrix} 1 \\ -1 \\ -1 \end{bmatrix}$$

The cyclic basis with generator Y is

$$Y = \begin{bmatrix} 1 \\ 0 \\ 0 \end{bmatrix} \qquad AY = \begin{bmatrix} 2 \\ 1 \\ 2 \end{bmatrix} \qquad A^2 Y = \begin{bmatrix} 4 \\ 5 \\ 11 \end{bmatrix}$$

1.13 The Canonical Forms

1 (a) A is already in both its rational canonical form and its classical canonical form.

(b) A is already in both its rational canonical form and its classical canonical form.

(c) To obtain the rational canonical form of A set $P = \begin{bmatrix} 0 & 1 \\ 1 & 1 \end{bmatrix}$. Then

$P^{-1}AP = \begin{bmatrix} 0 & -1 \\ 1 & 2 \end{bmatrix}$. To obtain the classical canonical form set $P = \begin{bmatrix} 0 & 1 \\ 1 & 0 \end{bmatrix}$. Then $P^{-1}AP = \begin{bmatrix} 1 & 0 \\ 1 & 1 \end{bmatrix}$.

(d) A is already in its classical canonical form. The rational canonical form of A is the same as the rational canonical form of the matrix of (c). To obtain the rational canonical form of A set $P = \begin{bmatrix} 1 & 1 \\ 0 & 1 \end{bmatrix}$. Then $P^{-1}AP = \begin{bmatrix} 0 & -1 \\ 1 & 2 \end{bmatrix}$.

2 (3) Both the rational canonical form and the classical canonical form of

$$A = \begin{bmatrix} -3 & 1 & 3 \\ 3 & -1 & -3 \\ -5 & 1 & 5 \end{bmatrix}$$

are obtained by setting

$$P = \begin{bmatrix} 1 & 1 & 1 \\ 0 & -1 & -1 \\ 1 & 2 & 1 \end{bmatrix}$$

Then

$$P^{-1}AP = \begin{bmatrix} 0 & 0 & 0 \\ 0 & 2 & 0 \\ 0 & 0 & -1 \end{bmatrix}$$

(4) The rational canonical form of A is obtained by setting

$$P = \begin{bmatrix} 1 & 0 & 1 \\ 0 & 1 & 0 \\ 0 & -1 & -1 \end{bmatrix}$$

Then

$$P^{-1}AP = \begin{bmatrix} 0 & -1 & 0 \\ 1 & -2 & 0 \\ 0 & 0 & -1 \end{bmatrix}$$

The classical canonical form of A is obtained by setting

$$P = \begin{bmatrix} 1 & 1 & 1 \\ 0 & 1 & 0 \\ 0 & -1 & -1 \end{bmatrix}$$

Then

$$P^{-1}AP = \begin{bmatrix} -1 & 0 & 0 \\ 1 & -1 & 0 \\ 0 & 0 & -1 \end{bmatrix}$$

3 $m(x) = x^2(x^2 - 7x - 2)(x - 1)$. The characteristic polynomial of A is $-x^4(x^2 - 7x - 2)(x - 1)$. The rational canonical form and the classical canonical form of A are both equal to

$$\begin{bmatrix} 0 & 0 & 0 & 0 & 0 & 0 & 0 \\ 0 & 0 & 0 & 0 & 0 & 0 & 0 \\ 0 & 0 & 0 & 0 & 0 & 0 & 0 \\ 0 & 0 & 1 & 0 & 0 & 0 & 0 \\ 0 & 0 & 0 & 0 & 0 & 2 & 0 \\ 0 & 0 & 0 & 0 & 1 & 7 & 0 \\ 0 & 0 & 0 & 0 & 0 & 0 & 1 \end{bmatrix}$$

4 $m(x) = (x - 2)^3$. The characteristic polynomial of A is $-(x - 2)^7$. The rational canonical form and the classical canonical form are, respectively

$$\begin{bmatrix} 0 & -4 & 0 & 0 & 0 & 0 & 0 \\ 1 & 4 & 0 & 0 & 0 & 0 & 0 \\ 0 & 0 & 0 & -4 & 0 & 0 & 0 \\ 0 & 0 & 1 & 4 & 0 & 0 & 0 \\ 0 & 0 & 0 & 0 & 0 & 0 & 8 \\ 0 & 0 & 0 & 0 & 1 & 0 & -12 \\ 0 & 0 & 0 & 0 & 0 & 1 & 6 \end{bmatrix} \quad \text{and} \quad \begin{bmatrix} 2 & 0 & 0 & 0 & 0 & 0 & 0 \\ 1 & 2 & 0 & 0 & 0 & 0 & 0 \\ 0 & 0 & 2 & 0 & 0 & 0 & 0 \\ 0 & 0 & 1 & 2 & 0 & 0 & 0 \\ 0 & 0 & 0 & 0 & 2 & 0 & 0 \\ 0 & 0 & 0 & 0 & 1 & 2 & 0 \\ 0 & 0 & 0 & 0 & 0 & 1 & 2 \end{bmatrix}$$

CHAPTER 2   THE THEORY OF GAMES

2.1 Examples of Games

1 (a) $\left(\bar{P} = [\frac{2}{3} \quad \frac{1}{3}], v = \frac{1}{3}, \bar{Q} = \begin{bmatrix} \frac{2}{3} \\ \frac{1}{3} \end{bmatrix} \right)$

(b) $\left(\bar{P} = [1 \quad 0], v = 1, \bar{Q} = \begin{bmatrix} q \\ 1 - q \end{bmatrix} \text{ for any } q, 0 \leq q \leq \frac{2}{3} \right)$

(c) $\left(\bar{P} = [1 \quad 0], v = 0, \bar{Q} = \begin{bmatrix} 1 \\ 0 \end{bmatrix} \right)$

(d) $\left(\bar{P} = [\frac{1}{2} \quad \frac{1}{2}], v = \frac{1}{2}, \bar{Q} = \begin{bmatrix} \frac{5}{8} \\ \frac{3}{8} \end{bmatrix} \right)$

2 (a) $\left(\bar{P} = [\frac{16}{23} \quad \frac{7}{23}], v = \frac{22}{23}, \bar{Q} = \begin{bmatrix} 0 \\ \frac{10}{23} \\ \frac{13}{23} \end{bmatrix} \right)$

(b) $\left(\bar{P} = [p \quad 1 - p] \text{ for any } p, \frac{1}{4} \leq p \leq \frac{3}{4}, v = -1, \bar{Q} = \begin{bmatrix} 0 \\ 1 \\ 0 \end{bmatrix} \right)$

(c) $\left(\bar{P} = [\tfrac{1}{2} \ \tfrac{1}{2} \ 0], v = 1, \bar{Q} = \begin{bmatrix} \tfrac{1}{6} \\ \tfrac{5}{6} \\ 0 \end{bmatrix} \right)$

3   Yes.

4   $A = \begin{bmatrix} 0 & 2 & -3 & 0 \\ -2 & 0 & 0 & 3 \\ 3 & 0 & 0 & -4 \\ 0 & -3 & 4 & 0 \end{bmatrix}$ Yes.   Find them.

5   $\left(\bar{P} = [\tfrac{1}{3} \ \tfrac{2}{3}], v = \tfrac{1}{3}, \bar{Q} = \begin{bmatrix} \tfrac{2}{3} \\ \tfrac{1}{3} \end{bmatrix} \right)$

6   Let

$$S_3 = \begin{cases} \text{if the card drawn is H, then fold} \\ \text{if the card drawn is L, then bet} \end{cases}$$

$$S_4 = \begin{cases} \text{if the card drawn is H, then fold} \\ \text{if the card drawn is L, then fold} \end{cases}$$

$$A = \begin{bmatrix} 0 & 1 \\ \tfrac{1}{2} & 0 \\ -\tfrac{3}{2} & 0 \\ -1 & -1 \end{bmatrix}$$

7   $A = \begin{bmatrix} -\tfrac{2}{5} & 1 \\ \tfrac{1}{5} & -\tfrac{1}{5} \end{bmatrix}$ $\left(\bar{P} = [\tfrac{2}{9} \ \tfrac{7}{9}], v = \tfrac{1}{15}, \bar{Q} = \begin{bmatrix} \tfrac{2}{3} \\ \tfrac{1}{3} \end{bmatrix} \right)$

9   For $n = 2k - 1$, the kth row of A is

$$\begin{matrix} & & & & & k & k+1 & & & \end{matrix}$$
$$\left[1 \quad 1 - \frac{2}{n} \quad 1 - \frac{4}{n} \quad \cdots \quad \frac{1}{n} \quad 0 \quad \frac{1}{n} \quad \frac{1}{n} \quad \cdots \quad \frac{1}{n} \quad \frac{1}{n} \right]$$

For $n = 2k$, the kth and $(k + 1)$st rows of A are

$$\left[1 \quad 1 - \frac{2}{n} \quad 1 - \frac{4}{n} \quad \cdots \quad \frac{2}{n} \quad 0 \quad 0 \quad 0 \quad \cdots \quad 0 \quad 0 \right]$$

and $\left[1 \quad 1 - \dfrac{2}{n} \quad 1 - \dfrac{4}{n} \quad \cdots \quad \dfrac{2}{n} \quad 0 \quad 0 \quad \dfrac{2}{n} \quad \cdots \quad \dfrac{2}{n} \quad \dfrac{2}{n} \right]$

respectively.

10   For $n = 4$ the matrix of the duel is

$$A = \begin{bmatrix} 0 & -\tfrac{1}{4} & -\tfrac{1}{2} & -\tfrac{3}{4} & -1 \\ \tfrac{1}{4} & 0 & -\tfrac{1}{8} & -\tfrac{5}{16} & -\tfrac{3}{4} \\ \tfrac{1}{2} & \tfrac{1}{8} & 0 & \tfrac{1}{8} & 0 \\ \tfrac{3}{4} & \tfrac{5}{16} & -\tfrac{1}{8} & 0 & \tfrac{1}{2} \\ 1 & \tfrac{3}{4} & 0 & -\tfrac{1}{2} & 0 \end{bmatrix}$$

For n = 5 the matrix of the duel is

$$A = \begin{bmatrix} 0 & -\frac{1}{5} & -\frac{2}{5} & -\frac{3}{5} & -\frac{4}{5} & -1 \\ \frac{1}{5} & 0 & -\frac{3}{25} & -\frac{7}{25} & -\frac{11}{25} & -\frac{3}{5} \\ \frac{2}{5} & \frac{3}{25} & 0 & \frac{1}{25} & -\frac{2}{25} & -\frac{1}{5} \\ \frac{3}{5} & \frac{7}{25} & -\frac{1}{25} & 0 & \frac{7}{25} & \frac{1}{5} \\ \frac{4}{5} & \frac{11}{25} & \frac{2}{25} & -\frac{7}{25} & 0 & \frac{3}{5} \\ 1 & \frac{3}{5} & \frac{1}{5} & -\frac{1}{5} & -\frac{3}{5} & 0 \end{bmatrix}$$

11 The matrix of the duel is

$$A = \begin{bmatrix} \frac{1}{20} & -\frac{1}{20} & -\frac{7}{20} & -\frac{10}{20} \\ \frac{4}{20} & \frac{2}{20} & \frac{2}{20} & 0 \\ \frac{8}{20} & \frac{1}{20} & -\frac{1}{20} & \frac{10}{20} \\ \frac{12}{20} & \frac{4}{20} & -\frac{12}{20} & 0 \end{bmatrix}$$

The solution of the duel is

$$\left(\bar{P} = [0 \ \ \tfrac{11}{13} \ \ \tfrac{2}{13} \ \ 0], v = \tfrac{1}{13}, \bar{Q} = \begin{bmatrix} 0 \\ 0 \\ \frac{10}{13} \\ \frac{3}{13} \end{bmatrix} \right)$$

16 $\tilde{A} = \begin{bmatrix} 0 & A & -E \\ -A^T & 0 & E \\ E & -E & 0 \end{bmatrix}$

where A is the matrix of G and E is the appropriate row or column vector all of whose entries are 1.

,.2 Maxmin and Minmax

(Example 2.1.1) maxmin = -1, minmax = 1
(Example 2.1.2) maxmin = 0, minmax = 1
(Example 2.1.4) maxmin = -1, minmax = 2
(Example 2.1.10) maxmin = -1, minmax = 1
2 (a) minmax = 2, maxmin = 1
3 $\min_{y \in \mathbb{R}} F(x, y) = -4x - x^2$ for all x. Thus $\max_{x \in \mathbb{R}} (\min_{y \in \mathbb{R}} F(x, y)) = 4$ at $x = y = -2$. However, $\max_{x \in \mathbb{R}} F(x, y)$ does not exist for any value of y.
4 (b), (d), (g).
10 (b) Every point (x, y) for which $y = x^2$.
(d) There are no extreme points.
(g) Every point (x, y) for which $x^2 + y^2 = 1$.

2.3   A MATHEMATICAL INTRODUCTION TO THE THEORY OF GAMES

6 (a) $a_{13} = 1$ is a saddle point.

   (b) $a_{33} = 3$ is a saddle point. (Note that $a_{31} = 3$ is not a saddle point.)

   (c) $a_{42} = a_{43} = -1$ are both saddle points.

13 The pure strategies

$$S = \begin{cases} \text{if I have } V, \text{ call } X \\ \text{if I have } X, \text{ call } X \end{cases} \qquad T = \begin{cases} \text{do not believe } V \\ \text{believe } X \end{cases}$$

are optimal strategies. The value of the game is 0.

2.4   Linear Inequalities

1 (b), (f), (h)

2 $\mathbb{K}(A) = \mathbb{K}(C)$ if and only if

$$C = A\begin{bmatrix} \beta_1 & 0 \\ 0 & \beta_2 \end{bmatrix} \quad \text{or} \quad C = A\begin{bmatrix} 0 & \beta_1 \\ \beta_2 & 0 \end{bmatrix}, \quad \beta_1, \beta_2 > 0$$

4 $Z \geq AX$, $X \geq 0$ has a solution if and only if Z is contained in the cone spanned by $\begin{bmatrix} 0 \\ 1 \end{bmatrix}$ and $\begin{bmatrix} 2 \\ -1 \end{bmatrix}$.

6 $Z = AX$, $X \geq 0$ does not have a solution for any positive vector Z (as well as for other vectors).

2.5   The Fundamental Theorem

2 $P = [\frac{2}{22} \; \frac{2}{22} \; \frac{3}{22} \; \frac{1}{22} \; \frac{14}{22}]$ is an optimal strategy for both players. The value of the game is 0.

6 The game with matrix B is the symmetrization of the game with matrix

$$A = \begin{bmatrix} 6 & 0 \\ 2 & 6 \end{bmatrix}$$

which has solution ($\bar{P} = [\frac{2}{5} \; \frac{3}{5}]$, $\bar{Q} = \begin{bmatrix} \frac{3}{5} \\ \frac{2}{5} \end{bmatrix}$, $v = \frac{18}{5}$). Thus

$$X = [\frac{6}{25} \; \frac{4}{25} \; \frac{9}{25} \; \frac{6}{25}]$$

is an optimal strategy for both players in the game with matrix B.

2.6 Relations among Games

1 Example 2.1.1 $A^* = A$, Example 2.1.2 $A^* = A$, Example 2.1.4
$A^* = \begin{bmatrix} 1 & 2 \\ 3 & 1 \end{bmatrix}$

3 The second and third strategies for each player are the essential strategies.
Thus $A^* = \begin{bmatrix} 0 & 0 \\ 0 & 0 \end{bmatrix}$.

4 (a) $\bar{P} = [p \quad 1-p]$, where $\tfrac{3}{5} \leq p \leq 1$, $\bar{Q} = \begin{bmatrix} 1 \\ 0 \end{bmatrix}$, $A^* = \begin{bmatrix} 2 \\ 2 \end{bmatrix}$

(b) $\bar{P} = [p \quad 1-p]$, where $\tfrac{2}{5} \leq p \leq \tfrac{2}{3}$, $\bar{Q} = \begin{bmatrix} 0 \\ 1 \\ 0 \end{bmatrix}$, $A^* = \begin{bmatrix} -1 \\ -1 \end{bmatrix}$

5 $A^* = \begin{bmatrix} 0 & 1 & 2 & 3 & 4 \\ 4 & 3 & 2 & 1 & 0 \end{bmatrix}$, $\bar{P} = [\tfrac{1}{2} \quad 0 \quad 0 \quad 0 \quad \tfrac{1}{2}]$ is the only optimal strategy
for player I.

$$\left\{ \begin{bmatrix} \tfrac{1}{2} \\ 0 \\ 0 \\ 0 \\ \tfrac{1}{2} \end{bmatrix}, \begin{bmatrix} 0 \\ \tfrac{1}{2} \\ 0 \\ \tfrac{1}{2} \\ 0 \end{bmatrix}, \begin{bmatrix} 0 \\ 0 \\ 1 \\ 0 \\ 0 \end{bmatrix}, \begin{bmatrix} \tfrac{1}{3} \\ 0 \\ 0 \\ \tfrac{2}{3} \\ 0 \end{bmatrix} \right\}$$

is a maximal set of linearly independent optimal strategies for player II.

12 (a) $A_{(2)}$ dominates $A_{(1)}$, and $A^{(2)}$ dominates $A^{(3)}$. Hence $A_{(1)}$ and $A^{(3)}$ can
be eliminated. In the remaining matrix the second row is dominated
by the first row and can be eliminated, leaving $[1 \quad 1]$.

(b) $A_{(1)}$ is dominated by $\tfrac{1}{4}A_{(3)} + \tfrac{3}{4}A_{(4)}$. $A_{(2)}$ is dominated by $\tfrac{1}{2}A_{(3)} + \tfrac{1}{2}A_{(4)}$.
Thus $A_{(1)}$ and $A_{(2)}$ can be eliminated.

2.7 Extreme Optimal Strategies

1 (a) $A' = \begin{bmatrix} 2 & 4 \\ 2 & -1 \end{bmatrix}$ yields $v = 2$, $\bar{P} = [\tfrac{3}{5} \quad \tfrac{2}{5}]$, $\bar{Q} = \begin{bmatrix} 1 \\ 0 \end{bmatrix}$

$A' = [2] = [a_{11}]$ yields $v = 2$, $\bar{P} = [1 \quad 0]$, $\bar{Q} = \begin{bmatrix} 1 \\ 0 \end{bmatrix}$

(b) $A' = \begin{bmatrix} 2 & -1 \\ -3 & -1 \end{bmatrix}$ yields $v = -1$, $\bar{P} = [\tfrac{2}{5} \quad \tfrac{3}{5}]$, $\bar{Q} = \begin{bmatrix} 0 \\ 1 \\ 0 \end{bmatrix}$

$A' = \begin{bmatrix} -1 & -4 \\ -1 & 5 \end{bmatrix}$ yields $v = -1$, $\bar{P} = [\tfrac{2}{3} \quad \tfrac{1}{3}]$, $\bar{Q} = \begin{bmatrix} 0 \\ 1 \\ 0 \end{bmatrix}$

(c) $A^{\cdot} = \begin{bmatrix} 3 & -3 \\ -2 & 5 \end{bmatrix}$ yields $v = \frac{9}{13}$, $\bar{P} = [\frac{7}{13} \quad \frac{6}{13}]$, $\bar{Q} = \begin{bmatrix} \frac{8}{13} \\ 0 \\ \frac{5}{13} \end{bmatrix}$

(d) $A^{\cdot} = \begin{bmatrix} 2 & 2 \\ 4 & -6 \end{bmatrix}$ yields $v = 2$, $\bar{P} = [1 \quad 0]$, $\bar{Q} = \begin{bmatrix} \frac{4}{5} \\ 0 \\ \frac{1}{5} \end{bmatrix}$

$A^{\cdot} = \begin{bmatrix} 2 & 2 \\ 5 & -6 \end{bmatrix}$ yields $v = 2$, $\bar{P} = [1 \quad 0]$, $\bar{Q} = \begin{bmatrix} 0 \\ \frac{8}{11} \\ \frac{3}{11} \end{bmatrix}$

$A^{\cdot} = [2] = [a_{13}]$ yields $v = 2$, $\bar{P} = [1 \quad 0]$, $\bar{Q} = \begin{bmatrix} 0 \\ 0 \\ 1 \end{bmatrix}$

2   $A^{\cdot} = \begin{bmatrix} 0 & 4 \\ 4 & 0 \end{bmatrix}$ yields $v = 2$, $\bar{P} = [\frac{1}{2} \quad 0 \quad 0 \quad 0 \quad \frac{1}{2}]$, $\bar{Q} = \begin{bmatrix} \frac{1}{2} \\ 0 \\ 0 \\ 0 \\ \frac{1}{2} \end{bmatrix}$

$A^{\cdot} = \begin{bmatrix} 1 & 3 \\ 3 & 1 \end{bmatrix}$ yields $v = 2$, $\bar{P} = [\frac{1}{2} \quad 0 \quad 0 \quad 0 \quad \frac{1}{2}]$, $\bar{Q} = \begin{bmatrix} 0 \\ \frac{1}{2} \\ 0 \\ \frac{1}{2} \\ 0 \end{bmatrix}$

$A^{\cdot} = \begin{bmatrix} 0 & 2 \\ 4 & 2 \end{bmatrix}$ yields $v = 2$, $\bar{P} = [\frac{1}{2} \quad 0 \quad 0 \quad 0 \quad \frac{1}{2}]$, $\bar{Q} = \begin{bmatrix} 0 \\ 0 \\ 1 \\ 0 \\ 0 \end{bmatrix}$

$A^{\cdot} = \begin{bmatrix} 0 & 3 \\ 4 & 1 \end{bmatrix}$ yields $v = 2$, $\bar{P} = [\frac{1}{2} \quad 0 \quad 0 \quad 0 \quad \frac{1}{2}]$, $\bar{Q} = \begin{bmatrix} \frac{1}{2} \\ 0 \\ 0 \\ \frac{2}{3} \\ 0 \end{bmatrix}$

$A^{\cdot} = \begin{bmatrix} 1 & 4 \\ 3 & 0 \end{bmatrix}$ yields $v = 2$, $\bar{P} = [\frac{1}{2} \quad 0 \quad 0 \quad 0 \quad \frac{1}{2}]$, $\bar{Q} = \begin{bmatrix} 0 \\ \frac{2}{3} \\ 0 \\ 0 \\ \frac{1}{3} \end{bmatrix}$

2.8 Linear Programming

2 The maximization problem is unbounded.

3 (a) The maximization problem is unbounded.

(b) $X = \begin{bmatrix} \frac{7}{5} \\ \frac{1}{5} \end{bmatrix}$, $\quad Y = [\frac{8}{5} \quad \frac{1}{5}]$, $\quad v = \frac{19}{5}$

(c) $X = \alpha \begin{bmatrix} 0 \\ 2 \\ 2 \end{bmatrix} + \beta \begin{bmatrix} \frac{10}{4} \\ \frac{3}{4} \\ \frac{3}{4} \end{bmatrix}$, where $\alpha + \beta = 1 \quad \begin{matrix} \alpha \geq 0 \\ \beta \geq 0 \end{matrix}$

$Y = [\frac{1}{3} \quad \frac{8}{3} \quad 0]$, $v = 12$.

4 If the farmer must grow the vegetables to feed his animals, then he raises 60 animals and 10 tons of vegetables for a profit of \$2,700. If he can buy vegetables at \$30 per ton, then he raises 70 animals, buys $\frac{70}{6}$ tons of vegetables, and increases his profit to \$2,800.

CHAPTER 3 FINITE MARKOV CHAINS

3.1 Examples of Finite Markov Chains

4 The probability that player I will lose is $\frac{3}{4}$.

5 The probability that a mouse initially in room 2 will enter room 3 is 1. The probability that a mouse initially in room 3 will enter room 2 is $\frac{1}{3}$.

6 If the rooms of the maze are ordered, 2, 3, 1, 4, then

$$I + A + A^2 + \cdots = \begin{bmatrix} \frac{4}{3} & \frac{2}{3} & 0 & 0 \\ \frac{2}{3} & \frac{4}{3} & 0 & 0 \\ \infty & \infty & \infty & \infty \\ \infty & \infty & \infty & \infty \end{bmatrix}$$

(Note that if the mouse is initially in room 2 and the cheese is in room 4, then the mouse may get no cheese. However, his expectation is still infinite.)

7 The probability that a mouse initially in room 2 will ever enter room 4 is $\frac{1}{3}$ The probability that he will enter room 1 is $\frac{2}{3}$.

8 2 minutes

9 $f_{51} = \frac{1}{7}$ $\quad f_{52} = \frac{2}{7}$ $\quad f_{53} = \frac{2}{7}$ $\quad f_{54} = \frac{3}{7}$

$f_{61} = \frac{6}{7}$ $\quad f_{62} = \frac{5}{7}$ $\quad f_{63} = \frac{5}{7}$ $\quad f_{64} = \frac{4}{7}$

$f_{12} = \frac{3}{5}$ $\quad f_{13} = \frac{3}{5}$ $\quad f_{14} = \frac{1}{5}$

$f_{21} = \frac{1}{2}$ $\quad f_{23} = 1$ $\quad f_{24} = \frac{1}{3}$

$f_{31} = \frac{3}{13}$ $\quad f_{32} = \frac{6}{13}$ $\quad f_{34} = \frac{2}{13}$

$f_{41} = \frac{1}{3}$ $\quad f_{42} = \frac{2}{3}$ $\quad f_{43} = \frac{2}{3}$

$f_{11} = \frac{3}{10}$ $\quad f_{22} = \frac{11}{18}$ $\quad f_{33} = \frac{6}{13}$ $\quad f_{44} = \frac{2}{9}$

10 The probability that the biased coin will be used for the third toss is $\frac{43}{72}$. The probability that the third toss will be heads is $\frac{259}{432}$.

3.2 Infinite Processes

1 (a) $\begin{bmatrix} 1 & -1 & 1 \\ 0 & 0 & 0 \\ 0 & 0 & 0 \end{bmatrix}$ (b) $\begin{bmatrix} 0 & 1 & -1 \\ 0 & 1 & -1 \\ 0 & 0 & 0 \end{bmatrix}$ (c) 0 (d) 0

2 $\begin{bmatrix} \frac{1}{2} & -\frac{1}{2} & \frac{3}{2} \\ 1 & -1 & 3 \\ \frac{1}{2} & -\frac{1}{2} & \frac{3}{2} \end{bmatrix}$ 3 $\begin{bmatrix} 1 & -2 & 0 \\ 0 & 2 & 0 \\ 2 & -4 & 1 \end{bmatrix}$

4 $I + A + A^2 + \cdots = \begin{bmatrix} 5 & -2 & 4 \\ 2 & 0 & 2 \\ -1 & 0 & 0 \end{bmatrix}$

10 One pair is $A = \begin{bmatrix} 0 & 2 \\ 0 & 0 \end{bmatrix}$, $B = \begin{bmatrix} 0 & 0 \\ 1 & 0 \end{bmatrix}$.

12 (a) $\begin{bmatrix} e & 0 \\ 0 & e \end{bmatrix}$ (b) $\begin{bmatrix} e^2 & 0 \\ e^2 & e^2 \end{bmatrix}$ (c) $\begin{bmatrix} 2e^2 & e^2 \\ -e^2 & 0 \end{bmatrix}$

3.3 A Mathematical Introduction to the Theory of Finite Markov Chains

(Example 3.1.15) 1, 2, and 3 are transient; 4 forms an ergodic set.

(Example 3.1.18) 1, 2, 3, and 4 are transient; 5 and 6 each form ergodic sets.

(Example 3.1.20) 1, 2, and 3 form an ergodic set.

2 (Exercise 3.1.6) 1 and 4 each form an ergodic set. The stationary vectors are

$$\begin{bmatrix} p \\ 0 \\ q \end{bmatrix} \quad \begin{aligned} p &\geq 0 \\ q &\geq 0 \\ p + q &= 1 \end{aligned}$$

(Exercise 3.1.9) 5 and 6 each form an ergodic set. The other states are transient.

(Exercises 3.1.10 to 3.1.12) The only ergodic set is the set of all states.

3 (a) There are two ergodic sets, {1, 4} and {6}; 2, 3, and 5 are transient.

4 (a) 3 forms an ergodic set. All other states are transient.

(b) There are two ergodic sets, {1, 3} and {2, 4, 5}. There are no transient states.

3.4   Periodic Transition Matrices

1 (*a*) The period is 4. If the states are ordered 1, 2, 5, 3, 4, 6, then

$$A = \begin{bmatrix} 0 & 0 & 0 & 0 & 0 & 1 \\ \frac{1}{2} & 0 & 0 & 0 & 0 & 0 \\ \frac{1}{2} & 0 & 0 & 0 & 0 & 0 \\ 0 & \frac{1}{2} & 0 & 0 & 0 & 0 \\ 0 & \frac{1}{2} & 1 & 0 & 0 & 0 \\ 0 & 0 & 0 & 1 & 1 & 0 \end{bmatrix}$$

There are no transient states.

(*b*) The period is 2. If the states are ordered 1, 3, 5, 2, 4, 6, then

$$A = \begin{bmatrix} 0 & 0 & 0 & \frac{1}{2} & 0 & 0 \\ 0 & 0 & 0 & 0 & 0 & 0 \\ 0 & 0 & 0 & \frac{1}{2} & 1 & 1 \\ \frac{1}{2} & 1 & 1 & 0 & 0 & 0 \\ \frac{1}{2} & 0 & 0 & 0 & 0 & 0 \\ 0 & 0 & 0 & 0 & 0 & 0 \end{bmatrix}$$

3 and 6 are transient states.

4 If the states are ordered 1, 3, 4, 2, 5, 6, then

$$A = \begin{bmatrix} 0 & 0 & 0 & 0 & 0 & 1 \\ 0 & 0 & 0 & \frac{1}{2} & 0 & 0 \\ 0 & 0 & 0 & \frac{1}{2} & 1 & 0 \\ \frac{1}{2} & 0 & 0 & 0 & 0 & 0 \\ \frac{1}{2} & 0 & 0 & 0 & 0 & 0 \\ 0 & 1 & 1 & 0 & 0 & 0 \end{bmatrix}$$

3.5   Recurrence Probabilities

1 4 drinks. Yes.
2 He can expect to win 9.
3 14 jumps. He can be expected to be at the bottom of the well 4 times
4 $F^{(1)} = A$

$$F^{(2)} = \begin{bmatrix} \frac{1}{2} & \frac{1}{4} \\ 0 & \frac{1}{2} \end{bmatrix} \qquad F^{(s)} = \begin{bmatrix} \frac{s}{2^{s-1}} & \frac{s}{2^s} \\ 0 & 0 \end{bmatrix} \qquad s \geq 3 \qquad M = \begin{bmatrix} \frac{3}{2} & 2 \\ 1 & \frac{3}{2} \end{bmatrix}$$

CHAPTER 4   THE THEORY OF GRAPHS

4.1   Some Examples

1   The missionaries and cannibals can cross the river in 11 trips in exactly
four ways:

$$(3, 3, e) \quad \begin{matrix}(1, 1, w)\\(0, 2, w)\end{matrix} \quad (3, 2, e) \to (0, 3, w) \to (3, 1, e) \to (2, 2, w) \to (2, 2, e)$$

$$\to (3, 1, w) \to (0, 3, e) \to (3, 2, w) \quad \begin{matrix}(0, 2, e)\\(1, 1, e)\end{matrix} \quad (3, 3, w)$$

2   There are two shortest paths of length 4:

$$(v_1, v_5) \to (v_5, v_4) \to (v_4, v_3) \to (v_3, v_7)$$
$$(v_1, v_5) \to (v_5, v_6) \to (v_6, v_3) \to (v_3, v_7)$$

4.2   Undirected Graphs

1

$$\{a_1, a_2, a_6, a_4, a_5\} \qquad \{a_1, a_4, a_6, a_2, a_5\}$$
$$\{a_5, a_4, a_6, a_2, a_1\} \qquad \{a_5, a_2, a_6, a_4, a_1\}$$

are the circuits of G_1 and also are the Euler lines from v_3 to itself.   There are
eight Euler lines from v_1 to itself.

4   (Fig. 4.1.8)   2   (Fig. 4.1.10)   4 $(v_2 \text{ to } v_3)$

4.3   The Vector Space of Sets

1   (a) $\begin{bmatrix}1\\0\\0\\1\end{bmatrix} = \begin{bmatrix}1\\1\\0\\0\end{bmatrix} + \begin{bmatrix}0\\1\\1\\0\end{bmatrix} + \begin{bmatrix}0\\0\\1\\1\end{bmatrix}$ (b) $\begin{bmatrix}1\\0\\1\\0\end{bmatrix} = \begin{bmatrix}1\\1\\0\\0\end{bmatrix} + \begin{bmatrix}0\\1\\1\\0\end{bmatrix}$

3   The minimal vectors of $\mathbb{E}$ are those vectors with two 1s.

$$\begin{bmatrix}1\\1\\0\\0\\\vdots\\0\\0\end{bmatrix}, \quad \begin{bmatrix}0\\1\\1\\0\\\vdots\\0\\0\end{bmatrix}, \quad \ldots, \quad \begin{bmatrix}0\\0\\0\\0\\\vdots\\1\\1\end{bmatrix}$$

is a basis for $\mathbb{E}$.   Thus dim $\mathbb{E} = n - 1$.

4.4 Matrices and Undirected Graphs

1

$$C = \begin{bmatrix} 0 & 0 & 0 & 1 & 1 & 0 & 0 \\ 0 & 0 & 1 & 1 & 0 & 1 & 0 \\ 0 & 1 & 1 & 0 & 0 & 0 & 1 \\ 1 & 1 & 0 & 0 & 0 & 0 & 0 \\ 1 & 0 & 0 & 0 & 1 & 1 & 1 \\ 0 & 1 & 0 & 0 & 1 & 1 & 1 \\ 0 & 0 & 1 & 0 & 1 & 1 & 0 \\ 1 & 0 & 0 & 1 & 0 & 1 & 1 \\ 1 & 0 & 1 & 0 & 0 & 0 & 1 \\ 0 & 1 & 0 & 1 & 0 & 1 & 1 \end{bmatrix}$$

2 The incidence and loop matrices of *G* are

$$B = \begin{bmatrix} 1 & 0 & 0 & 1 & 1 & 0 & 0 & 0 & 0 \\ 1 & 1 & 0 & 0 & 0 & 0 & 0 & 1 & 0 \\ 0 & 1 & 1 & 0 & 0 & 0 & 0 & 0 & 1 \\ 0 & 0 & 1 & 1 & 0 & 1 & 0 & 0 & 0 \\ 0 & 0 & 0 & 0 & 1 & 1 & 1 & 0 & 0 \\ 0 & 0 & 0 & 0 & 0 & 0 & 1 & 1 & 1 \end{bmatrix}$$

$$L = \begin{bmatrix} 1 & 0 & 0 & 0 & 1 & 1 & 0 & 0 & 1 & 1 & 0 & 1 & 1 & 1 \\ 0 & 1 & 0 & 0 & 1 & 0 & 1 & 1 & 1 & 0 & 0 & 0 & 1 & 1 \\ 0 & 0 & 1 & 0 & 1 & 1 & 1 & 1 & 0 & 0 & 1 & 1 & 0 & 1 \\ 0 & 0 & 0 & 1 & 1 & 0 & 0 & 1 & 0 & 1 & 1 & 1 & 1 & 0 \\ 1 & 0 & 0 & 1 & 0 & 1 & 0 & 1 & 1 & 0 & 1 & 0 & 0 & 1 \\ 0 & 0 & 1 & 1 & 0 & 1 & 1 & 0 & 0 & 1 & 0 & 0 & 1 & 1 \\ 1 & 0 & 1 & 0 & 0 & 0 & 1 & 1 & 1 & 1 & 1 & 0 & 1 & 0 \\ 1 & 1 & 0 & 0 & 0 & 1 & 1 & 1 & 0 & 1 & 0 & 1 & 0 & 0 \\ 0 & 1 & 1 & 0 & 0 & 1 & 0 & 0 & 1 & 0 & 1 & 1 & 1 & 0 \end{bmatrix}$$

6

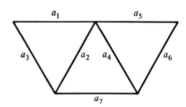

4.5 Directed Graphs

1

$$B = \begin{bmatrix} -1 & 0 & 0 & 0 & -1 & -1 & -1 \\ +1 & -1 & 0 & 0 & 0 & 0 & 0 \\ 0 & +1 & -1 & 0 & 0 & 0 & +1 \\ 0 & 0 & +1 & -1 & 0 & +1 & 0 \\ 0 & 0 & 0 & +1 & +1 & 0 & 0 \end{bmatrix}$$

$$L = \begin{bmatrix} -1 & 0 & 0 & -1 & 0 & -1 \\ -1 & 0 & 0 & -1 & 0 & -1 \\ 0 & -1 & 0 & -1 & +1 & -1 \\ 0 & 0 & -1 & 0 & +1 & -1 \\ 0 & 0 & +1 & 0 & -1 & +1 \\ 0 & +1 & -1 & +1 & 0 & 0 \\ +1 & -1 & 0 & 0 & +1 & 0 \end{bmatrix}$$

$$C = \begin{bmatrix} 0 & 0 & 0 & +1 & +1 & 0 & 0 \\ 0 & 0 & +1 & -1 & 0 & +1 & 0 \\ 0 & +1 & -1 & 0 & 0 & 0 & +1 \\ +1 & -1 & 0 & 0 & 0 & 0 & 0 \\ +1 & 0 & 0 & 0 & +1 & +1 & +1 \\ 0 & +1 & 0 & 0 & +1 & +1 & +1 \\ 0 & 0 & +1 & 0 & +1 & +1 & 0 \\ +1 & 0 & 0 & -1 & 0 & +1 & +1 \\ +1 & 0 & -1 & 0 & 0 & 0 & +1 \\ 0 & +1 & 0 & -1 & 0 & +1 & +1 \end{bmatrix}$$

The signs in any row of C and in any column of L may be changed.

2 $a_1 = \frac{1}{7}(3L_1 - L_2 + C_1 + C_2 + C_3)$ $\quad a_3 = \frac{1}{7}(L_1 + 2L_2 + 5C_1 - 2C_2 - 2C_3)$
$a_6 = C_4$
where

$$C_1 = a_1 - a_2 + a_3 \qquad L_1 = a_1 + a_2$$
$$C_2 = a_1 - a_2 + a_4 \qquad L_2 = -a_1 + a_3 + a_4 + a_5$$
$$C_3 = a_1 - a_2 + a_5$$
$$C_4 = a_6$$

4.6 Matrices and Trees

$$M = \begin{bmatrix} +2 & -2 & 0 & 0 & 0 \\ 0 & +2 & -1 & -1 & 0 \\ 0 & 0 & +1 & -1 & 0 \\ -1 & 0 & 0 & +2 & -1 \\ -1 & 0 & 0 & 0 & +1 \end{bmatrix}$$

There are 4 directed trees rooted at each vertex, 4 circuits, and 33 trees in the underlying undirected graph.

3   If the vertices of G are renumbered according to

$$v_1 \rightarrow v_1 \qquad v_2 \rightarrow v_4 \qquad v_3 \rightarrow v_2 \qquad v_4 \rightarrow v_5 \qquad v_5 \rightarrow v_3$$

then the resulting graph is identical to the graph of Exercise 2.

Subject Index

Index of Symbols

Till the Cows Come Home

Till the Cows Come Home

Sara Cox

CORONET

First published in Great Britain in 2019 by Coronet
An Imprint of Hodder & Stoughton
An Hachette UK company

1

Copyright © Sara Cox 2019

The right of Sara Cox to be identified as the
Author of the Work has been asserted by her in accordance
with the Copyright, Designs and Patents Act 1988.

A CIP catalogue record for this title is available from the British Library

Hardback ISBN 9781473672734
eBook ISBN 9781473672710

Typeset in Electra LH by Palimpsest Book Production Ltd, Falkirk, Stirlingshire

Printed and bound in Great Britain by Clays Ltd, Elcograf S.p.A.

Hodder & Stoughton policy is to use papers that are natural,
renewable and recyclable products and made from wood grown in sustainable
forests. The logging and manufacturing processes are expected to conform to
the environmental regulations of the country of origin.

Hodder & Stoughton Ltd
Carmelite House
50 Victoria Embankment
London EC4Y 0DZ

www.hodder.co.uk

Dedicated to my family, past and present.

Foreword

I really really enjoy talking. Anyone who's ever stumbled upon one of my radio shows will have no doubt about that.

I also love words and language. Maybe it's a hangover from my A level in English Literature – I do appreciate a clever turn of phrase. Or maybe this love precedes my years at sixth form – perhaps the seeds were sown in Dad's farm kitchen, my nana's front room or the pub where Mum ruled the roost as landlady: conversation, gossip, wry observations, banter – all swirling around me in a whirlwind of words as I sat quietly in the eye of the storm, soaking it all up.

Painting pictures with words is something I try to do on the wireless – conjuring up images that will hopefully transport the listeners to another place, whisking them away from their desks or the gridlocked traffic on the M6.

I love the buzz of making people laugh too. As the youngest of five I worked out pretty early on that the best way to get any attention from the grown-ups was to clown about. Not much has changed.

This book, then, is an extension of all the above. A big wordy pie packed with all the people, places and ponderings from my first twenty years on this glorious planet. I hope you enjoy tucking into it.

I wanted to tell the story of my upbringing on the farm because whenever I chatted to anyone about whether I could write AN ACTUAL BOOK I discovered my most vivid memories were always about the farm – wherever the sentence began it'd inevitably finish on a pony, by a tractor or in a field with some cows.

The farm is my dad's – he still lives there. I've always been a certified daddy's girl so it's been really rather lovely documenting our relationship, which, aside from my husband Ben and the kids, has been the biggest love story of my life.

Having said that, there's a little twist in the book. No, I'm not adopted (despite my four older siblings occasionally telling me I was found wrapped in chippy paper by the barn door), but writing this book has certainly helped me to unearth the person who I really have to thank/blame for making me who I am today. Typical – only I could write a book and surprise myself with its punchline.

The farm is the place where my story began and now it's part of my children's story too; Lola, Isaac and Renee love being at Grandad's to fuss over a newborn calf or clamber over the hay bales in the barn, just like I did.

They are another reason for writing this book. They just know me as Mummy – shouter of 'HURRY UP, we're now officially late!' in the morning and 'For the umpteenth time, BRUSH YOUR TEETH AND GET TO BED!' in the evening . . . and someone who talks on the radio and pops up on the telly.

I want them to know that as well as head chef and bottle washer I'm another person – a farmer's daughter, a former bullying victim, an ex-supermodel-cum-barmaid. I'm also hoping that if they know where I've literally come from then they'll understand better where I'm coming from when I nag them about wasting food, working hard, respecting their elders – all the stuff that punctuated my childhood (and therefore this book) and is a big part of who I am.

Now I'm older and approaching my thirties (OK, my mid-forties, whatever) I've become more fascinated by what – or rather who – came before me. I want to keep the memory alive of those people who shaped me but are no longer here. Through this book my own children can get to meet their maternal great-grandparents for the first time. I've tried to freeze-frame some moments from our family's history for my children (and perhaps theirs one day) to savour.

I think it's human nature that as we journey further away from our childhood, we yearn to return, to arc round, circle back towards the things that shaped us from the beginning – the people, the places, the food, the music. The land we lived on. For me it's my dad's farmyard, the streets where I played out, the lanes I drove down to my boyfriend's, the outdoor steps in faraway countries I sat on smoking with my best friend.

Like a thousand-piece jigsaw, all these places slot together to make me who I am. Each friend, each holiday, every family member has added a pencil stroke to the sketch of my life – the portrait of me as a person.

From idolising Lady Diana and several regrettable perms, to being spooked by Michael Jackson's 'Thriller' video, the polyester-clad eighties feature heavily in my story and it's a decade I remember with fondness and relish still today.

Looking back helps us to focus on who we are now. I don't feel I'm much different to the scrawny girl who was insecure and shy at school, but so secure at home nestled among my family that I was a confirmed show-off. Now I guess I'm just brave enough to do my showing off in the most public of ways, on the radio or telly.

Self-belief is a fragile, intricate structure, built up over years, getting stronger over time but still easily wobbled by a setback or criticism. How different am I now from the kid in the corridor praying the teacher arrives at the classroom before the bullies do?

Writing this book has also allowed me to accept my weaknesses and forgive myself for being bullied. I've felt frustrated at my seeming willingness to allow myself to be picked on, but as I've rifled through those memories I realise it wasn't my fault; I was powerless and scared – stuck between a rock and a hard-nosed thugette.

We should all look back and be proud of where we've come from and how we've grown. Occasional pluckiness should be applauded – from riding injuries to crossing continents pocketing adventures and a bit of cash along the way. And moments of fear should be met with compassion and kindness.

I hope you enjoy my jaunt back in time to Bolton, the glorious north – the place whose name runs through my very centre as if I were a stick of seaside rock.

Bolton is a town that's seen many changes through the years and has struggled to find a new identity since its traditional industries disappeared, but what makes the place is the people – stoic, friendly, practical, kind and hilarious.

Like any working-class town we know how to work hard and play hard. And I'm hoping this book will be a bit like a night in the pub, swapping stories about our upbringing – some funny, some poignant, some far-fetched. So pull up a stool, mine's a G&T, and let's begin . . .

4

1. One Girl and her Head

'Just jump!' So I did. It didn't go well. Let me explain. I was perched on the end of the loft platform in the barn on the farm where we lived. It was summertime, just before haymaking when all the lovely fresh hay would be brought in by an army of pals and helpers, fuelled by tea and ham butties.

Now, though, the barn was almost empty save for a few old bales knocking around on the dusty floor and a pile of half a dozen or so right next to the loft. These were to be my crash mat. All I had to do was launch myself from my position sitting up in the loft and push away from the wall, gliding majestically onto the scratchy landing below. Think 'falcon fledgling's maiden flight from its perch on a cliff'. It was a drop of perhaps eight feet, but to my seven-year-old mind the chasm below might as well have been the Grand Canyon. Or, seeing as this was Bolton, the 'Reet Grand Canyon'.

My older sister Yvonne was the one encouraging me to jump. I was grateful that she was playing with me so I was keen to impress. As the youngest of five I was used to being a walking,

5

talking booby prize to my older siblings. An unwelcome addition. A skin tag on the neck of fun. Like stabilisers on a bike I slowed them down and made them look uncool.

The unruly gang of scuffed-kneed siblings was led by Robert, the eldest, boasting a full twelve years on me. He was tall and bonnie with a snub nose and dark hair. He had a deep authoritative voice even as a teenager and was caring and sensible in equal measure. Next came the twins, a mere twenty months his junior. David – tall and slim, matching dark mop above a shower of freckles, and a practical nature but more shy than Robert. Dot – younger than David by two minutes and comfortable in the protective shadow of her older brothers, sensitive, willowy, with a sweet face and pale aqua eyes framed by fine mousey locks. Next was Yvonne, much younger than them but a full four years older than me – she wore her seniority with pride; long dark-blonde hair, olive-green eyes and porcelain skin, lips often chapped during winter so that in old faded photos she looks like she's wearing a smudge of red lippy. They were bigger than me – faster, stronger, busier. I seemed to stand rejected in their slipstream whenever they went about their business.

If they tried to sneak off on some adventure they'd edge away along the farm lane and just as they thought they'd escaped, Mum's loud, bird-startling voice would boom out, 'Take Sarah with you!' I'd emerge from behind her (while I still could before I outgrew her – Mum is tiny, a 4'11" pocket-sized, bespectacled badass, with an impressive bosom and cracking legs) and, with her cropped permed hair – blonde, auburn, hazelnut-coloured depending on the week – quivering, she'd holler instructions to them. She's a force to be reckoned with even now – a sharp tongue but a soft heart – you always know where you are with Mum and even if you're in the doghouse she won't let you languish there for long. Confident, funny. Really loves to natter on. This apple hasn't fallen far from the tree.

Back then though I was half girl, half forehead – a trait that was a target for classmates, both figuratively with a plethora of nicknames (slaphead, fivehead, fodder) and literally as they'd heartily slap it while running past, high-fiving my fivehead.

So the fact my sister had agreed to hang out with me and my forehead was a big deal. Sure, the downside was she wanted me to jump off high stuff to certain death for her own amusement, but at least I'd die knowing I wasn't a chicken. Turns out I wasn't a falcon either.

I took a deep breath. I jumped. I didn't push away from the platform, though, I sort of nudged my bottom over the edge, so instead of springing away from the wall I slid down it like a blob of custard dripping down a kitchen tile.

In fact I stuck so closely to the wall, hurtling down on my shoulder blades like sledge runners, that I didn't even make it onto the bales below but tumbled between them and the wall, like a lost mitten down the back of a radiator.

I looked up to see my sister's horrified face frozen in a grimace – part sympathy, part worry in case I grassed her up to Mum, and part admiration at how I'd managed a vertical human bobsleigh down a barn wall.

The good news was I didn't die and I didn't break any bones. The only evidence of my misjudged jump was the two red raw tramlines down my back that made me look like I'd been mowed down by a joyrider in a Silver Cross pram.

2. Making an Entrance

Growing up on my parents' farm in the seventies and eighties was like having my very own slightly deadly adventure playground, and us kids were often getting into scrapes.

Grundy Fold Farm is still there and my dad still runs it. It isn't vast – discard the picture you may have in your mind's eye of rolling hills stretching as far as the eye can see, like a giant's patchwork quilt. In a soft furnishing analogy it's less quilt, more fabric swatch.

It's a small tenanted holding. It used to be around forty acres but now is more like fifteen thanks to the surrounding housing estates, the semi-detached red-brick homes seeming to shuffle silently ever closer to the farmhouse, tarmac inching over the outskirts of Dad's fields, steadily nibbling away at its edges.

From the farmyard you can look out over the front field and imagine you're in the countryside but a quick glance to the left and you see the homes that now hug the farm lane, back-garden trampolines peeping over the fences.

Still, it is a proper farm and my dad's a proper farmer, as

were his dad and his own father before him. In fact, farming goes way back in our bloodline. A TV show about family history asked me to feature, to look at my forebears' role in the world wars, but after further investigation it became apparent that on Dad's side no one fought in any conflicts because as far back as records went the menfolk were listed as 'farmer' or 'butcher' – occupations excused from call-up because they provided food for the nation. My Cox ancestors brandished not a bayonet during wartime but a joint of beef.

My dad, Len, is still working hard on the farm. He's over six feet tall and gifted all us offspring his long legs. A face that's as handsome as it is lived-in, with pale-blue eyes, kind and crinkled, sitting underneath a nearly bald head that's tanned from being outdoors all seasons. He has a slight limp thanks to a badly broken ankle years ago, so he walks like he's got a bit of gravel in his boot. Driving the tractor, handling the cattle, lugging bales around and 'shovelling cack' (his words, not mine) have kept him fit and strong. His hands are the size of dinner plates and he's tough.

I've seen him smash through ice with his bare hands when the cows' water troughs have frozen over. It may or may not have been a coincidence that he did that once in front of a new boyfriend I'd taken up to meet him; a kind of 'Watch it, son. If I can do this to three inches of solid ice imagine what I could do to your neck if you dick my daughter about.'

The few boyfriends who were deemed good enough to get to the farm always looked meek and slightly pathetic, stood in my dad's mighty shadow, quivering in their borrowed wellies that looked as alien on them as kitten heels on a cow.

Spoken like a true daddy's girl, I know.

Like any family ours has had its dramas, and when my mum met Dad she was a petite nineteen-year-old and he was a divorced older man with three young children to look after: a boy of six called Robert and four and a half year-old twins, Dorothy and David.

Mum fell in love with Dad and with his kids, raising them as her own. Having kids now myself, I find that pretty bloody mind-boggling – it was a testament to how much she must've loved Dad (and how gorgeous the kids were too) that Mum didn't saddle up the nearest heifer and gallop off into the sunset. I'm glad she stuck around because a few years later they had my sister Yvonne and then, in December 1974, I put in an appearance.

Mum's story of my birth was less *Call the Midwife* and more *Carry on Through the Cervix*.

Being an awkward madam, I decided I wanted to rock a breach vibe and come out feet first, which, you don't need to be a genius to work out, stings a bit for the poor labouring woman.

Mum recalls how, with it being mid-December (Friday the 13th no less), the doctor who had been summoned to help had arrived straight from the hospital's Christmas party.

I like to think he'd been pulled out of the midsection of a conga chain and diverted to the labour ward, possibly with a few pissed orderlies still attached.

I also imagine his tie was wonky and he was still wearing a cone party hat held on with an elastic chinstrap. This was 1974 so I'm certain he'd have been wearing a swirly pattered shirt, possibly in orange, with brown flares. According to Mum he'd been enthusiastically putting the 'merry' into Merry Christmas and was half-cut to the extent that, once I was safely out, when Mum needed a little patching up, he had to close one eye as he struggled to thread the needle, tongue poking out of the corner of his mouth in concentration.

Knowing my mother, this last detail could be an embellishment, but she swears it's true. In fact, when I recently put it to

Mum that these details seemed a little over the top she even added to them, saying the young doctor had a bottle of Cinzano tucked under his arm.

My entrance into the world wasn't straightforward. Not only was I greeted at the door by a pissed man wearing a party hat like some rubbish seventies housewarming, but I was born with a dislocated hip and an allergy to milk.

Mum had to sit my four siblings down and gently explain that I was a 'special baby'. It's probably best they had a heads-up; I wasn't wearing the usual newborn attire of a cute Babygro but instead a contraption that was half dungarees, half tree-surgeon harness, with straps over my shoulders and round my waist and connected to my ankles. This look was topped off, literally, with my magnificent forehead, a bonnet stretched over the tip, like a tea cosy on a missile.

Despite being fixed by the clever doctors my legs have let me down on occasion. During times of extreme pressure and phys-ical exertion I'd often come a cropper. I'm no natural athlete and I maintain to this day a spring onion shape. Some women lament being apple- or pear-shaped but with my aforementioned acre of forehead plastered across my huge head and my lack of hips that whittle away down to skinny long legs I'm a dead ringer for a scallion.

I was never therefore very aerodynamic and struggled with running fast, my long legs going like the clappers with my gargantuan noggin bobbin about on top. From a distance I probably looked like I was running holding a helium balloon.

I dreaded sports days and struggled to find an event I could master. I sacked off the sack race and wasn't particularly smashing at the egg and spoon.

On one sports day I decided to go for glory. I was to compete in the 100m sprint. In fact, that's probably overstating it. I was five years old and the track was about twenty yards of a dusty sports field. As I stepped up to the starting line I looked sideways at my competitors, ready to take them down. I managed to block out the roar of the crowd (a few teachers and a handful of mums) so I could focus.

We took our positions and as the firing pistol rang out (whistle was blown) we surged forward, necks straining, arms pumping, legs flying.

At least that's what the other kids did. Not me. The pressure to perform and the flood of adrenaline through my skinny limbs meant that when the whistle blew, my legs gave way and I crumpled to the floor like one of Fred Dibnah's chimney stacks.

As the other kids streamed through the finish ribbon, I was marooned at the start. Mum came and gathered me up from the floor and dusted me down. Usually after an unsuccessful race your mum can say 'It's not the winning that's important but the taking part' but I didn't even manage that.

My school had two fantastic assets. The first was chocolate custard – a now rarely seen substance served at the same temperature as the earth's core and the same consistency as emulsion. The Farrow & Ball colour equivalent would be 'wet spaniel' or 'conker vomit'. The other good thing was the school's proximity to our farm.

The sports field that witnessed my start-line humiliation ran parallel with my dad's fields, separated by a line of prickly yew trees and a wooden fence.

Beyond that lay Dad's fields, split into wonky slices by trees and barbed-wire fences, the broad iron gates allowing passage from field to field.

Each gate swung out over a worn-away area. In wet times it

was a welly-snatching trap to navigate – all ploughed up by the cattle, congregating in the hope of being let into the dry or at the very least getting a few slices of sweet hay. In summer it became bald and dusty from the tractor and trailer driving through, compacting the hard earth.

Beyond the front fields lay four more, each with a name: the pond (containing a pond), the open cast, the lane, and Stopes, named after the road that ran alongside it.

From the school playground I could see the front fields and the barn, a vast light-grey brick block, its upper section made up of slats of pale wood.

The proximity of the farm was a source of both joy and pain; although it was torturous in that I was so close and yet so far away from home (just taking off and popping home for a swift half of Vimto was firmly discouraged by the spoilsport teachers), I still loved to be able to gawp out of the windows daydreaming and see home.

Not only could I see home, unfortunately I (and everyone else) could often smell it too.

In early spring Dad would take a huge orange tank, like a massive can of pop on its side, fill it with liquid slurry, hook it up to his tractor and take off around the fields, liberally spraying ample amounts of nutrient-packed but extremely whiffy poo smoothie.

The aroma would travel quickly and at playtime us kids were hit with the dense stench; a bit like the famous gravy advert, but as the wisps of scent danced through air towards their eager nostrils, instead of 'Mmmm . . . Bisto' they said 'Urgh . . . poo', quickly followed by shouts of 'Oi Coxy, that's your dad making that smell.'

I began to dread this time of year as I knew the kids would give me as much shit as my dad was spreading.

3. Benji and the Missing Ball

Masefield Primary welcomed me back recently to film a documentary there on friendship. It was lovely to sit watching the kids run riot round the playground, some completely oblivious to the cameras, others lobbing a ball towards us so they could coyly retrieve it while smiling shyly at the camera.

It would've been a great shot, me stood in the playground staring into the middle distance and saying wistfully 'I remember when all this was fields,' but it IS all still fields. Happily, not much had changed in over three decades, apart from a new wire fence cordoning off the playground from the grass beyond, which back in the day would've scuppered mine and Benji's fun.

Benji was a Lakeland terrier, a real jaunty, self-assured little fella who'd occasionally trot across the field from Dad's and swagger around the playground causing near hysteria among the kids.

Shouts of 'Dog! Dog!' would go up and I'd be packed off to grab him and return him home. Obviously for 'safety' reasons

I'd need to take a friend, so my chosen pal and I would escort Benji across the playing field like security ejecting a troublesome festival-goer. Dad would welcome us with open arms and crack open the ice pops. I've lovely memories of hastily munching snacks at the breakfast bar in the farm kitchen, knowing we'd have to leg it back to school sharpish.

Benji was gorgeous – just below knee height and compact, with small dark intelligent eyes and wiry chestnut-coloured hair on his face, legs and chest. He sported a superhero cape of black glossy curls along his back.

If he had had a superpower it would've been 'Mega ratter!' – the scourge of all whisker-twitching vermin daft enough to set paw on his territory. He was a working dog and was never happier than when snout-deep in rodent scent, like a wine connoisseur with their nose in a spicy Shiraz. He was often found wriggling his bendy frame behind bales of musty hay, on a one-dog mission to catch whatever rat was lurking in the crevices of the barn.

The 'Norway' or 'brown' rat population (which, puzzlingly, isn't remotely Nordic and isn't uniformly brown, rat fact fans) has been on the rise over the last thirty years thanks to warmer winters, which means longer breeding periods. With their fond-ness for merrily defecating near enough constantly, spreading diseases, damaging buildings and machinery and causing fires by gnawing through electricity cables, it's fair to say that rats really aren't high on the list of farmers' most favourite things, which, judging by my dad, I think begins with a brew at No. 1, then biscuits, then talking with his farmer buddies about cows; 'rats' is at the end at about No. 41,349.

We did our best to keep the buggers down and Benji took real pride in helping. Just as a rat is hard-wired to breed, to gnaw, to poop, this tough little terrier was bred to take them down.

At the height of summer when the barn needed ridding of the dusty old hay before the new fresh stuff was brought in from the fields, we'd gather to shift the bales that had lined the barn walls since the year before like the pastry casing of a huge square pie. The workers and friends who'd arrived to help with haymaking knew that the first morning would be taken up with clearing out not only the hay but its inhabitants as well.

On a couple of occasions I would perch on the edge of the loft next to my siblings (I wasn't tempted to jump – lesson learned) and as my dad's pals dismantled the remaining walls of hay like huge Lego blocks, we'd poke a pitchfork down the back of the bales, creating a kind of Indiana Jones-style obstacle course for the rats, who had to run the gauntlet down the corridor of prongs jabbing them towards the concrete floor. At these moments I secretly hoped I wouldn't feel the soft thud of furry flesh underneath my fork; that'd be too much. This was Bolton, not *Game of Thrones*.

As their empire steadily disappeared the rats would have to make a run for it through the vast barn doors and out to the open fields with Benji in hot pursuit.

More often than not they'd escape to freedom, but occasionally a cheer would go up when a slightly slower one was unlucky and Benji could be seen shaking his prize like a rag.

Looking back it all seems a bit gruesome and it was. It's a million miles from my own kids' lives. The only rodent they'll encounter is our hamster Cheddar; and our dogs Beano and Dolly, fluffy white Maltese ruled by the cat (Watson), might like chasing squirrels but if they came face to face with a rat they wouldn't know what to do with it. Beano would be just as likely to bake it a quiche as he would to break its neck with a few deft shakes.

By the time Benji died he had a few bits missing – a goolie and a corner of an ear had been lost in battle, as had a few teeth. If I have one resounding memory of that dog it's that if he worked hard, he played even harder.

He had his hobbies – he loved brawling, fast women and fast cars. Benji picked fights with any big dog that crossed his path despite being snack-size to them. It was like he was channelling every badass character Vin Diesel had ever played. He also was quite the romancer – if any bitch was on heat within a two-mile radius Benji would be up her garden path in no time.

His liking for fast cars, or at least chasing them, caused us no end of worries. On one side of our house was the farm proper, the buildings and fields, but on the other side, through a heavy, squawky-hinged blue gate, lay a housing estate.

Like a Bolton version of *The Lion, the Witch and the Wardrobe*, that gate was the door through which you'd step into a new world: not a snow-covered Narnia full of curious characters, but a vast estate with a fair few colourful characters. Hundreds of neatly kept small semis spread across a myriad of avenues and crescents. Little has changed; maybe the family Cortina and the kids' Chopper bikes have been replaced by Ford Focuses and scooters, but it was and still is a nice enough area. It was never rough – not mattress-in-the-front-garden rough.

Beyond the gate always felt like a different land; on the farm we were in our own little kingdom, we were safe, knew the rules and could gallivant around the yard, fields and buildings with no worries. But beyond the gate was the real world and this is where Benji would get into scrapes.

On the farm he barely batted a darkly lashed eyelid at Dad trundling past on his tractor and neither did he pay any attention to visitors pulling up in their cars, but once through the gate and onto the estate, Benji had an inexplicable and pathological hatred of vehicles and would pursue them, barking furiously. If it had an engine and four wheels it was fair game. His absolutely favourite quarry to chase through those streets was the ice cream van.

You've gotta admire this plucky terrier; he certainly had balls (well, one at least). Choosing a pastel-coloured ton of van as his mortal enemy, he'd set off towards the estate as the unmistakable strains of 'Greensleeves' chimed into audio range, announcing Mr Whippy's arrival.

If I'd managed to find Mum or Dad in time to cadge 50p I'd join the queue of kids, salivating at the thought of a lemonade or banana lolly, high on a heady combination of anticipation and the diesel fumes pumping out of the van's exhaust.

Benji would satisfy himself with sniffing around on a nearby lawn, acting as if the ice cream van was of no consequence to him. But when the crowds of kids had dispersed and the monster's engine roared and pulled away from the kerb, that was his cue to attack.

I'm not saying Benji wasn't a smart dog but to try to bite a moving wheel isn't the brightest idea. It's not only highly dangerous but must've hurt too – as the van trundled off Benji would be running alongside the rear wheel trying his damnedest to get a gobful of rubber tyre. He must've had gums of steel.

Luckily he never fully got a purchase on it or he could've become attached and whizzed round and round, thwacking road and mudflap in quick succession like some macabre wheel decoration. It was only when the van groaned from first to second gear and picked up speed that Benji's little springy terrier legs could no longer keep up and his nemesis would make its getaway, disappearing into the sunset; well, into Chester Avenue at least.

I like to think there's a real possibility that one or two of Benji's peggies are still lodged in a tyre lying in a scrapyard somewhere.

4. Cold Jubblys

Grundy Fold Farm lies just on the outskirts of Little Lever and Bradley Fold. The latter is quite remote. Dad knows a few farmers over that way. Narrow roads wind round sloping fields dotted with sheep. Stone cottages, the occasional tractor slowing the traffic and blind corners hiding horse-riders mark it out as being rural. It has its picturesque moments even in winter, its bare hedgerows framing the bleak but beautiful land. The former, Little Lever, is a place I often refer to on the radio as a 'charming little hamlet' as an in-joke to anyone who knows it. If Little Lever is a hamlet then I'm Dame Shirley Bassey. I love it though and I know my way round it better than anywhere I've lived since. I think the place where you ride your bike as a kid is the place you truly know back to front. I might not know all the street names but I know the footpaths, the ginnels and the short-cuts.

Residents refer to it as 'the village' but that's not accurate either – a village conjures up images of vicars, hanging baskets, duck ponds and thatched roofs.

What it is is a collection of sprawling estates and residential areas surrounding a high street that boasts my most favourite chippy in the world, a couple of chemists, shops and these days a massive supermarket, replacing the long-dead local minimarts, one of which was Hanburys. God, I loved that place, whose shelves were so cluttered it looked like one of those Channel 5 documentaries on hoarding.

Despite the aisle space being narrowed down to the width of a Weetabix by all the produce bulging from the shelves they still had little wonky trolleys, used mainly as mini cages on wheels to stop lil' darlings like me running riot and sliding a multipack of Blue Riband in among Mum's everyday essentials.

Recently while filming *Back in Time for Tea*, a BBC2 show about food and social history in the north, I discovered Hanburys had thirty-one stores in the north-west, eight of them in Bolton. They were first opened in 1889 by Jeremiah Hanbury and were among a handful of thriving independent supermarkets that could be found before the Big Four came striding into every town and swept aside all competition.

I find as I get older I care more and more about history, about where we come from, what went before.

If I'd told my twenty- or even thirty-year-old self about the beginnings of Hanburys I would've been met with the same blank stare I get from my own kids now when I point out a statue or interesting bit of architecture.

I first had an inkling I was heading towards middle age way back in 2005 while on a train to, ironically enough, a massive Radio 1 festival called the Big Weekend. Each summer the station would take the music line-up to a less obvious area than the usual big destinations. This year it was Sunderland's turn to welcome the circus to town. Tens of thousands of locals were allocated free tickets as world-class headliners rolled in for the weekend, hosted by all the Radio 1 DJs. They were huge events but as we approached Durham train station I wasn't struck breathless by the thought of the Foo Fighters, Kasabian or No Doubt – no, I was bowled over by the beautiful railway bridges with their high, narrow arches in warm yellow stone that framed the green hills behind.

Since then of course the signs have come too thick and fast to list or even notice . . . I now take pride in my full-blown middle-age symptoms and even took a puzzle book to do by the pool on holiday this year.

Back to the overstuffed aisles of Hanburys. In summertime Mum would treat us to Jubblys (stop giggling at the back), which were ice-pops encased in curious pyramid packaging made of the toughest cardboard ever created. I'm sure NASA would have some use for it if they realised its durability. Looking back, maybe Jubblys weren't that delicious and it was just such a mission to rip, bite and gnaw the packaging open that when the sweet, partly melted lolly eventually flooded your mouth the flavour was accentuated by relief.

Apart from Jubblys, Mum didn't really embrace too many food fads despite the early eighties being a hotbed of new, exciting convenience food. I'd watch the adverts on the telly for Findus crispy pancakes with their tagline of 'Every day has a different taste' and I'd think 'Yes! That's me! I want every day to have a different taste!' – the close-up of a slightly bobbly looking half-moon of pancake the colour of Homer Simpson, being sliced open to reveal its lumpy brown innards oozing out like some gravy mudslide, looked unbelievably scrumptious to my young eyes.

Much to my dismay, Mum insisted we ate proper food made from scratch and no amount of pleading would budge her. I absolutely love Mum's cooking though; her melt-in-the-mouth beef in red wine sauce was a thing of true wonder. Her Christmas lunch had mash AND roasties – potatoes two ways? What a woman.

Her broth was my absolute favourite though and she's now given me her old broth pan. Forget inheriting Great-Aunt Fanny's pearls, I was over the moon with my broth pan, which is silver with a few scorch marks and is big enough to bath an Alsatian in. The recipe involves some embarrassment as the main ingredient sounds like a sleazy bit of cockney rhyming slang – you need a certain confidence to walk into a butchers and ask for a ham shank without breaking eye contact. It's worth it though. The ham shank is so salty that if you were to throw it into the Dead Sea it'd gag and spit it back out, so it needs soaking. Once soaked you lob it into the pan atop a mountain of diced veg and boil the bejesus out of it, adding soaked soup mix along the way. The result is a greyish (stick with me) liquid with a flavour as deep as a mother's love. Each spoonful brings with it tiny treasures of celery, turnip, onion or lentil with a chunk of soft, stringy, pink ham.

If there is a national broth society I'm hoping they'd approve of the PR job I'm doing here.

Sadly, however, no amount of spin can convince my own family to eat it. After hours of me slaving over the stove, my face shiny with hope and condensation, only my eldest will have some and even then I suspect it's a sympathy slurp. No matter, I make it anyway and freeze it. Like salmon are driven to swim upstream to spawn, I make broth because it's in my genetic make-up.

It doesn't sound like much but this and her chilli con carne are the two recipes my mum has handed down to me and they were the two pillars that held up my childhood teatimes.

When I say Mum handed down the recipe for her chilli it's a slight exaggeration. After decades of eating the rich, dark chilli with rice, washed down with cold milk, I thought it was maybe time for her to pass on the secrets of its deliciousness. I was also about to appear with potty-mouthed chef Gordon Ramsay on his show *The F-Word*. The challenge was for us to compete to make the same dish, which would be blind-tasted and judged by a panel. I thought that chilli con carne would be perfect – who could beat Mum's home-made delight? The conversation went thus: 'Mum, please can I have your chilli recipe?' Now at this point I hoped for a hushed whisper, revealing a complicated list of rare spices sourced possibly from a rainforest, but instead she just said 'Oh, I've always used a packet mix, Schwartz is quite good.' I was genuinely gobsmacked that one of my all-time favourite meals had come from a packet. Seems Mum was up for convenience foods after all. Her secret trick was to put half a tin of baked beans into the chilli to soften the kick of the spice for us kids. Suffice to say I lost the cooking challenge and Gordon was almost lost for words that I'd committed such culinary crimes.

Despite the trauma I still make chilli with baked beans for my kids and they like it, whatever Ramsay reckons.

Mum's cooking skills came in handy on the farm as there were often extra mouths to feed, especially during haymaking, one of the busiest times on the farm, when there'd be a small army of helpers turning up to get the hay in. Arable farmers have for centuries watched the skies, trying to predict the weather and fathom its patterns. During haymaking my dad was at the mercy of the rain, which we get a lot of up north.

The far two fields in winter were knobbled and gnarled, but they transformed in the summer and grew lush long grass that towered above me. Once dried out by the warmth of the sun, it turned a beautiful golden and green hue and could be cut and left in long rows, ready to be gathered up by the metal fingers of the baler, swiftly knitted into neat blocks of compacted hay and then dropped out of the back like a huge mechanical goose laying oblong eggs.

For it to be cut and baled it had to be dry, which meant relying on the Great British Summertime. Farmers now use minute-by-minute weather apps to help save time and avoid disaster but Dad just had Michael Fish and luck.

Once Dad had had a couple of days of dry weather and the go-ahead from the weathermen it was all hands on deck in a race against the rain. Make hay while the sun shines isn't just a cute quip. Dad would rally the troops: my big brothers Robert and David plus a couple of their mates, including Andy with the metal-rimmed specs who had a look of Freddie Mercury, and Diddy, a lad from the village who was a right rum 'un. He reminded me of a pirate with his shining eyes, pierced ear and inky black hair. His nickname became the only name anyone knew him by, earned by him being what is technically known as a shortarse. There was also Mel, who helped out on the farm – a lovely man, quiet and considered, who Dad relied on; he sported a five o'clock shadow at 10 a.m. and had a head of dark curls with a matching tache. Then there was John, a family friend – bleached blond hair, toothy grin, daft as a brush – who was as much of a joker as he was a hard grafter.

John would be joined by Kim, his future wife, though neither of them knew it yet – then he could only dream of such riches. Kim had kept her horse on Dad's farm and was a similar age to my eldest sister Dot. She became a good friend of Dad's (now ex) wife. Kim was striking – tall and slender with bright hair the colour of pumpkins that fell in thick waves to her shoulders, and a dusting of pale freckles scattered over her nose. She was a grafter too, strong and athletic. She would muck in with the best of them and could cope with both the bales and the banter. John's face always lit up whenever he clapped eyes on Kim but it'd be another decade or so before she finally gave in to his charms.

Once the welcome cups of tea had been drained and the barn had been cleared and swept, haymaking proper would begin. Dad would be pulling a trailer with a tractor, following along behind the baler, with a couple of people picking up the bales and chucking them up to the men waiting on top of the trailer, balancing with their feet wide as if snowboarding. Chucking a bale upwards is most efficiently done using one smooth motion – lifting the bale first by the two lengths of orange twine holding it together, then a knee to propel it upwards momentarily onto your forearms, the momentum then helping the last upward movement as you push it up and away and onto the trailer.

The key was to get into a rhythm, using each body part like a footballer doing keepy-uppies. It was a schoolboy error to be seduced by the sunshine and wear shorts and a vest top – knees and arms would bear angry red scratches from the prickly bales, battle scars from a day's haymaking.

I was too little to gather the hay but I had the important duty of providing lubrication in the form of tea. As the crew toiled away I would make my way over the fields with a tray bearing cups of hot tea and coffee. It was hot, dusty, thirsty work haymaking, so I would feel like an important cog without which the whole machine wouldn't work. Each mug would be filled and then cling film stretched over the top; a genius idea, otherwise by the time I'd made it over the bobbly field most of the tea would've sloshed out onto the tray, like a small-scale *It's A Knockout* game.

There was plenty of banter, practical jokes and mickey-taking to lighten the load of so much physical slog. As well as the shouted instructions from Dad and Mel, laughter rang out across the field from this merry band of men – they were here because they liked Dad and my brothers, enjoyed the craic and were paid well, in sandwiches and maybe a turkey at Christmas. Mum would construct vast piles of butties ready for when the towering hay trailer came trundling steadily towards the barn for unloading – soft white bread, butter, ham, sliced in half, arranged on big dinner plates and devoured in seconds by the workers, like a swarm of locust through a barley crop.

I loved haymaking; it was one of my favourite times of year. The farm would be bustling, cars and vans bumping up the lane and pulling into the yard as the helpers rocked up. There was a real energy to the place, the feeling of folk mucking in together. The buzz of bringing in all the hay, the camaraderie and little me in the middle, fetching tea and trying not to get under everyone's feet. Once the barn was packed and stacked, fit to busting with new fresh hay and the massive blue sliding doors closed and padlocked, there was a real feeling of achievement. Knowing the hay was safely in and livestock could be catered for for the next year was something to celebrate; back at the farmhouse cans of cold beer were cracked open, more butties made, and the banter continued long after the sun faded over the newly bare fields.

5. Twelve Pots of Pond's

Away from the farm, on the weekends my eldest siblings would let their hair down at the local nightspot. Its name was Touch of Class but it had a few nicknames including 'Touch of Ass' or, my personal favourite, 'Touch of Crotch'. I never got to feel my feet stick to the dance floor of that particular establishment, though; instead I was down the pub. A family favourite was the Queen Anne – a vast red-brick building, a myriad of different rooms accessed along cream-tiled floors through dark wood and opaque glass doors.

These were the glorious eighties – decades before smoking bans, when parents thought nothing of sparking up centimetres from their offspring's faces.

In fact, so smoky were some of these pubs that kids got skilled at recognising their folks by their footwear peeping out below the dense fog of fag smoke. Once through the door of the snug the men became grey clouds wearing brown leather slip-ons.

My mum's mum, Nana, was the only woman who made having a fag seem remotely glamorous. Her packet of Benson & Hedges would be decanted into a silk purple paisley-patterned case with a golden clasp and she would hold her glowing ciggie between long fingers adorned with sparkling rings and always immaculately manicured fingernails, painted fuchsia pink. I have Nana's long nail-beds so they look decent even when cut to the quick – it's kind of comforting to think that once you're gone there are still whispers of you that echo through those you left behind.

Nana always smelled gorgeous, of face powder and perfume. Her golden-blonde hair was set into smart, slightly bouffant curls, framing a kind face with smooth skin and a dab of frosty pink lippy. Nana never washed her face with water, preferring instead Pond's cold cream. Each year on her birthday my grandad bought her twelve pots of the stuff to last her the year. It's not as showy as a dozen red roses but I think it's a thousand times more romantic.

Nana was pretty chic, often dressed in a floral number made at her dressmaking classes. I'd go along with her when I was really little. I remember her and her pals in the community centre, cutting out dress patterns, bent over sewing machines and gossiping over tea and fags.

Nana was a member of a few groups, including the fabulously named Lady Glades. Sure, in retrospect it sounds more like a feminine hygiene product than a secret society but that's exactly what it was, like the Masons but for working-class women. The Lady Glades gathered every Monday night in a room above the Queen Anne pub.

My grandad was in the male version of which the Glades were an offshoot. They had come up with their own female version as they weren't allowed to join the male one: the Royal Antediluvian Order of Buffaloes, or 'The Buffs' for short. Funny to think now that a group of middle-aged men in Bolton called themselves buffaloes. I like to imagine how the name came about in the first place, when the group was founded in 1822:

'OK, so settle down please, gents. Jeff likes Stag but there's too many connotations with that – there'll be no being stripped naked and tied to a lamp-post in this society, unless specially requested ha-ha, no, seriously though . . . Alf said Rhino but they're expensive – they charge too much – little joke there. Reg likes Buffalo cos like him it's heavy, hairy and has a slightly hunched back.'

'Last orders, gents.'

'OK, Buffaloes it is. Let's pop a "royal" at the front to make it fancy. Mine's half a mild if you're buying, George.'

Both groups were philanthropic organisations, raising money for charity, and each had a hierarchy and used badges, medals and sashes. I found the sashes fascinating – I'd only ever seen them on beauty queens, so seeing my nana slip hers on was like watching a grown-up playing dress-up.

But I suspect this generation that came through a war also enjoyed a bit of pomp and ceremony and were comfortable with order and rank. And liked getting together with their pals and having a drink.

On Remembrance Sunday I'd march with my mum, nana, grandad and the Buffs and Lady Glades. It being nippy in November, my grandad Vince would wear a massive dark-chocolate-coloured coat that I thought made him look especially buffaloesque. Maybe it was the pelt from an alpha buffalo he had to wrestle and kill and skin to gain entry to the Buffs . . . unlikely, in retrospect – buffalo weren't common in Little Lever and my grandad was only five foot in his stocking feet.

The procession of dozens of people of all generations, cagoules and perms fluttering in the breeze, solemnly wound its way through the village from the community centre, past the news-agents, a couple of chippies and the minimarket to outside the library, where the names of the village's war dead were carved on a huge stone plaque.

Grandad and his pals all proudly wore their medals on bright ribbons on the left breast of their coats. Some frailer than others, some in wheelchairs. I loved the feeling of us all marching for one reason, moving en masse.

I really got a sense of how important that day was, as the wreaths were laid and speeches made and then someone from the local church band played 'The Last Post' – a few bars of music that still has the power to move me to tears. It's always stuck with me and always will.

Years later, including during the time the tabloids would probably call the 'Party Girl Years', I'd always observe a minute's silence on both the eleventh of November and Remembrance Sunday, finding a quiet corner at work or pulling over in my car to bow my head and remember those who died in the horror of war so I could be free, glaring at the people around me who clearly weren't noticing the hour, the day, the month.

During the war Nana and her girl pals all worked in an aircraft-building factory. She didn't tell me much about it apart from she fixed the wings to planes. She made it sound pretty casual and well within her capabilities – popping a wing on a Lancaster Bomber, no trickier than hemming a pencil skirt.

My grandad Vince fought in World War Two but like many who served he never really spoke about it. Maybe it was more that I never asked, something I thought about too late.

As a nipper it's hard to imagine your elders as young people, in colour rather than in faded black and white. In my mind he was always just Grandad, the slightly grumpy one with a pot belly poking over his belt who would open the door to us with one hand while simultaneously turning down his hearing aid with the other. We didn't mind; we would just run to mither Nana. Grandad had his own armchair that no one else was allowed to sit on, a maroon velvet affair that he guarded like a swan guards its nest. If you dared perch on it for a moment when he'd popped to the loo, you knew to scram sharpish on his return. To be fair, Grandad's health wasn't the best – he had a vivid purple scar that ran from his chest and down his belly from his heart bypass and was often breathless and tired. But he never seemed frail to me – little yes, curmudgeonly sure, but never frail, and to the day he died he had a full head of shiny black curls and a large black tache – he looked like one of the old dudes playing dominoes in the background of *The Sopranos*.

Down at the Conservative Club he held court like a Mafia boss with his crew. He was naturally funny and quite the raconteur – he loved to make people laugh with his stories. Mum reckons I've him to thank for passing down the entertainer genes that have stood me in good stead in this crazy business we call show.

Vince had been a master baker (careful how you say that) but he had dreams to hang up his apron and take to the stage; he wanted to be a stand-up and even went to a 'comedy school' – night classes teaching the necessary skills to be a stand-up – but in the end he didn't get much further than writing down some of his favourite funny stories and jokes.

After he died Mum came across a small tan leather pocket-book full of his jokes. Because I love an emotional moment I savoured opening it, slowly leafing through the pages, hoping to maybe be brought to tears of both mirth and wonder at these incredible lost works.

The contents did actually take my breath away, not so much with their majesty but because . . . how do I put this? They were very much 'of their time' – written in the 1970s when nobody batted an eyelid about jokes about minorities cos it was just a bit of fun. I reckon if I'd told a couple of these gags to Bernard Manning even he might've thought I was out of order. Put it this way, at the time I was going out with a black guy and now I'm married to a Jew and I couldn't have shown the book to either of them.

When a few years ago a TV company asked me to take part in a history show called *War Hero In My Family* for Channel 5, I jumped at the chance. I was hoping to find out who he was before old age, deafness and a triple heart bypass moulded him into the Grandad I knew.

Old photos of Grandad as a young man showed him in his army uniform. He had more than a touch of David Niven, with a glint in his eye and a slight smile tipping up the corner of his moustache.

His army records revealed a mischievous lad who'd once, as a dare, nicked an officer's Land Rover and done a few laps of honour of the exercise yard with pals clinging onto the back, a joy ride during a joyless time. He was hauled out of the driver's seat and thrown before his superior, who handed out a fine and made a note of the disciplinary action on his record. It was amazing to be there seventy-odd years later looking out over the same exercise yard in which Private Vincent Fulford permanently blotted his copybook.

Our search led us to Italy, where, although obviously it was no longer wartime, I feared for my life a couple of times. My poor producer wasn't overly familiar with driving abroad, and inched along the dual carriageway, winding through the mountains in the style of a one-car funeral cortège with a slow puncture. We nearly needed a funeral as the Italian wagon drivers became incensed by her leisurely pace in our tiny hire car and roared up behind us in their trucks overloaded with fruit and veg, so close that we could almost smell the sweat forming on their furious brows.

We arrived in one piece at our location in the rolling hills near Naples. Once the scene of a desperate battle, now it was lush, green and peaceful, the only noise being birdsong. The idyllic surroundings made it even more difficult to imagine the horror that happened here in 1942.

It was my job to explain to viewers exactly what had happened – the Italians advancing on the British army in the valley below. How my grandad drove a makeshift ambulance – a Land Rover Defender – down into the battlefield and loaded up and ferried the injured back to safety.

I can't imagine how scary that would've been – bullets whistling past him as he drove back up the rough track with his fellow soldiers, his pals piled in the back as he bounced over the rough terrain, screaming in agony from their injuries, some already dead and others dying en route.

I really wanted to do the story justice and get across the nightmare that had unfolded in the beautiful valley below. I was to stand on the edge of a sheer drop and deliver my lines.

The only problem was that at that moment Grandad and I had more in common than any of the crew knew – as well as a penchant for performance my grandad also bequeathed to me a crippling fear of heights. I vaguely remember him lying on his back on the floor in their cluttered front room with his maroon-slippered feet up, when he had a vertigo attack, his little legs barely reaching the seat. I too have had to excuse myself from various social gatherings on charming roof terraces to avoid completely ruining the barbecue by flinging myself over the edge and onto, in my mind, the 'safety' of the pavement below, even if it's four storeys down. The fact that I would've, to keep with the Italian cuisine theme, splatted into a margarita pizza seems irrelevant to the vertiginous mind: the floor is where it's at. Only a surprisingly small voice in your subconscious whispers not to jump. Freaky.

So there I was, standing in the spot where Grandad bravely shuttled his friends to safety, and I could feel the drop behind me luring me backwards into its delicious void. To be fair, I was a good couple of metres away from the edge, but I still felt panicky, the palms of my hands seemed to have developed their own sprinkler system and my legs felt as sturdy as spaghetti. The cameraman looked puzzled as I kept taking fairy steps towards him and away from the abyss. If the director hadn't stepped in I probably would've kept going till my breath fogged up the lens.

I explained in a small voice how frightened I was and of course they let me move further away, though the view in the distance was even more distant than they'd intended, and I'm not sure the piece to camera had the same impact. It wasn't my most intrepid reporter moment. You've never heard Kate Adie say 'Ooh, can I just go and broadcast from behind that van – the gunfire really makes me jump?' or Sir David Attenborough refusing to film next to a giant African ants' nest cos they give him the heebie-jeebies.

At the end of the shoot the producer gave me a large brown envelope that I was to open on camera. As you'll know, with these sorts of shows emotional moments are encouraged, compulsory even. A sob, a sniffle, even a subtle lip-wobble would suffice. They set up the shot. I was to sit at the foot of a large tree and read the contents of the envelope.

The crew were speaking in hushed tones, clearly trying to create the right atmosphere so that when the director murmured 'Action' I'd feel able to openly weep at what I found, creating light entertainment magic as ex-party girl and 'Former Hellraiser' (copyright *Daily Mail* and Jeremy Vine's nickname for me) broke down in racking sobs.

The pressure was on.

I opened the envelope and peered inside. I was relieved – the sombre mood had for a second made me wonder just how bad things were. Was it a photo of Grandad wearing a French maid's outfit while cavorting with German officers? Did it, in a strange and unexpected twist, contain my surprise adoption papers?

No. It was in fact a letter from my grandad's group captain. I learned that after the war, when soldiers were discharged from service they were given a document that was somewhere between a school report and a CV to help them blend back into their former life on Civvy Street and rejoin the workforce. As I held the photocopied papers in my hand and read how Grandad had shown great bravery under pressure and had proved to be an honourable, hard-working soldier, I felt beyond proud.

Like farmers in a drought scanning the skies for rain clouds, the film crew watched my face, poised to zoom in as my emotional heavens opened and the first sweet drops fell. But my cheeks remained bone-dry. Usually I'll skrike at anything – news reports, songs, butter adverts – but like a kettle being watched that doesn't boil I couldn't just cry on demand.

I felt their eager faces watching me closely, so as I turned and delivered my final piece to camera I paused and stared thoughtfully into the distance and into a handy breeze, which at least made my eyes moisten a touch.

I came back from that trip with a fear of Italian HGV drivers and, more importantly, a fresh new respect for Grandad. I felt like I'd seen past the belly and hearing aid and met the brave, bright young soldier, popular with his fellow soldiers and respected by his superiors.

6. Bee's Knees, Ankles and Toes

Despite nearly dying at the hands of wagon drivers in Naples I have a huge respect for the much-maligned haulage community, thanks to me and my dad often hitting the road in his DAF lorry.

When you're one of five children and you're vying for attention not only with them but with a whole herd of cattle, time alone with your pa is so precious. As well as being a farmer my dad also drove a wagon. He was basically a man with a van but on a slightly bigger scale. He was freelance and had a rotation of regular clients who'd need their goods – from paint to tiles to garden furniture – transporting all over the UK.

Dad drove a modest flatbed DAF with a bright-blue cab. The back, as the name suggests, is flat, so the goods are loaded onto it on pallets and then covered with vast sheets, which are secured with long ropes thrown over and tied to hooks running along the underside.

I absolutely loved going with my dad in the lorry. I did it from the age of about seven, when I guess I was big enough to obey instructions but small enough to fit on the little bunk with him, up until the age of twelve-ish, by which time I was quite skilled at 'roping up'.

We'd rise ridiculously early. Or so it seemed. On winter mornings he'd wake me at 5 a.m. while it was still pitch black outside. Dad is a gentle soul and always woke me up really nicely, stroking my hair till I came to.

Mum on the other hand was much more vocal in her style, enthusiastically encouraging us to wake up by loudly shouting our names and telling us the time while whipping open the curtains and switching on lights.

Looking back of course I know it wasn't because Mum was a tyrant and Dad was an angel – it was because Mum was in charge of getting everyone up and off to school, and despite my efforts to be more gentle with my own kids, again this apple hasn't fallen far from the tree when it comes to mornings.

I try to coax my own offspring awake with kisses and whispered welcomes to a brand new day but with the clock ticking, dogs needing feeding and three of them to get up and out for 7.40 a.m. I usually resort to the curtain method. Recently I've had to try my very best to be more soothing, as my son banned me from waking him on account of my harshness and requested for his dad to be on wake-up duties instead.

On those dark mornings I really appreciated Dad's bedside manner. It didn't take much to get me up – not only was I excited to be going out in the wagon with him but I also saw myself as some kind of colleague or assistant on these trips and I didn't want to risk holding him up in case it jeopardised future invites.

Sure, being seven and feeling like a co-driver on a haulage job was totally normal. It felt like me and Dad against the world.

Traffic jams, middle-lane hoggers and tachograph frustrations held no fear for us if we had each other on these road trips. Like Thelma and Louise, if Thelma was a wagon driver with a bald spot and Louise was a scrawny nine-year-old in a hand-me-down polo neck.

I would get dressed quickly, partly due to keenness and partly cos it was always bloody freezing at the farm. Downstairs to the breakfast bar in the kitchen, feet swinging off the high stool as Dad dished up Shredded Wheat with hot milk.

And we were off. Well, once the wagon had rumbled into life, the heaters had kicked in and the windscreen mist had cleared. We'd bump down over the ruts and through the puddles on the farm lane and make our way through the sleeping village to the glorious M61.

All my school holidays would be spent out in the wagon with Dad. It felt like anything was possible on these trips – they were a real adventure to me and the world was our oyster. I imagine Dad saw them less romantically – loading up pallets of undercoat in Yorkshire and delivering them around Cumbria.

Dad is a quiet guy and so the deafening roar of the engine on these trips meant quality time together without the pressure to talk. I'd watch Dad and think he was not only the bee's knees but the bee's ankles and toes too. I loved stealing glances from the passenger seat as he drove this massive wagon. I thought he was the most handsome and brilliant man in the world.

We were higher up than most of the other vehicles on the road and I loved looking down from our lofty position. On long journeys with my own kids now the back seat looks like we've done a ram-raid on WHSmith – so many books, colouring pens, magazines and snacks to distract them from the hours on the motorway, as well as iPads and phones.

I didn't take anything to entertain myself in the wagon, though – I was quite happy to watch the world whizz by through a windscreen that seemed as big as a billboard.

I thought it was so cool how wagon drivers would flash their lights at overtaking lorries to signal that it was safe for them to tuck in in front of them, and in a thank-you for the courtesy the overtaking lorry would indicate left-right-left as they manoeuvred into position. It felt like a community – lorry drivers looking out for each other.

People sitting in the middle lane are the scourge of wagon drivers as the middle lane is their fast lane and if it's necessary to take evasive action and overtake a very slow wagon then it can be pretty dangerous if Clive and Claire in their Vauxhall Zafira are pootling along at sixty-eight in the middle lane, so engrossed with PopMaster that they're oblivious to the twenty tons of truck on their left needing to swing out to avoid a pile-up and potentially recycling the Zafira and its contents into a Turner prize entry.

I've not seen Dad angry much but middle-lane hoggers proper wound him up and I'm the same now – I never plonk myself in the middle lane and when pulling in front of a wagon I use my indicators in the left-right-left sequence favoured by HGV drivers as if deploying a secret handshake to identify myself as a member of their secret society.

I like to imagine they think 'The blood of a rich haulage heritage flows strong through yonder lass, she's one of us' and not 'Why's that daft cow in her Volvo indicating all over the shop?'

These trips with Dad were in the days before satnav, so Dad would smooth a bedsheet-sized map out over the steering wheel as we searched for an anonymous industrial estate tucked away somewhere in the drizzle of Yorkshire.

On the road with Dad we never really stopped at services except to use the loos. I was desperate to go to these brightly lit meccas of Big Macs and pick 'n' mix but they were seen as extortionate so instead we'd go to truck-stop cafés where ladies with upper arms the size and shade of partly defrosted legs of lamb, blue gingham tabards straining over their matronly bosoms, dished up sausage and chips and sloshed out our brews into white mugs.

I always felt quite like Sophie in *The BFG* in these places – everything seemed super-sized, from the truckers' bellies to the fry-ups they were wolfing as they leafed through the *Sun*. I was often the only kid in there, so I enjoyed feeling like a stowaway among the giants.

Sometimes we wouldn't have the fine dining experience of a truck-stop; instead Dad would pop into a shop and emerge with a pack of bread rolls and a packet of ham and, without cutlery (we were hardcore truckers, not picnicking at Downton!), he'd use his thumbs to create a crevice in which to poke a couple of slices of the ham. He was ahead of his time really. 'Len's Ham Pockets' could've been a thing if 'ham pocket' didn't sound like some horrible Shakespearean description of a bar wench's genitalia.

Whenever we were at a depot, collecting or picking up, I usually had to stay in the cab to dodge the wheels of the fork-lift trucks buzzing around. But if being confined to quarters in the cab was the dullest bit, one of the best bits was the occasional visit to a pub. ''Ey up,' Dad would say, 'do you fancy some pop? Let's go have a quick leg-stretch.' Zipping up his body warmer and smoothing down his hair we'd set off, me trotting alongside trying to keep up with his long stride.

The bracing walk took us from Dad's parking spot outside the depot we were delivering to first thing, through Whitehaven town centre, past the fishing boats bobbing about in the harbour, huddled together as if for warmth, and through the doors into the bright lights and cosy welcome of the local pub. After weekly trips delivering on this route for a few years Dad had become pally with the landlord, who bent the rules to allow me in. 'Alreet, Jeff,' Dad would say, 'pint of Guinness for me please and a lemonade for the ankle-biter.' There'd also be Smith's crisps, cheese and onion flavour, the green bag ripped right open by Dad and laid out like a sharing platter.

After a couple of drinks we'd head back to the wagon. I loved the sleeping arrangements. It felt a bit like camping. Behind the seats in the cab was a bunk that was about the width and length of a stick of chewing gum. Somehow all six foot two of Dad and me with my large head both squeezed onto it in sleeping bags, assuming the classic 'top to tail' position. It's a mystery how we got any sleep at all, but I imagine we were both knackered – Dad from map-reading and being angry at bad drivers and me from gawping out of the window – because after a quick 'Night night, sweetheart, love you' from Dad, we would both immediately zonk out.

The next thing I'd know it would be morning and Dad would have already wriggled noiselessly off the bunk and be busily going through his morning routine, accompanied by the strains of Derek Jameson braying and shouting his head off on Radio 2's breakfast show.

The cab of the wagon consisted of a driver's seat, a passenger seat and a big, hard plastic dome bit between the two that hid all the technical gubbins necessary for a truck to work, possibly involving spark plugs or something.

There was the thin bunk that ran behind both seats and, of course, a glovebox, full of tachograph cards.

That was it. But in my dad's hands this small space became like Mary Poppins's bag. From every nook, cranny and crevice Dad would produce stuff, like a plastic washing-up bowl for us to wash in – 'But hang on, Dad, we have no water!' – eau yes we do; he'd produce a big container of H_2O to soak the flannel and soap he'd just whipped out of a toiletry bag that he'd magicked from Lord-knows-where. Toothbrushes and toothpaste would be produced with a flourish, along with a fluffy towel. Finally Dad would dig out his electric razor and shave his greying bristles, gurning as he did so.

In my mind's eye it's like a movie montage flashback, Dad whipping out all this stuff, me watching all shiny-faced and goggle-eyed to the strains of Wham!'s 'Wake Me Up Before You Go-Go'.

The most regular run Dad and I would do would be the Cumbria one: Carlisle, Workington, Whitehaven, then round the coast to Barrow-in-Furness.

The last leg to Barrow involved Dad negotiating his wagon round roads that seemed to wind endlessly up and up with a stomach-bothering sheer drop at one side. I hated that road; with my fear of heights I always felt like we were inches from certain death – just one wrong twitch of the steering wheel, a violent sneeze perhaps, and us two plus wagon would hurtle over the edge to oblivion. Dad would never have admitted he was nervous driving this route but his face, set in grim concentration, spoke volumes. I always felt safe though.

Whitehaven was where a couple of tins of paint would always seem to fall off the back of the wagon and we'd then mysteriously acquire a massive haul of prawns, fresh from the sea and triple-bagged in plastic packed with ice to keep them cool all the way home.

When we arrived back it was my job to shell them in the

outhouse sink, snapping back their grey heads and gently tugging out the long black thread that was their intestinal tract. Like I say, these trips had ups and downs both literally and figuratively. The main up was the special time alone with my dad and the downs definitely included getting prawn poo down my nails when we returned home.

7. Seventies Tangerine

To get into the farmhouse you first went through the porch, which was added on in the nineties. This was painted white and housed a small loo cubicle so you could pop for a pickle without all the fuss of wrestling your wellies off. The porch was also a handy place to store mucky wellies and wet dogs.

From there and into the kitchenette through an ancient wooden door, made from wide planks joined vertically together and painted black with a cast-iron gate latch that had a knack to opening it to get in and on the other side a huge, cast-iron bolt as big as one you'd see on a castle door, to keep people out.

On the back of the door hung so many waxed jackets it was as though the door had grown its own shell from multiple layers of thick, dark-green skin. Behind the door was a dog bed, cosy next to the airing cupboard, a cork noticeboard beside the land-line that was hung on the wall, and in later years in a never-seen-before (or since) embracement of technology, a cordless handset with a metre-long aerial, like a radio control for a toy boat.

There was a small padded bench underneath the phone, a

perfect perch for yakking on the blower or taking off your wellies. The metal sink that Dad alleges I used to be bathed in was under the big square window looking out over the yard. The cupboard underneath the sink held all the medication for the animals, like something from the pages of Harry Potter – old-fashioned medicine bottles, huge syringes and brown glass ampoules jumbled up with plastic plungers, bandages and aerosols of purple antiseptic spray.

Next to the sink stood a tall freezer, often containing half of one of Dad's cows, and the top of which was the domain of Cindy, the dark-brown Burmese cat, who I loved. Like a feline version of Angie Watts, she was clingy and fiery. Depending on her mood she'd either be mewing loudly for your attention from her furry nest in the sky or would silently swat at the top of your head with a paw as you passed, just so you didn't forget who was boss.

When she wanted love she'd meow until whoever was nearby would eventually give in and stand still next to the freezer so she could climb slowly down onto their shoulders, using her front paws to inch down the vertical drop of the freezer's plastic door until both gravity and common sense suggested she was close enough to make the leap onto their upper back.

You'd have to brace yourself, for two reasons – not only was she pretty sturdy and heavy for a cat but for the first couple of seconds after landing she'd use her powerful claws to gain her balance; then you could set off to the sofa or wherever with Cindy hitching a lift too, her long brown tail helping her to balance, wrapping round your neck and across your face like a rogue handlebar moustache.

Cindy and our Jack Russell Sally were sworn enemies, so she'd always time her hitchhiking with Sally's absence and skedaddle the moment the dog bustled into the room.

*   *   *

Like most houses the kitchen of the farm was, and is, where all the action is, the heart of the home. Although the cupboard doors have moved with the times – from seventies tangerine to eighties pine to a more understated cream in recent years – the shape has remained the same, a square of units with windows looking out towards the coal shed and the small garden, the public footpath and gate beyond. As you enter from the kitchenette you get to the breakfast bar and wooden stools.

It's in this room that Dad knocks up a cracking roast beef and one of his 'home-grown' turkeys at Christmas. It's here that everyone gathered after haymaking or during calving season. It was in this room that Robert frantically shook the ketchup bottle without checking the lid was on properly, liberally spraying the walls behind him with the red stuff, much to our delight as we all watched agog from the safety of the breakfast bar. It's in the kitchen that a teenage David helped himself to a jar of yoghurt only to find out it was actually worming liquid for the calves that Dad had put in the fridge to keep cool, and it's into here that Robert came home drunk after a big night out and liberally spread pâté on some hot buttered toast before falling into bed. He was told the next morning that he'd actually snacked on the dogs' brawn.

Rivers of hot tea flowed as births, marriages and deaths were discussed here; all life was in that room and it was all debated over a couple of fig rolls at the breakfast bar.

Sitting in the kitchen never felt particularly private – in summer the door that led out across the concrete pad and to the coal shed was always open and people would often be popping their heads round, from the girls who stabled their ponies in the yard asking for the barn key, to the local policeman saying how-do to Mick the Tip – no, not a gangster from the Mafia who got his name because he liked to lure his victims to the local tip but a wagon driver who kept his tipper truck on Dad's land, bouncing up the lane every night, the chains on his truck clanging like a haulage ghost of Christmas past.

If the kitchen was a hub of activity then the lounge was a sacred place of relaxation, as this was where the carpet began and may the good Lord have mercy on the souls of those foolish enough to venture from the kitchenette to the lounge without first taking their wellies off. The lounge was where Dad could finally relax after a long day. In winter evenings a coal fire would be built and lights dimmed, and humans and animals alike would snooze in the soporific glow of the flames. The fireplace was surrounded by brick that travelled about two-thirds up the wall, upon which Dad's ornaments of porcelain Herefords grazed, a bull, cow and calf. I loved the porcelain calf and would always secretly play with it.

The lounge was a long rectangular room with an oxblood leather chesterfield sofa and matching armchair in front of the fire, with a telly the size of a skip at one end and a polished mahogany dining table at the other. It was at this table that I broke the news to Dad that I was a vegetarian and it was at this same table that I tucked into a leg of lamb a week later when the whiff of a cunning plan cooked up by Dad worked a treat and the smell of the roasting joint wafting in from the kitchen overwhelmed me and banished any notion of not eating meat.

New Year's Day buffets were spread on this table and home-work pored over, all watched by a fox's head that's still there today and is the focus of much macabre curiosity from my kids. It is a fox that met its demise on Dad's farm and is mounted on a wooden plaque with its tail hanging on a hook next to it, which to me seems in poor taste – bad enough you get shot and stuffed but then to have what was previously dangling off your rear end placed next to your face for eternity seems a little unfair, like having your own arse stuffed and popped up next to your head.

The fox obviously has some fake bits – glass eyes and plastic gums and teeth set in a snarl – though its expression is one of slight surprise, which I guess yours would be too if you were met with the wrong end of a farmer's gun.

Dad actually wouldn't bump off foxes if the red-coated buggers would just pop in to steal one hen for tea and leave the rest alone but, in a henhouse, foxes will decimate the entire coop and barely eat anything.

The upstairs was made up of three bedrooms and a bathroom where the shower used to go cold whenever one of the cows would take a sup from the automatically refilling drinkers; the house shared the same water supply as the cattle sheds and it was all connected – at least that's what Dad used to tell me. 'DAAAAAD! The water's gone freezing!' I'd holler and he'd shout his reply 'The cows must be thirsty, it'll warm up again in a minute.'

Us three girls used to share the biggest room and there were bunkbeds on one side and a single bed on the other. It's the room my husband and I sleep in now when we visit and it's tiny, although it seemed massive to my six-year-old self. It's in this room that I'd sashay up and down singing Gloria Gaynor's 'I Will Survive', coached by my sister Dot. Across the landing from where Dot would swing me over the stairs on the draft excluder was the boys' room.

The farmhouse always has a slightly wonky feel to it; the carpeted floors always seem slightly sloped and creaky, like the whole building has been pushed to one side by a giant hand and then hastily straightened. To be fair, one side wall of the house had to be knocked down and rebuilt at one point, the bricks removed to reveal the boys' bedroom, like an opened-up doll's house.

From the outside during my childhood the farmhouse looked like a child's simple pencil drawing – painted white, two windows upstairs, two down, a door and smoke coming out of the chimney. Only in winter obviously, and only at night-time. The only time I remember a daytime fire in the hearth was on Christmas Day and New Year's Day, and the rest of the time the house was – and still is – a bit parky. One winter there was a heavy snowfall and the house was absolutely freezing, so Rob begged Dad to let him light a fire, to which Dad responded by pointing at the three inches of snow on the roof and saying, 'See that snow, it's insulating the house so it can't be cold.'

These days my husband reckons it's colder inside the farmhouse than out but he's wrong – there's no such thing as a cold house, just the wrong jumper thickness. Dad used to be a firm believer that central heating made you ill by making you warm and cosy and – I imagine – a bit soft and therefore a more attractive target for germs. Not sure how the science of this plays out but let's just say Dad is still going strong and is to hardy what James Bond is to suave. My brother Robert has a fondness for made-up medical facts – for example, that long toes are a sign of intelligence (he himself has very long toes). I can see where he gets it from.

The house had a long double garage behind it that could've been a prop from a horror movie – made from wood and painted treacle-brown, it was always dark inside and was cluttered with junk and tools on the floor and old thick cobwebs hanging like grey garlands from the rafters. Its pitch-black interior meant that the bare bulb dangling from the ceiling couldn't throw enough light to illuminate the very rear, so it was as if the garage just dropped off into a black hole. The boys tinkered with their motorbikes in there but it was a self-imposed out-of-bounds area to me.

Beyond the garage was a small grassed enclosure that ran from the rear of the garden and along the tarmac path that led down to the estate. In recent years a small allotment area was created by my brother-in-law Gary, where sprouts sprouted and cabbages cabbaged with huge success – Dad reckoned the soil was so rich thanks to it being a former burial site for many of the farm's pets.

Before the deathly vegetable patch, my pony Gus would some-times be put in for a pick of grass, as would Leeroy the goat or the occasional orphan; not the curly red-headed all-singing all-dancing variety – no, Dad's mate farmed sheep and during spring we often adopted a little lamb after it was orphaned at birth or if it was one of twins and its mum didn't have enough milk for two. This was the stuff of dreams for a young girl – these little spares came to us to be hand-reared. At first it was pretty full-on – like any newborn they'd need feeding every few hours, including through the night. The lambs would be white, fluffy and the size of a big cat, eighty per cent skinny legs like pipe cleaners, with big doleful brown eyes, long pale eyelashes.

Their feed bottles would be the size of big glass pop bottles, with a long black rubber teat on the end. They'd suckle furi-ously and as they got bigger feeding them became like pulling a cracker that wouldn't crack, the lamb pulling one way while you'd try to cling onto the bottle.

The lambs would snooze in a big cardboard box filled with a couple of old jumpers before graduating to a spare dog bed, until eventually they were strong enough to be turfed outside. I'd read enough children's fiction to know exactly what was supposed to happen with these lambs – a strong bond would form, the lamb would follow me everywhere like a dog and people would comment in wonder at our special relationship, the likes of which they'd never witnessed in all their years dealing with sheep/children. The lamb would blossom into a sheep and we'd go through a rocky patch after a misunderstanding but then one day the sheep would rescue me by raising the alarm after I got wedged somewhere remote. Sheepy would then escape the chop and not become chops – and we'd head off into the sunset, hand in hoof.

What actually happened was I'd get terribly attached to these furry friends and one day they'd disappear. Just like that. Gone. Laters. TTFN. The clue could've been in the names my dad bestowed upon twin lambs who ended up in our care - 'Freezer 1' and 'Freezer 2'.

Still, I didn't ask too many questions of Dad as he's known for his no-nonsense answers, plus I'd grown up with the legend of Larry (there's more on this later) as one of our family folk tales, so instead I chose to believe they'd gone to live somewhere charming, possibly with a name like Honeysuckle Farm, where they had all the sweet grass they could ever need.

8. Nuzzle that Muzzle

Returning to the farm after a trip away with Dad always gave me a sinking feeling.

As we slowly bounced up the farm lane I knew that coming home meant I had to share my dad again – with my siblings, the farm helps, the cattle and also Dad's then wife.

By this time my parents had divorced – a drawn-out, traumatic time for everyone except me; the heartbreak was over my head as I was only six at the time. Even though the consequences rippled through our family for years – the hurt, the divided loyalties, the politics – I was cushioned from it by my siblings, especially my eldest sister Dot, who was always there with a hug and a gentle explanation of any drama.

By now I was living half a mile away from Dad in a little council house with my mum, stepdad Ken and sister Yvonne. The three eldest had remained on the farm with Dad and his new wife.

I still had my old bedroom at the farm and spent loads of time up there as it was only a ten-minute walk away and there was a pony who needed mucking out. Still, coming home in the wagon meant back to reality and the end of mine and Dad's adventure.

The pony that needed mucking out was the result of our fat little Thelwell-esque Exmoor pony Muffin having a dalliance with a local part-Arab stallion. This in itself was a marvellous yet puzzling thing, as Arabs are the equine equivalent of annoying people you see dancing in bars in Bacardi adverts – sexy, vibrant, beautiful things, big eyes flashing and slender hips swinging.

Muffin, however, was less white rum and more half a stout – she was no looker, bless her; I can only presume she had a great personality or the stallion had his sugar-beet goggles on.

Quite the opposite of a leggy filly, she had short, sturdy legs placed at all four corners of her like a small coffee table. Now imagine the tabletop had begun to give and sag under the weight of a large barrel that had been perched atop it and you've kinda got her silhouette, her low-slung belly almost grazing the long grass of her field. She was a bright bay colour and so unbeliev-ably stubborn that any mule who encountered her would've concluded that sometimes compromise can be a good thing.

She provided hours of fun though for Dot, who rode her endlessly round Dad's fields. Muffin was like a little tank – you could trundle around on her and no one got in her way, calves and ground-nesting birds scattering as she approached at a thun-derous trot. As long as you didn't expect to be able to steer her or adjust her speed she was a cracker.

The result of Muffin's dalliance with the stallion was a little bay colt with a white star on his face and one white sock on his back left ankle.

He was born in the early hours of 1 August 1979 and was named by Mum, who chose a moniker subtly inspired by his birthday – Augustus the first, Gus to his friends.

And he had many friends – Gus grew up to be a beautiful boy. He grew bigger than his mum and ended up 13.3hh. He was definitely sturdy, but slinkier than her thanks to his dad's genes, with a pretty face and huge dark eyes with full lashes.

He was super-affectionate with a personality as big as his arse, which was sizeable thanks to Muffin.

I truly believe that the relationship between a young girl and her pony is pretty unique. I think Gus was probably my first love – I spent so many hours whispering sweet nothings into his ear and nuzzling his muzzle, breathing up his nostrils, fancying myself as a bit of a horse whisperer. I'd clasp Polo mints in my lips and Gus would gently nibble them from me. Who needed real boys when I had this dark, handsome chap to cuddle, waffle on to, kiss and love? And Gus didn't want me to drink cider with him or try to put his hoof up my top.

When teenagerdom struck, I'm sure my dad felt this last point keenly. As members of the opposite sex began to catch my eye and I started to swerve slightly from Gus, my dad knew that keeping me interested in ponies would at least delay the inevitable.

I'd been riding ponies since the age of six. Dad used to take me to a local riding school. I loved it – each week you'd ride a different little pony for an hour, all of our charges following nose to tail round the edge of a large indoor school. We'd be put through our paces from walk to trot to canter, go over trotting poles and at the end of each lesson practise our balance exercises – the excitingly named half scissors and more advanced full scissors, where you'd have to swing your legs around like a shonky circus act without sliding off the pony and onto the dusty floor, and 'Round the world' – basically this involved taking your feet out of your stirrups and awkwardly shuffling round in the saddle till you had rotated 360 degrees, while hoping your mount didn't suddenly set off for a jog. Luckily the experienced little ponies saw it as a perfect opportunity for a quick standing-up nap.

On the way home we'd stop off for chips and I'll always remember the guy behind the counter giving me a single chip wrapped in paper. Either I had extraordinarily tiny hands or spuds back in those days were the size of rugby balls, because I remember holding this one chip, gripped in my fist, and it was the size of a Crunchie.

As well as the mammoth chips, my other main memory of that time is of my massive head almost thwarting my equestrian ambitions. My dad took me to a tiny saddlery to fit me with my own riding hat. Before then I'd popped on any old adult hat that was rattling round the farm – dusty affairs with no chin straps, the torn velvet faded by the sun. Now, though, with the introduction of Kitemarks and safety standards, Dad thought I'd better get a proper hat.

I loved and still love a saddlery – these days if I treat myself or my eldest daughter to some horsey gear the smell of leather from the tack takes me straight back.

There's something so special about a shop dedicated to one hobby or pastime, because not only will you find everything you could ever need for your passion all under one roof but you're only ever in there with other people who also love what you love. It's the same with all specialist shops – no one ever really goes browsing through the bait bins in a fishing shop – 'Ohh, I might treat myself to a pint of red maggots!' – or idly peruses the trombones in a brass instrument emporium. These are more than shops, they're hives for kindred spirits to gather and nod admiringly at a split cane bamboo fly rod, curved soprano sax or, in my case, new riding boots.

The shop I went to with Dad was tiny with high shelves, reachable only by one of those ladders that wheel along rails, heaving with everything from lead ropes to boots and brushes.

Despite having a nogginous maximus I manage to go about my daily business and, apart from occasionally knocking it on low-hanging shelves, I cope. The only time a big head becomes a real problem is when one needs to borrow or buy headgear.

The trouble is that my head is deceptively large and, as I've painfully learned, people aren't always willing to accept its true size.

So there's me, six-year-old Sarah, standing in the saddlery with my dad. The shopkeeper eyes me for a few seconds and then reaches under the counter, producing a box containing a beautiful velvet navy hat. She whips off the white tissue paper with a confident flourish and pops the hat on my head. It's too small and perches like a thimble on a beachball. She seems puzzled but nevertheless produces a hat one size bigger; again it's too small. She repeats this about four times and each time becomes more puzzled and more determined to make the hat fit, trying to wrestle it further down each time, forcing it over my increasingly tender scalp, my face reddening as it becomes perfectly clear I have an abnormally huge head. What next? Maybe Dad will sell me to a travelling freak show or the circus. I'd spend my days combing the bearded lady's face and my nights displaying my planet head to the crowds.

Luckily, out of nowhere (well, I imagine out of the back room) the lady's husband appears with a hat that he's clearly dug out from some specialist area as the box is so dusty. The huge hat slides on like it was meant to be. Relief floods the room – Dad hasn't had a wasted journey and I've got my Kitemarked, navy-blue-velvet-covered hat.

A repeat performance of this scenario occurred again about

twenty years later, this time with a motorcycle helmet in a biker shop in Essex with my then-boyfriend, who rode a very fast bike around very winding country lanes and wanted me to join him on the pillion. The scalp and my ego were just as bruised as the first time.

Riding was something I just loved. I always felt, and still do, that when I'm not riding regularly there's a piece of the jigsaw missing – I have that feeling that there's something not quite right, like Kevin's mum in *Home Alone* just before she realises she's left her youngest child behind.

Where there is light there must be darkness, and what goes up must come down. The darkness and the down came in the formidable shape of my riding instructor.

The shadow she cast was a big one in all senses. She was my stepmum's mate and they'd have a catch-up over a Mellow Bird's and a couple of fags before we went up to the field where she taught me. To be fair, maybe I was an easily intimidated child, but to me she was truly terrifying. I got the distinct feeling that, like the lost sibling of Miss Trunchbull, she loathed children. No smiles or chit-chat from me could melt the frozen Arctic tundra that was her heart. What she lacked in empathy she made up for in disdain. In jodhpurs and boots, her cropped bleached hair standing stiff in the breeze and holding a whip in one hand like a Disney villain, she'd stand in the field as Gus and I trotted in a circle round her.

When I struggled to get Gus to do as I asked, she'd holler instructions dripping with sarcasm, shredding any wisps of confidence I may have had in my possession.

I hated those lessons; they were my most miserable hour of the week. Seeing her thundering across the yard made my heart thump with dread.

One of the consequences of being born with a dislocated hip is that my legs are kinda wonky and no matter how many times she shouted at me I just couldn't keep my knees in the correct position – so one day she tied my stirrups to the saddle's girth with some baling twine, then tied my feet to the stirrups. Now, I'm no health and safety nut but even I knew that to be tethered to half a ton of animal with its own mind was pretty dangerous. Gus was a sweet pony but like all ponies had irrational fears and would occasionally take off if something spooked him (mostly carrier bags and buses). He was also a bit clumsy, and if he'd tripped with my feet tied into the stirrups I could've been crushed and killed.

Luckily none of that happened. But when I was untethered my knee was so sore from being held in an unnatural position that I came up with some cruel fantasies about my instructor from hell being sucked into a slurry pit or trampled by a thousand cows.

As I limped away from the lesson that day I think we all knew a line had been crossed. I must've said something to Dad because lessons with her soon stopped.

Riding should be fun, and I had to make up for lost time. I did just that on what were known as the sandy paths. Just a fifteen-minute hack down a disused railway line, these paths wound around a huge country park away from roads and civilisation. Clip-clopping past the stables and a couple of rows of cottages trussed up in flower-covered trellises and into the paths felt like entering a whole new world – the Kingdom of the Ponies! Like gangs of outlaws romping round the Wild West, herds of teen girls trotted their fat ponies round the paths, pausing only to catch their breath or smoke a mint Consulate. I knew a few of the girls but I was a bit of a loner – I've always been perfectly happy with my own company.

Plus, as a teen I loved it being just Gus and me – off on our escapades. I just loved the silence; the space to breathe, the only sound Gus's hooves clip-clopping along.

When I did hook up with a couple of the girls who I knew from the paths, bonds would quickly be made by shooting the breeze about each other's ponies and we'd light up ciggies and chat. Smoking in the saddle was the thing to do. It always felt like a ridiculously rebellious act.

Now of course I see it for what it was – just a ridiculous act. Trying to give off an air of nonchalance while wrestling a Clipper lighter out of your tight jodhpurs pocket and sparking up a fag wasn't easy. If there was a breeze you'd need to cup your cig as you lit it, meaning not only were you setting fire to something while sitting on top of an animal, you now weren't holding on to the animal either. Smoking is already bad for you without adding an extra edge.

Luckily Gus never took off with me during a ciggie break, not that he was averse to the odd gallop. Near the sandy paths was a huge hill, and we'd all hurtle up it. That was the plan anyway – what usually happened was we'd get to the foot of the hill and the other girls would set off at a canter up through the grass towards the brow of the hill, but Gus would get so excited at the prospect of going fast with his mates that he'd express that excitement the only way he knew how to – by bucking. Gus was a master bucker (on these occasions I used to call him a little bucker, or something similar at least).

He'd dip his head down low and put it between his front knees, instantly pitching me forward slightly, then he'd throw both his back legs up together and with some force. I did my best to lean back and stay on. I got quite good. If a vacancy had come up for a part-time cowgirl in a rodeo I reckon I would've been in with a chance.

It was like a real-life game of Buckaroo except I didn't have a plastic lasso, spade, sombrero or guitar. Once the bucking bronco sideshow had ended, if I was still in the saddle, Gus would gallop up the hill at a decent clip to join his mates. We'd arrive at the top pink-cheeked (me) and breathless (both of us).

Trying to avoid serious injury/death by either being thrown or setting fire to myself really worked up an appetite, so most hacks around the sandy paths ended with a trip to the fish and chip shop.

As the old saying nearly goes: 'You can take a horse to a chippy but you can't make it fit through the doors,' so Gus would wait on the pavement while I popped my head in and ordered a chip barm, a soft white circle the size of a CD crammed with hot, fat chips. With one hand holding Gus's rein I'd stretch with the other to pop my 50p on the counter. My usual fare from a chippy is chips and gravy but I wouldn't attempt that on horse-back – I may have been daft enough to smoke on board but I wasn't a lunatic.

I had some brilliant times with Gus – he was my best friend and annoying sibling rolled into one slightly roly-poly pony. When I got started in TV on Channel 4's *The Girlie Show* (if you didn't see it, it was a rather stuffy late-night political discussion show, no need to google it, just take my word for it, OK?) and had done a few interviews with the press, one paper, let's call it the *Maily Dail*, ran an exposé on me poo-pooing my working-class roots and insinuating I was actually a poor little rich girl whose daddy was landed gentry and I had had my pick of world-class ponies, golden unicorns and a candyfloss castle, or something like that.

The image of me as some Veruca Salt character trotting around Daddy's vast estate couldn't have been further from the truth. I was definitely lucky – I knew most girls my age would've killed for a pony – but I wasn't spoilt and this was never more apparent than when Bolton Horse Show was on.

It was held a few miles away on some land behind a mill. I think it's fair to say that most if not all the competitors arrived with their horses in a trailer being pulled by Mum or Dad in the family car. The few more serious contenders rocked up in a horsebox. I hacked there on Gus with my pal Vicky on her dapple-grey pony. Show days meant that everyone had to be smartly turned out – I'd be up at the crack of dawn, stomach churning with excitement, tiptoeing out to Gus's stable to give him the works. It was like a spa day for him; he was buffed, brushed, plaited and combed to within an inch of his life. I would just stop short of a vajazzle.

Poor Gus often didn't have much mane to plait as he suffered with sweet itch – I know it sounds like a seventies R 'n' B song but it was actually a skin irritation caused by allergies and midges that made him rub the skin around his mane raw, removing much of it in the process. We tried all sorts of lotions and potions and even put garlic powder in his feed, hoping the pungent garlicky scent of his sweat would deter insects from nibbling him, but it didn't work and he merely always smelled like he'd been out for an Italian the night before. In the end it was easier to hog him – closely shave all his mane off so he'd rock a look that was half yard brush, half Annie Lennox eighties buzz cut.

Still, with his hoofs slathered in hoof oil, leather tack all cleaned and coat shinier than a conker, he looked the part, which was just as well as he wasn't attending the show for his skills.

Despite there being loads of showjumping classes and cross-country, Gus and I chose to shun them in favour of the more beauty-based events – 'Prettiest Pony' and the classic 'Pony the Judge Would Most Like to Take Home'. This class was a thinly veiled attempt to make sure there was at least one class everyone could win. For 'Prettiest Pony' the pony had to on some level be cute, but 'Pony the Judge Would Most Like to Take Home' was open to interpretation – any ol' mule could mooch round the ring as long as they paid the couple of quid entry fee.

The reason we didn't attempt any proper jumping classes was more to do with my lack of skills than Gus's. I'd popped over many a little jump in Dad's field, made from fencing poles painted in pastel paint and propped up on empty oil drums, and Gus was pretty good. The only problem came when I had to remember my way round a jumping course. With my heart thumping in my ears and adrenaline gushing through every fibre of my body, I'd forget the order of the jumps. Like one of those news stories about a pensioner driving four junctions the wrong way along the M62, I'd get completely confused and overwhelmed. Each jump was numbered and the first two would be pretty straightforward but then it'd be a left or right turn to number three and this is where I'd fall apart, nerves shredding my memory of the course.

It was so humiliating when a crackly voice announced over the tannoy that you'd messed up – they might as well have said 'SARAH COX RIDING AUGUSTUS FIRST IS DISQUALIFIED FOR BEING A COMPLETE IMBECILE – PEOPLE PLEASE GATHER ROUND TO POINT AND LAUGH.'

They could've just flagged me down and discreetly informed me; that way I could've casually trotted out of the ring as if it was my choice because I'd just remembered there was a sale on at Tammy Girl. The humiliation stung enough to put me off for life.

Part of the excitement with riding is the unpredictability of it – but for all the thrills, what goes up must occasionally come down.

I've managed, in the main, to bounce on landing from errant equines with no real damage. However, as a twelve-year-old I found myself in the A&E department of Bolton Royal Infirmary. Eventually.

I was out riding with Vicky, a very gung-ho pal of mine who was also a kind of no-nonsense, jolly hockey sticks type, like a 1950s Guide leader crossed with Angela Merkel.

She kept her dappled grey pony on the farm, so we'd often go on a hack together, Gus skipping about excitedly at having a friend, trying but failing to play it cool, swinging his rear end around and generally acting like the Rizzo to the other pony's Kenickie.

We decided to head to the sandy paths. We could get there by one of two ways. The first was mainly roads and would mean negotiating the 'Seven Stars dip' – which sounds like an impressive astronomical event but was just the pub that stood at the lip of a huge steep drop in the road, a kind of narrow tarmac-covered valley. Mooching past the pub meant leaning back in the saddle to help your pony's balance as they negotiated the steep decline.

I was familiar with the sharp angle of this rollercoaster-esque drop as I'd ridden down it all my life on my sister Yvonne's Chopper bike.

From the age of around six, with scant regard for my personal safety, Mum used to let me get a 'backie' from Yvonne, and it was literally a backie – we'd travel back to back.

Yvonne would be facing forward, face set against the wind in a determined grimace, mousey long hair streaming behind her, flared denim dungarees flapping as we hurtled down into the abyss of the dip, sun glinting off her white knuckles as she gripped the big Harley-Davidson 'ape-hanger' handlebars and steered the Chopper round the gravel-filled potholes. There was me facing backwards – head jiggling with the vibration, knees tucked up to avoid the wheels – luckily the L-shaped saddle of the Chopper meant I was able to cling on to the vertical bit of the seat like a koala on a tree trunk.

At the bottom of this plummeting drop was the entrance to a builders' merchants, which meant that as I hurtled backwards down the hill there was often a lorry or tipper truck following closely behind, bearing down on us as my sister negotiated the grids and ruts at full pelt.

It was these lorries that played on my mind whenever I considered which route to take on horseback. Gus wasn't bad with trucks usually, having got used to Dad's being parked outside his stable most nights. His greatest nemesis were double-decker buses; the quiet swooshing noises they made freaked out his little pony brain.

However, the trucks from this builders' yard were operated by young drivers who I presume thought it'd be character-building for young riders to have a few tons of metal roar up behind them as they trotted along the road.

I decided on that fateful day with Vicky that the disused railway tracks would be a better and safer bet. This route ran for a couple of miles from near enough the bottom of my dad's farm lane all the way to the entrance to the sandy paths.

The railway lines were really overgrown and actually quite picturesque – lots of bushes and greenery with the odd glimpse of wooden sleeper or track peeking through the weeds and the nettles.

There was only the occasional hazard. One of the best things about riding is you don't have time to think about any of your own worries because you spend the entire time thinking like a horse.

To the untrained human eye it may just be a carrier bag

caught on a bit of barbed wire fencing, but when you see it with your pony eyes you know full well it's actually a small monster, fluttering in the breeze and ready to fly at you and eat your face.

Hear a scrambler motorbike in the distance? Sure, it's a slight annoyance to our human ears but to our pony ears it heralds the arrival of a terrifying roaring dragon, again who could try to eat your face and possibly bite your bottom. Thinking like a pony is rather lovely because it helps you to subtly brace for any potential upcoming drama. I never went too method, though; me wearing a bridle and pooing on the road would've been a step too far.

This day on the disused railway seemed to be particularly hazard-free. It was bright and warm and only the sound of birdsong filled the air. It was the sort of sunshiney, woodland-creatures-smiling-at-you, dickybirds on your shoulder morning mainly found in fairy tales.

Once away from the road we pushed into a trot, which became a canter, which became, out of the blue, and entirely decided by Gus, an emergency stop.

My pony brain was too busy looking ahead for hazards to be able to see this one coming. If I had been approaching a jump I wanted him to clear, a stream I expected him to cross or even a path with unfamiliar zigzag road marking I could've braced for a stop. But this? This seemed to be an emergency stop made entirely from mischief. I truly believe he did it just to see what would happen.

What happened was this. Gus stopped but I didn't. The speed we were travelling and the surprisingly sudden nature of him slamming on the brakes meant the momentum kept me travel-ling forward for what felt like a long time. Going by the double-decker bus, the measurement tool popular for record-breaking skyscrapers, blue whales etc., I reckon I kept floating through the air for at least the length of one bus.

When I eventually landed, I managed to take the full impact of the fall on my left elbow. Yup, when hurtling through the air, always best to choose the pointiest boniest bone to land on.

I remember sitting on my arse in the shrubbery, looking up at Vicky, who'd turned back and was now looking down at me from her pony a few feet away. Both rider and mount were wearing confused expressions, their eyes wide, heads cocked quizzically to one side.

Gus meanwhile took the opportunity to grab a few mouthfuls of grass. Behaving like a complete dickhead had clearly given him an appetite.

The art of falling from a horse can be mastered – the tuck and roll is the method professional jockeys use when they come off their charges among the clatter and chaos of dozens of horses' hooves hammering past you on the racetrack. I'm certain 'Try to land on your elbow' is not part of the tuck and roll because it bloody hurts. I struggled to my feet and grabbed Gus's reins with my right hand and somehow managed to swing my sorry ass back into the saddle. We were only a few minutes from the farm, so I pulled Gus round ready to head back to the safety of the yard. Vicky had other ideas; it was a lovely day after all.

Combining all her medical training with her impatience at the hack being hampered by my dramatics, she said 'You OK to carry on, yeah? Reckon you're just a bit bruised and it'll feel better in a bit, let's just do a quick hour and we can cut back if it's still sore,' at which point I should've replied 'No thanks, because I may well vomit with pain down my horse's neck any minute,' but being a young teen and not yet equipped with the confidence to vocalise what was best for me, even when in agony, I replied meekly, 'OK, well, can we go slow please?'

I was within sniffing distance of Dad's farmyard and there I was, willingly trotting away from help. I didn't want to be a wimp, but it can't have been much fun, a tentative shuffle around the railway lines with a girl quietly whimpering over every lump and bump.

At last we made it back to the farm and I slid out of the saddle gratefully. I found my dad, as usual tinkering with something, trying to fix a bit of tractor. I didn't want to panic him so I calmly told him what had happened.

Dad looked at me, ruffled my hair with one of his huge hands and reassured me that I'd 'Be alreet', and to get Gus 'fed and mucked out'.

Not the usual medical advice admittedly. In agony? Faint with pain? Go lug around a couple of feed buckets and pick up some pony poo – it works a treat for all injuries! Available from NO major pharmacies because IT'S MADNESS!

It wasn't what I was hoping for, to be honest. But I didn't protest; I think I just trusted Dad's judgement implicitly, plus with him I was always conscious of not wanting to seem soft or make a fuss.

By the early evening, having mucked out and fed Gus literally single-handedly, I was feeling terrible and my left elbow was throbbing sickeningly; a dull thud that reverberated up my arm towards my throat, making me bilious. I remember sitting on the big oxblood chesterfield couch in the lounge feeling woozy and hoping someone would notice this pale, unusually quiet child and take pity on me. Luckily they did – Dad suddenly realised that even for me, pretending to be in pain for six hours just to get attention was a bit much, and decided I needed to be taken to hospital.

Dad didn't take me himself – it was calving time and he

didn't dare leave one of his prize cows that was showing signs of going into labour. Given the choice between his youngest offspring and a bouncing baby Hereford there was no competition – I wouldn't fetch as much at market.

I eventually made it to A&E thanks to my stepmother's dad giving me a lift. I arrived home from Bolton Royal Infirmary with my arm in a foam sling and a confirmed fractured humerus – I'd broken my funny bone. It was, ironically, no laughing matter.

9. Spiral Permed Butterfly

It was around this time that I started a new school. I was twelve years old and had had to leave Bedford High, the school I liked and the friends I loved to become the new girl at a secondary school back in Bolton. For a couple of years we'd moved a few miles away but we'd now returned to our home town. My very first day was after the Easter holidays in the second year of 'big school'. Although the pupils had only been there eighteen months it was, in their short lives, plenty of time to make friends, form bonds and subconsciously create tribes – the swotty, the sporty, the rebellious. It wasn't easy to infiltrate an already established school hierarchy. It didn't help that my arm was still in its foam sling after my riding accident. It wasn't a great look, to be honest – a plaster cast earns gasps of excitement as people excitedly crowd around you, pens at the ready to joyfully graffiti it, but a flesh-coloured foam strap that had become a bit manky and bobbly on the edges wasn't really the focus of any envy.

Unfortunately, my ear-length, permed bob (yes, there's such a thing), lanky legs with three pairs of long socks scrunched down round the bottoms so I looked like a foal with heavily bandaged ankles and my general lack of confidence didn't make up for the un-coolness of the sling.

No worries, I thought on my first day, I'll just keep my head down and once my arm is fully recovered I will fling off my sling and emerge like a spiral-permed butterfly from a chrysalis.

This plan was slightly derailed when the headmaster spotted me at lunchtime struggling to carry my tray of food one-handed and, before I could protest, insisted he carried it for me and found me a seat.

It's easy to forget what an important figure the head is. A bit like when a police car pulls up behind you on the road and you automatically start driving like you're on your driving test, hands at ten-to-two, sitting up straight, hiding your shooter and cocaine in the glovebox – OK, not really, but you know what I mean, you don't fully relax until they've driven off.

With the head teacher everyone eyes them and tenses until they've glided past. So it was on this day; as the headmaster strode through the canteen holding my tray of chips and beans aloft, everyone was watching him and the girl with the knobbly knees following behind, her cheeks burning bright red at the unwanted attention.

I may as well have been wearing an A-board saying 'Wanted – full-time bully. Must be able to work individually or as a team'. The post was soon taken up by two girls who were to make my school life miserable for the next couple of years. Like cartoon baddies, one was a big oaf who towered over the other – a short, scrawny, mean-faced girl with long thin hair and a pinched expression. Let's call them Big and Little.

The bullying, as is often the case, ranged from so subtle it was barely noticeable to others – snide whispering, passing notes, elbowing me as I made my way past – to more physical, humiliating stuff like shoving me or tripping me up so I'd land with a bang on my knees. I went to speak to the teachers about what was happening but Big, although not the brightest, was popular with the teachers because she could be quite a fun, nice character when she wanted to, so I was dismissed as the new girl not settling in and being a little oversensitive.

I had friends but they were too afraid to stick up for me in case the bullies turned on them. What was worse was that the bullies would often be charm personified to my friends so I felt truly alienated. It's a tried and tested formula: be so nice to your target's friends that it's difficult for the friends to defend the target fully. Secondary school politics were fragile and it was a minefield to negotiate. As much as I'm sure my friends sympathised, they didn't want to put themselves in the line of fire by defending me too stridently.

My two best friends at the time were Sakira and Firoza. I'd latched on to these two pretty soon after starting at Smithills School, no doubt drawn by Sakira's mischievous dancing eyes and bright toothy grin and the shy smile of Firoza with her big brown eyes, soft, round shape and a contagious giggle (hiding a wicked sense of humour) never far from her lips.

Sakira was actually quite sassy, so the bullies would try to befriend her, and Firoza had such a placid and serene nature that to be horrible to her would surely incur the wrath of the gods.

Rather than defend me, they just listened to my woes. 'God I hate Big, such a stupid cow,' I'd whine and they'd agree. 'She's a silly bitch,' Sakira would hiss angrily, nodding in agreement as I recounted the injustice of being heckled and shoved in the queue outside the science lab.

Sakira would sometimes make a joke to distract Big from picking on her prey and that would work, temporarily earning me a reprieve from the bullies' attentions as they switched their focus to whatever teacher Sakira had made a wisecrack about, but at other times she'd be a bystander to the abuse, unable to intervene. It was a calculated stance – she had to choose her moment to defend me carefully without winding Big up or humiliating her in front of her rapt audience (usually of one – Little) and becoming her new victim. Big liked Sakira but, make no mistake, if Sakira pushed her luck too much Big would flex her bully biceps in her direction.

Sakira was from a big, loving family. Her mum was a quiet, calm lady who I'd see after school preparing the evening meal, cross-legged on their living room carpet peeling a mountain of vegetables onto a layer of newspaper. Her dad, though they were really close, was pretty strict about who she hung around with and I didn't quite meet the criteria for the sort of girls he wanted his daughter to be pals with – this girl with the big perm and big voice. Sakira was very funny and we'd often be giggling and roaring with laughter. It was usually about completely daft, childish stuff but I suspect her dad thought we were plotting about boys and bad behaviour. She asked a few times for sleepovers at mine but permission was always denied; she'd call him at work, slipping into Urdu as she tried to convince him she'd be on her best behaviour, but it was no use.

So although I was close to Sakira we never got to hang out much after school. We still managed to have fun at school though; at least, we did when the bullies weren't around.

Anyone who's been bullied knows it can become debilitating and all-consuming. It was one of the first things I'd think about every morning before school. There would be those lovely nano-seconds before I was fully awake, barely conscious and still blissfully ignorant, but then with a thud I'd be down to earth, remembering the horrible treatment that awaited me. I'd feel instantly deflated before the day had even begun.

Mum tried her best. 'You OK, sweetheart?' she'd ask as I ate my Weetabix at the little round white wooden table in our tiny dining room in the terraced house where we lived. I'd sullenly watch the portable telly on the sideboard, Anne and Nick warbling away happily as I forlornly spooned up my soggy cereal. 'I'm fine, I just hate Big, she's such a cow to me.' I could never really say the 'b-word' – bullied. It would seem like an admission of weakness and I felt such a failure being a victim. Pathetic. Mum was really busy and worked so hard – I felt she had enough on her plate without my woes adding to her load. Occasionally she'd notice me looking tearful in the morning, but I imagine she presumed I was just tired and in the case of Big she just thought it was the usual teen squabbles, as I didn't point out exactly how bad things were. To quote one of Mum's favourite lines, 'I'd be in a tent on Blackpool prom if I could read minds' – she had no idea how much I was suffering because I suffered in silence at home. I'm a little like that still; for someone who talks for a living I'm very good at keeping schtum and internalising my worries.

Very occasionally there'd be proper tears in the morning and Mum would let me stay off school; at those moments it'd be like the valve had been loosened on Mum's pressure cooker and all the negative energy would fizz out and evaporate.

Mum would leave me at home with reminders of the chores I had to do, which I'd immediately put off to settle down for a day on the squashy foam sofa bed we had in the lounge. These would be 'double *Neighbours* days' – the ultimate in sloth-like behaviour, watching both the lunchtime and teatime showings of the same episode from my horizontal position inches from the screen.

I couldn't be off every day though, so the next morning I'd get the bus with the same feeling of dread rumbling in my stomach – it was pretty exhausting worrying about walking down a corridor in case a foot or elbow came flying out at me, because the school day in general is walking down lots of corridors (walking, mind – NOT running, something the teachers were obsessed with preventing).

If I could travel back and have a word with myself I would probably tell the twelve-year-old me to retaliate. To push back. I was a strong girl from lugging buckets of water and bales of hay and probably could've battered them. But I didn't because ultimately I was a bit of a chicken and not naturally violent and they took full advantage. While I was at it I would probably also tell myself to maybe rethink the spiral perms.

Both the perms and the bullying pockmarked my early teens. I didn't confide in Yvonne as she was away at uni and my older siblings weren't around as much – they were all married off. When I did see them the last thing I'd want to do was prattle on about my woes. Eventually Mum talked to the school but I hadn't fully accused anyone of anything, so it was difficult for her to go in there and read the riot act. It meant that they didn't take her concerns seriously and just fobbed her off, promising to keep an eye on me and reassuring her I just needed to settle down.

I never would've guessed that I would be eventually rescued, in a roundabout way, by Mr Whippy.

One day at breaktime the head of my year looked out of his office window and witnessed a kerfuffle in the playground – Big was holding my arms behind my back as I struggled while Little shoved a cone of ice cream into my face. I remember I tried to struggle free but couldn't, so I gave in, staying still just to get it over with. Having seen Little mush the ice cream into my face, it coating my chin, my nose, getting into my eyes, as other pupils shrieked with thrilled shock at what they were seeing, he finally believed that I was indeed being bullied. He apologised to me and Big and Little were hauled into his office, warned off me and put on detention. Other teachers were tipped off. It did the trick and the bullying ceased, but I've never eaten ice cream since. Ha, not really. I joke but it had been a truly horrible time and, with the bullying finally over, at last I felt like I could breathe, like a breeze block had been lifted from its resting place on my chest.

I'll never know if Big realised how hellish she made those couple of years. But I do know that Little seemingly had no clue that she was my bully, because years later she texted Radio 1 when I was hosting the breakfast show, wanting a 'shoutout'. I was flabbergasted.

The text she sent was a fond one, as if we were long-lost school pals. Didn't she realise the tears I'd shed thanks to her? The anxiety she caused, the humiliation, the dread, the sadness? She got her shoutout all right. Though not the one she was expecting. I relished dedicating an entire link to reminding her of her horrible treatment of me. Was revenge sweet all those years later? I'm not sure. I'd waited fifteen years to get a voice loud enough to answer my tormentors. Was outing her on the radio to millions of people tantamount to bullying . . . ? Probably. Did I care? Not really. I wouldn't do it now but back then, in my late twenties, I was much more hot-headed and I suppose still hurting.

Perhaps as a response to the bullying, I began to assert my authority in other ways and embarked on a short rebellious stint at school. But my one attempt at wagging a class ended in disaster when I decided to skip a lesson and spend my illicit freedom hanging out in the school toilets with Sakira and Firoza. After about ten minutes we were busted by a teacher. How on earth did she know we were there? Why was our cunning plot halted in its infancy? We didn't have long to wait to find out because the teacher told us: she'd spotted our silhouettes leaning up against the big frosted window in the girls' loos. Our hearts clearly weren't in this wild behaviour. She didn't even punish us, just sent us on our way to class.

Life at secondary school had some welcome distractions, mainly playing in all the sports teams despite me being the least talented member of each squad. In netball I was GD – Goal Defence. Anyone who knows the game knows the court is divided into three sections, with restrictions on which thirds you can enter. Most players can enter two zones, but Goal Defence is allowed in only ONE. Which spoilsport (literally) thought of a game that restricts its participants from using most of the court? It's like being given a delicious trifle but only being allowed to eat the jelly. There's only one position that can run free and gazelle-like around all three zones – Centre. It seemed to me that there was always a particular type of girl who got to be Centre. Glossy-haired, bouncy, not knock-kneed and goofy. I think I can spot a Centre girl even now. Gabby Logan, Myleene Klass, Susanna Reid, Alexandra Burke; they are the Centres, determined, gritty, dedicated – they get the jelly, the cream AND the custard.

Still, I made the team. I enjoyed standing stock-still, marking my opponent, arm in the air and fingers outstretched like I was feeling along a tall shelf for some door keys.

I also didn't excel at hockey. I can remember standing on an all-weather hockey pitch in deepest winter and glancing down to see my legs had been replaced by two sticks of corned beef, pink-and-purple-mottled. My hands were frozen to the hockey stick, it was dark and I could see my breath as I panted, chasing the ball in the white glow of the floodlights.

I loved it though. I rarely scored, whether it was netball, hockey or running a rounder in rounders. Maybe I just enjoyed the smell of sweaty changing rooms, the rumble of the school's clapped-out minibus or wearing an elasticated bib.

Maybe it was that I felt like I belonged. I'd always felt loved at home but being younger than my siblings I never quite made full membership of their club. Humans are naturally pack animals, even if we occasionally enjoy our independence; the feeling of being in a clan at school was one I treasured – perhaps the spectre of bullying, which potentially could've raised its head again, made me feel that safety in numbers was the wisest choice. Even if I was a bit of a butterfingers during netball, misfired at goal during hockey or rarely scored a rounder, the teams still accepted me.

If my team track record was mediocre, my actual track record was abysmal. To look at me back then, with my long coltish legs making up three-fifths of my body, you'd think I was built to run. Nature had other ideas and had also gifted me slightly knock knees, and the ensuing nicknames knocked my confidence so much that I hated even walking in front of people, let alone running. Of course, when you're fourteen you may not have the emotional intelligence to explain this to a tough Glaswegian PE teacher with a tight black perm and a mouth set like a pale pink hyphen. So what I did was a sulky half-run, half-galumph round the athletic track, like a wildebeest whose tranquilliser dart was just starting to take effect. It wasn't appreciated by the teacher and I was a constant source of frustration for her. Sorry miss.

10. Nailed It

I was a toddler with no toddle, thanks to being hypermobile – a fancy-sounding way of saying a bit wobbly. My joints – from elbows to knees to ankles – can bend in all kinds of unnatural and uncomfortable-looking ways and, when I was little, even getting upright was a struggle.

The only way I'd sit to watch telly was on my bottom with my knees and thighs together flat on the floor and each shin folded out at my side. I had to sit like this in school assembly too as forcing my gangly legs into a crossed position used to really hurt – after a couple of Bible stories, some announcements and a rousing rendition of 'All Things Bright and Beautiful' unlocking my kneecaps from that unnatural shape was agony.

It wasn't all negative – my party trick as a child was to grab my left foot with my left hand and pull it sideways and then backwards towards my spine, arching my torso left till I could look over my left shoulder and touch my nose with my big toe, to howls of delight and disgust from my friends.

Being hypermobile may have been handy for those moments

TILL THE COWS COME HOME

when it was absolutely crucial to touch my face with my foot but it wasn't very helpful when it came to walking.

At around twenty months, a time when most kids are walking, I was struggling. I would pull myself up to a standing position but my knees – unsupported by my plasticine ankles – would give way and I'd crumple to the floor, my nappy-clad bottom hitting the deck hard.

Watching my frustration gave Mum a brilliant idea. She knew I was ready to walk; I just needed a little support – literally. The helping hand arrived in the form of some little red wellies, my new best friends.

They came up to just below my knees and were of sturdy enough rubber that when my skinny legs weakened and wobbled, they'd provide just enough scaffolding to help me remain upright and grow stronger and more confident. In no time I was toddling about the farmyard, thrilled. I wore them all the time, often pottering around in just my nappy and my red wellies – I imagine I looked rather swish.

My own three children were late walkers too, having inherited my ankles. I knew from my own experience why they were struggling to walk and so at around twenty months I popped a pair of wellies on them and, hey presto, they too were walking within a day.

Once I was fully upright, my troubles really began. My legs picked up a fair few knocks from various exploits on the farm – one scar on my knee remains to this day. I was thirteen and Dad asked me to carry a crate of chickens from the shed to a new coop. The crates were about the size of a bale of hay and awkward to carry, made of heavy-duty brown plastic with slots for holding on to them. There's no way I could've managed by myself, so one of the farm helps took one end and I the other.

It was dusk and disgusting weather – blowing a gale with a helping of sideways rain – so over my jodhpurs I was wearing my waterproof trousers, which were as glamorous as they sound, but these huge navy rubber affairs did the job and kept me nice and dry.

As we slowly made our way through the gate into a little enclosure I slipped on the sopping wet grass and pitched forward, letting out a yelp. Not wanting to drop the hens, I clung on and landed on my right knee with a thump. My kneecap connected with a semicircle of smooth rock that was poking out of the grass like a discus that'd been thrown with such force that most of it had embedded itself in the earth.

My knee was sore but I staggered to my feet like a spurned lover after an unsuccessful proposal and eventually, in the howling wind, we decanted the clucking hens into the warmth of their new coop.

Job done, I went about my other chores with my knee slightly throbbing. I fed and watered Gus and as I pushed a wheelbarrow of muck up to the midden through the rain I realised that my faithful waterproof trousers had obviously suffered a tear when I'd fallen and the rain had got into them – my right leg felt very wet.

I leaned the barrow up and made my way towards the house. In the dim light of the kitchenette I wearily pulled off my wellies and peeled down my rubber trousers to reveal my jodhpurs, now looking rather snazzy as they weren't just cream but patterned red too . . . The wet I'd been feeling wasn't rainwater but my own blood, gushing from a pumpkin-seed-shaped gash on my kneecap. The blood had soaked through my jodhpurs and it looked like I had a bright scarlet bandana tied round my leg.

To this day, every time I shave my shins I spot that little

battle scar on my knee and feel strangely proud that despite everything that happened, I didn't drop those bloody hens.

Minor injuries were common on the farm. One day I was playing, as you do, on a huge pile of planks that had been stacked up against the side of the cattle pens and which Dad obviously had plans for. They were 'pre-loved' and thus had lots of nails poking out of them; huge, gnarly nails. The kind of nails you'd imagine Hagrid would have rattling round his pockets.

See if you can work out what happens next, like on a TV bloopers show where the host pauses the footage and you have to guess the outcome after the adverts. A five-year-old girl is playing on a nail-strewn plank pile while her dad shouts at her from his perch on the tractor, 'Get down or you're going to get hurt!' I, being five, obviously ignored my dad because a) these planks were tremendous fun, and b) Dad's control over me was limited as he was in the process of steering a tractor and moving silage bales.

I carried on merrily skipping up and down the planks until the inevitable happened. I went flying. As I landed I did what comes naturally – I put my hands out to stop myself. The nail that punctured the fleshy part of my right hand, the meaty bit below the thumb, pushed through with such ease and poked out the other side so cleanly, like a skewer through a chicken drumstick, that it almost looked like something from a joke shop. No one was laughing though. I instinctively pulled my hand up and off the nail and stared at the hole it had left in shocked silence. Then I snapped out of it – as the blood began to pour from the wound – and my screams echoed around the yard. Dad leapt off the tractor and came running over, his face growing paler with each stride. He reached me and immediately gave me . . . a massive bollocking.

That's what happens. When you ignore your parents' frequent warnings, then get injured, their frustration morphs into a brief, hot, fury.

Dad rarely got angry, so I must've really shaken him up. I was whizzed to Bolton Royal Infirmary and was patched up. I remember having to have a tetanus jab in my bum cheek and trying my best to punch the doctor away as he approached with the needle. I'd like to apologise, thirty-eight years later, to that no-doubt overworked, under-appreciated young NHS medic for being, in all respects, a pain in the arse.

11. Geese and Guns

Growing up the youngest of five meant there was always someone around to mither. Someone who'd hopefully let me tag along. I love being part of a big family – we're a clan, identified by our long lanky legs, toothsome grins and blue eyes. The kitchen, with its classic 1970s bright orange units, was always a whirl of activity when I was little – I'd stand in the eye of the storm as the rest of the family swirled around me, shoving past to put the kettle on, Mum rattling pans on the stove, Dad having a hasty brew, flicking through the *Farmers Guardian* or opening post. ('You can have the bills if you like, Sarah.')

There'd be various siblings entering and exiting with buttered toast or a cadged biscuit hanging from their lips. Robert and David didn't hang around – they always had something to tinker with or jobs to do around the farm. Dot was a calmer spirit, often quietly going about her business, helping with housework, tickling up. You'd find Yvonne in a quiet corner of the lounge curled up with a book.

I miss them now. I live 200 miles away in that there fancy

London where the streets are paved with sashimi and I miss being able to call on my older siblings for a quick brew and – let's be honest – free last-minute childcare and some DIY. Whenever my eldest brother Robert comes to visit with his wife Sarah-Jane he's barely through the door before he starts oiling a hinge here, bleeding a radiator there, and she's reading a largely ignored child a story while popping the kettle on.

Robert and I are really close. Because he's thirteen years my senior I've always been slightly in awe of him because he's eternally had an answer for everything – he managed to seem wise even as a teen. He's a fixer – he can turn his hand to anything – and has always been driven, from his first business selling mice and bunnies to owning various companies and now selling his art, thanks to a previously untapped reserve of artistic talent that means he can paint. So he's a do-er.

He's also a looker – in the eighties he became a bit of a gym bunny and loved a sunbed – back then lightly grilling oneself to a crisp on a sunbed tucked away in a spare room was all the rage, blinding white tubes mere millimetres from your face pumping out UVBs and Cs while you prayed your mate's mum didn't barge in and find you horizontal, sweaty and naked apart from a pair of tiny foil goggles.

Robert had it all – he was tall, tanned and muscly with a mop of jet-black hair and a bright white smile. He looked just like Dad if Dad had been put through some sort of heart-throb machine. My friends all fancied him like mad and would swoon whenever he popped round to Mum's and I'd kiss him hello. 'You're so lucky you get to kiss him!' they'd squeal – which was kinda weird; he was my big bro and, as much as I loved him, to me he was no Morten Harket.

Robert always had a knack of talking like he was incredibly knowledgeable about his subject, even if occasionally he'd have little evidence to back up his wise musings. Whenever he was in charge he'd not allow us to have the radio on because he claimed we'd get indigestion cos we'd 'eat to the beat' – something that still makes me chuckle to this day. I'm sure he just didn't fancy having the radio on, so he came up with this fantastical logic to be able to persuade us. He was good at making up facts to suit him, like his medical fact about long toes like his being a sign of intelligence.

It wasn't just my friends who thought he was gorgeous. Robert also attracted the attention of a goose. Geese are brilliant at seeing off intruders or potential threats because they're so ballsy – I was always wary of them, but when I was about seven this one gander called Max, behaving not at all like a typical fierce goose, developed a deep and all-consuming obsession with Robert.

It would follow him round the yard like a feathery shadow. At first it was funny but the novelty quickly started to wear thin and Robert attempted to avoid it. Whenever the coast was clear he would venture out and suddenly the goose would appear from nowhere as if it'd been hiding somewhere on surveillance. It'd waddle after him until it fell into step alongside him, all casual like. One boy and his goose. Robert eventually succumbed and just let the goose hang out with him. At the height of the affair, Robert would be on his back on an empty feed bag under his car, and while he tinkered with the engine the goose would settle on his lap.

Robert has the best stories about growing up on the farm – stuff that happened when I was still knee-high to a heifer.

Like the one about Dad's mate Brian, who was a 'real seventies copper' with a fascination for high-powered weapons, which he'd bring up to the farm to show to Dad.

One day he brought a machine gun with him, an actual machine gun. Rob was duly dispatched to go and 'borrow' one of our sister Dot's dolls, which was then tethered to a fence post in the field before Brian fired off dozens of rounds at it, machine-gunning the doll to smithereens.

Rob himself had guns and pistols from a young age because, as he explains now, 'There's always something to shoot on a farm.'

One time a bloke called Billy who lived down the footpath from the farm agreed with Dad to swap his .410 shotgun for six large trays of eggs. Rob was sent off to do the transaction, carrying the eggs down the overgrown path and walking back with the shotgun, an old thing that had been painted grey. Robert stripped it back, varnishing it till it was like new. Around the same time, Grandad Cox passed away and Dad ended up with an old box of bullets, likely from the war years, all different sizes. Realising none of these fit his new gun, Rob decided to have a bit of fun with them; he waited till everyone was out and built a small campfire next to the garage and put a handful of bullets on the fire to 'see what would happen'. Well, what happened was of course that the bullets, containing gunpowder, exploded.

Robert legged it into the garage and hid. It was only once everything had fallen silent, apart from his heartbeat thudding in his ears, that he saw the new small hole a few inches away from his head in the garage wall and realised how close he'd come to killing himself that day.

Always the entrepreneur of the family, when he was a teenager Rob bred mice in the coal shed. The bottom half was piled with coal and the top half was his home-made mousearium. It was a pretty snazzy pad for the mice – multi-level, linked by wooden ramps and tubes and enclosed by chicken wire. I remember standing on my tiptoes and peeping into the labyrinth of this rodent penthouse and thinking it looked quite magical. Ironic that his pets were kept in the mouse equivalent of P. Diddy's Miami mansion while their cousins the rats were being hunted and hated up at the barn. Rob sold these mice to pals and to Mittens, the pet shop, and before long he progressed to breeding rabbits for the same shop, making a few bob.

As well as being an astute businessman Robert was also quite the inventor. One day he made stilts by getting thin oblong lengths of wood and nailing tyre stops onto them. Similar to doorstops, they were wooden wedges shaped like cartoon cheese. He fastened the underside to the wood with the thin end of the wedge pointing down so his feet could stand on the square surface at the top. Ingenious. They weren't very high – when he eventually managed to get up he was perhaps two feet off the ground. We all trooped out into the garden for the demo. He could get up OK, but staying up and moving proved difficult. He'd underestimated the weight of the wood, and lifting these chunks of timber to take a step while remaining balanced was too tricky.

After numerous failed attempts, Robert's audience, by now bored, melted away. I popped back into the kitchen to carry on looking at the back pages of my dad's *Farmers Guardian*; I loved drawing biro rings round my dream horses in the classifieds, getting lost in a world of if-money-was-no-object fantasy ponies.

Something caught my eye. Robert seemed to have mastered the stilts – once, then twice past the kitchen window; he'd cracked it! What I didn't know at that point was that, in order to solve the falling off the stilts issue, he'd tethered his wellies to the actual stilts and, using the side of the house, had managed to get his feet back into the wellies, pull himself vertical and was now proudly strutting his stilty stuff around the lawn.

On his third walk-by there was a stumble that uprooted him. He started to fall and I remember watching and wondering why he wasn't jumping off like he'd done before. Instead he just fell, almost in slow motion, past the window. No arms flailing, no drama, just a silent fall, like a tall skinny tree being felled right outside the window. I imagine that was the exact moment he realised the main disadvantage to tying wellies to stilts. He landed with a crash, gashing his forehead on the corner of the garden wall.

When he wasn't being seduced by geese or crashing his stilts, Robert was very popular. His birthday parties were always wild affairs and I thought he was very cool – his mates were shaggy-haired lads in skintight double denim and girls with blonde perms and short skirts, all of them tumbling out of the tent put up in the farm garden, mysterious potent-smelling smoke billowing out after them.

From about the age of fourteen Robert loved motorbikes and would often bounce across the fields on them. My other brother David followed in his tyre marks and soon the two of them were obsessed with this new hobby. Robert started with a little Puch moped, which he soon tired of and sold to David. Makes me laugh that he didn't give it to him but sold it to him – both lads obviously knew the value of a quid! Plus Rob needed the cash for his next bike. This one is the one I remember the most vividly, probably because of the colour. Rob bought an old Lambretta scooter from his mate Warren for £15. He had great plans for it – he chopped off all the metal panels and painted the thing lime green and used it as a scrambler.

By sixteen he had a Yamaha FS1-E 'fizzer' – a bike that was small and spindly with big wheels but thin tyres that looked and sounded like a mosquito – that he'd razz off to school on.

Both lads would let me ride on their latest bikes, between their legs on the narrow seat, and I'd just perch there like a wild-haired mascot between their outstretched arms as they steered the bike up the farm's gravelly lane, avoiding the potholes.

David was tall and gangly with legs up to his armpits – a trait that connects all the Cox siblings. As a teen he had a mop of hair as dark and shiny as liquorice, a bright toothy smile and playful eyes sheltering under more serious brows.

We had a slurry pit, two cavernous underground tanks about ten feet deep that held thousands of gallons of liquid muck from the farm, ultimately intended for spreading on the fields. The pit had an old wooden cover and when David was twelve he fell through it. Robert saw the cover give way and David disappear into the darkness beneath. David would've drowned if it wasn't for Robert – he ran and screamed for Dad, who also came running, grabbed a long wooden ladder and lowered it into the slurry, scrabbled down it and managed to somehow grab his son from the pitch-black stench. Beyond terrifying. I was a baby at the time, so I don't remember this, but I'm crying as I tell the story. Thank God I didn't lose my lovely brother that day. Dad replaced the broken cover with a new steel manhole cover and no doubt relived the horror of what could've been again and again.

David doesn't like a fuss and when I asked him to describe the whole experience, he simply replied 'Smelly.' He said it wasn't the slurry pit that was his worst experience; his most frightening moment growing up on the farm happened when he was fourteen and I was two. David was in the garden, cutting the grass with old shears. I toddled up to him and grabbed at the shears and my tiny fingers got caught in the closing blades. I screamed. David froze. Luckily they were rusty and blunt so no real harm was done. It's touching that from all the dramas on the farm, including his own near-death experience, it's that one, when he momentarily thought he'd hurt his baby sister, that springs to his mind as the worst.

Both Robert and David always looked out for me when I was little and I remember them being gentle and kind to their kid sis. Thanks to the large age gap I was no competition and didn't have the wit or inclination to wind them up. I think I was seen as a harmless enough little thing who just occasionally got under the feet, like a cat.

There was only one time when they really upset me.

Pippi Longstocking was my beautiful rag doll. I was only six when I got her and she was similar in height to me. She had yellow wool hair tied in two neat thick plaits and huge round bright-blue stitched-on eyes with long black stitched-on lashes, and wore a matching pink floral cloth hat and sundress. Physically she was similarly built to the Cox clan – her legs took up four-fifths of her and it was these limbs that my teen brothers were attracted to. Not in a weird way – she was a rag doll. No, Robert and David would occasionally each grab one of my darling Pippi's feet and use her for a brief tug-o-war. This horrific practice would usually stop as soon as I clocked what was happening and would have the screaming abdabs until they stopped and I could snatch poor Pippi from their grip.

One fateful day things went too far – I screamed, I shouted, but my brothers continued with their tug-o-war, deaf to my pleas of 'GET OFF HER!' Both boys' faces were strained and pink with exertion and I realised this was no longer about Pippi or me – this was about them and no one was stopping till the tug-o-war was won.

Pippi's face flew back and forth, her plaits whizzing around, her blue stitched-on eyes looking wider with shock as the boys kept pulling and I screamed and shrieked, trying to rescue her. Something had to give, and what gave was Pippi. Or more specifically Pippi's lady area. She split right up the middle. White cotton woolly stuffing immediately started to peep out of the split seam. When I was little and attempting some unlikely gymnastic manoeuvre or eccentric dance move Mum would warn me that I'd 'split my difference', and that's exactly what happened to Pippi. Both boys snapped out of their hypnotic tussling, immediately dropped her and legged it out of the bedroom I shared with my sisters, hoping to make themselves scarce before I could holler the standard distress call; a double-syllable 'MU-UUUUM!'

Mum tried some emergency surgery and stitched up poor Pippi but she was never the same again.

I myself was treated more like a souped-up doll, rather than a little sister, by my sister Dorothy. David's twin, so eleven when I was born, she was all light-brown permed curls, large green eyes, a shy unsure smile and a quiet voice that faded before the end of every sentence as if she had little faith in the words that were coming out. She was tall with a willowy frame and, back then anyway, flared jeans. She cuddled me and fussed over me like you do with a favourite doll, but also did a lot of throwing me around.

One of our favourite games involved a draught-excluder snake. Back in the seventies and eighties, it seems now, instead of making doors that actually fitted into the frame someone decided it'd be easier just to ensure that every house had a few (usually home-made) draught excluders. These were long, thick, stuffed fabric cushions (usually, due to the shape, either styled as a snake or a sausage dog) that lay dolefully on the floor at the bottoms of doors to block the gaps left by shoddy workmanship, watching you with their button eyes.

Of course to kids these were just extra toys – at my nana's I'd always be faffing about cuddling the draft excluder, much to the annoyance of Grandad, who'd strongly insinuate that the entire room of adults were experiencing the first symptoms of hypothermia because I was preventing the excluding of the draught by messing with Sid the snake, who was made from conker-brown velvet so was very huggable thankyouverymuch.

We had a blue Aztec-patterned snake at the farm (Ooh, is that the sound of pencils being sharpened down at *Elle Decor* I hear?) and it was with this that Dot and I came up with a fantastic game. She'd loop the draught excluder round my chest and under my armpits, then, standing behind me, hold the snake's head and tail tightly in her hands. Looking back I'm pleased she held it tightly because once the snake was in place we'd manoeuvre to the top of the staircase and she would swing me out over the stairs. For a brief moment I'd be airborne, squealing with delight as I flew through the air, the steep wooden stairway my only cushion if Dot lost her grip. Luckily she never did.

The only time she mildly injured me was when she dropped me on my head. She used to like holding me upside down by my ankles so my head was just a few inches from the cold tiles of the farm's kitchen floor. She'd then sort of dunk me up and down like a digestive biscuit in a cuppa, as if trying to shake change out of my pockets.

I know as you're reading this you might be thinking 'Jeez Sara, if this is how the sibling who loved you the most treated you, what happened to you at the hands of the others?' but I loved it. I'd giggle as my face flushed and the blood ran to my head.

On this particular occasion there was only one voice that

could be heard above our whooping and laughing and that was Robert's: 'Dot – you're going to drop her on her head, put her down.'

Sure enough, at the climax of what was to be our last game, Dot's grip loosened and she dropped me on my bonce. I remember it really hurt. She didn't let go of me completely – she just misjudged the distance between my head and the floor and therefore the depth of her dunk.

Even as I screamed I became aware that to make a fuss would get Dot in trouble and result in a severe case of told-you-so's from Robert. I may only have been six but already I knew about loyalty and so stopped screaming and instead laughed hysterically despite the pain. I don't know if they realised I was trying to protect Dot, or if they just presumed my cackling was due to serious concussion. Either way it did the trick and Dot escaped a serious scolding from our big bro.

Even though he would get frustrated with Dot, Robert loved her to bits. When they were a little older Dot used to do the vast majority of the cleaning in the house – even in family photographs she's scrubbing, wiping and polishing like some kind of Boltonian Cinderella. But one day she got herself into a bit of a pickle. She'd been cleaning the brand-new avocado-coloured bathroom and matching carpet – it really was a sea of milky green in there when it was new, although the carpet faded over time until eventually it looked like a thin coating of lichen had grown over the floor. Dot accidentally dripped some bleach on the carpet round the bottom of the loo. I can imagine her poor face as before her very eyes big spots of the carpet's rich avocado shade started to disappear and in their place blossomed splodges of pale sickly yellow as the bleach ate its way through the pigment. Dot panicked, knowing she'd be in so much trouble for wrecking this new and luscious mushy-pea-shaded carpet, and did what we all did (and still do) if there's an emergency – she went to find Robert.

Robert is, and already was back then at nineteen years old, what is called in Yiddish a 'mensch'. One of life's good guys, solid, kind, reliable, the kind of 'drop everything' bloke who is really good to have on your side. So in this emergency, Dot found Robert and he got his thinking cap on.

On surveying the damage, he asked Dot if she had any green eyeshadow. Luckily, this being the early eighties OF COURSE she had green eyeshadow. In fact, she had many shades of green to complement the various shades of frosted pink lipstick she sported back then. Some might say that pale pink and green isn't a good combination and you know what? They'd be right. But that's the eighties for you: bold fashion statements and avocado bath suites.

Robert sprang into action and used the palette of green eyeshadow to cover the bleach marks, perfectly blending Fir Forest with Parsley Punch until eventually the marks were no longer visible. It was a triumph and Dot was thrilled. She kept an eye on those spots and touched up the area whenever the eyeshadow faded and the pale bits started to peep through.

Fast-forward a fistful of decades and after various successful career moves, Rob is now an artist whose work sells well to his numerous fans. He creates mainly palette-knife oils on huge canvases, but I like to think that it was crouching by the loo back then, mixing Dot's eyeshadow palette and dabbing it on the bleach spots, that ignited the art impulse in him.

If Robert was the MD of us kids, Dot was head of the care department. She always looked after me and was my constant defender if I was under people's feet, being noisy or annoying.

During times of family strife she'd shield me from the fallout until the storm blew over. She always had time for me.

Dot was the one who taught me how to ride a two-wheel bike. Let's just say I wasn't a natural Bradley Wiggins – up until the age of nine I couldn't ride a bike sans stabilisers. The final reckoning came for me on a camping holiday – all the kids on the site were whizzing around in a gang on their hired bikes giving it the full *Goonies*, but there was me at the back of the pack on a tiny bike with stabilisers, my knees up to my ears as I pedalled clumsily after the others. I had to learn to ride a proper bike, but I was terrified of falling and didn't trust anyone to help me learn apart from Dot. Eventually she stepped in to make sure that I would be on two wheels by the time I was into double figures. It only took an afternoon at the local park, wobbling along the paths with me screaming 'Don't let go,' until eventually of course she did let go long enough for me to wobble off on my own.

It was for Dot's wedding that I got my first perm. I was twelve and on bridesmaid duties, and what better way to celebrate a wedding and top off a pale blue pouffy satin dress than with a perm? No ordinary perm, mind – I wanted to channel my favourite pop superstar and hairspiration Whitney Houston. I dreamed of having a spiral perm so I could recreate her head of bouncy sun-kissed caramel curls with my own fine, mousey brown hair.

After months of intense campaigning, Mum finally folded and agreed I could perm my hair and debut it at Dot's nuptials. She took me to her hairdressers in the village and I was encased in a plastic cape, my head slathered in perming solution so potent that I took shallow breaths to avoid inhaling too many fumes, and deposited among the old ladies under the huge dryers to wait patiently for the magic to happen.

Eventually I was rinsed off and waited with anticipation to

see a north-west Whitney emerge. But no curls tumbled down from the soggy towel, just a damp mass sat on my head – no matter! Surely once dry its full majesty would be revealed.

The hairdresser set to work with her hairdryer. Unfortunately, someone had missed the memo about which strong female eighties icon I wanted to resemble; with my jawline-length hair blow-dried not into springy corkscrews but big, blowsy curls, rather than Whitney from New Jersey it was Deirdre Barlow from *Coronation Street* staring wide-eyed back at me from the mirror.

My poor sister now had for a bridesmaid an awkward pre-teen with a terrible perm. On the big day I felt so mortified at being the centre of attention with my terrible middle-aged barnet that I was too nervous to actually smile; to stop my teeth chattering with nerves I clamped my jaws together in a determined grimace, creating the kind of underbite favoured by Les Dawson's Ada.

Still, it was a special day and I was thrilled to be part of it. Dot was marrying Gary, a nice lad from the village with dark curly hair, a snub nose and a brilliant sense of humour, would play-fight with me and let me duff him up, so he was a-okay in my book. And he and Dot had kept me tucked firmly under their wings since the beginning of their relationship – from letting me sleep on a blow-up mattress next to their bed on their first night in their first home, thus extinguishing any frisson of first-night passion, to my stint clutching a bouquet and trundling up the aisle behind my sister.

In fact, I like to think their marriage has stood the test of time for thirty years precisely because it had to strive to get over its slightly shaky start with a clingy little sis who was also the world's most awkward bridesmaid.

The sister closest to me in age – and closest in proximity, as we were occasionally forced to share a room – is Yvonne. The friendship closeness took time to nurture and grow thanks to the age and proximity closeness. We'd fight like cat and dog. I was, I think, an irritant to Yvonne with my annoying habit of trying and often succeeding to be, the centre of attention by making people laugh. Being the youngest is easy – it's the plum position in any family, I think – and it was a role I fully embraced. By the last child the parents have grown tired of discipline and control and the care given to the little one is similar to how you'd treat a cactus: feed and water it occasionally but mostly leave it to its own devices.

Being the youngest gives you licence to really rock the ol' cute card. Habits that in older siblings are seen as petulant and annoying, like sulking, are, when looked at through the kaleido-scope of kids who are the baby of the family, cute and endearing. I was famous for sticking out my bottom lip, a super-sulker – the Incredible Sulk, Grumplestiltskin – but instead of being told off for being a mardy arse it was 'Ooh be careful, you could trip over that bottom lip!' – all fond chiding. At family meals I'd perform rubbish impressions of the then prime minister Maggie Thatcher: 'Mum, please pass the sprouts,' I'd say in my best Iron Lady drawl and my mum and nana were enraptured, applauding wildly as if on the front row of a sell-out Broadway show.

No wonder Yvonne found me annoying.

Before I go any further I should make it crystal clear that Yvonne and I are now very close. She does still have the ability to make me grit my teeth in fury with a single seemingly innoc-uous comment – I call it top-bunk mentality. In my mind she was always top dog, the boss, the one to be feared/avoided, so I'm particularly sensitive to any criticism from her. Whenever we argued she'd win, because I'd get too emotionally involved and would muddle up my words. It was usually words I'd heard her say that I would then put through my brain-mangle until they were misshapen and spit them back at her, much to her amusement – 'You're so IGRANONT,' I'd shout triumphantly.

In adulthood, we became pregnant at the same time (not literally to the second, that'd be weird), and from then on our relationship defrosted and re-formed as a strong sisterly bond born out of our sudden similarity – becoming mums for the first time was a major shared experience for us. And I love her very much.

As a child I always demanded a lot of fuss and love, not just with epic sulking or Maggie Thatcher impressions, but from the moment I arrived with my wonky legs. No doubt Yvonne, then four, thought I was a right pain with my dislocated hip, stealing her thunder.

Loads of family photographs show Yvonne and me on Christmas mornings, looking tired and dishevelled having leapt out of bed before dawn. We're posing proudly with our favourite new toys, our matching light blue nighties clashing with the swirling orange seventies wallpaper behind us. We have the same mousey hair – 'fine but lots of it' is how hairdressers have often described it, which seems like a backhanded compliment – and in those photos we both look so pale, almost translucent – like ghost girls clutching their Tiny Tears dolls. We have always looked very similar – spooky twins in our long-sleeved polyester numbers – but we were always very different in personality. Yvonne has always been slightly quieter, more reserved. The Cannon to my Ball-ache.

We were thrown together when our family lived for a while in a two-up two-down in Boothstown, a little place that squats by the East Lancs Road just outside of Manchester. Our house was tiny, a little terrace attached to the Conservative Club that Mum and Ken ran, so Yvonne and I shared a room, her luxuriating on the top bunk and me roughing it in the lower. We considered trying to split the room in two so we each owned a side, strictly out of bounds to the other, but once the bunkbeds and chest of drawers were in, the floor space that remained was roughly the size of a baking tray, barely worth sharing. It was more of a cell really, and if either Yvonne or I had fallen in with the wrong people and became hardened criminals it would've been good practice for a ten stretch.

Trying to squeeze past each other to get in or out of the room was like something from *The Crystal Maze* – 'Left a bit, no, you go back and I'll step here, go right, right! Yes, we've got the crystal!' – not helped by that fact we were both blessed with long limbs and a lack of coordination. No wonder our tempers frayed – we were literally living on top of each other in our bunks and we used to have some horrible fights. Poor Mum would often threaten to 'Bang our bloody heads together' but luckily Mum is the size of a Borrower, so she was never physically able to carry out that threat.

Yvonne's frustration with me was compounded by my irritating habit of using her stuff without asking and then promptly breaking it. She always looked after her stuff – she was always more responsible and conscientious than me. She'd cherish her things and treasure them, so no wonder there'd be hell to pay when ol' sticky mitts Saz would come along and help herself, like a magpie in corduroy trousers.

The event that best sums up the difference in personalities between Yvonne and me happened every year at Easter. We'd get about seven Easter eggs each from various family members and by 11 a.m. on Easter Sunday all seven of my eggs would be open, some partly devoured, others polished off. It was carnage. Shards of chocolate poking out of the packaging, shreds of foil littering the immediate area. Like a fox in a henhouse, I couldn't just eat one egg at a time; I'd tear into every box in an orgy of consumption. We kept them displayed (do come in, admire our display of Cadbury's finest) on top of the upright piano, and mine looked like a crime scene.

Yvonne, however, would carefully open one egg, eat some of it, then neatly replace the foil so the egg still looked untouched. Like Ferris Bueller arranging his duvet so it looked like he was still in bed, Yvonne would cleverly reshape the foil so it looked like there was still intact choccie underneath.

She approached the eggs in the same cool, considered way she approached life and it drove me mad. It always seemed like Yvonne's eggs lasted for weeks whereas mine were gone in a few hours.

This analogy isn't really to suggest I'm wild and unbridled and she's stiff and controlled, but for sure she's much more considered and I'm more slapdash, so growing up there were bound to be clashes.

Despite our arguments we did love each other; we just couldn't live together, like Syrian hamsters.

When Yvonne was around fifteen she took herself off to stay with a cousin in London and do some work experience as a hospital cleaner. This now feels very 'of its time' – the eighties were clearly pre-health and safety and insurance concerns – 'You're fifteen? OK, grand. Here's a mop, go sort out that large pool of bodily fluids over there near that naked old man!' – but thanks to a mixture of adventurous spirit and, no doubt, annoying sister, Yvonne booked her train ticket to head down south and get stuck in.

I was eleven and to me London might as well have been Mogadishu – my only knowledge of it came from *The Bill*,

Minder and new soap opera *EastEnders* and it seemed to me that London was mainly made up of criminals, con men and large-haired, loud pub landladies. I was genuinely worried for Yvonne's safety.

In the days before she left we started to read Shakespeare together – I don't know why. I think we rather fancied ourselves as modern-day Little Women like in the books, huddled together reading by the light of a crackling hearth. Although in reality we were both squished into Yvonne's top bunk reading aloud by the glow of a 40-watt bulb. Yvonne would hold the *Complete Works* in her hands and we'd read all the different roles. I absolutely loved it.

Maybe she did the Shakespeare thing because before she left for the big wide world she wanted to do a bit to educate this knotty-haired gobshite who'd blighted her life so far. All I know is that when she did leave I was in a real state. Even the thrill of acquiring her top bunk for the summer did nothing to quell my fears. I imagined all sorts of murderers and ne'er-do-wells lurking in the shadows of London, waiting to harm my sister. I convinced myself she'd be hurt in some way and I cried myself to sleep for the first few nights she was away.

I think Mum was as touched as she was incredulous.

Of course Yvonne returned in one piece but I could tell shaking off the shackles of small-town life had had a profound effect on her. It was the beginning of a decade-long case of intermittent itchy feet – not a fungal infection but the desire to travel. Not long after London she acquired a short bob hairstyle, cut sharply like the girl from Swing Out Sister, and, in the spirit of their hit single, she was ready to 'Break Out' and discover the world. Me, on the other hand . . . never mind spreading my wings; I curled them round me, perfectly happy to keep things as they were, always happier as a big fish in a small pond.

Yvonne was really close to Mum and I in turn could wrap my dad round my little finger with a killer combo of cuteness and comedy. Whenever Dad, Yvonne and I were together I'd be jazz-handing all over the shop in a bid to hog the spotlight. My sister is a talented pianist and had lessons every week. Dad used to take us and while she was learning her arpeggios with a local lady, called Mrs McCartney, who had a dark-brown bob and an extensive range of flowing shin-length skirts, we'd scarper to a nearby park and I'd get Dad to myself.

I had lessons too, but for a very short time because I lacked Yvonne's dedication and patience. I also wanted time with Dad more than I wanted to squeeze up on a piano stool next to Mrs M in her stuffy front room. Swings and slides or scales? No-brainer.

Yvonne has always worked hard for things, from Grade 8 piano to A levels to a law degree, whereas things have often fallen in my lap. And if it wasn't for Yvonne and her fabulous brain, I wouldn't be where I am now. It was while I was visiting her in Paris, where she was studying, that I first got scouted for modelling.

12. Red, for the Winner

Before the heady days of modelling and the glamour of the French capital, though, I was often busy helping to beautify bevies of a different kind of supermodel, of the bovine persuasion.

Agricultural shows are amazing events showcasing the many different sides of farming, rearing animals and general country pursuits, and they were the highlight of our summers.

In vast white marquees folk hoping for a rosette would display everything from award-winning Cheddar to prize pumpkins, bantams, bunnies, pigs and pigeons – all churned, grown and reared with love and dedication. Spectators sitting in the temporary stands would watch showjumping, sheep-shearing and motorbike displays. Throw in stalls selling everything from wellies to woodcraft to waffles and it was heaven. To me these were the best places on earth. They were the Glastonbury of the farming world, where instead of Ed Sheeran on the Pyramid Stage you'd watch a Gloucester Old Spot boar taking home a trophy in the pig ring.

In the build-up to show days Dad would spend a lot of time

bathing the cows. Well, I say bathing, but although I love the idea of a cow leafing through a *Take a Break* in a bubble bath surrounded by tea lights, the reality was far less luxurious and I guess technically more of a shower, as he would use a hosepipe and maybe a bucket.

The shows all happened in summer, so luckily for the cows the weather would be warmish. 'Up north warm' though, which really just means no sleet.

Dad would pop a rope halter on the cow or bull and tie it to a metal ring attached to a wall, then the fun would begin, hosing down the beast with lashings of Fairy Liquid and lots of rubbing and scrubbing until the white bits were sparkling and the brown bits glossy and squeaky clean. The cows on the whole didn't really seem to mind – there might've been a bit of skipping about from the younger animals but I reckon most kinda enjoyed having their crusty mucky bits scrubbed and soaped up. The prep continued with tools from the show box, a big wooden chest filled with hairspray, brushes and combs, much like a Kardashian's weekend bag.

Added to that were the blocks of chalk to rub on their ankles, the bright white lead ropes with gold clips and the leather headcollars all polished up. The humans needed to be well turned out too; anyone entering the show ring with an animal had to wear a white buttoned-up coat, like a junior doctor on *Holby City*.

Everything would be loaded up and off we'd rumble down the farm lane, the cattle wagon packed fit to bursting with cows, calves, the big bull, show box, bags of grub (for humans and cattle), hay, camping stove, sleeping bags and a couple of Z beds – because when night fell the straw would be swept to one side and the main part of the wagon magically transformed into our cosy accommodation.

Then would follow three days of the show – hours and hours of hanging around interspersed with short bursts of manic preparation and intense, nerve-wracking turns in the ring.

I'd watch Dad on his stints in the ring and my stomach would churn with nerves for him. This wasn't just about a bull catching the judge's eye – this was business, and reputations were at stake. The more prizes Dad's bull won the more its offspring could be sold for and the more Dad could charge for the bull to impregnate their cows. Each breed had a small and close-knit community of farmers supporting it, so word would travel fast about what had happened at the show, good or bad.

Dad never looked nervous in the ring but I know the adrenaline would've been flowing. In the last half-hour before the class, he would become quieter, more serious. He'd check the cattle over and over again to make sure they were pristine. He'd nip to the loo and comb his hair, straightening his dark-green Hereford-patterned tie, smoothing his white coat. His slender face would be still, just his light-blue eyes darting around as he performed last-minute checks on the animals. Then it'd be time to head to the ring, spectators scattering as Dad steadily made his way through the crowds. 'Mind your backs, please,' he'd shout, and families would glance round to see half a ton of beast steadily bearing down on them. One of us would be bringing up the rear, literally, holding a handful of straw in case the cow or bull decided to poo en route. If it did their tail would be held out of the way to avoid getting muck in it and we would wipe the cow's arse with the straw.

In the ring, everyone looked so serious, faces set in concentration as they stood their animals well and made sure they held their heads right. Dad would manoeuvre his massive bull round the ring so the judge could watch how he walked. Dad's eyes would rarely leave the judge's face as he watched for signs of approval in their expression. The crowds of families watching would be near enough silent, kids in buggies munching ice-cream cones and old folk leaning on their sticks, all rapt, seeming to hold their collective breath as the air crackled with tension. Then, when the judge pointed at the winning beast, the crowd would erupt into cheers and applause.

I remember feeling so proud watching Dad in the ring – I've welled up just writing about it.

Now I have my own family I know the importance of having different hats. I'm always me but I have my mum hat, work hat, friend hat – different parts of my personality given expression in different roles. I think I loved seeing Dad in the ring because he wasn't wearing his hassled-dad hat or knackered-wagon-driver hat; here he was in his prize-cattle-breeder hat, skilfully showing off his animals and taking pride in all the hard work it had taken to get them up to that calibre.

Dad loved these moments in the ring. Relief and pride would wash over his features as he stepped forward to collect his rosette. He would be particularly proud when it was the best colour rosette of all – red, for the winner.

During downtimes at the shows, Dad would mainly banter with his farmer mates. I loved watching him chatting so animatedly with them all; he seemed so far removed from the busy Dad juggling everything on the farm. I'd watch as his face would darken when discussing serious stuff and then suddenly light up, his blue eyes crinkling at the edges as he cracked up laughing with his buddies. The one usually holding court was another Hereford man called Jack Henry, a small wiry fella who had a precise, pointed black beard and moustache like Guy Fawkes and merry dark eyes that would twinkle as he talked. He always took the mickey out of me and I loved him. His wife was called

Win and she was a petite bird-like lady with neat pale-blonde curls, as glamorous as a lady could be in a white pharmacist coat holding onto a massive cow. She was quick to roll her eyes at Jack's chat but you could tell they were very much in love.

At the end of each long day Dad and his pals would pull up a few bales of straw and someone would produce a bottle of whisky. I'd hang out a bit, ear-wagging their chit-chat before cosying up in my sleeping bag on the camp bed in the cattle wagon.

Thousands of people would visit the shows, crowds filing through the huge marquees that were the animals' home for the duration, pausing to admire the sheer awesome size of a bull or pointing out a calf to the kids. I loved nothing more than busying myself around the cows, wearing the world-weary expression of a kid whose life's work was to make sure the white end of a cow's tail was backcombed to within an inch of its life and brightened with chalk. In clouds of Elnett hairspray and chalk dust I'd toil away, loving having a few passers-by stop to watch me skilfully coif the tail into a replica of Bet Lynch's bouffant hair circa 1986.

We kids not only helped with keeping the cows pristine; we were also allowed into the ring to show the cattle. Usually Yvonne and I would be given a calf each to hold, although the sensible-seeming logic that the smaller the animal the easier it is to handle was kinda flipped on its head with Dad's cows.

The massive stock bull, all three-quarters of a ton of prime beef, was surprisingly easy to deal with, probably because he was so massive he couldn't be arsed to summon up the energy to try to run about. Apart from the odd sidestep and guttural roar at a rival beast (the bull equivalent of saying 'C'mon then, you want some, do ya?') the big fella was pretty quiet and docile.

But a calf weighs the same as your average twelve-year-old human when it's only a couple of months old.

When it's so young, a calf is always shown alongside its mum, so you'd think – or hope – it'd just want to stick close to her. However, like a toddler being dragged round the shops, a calf in a show ring would often get bored and want to wander off and have fun, and those Bambi-esque limbs were deceptively strong.

That's exactly what happened to Yvonne at Bury Show one year. She was leading a young bull calf when suddenly it decided to make a bid for freedom. We were all holding an animal of our own, so we couldn't really give chase to help pull up this high-speed Usain Bull. We just watched helplessly. Yvonne momentarily tried to dig her heels in and stop the calf from gaining enough momentum but it was too late; the bull calf took off, pulling her along with it. She should've let go but some weird instinct must have kicked in – she clung grimly on to the lead rope as spectators, families and stallholders all leapt out of the way of the runaway calf with the girl attached to it like a rag doll tethered to the tail of a kite.

When she was unable to keep up any longer she fell and was pulled along instead, bouncing over the grass on her front like the speedboat she was waterskiing behind had run out of lake. Eventually the bull calf ran too close to a marquee and Yvonne got caught on one of the guy ropes, forcing her to let go of it and come tumbling to a halt as the calf simultaneously got bored of its adventure, slowed to a trot and stopped a few feet away.

Yvonne was fine – just a bit bumped and bruised. She went up in my estimation that day, with her display that would've impressed the most hardened Hollywood stuntwoman.

*   *   *

Runaway calves weren't the only excitement to be had at the shows. Dad and his mates would get involved in a little competition to guess the age, sex and breed of cattle. Easy, I hear you say – look at the size, the colour and the dangly bits. The twist, though, is that these weren't live animals – they were butchered, cooked and served in little chunks on numbered platters, like the samples they put out on supermarket deli counters.

Dad and his chums would chew the beef and then chew the fat over which animal they thought it had come from. 'Hmm . . . what do ya reckon, Jack?' Dad would say. Between these two great friends there was always a healthy dollop of rivalry, whether it was serious, like in the ring, or a bit of fun as in the meat tasting. Jack would eye Dad: 'Eeh, I'm not reet sure, Len . . . tricky one this,' he'd reply, squinting thoughtfully for effect, and then suddenly, eyes widening in a Eureka! moment, he'd scribble down his answer on the printed-out sheet.

I was always flabbergasted that they could get even one right but more often than not they got most of them; apparently from the taste, firmness and texture they could tell a lot about the beast the meat had come from, stopping just short of eye colour, star sign and hobbies.

Every show ends with the Grand Parade, which is basically aversion therapy for anyone with a health and safety addiction. On paper it sounds dangerous and in reality it was always insanity. As a celebration of the day's success each animal that'd won its class was taken into the main ring to do a lap of honour for the watching crowds. Considering there are dozens of classes, that meant dozens of animals all jostling to get into the ring; all creatures great and small: prize rams, ponies with their rosette ribbons fluttering, loads of cattle – it was a bit like Noah had parked his ark nearby but the memo about only two of each species hadn't got through.

I'd inevitably be holding on to a calf and you'd end up getting carried along on the swell of people and cattle. Not only animals but horse-drawn carriages and tractors would do a circuit too, like the Wacky Races. All of this made more hazardous by the fact that nearly all the farmers would be slightly delicate after the revelry on the bales the previous night.

During the laps of honour someone would announce over the squawky tannoy each animal's fancy long show name and its accolade. Much to Dad's annoyance, they'd often mispronounce his bull's name. He was called Lowesmoor One Fatham but they'd drop the 'th' sound of Fatham and call him 'Fat ham' – a name suitable for a prize boar maybe, but not a bull. I would see Dad shake his head in annoyance.

Grand Parade completed with, miraculously, no serious injuries, we'd finally load up the animals and get on our way, joining a long line of cattle trucks, trailers and horseboxes trundling off home.

13 Turkey Spaceship

If summers on the farm were all about the shows then winter
was all about the turkeys. Across the yard from the farmhouse
were two large, long wooden sheds. To get to them you'd have
to run the gauntlet past Radge, a rescue Alsatian that Dad had
as a guard dog tethered on a long chain near the sheds. I don't
know where he sprang from, but he was one of many strays that
found a home and a role at the farm.

Radge was not blessed with good looks – he was scrawny
with a slightly hunched back and had a touch of the hyena
about him with his wild eyes and slathering mouth. He was also
always filthy and wet, even on warm dry days.

What he lacked in looks he made up for in enthusiasm, and
whenever you'd walk within range of him he'd bound up and
fling himself at you, risking a garrotting if he misjudged the
length of his chain. (Radge had an insatiable appetite for affection and also for neighbourhood cats, hence the chain.) It was
like being scrubbed on the legs by a stinky yard brush, but still
I loved him.

Once past Radge you'd reach the oblong sheds that had housed chickens when I was very young but which, in later years, were turned over to the turkeys Dad raised in the winter months. The huge wooden doors would slide open to reveal dozens and dozens of young turkeys that were spending the last weeks of their short lives on the farm so my dad could fatten them up in time for Christmas.

The turkey shed was warmer than a hospital ward and really stuffy, lit by glowing yellow heat lamps that cast massive monster-turkey shadows on the walls. The shed was lined with sheets of silver insulating foam, which made the whole place feel like you were in a spaceship made on *Blue Peter*.

The turkeys would mill about chunnering among themselves, making a low warbling noise. They had plenty of room and were pretty content but very nervous, sensitive to sound and sudden movement, so you'd have to walk among them like Tom from *Tom and Jerry* on tiptoes past Spike the bulldog.

They were hefty birds and if anything startled them they'd keel over, their little hearts already under strain from being over-weight. The heart-attack victims would be boiled up for the dogs, semi-plucked and bouncing around in a massive boiling bubbling vat of greying water, scum gathering on top, smelling like death.

Dad would mostly take orders for them, but the odd house-wife would come up to the farm and Dad would offer to take her to the shed and wind the woman up, saying 'Come and point out the one you want so I can put a ribbon round its neck.' Women have always loved Dad; the ol' strong silent vibe he gives off makes women keen to get his approval, to make him laugh, to get him to drop his quiet, serious demeanour. I'd watch ladies go all gooey around him, laughing at his teasing, secretly thrilled that he was winding them up, his bright-blue eyes shining mischievously.

Slaughter time and the laughter ended, though I imagine it was marginally worse for the turkeys than for Dad.

Living on a farm, you have to be pragmatic about the use of animals for food. I never really had any moral wrangling over it. I remember asking where various meats came from and the answer was always a straightforward 'That's pig/cow' or whatever. I'm the same with my own kids; I answer honestly and it's up to them to decide how they feel about it. My eldest daughter Lola is now a vegetarian, so maybe I've been a bit too honest.

When I was tiny, my older siblings had a much harsher introduction to the food cycle, when (so family legend goes) they asked at the dinner table where their pet lamb Larry had got to and Dad replied, 'You're eating him.' Dad farmed in a fair way, though – he always made sure the animals had good living conditions and were treated with respect before they met their end.

A stable was used as the killing room, turkeys held firmly by the legs upside down, their heads sliding between two flat metal rails, which had an electric current running through them – this would stun the birds unconscious. Dad would then neck them, twisting their heads between his fingers in a quick, deft movement, before placing them head first into metal cones. The body would be encased tightly in the cone to avoid bruising to the flesh, because the bird would continue to twitch and writhe even after death, as all the nerves would still be firing – it also avoided broken wings. Their purple heads would dangle out of the opening at the narrow end of the cone. Plucking the birds was my job, and it was almost comical. After being dipped in boiling hot water, a machine that was basically a cylinder of rubber fingers would spin at high speed, as I held on to the chicken's feet like they were the reins of a runaway horse and carriage, and the birds' carcasses bounced over the whirling fingers.

Writing this shines a light on the darker side of eating meat but it's what has to happen if on Christmas Day you want slices of succulent turkey alongside your sprouts and roasties. As much as I like the idea of animals living to a ripe old age and dying of natural causes while surrounded by loved ones in a sun-kissed meadow, and only then being used for meat, that just ain't the truth.

At Dad's it was a small-scale operation but an efficient one, and done as humanely as possible. There was a shed opposite the house where he 'cleaned' the turkeys, as he called it. Hour after hour Dad would stand in that freezing shed sporting a big cream rubber apron, smeared with blood like macabre toddler finger-painting. He'd use huge knives to slice off heads and feet before pulling the entrails and innards out of the turkeys like a magician conjuring endless ribbons from his sleeve, and neatly parcelling up the giblets in a little plastic bag. Next to him on the floor was a silvery metal bucket of doom containing all the cut-offs – a grim pile of pastel-shaded organs, pale lilac intestines, rose-pink stomachs, papery faces gawping blankly up at you.

I always felt sorry for Dad, whose back must have ached from standing so many hours in his wellies on the cold concrete floor. It was a smelly, monotonous task and the one part of farming he didn't relish at all, which is saying something cos – much like parenting small children – a lot of farming is monotonous and smelly.

Turkeys bagged, collected and paid for and Christmas could commence. It was always special, but it was super-special if we got a bit of snow because then Dad would hook up a plastic sledge to a rope connected to the back of the tractor, and with a shout of 'Right! Hold on, kids!' he'd tow us around the undulating front field at a breakneck top speed of 3mph (obviously it felt much faster).

Looking back I don't think about the safety of sliding along behind a huge beast of a vehicle or the diesel fumes we probably inhaled; I think of Dad's face as he laughed, steering the tractor, half turned round in his seat to watch us as we screamed and squealed all round the field.

Health and safety, or lack of, is a recurring theme in this book. It's a miracle we survived the seventies – if we weren't second-hand smoking we were sliding around in the back seats of cars without seatbelts driven by adults with relaxed views on drink-driving. Add to that the extra excitement that comes with being on a farm surrounded by unpredictable animals and reversing tractors, and the Cox clan were lucky to escape unscathed.

There were a few near misses.

We had a lovely old dog called Spud, a lurcher crossed with a collie. He was huge and black with a dickie bow of white fur round his neck and a white tip to his tail like a paintbrush that'd been freshly dipped. He had the long gangly legs of a lurcher and a taste for smaller family pets – so he was also kept on a chain. He guarded the front of the house and would always demand a cuddle when you went in and out, sort of like an affection tollbooth.

He'd lean up against me and nearly knock me over, the top of his bony head reaching to my chin. He was getting on a bit and by the time I was eight he looked like one of the Rolling Stones, his hip bones jutting out and his fur greying. In his day he had been a fierce guard dog but now his eyesight was failing and he was half deaf.

It was this that created a problem one day when Dad was at

the top of a very tall ladder at the front of the house fixing a bit of guttering. Dad's friend Mel, who often helps out on the farm, was supposed to be holding the bottom of the ladder but had been distracted and left his post. Spud was on his chain as usual and mooched his way over to investigate what Dad was up to. Dad clocked this right away and started to shout at Spud to deter him from coming closer.

Unable to see Dad and only able to hear him faintly, poor confused Spud, like the world's oldest morris dancer circling a maypole, walked slowly round and round the base of the ladder looking for Dad, dragging his chain behind him.

Dad started to shout louder, realising that if Spud gave up his search and started to make his way back to his bed in the shed he'd pull the ladder with him.

Spud's chain was getting shorter and shorter as he wound his way round. Luckily, hearing Dad's shouting get louder and more panicked, Mel legged it back across the yard and grabbed the dog, turning him steadily round and walking him back the other way, eventually untangling the bottom of the ladder – but not before a small crowd of us kids had gathered to supportively laugh and point.

Dad did eventually see the funny side, once he was safely back down the ladder feet first rather than face first.

14. Bovine Sally Gunnells

One of the commandments of living on a farm is 'Thou shalt grab yonder wellies, come rain or shine, when commanded to round up escaped cattle.' Usually after I'd fed, watered, mucked out and tucked up my pony Gus for the night, swept the yard, peeled off my soggy joddies and was lounging in my jammies in a prime spot on the couch in front of a roaring fire with a couple of biscuits and something good about to start on the telly . . . THE CALL WOULD COME.

The phone would ring. If it was after 8 p.m. it'd never be good news – no one rang for idle chit-chat at that time – so as soon as the first ring trilled out from the big cream phone hung up in the hall we'd all look at each other with dread.

Dad would haul himself out of his chair grumbling and head towards the phone in the kitchenette. We'd be straining to listen to his replies – 'Where about? How many do you reckon? Are they near the road? OK, love, thanks for letting me know,' then once he'd put the phone down, the cry would go up 'Cows are out!' and it'd be a mad scramble to get wellies on and head out.

Like the bell ringing at a fire station or the Bat Signal being shined into the air, we'd all spring into action. Maybe Dad should've used a more exciting alert, some sort of klaxon or a cowbell.

Either way, we'd trudge through the fields to find the cows, who would be having a whale of a time, having pushed and leaned on a fence post until it gave way or found a weakness in the barbed wire thanks to the handiwork of some of the charming local teen vandals who often amused themselves by dicking about with Dad's fences.

The cows usually didn't actually go too far, none of them having the energy or inclination. Let's just say they lacked an adventurous spirit. They'd only roam onto the park that ran parallel with the bottom boundary of Dad's land or onto my school's fields, where they'd be spotted by a local, who'd ring the farm.

Depending on the weather the ground could range from squishy to knobbly to slippy, making it tricky to run after cows in the dark wearing wellies and your nightie.

The usually docile cattle could suddenly turn into bovine Sally Gunnells, skipping away at high speed and dodging past us in a zig-zag like they were scoring a try on the rugby pitch.

Us kids and Dad would be spread out in a circle like some weird dance performance, arms outstretched like scarecrows, slowly closing in on the cattle, hoping to shoo them back towards the field they'd come from. After a bit of pratting about the cows would eventually get bored of their new game and make their way back home whilst Dad patched up the fence.

We'd all pull our wellies off and flop back onto the couch, knackered, rosy-cheeked and strangely exhilarated by the adrenaline rush of an impromptu cattle drive.

*　　*　　*

Occasionally a call would come to say that something altogether more sinister was happening – there was a dog in the field with the cows, calves and expectant mums.

This was a less jolly affair – even the sweetest family pet who may by day like nothing more than a belly tickle and snooze on an armchair can undergo a transformation when in a field with a bunch of frightened cows and many hark back to their wolf ancestors, becoming slavering predators, like when the Hulk gets cross then can't remember anything after.

A dog suddenly appearing in the field barking and nipping at the ankles of the animals causes a mass panic among the herd, which is particularly bad for the cows that are with calf. Any woman with kids knows the last thing you want to do when you're heavily pregnant is run around a knobbly field in the dark being chased by dogs. You'd much rather lie on a squishy sofa eating custard creams. Although that obviously wasn't an option for the pregnant cows, having a relaxing time in the fields, chewing the cud and snoozing, was important. There'd also be new calves in the field that were at risk of injury from the dog or being trampled by the running cattle.

When Dad got a call about a dog that'd be the cue for him to go and unlock his rifle cabinet. Dad, like lots of farmers, used to have a gun for clay pigeon and rough shooting and had a licence to use it to protect his livestock.

That'd be my cue for hysteria. Weeping and begging Dad not to shoot the dog, I'd try to block his way and make him promise not to kill the family pet currently stalking the poor cows out in the field.

Once, I remember it was my job to run up the lane to the top field and swing open the big iron gate so Dad could drive through in his jeep and head up towards the cows and the dog. Instead of wrestling the old frayed blue rope off the tombstone-esque gatepost I stood there in the glare of the head-lights, refusing to open the gate, arms folded and face set in a determined grimace. It was vital to get to the dog before it did any damage, so this protest was probably pretty annoying. Dad jumped out the jeep, gently but firmly moved me out of its path, swung open the gate and roared off in the jeep, shouting 'I'll only pepper its bum.'

This charming phrase meant that he would only fire once the dog had scarpered, so it'd get nothing worse than a few tiny bits of shot in its rear end, hurting it just enough to deter it from returning.

What usually happened was Dad would get to the cows and find that the dog had gone. Or sometimes he would indeed just pepper its bum – I don't know if he meant to miss but let's just say that Dad wasn't always the best shot.

15. Death Row Gameshow

On a few occasions Dad would take me to the cattle market when he had a couple of bulls to sell. I loved it. Bustling, huge sheds with corrugated metal roofs containing rows and rows of pens holding all manner of livestock from sheep, pigs and cattle to geese, goats and hens.

Animals being loaded and unloaded from wagons; market workers and farmers shouting, watching the auctions, queueing for a polystyrene cup of tea. Men with red, weather-beaten faces as creased and crinkled as old leather satchels would gather round the indoor rings. These were metal enclosures with a gate on either side that could operate like a one-way system – livestock in one gate to be seen and bid on and out the other side once sold, hopefully off to a new farm and not the abattoir.

Dad always hoped for the former – a farm only needs one big stock bull, so he'd need to offload any spares. They were still good-quality young bulls with a great pedigree for showing and breeding, so ideally Dad would want them to go off to a new farm to service their own bevy of bovines, like a hoofed Hugh Hefner.

To help raise the stakes and stop his bulls becoming steaks, Dad would get me to walk them round the ring. This was a very different experience to showing cattle. It felt like some death row game show – 'Ladeeeez and gennelmen, will this young girl impress the crowd and save this beast from certain doom or will the farmers make mincemeat out of her . . . AND the bull?'

Either way, our plan often worked. I'd walk the bull on a rope headcollar round the ring and the auctioneer, before launching into his special hyper-speed babble as the bids began, would do a little spiel on the microphone about how easy the bull was to handle as 'a nice young lady is leading it round'. I would try to make light work of encouraging this slightly freaked-out animal to walk smoothly round the ring, smiling beatifically if it skipped about or misbehaved. ('C'mon, you schmuck! I'm trying to save your life here!')

Let's be clear here, by the way, this was when I was fourteen, with a terrible wavy bob just below my ears, boobs still a distant dream and registering about 250 on the awkward scale. I wasn't like one of the girls they used to have at the boxing, sashaying round the ring in between rounds using my cute looks to sell a bull.

The audience of mainly men would cross their arms on the gates and lean their chins on their forearms, burying their mouths in their jacket sleeves, possibly to hide their expressions so they could play their cards close to their chest about what beasts they were interested in. This left a narrow strip of face available between their nose and the pulled-down peak of their flat cap. Nevertheless, they still managed to bid, communicating with the auctioneer with the most minuscule of movements – a wink here, an eyebrow-raise there or the most minute twitch of an index finger were enough to place a bid. People joke that at an auction you could sneeze and end up accidentally purchasing a painting or expensive vase. At the cattle auctions your pupils contracting could buy you a cow.

16. Pine – the Wood that Powered the Eighties

When we left the farm we moved about half a mile away and lived in a red-brick three-bedroom council house on School Street. I was about six and a half when we left – the kind of age where you don't really question the whys and whats, you're just taken along for the ride. I can't remember feeling sad about leaving the farm – I had the priorities of a six-year-old and I'm sure the promise of my own bedroom was enough to quell any lingering unease about why we were moving home. Over the following weeks, months and years I'd gather enough information to piece together what had happened: ear-wagging on whispered conversations between Mum and Nana amid clouds of cigarette smoke; hearing curt greetings between my parents when Dad dropped my sister and me off at Mum's that occasionally escalated into shouty arguments – the frosting of civility still too fragile and cracking under the weight of Mum's heartbreak. It was a sad time, especially for Mum, but I was sheltered from a lot of it. Before I knew it, Mum was in love with a man called Ken and we were suddenly a shiny new little family and

Mum was much happier. Dad was with his new wife and we all had to readjust, now the dust had settled slightly. It wouldn't fully settle until twenty years later when I saw my mum and dad chatting warmly at my sister Yvonne's wedding. It took both the other partners to be gone – Ken through death and Dad's wife through divorce – before Dad and Mum could allow the old wounds to finally heal enough for them to become friends again. It was a surreal thrill, seeing the two people responsible for my existence talking together after all that time.

Life on School Street turned out to be a golden time. It was a quick walk from the farm – you could nip out of the farm gate and, depending on the season and therefore how overgrown it was, usually go down the little footpath that hugged Dad's paddock and continue along the side of Masefield School and out onto Masefield Road, which had neat little semi-detached houses on the left and the Lorival factory on the right, although you couldn't see it thanks to the high ivy-covered railings. First right and onto School Street and our house was the second on the right.

We lived at number 69 and I loved this house for many reasons: it had steep steps at the front that led up to the garden path, so to me the house seemed like a mansion on a hill; my nana and grandad lived just over the road at number 32; the back garden seemed huge – it was long and narrow with big leafy trees lining each side and, at the bottom, a rockery, something that was very popular in the eighties (I mean, rocks half-embedded in the earth and clumsily painted white; not surprising they were so popular); and my bedroom had the most amazing wallpaper – of sweet rosy-cheeked cartoon cowgirls riding their plump, peachy ponies, all in faded golden hues.

The house was always warm and cosy – in the lounge thanks to the gas fire and in the kitchen thanks to Mum's cooking. The window overlooked the back garden, not that you could look at it through the steamed-up windows created by a lack of ventilation and Mum's enthusiastic use of a pressure cooker.

I always found Mum's pressure cooker faintly terrifying, a huge silver vessel with a thick, screwed-on lid that was more 'engine part that's fallen off a spacecraft' than a pan. It squatted on the kitchen counter, hissing and spitting like a cornered cat and generally giving off the impression it was going to explode at any minute, blowing the entire kitchen and me all the way to the rockery.

From the pressure cooker and the slow cooker came a constant flow of delicious meals, from her melt-in-the-mouth beef in red wine sauce to chicken stew.

Our kitchen was dominated by the table, which was like a bench from a picnic area and made of pine – the wood that powered the eighties. It wasn't the pale untreated pinewood that's more popular today; our table was glossed to the colour of Irn-Bru and had large black knots in it. It was like eating our tea on a reclaimed wall from the inside of a Swedish sauna.

The front room was oblong with a big square window looking out onto the street and on the opposite wall a dark wood upright piano. Yvonne needed to practise, and before Mum could get round to buying a keyboard for her, Grandad heard of a friend of a friend who had a second-hand piano to sell, so there it stood, massive, dominating the relatively small room. There was always piles of stuff on the piano. Like an exercise bike that becomes a hanging rack for clothes, the piano became like a large shelf, covered with books, unopened post and general clutter. When the lid was open there'd often be one of our cats on it, casually stepping on the keys and accidentally playing a disjointed, mournful tune.

The squishy, pale-blue velvet sofa was opposite the gas fire and angled towards the telly in the corner. This was my spot, sitting on the floor, leaning against the sofa among a jumble of legs, my face shining with heat from the fire. It was in this position in this very room that I witnessed many historical moments, from Michael Jackson's 'Thriller' video when it was debuted on telly (my cousin Anthony came round specially to watch it with us and ape the dance routines), Bucks Fizz's skirt-ripping triumph at Eurovision and the Royal Wedding.

I was especially excited by the Royal Wedding. My mum and my nana absolutely loved the royals – every Christmas Day we'd watch the queen's speech and we couldn't talk or mess about in case a single syllable was missed. Mum and Nana would be nodding and murmuring their agreement with the queen, as if they were having a gossip with Liz over the back fence. Whenever the royals had troubles, Mum and Nana would discuss them as if talking about members of our own family, passing judgement on some of the seemingly sloppier members but always having sympathy with the queen.

The engagement of Charles and Diana took over the nation and everyone in our house was swept along with the excitement. I decided Diana might need some help with designing her dress, so I drew a picture of her in a huge pouffy frock with diamonds on and sent it to the Palace. I got a reply within a couple of weeks, on expensive, stiff, creamy-coloured paper with a red Buckingham Palace insignia on the top and on the envelope. I was beyond excited. Grandad, who seemed to have a lot of 'friends of friends', found another friend of a friend who worked at the local paper, the *Bolton Evening News*, and aged six I had my first taste of mega-stardom – I appeared on its smudgy black and white pages, proudly holding my letter from the Palace. In it the secretary to Lady Diana thanks me for the drawing. There I am, hair neatly pinned down in a side parting, forehead shining and grinning a shy smile. Pleased and proud as punch.

In the end Diana shunned my designs and opted to work instead with David and Elizabeth Emanuel, which I took on the chin. The dress was magnificent in a very eighties-princess way and I secretly thought that she'd been inspired by my pouffy big number.

We soon had our own wedding to plan, when Mum decided to marry Ken. He'd been her childhood sweetheart and once Mum, Yvonne and I had left the farm they rekindled their love. Ken was younger than Mum (only by a few months but she still took great pleasure in claiming she had a toy boy) and had been a bit of a wild lad. He had mousey-brown hair in a flick and a matching mousey moustache, olive skin and a gap-toothed smile. He loved to wear his keys attached by a ring to the belt loop of his jeans. As time went on and they became tenants in pubs the number of keys increased – cellar door key, pool table key, front door key, jukebox key, cigarette machine key, till key – all different shapes and sizes of keys hung from his belt in a massive bunch the size of a boxing glove; you could hear him coming for miles as he jangled along the corridors of the pub.

When we were all thrown together at School Street, Ken was kind and fun and his bad-boy rep melted away when he fell for Mum. He must've really loved her to take on not only Mum but six-year-old me and ten-year-old Yvonne.

My main memory of their wedding day was that for some reason I rebelled and refused to hold my hands in a ladylike manner for my individual photos. The photographer tried to cajole this seven-year-old bridesmaid to lay one hand gracefully over the other and then lean my head on them as if daydreaming some whimsical thoughts, but to no avail – there are photos of me with my chin resting firmly on two tightly clenched fists like a chess player trying to psych out an opponent.

Our dresses were beautiful and I loved them. We went to a local dressmaker and in the build-up had to endure many an awkward fitting, standing in vest and pants in the dressmaker's front room. The dresses were floor-length and made from floaty chiffon and were cream and pale green; I looked like a walkin' talkin', fist-clenchin' summer meadow.

In the photos Yvonne and I look our usual milky-skinned, pale selves, like we're haunting the photo ('in this example you can just make out the spirits of two little bridesmaids who choked to death a century before in the great confetti disaster of 1889'), and Mum looks serene in a satin cream number with headgear that looked part lacy pancake, part 1920s pilot's leather helmet.

The night before the big day I couldn't sleep, not from excitement but from mild agony. For some unknown reason my mum had snubbed the use of heated hair appliances (despite the eighties being the decade of the hot brush and crimpers) in favour of going back to Victorian times, and had created ringlets for us using 'rags'. These rags were metre-long, narrow strips of creamy fabric that were possibly the shredded bedsheets from a Crimean War field hospital.

Each 'rag' was folded in two and tied round a small section of the hair at the root. It then entwined, wrapped and encased the hair so I ended up going to bed with what looked like huge, white, cotton dreadlocks. They weren't the comfiest – whichever way you leaned your head on the pillow it felt like a small elf had grabbed a fistful of your hair at the scalp and was pulling it with all the strength its little elf arms could muster. Horrible. Still, the result was lovely and I had enough ringlets to make a Victorian weep with envy.

Being fitted for the bridesmaid dress served me well for the next big social event that lurked round the corner. For the Christmas play my primary school had decided to put on that Christmas classic *The Wizard of Oz* and I had been awarded the role of Toto, Dorothy's dog. I was very happy with this and, as an aficionado of the film, I knew I could do justice to the role of the small, sassy hound.

I would need an outfit, of course, in order to truly look the part.

Mum consulted the classified ads in the back of the *Bolton Evening News*, which was a bit like an eighties Amazon except everything was second-hand. Whatever you needed, you'd have a good chance of finding it in the back pages. Adverts selling second-hand skateboards, mattresses and food mixers sat alongside more for cars, cats, violins and walking boots.

Some items were 'free to a good home' and the rest fell into one of several categories: 'Under a Fiver', 'Under a Tenner' and all the way up to, if you were feeling flush, 'Under £100'.

Mum scoured the small ads and sure enough she found a second-hand dog costume. The lady who was selling it lived on a local estate and so one day after school we headed to her house to view the goods.

In the living room, as her three kids lounged on the sofa watching *Emmerdale Farm*, the lady produced the outfit with a flourish. It was conker brown, with shaggy fur and floppy ears. Embarrassed by having to change in front of the other kids, I crouched down behind the sofa and remained crouched there in my vest and pants until Mum held out the outfit, helped me step into it and zipped me up the back. I popped up the hood and that was it. It fit like a dream and I WAS Toto.

The play went down a storm. I remember building up my part, sliding around on my knees and enthusiastically yapping. The watching mums and dads started to laugh at this very vocal Toto and the more they chuckled the more I barked.

17. How Much for the Donkey?

We had real dogs to look after at Mum's and they came in a variety of shapes and sizes. Gemma was a beautiful setter-cross pup with a black, glossy coat, liquid brown eyes and the waggiest tail you'd ever seen. She also had a penchant for buggering off if someone accidentally left the front door ajar. One morning she escaped and Mum searched all day, only coming home when it went dark. Later that evening we got a call. Someone had found her dead by the side of Seven Stars Hill; she'd been hit by a car. We were absolutely distraught. I remember Mum crying. It took us all a long time to get over that.

We begged and pleaded for another puppy but I think Mum decided that the way forward was to get an older rescue dog that wouldn't have the energy to scarper and get squished. She contacted her mate Margaret from the local RSPCA and gave her a heads-up that we were on the lookout for a new four-legged friend.

Therefore, without any prior discussion with her family, Mum set in motion a repetitive chain of events that went like this:

We would rehome a middle-aged, whiskery dog that would last an average of five years. We'd of course absolutely love the dog but it became normal for us that we'd come home from school to see the dog's collar on the white circular dining table, which was the sign that they'd shuffled off this mortal coil.

After the tears, the sobbing and snotty hiccups had ceased we'd blow our noses and then beg and plead with Mum for a puppy.

Somehow Margaret would get wind of the latest pet vacancy at ours and would ring Mum with details of yet another whiskery old dog that desperately needed a home.

Each time she used the killer line 'Honestly, Jackie, he's lovely, and if we don't find him a home he's getting put down in the morning.' Unsurprisingly, that always clinched the deal. We'd end up with another death row dog.

We steadily worked our way through a variety of mutts, beginning with multicoloured Merl, named after the wizard Merlin, a knee-high mongrel who looked like he was fashioned from a thousand carpet offcuts. He had a black head, a white chest and greying ears and beard; his back was a brilliant mixture of brown, more grey and undercoat of black; and there were bright ginger layers all round his back legs. His coat was shaggy and rough and he had bright, smiling eyes.

He was no trouble and was Ken's best mate, following him round like a slightly shambolic shadow.

When we inevitably found Merlin's collar on the table we started begging as usual for a puppy, but I swear as soon as we said the word the phone rang. Yes, it was Margaret. Old dog . . . lovely personality . . . execution planned for tomorrow . . . blah blah.

So we then opened our arms to Scooby, a short stocky black terrier with a white tip on his tail, a white star on his chest and four white socks. He looked like the result of a romantic liaison between a rogue corgi and a Border collie.

Mum could be a sucker when it came to cute animals. There was an occasion at a local funfair when I almost persuaded her to buy me a donkey. I'm not even kidding. It was dusk and we'd been having a whale of a time, but now the fair was winding up. I spotted Holly, a small grey donkey, tethered to the side of the owner's horsebox on a break from giving rides. I went over to stroke her and all I can remember is feeling an instant connection with this donkey. I was convinced she was communicating to nine-year-old me, telling me through her snuffles and grunts that I was her saviour and could take her away from this life of drudgery. I looked deep into her dark-brown donkey eyes and knew we were meant to be together.

Mum came over and I explained all this. That I loved Holly (her name was stitched onto the browband of her bridle, by the way; I'm not psychic), that we were meant to be together. I don't know what I expected Mum to do; swagger up to the owner like Veruca Salt's dad, pull out a wad of fifties and declare 'Alreet, sunshine! How much for the donkey?'?

What she actually did was explain that we couldn't take Holly home. She was always the tougher parent – the eat your greens/ do your homework/get to bed parent. 'Darling,' she said gently in her broad Bolton accent, 'We can't take the donkey, where would we keep it? No, love, now come on, say goodbye to the donkey and let's head home, eh? Do you want some candyfloss?'

At this point I calmly accepted the fact that it wasn't meant to be and we went home. Not really! At this point I went into meltdown. I threw the biggest paddywack the world had seen. I cried, I wept, I begged, I sobbed and snotted and screamed and sweated and gulped and threw my arms round the donkey's

neck and made her fur soggy with my hot tears. I hyperventi-
lated, hiccupped, whimpered and wailed. It was actually pretty
out of character for me. I'd learned from an early age that a
tantrum achieves not what thou doth desire. In fact, it was so
unlike me that Mum was compelled to go and ring Dad from
a payphone and explain the state I'd got myself into. She asked
him if we could get the donkey and if it could live at the farm.
To which came the resounding reply 'No. Absolutely not.'

Once he'd said no, Mum seemed to snap out of the spell I'd
cast with my hysterics and broke the news to me and dragged
me away from Holly and into the car.

As always, given a bit of literal and emotional space from
the incident, I now realise that when I decided to fixate on the
donkey it was just stood there minding its own business. It wasn't
trying to communicate with me. There was no connection. It
couldn't move – it was tethered to a wagon – and I was just a
curious distraction from the boredom of hanging around in the
cold.

I'd eaten my own weight in sugar and my hysterics were most
likely a result of a crashing candyfloss comedown.

I have warm memories of life with Mum and Ken at School
Street; they were happy times. We didn't have much, and
although it wasn't quite 'There were fifteen of us living in a
crisp packet by t'side of the road but we were happy', I was
aware, usually from ear-wagging on Mum and Ken's conversations, that there wasn't money to spare.

We were lucky though – we still had holidays, driving down
to Cornwall for camping trips with my 'Auntie' Carole (actually
Mum's best friend) and 'Uncle' Trevor.

Mum and Carole had been friends since they were teenagers,
both of them petite and back then having the same dark bobbed

hair and an obsession with the Beatles, even wearing stockings with a pattern of tiny beetles – of the insect variety – scuttling up their calves.

Uncle Trevor, a short, handsome man with inky black hair and tache who died far too young, loved to fish and on one holiday took Ken out on a day trip with local fishermen. Trevor also loved to cook and I will never forget the smell or the taste of freshly caught Cornish mackerel being grilled over the hot coals of a barbecue on a sunny but windswept campsite. It's no exaggeration to say that although I've not exactly dedicated my life to finding mackerel as delicious, every time I order it in a restaurant I have Uncle Trevor's mackerel in mind and I'm always slightly disappointed that it never quite matches up. Is it that the fish all those years ago was a once-in-a-lifetime dish or have my tastebuds dulled at the same rate that my memory has become more rose-tinted?

We had a couple of trips like that, the grown-ups supping little bottles of imported beer round the glowing embers of the barbecue while Yvonne and I ran feral with our 'cousin' Laura, our hair knotting in the sea air.

One year they rocked up with a trailer tent, which to me might as well have been a spaceship. I was in awe. Up until then, our family had observed the five commandments of camping: 1) Thou shalt be missing a vital pole; 2) thou shalt curse the tangled-up guy lines and wonky tent pegs; 3) thou shalt argue with thy loved one during the putting-up of the tent; 4) it will start to rain/go dark during the tent erection; 5) thou must not crack open a small imported beer until the tent is up – it doesn't help.

No wonder then that, when Uncle Trevor's Sierra rolled serenely over the grass of the campsite and he reversed his trailer tent into their pitch, I was intrigued. We all gathered round while Trevor proudly demoed his new purchase. In my memory he pressed a big red button and, Wallace and Gromit invention style, the top of the innocuous-looking trailer sprang open and out exploded a ready-made massive tent, like a great big blue fabric vomit, where it settled, all ready and upright on the grass. I know now that there was a bit of prep and the odd tent peg to be hammered in, but to me then it was magical.

It was thanks to those holidays that I was nibbled by the camping bug and I still have the compulsion. These days I make my young family endure a few nights under canvas at a family-friendly music festival every year.

As festi time approaches I spend hours raiding the attic and garden shed for all the camping gear, stacking it all up and ticking items off my checklist, then my husband spends ages packing, squishing and brute-forcing bits of kit – gas camping stove, foldaway table, pillows, inflatable mattresses – into my car, then crowbarring the kids into the remaining gaps. We then drive to Dorset, to Camp Bestival, a weekend-long festival held in the magical shadow of Lulworth Castle.

Once we've parked up we unload all the gear and cart it all to the campsite, where I proceed to fall foul of each of the five camping commandments. It's also a tradition that I then almost ruin the weekend before it's started by shouting at all my children in turn, for getting under my feet/mithering/booting a footie full pelt at the half-erected tent/asking for a snack just as I'm on my knees wrestling a metal peg into submission.

In a neat reflection of the shift in attitude around women's and men's stereotypical roles in society, it's me who loves to put up the tent, usually with the help of two strapping youths I've flagged down. I love the fact that some fabric and poles can be brought together to make a place good enough to sleep in. All it takes is teamwork, a bit of elbow grease and some swearing.

I'm in my element camping; I love the sound of a zip in the morning, I love poking my head out of the flap, surveying the skies, hoping to see blue not grey. I'm at my happiest frying an egg on my camping stove, sitting on a picnic blanket, supping a plastic mug of tea.

A few years ago at Bestival we didn't camp, and instead hired a little motorhome all ready and parked on site, but it wasn't the same. Plus there was a slight hiccup with the route. Instead of typing Lulworth Castle into the satnav I put Ludlow Castle and slavishly followed its directions like a fool. I just glanced at the initial big map of the UK that pops up on screen before returning to the road-by-road directions and it looked about right – we were heading west so that'd do, I thought. But in the end it definitely didn't 'do'.

Instead of heading south-west to Dorset we drove north-west towards the edges of Shropshire. As we were nearing Ludlow I started to get a slight nauseous feeling in the very pit of my stomach. That bit that churns whenever you're late collecting a child from school or when you're six and you've smashed a dinner plate.

I was looking out for the usual yellow AA event road signs in vain; there were none, of course, because we were about to arrive at our destination – ONE HUNDRED AND FIFTY-FIVE MILES AWAY from where we were supposed to be.

Two miles to go and still not a single telltale sign of a nearby festival – no other cars packed with gear and kids, no increased traffic, no teenagers walking the last couple of miles on foot with their rucksacks stuffed.

It was at exactly the time that I started to get the sinking feeling that something had gone horribly wrong that the kids' tempers started to fray; they'd been cooped up in the car for four hours and were desperate to get out. 'Don't worry,' said

Ben cheerfully, 'We're there in a few minutes.' Little did he know of the horror that was about to unfold.

The satnav informed us we had reached our destination and we pulled up as instructed. But where were the security checks, the crowds, the cars, the funfair rides and huge circus tents in the distance? They weren't here. What was there was a castle and, before it, a street deserted save a couple of old dears heading into a tea shop.

I then uttered what would have been the standout line if this had been a movie, the kind of thing like, 'We're gonna need a bigger boat' and 'That's not a knife, this is a knife' – I turned to Ben and said, 'We're at the wrong castle.'

As the kids squabbled, unaware, I quickly corrected the satnav to Lulworth Castle; as it started to calculate the distance and time my head was spinning with wild thoughts and hopes – maybe we've just gone wrong by ten miles and before we know it we'll be sipping a beer by the motorhome, chortling at my minor mistake.

But my mistake was major. This became apparent when the satnav announced that our estimated time of arrival at Lulworth would be gone 11 p.m. It was currently 6 p.m.

The kids were so shocked they took it surprisingly well – they fell silent and must've sensed that this wasn't the time to whinge as Mummy was fully aware what a humongous mistake she'd made.

Ben also just murmured quiet words of support as I turned the car round and headed away from Ludlow and towards Lulworth. I gripped the steering wheel, teeth clenched, and bedded in for a long journey.

That was a low point. As was stopping off at a service station to change the children into their pyjamas in the grotty loos, hopscotching over puddles of pee.

It was at the services when I was queuing for a coffee that I got the phone call from a lovely lady from the festival reassuring me we would definitely be allowed on site no matter how late we arrived. She was so kind and I was so relieved that we wouldn't be missing the first night on site, another burden of guilt I couldn't bear on such an emotionally charged day, I burst into tears of gratefulness. As I sobbed, clutching my cappuccino, a lady's kind face hove into view and, seeing my tears, she offered . . . not a clean hankie to dry my tears but a Starbucks takeaway cup and a biro, and asked me for my autograph.

Many hours later we arrived at the festival gates and Ben unfolded his six-foot frame out of the human origami shape he'd performed in the back seat in between the two smaller sleeping children. We unloaded bags and kids into the motorhome and flaked out.

The next morning was bright and warm and our troubles of the day before melted away. The rest of the weekend was brilliant. Word spread round the festival staff about my monumental mess-up and it was possibly the best anecdote of the weekend and has gone down in history as part of the Camp Bestival legend. Every year since, Rob, the bloke who owns the festival, always quips as he passes me happily frying eggs outside the tent, that at least we 'Got the right castle this time.'

Back in 1980, decades before I knew how it felt to try your best to make your kids happy, I was the kid, six years old, and although things were a bit of a struggle money-wise I was cushioned from any stress by Mum's love and my own blossoming social life on the street.

Opposite our house was the 'half-circle' – smaller than a precinct but bigger than a regular pavement. Where School Street sprouted off to the right from Masefield Road on the left there were two houses marooned together in a no-man's-land, being half in each street. In front of these houses was the half-circle.

This was where we all played out. There was a boy called Paul who split open my head by lobbing a rock at it, a neighbour called Claire whose dad had the first official goatee beard I'd ever seen, and Neil who lived further down the road, whose dad had a penchant for argyle jumpers and who would come out to call him in for tea by standing at the garden gate and shouting at the top of his voice 'NNNEE-IIL', at which point we'd all fall to our knees on the cold pavement. That joke never got old.

I especially liked Claire because her fridge always had a can of that disgustingly delicious aerosol cream in it and I'd flip my head back like a pedal bin so she could fill my gaping gob with big sweet clouds of the stuff.

She'd also get a massive tub of margarine (a whopping catering-sized vat, big enough to bath a spaniel in) and scoop out a teaspoonful of marge, then roll the marge in the sugar dish and pop it in her mouth. A ballsy attitude to nutrition – not even bothering to eat food laden with sugar and fat, just rolling some fat in some sugar and chowing straight down.

The half-circle was our domain. If you crossed the road and went down a wide path between the houses you'd get to the park. It was the park where my sis Dot taught me how to ride my bike and where I once went when I decided to run away and start a new life after being told off. I packed three malted milk biscuits, an apple and some clean pants. I hid under the slide till dusk, then got cold and bored and went home for tea. Mum hid her relief very well and played it cool. Either that or she hadn't even noticed my dramatic exit.

18. A Case of
Leprosy Perhaps?

Mum was always busy; she worked hard, often holding down a couple of jobs at a time.

I think a combination of her being busy and me wanting attention, coupled with my penchant for drama and play-acting, led to a really peculiar habit of mine that showed itself a few times when I was around six.

As I write this now it sounds completely mad and, to be fair, it was. But it did show a great deal of imagination and an impressive determination to avoid school.

On a regular morning, and it must've been winter because I remember it being dark outside, I would be sitting happily at the kitchen table/picnic bench eating my Ready-Brek. Then as soon as Mum left the kitchen I would pretend to have fainted. With no audience to witness it, I would simply lie down on my front on the cold lino and carefully arrange my limbs in a classic police chalk outline shape – head turned to side, one arm bent at the elbow, fist near face, other hand bent behind me, legs split in a running position. I'd clearly seen too much *Columbo* as I had perfected the 'how people land when they've been felled by a bullet/candlestick/desire to skip school for the morning' technique.

I would then close my eyes and breathe calmly and deeply to give the impression that I was unconscious. So far so normal – NOT!!

Mum would then re-enter the kitchen to find her youngest daughter laid prone on the floor and would let out a small yelp before gently gathering me up in her arms and sitting me at the table, as I started to 'come round'.

She'd always grab a blanket from somewhere (I don't know where from; this was way before the high-falutin' days of 'throws' – maybe she kept a special 'fainting blanket' to hand) and I remember nestling in her arms as I bravely fought my way back to full consciousness.

I have a daughter of similar age now and I'm pretty sure if she decided to do an Oscar-worthy performance of 'Girl faints unexpectedly' I'd smell a rat.

However, having just spoken to key witness number one (my mum), she's confirmed that she totally fell for my act. In fact, so convinced was she that I was actually fainting (remember, this happened more than once), she took me to see our family GP, the legendary Dr Parikh, a middle-aged Indian man whose manner was a grumpy mix of suspicious and slightly bored in equal measure.

I remember him sighing a lot during appointments, as though he was frustrated and had something more pressing to attend to than rashes, lumpy bits and listening to our wheezes and splutters.

It's as though his mind was elsewhere or he wished someone would at least come in with something a bit more exotic – a spot of leprosy perhaps – something to get his teeth into and test the knowledge he'd gained while slogging away at medical school.

Anyway, Mum would explain how she'd found me unconscious on the kitchen floor and he'd eye me sideways from
underneath his thick dark lashes and shiny black hair-flick. He
always wore a grey suit and I remember thinking he was so wise
and important – the font of all medical knowledge, if a slightly
bored and suspicious font. He'd check me over, but I imagine
now that the whole time he knew there was nothing wrong with
me apart from a severe case of dramaqueenius skiveus maximus.
He knew I was faking it, wasting his time when he could be
doing something much more interesting like being a secret agent
or whatever it was he seemed to have his mind on.

I can't really remember my motivation for these faux faints
but I'm guessing the cuddled-in-a-blanket-missing-school vibe
was one I embraced.

If Dr Parikh thought these episodes were a tad OTT he was yet
to experience the full potential of my imagination.

Having had two or three fainting trips, my six-year-old self
obviously decided I needed a new act. And so it was that in the
middle of a lesson at Masefield Junior I felt compelled to put
my hand up to get the teacher's attention. The teacher looked
at me, said 'Yes, Sarah?' and I replied 'Miss, I can't move my
legs.'

What is fascinating is the absolute lack of planning that went
into this; no thought of the consequences or why one maybe
shouldn't fake paralysis for kicks. I didn't stop to think if it'd get
me in trouble, I didn't plot how I would make it work, I just
said it. It also illustrates the chutzpah six-year-old me had.

Once I'd alerted the teacher to the 'fact' that I couldn't move
my legs, there was no going back: 'No, miss, I can't wiggle my
toes – I can't feel my legs at all. Up to where, you say? Er, up
to my waist . . . No, I definitely can't stand up.'

The teacher went to fetch our headmistress, the no-nonsense Mrs Edge, but even this formidable woman, all below-knee tweed skirts and neat, cropped salt and pepper hair, softened on seeing this poor stricken child sitting helplessly in her chair while her classmates gawped in wonder and awe.

Unsurprisingly, the teachers didn't try to force me to stand – I was clearly unable to BECAUSE WHAT SORT OF CHILD WOULD FAKE PARALYSIS?????

Mum was called to the school and she came to collect me. I remember being carried up the garden steps by her and a teacher, sedan-chair style, one on each side, holding me under my bottom. I was put on the pale-blue squashy sofa to relax while Mum rang the doctors. An appointment was made and back down the steps I was carried and we headed off to see Doctor Parikh.

Mum managed to carry me into the waiting area, then into the doctor's room. Dr Parikh at last had an exciting medical mystery to get his teeth into – the unexplained and sudden paralysis of a young girl. Except within seconds of me being carefully plonked on the plastic cushion of the chair in front of him he knew it was an elaborate scam. He checked me out and told Mum there was no physical reason why I couldn't walk. Mum, boosted by this, asked me to try to walk. To my eternal shame I took a few faltering steps, wobbling and unsure, like a real miracle had just happened before their very eyes. What was a real miracle was that I didn't get a wallop from Mum across the back of my (now recovered) legs for lying so outrageously. She was just so relieved that I was OK, and says now that she remembers thinking she had to laugh as I skipped across the car park to the car, magically cured, not even bothering to keep up the frailty act.

19. The Ghost of Toulouse-Lautrec

There was an annual event where my taste for drama came in handy: Bolton's legendary carnival. Every summer different pubs would design a float for the carnival and it would be all hands on deck as each pub competed to produce the best float with the most exciting, fun get-up for the cast of characters riding it. The floats would then take part in a procession alongside performers on foot performing circus tricks, and dancers in amazing outfits.

There would also be the May Queen, May Princesses and May pageboys, who I was completely fascinated by. In a tradition apparently dating back many centuries, the May Queen would lead the procession, officially signalling the beginning of spring.

I don't know how the royal party travelled along the route in the Middle Ages (or how they travel now, for that matter), but in 1982 the Little Lever Carnival Queen mainly travelled on the bonnet of a Vauxhall Cavalier. Imagine the elegance: a car decorated with dozens of white paper carnations, stapled to a white bedsheet that was tethered to the wing mirrors and then stretched over the front, held taut by trapping the front edge under the bonnet. The Queen would be sat on top of the sheet, legs tucked demurely to one side of the bonnet, wearing a beautiful frock – usually the second outing for a confirmation dress – a tiara perched on her head and wearing a satin sash.

Following behind her on other slow-moving cars were the rest of her entourage. Oh what I would've done to be crowned a Carnival Queen. Unfortunately, the chosen queen earned her coronation by racking up the most attendances at church and Sunday school, so I never had a chance. We only went to church on special occasions – when family members were hatched, matched or dispatched. I went to Sunday school twice when I was very small – I loved it the first time as we sat at little tables colouring in. The next week the teacher kept banging on about Jesus and I found it really boring – I refused to go again and that was the end of my Sunday school education.

Our float was from the Queen Anne pub. The carnival committee, which included the Lady Glades and the Buffs, would decide on a theme (themes varied but all required the participants – by day car mechanics, cleaners, supermarket workers – to dress in flamboyant costumes and dance around on the back of a wagon) and the locals would cobble together costumes, often with incredible results. Everything from hats to headdresses and floaty frocks to fake facial hair would be made, bought or borrowed.

The float itself was usually a flatbed wagon – a cab fixed to a trailer, driven by owner Sam at strolling speed along the carnival route with spectators waving and cheering from the pavement, as the convoy of brightly decorated lorries trundled past. From my position on the float, I'd wave and grin my gap-toothed smile at the crowds; I loved it. My sister Yvonne and I loved the outfits and it was always a real family affair with Nana, Grandad, Mum and Ken and our cousins Anthony and Tracy, the kids of Mum's late sister Vivien (a lovely, vivacious woman who'd died tragically young from an asthma attack) joining us aboard too.

The floats would do a two-mile circuit of the village and end up back at the pub, where everyone would hop off and celebrate what a brilliant day it had been.

Waving at the crowds (made up of the local population of the village) meant you were potentially making a complete buffoon of yourself in front of ex-lovers, your kid's head teacher, former bosses, people you hated from school – so unsurprisingly some Dutch courage was needed; riding on the carnival float therefore became quite a boozy affair – not that I realised; I just thought everyone was having a fabulous time, not getting gently sozzled.

One theme that allowed there to be an actual bar on the wagon – while most of the characters fell off it – was the Moulin Rouge float of 1981. Dressing up a lorry to recreate the legendary late-nineteenth-century Parisian cabaret club was no mean feat but they did it; gold and red fabric was draped along the edges of the wagon, the famous three Moulin Rouge wooden windmills were recreated with a bit of MDF, and a few large wooden beer barrels were stood on their ends and secured to the floor, creating a safe surface for drinks.

Bookending the whole scene was an upright piano at the tail end of the wagon and a small but apparently amply stocked wooden bar at the front, nailed into position behind the cab. My sister Yvonne and I, along with all the other young girls, wore can-can, multi-tiered dresses in rainbow hues, knocked up by the sewing-machine whizzes at Nana's dressmaking group. To a seven-year-old girl a long dress is a big tick anyway, but add frills and a feather in the hair and that right there is the dream costume thankyouverymuch.

My nana was in a slightly more grown-up can-can dress in black and red, with a lace choker embracing her slender neck. Randomly, my mum sported a black eyeliner beauty spot just above her lip and a French maid's outfit, the sort that is popular for fancy-dress fun between consenting adults.

I imagine there was a shortage of can-can dresses and so the theme was broadened to include any French-leaning outfit – there were a couple of chaps who were wearing Breton striped T-shirts and a string of onions. My Grandad Vince, however, surpassed everyone with his Toulouse-Lautrec. Added to his moustache was a beard, round wire-rimmed specs, derby hat and overcoat to recreate the look of the famous French artist. The *pièce de résistance* was Grandad's determination to make his interpretation historically correct. To get across Toulouse-Lautrec's diminutive stature he attached shoes to his knees and knelt on them so his overcoat came almost to his false feet.

This is hilarious for two reasons – 1) the kneeling-on-shoes thing will always be funny, and 2) my grandad was pretty little anyway and if the life-size ghost of Toulouse-Lautrec had appeared that day he'd have towered a good two inches over Grandad Vince – so there was really no need to make himself smaller and spend two hours shuffling around on his knees on a slow-moving wagon.

Another year, the theme was 'Turkish Delight' and the whole wagon was covered in yellow-painted cardboard to depict sand dunes with plywood palm trees dotted around.

I've always thought that theme came about after about thirty-two miles of sheer, dusky-pink tulle and salmon-pink satin somehow fell into the laps of the carnival committee and they leapt at the chance to use it. I imagine they worked through a number of pink themes – Inside the womb! Flamingoes! – before thankfully settling on the rather vague Aladdin-meets-*Arabian-Nights*-via-Bolton.

Every person involved was mummified in the pink tulle and satin – the men and boys wore harem pants made from it, with matching little cropped waistcoats revealing bare chests and turban-esque headdresses made from twisted tulle, while the women and girls wore more of the fabric fashioned into impro-vised bikini tops with harem pants and a strip of fabric round the forehead.

We had a ball – Grandad's moustache was waxed and curled up at the ends, giving him a touch of the 'Genie's' and both Nana and Mum managed to look impossibly glamorous. Ken, with his olive skin and boasting a beer belly at that point in its earliest infancy, managed to get away with the cropped waistcoat/ bare chest look better than most.

I'm not sure where the booze was stashed that year – reports suggesting it was inside a large wooden camel are unconfirmed. I do know that ninety per cent of the adults on board smoked, so they were playing a dangerous game, wrapped in so much flammable fabric atop a wagon with a full tank of diesel; one stray spark from a Regal Kingsize and we'd have all gone up in a fireball visible from Turkey itself.

I don't think health and safety was a thing in the days of our carnival. This was very apparent the following year; the theme was the sedate-sounding 'English country garden' but the carnival itself was anything but sedate.

The costumes were less bonkers than in previous years as we merely had to recreate a quintessential English garden party – imagine a *Downton* picnic on a lorry and you're there. All the women and girls had long, floral frocks on and luckily Yvonne and I still had our green and cream flowery bridesmaid dresses, so we wore those topped off with wide-brimmed flowery sunhats.

Mum and Nana wore long dresses too and the men togged up in smart suits and top hats. Casual spectators would've drunk in the wonderful sight of the Queen Anne float approaching, bedecked in hundreds of multicoloured scrunched-up paper tissue roses, the party attendees sitting at white wrought-iron tables sipping 'tea' from fine china cups poured from delicate teapots. However, it wasn't really tea they were sipping but vodka! Outrageous. Getting smashed in plain sight. Wouldn't have happened on the Dowager Countess's watch.

This year we were appearing at the main Bolton carnival, so we were all very excited as it was a bigger affair and would take place a couple of miles away. The other difference was that we weren't on Sam's usual flatbed wagon, so instead of Sam driving it was a different bloke and a bigger, longer, curtain-sided lorry. These are the wagons with a roof held up by columns along the side and huge, thick plastic curtains that can be pulled along and fastened into place, creating a lorry that can protect its load from the elements.

Of course, on carnival day the curtains were open (after all, this was a charming village tradition, not people-smuggling) and were tightly concertina'd to the columns, acting as trellises for the dozens of roses wound round them for the occasion.

The carnival was fantastic – we waved, the crowds cheered, the grown-ups drank their special tea. It had been a spectacular debut for the Queen Anne.

Our new driver manoeuvred the float towards home and, as he did, we felt the first few spots of rain. It didn't cause any panic – this was Bolton, after all, where precipitation is a familiar companion – but then the rain started to get a little heavier. As it got heavier, the lorry went faster and the screams from us got louder.

The driver, on seeing the rain first spotting and then streaming across his windscreen, must've panicked and decided the best course of action was to get us home as quickly as possible. Either that or he wanted to get back home himself in time for *The Generation Game*.

Unfortunately, he simultaneously overestimated how much we disliked the rain and underestimated how difficult it is to remain upright on a fast-moving wagon, especially when you're half-cut from drinking vodka from teapots. He put his foot down, unable to hear our shouts over the roar of the wagon's engine. The sedate pace of Sam and his flatbed was a distant memory as the new guy hit an estimated top speed of . . . 20mph. We were all clinging on for dear life. What a sight it must've been – a bedraggled tea party lurching around on top of a wagon, bonnets flapping in the wind. Once we got our balance we dug in and clung onto whatever was nailed down, accepting that the new driver just wasn't quite au fait with the etiquette of driving a float.

Then we noticed he was manoeuvring towards the right and was actually pulling onto the dual carriageway. We all started screaming again and some of the men managed to bang on the back of the cab, but to no avail.

Luckily, once on the dual carriageway, the driver didn't go any faster, instead crawling along in the slow lane. Once we'd exited the dual carriageway, after much thumping on the cab and shouts of desperation from some of the women, he finally pulled over. After all that drama (combined with several gallons of booze), some of them desperately needed an emergency pee stop. The driver managed to find a big brick wall to stop next to and, windswept and traumatised but thankfully in one piece, several of the women clambered down from the wagon and over the wall. Squatting, they hurriedly tugged down their bloomers, wide-brimmed hats askew, giggling uncontrollably at the crazy predicament they'd found themselves in – and no doubt high on the adrenaline of surviving and the relief of emptying their bursting bladders. The men followed suit and then we all slowly trundled back home on the wagon.

20. Gumption

My grandparents were always organising fun stuff – they were after all the war generation and relished life, working hard and playing hard. They were among the millions of sun-seekers who took advantage of the introduction of cheaper package holidays in the seventies and jetted off every year to the Costas. I remember Nana always looked even more glamorous after a stint in the Spanish sunshine, her bronzed skin setting off her light-honey bouffant, her gold chain glistening on her tanned décolletage.

Buying gifts on holiday for the people back home was a real thing when I was growing up – we're not just talking a bottle of ouzo for the lady who minded the cat, but souvenirs for all the family. My grandparents would return laden with gifts for everyone: white T-shirts with hems shredded into long tassels round the bottom and the sleeves, which Yvonne and I considered to be the absolute height of fashion. Souvenir ashtrays, bottle-openers and miniature straw donkeys all made the journey home as well as duty-free Bell's whisky and cigs for Mum and

All the siblings together. For some reason I seem to be wearing an old lady's shower cap.

Here come the girls – cousin Tracy, Sis Dot, Yvonne. Me in tartan either giving a thumbs up or subtly picking my nose.

First day at work on the farm aged 9 months.

SELECTMATIC 880

Clearly very pleased to be cuddling up to dad.

No Yvonne, you can't blow out the candles or make a wish until we've taken an awkward photo.

Mum and Merl th
dog having a curl-o

Dad not letting a broken ankle or health and safely concerns get in the way of driving half a tonne of tractor.

Nana showing Yvonne
a thing or two about
how to sit like a lady.

Late stepdad Ken who made my mum's smile as wide as his shirt collar.

You can tell from my hesitant smile that I'm already regretting the wonky plastic earrings.

arling mum and my wonderful Nana.

Small but fearsome. Mum in full kickass landlady mode.

Sara pushed aside any thoughts of wild teenage parties and got on with brushing the cow.

Baby cow.

It's behind you!

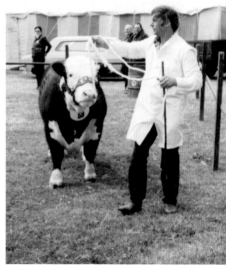

Moments later the wagon mowed down the bull. Not really.

Dad wowed the show crowds with both his bull and his impressive sideburns.

Carnival king Vince with
his princess Jackie.

Only Nana was pleased
with the amount of
fabric she had been
given for her costume.

One naked flame and
we would've gone up
in seconds.

Photo booth
hilarity with Joanne.

Gus looking very handsome,
hoping to distract onlookers
from my inappropriate t-shirt.

The art of steering a horse
using only a rainbow jacket
and heavy fringe

On Muffin with a very
leggy baby Gus.

Where are the breaks?
On Martini.

Ani and I trying to pull off the tricky Victorian dicky bow and espadrille look.

Beyond 80s : The perms, the patterns, the dungarees.

e embarrassing time Joanne and I our hoop earrings tangled together Ibiza.

What has a furry coat, pouts and wears trousers higher than Simon Cowell? Me in 1990 clearly.

'I've not got your bobbled beret, honest.'

Clare and I modelling in south Korea, one of us has probably just said something rude.

Ken. Holiday shopping became akin in stress levels to Christmas shopping, completely counteracting the whole point of getting away from it all and relaxing. One year before leaving for a week in Cyprus with Mum and Ken, Dad fished around in his jeans pocket and pulled out a crumpled fiver with a flourish. To my teenage self, these spends – once changed into the local currency – were to be used to buy my father a magnificent gift. I treated it as if I was a buyer for a billionaire businessman, on a mission to source Cypriot artefacts. Sadly, my time on holiday was slightly tainted by my stressful search for Dad's present. The hotel gift shop didn't sell anything he'd like and in the end, in last-minute desperation, I spent the whole lot on a bootleg copy of A *View to Kill* from a street vendor.

Both home and away my nana and grandad found ways to have fun. By 1980 Grandad was working as a forklift truck driver at Dobson and Barlows, a local engineering firm. With his friend Bob they set up Dobbies, a social club they started in the factory canteen so the workers could get together after clocking off.

The factory is long gone now but Dobbies is still going strong. It's no longer in the canteen but is a thriving members-only club with a venue of its own.

One of the biggest events Grandad and Bob set up for the club was the annual treasure hunt, which was hugely popular, with around forty teams taking part.

The treasure hunt involved several generations squeezing into the family car and setting off on a route around Bolton and out into the surrounding countryside of the Rossendale district, searching for answers to mystery clues. Grandad and Bob devised the whole thing and it was against the clock; they handed the competitors a printout that would be clutched by the co-driver as the cars screeched over the gravel of Dobbies' car park and out onto the road.

They were supposed to be a fun-filled activity for all the family, and that's how they started, but soon frustration would snowball into a stressful car chase around Bolton, gripping a crumpled sheet of A4 in your hand, shouting at the bored, hot kids in the back, being impatient with the grandparent also in the back, and the co-driver barking directions to the driver while trying to solve clues.

It's fair to say these treasure hunts were a test not only of the teams' intelligence, imagination and gumption but also of the marriages of the competitors. Uncle Trevor and Auntie Carole also used to take part and Mum now jokes that they came close to divorce at least twice during the event.

The clues would be pretty cryptic, occasionally involving foreign languages, literature, mathematics, history or geography, and led the players to a wide range of places where the answer would hopefully be found – a shop sign, the name of the vicar written on a church's noticeboard, a house number or type of tree.

The entry fee was a pound per car and the prize for being the first back to Dobbies with the most correct answers was all the entry money. That could be around forty quid, which of course in 1980 wasn't to be sniffed at – it could've bought you a three-bed house with change. I think the main prize, however, was pride; although I wouldn't know as we never won it – most of the cars would already be in the Dobbies car park by the time we returned, and the occupants would be on their third celebratory pint and the second helping of pasties and peas.

I guess it was a 1980s version of geocaching, but instead of using satellites and global positioning systems to find hidden containers anywhere in the world we used our wits and a rickety family motor around Bolton.

21. Cowgirl

Despite not living on the farm any more it was – and still remains – an important piece of the puzzle for me. As a teenager it was the place I'd escape to, to tell Gus all my troubles. For a while I'd stopped riding but I always came back to it. Even now if I don't sit on a horse for a while I get the urge to – the equestrian form of really craving a KFC every once in a while.

While Gus and I were on a serious break, Dad decided to loan him out to a local girl so he'd be ridden and get all the fuss he deserved. After a few months I started to find this really hard and I was desperate for him to be officially mine officially.

Christmas morning when I was at the farm there was a weird energy in the room, like Dad couldn't quite relax. In retrospect there was the air of an undercover operation, Dad and his wife communicating just with eyebrows and shrugs. Then Dad popped outside and a few minutes later there was a knock at the door and lo and behold there was Dad holding Gus, who had a red ribbon tied in a bow round his neck and a new hairdo, clipped up like he'd joined the Marines, dark brown coat shorn to reveal a velvety liver-coloured undercoat. He was sporting a new rug – navy blue with red piping round the edges and his name stitched on the bum in big red letters. It was a lovely surprise and I was so happy he was mine again.

There was no stopping us. Gus and I once again ruled the sandy paths and although things at school still weren't going smoothly, riding and school, both such big elements of my life, were about to combine in the most fantastic way.

In the third year of school every pupil had to take part in work experience. This involved spending time out of school and pretending to have a job, which generally involved looking busy, clock-watching and finding lunch break and home time the best bits of the day, so really preparing you for the real adult workplace. From filing and photocopying in an office to donning a hard hat and making tea on a building site, the jobs on offer were varied and sometimes depended on what your parents did for a living – if Mum was a dentist or Dad an accountant you'd maybe be able to do a stint working for them, including cushy transport to and from your new job in your parent's car.

I didn't need a lift to my work experience though; I provided my own wheels thankyouverymuch. Well, not actually wheels but hooves. One of the jobs available to me was to work at the local holiday kennels and cattery. It was situated on the country lane between the notorious Seven Stars dip and the sandy paths and I jumped at the chance to work there. Scooping up cat poo and walking dogs – what's not to love? This Holiday Inn for pets was fronted by a neat bungalow behind which was a square courtyard with each side lined with long, open-fronted sheds split into separate kennels by chicken wire, each with a bed and feed bowls in there for whatever dog or cat was holidaying there at that time.

One of the kennels had been converted and refurnished as our staffroom, which was where I hung out with the other young girls who worked there, eating our lunchtime butties and gossiping about boys. Well, they gossiped, I mainly listened wide-eyed to their chat about boys and their various sexual adventures. I often nearly choked on my ham barm.

Clearing up the cat poo wasn't the most fun but I absolutely loved walking the dogs. We'd take them out to the disused railway line that ran along the back of the kennels. We kept them on their leads to avoid losing them and occasionally it was the dogs who'd walk us, especially the big breeds. I remember once a pair of big, ginger Rhodesian Ridgebacks continually taking off with me, only pausing when I'd managed to dig my heels into the gravel or they stopped to sniff an interesting bush.

The absolutely best bit about the job though was that I rode Gus to work every day. The owners had a small paddock in front of the bungalow, next to their front garden. Dad knew them and they kindly agreed to let me keep Gus in there each day while I minded the moggies and pampered the pooches.

Saddling up for a day at work was the best – the closest I've come to being a cowgirl (my ultimate career ambition). All week I stayed at Dad's and every morning I'd put Gus's tack on, climb aboard and head off to the kennels with my lunch in my backpack.

There was something so gloriously old-fashioned about this, with serious whiffs of *Little House on the Prairie*. It was a wonderful way to start and end the working day and, hand on heart, if I was able to ride a horse into work and back every day now I'd absolutely love to. I can see it now, the outside of Radio 2 in central London looking like a saloon bar in the Wild West with a wooden post with various charges tethered to it – Vanessa Feltz's glamorous palomino pony with the flowing mane, Ken Bruce's stocky cob munching hay next to Jeremy Vine's leggy chestnut gelding.

Finishing my work experience at the kennels and getting back on the school bus was a sad jolt back to reality. To this day, every time I'm up north and drive past that place I smile to myself and remember how special it was. I don't know if the owners ever knew just how much it meant to me to ride Gus to and from their place every day.

Working at the kennels wasn't my only job as a teen. In between getting fired for dressing inappropriately while glass-collecting and becoming the world's most surly barmaid I managed to secure a Saturday job at Scotts Sports in Bolton town centre. Lord knows how. I can't remember the interview; I imagine there was just a shortage of grumpy sixteen-year-olds on the day they needed to fill the position. I must've shown enough enthusiasm for the gig. That was always a speciality of mine growing up – Girl Guide keenness only matched by the Famous Five en route to solve a mystery, even if that keenness would dissolve quicker than a sugar cube in lashings of ginger beer as soon as I'd been required to roll my sleeves up and get properly stuck into any activity. From helping Mum in the kitchen or garden to hobbies and work, my enthusiasm would always wane, and my job at Scotts Sports was no exception.

Every weekend I would get the bus into town and make my way over the pedestrianised precinct to the double-fronted store, over the way from the Early Learning Centre and round the corner from the glorious golden arches of McDonalds. My role was as head of the ladies' department, which wasn't as highfalutin as it sounds. The 'ladies' department' was basically a rack of women's jogging suits and a few bras. A lady would come in looking for some leisurewear and I would mooch at a snail's pace to rummage around for a size 12 in dusky pink or whatever she wanted in the stockroom downstairs before climbing back up the stairs many minutes later with her goods.

The stockroom was where I spent the vast majority of my time, hiding as the hours inched by, lazing around, flopped on the stacks of polythene-wrapped leggings and hoodies. When I wasn't blatantly avoiding work completely I was trying to look busy, rejigging the boxes of Reebok pumps. Of course I know now that the time would've gone much faster if I'd just thrown myself into my role at the shop. I also know now that when I'm confronted by a teenage assistant while out shopping not to expect too much, because if they've graduated from the Sara Cox School of Retail they'll not give a monkey's.

My retail career quickly segued back into one in hospitality – I had experience as a glass collector, after all, so I found a job working at a nightclub in town. At sixteen I was too young to work behind the bar, so I was the guest-list girl; just inside the door of the club was a sort of wooden lectern and I'd perch on a bar stool behind it, clutching a clipboard. The guest list would be attached to the clipboard and it was my job to check people's names against my printed-out list. I wasn't that busy, to be fair – this was Bolton, not Ibiza, so there weren't exactly queues of revellers desperate to get in. The Saturday night would rumble on; my hours were 9 p.m. till 2 a.m. and for the last half-hour of my shift I would largely be entertained by the security staff encouraging troublemakers to leave the premises. The club itself was up a flight of stairs just next to my seat, and by 1.30 a.m. the bouncers were living up to their names by literally bouncing people down to the bottom of the stairs, where they'd land in a sozzled, angry heap. Their girlfriends would come screeching down the stairs in a clatter of high heels and drag them the last few feet to the door and out onto the pavement.

It was a job where even I questioned if I was old enough to be doing it; I was only sixteen. Then again, I'd been around pubs all my life – from being a regular in the Queen Anne with Nana and Grandad to living above the Pineapple (more on which later) – and had seen more bar brawls than your average Wild West saloon. And I'd started drinking in pubs aged fifteen, using a borrowed driving licence to gain access to the hostelries of Leigh in Greater Manchester with my best friend Joanne where we'd guzzle Taboo and lemonade, syrupy drinks sweet enough to make your eyes fizz. Even so, the scenes at the end of the night at the club were more out of control and extreme than the boozy squabbles or testosterone-fuelled fisticuffs I would witness at the Pineapple and I was glad to leave my position at the idiot's landing spot after a few months.

22. Cool as Fook

Even though I'd moved from Boothstown back to Bolton for
Mum and Ken to run the Conservative Club at Astley
Bridge when I was twelve, I remained really close to my friend
Joanne Lloyd, who I'd met at junior school there. Joanne lived
in a busy bungalow with her two older brothers and her parents,
who ran the family business, a maggot farm. I was never really
sure about the methods of maggot farming; it was all a mystery
to me. I just used to see the odd tray of maggots outside Joanne's
door, sometimes dyed bright pink, which Joanne couldn't resist
performing her special party trick with. She'd lay her hand flat
on the surface of the six-inch-deep pile of maggots, thousands
of tiny creatures that looked like oozing, overcooked rice, and
they'd wriggle and writhe en masse over the back of her hand
until it slowly sank under their weight and disappeared under
the squirming sea. It used to horrify me – which of course was
half the fun.

Joanne's mum Gill was tall with a neat halo of mousey permed
curls and a wide, full-lipped mouth and dad Fred was little,

with permanently tanned arms up to his rolled-up shirt sleeves from being elbow-deep in pinkish maggot dye. He was ruddy-cheeked and was rarely without his flat cap, hiding his thinning black hair.

They both spoke in broad Mancunian accents that swung and stretched every other syllable – 'Jo- WAAAN' – and the entire Lloyd clan were noisy and shouty and I loved them.

They struggled to communicate at usual volume, shouting and hollering at each other even when in close proximity, partly I think due to Fred's less than perfect hearing. The bungalow was chaotic – in the big kitchen, random empty maggot crates battled for floor space with occasional saddles, wellies and the gangly legs of their whiskery lurcher, spread out having a nap on the lino.

There was always a subtle smell of the maggot farm in the house, a kind of warm, offal scent that was masked only partly by the mouth-watering whiff of crispy bacon that seemed to be constantly frying – Fred loved the stuff and his favourite thing was to pair it with beetroot in big butties.

I used to go round there and get swallowed up in the madness; Joanne's brothers – John, with his golden-blond hair, wide mouth and light-blue mischievous eyes like Joanne's, and the eldest, Anthony, dark-haired and handsome – would be forever banging and crashing round the house.

They both went straight into the family business after they left school, and worked hard and played hard too. They were having a new home built on a small farm nearby and in the meantime it felt like the bungalow might burst. Every surface was piled high with cleanly washed laundry, dinner plates, used teacups and horsey paraphernalia – bridles, saddle soap, riding boots and leather girths – and when I was there after school the lads would bowl in from work, wolf their tea in minutes, then rummage through the towers of folded clothes that formed a fabric skyline along the kitchen units, trying to find the right outfit to wear for their evening out, before slamming back out through the kitchen door and into the night, like two rougue-ish whirlwinds.

Joanne's mum used to ride when she was little and was determined for Joanne to follow in her footsteps, so she'd got Fred to knock up a little stable, tacked to the side of the bungalow like a bumbag round its waist and painted white. It was essentially a tiny lean-to that was home to Joanne's little grey pony and took up the area at the top of their cramped driveway, onto which was also squeezed an old blue horse trailer and their battered tan-coloured Mercedes estate. Joanne soon outgrew the pony and a new stable and field were rented a couple of miles away for her upgrade, a beautiful piebald gelding called Apache.

He and Joanne were quite a team and competed most weekends at horse shows, eventually qualifying for the Wembley Horse of the Year show, the Wimbledon of the showjumping world, when Apache was the ripe old age of twenty-one. Joanne was my riding hero; she was brave and gutsy and dedicated – trophies, rosettes and medals jostled for space on the walls alongside framed pics of Joanne and her ponies in mid-air, flying over brightly painted jumps, her plaited hair trailing behind her.

Joanne was always more daring than me. Whether it was on the netball court or whizzing downhill near some local lock-ups in our navy suede roller boots with the rainbow on the ankle, she was always faster, braver and stronger.

She was shy though and would blush bright strawberry-red if put on the spot. I was more confident socially, so we made a good team. We were also partners in crime and one time back when we were ten we were particularly naughty.

After a busy morning balancing the books, setting up the bars ready for opening, taking deliveries – rolling full barrels from the brewery's dray wagon down a massive wooden ramp into the cellar below the bar – and collecting the empties, stepdad Ken would come home from Boothstown Conservative Club. His commute was quick – he'd nip in through a door just past the snooker table, like exiting one world and entering a new one, a mustachioed Alice in Wonderland – from the excitement and hustle and bustle of the club and into the little two-up two-down that was latched onto the side like an after-thought. He'd get a brew and the jangling bunch of keys on the belt loop of his jeans would fall silent as he settled into his scuffed-up velvety armchair, where he'd flick through the paper and fill out a Spot the Ball coupon.

Inevitably tiredness would get the better of him and he'd snooze off. This is when ten-year-old me, with Joanne as my wing-woman, would creep silently towards him and, moving slowly, inching forwards, soundlessly extract two cigarettes from his packet of John Player Special that sat with a lighter on top on the unit next to his armchair. We'd replace the packet and slip the ciggies up our sleeve, then hotfoot it through the kitchen and out the back door, thrilled and scared in equal measure at our daring and our illegal bounty. We'd make our way to our junior school, which was closed for the school holidays, and sit concealed behind a low wall in the deserted playground. Before lighting the cigarettes we'd each put on a woolly glove (we shared a pair) to protect our fingers from the stink of the fags in case we were busted by my mum and our digits sniffed for evidence. We puffed on the ciggies, obviously hating the taste and coughing and spluttering but for some reason thinking it made us ever so grown-up and sophisticated. We'd leave the playground and eat a fistful of Polos and, having removed the gloves, we'd lick a mint and rub it onto our fingers as backup in case the smell had seeped through the wool.

All that effort to smoke a filthy fag!

When Joanne and I left junior school we started at the local comp, Bedford High School. Her brothers were still there then, holding court on the top deck of the bus. There was always a pecking order on the bus – the ultimate seats were upstairs at the back but it took years to work your way up to the heady heights of that position – you had to start downstairs with the mums, buggies and pensioners and gradually try your luck getting a seat upstairs, occasionally clambering up the terribly designed spiral staircase and peering over the handrail at the top, praying there'd be space in the very front seat so you didn't have to retreat and do the walk of shame, lurching back down the stairs to the mums 'n' buggies bit.

Upstairs very front was brilliant for two reasons – I used to love the slight vertiginous effect, looming over all we surveyed like Leo and Kate on the front of the *Titanic* but without the imminent disaster bit. It was also far enough from the kids at the back to be out of range of any missiles/insults they might fling. It was never too intimidating though and felt very much like the natural food chain on the bus, within which we first-years were definitely the snack-sized minnows at the bottom. To get attention from the fifth-formers was not good; at eleven years old the aim was to silently observe the wild behaviour of the older kids without them noticing you, so you'd slink quickly to your seat at the front and essentially melt into it to avoid catching their eye. From there you could ear-wag gleefully at their shenanigans, shouting, showing off and puffing on fags, without the risk of an empty Cherry Coke can ricocheting off your head.

Thanks to Joanne's older brothers, who ruled the back seat,

we were given a leg-up and allowed, even as first-years, to sit in our spot upstairs at the front without too much trouble; bus seat nepotism was alive and well.

I had some great friends at that school – Olivia, whose mum bred West Highland Terriers and who was Amazonian with dark curls and a pretty snub nose perched on a round olive-skinned face; Joanne number two, who had white-blonde hair scraped into a high ponytail and was the first of our gang to get a bra, and red-headed, slender Lisa. I was happy there; I had good pals and most importantly I had Joanne number one. So you can imagine how devastated I was when Mum told me we were to leave Bedford High, the top deck and Joanne after the first term.

Mum and Ken wanted to move back to Bolton to run Astley Bridge Conservative Club, going where the work and opportunities were. When you're little you don't really have any choice in where you go; you just have to follow your clan wherever they roam. Of course, the grown-ups have their reasons, and as the child you just have to follow them; and before you know it you're the new girl at school all over again with a cracked funny bone and the head teacher carrying your lunch tray.

We were reassured that moving back to Bolton was going to be a good thing – we'd be about four miles from the farm, so closer to the rest of the family, and Astley Bridge Conservative Club was bigger and a new exciting prospect. We moved after three happy years in Boothstown, in the January of 1988 when I was thirteen and in the second year of big school. I was devastated to leave Joanne but I was happy to be back in Bolton. We'd also be trading in our two-up two-down for a sprawling three-up three-down; a wee terraced house not attached to the Conservative Club, as in Boothstown, but over the way, meaning

Mum and Ken could keep work and home more separate than it's possible to with just a paper-thin wall between you and your customers.

I didn't know it then, but we'd stay in that area for the next decade or so, which for us was a long time. Looking back it's clear that from quite tough times at School Street, when Mum had done lots of casual cleaning and bar work to make ends meet (and Ken struggled to find any work at all), there's a clear upward arc as Mum and Ken grafted and grafted to improve all our lives. With each career move came more responsibilities, more ambitions for the places they managed, bigger homes to live in and more spare cash to spend on nice things.

It was hard though – I loved Bedford High School and was gutted to leave. I would miss my form teacher Mrs Pimlott, who was one of the best things about that school. Moving up from juniors to secondary is always pretty intimidating but Mrs Pimlott had been the perfect person to oversee that transition and had eased us from little to big school with her kindness. She had a name and a fashion sense that were more Hogwarts than north-west state school, sporting long, silky flowing skirts in swirling fuchsias and lilacs, teamed with acid-bright angora cardigans; she was like a tall, slender exotic bird. Her hair was cropped, spiky and peroxided and she had huge eyes with lashes of bobbly mascara so she looked like someone had squished two massive spiders on her face. She always seemed to be in a rush – she'd swirl into a classroom like she'd been blown in on a brisk breeze like an autumn leaf.

Even now, if my mascara goes lumpy I think of Mrs Pimlott. I loved her because she was kind and funny.

It's amazing the imprint teachers leave on you, good or bad; I was only at that school for a couple of terms but Mrs Pimlott crosses my mind regularly – as does, for different reasons, my maths teacher Mr Jones, who I still blame for my fear of numbers. He was strict and sarcastic, with cropped deep-red hair matching his ruddy complexion and maroon V-neck jumpers – he was like a walking talking Dulux colour chart for reds. He was left-handed and wrote on the blackboard with such vigour that his entire body moved. His right hand would ape his left, as if writing the exact same thing as the other on the board with invisible chalk.

He wasn't a tyrant or cruel, he just went so fast and was so scathing that I quickly fell behind, too intimidated to ask for help when I didn't understand something. My hatred and fear of maths was set; I couldn't do it. Maths is like a language – you just need to work it out to be able to speak it fluently – but in my very first year at secondary it became an indecipherable foreign tongue, and has remained one. We all knew that to be allowed to study A levels we had to achieve the holy grail of C or above in GCSE maths; it took me THREE attempts. By the final time I felt like that scene in *Elf* where Buddy is in the classroom towering over his classmates.

Teachers aside, Joanne and I were best friends and there was no getting away from the fact that I was about to move seven miles away from her. This resulted in me spending a lot of my teenage years on buses – I would catch one to Bolton town centre, then another one on to Boothstown for a thrill-filled night hanging out on the precinct in the drizzle with Joanne and our other pal Lisa.

This was 1989 when Manchester was the centre of the universe for bands and fashion and coolness. Until now music for me had been mainly what was played at family parties and get-togethers – my musical memories up until my teenage years are just random moments, scraps and snatches of songs providing the soundtrack to Cox clan moments. My Uncle Stan, the widower of my mum's sister, jiving with Mum to the strains of Bill Haley's 'Rock Around the Clock'; dancing to Status Quo at my older siblings' birthday parties at the farm (a sea of tight double denim swaying and swinging to 'Rockin' All Over the World', thumbs firmly hooked in jean belt loops); hanging off Yvonne's coat-tails as she played her Prince CDs; and, one of my earliest memories, my eldest sister Dot teaching me the words to Gloria Gaynor's 'I Will Survive' – even though I was only three or four she'd have me perform it in the bedroom we shared, a pocket-sized diva sashaying up and down the full length of the room in my lilac nylon nightie, giving it some welly for my audience of two (Yvonne was usually there), me channelling Gloria as she sang of triumph over heartbreak.

The whole Manchester thing was different though. This was something I could own, belong to, be part of the tribe. Homegrown talent, from bands like the Happy Mondays, Inspiral Carpets and the Stone Roses to clubs like the Hacienda and fashion labels like Red or Dead and John Richmond, flew the flag for the north-west and Manchester swapped the 'n' for a 'd' and became Madchester. We all wore Joe Bloggs jeans and Inspiral Carpets T-shirts with the cow face logo mooing 'Cool as Fook'.

We loved having our melons twisted and the band James implored us to 'Sit Down'. We all wore Kickers boots and chunky black and white British Knights hi-top trainers, as clumsy and clumpy as strapping a couple of baby's car seats to your feet.

We girls sported spiral perms with fringes structured into a stiff vertical wave, cresting atop our heads. Impressive, slightly crispy monuments secured with layer after layer of hair-dryered Rock Hard hairspray. The boys cultivated hirsute 'curtains', their hair hanging limply down each side of the face as if they were wearing an open pamphlet like a hat, the spine stapled to their skulls along their centre parting.

Joanne and I loved going out – perms piled up and pins on our heads, we'd head into the centre of Leigh, a local town, and hit the pubs there, taking advantage of the rather lax attitude to the alcohol laws back then. Bouncers were likely to let you in if you walked up confidently enough and smiled. I became good at just strolling into a pub as if it was the most natural thing in the world – after all, I was a sophisticated eighteen-year-old, wasn't I? Err nope, you're sixteen, clutching a driver's licence borrowed from your mate's brother's girlfriend. Almost every time I fooled them with my faux confidence – either that or the bouncers were momentarily incapacitated by the cloud of Body Shop Dewberry scent I insisted on dousing myself with. We'd drink Taboo and lemonade – a drink sweeter than a thousand fluffy kitten gifs – thinking we looked sophisticated. If we wanted to switch it up we'd drink Mirage and lemonade, which was a bit dryer. We'd stand with our drinks, swaying about to whatever the local DJ on the decks was playing, getting gently sozzled.

It wasn't a pub crawl, more a pub loiter – we were so relieved to be let in that we didn't want to push our luck by leaving one place and trying to get into another in case we were refused.

We also went to clubs, cadging lifts from Ken on Saturday nights to places like Angels in Burnley, Monroes in Blackburn and Pleasuredrome in Farnworth.

The latter was a vast open space with all the charm and intimacy of a Sports Direct off the A41, and the two others had speakers as big as garden sheds and no air con. In Monroes there'd be water dripping down from the low ceiling, landing on your face, down the back of your neck, on your head. Except it wasn't a leak, it was condensation from the hot clubbers dancing in the dry ice and strobe lights. It was recycled raver rain, a sweat storm. But we didn't care – we were there for the house music, we were cheesy quavers, we were ravers.

We loved house music with piano popped on top of thumping bass – I bought N-Joi's 'Anthem' on 12-inch and played it non-stop on rotation with Nomad's echoey 'I Wanna Give You Devotion'. We followed our favourite – DJ Welly – and we saw 808 State play a gig at the G-Mex in Manchester with MC Tunes.

I wore my raver uniform: red Puffa coat with a Vicks Rave-a Rub sweater, denim shorts and thick black tights, the look bookended with my long perm and Kickers boots.

At chucking-out time we'd fall out the club's front doors and into the cold, faces pink with exertion, ears buzzing and hair wet with sweat (some our own, but mainly other people's) and look for a lift home.

These clubs were massively popular in the north-west and there was always a good chance there'd be people there from Bolton and somehow we'd find them.

The raving community was kind – we were part of a movement that believed there was no space for bad vibes, just good music and a coming together.

Having said that, my blood does now run slightly cold at the thought of Joanne and me piling into a stranger's car to get a lift home, slamming the door of the 'scooby-doo' (XR2) behind us and putting our sixteen-year-old lives into the hands of lads only a year or so older than us. I don't mean we were in danger of being attacked, more that after a night of clubbing maybe a teenage boy wouldn't be the best driver. But we always got home in one piece, possibly thanks to good luck more than good judgement.

Occasionally we'd venture with some pals to Blackpool, the

Vegas of the north. We'd go to the Zone, which was on the seafront and always had a huge queue to get in. Back then the longer the club queue the more I wanted to go. Which is in sharp contrast to now. I can't queue at all now. I'm allergic. My indignation swells and my breathing becomes shallow. In London you see fools queuing around the block, waiting for a table at the latest Vietnamese café because their pho is to die for. Nah, you're OK, I'll go home and make an omelette. Eateries are now essentially the new clubs, people desperate to get into the latest place.

Back in 1990 though it was all about the clubs – and we'd think nothing of queuing outside of the Zone in Blackpool for an hour or so, huddled together for warmth like penguins. We'd go into a vegetative, meditative state, inducing in ourselves an out-of-body experience to escape the reality of the chill that was seeping into our bones, the brutally cold wind from the Irish Sea whipping around us.

I lost a pale pink Kickers boot at an Inspiral Carpets concert at the now defunct GMEX venue. It was July, the hot summer of 1990, and I was sixteen. Joanne and I were driven to Manchester by Mum and we ran from the car, up the wide steps and through the huge glass doors and through to the gig. We were so excited, but alas we found ourselves at the back of twelve thousand equally excited fans, bobbing about, curtains swinging, the heady scent of Lemon Hooch filling the air.

We set about slowly pushing ourselves through the crowd, steadily working our way towards the stage, but a few metres from the front we were blocked by an impenetrable wall of Manc teen. Our method of apologising sheepishly and looking a few feet in front of us as if trying to reach invisible friends just up ahead stopped working. We could see our heroes now

– with one determined push, surely we could get to the front? This was when I came up with, to quote Baldrick, a cunning plan.

If I utilised all the skills I was currently learning in Drama GCSE then maybe I could feign illness – after all, I had form from my Dr Parikh days. Obviously no one wants a teen with a spiral perm vomiting on them at a gig, so sure enough, with Joanne playing the role of 'concerned friend', one arm round my shoulders, the other gently tapping people on the back and gesticulating at me, mouthing the classic line 'Let us through, she's gonna be sick,' we started to push through the throng.

It was working!! The crowd started to part as people got the gist of what was happening. It was packed near the front but somehow, even while being carried on the current of the mosh pit, flowing right or left a couple of feet, pushed by the swell of the crowd, still a path to the barriers at the very front was opening up. I was doing my best bilious grimace – cheeks puffed out, eyes slightly rolling, my tongue lolling from my mouth occasionally – while Joanne gently ushered me to the front. People were looking really concerned. Not only were they moving out of our way but now they were pointing at us, obviously signalling to someone that we needed help. Oh no. This had gone too far. We felt a backfire approaching. We'd done so well to get to the front but as we neared the barriers just in front of the stage, we saw that everyone was now signalling to the security staff, who were standing there waiting to help us. We managed to glance up at the band at close range for possibly two seconds before the big security guys in luminous jackets grabbed an arm each and hauled us over the barriers, plucking us to safety. To add to the humiliation, as we were birthed out of the crowd, the suction as we emerged pulled off one of my boots. Security delivered a mortified Joanne and a limping one-shoed me to the St John Ambulance set up at the side of the stage in a small marquee.

After ten minutes of politely sipping water from a plastic cup and reassuring the lovely ambulance crew over the deafening din of the gig happening mere metres away that my nausea had subsided, we were released like rehabilitated otters from a wild-life sanctuary and shuffled off forlornly to resume our position at the back of the crowd. By some miracle, the Kickers boot had been spat out by the undulating crowd and was duly returned to me by security.

I'm pretty sure there must be some ancient Chinese proverb along the lines of 'She who hopes to gain from feigning sickness is doomed to lose face and shoe.'

Joanne and I were thick as thieves. She would come on family holidays with us; Yvonne was always either at uni or hiking in the Himalayas, so it'd be me, Mum, stepdad Ken and Joanne.

Mum and Ken worked long hours – our accommodation was often connected to the licensed premises they ran, so there wasn't much escape from the hubbub; bar deliveries would arrive just as we'd sit down for tea, the hours were long and spent dashing around and Mum and Ken were permanently stressed and knackered. No wonder they grabbed the chance to seek out some sunshine.

It was long before the convenience of online booking and Mum and I would make the pilgrimage to Bolton town centre, sitting patiently at the travel agent's desk while she tapped away at her keyboard, squinting at the screen of her monitor, which was in those days the size of a shipping container. It was an exhausting experience that took hours – we'd ask a question – 'Is there a sandy beach?' and she'd launch into another eighteen minutes of continuous typing and bashing of the return button, peering at the flickering screen, grimacing, ho-humming, then eventually answering 'Yes. Well, it's four kilometres away. And shingle.'

After what felt like days she'd eventually come up with the flight itinerary and accommodation.

Well, kind of. Mum and Ken used to plump for the sort of holidays that had an exciting element of surprise – we never knew where we'd be staying, just that it was self-catering and guaranteed to be three-star or above.

So it was essentially a lottery and where we ended up was in the lap of the Greek gods. Or possibly the Majorcan ones.

We weren't well off, so if we wanted to go on holiday some-where hot we could only really afford this kind of deal, where apartment owners and holiday companies found a clever way to make sure their accommodation wasn't left with any empty units – by slashing prices and bunging those brave enough to take a punt wherever was available.

I loved that about Mum and Ken – they were pretty gung-ho about the whole thing. We still laugh now about the transfers from the airport to the apartments, peering nervously out of the windows of the coach. Occasionally there'd be a cement mixer beside a half-built reception area and you could feel everyone hold their breath as the rep from the travel company held aloft her clipboard, grabbed the crackly microphone and announced the names of the holidaymakers who were doomed to stay at Los Apartamentos Derrelicta. We'd all be silently praying our names wouldn't be read out, and would sigh with relief as the coach pulled away, leaving the unlucky ones standing with their suitcases in a cloud of diesel fumes and brick dust.

We were fortunate and always seemed to end up in nice enough apartments – 'basic but clean', as Mum would say.

One year Joanne and I noticed an army of ants marching

along our balcony and making their way in single file through our bedroom towards the kitchen. We came up with an ingenious way to stop them – Mum came in to find Joanne and me on our hands and knees by the patio doors, giggling like mad, wielding a tube of toothpaste each as we squeezed out along the linoleum floor a minty barrier to stop – or indeed stick – the ants in their tracks. It was like an ant-size version of the Great Wall of China, a fluoride-filled fortification that remained there for the rest of the holiday.

Ant infestations and cement mixers are a world away from some of the lovely five-star places my own kids have stayed at. Although having said that, my kids don't live in a luxury bubble when it comes to travel – they've loved a half-term break in a slightly mouldy-smelling, three-star, dog-friendly place on the Isle of Wight, been constantly drizzled on in Devon, nearly blown away in Cardigan Bay, and their favourite hols have probably been cheap and cheerful all-inclusives, running around in the sun with a gang of their newly made mates, smiles bright pink from the non-stop strawberry slushies and wall-to-wall chicken nuggets for dinner in the buffet restaurant.

Going away with Joanne was brilliant for me, and easier for Mum and Ken as she was a playmate for me. Though there were rules about how much we could actually play – when we went to Ibiza we were fourteen years old, too young to go to the proper clubs but too old to hang out with Mum and Ken, so we were allowed to frequent the local strip of bars a few minutes' walk from the apartments.

We felt terribly grown up and would carefully tether our perms up into ponytails high on top of our heads in the classic 'pineapple' style. Our curls would be looking particularly blonde because every day we'd sit in the blazing sunshine trying to lighten our locks with lemon halves – skived from the barman at the pool bar – which we'd squeeze directly onto our hot, pink scalps. We had to keep our heads dipped slightly back and didn't dare look down in case the juice ran off our foreheads and into our eyes, and we would gather an extensive collection of lemon pips and pith that got tangled into our hair. Not to mention the constant threat of wasps, circling our heads, attracted by the sweet citrusy smell.

Once we'd washed away the fruit from our scalps, we would slip on our floral culottes and Kickers and hit the town.

This was during the Italia 90 World Cup, which captured a lot of people's imaginations – it featured the infamous England semi-final match when Gazza burst into tears at being booked and the team crashed out on penalties against West Germany – so wherever we went big screens were showing the matches. It was also at the very height of the Madchester era, so whoever we chatted to – always people our age, seemingly also marooned between their parents and the real clubs – I'd tell them I was from Manchester. Straining to be heard over the cheesy Euro-house music at the half-deserted, neon-blue-lit bars, I would revel in my 'Mancunian' status, proud to be able to say I was from the current capital of cool. I'd even try to slip in a bit of a nasal Manc twang to add to the effect. Sadly, Joanne would always helpfully remind me in front of our new friends that I wasn't actually from Manchester, I was from Bolton.

She did it from pure mischief, to wind me up and bring me down a peg or two. In fact, though, since then Bolton, which historically was part of Lancashire, has been reclassified as Greater Manchester, so I was right all along – in your face, old pal!! I'm from Greater Madchester, cool as fook!

23. Molten Hot Chips

I was always very happy to be going abroad on holiday, apartment lottery or not, because it'd be fair to say we had a wide variety of holidays in the UK that . . . varied in their fabulousness.

I can't remember ever holidaying with Dad, although before I was born he and Mum had trips with the eldest kids to Venice – there are lovely faded photos of the twins Dot and David, aged about five, looking very sweet in matching pale-blue dungaree shorts, chubby hands holding ice-cream cones, in some beautiful Venetian plaza.

I'm guessing that when three became five and then the farm became busier, holidays weren't a priority – farmers rarely get to go away and leave their land and livestock. I've never lain on a hot beach with Dad or paddled in the sea. The only pool I've swum in with him isn't an infinity one overlooking a Caribbean island, or even at a waterpark in the Med, but at Bury Baths.

*   *   *

Most Saturday mornings when I was little Dad would take Yvonne and me swimming. Although it's not really accurate to say we went 'swimming' – really we went 'clinging' as we spent approximately eighty per cent of the time clinging to Dad, our arms wrapped round his neck like we'd just been tossed from the deck of a ship into a churning sea and were going to drown at any moment.

Of course now, thanks to my own children, I know how intense it is to be clung to like that, trying to peel their skinny little arms from round your neck; and I empathise with Dad when I remember him also trying to loosen our vice-like grips from round his windpipe. I was never very confident in the water but I also think I relished being attached to him like a whelk; it was a chance for me to slightly regress and be carried by him, be his baby again. I felt protected and safe in his arms.

The pool was always busy, full of screams and splashing. There would be other dads and their kids – maybe they were all the Saturday fathers – bobbing around in the pool like a big broken-family soup, trying their best to spend quality time with their kids. You see them still on a Saturday at the cinema, the park and Pizza Express, studying their offspring's faces, wondering how they got so grown up, mourning the daily mini-milestones they're missing.

Luckily for us we were, Dad included, always at the farm so he was never just a Saturday father. He just took us swimming as part of our family routine. I loved being with Dad and though he was never overly demonstrative, never overly cuddly or gushy, he showed his love for us by letting us half-throttle him at Bury Baths every Saturday morning.

The cherry on the cake was the paper cone of molten-hot chips we always got from the café afterwards, which we (famished in the way only swimming can make you) devoured, burning our tongues. We would wash them down with vending-machine hot chocolate, our sopping wet hair hanging in rat-tails down our backs and making our sweaters soggy.

The closest we came to a holiday in those times was visiting Dad's friend Jim on the Isle of Arran. The three of us, Dad, Yvonne and me, set off at dawn and rumbled up towards Scotland in Dad's wagon, which was fully loaded up with a tower of straw on the back for Jim and, in the cab, a tower of ham butties for us, all wrapped in foil.

Like Hansel and Gretel we left behind us a trail of bright yellow straw, fluttering along the motorway, loosened and freed from the bales as the load was buffeted by the wind as Dad roared along the M6.

We drove all the way up through Cumbria, skirting Dumfries, bypassing Glasgow and taking in the views of the Trossachs National Park, until we arrived eight hours later in Ardrossan. It was summertime and still light as we passed the little fishing boats bobbing around in the port and finally arrived at the more industrial ferry terminal.

The trip on the ferry was exciting – Yvonne and I sat in hushed reverence so Dad could concentrate as he edged the wagon and the tower of straw onto the ramp and into the belly of the ferry before we clambered down from the cab and he jumped down, his work boots landing with a clatter on the metal floor, hitching up his jeans at the same time. We cadged a can of Lilt and some crisps at the on-deck cafe from Dad, and docked at Brodick just under an hour later, Yvonne and I trans-fixed by the grey-blue water of the Firth of Clyde rolling by.

We unloaded the wagon and wound our way through Arran's country lanes, hugging the coastline, brushing against the hedge-rows, no doubt depositing some of Jim's straw in the odd low-hanging tree branch. We arrived at Jim's and, before the straw was unloaded into the barn, the kettle was put on for a journey debrief and general catch-up. We spent the night there, then next morning it was up and at 'em. Jim took us in his jeep with his collie dog up to the top fields to check out his herd of commercial cattle, which means a mixture of beef breeds bred for fattening, as opposed to seed stock like Dad's, which is a herd of one breed that is nurtured to a high pedigree standard.

With a few bursts of barely decipherable instructions Jim had the dog fetch the cattle closer so Dad could check them out. It was a bright but blustery day and as we walked across the field I realised we were on a clifftop, looking out beyond the cattle to the rolling sea below. I found it amazing that these cattle grazed so close to the cliff edge, just a couple of strings of barbed wire separating them from a sheer drop and certain death.

We headed back to the farmyard and, before setting off on our return journey, drove in Jim's jeep down to the coast and watched seals lounging on the rocks in the distance. After our whirlwind visit it was time to set off, back on the ferry and towards home, which was obviously quicker because the wagon was empty and, as my dad was fond of saying, it's all downhill from Scotland.

I returned to Scotland a few years later with Mum, Ken and Yvonne. It was August 1987. I know this because it was the month that Yvonne turned seventeen and therefore gained her provisional driving licence. It was decided that the best course of action to mark this milestone occasion would be to embark on a very long car journey, with Yvonne using the high roads and low roads of Scotland to practise her blossoming skills at the wheel. Ken would drive for the majority of the journey, then once off the motorways Yvonne would whip out her magnetic L plates, whack them on the car, and the adventure would truly begin as she attempted to negotiate the smaller B roads.

Our weapon of choice was Mum and Ken's Renault 12. A car that if you saw it now, it'd be vintage, charming and slightly kooky. It'd be driven round Brighton by a girl wearing a polka-dot headscarf and her hipster boyfriend with a beard, both in their ironic cowboy boots. They'd be sipping turmeric lattes and listening to Beck. But in 1987 it was just a regular, and occasionally rubbish, car.

I'd always had an issue with the way this car looked – the shape was just strange. The bonnet looked a little too long and was flattened at the front, like a harmonica that had been accidentally stamped on. Moving round to the rear, the back end sloped downwards at a horrible angle like it was giving up on life. It looked like it had been fed, rear bumper first, into a car crusher but the owner had had second thoughts and dragged it out again.

I think I'd harboured ill feeling about this car since the day it bit me – a few years before, when I was eight, we were loading up the car to head out on an errand. I was round the back of it, generally getting under Mum's feet as she was busy wanging on at Ken about something or other. We were running late as usual – looking back we often resembled the family from *The Fast Show*, the ones holding lots of bags or equipment and always legging it through airports, bus stations and shopping centres yelling 'Sorry! Excuse me!'

We were almost ready to set off and I was still leaning on the back of the car when, in a hurry, Ken slammed the boot shut with considerable force. Unfortunately, my finger was still there. To be precise, my wedding-ring finger was now trapped in the boot of the car. I did what anybody would do in this situation, and screamed. A blood-curdling screech of a scream; the sort of scream that awakens every primeval instinct in a mum and makes them drop whatever they're doing and run to their offspring. On seeing my finger Mum also started screaming. All of this wasn't really helping Ken, who had the unenviable task of finding the key to unlock the boot of the car. You see, for some unfathomable reason, in those days boots locked automatically when you closed them and you needed a key to open them. For any whippersnappers reading, this was centuries before it was possible to unlock cars remotely with a fob and millennia before the magic of a keyless start.

What we had on our hands (or my finger) was a perfect storm. The automatically locking car boot and Ken's penchant for keeping dozens of keys on a ring attached to his belt loop meant that, as Mum and I screamed and his stepdaughter's finger went on being mangled, he went panic-blind – jangling the keys madly, frantically searching for the small boot key but he couldn't see it. The more we screamed, Mum helpfully shrieking 'UNLOCK IT!', the more the keys jangled. It felt like I was trapped for ever but it must've been only a few seconds until Ken found the key and, with shaking hands, opened the boot and released my poor rapidly purpling finger.

I lost more than just respect for the car with the stupid automatically locking boot that day. I also lost my fingernail.

Fair to say from then on I hated that car. Still, I sometimes didn't fancy walking to school in the rain so I would occasionally harangue Ken into giving me a lift, but because I didn't want any friends to spot me arriving in this misshapen motor, with breathtaking cheek, summoning up all the ungratefulness only a child can muster, I'd duck down out of sight in the passenger seat, pretending to tie my shoelaces, before doing an SAS commando roll out of the car and onto the pavement.

Fast-forward a few years and we were loading up the Renault 12 outside our little house in Astley Bridge preparing for our Highland adventure, the entire car interior looking like a shambolic version of Tetris.

A neighbour who lived a few doors away on the neat row of tiny terraced houses was a car mechanic; Jimmy was a stout, muscular man with bright green eyes, a booming voice, broad Bolton burr and a scar where a cleft lip had been fixed, which just added to his charm.

He had a nice wife and a couple of preschool kids who were often shyly hiding behind his legs when he knocked to chat to Mum and Ken. Before our epic journey Jimmy checked out the now slightly decrepit Renault and gave us the all-clear for going to Scotland in it.

We set off late afternoon, nosing our overstuffed car into the Friday traffic, heading north. Mum and Ken had planned to head to a static caravan park in Fort Augustus, just south of Loch Ness, some 360 miles from home and a six-and-a-half-hour journey. We didn't want to attempt the journey in one fell swoop – we were in no rush and we didn't want to push our luck with the Renault – so we scheduled a stop in York, a mere ninety minutes away from home, arriving as darkness fell at a B&B bungalow run by an elderly couple, every inch of every wall covered in floral wallpaper and framed family photos and the faint smell of lamb chops.

Next morning we hit the road early, Ken at the wheel, Mum juggling her many roles – co-driver, map-reader, peacekeeper, caterer ('Another custard cream, Ken?') and general commentator, in which role she critiqued selfish manoeuvres by other road-users, mused on the purpose of interestingly shaped buildings in the distance and pointed out roadkill ('Aww . . . Badger').

She was also the DJ in charge of the entertainment, which was mainly Radio 1 – Mum absolutely loved Simon Bates's mid-morning show with its emotionally charged feature 'Our Tune', where Simon's usually light-entertainment voice would drop to a deep baritone as he read out listeners' stories, adding gravitas and solemnity to the tear-stained tales, which always featured heartbreak, occasionally death, sometimes unrequited love and – if it was a really juicy one – all three.

When Mum wasn't sniffling along or openly sobbing to heart-wrenching radio features she'd pop on a tape – and the song that soundtracked that entire trip was Queen's 'Another One Bites The Dust'. I will always remember it playing as Yvonne and I lounged about in the back seat, all tangled limbs and Vimto smiles, occasional squabbles breaking out, Mum shifting round in her seat to referee if things got too physical. That song would be blaring out with Ken on percussion in the driver's seat, beating out the drum rhythm on the steering wheel as he picked his way through the winding country lanes. The more we squab-bled, the harder Ken hit the steering wheel, beating out the 'Dum dum dum – another one bites the dust . . .' in time.

The journey up to Fort Augustus was painstakingly slow because August 1987 was wetter than average for Scotland and indeed it felt like for the entire week we had a cartoon raincloud following us, hovering just above the Renault wherever we drove. The car was steamed up and cramped, the roads were treach-erous and there were two teenagers folded up in the back, by turn reading, snoozing, arguing over Travel Connect 4, munching biscuits and complaining of boredom. I wondered what in Ken's head was the 'one' in 'Another One Bites The Dust'? His freedom? His prime? Happiness? Patience?

Mum and Ken had given up smoking a few years before but Ken had struggled and so had switched from cigarettes to thin Hamlet cigars, in the beginning just smoking a couple a day – surely better than a twenty-pack of cigs, he reasoned – but inevitably his intake crept up to a dozen or so.

I always worried about their smoking but of course at thirteen you can't always find the right words to vocalise your worries and concerns, so instead I'd just grumble and gripe, fanning the air in front of my face at every opportunity.

The switch from cigs to cigars always seemed like madness to me – I thought that if anything, cigars, even thin ones, must be worse than regular cigarettes. Like saying 'I'm giving up juggling knives, it's too dangerous – instead I'm going to juggle live grenades.'

Every hour or so, the stress of the journey would take its toll and we'd pull over for Ken's smoke break. If the rain had eased to a light mizzle we'd all get out for a stretch and to lean against a soggy lichen-covered boulder and take in some of the beautiful views.

Yvonne would occasionally get a turn behind the wheel, which considerably slowed down what was already a pretty sedate journey. Ken would go in the passenger seat to guide her and Mum would be demoted to the back seat with me, which was actually quite nice as I got to cuddle up to her and make myself comfortable on her ample jumper bumps.

Despite getting Mum to myself in the back, it was during the stints of Yvonne driving that I would be at my most grumpy because we couldn't have any music on and we couldn't talk in case we distracted her and she sent us plummeting off a mountainside.

I was jealous because she was getting lots of praise and was the centre of attention – a position I was used to occupying.

With every correct indication and snippet of praise from the front of the car I'd roll my eyes and huff and puff, muttering my displeasure at the snail's pace and lack of Radio 1. I remember thinking 'I hope she crashes' – alarming to think a car accident involving my loved ones seemed preferable to the torture of my sister getting a bit of a fuss.

The car was so steamed up I couldn't make out the rolling green outside; it just looked like there was a bright green tea towel flapping around outside the window. After what felt like an eternity we arrived at our static caravan on a site in Fort Augustus, where moods lightened even as the skies above darkened with a fresh batch of good Scottish rain to pelt down on us.

The caravan was lovely – furnished in fifteen shades of beige – but the whole week was so wet we barely had the chance to unpack the swingball. It all went by in a bit of a blur – I just know it involved a lot of Connect 4, rainstorms, more of Yvonne's driving practice, scampi and 2p penny-pushers at the campsite's small arcade.

With some relief on all our parts, our final day rolled round and we packed up the Renault.

Mum wanted to take a different route home and stop off at Fort William. Which was where the car, which had until then been performing like a dream, decided to give up the ghost. No amount of coaxing could bring it back to life and Ken, after half an hour of looking at the engine, scratching his head, giving something a waggle and then shouting to, um, to try the ignition, finally admitted defeat, slammed shut the bonnet and went to call the RAC.

And so it was we arrived back home in style at 3 a.m., the entire row of terraced houses illuminated by the blue lights of the RAC wagon; all of us were snoozing, squished into the huge warm cab, the little red Renault strapped onto the back like a felled deer on a hunter's truck.

The next day our mechanic neighbour Jimmy called round and Mum thought she'd pull him up. 'Hey, Jimmy, I thought you said the car was all right?' 'I said the car would get you to Scotland,' he replied. 'I never said it'd get you back.'

24. Bob and the Beanstalk

On the farm there was always a new arrival to look forward to meeting. Waifs and strays would rock up – a flock of twenty guineafowl one day, who landed in the yard and decided to make it their home, sleeping in the rafters of the cattle sheds, waddling quietly around the farm during the day, chattering quietly among themselves like a group of day-trippers round a National Trust property. Another stray was the appropriately named Max – who the vet said was the biggest Alsatian he'd ever seen. He was malnourished and mistreated when he first arrived, in need of some serious TLC. He had a shiny dark coat along his back, with thick, brush-like blond fur across his broad chest that continued under his belly and along the underside of his always-wagging tail. He was huge, with paws the size of side plates and a massive head, out of which lolled a wide tongue, the colour of boiled ham and the shape of a size twelve shoe insole.

He was physically massive but even his impressive size was dwarfed by his personality – Max was one of the most loving dogs we ever had. He was a rescue dog and had come to us weighing roughly half what he should've done. Whatever sins past humans had inflicted on him, he obviously didn't blame us for he was the sweetest soppiest hound ever, always on the lookout for affection. In fact, he was so soft that I wondered if maybe his past owners had maltreated him for not shaping up as a guard dog; he looked the part, though, and definitely sounded it too, with a ground-shaking woof and an intense stare that quickly melted into a dreamy-eyed gaze whenever he recognised a familiar face. When I'd arrive at the farm he'd come lolloping over and nearly knock me down by leaning his full weight on me, wanting a cuddle. He was always panting and a long stalactite of drool was often suspended from each side of his huge mouth and shiny black lips. We knew to duck and cover to avoid being splatted whenever he shook his head, as the dribble would cast out from him in long gluey strings.

The farmhouse was often freezing inside until the fire had been lit (in fact, before we had central heating, frost would settle on the inside of the windows), and Max made an excellent furry duvet to cuddle under on the floor while watching telly. When he'd get up you'd be left covered in a thin blanket of his moulted hair and possibly a couple of his calling cards – damp spots of drool.

Max formed a partnership with the Jack Russell Sally, who was all white except a choc-chip-coloured patch on her tail and a caramel-coloured ear. Sally was supposed to be a tough little farm dog but since she had arrived as a tiny puppy, able to curl up neatly into the size of a bagel, she had been treated like a princess. She loved nothing more than being in Dad's lap, whether he was driving about in his bashed-up old Daihatsu jeep or sitting in his leather armchair in front of the fire.

Sally, however, was no princess when it came to ratting and

loved snuffling after them, like her forebear Benji had. Together Max and Sally made the greatest partnership since Scooby Doo and Scrappy – and they were similar in proportions. Sally was fearless when it came to protecting the house – whenever a car pulled up into the yard she'd scramble up and at 'em, yapping her head off and obviously feeling especially brave because always following a few feet behind would be the biggest Alsatian the vet had ever seen.

The vet was Bob Potts. Of all the characters who came up to the farm, Bob was one of the most memorable. He was a large man, with a bulbous nose and a windburnt face, a close-knit web of spindly purple veins blanketing his ruddy cheeks. His small eyes were bright and sheltered beneath huge eyebrows like two black marbles under a couple of pan-scourers. His sideburns were wire wool, as was the hair that stuck out from the rim of his flat cap. Bob Potts spoke in the broadest Bolton accent, thick with dropped aitches and dialect. He also talked very loudly, in a booming voice people usually reserve for very old deaf ladies or foreigners. Being within chatting range of him was like standing behind the engine of a 747 taxiing for take-off.

Considering he spent the majority of his time at the business end of cows Bob was exceedingly well turned out, always in an olive-green trilby and tweed suit tucked into wellies – topped off, if he was planning on doing a bit of bovine puppetry, with a big green rubber apron and a disposable plastic opera glove going up beyond the elbow. The outfit sounds a bit *Toad of Toad Hall*, I know, but he was a big enough build and personality to carry it off.

'AYE UP, LEN,' he'd holler at my dad, and I'd hide behind the kitchen door, peeking through the crack of the hinge, fascinated and terrified in equal measure. He'd always clock me and try to speak gently but it was physically impossible for him.

His big red face would lower to mine and he'd boom 'WHO'S THIS THEN?', stray whiskers from his sideburns twitching as he spoke, me frozen to the spot. I remember one winter he pointed out the crusty sleeve of my dressing gown, which was stiff with dried snot – for a few nights I'd been cleverly utilising the right sleeve as an emergency hankie and it now resembled a piece of pale-green armour, a vambrace of germs. Mum came to my rescue and ushered me away, and once out of sight she whipped off my dressing gown and bunged it in the wash.

Whenever Dad said he'd have to call the vet Mum would secretly hope it would be Ian, one of Bob's partners in the vet practice, who according to Mum was tall, handsome, blond and always wore Old Spice aftershave. I love the idea of Mum being one of the many farmers' wives who'd quickly run a brush through their hair and sashay across the yard in the hope of seeing (and smelling) Ian, only to be met by the more senior partner Bob and his bristles.

To me, Bob was like some giant who'd clambered down from a beanstalk, but to Dad he was a friend and a loyal, dependable vet. When you run a small farm with a few dozen head of cattle it's a business but it's also a passion. Dad knew all the cattle by name, not just by a number on an ear tag. He cared for each cow or bull and wanted to keep them well. Bob knew this and would always do his best to help out Dad, rattling up the farm lane in his green Land Rover Defender at all times of day and night.

According to Mum, one windy day in March 1974 he'd been up to give one of the cows a check-over and a jab and, as he headed across the yard, he spotted Mum. Bob took one look at her and said in his usual broad, sonic-boom-bothering voice, 'BY 'ECK, JACKIE, YOU'RE IN CALF AGAIN!', which shocked

Mum because only that morning she'd made an appointment with Dr Parikh as she suspected she was may be pregnant. And indeed, despite her not showing at all, Bob turned out to be right – she was six weeks gone with me. It seems somehow fitting that the man who was responsible for Dad's prize cattle was also the one to detect the existence of the last offspring to be added to Dad's personal herd.

25. Feather Carpet

In the same way that downing a few pints at a family barbecue before driving your kids home sans seatbelt, them sliding around in the back of the car while you chuffed on a delicious fag in the front, became outmoded as we became more educated about the dangers, then so did many farming methods. Dad had to move with the times and by the early eighties it was becoming increasingly obvious through the media and people's shopping habits that keeping battery hens was no longer the done thing. Along with boys getting perms and the Reynolds Girls, battery-hen farming was soon to be consigned to the eighties rubbish bin for ever more, and quite rightly.

Before the general population turned to the belief that it was the basic right of animals to be able to live a life free of distress and pain and be able to turn around in their living space, egg farmers kept their hens in battery cages. Different times, as they say.

Dad's laying hens were kept in the long wooden sheds that were turned over to the Christmas turkeys after the egg business ended. These sheds were narrow, squat buildings about a hundred metres in length and looked like the wooden dorm blocks you see in army barracks in old war movies. Inside the sheds the 'batteries' were stacked three high and ran the full length of the shed in four long rows. Seeing as each battery was essentially a wire cube with a floor the size of a tabloid newspaper and there would be three hens in each cage, that meant a lot of chickens.

The first thing that hit you when you slid aside the huge wooden doors on their runners was the smell. Hen muck has a sour scent to it and it was carried through the air on a gently swirling snowstorm of dirt, dust and tiny feathers. To this day if I'm out in the sticks and get a whiff of a chicken farm as I'm driving along, my nose wrinkles and I know what it is right away. After the smell came the deafening noise. Hundreds of hens make a racket and it wasn't a contented hubbub, more of a pained cacophony.

Of all the spooky, smelly or noisy places on the farm, I found the hen sheds the worst. It was a real attack on the senses – the noise, the stink, the dust. Walking among the stacks of hens was so scary; the floor a springy carpet of compacted reddish-brown feathers and muck, their beady eyes watching you, the constant moving as they lurched and flapped, trying to get purchase and stand comfortably on their wire floors.

Underneath each cage was a metal shelf like an in-box on an office desk and the droppings would drop down between the bars and land there. The floor of the cage slanted downwards, allowing any eggs to roll to the front of the wire and through a small gap, where they came to rest in a curved metal lip, ready to be collected – which was another job that really scared me; I'd have to close my little fingers on a still-warm egg and try to whip it away before a hen strained to peck at my soft knuckles. The design of the crates was actually genius in some ways but, as we now know, the hens weren't living, they were merely existing.

Back in the seventies and eighties though, attitudes to animals were very different. They were seen by consumers and farmers alike as commodities to be used, discarded and replaced and Dad was only one of thousands of people who farmed this way.

Changing times and a broader awareness of the hellish effects on animals means that, in the UK at least, battery, veal crate and sow stalls have all been resigned to history. Just as women are now allowed to vote and we don't shove kids up chimneys any more, we now know more about animal welfare and allow hens to live and lay in much better conditions.

26. Genie from a Bottle

I look back on most of the farm dogs with a warm glow of love. I have happy memories of being tiny and cuddling huge Spud, the lurcher/collie cross, my arms slung round his neck, face buried in his thick black fur with the garland of white round his neck – him wolflike, the Akela to my scruffy Mowgli.

Sally snoozing on my lap, Dad's gun dog Brin, a beautiful, sleek short-haired German pointer, docile and sweet, resting his heavy head on my lap wanting some fuss.

Even Radge, caking me in stench every time he threw himself at my legs, desperate for affection, and, most recently Jed, Dad's loyal black Lab with a tail so waggy it looked like it was wagging him.

There is one dog, though, who holds no happy memories for me. Nero was a rescue dog who arrived at the farm looking in good shape physically but who, we were soon to learn, hadn't recovered mentally from the abuse he'd suffered at the hands of his former owners.

Nero was a Rottweiler, 140lbs of muscle with a shiny black

coat and a tan muzzle, neck and eyebrows. He was chained up in the yard and was one of our 'second-chance dogs' – we had a few over the years for whom the farm was to hopefully be their second chance at happiness. His backstory was pretty grim, according to his rescuers. He'd been abused horribly. He was owned by criminals who wanted to use him to attack the police, so they used to wear fake police uniforms and beat him, instilling in Nero a deep hatred of anyone in a uniform. Whenever our local bobby used to pop up to the farm Nero would go mad. He'd be wrestled into the porch but even then he'd throw himself against the door and the window, trying to get to him. He'd bark furiously, his breath misting up the glass, drool smudging the window as he shoved his face against it.

He was a damaged dog and we were aware we had to treat him differently – apparently big dogs who are troubled don't like to be stared at directly in the eyes as they see it as a challenge, a confrontation. I was always scared of accidentally locking eyes with him, as if he was a Hollywood star and me a mere assistant.

When he was on the chain I'd make sure I'd stay beyond its reach as I went back and forth filling up water buckets, always keeping myself aware of where he was.

Avoiding him was harder in the house, where he obviously wasn't on a chain, but I just tried to stay out of his way.

One chilly afternoon, though, I'd ridden Gus and done my chores so decided that I deserved a little break. Dad was mooching about in the yard, so I was alone in the house. I helped myself to a couple of malted milks. These were the biscuits that we always had at the farm, small sweet rectangular things showing a raised scene of a biscuit cow and calf grazing in a biscuit field. I always particularly enjoyed that the biscuits were a confectionery representation of Dad's fields.

I was so engrossed in getting to the sofa without the biscuits sliding off their plate that I didn't even notice Nero was lying in the corner of the living room. It says something about the deliciousness of malted milk biscuits that they stopped me from spotting a whopping beast of a dog just a few feet away.

I flopped onto the leather sofa, still watching the plate of biscuits, and swung my legs up onto the seat next to me.

Like a canine genie from a bottle, Nero appeared from nowhere.

He was massive but it wasn't his size that intimidated me, despite his head alone being the size of a Nissan Micra. I was used to big dogs – Max and Spud were definitely bigger and I'd cuddled and fussed over them, but Nero was to cuddling what loose bowels are to scuba-diving. It wasn't his size but his fearsome reputation – I sensed from Dad that he didn't entirely trust Nero and there was a general consensus that the dog was very much a work in progress.

So there he was, wide as a bison, big brown eyes staring intently at me, then at the malted milks. His glare switched from the biscuits and back to me several times as if he was weighing up which to eat first, like scones on a cake stand. Any other dog would've been gently shoved away or I would've shared a bit of biscuit with them, but with Nero I was frozen to the spot like someone had stitched the rear of my jodhpurs to the chesterfield. I quickly ran through my options – scream for Dad, make a run for it, shout at Nero, ignore him and just eat my biscuits – but each one ended with the same conclusion – I'd be torn to pieces by a furious Rottweiler.

Meanwhile, Nero intensified the stand-off by inching slightly closer to me so I could feel his broad chest pushing up against my legs. And then the worst bit. He made a deep, low noise – a growl that came from the very pit of his soul. It wasn't loud and was in fact more petrifying for its quietness, like it was our little secret. The teachers I always feared were the ones who'd tell you off with an intense hoarse whisper rather than the shouty ones.

In the end I decided I had to escape and in one smooth move threw the plate and with it the biscuits onto the carpet. As Nero moved towards it I leapt over the back of the low sofa, out the door and into the kitchen, shutting the dog and the malted milks in the living room.

I was so relieved – and seriously shook up. I'd grown up around big dogs, animals that given half a chance would literally make a meal out of local cats, rats, other dogs, but I always knew they'd never hurt me.

In fact, the only other animal that I'd been really wary of on the farm was Leeroy the giant goat. What a force they could've been if they'd teamed up: the goat and the Rottweiler, the Phil and Grant Mitchell of the farmyard, demanding malted milks with menaces.

If we had our suspicions that Nero was slightly unhinged then sadly we were to be proved right. One day Dad was up in the field with Nero, fixing some fencing, when he saw a Land Rover bumping up the farm lane. It was Jack Henry, Dad's good mate, popping up to the farm to have a brew and look at Dad's new bull. Spotting Dad, Jack stopped the Land Rover, got out and walked up the field to meet him.

Dad blames himself but I think it all happened so quickly it would've been impossible to stop. As Jack approached, Nero's hackles rose up all along his spine like a mohican and he stood across Dad, blocking Jack off and – he thought – protecting Dad from this stranger. Before Dad could stop him, Jack reached over Nero to shake hands and Nero launched himself at Jack, tearing at his wrist. Dad managed to get Nero off, kicking him away, but an ambulance had to be called and poor Jack taken off to be patched up. It was a shocker. Nero had to be destroyed and Dad's wife was devastated. I think we'd all hoped Nero could be fixed, but the legacy of his abuse shadowed him for the rest of his short life.

Rottweilers as a breed have often been the subject of press campaigns over the years but, as far as I'm concerned, a dog isn't born bad, it's made bad. Having said that, would I ever have a Rottweiler myself? Nope. Nero left his mark on me for sure and I'm lucky I escaped physically if not mentally unscathed.

27. Fern Britton was a Cow

Starting from long before I 'helped design' Lady Diana's wedding dress, there had been a long line of doe-eyed Dianas on the farm – of the bovine persuasion. Dad bred Hereford cattle, a breed native to the UK, from – spoiler alert – Herefordshire, where the breed evolved. There are mentions of Herefords by agricultural authors as far back as 1600. The breed is thought to be founded on the draught ox descended from the small red cattle of Roman Britain, mixed with a large Welsh breed popular along the border of Wales and England. Love that a fruity Welsh loverboyo used to pop over the border to romance the hot red herd, or some slinky scarlet lady cow scarpered to Wales to find her big, as they'd say there, 'golygus' bull.

Herefords are a gentle breed, not highly strung; they're calm, kind, solid, long-lashed and bonnie. If Fern Britton was a cow she'd be a Hereford.

They're so placid in fact that as a little girl I would often be plonked on Dad's prize bull after a successful showing season.

In the garden, with the bounty won that summer – cups, trophies, rosettes, all laid out in front of the bull – a photo would be taken as a memento of the family's achievement. For the last pic, for a bit of fun, I would be sat on the bull's back, grinning over a ton of twitching, muscular prime pedigree beef.

However, before the heady days of rosettes and accolades, in the early sixties when Dad moved on to Grundy Fold Farm from his dad's farm, he had a 'starter pack' of cattle – a mish-mash of commercial breeds, good for beef but not for the show ring.

Once Dad had settled on the farm he decided he wanted to start a proper pedigree herd, so he took himself off to a Hereford sale over the Pennines in York. Mum put a call in to the school and my brother David was given special dispensation from the headmaster to have a day off to accompany Dad to the sale – after all, Dad was purchasing what would be the beginnings of a herd that one day David would take over, so it was seen as work experience.

I don't think that'd wash these days. Taking one of my kids to Radio 2 with me because they fancied being a DJ one day would probably not get past the head teacher.

That day Dad came home from the cattle sale with two heifers. These were the seeds from which his whole herd would blossom. Two Eves who just needed an Adam. The Adam came in the form of a young bull Dad bought privately from a farmer called Sam Woods, who became a trusted seller. Sam, who lived in Blackburn, brought over the young bull, a fine polled Hereford lad called Halesdown Dermot, and the beginning of Dad's first herd sparked to life. Dermot was introduced to the heifers, and I imagine Dad lit a few scented candles, popped on a Barry White LP, and left them to it. The heifers were sisters and both were called Diana.

Diana 1 and Diana 2. So not confusing at all. All pedigree cattle have by law to be registered before they're sixty days old, so they were already both named (after their mothers) by the time Dad bought them. There were three people in this marriage and that suited them just fine. Dermot (like my radio buddy O'Leary) was obviously a charmer and he set to work creating 'Masefield Herefords', the herd being named not after the farm but after the area.

Masefield Herefords at its height would have forty cows with calves at heel plus the main stock bull, as well as maybe a dozen or so young bulls at any one time. Although it's a more modest size now, the herd is still going strong today, half a century later.

Hereford farmers name their animals alphabetically, each year bringing the next letter, like car registrations. It's always exciting when a letter comes round that you can really do something with – out there somewhere there's a young bull called Isaac and a cow called Lola, named after two of my kids. Let's just pretend they're out there anyway, and not in a burger or Bolognese.

Along with the cattle there were pigs. Dad had dozens of Large Whites – huge white pigs with pretty, slightly dished faces and big pink erect ears. I was only young when Dad had the pigs and the sows seemed to me the size of a two-seater sofa. My main memory is of sitting on the barn floor in my dungarees, my bottom cushioned by a thick bed of clean straw, clutching tiny squirming piglets on my lap, their sweet little pale snouts the colour of strawberry milkshake.

I remember becoming vaguely aware of what their final destination was when Dad loaded them up in the wagon a couple of months later. I felt so sad for these wee pigs – they were no longer tiny and as cute, but I felt the injustice of their premature deaths keenly. 'Dad, pleeeeease don't take them,' I begged. Dad didn't want to leave me at home, getting under Mum's feet, moping around and sulking, so he took me to the slaughterhouse with the pigs. I must've exhausted myself with my histrionics because I'd fallen asleep before we even got there. All I remember is waking up chilly, with a rumbling tummy, in the silent, cold cab of the cattle wagon parked back home. After this trip Dad wasn't quite so keen to, in Iron Maiden's words 'Bring his daughter to the slaughter' and I never again asked if I could accompany him.

28. Up in Smoke

The farm was a busy, bustling place. From spring to mid-autumn (if the weather was kind) the cattle would be out in the fields. Summer was all about keeping one eye on the pregnant cows for calving and the other on the skies for haymaking.

All year round the pigs and laying hens kept ticking over. Winter, when the weather turned bad, was when the cattle came in. The wet fields would get churned up by the herd standing at the gates, mooing mournfully, protesting at the wet, wanting to come in to the dry. Once in their pens the hard work really began. Feeding up happened twice a day and the work of mucking out all those different sheds was pretty much constant. Add to this the odd casualty that called for a visit from Bob the vet, plus foot-trimming, worming, occasional escapes . . . running a farm isn't a job, it's a vocation. A challenge. Twenty-four hours a day, seven days a week. Plus us five kids running round. Busy with a capital insane.

*   *   *

Dad had never been afraid of graft, though, and he worked to build up his herd over the following years until a cold November night in 1981, when all his hard work went up in flames. Literally.

It was a Saturday evening. I was living at School Street by then and so at the farm was just Dad and the older kids. Everyone had just finished their tea and was settling down in front of the telly for the night when there was a hammering at the door of the farmhouse. It was a lad called Ian, who was the brother of one of Rob's pals, raising the alarm – the barn was on fire.

Everyone scrambled into their wellies and ran up top to the barn. I can only imagine the panic as they raced across the garden and turned a right onto the lane to run up past the main cattle sheds to the barn. It was a beautiful old traditional Dutch-style building, with a high roof made from old timber and the lower sections from huge long logs, sleepers reclaimed from railway lines. The main body of the barn was full of the hay that had to last until late spring, when the cows could at last kick up their heels out in the fields and have a pick of grass with the warm sunshine on their backs.

The fire had to be stopped. At the nearest side as you approached, underneath the loft, were three stables, two of them containing young bulls and the third holding Muffin, my pony Gus's mum. With Dad shouting instructions, he and the kids swung open the stable doors and shooed all the animals out. Muffin clattered onto the lane, desperate to escape the flames, which had begun to lick around the old timber at the rear of the building.

There had been a huge blaze that afternoon in Bolton Town Hall, which was connected to the historic Albert Halls, and so all local fire crews were in attendance there. No regular engines were available and eventually two army engines, Green Goddesses, roared up the lane and fought the flames. They managed to put it out in minutes. The damage was minimal; just a few bales were damaged.

Dad and the animals had a lucky escape that night.

The next night their luck ran out.

The local lad who the police eventually collared for the crime had returned. Nobody knew why. It wasn't a grudge against our family – his brother was pals with David. So was it a fascination with fire? He never gave a reason.

Whatever his motivation, the bastard set light to the barn again.

It took hold. There were two terrible differences now from the evening before: it was a blustery night and the wind was against them. It fanned the fire, blowing over the fields towards the barn, pushing the hungry flames through the bales, the heat rampaging at speed, eating up the wood, devouring the straw stacks, heading up to the roof.

It was also later in the evening than the previous night, so no one was out to stumble upon the fire and raise the alarm. It had time to burn.

By the time Dad knew about it the barn was fully ablaze, flames roaring up into the pitch-black winter sky. In no time four fire engines were speeding up the bumpy lane and firefighters spilling out, ready to do battle. But this time, although the animals again escaped, the barn couldn't be saved.

Dad was helpless. He had to stand there and watch as his beautiful barn burned.

Inside the barn was all of Dad's hay and straw he'd harvested in the summer. It was his only means of feeding and bedding his livestock and it had all gone up. A whole winter's worth of food up in smoke. Balancing the books on a farm is a delicate business as it is; having such a valuable asset wiped out like that was absolutely devastating. The firefighters had to stay for a week to keep dampening down the fire. The hay was good quality and dry, so it kept reigniting. They cut open and spread out the blackened bales, dragging out the hay and straw over the gravelly lane, thinning it into a still-smoking carpet. It smouldered for days, the acrid smell drifting over the fields and the estate.

A handful of cattle who were temporarily homeless kept having to be shooed away from the thick blackened tapestry of hay; it was ruined, so they weren't eating it; they kept lying on it because it was so warm.

The barn fire nearly finished Dad's business off and his career as a farmer was almost snuffed out before it had even begun. He had to buy in fodder for the cattle to replace the hay lost in the fire. My brother Robert remembers this time vividly, the sounds and the smell of the fire – awful stuff. He says it was lucky that Dad had lots of hens – they ate only eggs and chicken for about six months after the fire. It was only because of those hens that Dad managed to feed the family through the winter.

If the kitchen was the heart of the farmhouse, then the barn was the beating heart of the farm – it's the thing that stands solid, crucial to the survival of a smallholding. It feeds and homes the animals. It is storage and it is shelter. It's hard to comprehend and it's hard to forgive when someone is callous and cruel enough to rip out the heart of a place like that.

Eventually, after a nerve-shredding wait and lots of investigation, the insurance paid out and Dad was able to start piecing together the barn area. The blackened skeleton that remained was removed and in its place was built a more modern structure, half grey breeze block, half pale tinder slats, with huge bright blue sliding doors. The new barn of course lacked the charm of the original, but I have lovely memories of playing in it with my buddies, lounging on the bales gossiping with my pony-riding pals. It's still there today, breezy and well lit, and it echoes with the laughter and banter of over thirty haymaking seasons and with the squeals of my own kids from the times they've spent in there, leaping from stack to stack of the bales.

29. Martini –
Shaken and Scared

Until Gus was mature enough for me to ride, Dad acquired a small, pretty mare called Martini. She was dun – pale beige with a flowing, long, darker mane and tail. Like with most of the animals that weren't bred on the farm, I don't know where she came from but I heard rumours that Dad's mate's daughter couldn't handle her so Dad got her for a knock-down price. I presume that I was expected to get in the saddle and show Martini she didn't shake or stir me; basically I was to be the one to take charge of this little asbo pony and knock her into shape.

Problem was, I was six. I also wasn't brave enough to act as some sort of screw in a horsey borstal. Martini and I got off to a shaky start; she wouldn't go forwards or backwards – in fact, she wouldn't move at all. When she did decide to move it was usually sideways, always an alarming manoeuvre on horseback unless you're deliberately doing a jaunty dressage display. The ponies at the riding school were like being on a merry-go-round, happy to just follow the bum in front, so to be confronted with a pony with attitude was alarming. I was scared of her – in the world of ponies she was no Shire (and retiring type, lol) but I was only a wee scrap of a thing and she was stroppy and, compared to me, massive and strong.

All these issues were taken into consideration and it was decided that the best course of action would be for me to have a few lessons on Martini to slowly build up my confidence so we could begin a new and everlasting friendship. Only joking! Instead it was decided that I should be locked in a field with her.

Yes, classic. Read any respected psychology studies and they'll clearly state that the best way to deal with something you're frightened of is to be locked in a field with it.

I'm surprised Relate don't just welcome warring couples into their therapy room and then show them out of another door leading straight to a paddock, in which they'll be locked until they've sorted out their differences.

Anyway, I was plonked into Martini's saddle and led into the field. Once there, my dad's wife swung the big gate closed, bolting it shut.

If this was a Hollywood movie, Martini and I would've stayed in the field and there would've been tears (me), bucking (Martini), falling off (me), snorting (Martini) and determined grimaces (both) – then after a montage of all this, possibly to the *Friends* theme tune, I would've horse-whispered the heck out of Martini and we'd be best friends, galloping up to and jumping over the locked gate, sailing over in slow motion.

What actually happened was Martini and I remained parked in the same position, right next to the gate where we'd been deposited, and I wept. I cried my eyes out. I was cold and frightened and it was getting dark.

Some kids are fearless on a pony – put them on anything and they'll cling on, boss it and eventually win it over. I just wasn't that kid though. I hated falling off and I found it scary when a pony played up.

I felt a complete failure, like I'd let Dad down. I knew I'd messed up but to me it wasn't altogether clear how it was my fault; I was scared of a pony I hadn't asked for and was now being punished, which to my six-year-old (and to be honest, 43-year-old) mind seemed rather unfair. Anyway, I busied myself with the kind of gulpy, hiccuppy crying that small children favour for what felt like for ever but was probably less than half an hour.

Eventually my dad's wife came stomping out, dark permed hair quivering with fury, face thunderous, naturally disappointed that the cunning plan hadn't worked and that locking a frightened child in a field with a pony hadn't turned out to be the best way to resolve confidence issues.

Martini's time on the farm soon came to an end and she left as promptly as she'd arrived, hopefully to a braver kid who'd do her justice.

Over the years my confidence in the saddle has fluctuated but the highs – hours with my best mate Gus, as an adult competing in charity horse races – and the literal lows – facefuls of dirt and a couple of fractures – have all been worth it.

At the age of fifteen, I had outgrown Gus; the famous Cox legs had truly sprouted and my limbs dangled way past his rotund belly. If I wanted to keep riding him I'd have had to wear roller skates so my feet didn't drag along the ground. There's something peculiarly humiliating about riding a pony that's too small; you feel cartoonish and ridiculous. It probably also reminded me of being the tall girl on the little bike with stabilisers. It's a similar feeling to when you outgrow your trousers and they become half-mast.

The answer was Chantel – a beautiful bay mare, tall, elegant and sweet-natured, with the build and breeding of a racehorse but the personality of a yoga teacher. I arranged to have her on loan for a while, but our time together was to be sadly cut short.

One of the worst times of year for riders is the couple of weeks leading up to Bonfire Night, when fireworks go on sale. As you can probably imagine, small explosives and horses mix as well as fun runs and minefields, so I was always wary in the build-up to the fifth of November. Young ne'er-do-wells were fond of setting off fireworks at all times of day and night and the loud bangs echoing round the skies were terrifying for horses and their riders. Still, poor Chantel couldn't be confined to barracks for the whole of October, so one afternoon we set off from the farm, squeezing through the blue iron gate and down the path and into the estate. We'd barely been out for five minutes when someone let off a banger in a nearby street. The loud crack that it made either terrified Chantel or she thought it was a starter pistol because she set off at high speed away from the noise, along the road. Mercifully the road was quiet but I had no idea what was round the corner. At any minute a car could appear and we could career into it. And Chantel was carting me off towards a T-junction that opened up onto a busier road. I had to stop her.

I can remember the sound of her metal shoes clattering along the tarmac. When a horse gets the bit between their teeth and decides they want to leg it, you can't just pull on the reins and pray. You have to take emergency action. There's one surefire way to stop them, though it's not without its risks. If something is running really fast in a straight line the best way to slow them down is to turn them in ever-decreasing circles until they've nowhere left to go and have to stop.

Problem is, turning a racehorse-sized mare in a narrow road is a bit like turning a high-speed conga line in a corridor; it's easy to crash. Which is what Chantel and I did. I quickly grabbed the right rein close to her bit and, using all my strength, pulled on it and hauled her round to the right; we managed to circle in a clockwise direction and she slowed down, but then in all the drama, as she was coming to a stop, she tripped up the kerb. She stood there panting, her neck sticky with sweat, but we were in one piece at least, and I hopped off and led her slowly back to the farm. We'd soon discover she was so lame she had to be retired from being ridden.

I've never enjoyed a firework since. OK, not really.

Riding stayed with me throughout my teens and, in various forms, up to the present day. Even while waiting to hear if I'd got my first job in telly on *The Girlie Show*, I was working as a groom for a married couple in Bolton called Mark and Joanne – he was my optician and a keen showjumper and competed every weekend while Joanne cheered from the sidelines. My duties included loading and unloading the horses from the wagon, tacking them up and untacking them – just general grooming duties, but I loved it.

I think they slipped me a few quid but I didn't do it for the hard cash, I did it for the thrill of the whole thing – watching the classes, being around the camaraderie of the shows and smelling of sweet, sweaty horse when they'd drop me off home after a long day helping out.

I'm sure it was during an eye test that our shared passion for riding came up, and I'm pleased it did. I was in flux at the time, managing to combine the two careers of mediocre model and slightly sulky barmaid while approaching the brow of an exciting rollercoaster-ride few years, and this would turn out to be my last chance for a while to get down and dirty with my four-hoofed friends. I remember Joanne telling some of the other riders at the shows that I wasn't going to be grooming any more as I'd got a job in showbiz and them laughing, fascinated that the quiet knock-kneed girl sweeping out the horsebox was going to be on the telly.

30. Leather, Lace and Fishnet

When I was seventeen, running in parallel with my life with horses was real life, and the move from Smithills Secondary School to Canon Slade School's sixth form was looming. I was about to switch from being a small fish in a big pond to being in a slightly smaller pond, but any pond is intimidating to a fish with self-confidence issues, so my nerves about starting this new chapter were off the scale.

As well as worrying about making friends and fitting in, I was in a spin of contradictory emotions about the fact that at my new school we could wear our own clothes. So mind-blown was I about the heavenly release from the constraints of uniform that I absolutely could not get my head around this. The dress code at the sixth form was actually a little stricter than your average college as it was connected to a Church of England school and had an overall much tighter grip on its pupils than my old school ever had. It was decreed that jeans wouldn't be allowed, or leggings. I was soon to discover they upheld these uniform rules so fervently that I imagined they thought denim was the devil's work and, to be fair, with the parallel jeans I used to sport, like two wide windsocks completely concealing my Kickers boots, giving the impression I hovered rather than walked, they weren't far wrong. As for leggings, I think society would benefit as a whole if they were outlawed nationally.

With my faithful jeans and black leggings no longer options, I floundered. I wanted to look smart and stylish, so went shopping for some day-to-day A level essentials, flashing my Clipper Card and travelling not to Rodeo Drive, baby, but to Bolton town centre. I hit the shops (Etam, Mark One) and, after refuelling at Greggs, bussed it home, pretty confident I'd sorted out my look.

The first day rolled around and Mum ran me to school as I was sick with nerves – my history of being a new girl was riddled with negative experiences and I hated the prospect of having to make a first impression to dozens of fellow students that could potentially set the tone for the rest of that first term. The only consolation was that, even though the majority of pupils would've come from the fifth form of the school into its sixth form, today was still a nerve-wracking transition for everyone.

Back then Canon Slade school was a huge creamy-brick building with a squat two-storey sixth form tagged on at the end like an afterthought.

As I walked towards the double doors of the sixth form I looked down at my outfit and was suddenly racked with doubt. Not surprising really – for reasons that to this day elude me, I'd settled on a pair of Baileys-coloured slacks and a cream buttoned-up cardigan with pale pink and blue chunky flowers knitted onto the front and round the neck. The effect was less cool college kid and more born-again Christian working part-time at the library.

As I was new to the school, aside from a couple of Smithills lads the first day of sixth form was literally the first time my new schoolmates had clapped eyes on me. The first impression of me must have been baffling – who was this Sunday school teacher who'd been shipped in from 1950s Alabama?

I walked into the common room and although it wasn't quite like a stranger walking into a rough locals' pub – jukebox falling silent, pool balls stopping on the ripped baize – it felt like it. As the door swung shut behind me the membrane that had caused my fashion-blindness fell away from my eyes and I could see clearly and instantly that I looked all wrong.

I was met by a vast array of better dressed pupils: in the goth corner was a sea of black miniskirts and thick tights, baggy jumpers and Doc Martens; the drama gang wore floaty ankle-length floral skirts, polo necks and an air of importance; and the lads wore the kind of pastel-coloured chinos that are favoured now by David Cameron on his Devon hollibobs but which back then were bafflingly passable for teenage boys.

Safe to say, the next day I upped my game and stopped dressing like a florist from *Midsomer Murders*.

In fact, I moved a little too much in the opposite direction and a few weeks later I was hauled before the head of sixth form for flouting the uniform code. Mr Dyson could've been Mr Bean's older brother – he had dark bushy brows and rubbery lips that, when displeased, would contort and twist like they were trying to wriggle free from his face.

With his wild wiry hair, greying at the temples, he always looked dishevelled, as if instead of travelling to school by car or bus like the other teachers, his preferred method of transport was hurricane. He always seemed to wear the same grey checked suit that looked like he'd buttoned it up wrong, with his tie skew-whiff and tattered too-long trouser hems catching the bottom of his battered brogues. Ironically, though, he was a stickler for smart uniform.

For all I know, away from school Mr Dyson was a lovely

bloke – maybe he was a popular member of a pub quiz team, collected toby jugs, enjoyed charades and cooking Mexican cuisine for his family, was keen on yoga or his favourite film was *The Commitments*.

To me, though, he was just my head of sixth form, devoid of personality, compassion or human emotion. A teacherbot and my arch-nemesis for the couple of years I was at Canon Slade.

After a couple of run-ins about my trousers being too tight he was constantly on the lookout to check I'd fallen in line with the dress code. Whenever he'd encounter me in the corridors he'd quickly and surreptitiously scan me for uniform contravention. One day I was wearing ski pants, a curious fashion long since consigned to history's style sin bin. They were fitted trousers, often with an elasticated waist, and at the end of each leg had a thick fabric stirrup worn under the arch of the foot to keep the trouser leg stretched and taut. I've never been skiing but I imagine the foot stirrup is the only similarity between ski pants and actual skiwear, as the flimsy black numbers I enjoyed wearing in the early nineties surely weren't suitable for any kind of sport. Except perhaps darts.

Mr Dyson's eyes settled on my ski pants and he ordered me to his office for wearing leggings. Once inside, after the usual lecture about how rules are rules, I played my best hand; I popped off my shoe, exposing the fabric stirrup that was to save me. A simple strip in a stretchy polyester/Lycra mix was all that lay between me and a right rollicking.

Triumphantly I flashed the bottom of my trousers, showing that I wasn't actually wearing leggings at all but SKI PANTS! In your face, Dyson! Mwahaha.

I watched him gaze at the stirrups and waited for his face to crumple as he realised that on this occasion he was beat. Maybe he'd use his hankie as a surrender flag?

Instead he said, 'They're still leggings,' and sent me home to change.

Despite my battle with Mr Dyson I did enjoy my time at sixth form. I liked my chosen subjects, English Literature, English Language (we were convinced our two English teachers were secretly in love, Mr Jones and Mrs Smith; he had big square specs, matching mousey beard and floppy hair), Geography and Theatre Studies. Especially that last one. In my second year a new pupil arrived who was so good at drama you could see the teacher's eyes shine with excitement; like pirates opening a treasure chest, they knew they'd found something special. And they had; the girl with the broad accent, white-blonde hair and penchant for blocky heels and chunky knitwear was Maxine Peake, now an award-winning actor of both stage and screen. I like to think that I can take some credit for her career success – surely watching my faltering performance of a *Medea* soliloquy inspired her to try even harder.

During my time at Canon Slade I started to wear glasses, having obviously inherited not only my mum's knobbly knees and jumper bumps but also her less-than-perfect vision. I was actually quite thrilled that I was officially slightly short-sighted; during my 'faux fainting' years I'd always tried to cheat at colour-blindness tests, pretending I couldn't make out the number nine or the letter T written in different shades to the background colour. I was desperate to be officially declared medically lacking or different in some way – oh to be diagnosed as having something as exotic and exciting as colour-blindness! It never happened, though – the opticians always saw through my lame attempts to fake it. I clearly wasn't the first kid – and probably not the last – to try to fool them.

Imagine my excitement when a decade later it was revealed that I needed to wear glasses when I needed to see the whiteboard in class and for driving. A lovely tortoiseshell pair were picked out and I was absolutely thrilled. The only issue was that my mum came with me and with the two-for-one deal on offer she thought it'd be a great idea if, like her, I had one pair with 'react-to-light' lenses, which automatically darken when subjected to bright sunshine but when indoors remain clear. The only problem was that they never darkened quite enough to look cool like sunglasses and they took ages to lighten up when I came back in from outdoors, meaning that when I lost my regular pair and had to wear the react-to-light ones, I spent the majority of my time in lessons looking like a Roy Orbison tribute act.

Despite my unusual look, I had some really lovely pals at sixth form. Take Sonja, who was fully obsessed with Madonna and lived in a huge, higgledy-piggledy three-storey house. Sonja's room was on the third floor so her mum, to save her voice, kept a big brass handbell at the bottom of the stairs to ring when she needed Sonja to come down to the phone when one of us rang her.

Sonja with her poker-straight, long dark-blonde hair, who would pause and widen her eyes while pursing her lips to add extra va-va-voom to the nugget of juicy gossip she was about to impart, the facial equivalent of rolling out a red carpet for her salacious titbit to make its grand entrance.

And there was lots of gossip – Sonja came out in the first year, which we all found extremely exciting. We all threw ourselves into the big gay adventure that followed. Like Mickey Mouse in *Fantasia*, Sonja was at the centre while we all swirled around, engulfed by the excitement of her love life – late-night phone calls, bitter break-ups and passionate make-ups. We loved it. We'd all troop off to Blackpool to drink in gay bars with her, feeling ever so worldly. We loved the drama – listening, rapt, in the common room as she laid out all of her complicated affairs on the communal table and painstakingly untangled them for us like dozens of balls of wool.

Sonja's audience was made up of myself and Rebecca, who had huge brown eyes, a gap in her front teeth and dark wavy hair that reached down to her bottom. She worshipped Prince and I still can't hear 'Diamonds and Pearls' without thinking of her.

There was also Louise, crow-black, poker-straight hair, a penchant for thin black velvet chokers and as sarcastic as she was straight-talking. Think Wednesday Addams with excellent liquid eye-liner.

We'd all go to 'Flesh' at the legendary Haçienda on a Wednesday night. By now it was 1993 and the halcyon days of the club were behind it, but Flesh was still special – an intox-icating mix of gay men, lesbians, drag queens and party girls. We were the latter, drawn partly by the music, the style, the crazy crowd bedecked in leather, lace and fishnet, but mainly because we felt safe there; it was less intimidating than in the more alpha-male days of old. One of the night's promoters, Paul Cons, has been quoted as saying 'This isn't Madchester any more – it's Gaychester.'

Flesh was fun, inclusive, friendly. It throbbed with personality and slight craziness. I could wear what I wanted, even be, as the *Sun* newspaper would say, 'scantily clad' without feeling like it was a meat market – once we'd run the gauntlet past the slavering bulldog bouncers.

One week as I danced I was approached by an older girl, who slipped a flyer into my hand – she was scouting for people to be in the audience of *Dance Energy*, a 'yoof' music show on BBC2 hosted by a DJ called Normski. I went along to the recording, so somewhere there is footage of an eighteen-year-old me dancing self-consciously in the background in a starkly lit studio wearing my faux-leopard jacket.

I absolutely loved the girls I was friends with, but I found them complicated. The sands of friendship would always be shifting and from day to day you'd never know where you stood. I'd literally wave goodbye to my friends on a Thursday afternoon and we'd be all smiles and everything hunky-dory, then the next time I'd see them, on Friday morning, there'd be a frosty silence shrouding their table in the common room and they would be cool with me, exchanging knowing glances with each other. It was like starring in a soap opera where no one ever gave me the scripts.

Each day I'd walk in not knowing whether I'd be met by smiles or silence and I found it mentally exhausting. I'd been put through the mangle by the bullies at secondary school so I was extra-sensitive to animosity. Of course these girls weren't being horrible on purpose – everyone at this age was just busy working out who they were and what they were about and that involved sometimes nudging the weakest chick from the nest if they needed more space. They certainly weren't bullies, just helpless passengers clinging on a runaway chariot of their own hormones.

And I was weak – I always felt frustrated by my cowardliness, for not being able just to blurt out how I felt, stick up for myself, push back a little when people tried to intimidate me. Instead I shrank away from confrontation. I've always been familiar with the feeling of pushing my head into the sand. This is what I did with the girls – I didn't fit in with them so I slowly melted away, out of their grasp, away from their drama.

31. Plaque in the Attic

Luckily I was put out of my misery by meeting Ani, a lad in the year below me who became my first love. Ani was many things – charming, hilarious, witty and kind – and he had what I saw as the perfect family which, I'm now slightly ashamed to say, I was very in awe of and envied. I was happy at home but things weren't always easy; Mum was busy and stressed and life above the pub wasn't the most relaxing. By comparison, Ani's house was a stone-covered cottage down a country lane and his dad Jeff was an art teacher, all slightly wild grey wavy hair, slouchy jeans and Birkenstocks, hippy vibes with a bit of edge. Jeff loved to hang out with us in the kitchen, taking the mickey, cracking gags while tall gangly Ani warmed his arse on the Aga.

My mum used to laugh because customers at her pub would say to her, 'Hey, Jackie, I saw your daughters earlier, walking hand in hand.' Yvonne was away at this point, so she knew the 'daughter' in question was in fact Ani. He had straight dark-blond hair that fell below his shoulders and from the back, with his long legs and willowy figure, could easily be mistaken for a girl.

His face, though, was handsome, with bright, kind eyes and full lips, which he often gently chewed on while chatting, as if mulling over his next quip.

He was a really lovely lad with a lovely family. His mum was a petite yet formidable woman called Val, who had sharply bobbed hair and was a lawyer. She was faintly terrifying because she seemed to be able to look into my very soul whenever she asked what we'd been up to. I thought she was awesome though; she was always kind to me and the deep love she felt for her sons always shone from her, even when she was interrogating us on what time we'd got in and where we'd been. I thought at the time that they were the perfect family, even though we all know that doesn't actually exist.

Ani – a nickname acquired at birth when his toddler brother Jeremy couldn't pronounce Antony – had a bunch of brilliant friends and they quickly became my friends too. I'd found my tribe. Boys seemed a lot less complicated than girls and I loved being one of the lads. You'd never meet up after college to find that one of them was inexplicably not talking to you. The worst they'd do was nick your seat when you went to the loo and tell you when you came back that you'd been 'taxed'.

There were about eight of us who hung out and, from sixth form to when I eventually left for London at twenty, they were a big part of my life. The lads were hilarious, sarcastic, daft and strangely caring in a gruff way. There was a uniform – baggy jeans and Karrimor or Berghaus Gore-tex jackets, the latter designed for mountaineering, though the only expeditions they were used for were hanging around the chippy and scouting for magic mushroom patches on the playing fields.

No one really went by their given names, instead preferring a bastardisation of their surname – Braddy, Middy, Briggy, Fitz

– and Fads, who answered to this name even though it stood for 'Fat and dopey shit'. He wasn't particularly fat (although he did, as my dad would say, 'Have enough cover on him') and wasn't dopey, more affable. I think it speaks volumes about his lack of insecurity that he not only answered to the name but found it humorous.

We looked out for each other and smoked, played *Street Fighter III* and listened to the Kinks and the Orb in the attic of Phil's massive house in Bolton. Phil had dark floppy hair framing huge round eyes that looked too big for his slender face and he had a fine patter in hippy-leaning pseudo-philosophy. He paid homage to those years in the attic by getting the *Screamadelica* album cover tattooed on his upper arm – we played that Primal Scream record to death.

Phil's parents very much took the view that if their son was going to get up to the usual teen behaviour they'd rather he did it under their roof. And they let all his mates in too.

I spent so many hours in Phil's attic with the lads that there should be some sort of plaque there commemorating the formative years that occurred under that roof among a gang of pals, laughing and nobbing about.

I'm still grateful to those lads for, if not quite taking me under their wing, at least letting me be part of their crew. Growing up on the farm I was always comfortable in the company of men – from the lads helping out at haymaking to my big brothers and their friends to trips with Dad in his wagon, chatting to the geezers at the depots – I admired their uncomplicated friendships with each other, the easy conversations, the relaxed banter. It's no real shock that after slipping away from the dramas of my girlfriends I felt relieved to be hanging out with Ani, Phil and the rest of the lads.

*   *   *

My relationship with Ani lasted about three years and could be tempestuous – I loved the bones of him but I was immature, despite being a year older than him, and I think I found this 'love' lark a little overwhelming. I was insecure and jealous, easily threatened if he spoke to other girls, even though his behaviour was always exemplary. I was just daft and young. For a huge part of the relationship, my sulks aside, there was a lot of laughter and fun times. He was a brilliant, gentle, good lad and the perfect person to do some growing up with; together we trod that awkward path between teenagerdom and adulthood, navigating the bumps along the way. I consider myself very lucky that my first love was such a sorted person.

32. Molehills

During this time I lived about a quarter of a mile away from Phil above a pub called the Pineapple Inn. It was a stone building on the corner of Blackburn Road and a massive dual carriageway called Moss Bank Way. I've no idea why it was called the Pineapple but I always thought it was a strange name for a pub.

The public area was on the ground floor and consisted of two rooms – the smarter room, with old-fashioned table lamps on the deep windowsills and floral-covered seats, and the games room. The bar was in the middle, separating the two. There was a narrow corridor that ran between the two rooms and, running off that, the doors to the backyard and the loos, which were opposite the public coin-operated phone.

Mum and Ken were tenants in the pub, employed by the brewery who owned the bricks and mortar. Mum was landlady and Ken was the landlord – he was mild-mannered and definitely the good cop in there while Mum was, not bad cop but definitely badass cop. Lads who were pratting about and generally misbehaving would quiver as the small shadow of Mum fell across them. Everyone loved my mum and the minute she told someone to behave, they did. She may have been belly-button height to most of the big bruisers but they still did as she told them.

The games room consisted of a pool table in the far corner next to the dartboard and next to that the fruit machine and jukebox. Men would come in here for a pint on the way home from work and it was fine if they had mucky work clothes on as this room wasn't posh like the other, 'best', room. There were maroon faux-leather banquettes with matching seats, wipe-clean in case grease, booze or blood got spilled. The tabletops had a shiny dappled bronze effect, with thousands of tiny indents on the surface, perfect for collecting drips of bitter or tiny flakes of fag ash. These of course were the days of smoking in pubs. People complain that now pubs smell of stale beer and farts, but at least that can't give you cancer. Well, I don't think so. Fart cancer sounds bad.

Each table had a large tin ashtray gifted by the breweries, emblazoned with 'Tetleys' or 'Guinness' or whatever. By the end of the night, if it was busy, they'd be close to overflowing, piled high with little molehills of grey ash and dozens of cigarette butts, some glossy with transferred lipstick.

When I was working a shift at the pub one of the jobs I hated most was 'doing the ashtrays' – I'd walk around with a little red plastic bucket and a paintbrush. I'd decant the contents of the ashtray into the bucket at arm's length to avoid the grey cloud that would billow up into my face, then whizz the paintbrush around the tray to dislodge any last traces of ash so they could be piled up and taken to be soaked in the big metal, trough-like sinks. At the end of each shift all the glasses were loaded into the massive glass-washers behind the bar, whose scalding hot steam would rush up into your face; the jukebox and fruit machine were unplugged and fell silent and dark. I loved the pub when it was closed; it was like a sleeping beast – dark, calm, subdued. It was so spacious and as I unstacked the clean ashtrays and put them out on the tables I'd always play the same song on the jukebox – Bjork's 'Violently Happy'.

When the pub was in full swing, though, all the action happened in this more relaxed games room, the pool table and the customers being looked down upon by framed black and white photos of former Bolton Wanderers players, whole squads surveying them like the ancient ancestors looking down in *Mulan*.

If you could make your way through the smoke to the jukebox you could put on a tune, but only if you beat Wayne to it. Wayne was one of the regulars – he was a heavyset biker dude, complete with a ZZ Top beard that looked out of place on his young face, which was topped off with a red bandana tied on his head. He sported pale jeans and a black leather waistcoat over a T-shirt.

Wayne's favourite song on the jukebox was Meatloaf's 'I'd Do Anything For Love'. Inexplicably, the version on the jukebox was the album version, which was a face-melting twelve minutes long. He'd put his three plays to good use by selecting this song for each one, which resulted in an ear-spanking thirty-six minutes of Meatloaf. You'd always know when Wayne was in because 'I'd Do Anything For Love' would be blasting out of the jukebox; it was his walk-on music, as well as his sitting-down-drinking-beer music, popping-to-the-loo music and walk-off music for when he slipped into his biker jacket and left the pub. He certainly got his quid's worth.

Whatever was playing on the jukebox had to battle to be heard above the general pub hubbub that increased with every hour and every unit of alcohol consumed. There were some real characters in the pub, like Dave the stonemason, who by day fashioned gravestones from big blocks of granite and by night enjoyed a banter and a pint of bitter. He was a big bloke, ruddy-faced with clouds of grey hair and matching tache. On his arm was always his lovely pocket-sized partner Freda, small and curvy like a Russian stacking doll, with a big laugh and tight silver perm.

Next to them were Michael and Marlene, a couple who I thought were very glamorous; she was well-spoken, in her fifties, with expensive-looking blonde hair cut in a flattering short bob. She was very stylish and had a touch of the Helen Mirrens about her. Michael was a few years younger and his hair was receding, but he still had a decent covering of tight auburn curls that ran in a wide strip round the back of his head and a runway that carpeted the top of his head. He was a trim bloke and an alpha male – chest thrust out like someone had just put an ice-cube down his back, fond of wearing a lambswool V-neck in maroon or baby blue set off by a fine gold chain round his neck and a gold Rolex. He was a bit of a twinkly-eyed rogue – a former bad boy, past his battling days but a good bloke to have about if you needed an extra pair of hands or, rather fists. He had a gift – he could always sense when there was discord in the air. The secret signal to the rest of us that trouble was afoot was when Michael calmly took off his watch and put it on the bar so it wasn't damaged in the anticipated melee.

The Pineapple wasn't generally rough, to be fair, but there'd be occasional trouble. Too much booze combined with the posturing of the young and stupid sometimes led to fights. One morning Mum was wiping down the radiator next to the pool table, which would normally be the cleaner's job. I asked her what she was doing and she replied matter-of-factly, 'They were bloody battling last night and blood got splashed on the bloody radiator' – she said it in the casual manner of someone discussing the weather at a bus stop. That was a bit of an eye-opener – poor Mum having to do that, plus some poor bugger was waking up with a radiator imprint on their face.

There'd also be parties sometimes, big, rowdy, drunken affairs that meant loads of people and new faces flocking in to wave off a mate. But they weren't for a pal who was emigrating or off on a gap year backpacking round Oz; they were for someone who was off to prison. Their date for sentencing was imminent and they would be having a pre-nick knees-up. Inevitably we'd see the same crowd again, usually a few months later, when their mate was released.

Twice there was fighting during my shift and Mum shouted at me to call the police – usually Michael, Ken and a couple of locals would have managed to shove the troublemakers through the front door and out into the street, bolting it shut behind them, but then the idiots would hang around outside, trying to boot the front door in or threatening to put the windows through.

I can't remember feeling scared – I found the process massively exciting.

999 operator (serious and professional): 'Which service do you need?'

Me (squeaky and secretly thrilled): 'Police please!'

I was harbouring some small shreds of ambition to join the force (mainly because I wanted to be in the mounted division – another bid at finding a career that involved me sitting on a horse for large parts of it) and to me just making this call made me feel a step closer. I mean I was literally part of a live police operation!

Police operator (gruff and calm): 'You're through to the police, what's the situation please?'

Me (excited and pumped but now fully aware that I'm basically IN THE POLICE so trying to rein in the squeakiness):
'Yes, there are three people trying to kick down the front door
of me mum's pub . . .'

Within minutes we'd hear the sirens, the lads outside would
usually scarper and the games room would be bathed in a blue
glow as the flashing lights shone through the gaps in the curtains.
The local constabulary would be there, rounding up any
remaining idiots who were either too drunk or too stupid to run
off, or both. I used to love the police officers. They'd come in
for a statement from Mum and Ken, pink-cheeked with adrenaline from sorting out the troublemakers outside in the cold,
and I'd hang about, giving the bar an extra wipe down, ears
wagging.

Once the police had left, there'd be a lengthy lock-in for a
major debrief where the drama was discussed over a few bevvies
until the wee small hours. Discussed and perhaps built up a bit
– by the end there had probably been thirty-five assailants, six
helicopters and a samurai sword fight on the pool table.

On the whole, though, the Pineapple was just a regular pub.
These few thrilling instances of sirens and shouting aside, most
of the noise came from the frantic thumping of the thick end
of the pool cues against the floor in celebration of a good shot,
a rumbling noise that could be clearly heard upstairs in our
living quarters.

The pool table was put to good use by me and my friends; on
occasion we'd sneak in after hours to play pool and eke out the
last minutes of our night out while Mum and Ken slept upstairs.

Through a large black door next to the pool table was the
entrance to our home; a cold, sterile white corridor led to our
back door and the stairs that took us up away from the smells

and (most of) the sounds of the pub and into the peace of the upstairs living quarters. The set-up was, to say the least, bizarre, as if each room was a dice that'd been shaken in a giant's hand and tossed onto a table, settling in a higgledy-piggledy pattern.

Mum and Ken's bedroom didn't have its own entrance, so to get to my bedroom I had to walk through theirs. The only bathroom was an en suite that was also attached to their room, meaning they had no privacy at all, their supposedly private inner sanctum racking up roughly the same footfall as Oxford Street on the last Saturday before Christmas. This was when I was between the ages of seventeen and twenty and my social life was in full bloom, so I'd often be coming home at all times of night, walking tipsily through their bedroom as they slept, clattering around in the bathroom.

Being a parent now, I can see how annoying that was. It's annoying enough having to shoo my kids out of my bedroom and they're pretty small, but great big galumphing teens trampling through my room? No thanks.

There was a spare box room that served as an office and had a sash window, and on sunny summer days I would climb out and onto the flat roof to sunbathe. Like a burglar obsessed with soft furnishings, I would squeeze through the open window with armfuls of cushions and towels to lie on. I'd lie there on the roof of the pub, which was about half the size of a squash court with a two-foot-high wall running round the edge, and sunbathe like I was kicking back in the French Riviera, never mind the traffic of the major dual carriageway rumbling past a few feet below as I lay there in my bikini. When the sun went in, so did I. I'd always have to shower because my skin would be covered in tiny black flecks of dust and pollution from the traffic roaring by below that had settled on my skin, sticking to the slick of Mum's leftover Hawaiian Tropic I'd slathered on.

237

The lounge was at the front of the building and looked out through big sash windows over the traffic outside, and it was here I'd lie prone on the sofa, psyching myself up for a shift at the pub. I'd work on a Sunday night, starting my shift at 8 p.m. and – thanks to Sunday opening hours – finishing at 11 p.m., getting thirty quid from Mum for my efforts. In one way it's kinda jammy to live above your place of work – no schlepping around in the cold, no missing the bus to work, no shelling out for taxis home – but the downside was you could always hear the sound of the pub. I know this made it difficult for Mum and Ken to switch off. I, on the other hand, was so good at switching off that my problem was switching back on again. I'd lie on the sofa on a Sunday evening until one minute to eight; if I heard the beginning of the *Heartbeat* theme I knew I was late and I'd have to scarper downstairs before the phone trilled – Mum, calling from downstairs to chivvy me along.

I didn't really blossom as a barmaid. I hated having to work, though of course I loved the cash. I could and still can pour a decent pint and was pretty good at the maths, I just couldn't be bothered to make an effort with the social side – the banter with the customers that Mum was so good at, taking an interest, making small talk. Ironically, I'd probably make a brilliant barmaid now – I love talking to people and I'm naturally nosey, a combination that means I can talk for hours on the radio. When I've been asked how I sound so cheery when I'm broadcasting early or late, live on the wireless, I often put it down to Mum's example – no matter what was going on in her personal life or if she was tired or fed up, she'd have to put it behind her, pop on her lippy, run a comb through her hair and get behind that bar, and that's what I do on the radio (but without the lipstick – and sometimes without brushing my hair).

Most of the customers found me quite stroppy, and a few used to ask me outright why I was so miserable.

To them I was this studenty-looking girl, different from them in my vintage striped polo necks and tight cords. My hair was long and straight, not permed, and I wore no make-up, unlike the women who had made an effort and put their face on to come out for a Sunday night in the pub.

It didn't help that I've always had what's now called a 'resting bitch face', so even when I'm happy as not only Larry but his equally jubilant brothers Barry and Harry, if I'm not grinning then my mouth naturally curves down slightly, making me look like I've just smelled some very off pâté. My whole life I've been helpfully advised by builders and random men (always men!) to 'Smile' because 'it might never happen' when I've dared walk past them without a rictus grin plastered on my face. My sister Yvonne has the same facial issues, which didn't help our already fractious relationship because we'd both think the other was giving them a dirty look. I've got into arguments at school about it, all because I naturally look so miffed. When I start working with a new person now I always warn them I'm not furious, it's JUST MY FACE. The irony is that on the whole I'm not a glass-half-empty but a glass-overfloweth kinda gal. I'm annoyingly positive. Of course, I didn't know all this about myself at the age of seventeen as I stood there pouring pints of mild behind the bar, and the customers didn't know it either. It was just shyness and immaturity, but I guess it came across as arrogance, and to them I was simply a mystery wrapped in a snooty cow.

33. Grey Mafia

I should've been used to being behind the bar – I'd already worked for Mum for years. I started back in Boothstown, where Mum and Ken ran the Conservative Club. Being about ten years old, I was only back-room staff. They had a big upstairs room where people would throw wedding receptions and retirement parties and, as the guests couldn't live on warm Chardonnay alone, Mum would do the catering. I loved helping her.

I would spend hours finessing a large plate of crudités and salad; I was famous (in our house anyway) for the intricate arrangements of cherry tomatoes I would produce on huge foil platters, creating fabulous patterns with fresh veg, a mandala made entirely of your five-a-day.

I loved being Mum's sous chef and going with her to collect the hot pasties from Carrs, a famous Bolton pastie shop. We'd go in Mum's little Austin Metro hatchback and she'd put the big plastic tray of pasties in the boot. The heat from the pasties would steam up all the windows of the little car and the incredible smell of the buttery shortcrust pastry, potatoes and beef would fill the car so it was like we were driving along in a pastie on wheels.

I would wipe the drool from my chin with crossed fingers, hoping for a cat to run out, causing Mum to slam on, hurling the pasties forward and breaking one in two so we could have half each, but it never happened. It was torture.

Mum would dish out the pasties with steaming pale mushy peas. I still love Carrs pasties; I always have some in the freezer and served them at my wedding reception a few years ago and they went down a storm.

From Boothstown we moved to Astley Bridge in Bolton, where Mum and Ken took over as steward and stewardess of the Conservative Club there. I used to work there too, promoted from tomato artist to the dizzy heights of glass collector.

It was during these years that I used to joke that I was raised by Post-it notes. Every afternoon when I got home from school, Mum and Ken would be at work and in their place would be a paper trail of yellow squares instructing me what I needed to do.

They were always in Mum's big writing – she writes as she talks: loudly. From chores ('Sarah please wipe up' – stuck to the draining board) to instructions ('please feed dog') and the nicer ones ('Chilli for tea, zap in mic') and then at the end of this yellow informational bunting 'Luv you' – 'love' spelled in that colloquial way that she still does now, much to the amusement of my kids. As a working mum I now fully appreciate the luv she showed by leaving those notes. Even in absentia she was still looking after me. She worked really hard but she made sure I didn't come home to an empty, cold house. It was warm and lit against the winter dusk and those notes showed she loved me and she was there with me, in paper form. Now I do the same for my teenage daughter in the modern version of Post-it notes, texts. Her phone buzzes with my maternal love – checking what time she's home, dinner instructions, homework nudges, signed off with a 'Love You'.

Mum and Ken worked hard at Astley Bridge Conservative Club, which was a red-brick cube with a blue stripe round the top and one round its middle, like an iced cake. The building squatted beside a flagpole just a few metres away from our front door. The only thing separating it from Moss Bank Way, a busy dual carriageway, was its square car park, tucked in the front of it like a napkin.

Mum and Ken ran the club on a day-to-day basis, serving behind the bar, keeping the cellar ticking over and doing the books, but it was actually ruled with an iron fist by 'the committee'.

The committee were a gang of elderly men who were retired and clearly missed the cut and thrust of industry. Instead of spending the autumn of their years travelling, playing bridge or trying Pilates they instead became a kind of grey mafia, wielding their power over my mum with all the vim and vigour they should've saved for a crown green bowling ball.

Like the mafia they were all well turned out, in smart grey suits, and had glamorous septuagenarian wives on their gold cufflinked arms, who were the twinset and pearls crew. Also like the mafia, they revelled in their notoriety and could be ruthless smiling assassins, the cabaret lights glinting off their dentures as they grinned at you before stabbing you in the back.

Mum and Ken worked all the hours God sent at that place but never felt they got the recognition or respect they deserved from the committee, who would often override their decisions. The customers and staff all loved Mum and Ken, though, and that's what mattered to them.

*   *   *

Downstairs in the club was a large, smart room with plush seats and pale soft furnishings, and the sports room, which was huge, containing two full-size snooker tables, always with clouds of ciggie smoke hanging over clusters of men gathered around the tables in pastel V-neck jumpers.

The rooms were separated by the hallway, from where you could walk up the sweeping blue swirly-carpeted staircase past a huge gold-framed photograph of Queen Elizabeth II (the 1960s glam brunette version) and into the upstairs concert room.

This was where all the glitz and glamour happened. During the week it lay quiet and still but at the weekends it sprang into life. It was a large oblong room with all the seating favouring a large stage, painted black, at the end with a well-sprung dark wood dance floor in front of it and a long bar running along one side and windows along the other.

Weekends it would be packed with customers dressed up to the nines; women in their best dresses and jewellery, hair blow-dried to glossy perfection, lipsticked and powdered to the max. There really was an air of old-school Hollywood glamour about the place – sure, a lot of the customers were of a more impressive vintage than any bottle on the wine list, but they could still turn on the glitz come Saturday and Sunday nights.

The man in charge of making sure everything went with a swing was Paul the compère; in his early fifties, he had hair dyed jet black and sported a shiny black shirt that was tucked in at the bottom to gently wrestle his slight paunch into submission, and unbuttoned enough at the top to show a suggestion of manly chest hair to thrill the laydeez.

On Saturdays and Sundays before the evening's live entertainment commenced, Paul would wheel an old-fashioned bingo ball out onto the stage. It looked like a wire globe with a handle to turn it at one end and multicoloured balls in the middle.

Everyone would produce the bingo cards they'd purchased on the way in and it was eyes down for a couple of games before the cabaret began.

I thought Paul was awesome – he commanded the stage, striding up and down, warming up the audience before the other acts came on. He was a TV repair man by day but by night he was a star – a club singer and minor celebrity in his own right. Paul always had a touch of the Vegas about him – Astley Bridge Conservative Club was his Caesars Palace.

Paul always included Bruce Springsteen's 'Dancing in the Dark', The Searchers' 'Love Potion Number Nine' and Elvis's 'Suspicious Minds' in his set, accompanied by the house band, a couple of blokes on keyboard and drums.

Paul would break a sweat by the first chorus, face red, veins in his forehead throbbing with the strain of conjuring up the Boss and the King in one set.

Once Paul had done the opening song and patted down his brow, he'd introduce the first performer – maybe a female singer in a sparkly trouser suit and big hair, singing everything from Tyler to Turner, a local stand-up telling jokes about the missus and her mum – 'Take my mother-in-law – no, really, take her, please' – or a magician pulling stuff out of a hat.

Refreshment in between sets could be found at the bar, where Mum and her staff did their best to keep the booze flowing. Sometimes Mum would do pasties and peas to help soak up the drinks. By the end of the evening the room would be swinging and a lot of couples would take to the dance floor as Paul closed the evening, busting a gut to give it some welly on stage with another couple of big songs before the main lights went up, pint pots and Slim Jims were drained and the audience wobbled unsteadily down the stairs and out to their waiting minicabs.

Peter Kay always says his toughest audience is playing in

Bolton, his home town – everyone loves him but the funny stuff he says is almost too familiar to a local audience. I know what he means – cabaret nights at the Connie Club were basically the Phoenix Club.

My job at the Connie Club, because at fifteen I was too young to work behind the bar, was stacking the empty glasses in great swaying towers in my arms, squeezing between the seats at the weekends when it was busy, trying to appear invisible to avoid a stand-up making a wisecrack about me.

I enjoyed it – I loved having a few quid in my pocket and loved watching Mum run the upstairs bar like a conductor directing an orchestra. She kept everyone happy, customers and her bar staff, darting around, a blur of streaked permed hair and coral lipstick. At the weekends she was in charge of the big bar upstairs while Ken ran the downstairs rooms. Mum knew all the customers by what they drank, like Dad might tell his cattle apart by a dark patch of fur near their eye or a very curly face. 'You know, Jim,' she'd say, 'half a mild,' or sometimes it'd be 'Jenny, frizzy hair, G&T.' I worked hard, the glasses smelled of stale booze dregs and the main room upstairs was smoky and noisy. On cabaret nights it was always rammed and I'd spend the whole night running back and forth with armfuls of glasses, loading and unloading the glass-washing machines, replenishing the shelves with clean glasses.

My undoing here wasn't to be my work ethic but my dress sense. Or, some may argue, lack of sense when it came to how I dressed. I had the attitude of a regular fifteen-year-old, wearing my budding sexuality with a casual nonchalance. I had long slim legs up to my armpits and long dark-blonde hair down my back but wasn't really aware of how I looked. The bullying at school had undermined my confidence and I didn't realise that I was blossoming into a pretty young thang. This fact hadn't escaped the notice of the committee wives, though, and my mum was convinced that they had a hand, or rather a voice, in my dismissal.

For my part, I didn't see anything wrong with the outfits I wore but this was the reason the committee gave when they went over my mum's head and sacked me. My offence? Wearing hot pants. It remains one of my favourite sentences – 'I was fired for wearing hot pants.'

The committee dealt with it by informing my mum after a meeting that I was no longer employed as a glass collector. I love that nestled among the usual business to be discussed in the meeting – profit margins, proposed renovation of ladies' toilets – were me and my hot pants.

I loved hot pants and used to sport them with thick black tights, chunky shoes, a white shirt with a black sweater over the top and shirt cuffs pulled out in the dandy style that was all the rage in 1989. I thought I looked marvellous but I probably looked like a weird Dickensian pageboy.

What was the crime exactly? Malicious flashing of thigh? Distraction by hot pant? I did my job well enough but I imagine the committee wives with their crispy duck décolletages didn't appreciate someone on the cusp of fruitful womanhood bending over tables in front of their old husbands and jeopardising their blood pressure. I was completely oblivious to all this; I was just trying to reach the empty glasses.

We laugh about it now but back then Mum was fuming at the treatment we got from the committee and it soured her relationship with them beyond repair. A year or so later we left Hotpantgate and the committee behind when Mum and Ken decided to go for tenancy at the Pineapple pub, approximately 100 yards away, where I would go on to become the world's sulkiest barmaid.

34. Running on Fumes

It was just before we left the Connie Club that I started to learn how to drive – after the Scotland débâcle we'd finally scrapped the Renault 12 and were now the proud owners of a shiny second-hand Austin Metro in navy blue. This was to be my ticket to the highways and byways of Bolton, my ticket to freedom.

My teacher was a man called Derek who also drove an Austin Metro, which was clearly the car to be driving in the very early nineties. Derek was a nice enough chap who made cheesy quips and laughed too loudly at his own jokes, his guffaw escaping from underneath his neat black moustache. He would pull up outside our terraced house and toot the horn of his little red car, which was plastered in L-plates and had paintwork decorated with ads for his driving school. Every time I got in the car he'd reach in his top pocket for a little tin of breath mints, which were small, round and white. Barely able to suppress his chortling in anticipation of the joke he was about to tell, he'd hold out the mints and ask, 'Do you want an E?' Without fail I'd

247

politely decline and he'd burst into fits of laughter before popping a mint in his mouth and needlessly reassuring me they weren't actually Ecstasy tablets.

It was the rave era and Ecstasy was occasionally hitting the headlines and this was Derek's peculiar attempt to be down with the kids. Still, he was patient and relaxed and, recreational drug quips aside, he was a good teacher.

Mum was keen to help me pass my driving exam because the fewer the lessons and the fewer the tests the less it all cost. Plus once I could drive, just as I'd be free of having to beg her for lifts she'd be free of being begged to run a one-woman taxi service. Thanks to Dad I'd been able to drive a tractor since the age of twelve, so I think I had already fine-tuned my spatial awareness and confidence with manoeuvring a vehicle. Plus a David Brown tractor was massive compared to a Metro, so how hard could it be?

In fact, not very hard as it happened. I loved driving and I'm forever grateful to Mum for her patience, bravery and generosity in donating both her time and hard-earned cash to the cause. It seems completely mad to me that anybody would be willing to sit in the passenger seat as their teenage offspring takes charge of one and a half tons of metal, commandeering it around the roads among other cars filled with people, but that's what she did and what loads of parents do – and what I'll be doing at some point. Eek.

So Mum and I would set off for a twenty-minute pootle in the Metro – out onto the dual carriageway to do a circuit of Astley Bridge, Mum only occasionally slamming her own right foot down on a non-existent brake pedal in the passenger foot-well, her instinct to survive overriding the fact she was in the wrong seat. We'd return home to Manley Terrace and Mum had a wee victory catchphrase she loved to yell as we turned in to the street – 'Slowly into Manley!' she'd exclaim as we rolled round the corner and came to a halt outside the house.

Dad used to let me have the odd go in the jeep on the lane, where I'd bunny-hop the car along because I couldn't master the clutch control; he'd claim I didn't have petrol in it but 'kangaroo juice' because of the big uncontrollable leaps we'd make. The tractor and the jeep, but mainly Mum and Derek's hard work, meant that after eleven lessons I passed my test. I was so thrilled. I loved driving and still do; I've always loved the open road. Maybe I link long journeys back to Dad and the only time I got him all to myself, when we were out in the wagon.

When we go up north or drive for a UK holiday now, it's me who does the majority of the driving. Once the kids are set up with their own in-car entertainments, I pop Jeremy Vine on and I'm off. With my wagon-driving genes, I want to get as many miles under my belt as possible before stopping for anything as indulgent as a loo stop or leg-stretch, though eventually I have to give in to the demands of my family. It's on those journeys that I always think I could do with one of 'Len's Ham Pockets'.

Once I'd passed my test I was allowed to use Mum's Metro. All I had to do was follow the three golden rules – keep the car tidy, absolutely no smoking in the car, and always keep the petrol topped up.

Rule one was regularly flouted, rule two only very occasionally, but sadly rule three, due to lack of funds and common sense, was impossible to obey.

There would be many times when I'd get a rollicking because I'd left the car, in Mum's words, 'running on fumes', which was bad for the engine and for Mum's mental health. I got away with a lot. I only hope there are some crumbs of comfort for Mum in the fact that all this annoying teen behaviour is now being inflicted on me by my own daughter – yes, it's started already, the mess, the 'borrowing'/losing of my stuff. The universe is dishing up a steaming hotpot of *Schadenfreude* for my mum now as I'm left picking up after my offspring and picking my battles. As Elton John sings, 'It's the circle of life'.

Despite me making it run on fumes the Metro still managed to get me to Ani's house one New Year's Eve, when sadly it was to meet its demise.

On my way home to get ready to go out and celebrate, I turned right out of a small road. It was dusk, light was fading and the light-coloured car seemed to come out of nowhere; I just remember that in the split second before impact we had eye contact. I noticed she had her hair neatly wrapped in big Velcro rollers. She walloped straight into the driver's side with an ear-splitting crunch, pushing me and the Metro a few feet up the road. I was lucky for two reasons: she wasn't going fast and she ploughed into the middle of the side of the car, so the stronger bit between the front and back doors took the full force of the impact. If she'd hit the driver's door it would've been a different story.

I managed to get out of the Metro but it was a write-off – the roof had buckled upwards when the impact had crunched into the side. I felt terrible for the girl, who was en route home from the hairdressers prepped to hit the town for New Year's Eve. She and the police were very nice; it was my fault but it was an easy mistake to make when visibility was low. Mum arrived and burst into tears when she saw the state of her car – it did look pretty bad. I didn't even get shouted at; she was just relieved I was in one piece. She hugged me so tight I'm sure she dislodged a kidney.

It's moments like these with Mum that have made an

indelible mark on me and my parenting style. My kids know I shout, I snap, I rant, I grimace, but they also know that all I want is for them to be OK and that when it comes down to it, they're my number one. Mum worked hard and loved us so much and I feel more and more, as my kids grow, that as close as I was to Dad it was Mum who was always there. Like you don't notice air but you breathe it, Mum was always there even if I didn't always notice.

By night, barmaid, by day, average A level student. I got four A levels and left sixth form, but not before the leavers' do at a local hotel. Everyone wore their best frocks and black tie. It was to be a huge bash and I wore a long white John Richmond dress that was V-neck, clingy and sheer, like a T-shirt that had got mangled in the wash and ended up ankle-length and with a split. On my feet I sported some red espadrille wedges from Red or Dead, one of the coolest labels of the time. I felt lovely and thought I looked OK too. I wore my straight, dark-blonde hair down and just a bit of make-up. When I look at pictures from that night I think I look gorgeous but I was never terribly confident in myself – I still hung out with my crew of lads, was more comfortable in jeans and trainers, and the marks left on my psyche by bullies in the years before were deep and ingrained, so I didn't trust people easily, especially some of the girls at sixth form, who always seemed to have a snide remark bubbling just under the surface of their fake lip-glossed smiles.

I was right to be wary; some of these girls had come up with a mini awards ceremony that was to happen during the evening – it had the usual prom king and prom queen awards but then they announced the next award – the 'I love myself, somebody has to' award – oof! I thought, that's a bit cruel. I wouldn't want to be on the receiving end of THAT accolade. Then they read my name out and everything went a bit slow-motion: the look of anger on Ani's face, people's heads swivelling to see my reaction as my face burned with humiliation. I remember my head swimming as I went up to the stage to collect the award that nobody wanted; I was mortified – if they thought I was vain and loved myself they couldn't have been more wrong.

Making my way back from the stage compounded all my feelings about sixth form and the cruel, shallow, bitter clans of girls who could make people feel rubbish about themselves with a click of their French-manicured nails. If that's what they needed to do to validate their own feelings of lack of self-worth then they were the weak ones, not me.

Many years later when I became the 'Pretty Polly girl' – a modelling contract with the hosiery firm that in the nineties was quite a big deal – I felt like it was a bit of a fairytale ending, this ugly duckling becoming a 30-denier swan; it was a big two fingers sticking up at the bullies who made fun of me for everything from having long skinny legs and big lips to daring to look nice at the leavers' ball. Here I was suddenly being celebrated for the things that I was previously mercilessly teased for – not bad for a rubber-lipped crazy-legged crane who loved herself because someone had to.

35. Rhymes with Carer

I waved goodbye to sixth form in 1993 aged eighteen. I wondered what to do next. I had the whole summer to decide. My sister Yvonne was studying for a degree but I didn't much fancy following in her footsteps – from what she said she was constantly skint and she had holes in her shoes and there was fifteen of them living in a biscuit tin eating potatoes. Or something like that. I flirted with the idea of being a journalist and if my arm had been twisted I probably would've tried to study media somewhere.

Luckily fate stepped in and my plans changed, thanks – ironically enough – to Yvonne. Despite not feeling particularly inspired by my sister's uni experience, I wasn't daft, so when she went on a placement as part of her French degree to work in Paris and when she invited me to visit, I leapt at the chance.

I flew to Paris Charles de Gaulle and found my way to Yvonne's place, on the third floor of a narrow grey brick building in Montmartre. The apartment was too small to swing a *chat* in but I loved it. High ceilings, cream walls with ornate cornicing, wooden floors and a tiny wrought-iron balcony pinned on the front like a nose piercing.

This was the first time I actually started to realise there was life beyond Bolton; up until then Yvonne's need to backpack, InterRail and see the world was beyond my comprehension. Bolton had everything you could ever wish for, I'd thought; why would you want to leave? It was massive, all our family was there, it had a Body Shop and a Greggs and a drive-thru McDonalds was planned for nearby. It was Utopia.

During her uni holidays Yvonne was always off – from teaching English in Russia to travelling round China and Hong Kong, trekking in Nepal and surfing in Hawaii while I stayed in Bolton, working at the pub and hanging out in Phil's attic.

So Paris was a real eye-opener. Yvonne had a job to go to every day, working at the French Electricity Board, leaving me in the apartment armed with a door key and a metro map. This was before the days of mobile phones – nowadays you could pop any destination into Google Maps and your iPhone could basically be your nanny and tour guide all day, but back then I really was on my own.

I set off into the streets of Paris with only a very light smattering of GCSE French and a gung-ho attitude. Every morning I popped to the same patisserie for a pain au chocolate to eat en route. I insisted on ordering it in my pidgin French and every morning the same stroppy woman who served me replied in cross English, fury etched on her brow as if I'd just spat on her best blouse. But I loved the whole experience. Munching my breakfast, I'd set off on my adventures – art galleries and mooching along the banks of the Seine. None of it fazed me and I began to think maybe there was something in this travel lark. In Paris as I strode along I could've been anyone, not just a sulky barmaid.

At night-time Yvonne would whizz me out on the town and

we'd go for dins and dancing. At this distance from home she and I reinvented our relationship; we were no longer squabbling sisters but friends hanging out. In cabs Yvonne would direct the cabbie and get chatting to them and invariably they'd think she was French-Canadian – her grasp of the lingo was so good they presumed it was her first language. I remember feeling proud as punch of my sophisticated sis. She took me to her favourite restaurants and made me taste mouthfuls of beef that melted on my tongue, washed down with deep red wine. We blagged our way into Les Bains Douches, the city's coolest nightspot, where we danced – and shared a drink, seeing as they were fifty pounds a pop.

I was only there for a few days but it was life-changing. I felt like a veil had been removed from my eyes and suddenly I could see the big wide world out there that was mine for the taking – all you needed was a couple of quid for a train ticket and enough chutzpah to find your way around a bit. Then, on my last day in Paris, something happened that set me on the road to where I am today, though of course I didn't know that at the time.

I was out window-shopping, wandering around the city in the late spring sunshine. Career-wise I had no idea really what I wanted to do; I was ready and willing to be blown in any direction, like a feather. Fate happened to waft me into a small boutique, where I looked at the rails of clothes I couldn't afford. I felt the shop assistant sidle up to me from his perch behind the till.

He looked at me closely – he was a petite, slender guy with a huge flick of peroxide hair over one eye. In a strong French accent he asked "Ave you ever thought of being a model?'

Now, that line has got to be up there on the 'Lines that should make alarm bells ring' list, alongside 'Will you help me find my puppy?'

I fought the instinct to karate-chop him in the windpipe as he didn't seem remotely predatory – he hadn't exactly whipped a leopardskin rug out and asked me to get naked on it.

What he did whip out was a business card with the name of a Parisian model agency on it. He told me to call them as, according to him, I "ad a very strong look for ze moment".

I must confess I was surprised as I wasn't swimming in self-confidence, but, like a lot of girls, I'd occasionally wondered to myself if with my slender figure and decent height I'd be able to model – the nineties were the decade of the supermodel and I rather fancied not getting out of bed for less than ten grand thankyouverymuch.

Every year since I was about thirteen, for my birthday treat my mum and I had gone to Clothes Show Live at the Birmingham NEC when the TV fashion juggernaut rolled into town every December. It was the ultimate girls' day out and we loved it. The huge aircraft-hangar-sized exhibition space was taken over by clothes stalls, shoe shops, hairdressing demos, catwalk shows and modelling agencies scouting for the next big thing.

Models One, Select Models – all the biggies – had stands with people holding clipboards, subtly scanning the crowds of girls mooching around, hoping to spot a cheekbone you could slice cheese on or someone sporting an overstuffed sofa where their lips should be. These scouts were looking not for ready-made supermodels but for potential; in the right hands that gangly fourteen-year-old with the knock knees and lank hair could be on the cover of *Elle* and the catwalks of Milan within six months.

I was that gangly fourteen-year-old at the NEC but I was clearly yet to blossom even to the potential stage – no matter how many times I walked past the scouts they failed to spot my model good looks! I cringe now as I remember trying my best to nonchalantly sashay past the Models One stand, bracing myself, convinced that any second I would feel a hand grab at my arm as a clipboard-wielding person with an edgy haircut and a flick of eyeliner would stop me to pluck me from my Bolton life and catapult me to fame and fortune as I became the next Kate, Naomi or Linda.

If I'd told Mum she probably would've marched me up there, blinded as she was (and still is) by maternal love, believing I'm beautiful. Even now when I ask her opinion on an outfit she still proudly proclaims, as she's done since I was a teen, that I'd 'Look good in a bin bag.'

So this Parisian encounter was really rather exciting. To me the chance to model wasn't exciting because I was particularly passionate about fashion – rather, it would give me something to do other than become a university student.The clock was ticking on my summer and I'd need to decide what I wanted to do; a contract with a modelling agent could provide the perfect distraction for a couple of years.

I rang the agency that evening and it was decided that as I was leaving Paris the next morning I would go to their sister agency back home. MMA – Manchester Model Agency – was situated on the second floor of a beautiful Edwardian baroque building on the stylish St Ann's Square. The agency was run by an incredibly glamorous woman named Mavis who had a fondness for dusky pink, its hue liberally slathered over all the walls and soft furnishings in the waiting area. She had poker-straight, creamy blonde hair, with a fringe framing twinkly bright eyes. She was in her early fifties and behind the sweet, candy-pink-lipsticked smile she was clearly a tough businesswoman, managing to run her independent agency despite competition from other huge local agencies.

She wore half-moon glasses on a gold chain round her neck

and she'd pop these on her nose in order to peer closely at whomever she was chatting to. You always felt like she was scanning your face, weighing up strengths and weaknesses in your features. It took me back to being in the ring at the cattle shows, except this time it was me being studied, not the cow at the end of the rope I was holding. Like the cows, I would also be judged against the specimen next to me – was she thinner, taller, more eye-catching, more unique? The best of breed?

The first time I met Mavis, I waited with sweaty palms until she came out, introduced herself, and had a good nosey at my face. She thought I had potential and said she would allow me onto her 'modelling course', which involved me paying £150 for make-up and beauty tutorials and a photoshoot, the pics from which would be the beginning of a portfolio that could then be used to get work. I'd read enough articles in *Just Seventeen* to know that paying for a shoot was actually not necessary and that a reputable agency wouldn't expect any money up front – but I didn't care; I liked Mavis and found it all very exciting, plus she'd recoup the costs of the shoot from my first earning as a model – I mean it was fine, £150 wouldn't make much of a dent in 10k anyway.

The shoot was set up with a local photographer and as instructed I duly turned up to his Manchester studio clutching a bag of various 'looks' – including a clingy black party dress that we shot in a dimly lit studio like I was in the spotlight on stage at a smoky jazz club with my hair in big, bouncy curls, and a more casual striped polo neck and velvet trews combo that I wore while being photographed leaning nonchalantly on the railings of a bridge stretching over the Manchester Ship Canal, staring moodily down the lens.

After a long day of posing for pictures we had enough to put together a composite card, a postcard featuring several shots of me in various poses and moods – cheeky grin to sultry sulk – plus my name and measurements. These cards could then be posted out to potential clients and left behind at castings. The photos were also put into my portfolio – a big black book that would soon hopefully contain magazine pages of my photoshoots but for now just had my test shots in it.

The day arrived when I was to go in and collect my composite cards. I was well excited. The photos looked great to me, black and white shots arranged in a little collage with my measurements written in white on a black background in the bottom left corner: bust 34, waist 23, hips 34 and, above that, my name in bigger letters. Except . . . hang on. That's not my name. My name was Sarah. With an h. Rhymes with carer. But the name in gleaming white print on my new, very expensive cards was Sara, no h. My h had been haccidentally dropped.

I knew there was no way I could afford to get them redone. Of course, nowadays, being all growed up and knowing my rights, I would simply demand they be redone with my name spelled correctly as it was their mix-up. But at eighteen I just accepted that I would now be known, for work anyway, as Sara Cox.

I'm pleased in retrospect that I was too young and daft to get the spelling corrected. I still work under the dropped h and my maiden name now. There's something rather lovely about there being two of me. There's the Sarah that my family and friends know, which means the Sarah written in birthday cards from my mum is different to the Sara that's occasionally fallen prey to hideous articles in the *Daily Mail*. It's easier to distance myself from any hurtful comments if it's work Sara instead of me, mummy, wife, daughter, sister, friend Sarah. So I owe a big thank you to whichever Manchester printer managed to spell my name wrongly all those years ago.

36. Pretending to Fish

So I was now Sara Cox, model. I had my cards, my portfolio and a landline. I was good to go. Watch your backs, Kate et al. All I needed to do was wait by the phone above the Pineapple pub and attend castings when the calls came in and I'd be on my way.

Our phone line, sadly, was shared with the payphone in the pub, so sometimes when I was talking to MMA I'd get a half-cut customer downstairs interrupting, all confused because they'd be trying to order a taxi into town and could hear people talking on the line. No matter. I rang MMA most mornings, checking in and asking if there were any castings going on. Usually the answer would be a resounding 'Nope' from the booker Georgina, a pretty glossy-haired brunette who manned the phones and was Mavis's right-hand woman and all the models' protective big sis.

Sometimes I'd pop into Manchester and swing by the agency instead. It was easy to spot models once I'd become one of them; they'd always be smoking and carrying a portfolio in their arms – they're an awkwardly big size, too big to go in a regular bag

– and would be wearing black, straight hair not troubled by perm and face largely untroubled by make-up.

It seemed as I walked through Manchester that every third girl was a model, and that was the problem; the city had a handful of big agencies, a lot of models and very few jobs. It was hugely competitive. Manchester is a very cool city but it wasn't the capital of the fashion world – that was the actual capital. Unable to just uproot and move to London, I joined the dozens of girls at every casting. Sometimes it'd be a mixed casting with the male models. We'd all be herded into the back room at MMA and, once cocooned in the dusky pink womb with the dusky pink carpets and walls to wait, it always struck me how different the male and female models were. All the girls would be clustered together in little groups talking about how fat, short or spotty they were. Bear in mind the average size was an 8, the average height 5' 9" and the skin was the dewy fresh stuff usually sported by your average nineteen-year-old.

Still. The girls loved to put themselves down. The boys, however, jostled for position in front of the row of illuminated mirrors in the back room, slicking down their hair, pouting at their reflections and generally loving themselves. To quote my mum, if they'd been made of chocolate they would've eaten themselves.

They had no insecurities, unlike us girls, who held our own self-criticism Olympics every time we gathered together.

The casting would begin and one by one we'd be called out to walk down the pink-carpeted catwalk in front of Mavis and the client. Again, faintly reminiscent of the cattle sales I'd frequented just a few years before, except instead of wellies I'd be wearing pointy ankle boots.

It was always slightly awkward because the catwalk must've only been about two metres long. It was a T-shape, so you'd walk out and turn right along the top of the T, then hang a sharp left and head down towards the end, but it was so short you always felt as if you might walk right off the end – more gangplank than catwalk.

The lucky ones would be shortlisted and the rest would be dismissed, heading back out into the Manchester drizzle, clutching our books and sparking up a consolatory ciggie.

My first job was for the cover of not *Vogue* but a travel brochure for a local firm. I went to a tiny studio on the outskirts of town, where I was slathered in fake tan, my hair gelled back off my face as if wet and heavy eyeliner applied. I had to wear a foul, Versace-style strapless one-piece costume that was patterned with big gold swirls. I was instructed to pose with my hands on my head as if I was being held at gunpoint and to close my eyes as if luxuriating in the glare of the hot Mediterranean sun rather than shivering in a draughty studio in Ancoats. The photographer was middle-aged and grumpy, barking instructions at me and clearly frustrated by my amateurish attempts to pose and pout.

I remember getting home that night knackered; holding an unnatural position while trying not to shiver was surprisingly tiring. It wasn't a very jolly or glamorous introduction to the world of modelling but that was probably for the best – it was at least a realistic introduction to modelling. I got paid maybe £80 for that shoot, which to me was riches, at least three shifts at the pub. When the brochure came out I still put it in my book – a cover is a cover, after all – but it wasn't the best; I didn't look like a hot Spanish señorita but more like I'd popped my mum's old cossie on and smeared gravy browning on me.

Other scraps of work came, including me prancing around in pigtails and checked shirts, pretending to fish by a pond for a fashion shoot in *My Guy* magazine, but on the whole I lounged around at home waiting for the phone to ring.

I drove Mum barmy; she'd come upstairs from working lunch-time in the pub and I wouldn't have lifted a finger all morning. Not one dish would be washed – instead I'd be watching telly and painting my nails, because of course a spot of self-maintenance is hugely important when you're a top model. 'Sarah, can you wipe up please?' she'd ask through gritted teeth. 'I can't, Mum,' I'd explain slowly, waggling my fingers at her, 'I'm DOING MY NAILS.' I'd say it as if it demanded the same understanding as 'Sorry Mum, I can't, I'm on a conference call with the UN about solving world poverty.'

Mum was surprisingly patient with me – this overgrown teen-ager lolling about like a long-limbed sloth while she continued to zip around like, as she used to say, a 'blue-arsed fly'. She'd remark, 'If you were any more laid-back you'd be horizontal.'

I was still working some shifts behind the bar, so at least I wasn't cadging money from her – I was earning enough to get into Manchester occasionally for castings and a mooch around Afflecks Palace, a grand building containing a slightly musty-smelling, multi-level indoor market, housing stall after stall of fabulous stuff and tantalising tat – a mixture of new young designers, second-hand clothes, entire stores stacked with indigo blue jeans and stalls selling crazy shoes in experimental materials like spiky black rubber stilettos and cherry-red platforms with smiley faces or glow-in-the-dark plastic bows on the toes, more like works of art than footwear.

Afflecks Palace was a Manc mecca. Like goths to a flame, students, indie kids and floppy-haired ravers made the pilgrimage from all over the north-west, collectively swooning at how cool it was – the exposed pipework, the sleepy-eyed stallholders sucking on roll-ups, the facial piercings, poster shops, display cabinets of bongs and poppers, vinyl stores, tattoo parlour, the mosaic on the wall outside stating 'On the sixth day God created

Manchester' – all of it was achingly cool, something that always made and still makes me feel completely out of my depth. The cooler a place was the more I used to feel myself shrink into my shoes. I've always had trouble blending into my surroundings if those surroundings aren't a farm or cattle show, and both those things are admittedly pretty niche.

I've always been more comfortable in my scruffs and my footwear of choice is wellies, trainers or boots over a heel any day. In fact, I've never been able to slink about in heels – I can't physically stand up or walk in them. I'm convinced I'm missing the gene that allows one to sashay in a towering heel. Instead I walk with the wide-legged gait of a cowboy who's had his usual boots nicked and has been forced to pop on a Jimmy Choo. The fact I chose to stick with modelling for two years – the one career where you're absolutely required to dress up and walk confidently in heels – proves just how determined I was to avoid getting a proper job.

Although I still can't walk the walk I'm getting better at talking the talk and not being intimidated; I can now throw my shoulders back and meet people's gaze with confidence – largely because one of the joys of growing up is you don't give as much of a monkey's about what people think of you, and also you're a little more cynical about the perception of cool. I now don't think anybody is all that 'cool' – once you've reached your forties you're wise enough to know we're all racked with the same insecurities and anxieties. Everyone is blagging it, just running around like ants in an anthill.

Back then, though, at eighteen, I always felt awkward, like a baby giraffe in a Puffa coat. I'd stroll around Afflecks Palace, clanging up and down the industrial metal stairways, my pub money burning a hole in the pocket of my flared jeans.

I bought quite a few interesting pieces there, and by interesting I mean hideous and by pieces I mean abominations to fashion – like a battered, dark-brown leather men's blazer with a paisley lining that came with a Ventolin inhaler still in the inside pocket. I wore this with pale denim parallel jeans – I basically dressed like Terry from *Minder*. Another beauty, purchased during my college years but still worn, was a second-hand faux fur leopard-print jacket that had alternating panels of tufty fake fur and flaking leather. It was tatty but I liked it. I must have looked like Bet Lynch after she'd been run over by a council gritter.

At one point during my A levels I lived back at the farm for a few months with Dad and his now wife and used to drive Dad's banged-up Daihatsu jeep – a deep-green bone-rattler with no windows in the back and a bench seat running along the sides in the rear, a real farm vehicle. I used to go and collect Ani and our mates and squeeze about six lads into that jeep, occasionally moving a bag of cattle feed that Dad had left in the back to make room.

I remember relishing the funny looks I'd get at traffic lights when the occupants of the car next to me would spot not a flat-cap-wearing farmer type but an eighteen-year-old girl in a leopard-print faux fur coat and round, pale-pink John Lennon glasses. I loved that jeep and I only mention that it was banged-up because it was me that banged it into the disused milking shed when reversing at high speed out of the farmyard one night after a row, fuelled by diesel (the jeep) and teenage angst (me). Sorry, Dad.

I only managed a few months at the farm before returning to my mum, tail tucked firmly between my legs. I loved being with Dad, I liked making a fire every night and didn't even mind the extra housework chores I had to muck in with – but it wasn't meant to be.

The old two's company, three's a crowd thing really resonated at the time. I for one loved living at the farm and for the first couple of weeks it was great, but then the cracks started to appear: some were subtle (frosty silences, being ignored) and some not so much (Lulu's latest hit single 'Independence' being played loudly and repeatedly on the battered hi-fi in the lounge).

I tried to be a helpful, positive member of the household – essentially I kept my head down – but it proved too much and Dad conceded that it was best if I left. I agreed – I loved Dad so much but I could see he was sad to be caught in the middle of the two women in his life and I didn't want to cause him any more grief. I moved back to Mum's.

My mum wasn't remotely surprised that I came back to my cushy life with her – you don't know what you've got till it's gone and all that. I returned to her bosom and lapped up her unconditional love, Post-it notes, central heating and packet chilli con carne, perhaps fully appreciating for the first time what side my barmcake was buttered on.

I'd always been a fully paid-up member of the daddy's girl club but things in life aren't black and white. As a teenager I think I viewed my family members as you would characters in a film – there were goodies and baddies. Dad was always the goodie and Mum, a lot of the time, though my guilty fingers falter over the keyboard as I type, was the baddie: setting rules, daring to ask for help around the house, demanding a little appreciation. How very dare she.

Now I see it all very differently because I've learned that in life there are no T-Birds and Scorpions; nothing is as simple as the Rebel Alliance and the Empire. What I gradually learned as I grew up was that nobody is perfect: the people I idolised – Dad – had their faults and the ones I clashed with – Mum – usually had my best interests at heart. I think you only begin to accept people and their insecurities, bad tempers, vulnerabilities, frustrations and sadnesses after you've lived a bit and realise that none of us really know what we're doing.

I always presumed my parents had all the answers – they were adults! Their actions should be controlled and considered, not influenced by emotions or anxieties – they were my parents!

Of course, now I'm a mum I realise that when you become a parent there is no miraculous moment where the brainwaves alter and a switch in our soul is flicked from floundering young person to responsible, 'together' adult. I was just the same old me, picking my way through the puddles of life, only now with a newborn baby in my arms.

The only thing that changed with me when I had my first child (beyond of course the whole feeling-love-stronger-and-more-potent-than-the-whiffiest-Christmas-cheese thing) was that suddenly I would cry at sad newspaper stories and gravy adverts.

And I realised that it's not advisable to spontaneously jump on a bouncy castle post-partum without first nipping to the loo.

Over my teenage years the picture I'd held in my head of my siblings and parents began to change, the outline of each person shifting and fading, the people I love and their personalities blending in to one another like a smudged pencil drawing – who was on my side, who I trusted the most – and now they're just my family and, I imagine, as complicated as all other families – loving, often hilarious, occasionally irritating, but always loyal.

My modelling career didn't exactly take off.

I did a lot of shifts at the pub, interspersed with the occasional casting. I did a summer photoshoot for *Just Seventeen*, wearing cut-off dungarees and checked shirts, wielding rods and nets, perching on a wicker picnic basket as if I was off for a day fishing by a river. I was happy, though. Magazine fashion shoots barely paid sixpence, plus the agency would take their twenty-five per cent too, but they were fantastic for building up your portfolio. Work bred work; the more evidence of different shoots, the more 'looks' you could demonstrate, the more experienced you became. Mavis had explained to me during our first meeting that I had an 'editorial' look, which meant a unique, more quirky look. You were deemed to have either an editorial or a 'commercial' look, which was all teeth and eyes – think toothpaste adverts and girls next door. Editorial was seen as more high fashion – staring moodily out from the style pages of glossy magazines, stalking up and down catwalks, big-money ad campaigns, *Vogue* covers.

Those of us who were declared editorial were quite smug about it, but it was the commercial girls who'd make cash – big-money jobs for shampoo, cars, breakfast cereal. It was unlikely us editorial girls would make our millions; in fact, we'd be lucky to make our thousands.

It's the lucky few who get a spectacular breakthrough due to sheer luck or from just being so striking and perfect to photograph – the Mosses, the Campbells – then there's the layer underneath – the girls who eke out good careers by constantly working, magazine shoots and catwalk work, on rotation round all the big shows every February at New York Fashion Week, then London and Paris.

They make good money for braving the whirling, non-stop fashion juggernaut – racing from one show to the next, face and hair being scrubbed clean in between each one so they are a new, fresh canvas for the next designer's whims and follies.

The rest of us realise fairly early on that it's not really a viable long-term career. If modelling was a crème brûlée then the crunchy, delicious caramelised bit at the top would be the models with a successful, sustainable 'buy a flat in London and have savings' career and the rest of us would be the custard.

Hopes of *Vogue* covers and private jets to Bajan bikini shoots evaporate within the first couple of months, when you keep finding yourself squished into a clammy room in Manchester with two dozen equally pretty girls waiting to walk out onto a tiny catwalk to be barely glanced at by a couple of bored-looking clients.

The good news is, once we'd got over our delusions of grandeur the real fun began. Most of the models I knew just did it to escape real work, to get drunk in exotic faraway places, to avoid university and escape their home towns. After all, that's why I did it. I'd had vague and varied career ambitions that had been mulled over and mostly discarded: vet (too much studying, not clever enough), mounted police (not brave enough, bad at running, lifelong fear of knives) and journalist (this one I still harboured, loosely believing that one day I'd return to a place of education to study media and journalism); but for now I was very happy to spend a couple of years treading water before being forced to grow up.

The biggest photographic job I did in the UK was for Boots tights. They were looking for someone with very slender hips rather than any hint of womanliness and from the front my hips had about the width and shapeliness of a stick of gum. They wanted to shoot for some hosiery packaging – imagine that! My wonky legs gracing the packages of tights in every Boots store! Rows and rows of tights with me on them. Beyond exciting. The new line they were launching were 'suspender tights', which sounded like something a structural engineer might use on a railway bridge and were about as sexy.

They wanted to save ladies the faff involved with attaching stockings to suspender belts by making tights with areas cut out on the thighs, revealing flesh and creating the illusion that they were complicated lingerie.

If you were in a dimly lit room and squinting a bit, it would almost certainly look like the lady was wearing stockings and suspenders. It'd only be once you approached/flicked on a light that you'd realise they were tights with bits missing and the woman would be revealed to be a lazy slut too slothful to bother to attach real stockings to a suspender belt. Sure, hooking up the real thing is akin to a *Krypton Factor* puzzle and you need to remove a vertebra to twist adequately enough to fasten the rear ones, but any women worth her salt knows it's worth it for the pleasure of feeling like you're wearing a racehorse's bridle round the top of your legs.

On these charming suspender tights' packaging would be just a close-up, not of my face but of my pelvic area. But that didn't stop my mum buying a few pairs as souvenirs of my stratospheric rise through the ranks of hosiery modelling. These tights never really took off; in fact, I imagine Mum was probably responsible for about forty per cent of the overall sales.

I also modelled some over-the-knee socks and my face WAS on the packaging for those! Although unfortunately my face looked weird, gurning sideways at the camera as I leapt over imaginary puddles in my long socks, miniskirt and blazer, clutching an umbrella. Still, my loyal mum stocked up. The lady at the till in Boots must've wondered why this petite lady needed so many socks.

Little jobs like these were lovely – it was undeniably thrilling that my pelvic area was such a hit (maybe I'd be recognised in the street and fans would want a picture with my hips?). They were also physical proof that I was working and enjoying some small successes. Dad loved my modelling photos; he was very proud – for years he had one of my composite cards in the side window of his wagon in the spot that was reserved in other wagons for Linda Lusardi.

Hilariously, I modelled a couple of times on *This Morning*, which was incredibly exciting but beyond nerve-wracking. With hindsight I now know why – when you're a model you're not there to talk, or have an opinion; you're not even mic'ed up, so even if someone asked you a question your answer wouldn't be heard. There's something unsettling about being muted. I was like the Little Mermaid after she's tricked into giving her voice to Ursula. Silently nodding, smiling, beatific.

When I started working as a television presenter I loved the freedom to chat, exclaim, laugh. I think that's why I talked so much, blurting out whatever came into my mind – my gag was gone, my voice and therefore my opinion, sense of humour and personality reinstated.

My modelling stints on *This Morning* happened during the golden years of Richard and Judy. We'd be herded around backstage by the model wrangler, whose job it was to get us organised and into our outfits. It was always quite a stressful gig; live television obviously put a bit of pressure on all the production team and this clearly seeped down to the women who had to get us sorted. They seemed constantly frazzled and ordered us about.

We'd then be cued on individually to stand there like a lemon while the fashion expert and the hosts chatted about the clothes.

Years later whenever I went on *This Morning* as a guest (WITH A MICROPHONE!!) plugging a TV show or charity campaign I'd tell Richard and Judy my tale of modelling on the show; they tried a few times to find the footage but luckily it never surfaced.

I used to joke on my Radio 1 show that Richard and Judy were my fantasy parents; in response when I went on their Channel 4 chatshow they surprised me by bringing on set a pony as a pretend present for me (thus only confirming their status as dream parents).

The floor manager had told me sternly beforehand that under no circumstances should I get on the pony because I didn't have a helmet on, which resulted in an Alan Partridge-esque stand-off on live telly with Richard shouting at me to 'GO ON!! GET ON THE PONY' while out of shot the poor floor manager stood shaking his head furiously at me, mouthing 'NO! DON'T!' Richard and Judy perfected that 'anything can happen' controlled chaos during their live shows and that's why I love them.

The modelling stints on *This Morning* weren't quite as crazy as that, though I was very unconfident. My body language didn't scream 'hot model', more 'cleaning lady who's accidentally come through the wrong door and stumbled onto the set and is trying to style it out'. I was an unsure, round-shouldered rabbit in the headlights.

One time on *This Morning* it was a 'Summer hats special' and to ensure all eyes were on the headgear we all had to bring along a black catsuit to wear. With the wide-brimmed summer bonnet atop my skinny body clad all in black I looked like a quivering floor lamp.

The irony wasn't lost on me that after an entire childhood being mocked for how I walked, to the extent that I didn't even like walking ahead of other kids en route to school, I was now expected to walk for a living in front of thousands. My crushing self-consciousness about how I moved meant I never cracked the knack of the proper model walk; despite the best efforts of model friends to teach me, the hip-swinging, determined stomp of the catwalk models was always out of my grasp. I walked instead like a wooden puppet whose strings were being held by someone in the throes of a sneezing fit.

I was good at standing still though (Go me!) and on one job I was even required to sway slowly and it was a triumph.

There was a casting at MMA for a pop video and I was chosen for the gig along with three other girls from the agency. We were pretty excited about this because the band was OMD, who I knew had had quite a few big hits in the eighties. Of course, now I've interviewed them and played their music loads on my *Sounds of the 80s* Radio 2 show, but back then I just knew they'd once been a pretty big deal – they were definitely from the cooler end of eighties pop, while I was busying myself with Wham! and Bros.

On the day of the shoot we arrived at the studio in a warehouse in Manchester and were told we were to be nymph-like creatures, hair pinned up, wrapped in long lengths of cream fabric that looked a lot like dust sheets. With youth on your side, it seems, you can even rock the 'protective covering on a statue during renovation' look.

We were to stand with the sides of our faces pressed against the wall as if trying to ear-wag on the neighbours' arguments through a thin wall. Lead singer Andy McCluskey was a bit grumpy that day and didn't seem to enjoy the shoot – it's worth saying here that whenever I've met him since he's been very nice, so maybe he was having a bad day. Also, video shoots are by their nature pretty repetitive: Andy's job was to sing along to the song again and again as we hugged the walls next to him.

It was nippy wrapped in just a sheet but once under the

lights, like newborn chicks under heat lamps, we felt warm again.

At the end of modelling shoots and indeed now, when I do photoshoots to promote a new show, it's always a joy to slip back into the comfort of my own jeans and jumper after being scantily clad, especially if everyone else on set – photographer, photographer's assistant, client, make-up artist, PR person – is fully clothed.

There's a certain vulnerability that's created by having on less clothing than everyone else and in my modelling days that was magnified by also not having any choice in what I wore. When you're a model no one cares if you're not keen on a skirt you're wearing or if you think the foundation that's been slapped on your face makes you look like Gollum's sister; you're there as a clothes horse, a blank canvas. On shoots, I'd feel equal to them again and able to have a voice, an opinion, only once I'd got fully dressed.

Obviously these days I do have a say. Hurrah. I only wear what I actually like and my definition these days of a 'scantily clad' shoot means I've maybe unbuttoned my cardigan a bit.

37. Three Thin People

Spring 1994. I was nineteen and was offered the chance to work in Milan.

I can't remember a lot about the place – I was there for a month but it was such a whirlwind that just a few memories rise like bubbles to the surface.

We were flown over by a sister agency of MMA and I shared an apartment with a few other girls. I was pretty naive and remember being shocked that we had to head out alone and find our way around the city to get to our castings. Luckily, thanks to Yvonne and my Paris trip I had some experience of navigating a new place armed with just a street map and a few quid in my pocket, although this time I was lacking even a GCSE-level grasp of the language. I did pick up a few phrases – hello, how are you, goodbye and f-off bastard.

The latter was basically the main one we needed while trooping around Milan, map in one hand and portfolio in the other. To say the men we encountered there were perverts is a little like saying a Premiership footballer is on a decent wage.

It may have changed now, but back then it was as if my visit coincided with the annual month-long sex pest convention. Wherever we walked in our little gang of girls, grown men reacted as if they'd been in solitary confinement and not clapped eyes on the female form for decades. They'd look at us like a starving dog would look at a T-bone steak. In the street, on the tube, they would make this strange hissing noise, the same one you'd make to get the attention of a cat. I don't know how any woman was supposed to react to being literally cat-called like that – 'Oh, how perfectly charming! You're hissing at me and clicking your fingers – we simply must have dinner some time.' We found it ludicrous and insulting.

There was no support from the sisterhood when we were getting blatantly ogled or harassed. Women on public transport didn't offer empathetic glances, they just glared disdainfully at us. I like to think if a man was being openly predatory towards a clutch of young ladies on the London tube now I'd at least, as a 43-year-old mum, keep an eye on the girls and make it obvious I disapproved of the man's behaviour.

When we were in groups of two or more and getting unwanted attention we'd use our new Italian phrase and tell the pervert to F off, but on the whole we hurried on and kept our heads down.

I imagine we were easy to identify as models thanks to us being foreign, tall, slim and, oh yes, holding our portfolios. Maybe models had a reputation as good-time girls and so were thought of as easy targets for sleazy men.

We were certainly used as a commodity; a few nights a week we'd be picked up by one of the bookers and taken out first for dinner with other people from the agency and then on to a club. I remember at dinner, even though I smoked, being disgusted at how everyone would puff on fags as the food was served and between mouthfuls, the smoke from their cigarettes constantly curling up from the ashtray. The bookers talked loudly to each other in Italian while the English models chatted among themselves and sipped the wine, hoping to appear grown-up and sophisticated.

Of course, a free dinner and a night out was not to be sniffed at for a nineteen-year-old girl but I soon sensed a darker side to it and started to feel uneasy about these nights out to clubs; we weren't there for our own entertainment, we were there as decoration for the club whose clientele was men old enough to be our dads.

We were just young women shipped in to make the place look pretty. The model bookers obviously had some sort of deal with the clubs that guaranteed they'd be treated like kings in return for furnishing the VIP area with models. I never felt in danger in the clubs – nothing was expected of us except to have fun and look the part – but I remember, as my trip neared its end, how I started to hate going out and to resent being one of the many pretty pawns in their shallow little game.

That month I spent in Milan felt hard; here I was, a model in an exciting city, one of the fashion capitals of the world, and all I wanted was to be at home on the flowery couch above the Pineapple pub, Mum shouting at me for not wiping up the dishes.

It was my first time working away and I was terribly homesick. I missed Mum, I missed Ani and hanging out with the lads. Everything was so (unsurprisingly, of course) foreign and I longed for the very streets I had thought I wanted to escape: walking up Blackburn Road to Phil's house, popping over the road to the massive Asda near the Pineapple, driving to Ani's to go for a drink by the log fire at the Bob Smithy pub. I should have felt slightly at home at least with the weather in Milan – it was very Mancunian and poured down almost every day, grey skies sluicing the streets with rain, soaking me on the way to my castings.

The castings themselves were depressing; each morning the agent would ring with a list of clients and locations, which I'd scribble down. The mornings were all a little bit like an episode of *The Apprentice*, with me looking frantically at maps of the city and public transport routes, working out the most logical way to get to all the castings, sometimes as many as six in a day, sprinkled liberally all over the city.

I'd take the metro, then a tram, then a bus to a grey miserable street. After walking around a bit in the drizzle trying to locate the building number, I'd schlep up four flights of stairs into a room containing three thin people dressed in black, with angular haircuts, smoking at a table. The people wouldn't even look up from their espressos; one would just thrust out an open hand, into which I'd slot my book. They'd flick through, looking at maybe three or four of the shots before mumbling 'Grazie' as they handed it back. That would be it. I'd slink out of the room and start my journey to the next casting.

Lunchtimes were equally disappointing. I was in a country famous for its love of good-quality, locally sourced fresh food and here I was, queueing in McDonalds for a Filet-O-Fish. What can I say? When you're nineteen and homesick you seek out the familiar. But it didn't taste the same as at home, and their McDonalds even served salads, proving to me once and for all that Milan was just plain weird.

Once back at the apartment I couldn't even escape by sinking into the sofa and gawping at the telly, because – surprisingly, I know – all of the channels except one were in Italian. The exception was MTV, which I watched endlessly – a background blurb or, if you actually watched it, a rolling visual and audio assault of candy-coloured ads and shouty hosts mixed with all the biggest nineties non-stop hits, interspersed with repeats of *Beavis and Butt-Head*.

I was pleased to return home after that month, no extra money in my pocket but a few new pictures in my book, which was at least to be helpful for getting my next gig.

38. Too Big for Japan

In August of 1994 there was a big casting for a contract in Japan. I'd always wanted to go to Tokyo – it had a reputation as a top destination for models, with quality work and great money. I was super-keen and the casting seemed to go well; I came out onto the mini catwalk and sashayed about in my usual 'accidentally walked through the wrong door' style, channelling Mrs Overall.

I wasn't worried. I knew Japan was all about the photographic work and wasn't catwalk-centric. I also knew my portfolio was quite strong. The Japanese agent loved my look, liked my book, but there was one stumbling block. Or rather two. My breasts. In Japan big boobs just wouldn't work – they liked their fashion models slim as reeds (check!) with narrow hips (check!) and flat chests (fail!) – so I was denied the chance of a Tokyo trip. The only positive part of this rejection was that I've been able to state ever since that 'My boobs were too big for Japan,' as if my mere presence in the country with such enormous breasts would cause natural disasters on a scale never seen before, my knockers disrupting equatorial weather systems.

However, there was a silver lining in the shape of South Korea. MMA had recently connected with an agency in Seoul called Classy and told me there could be an opportunity to work there.

It was explained to me that using western models in South Korea was a complicated business and although the agency was above board, there were laws regarding how many western models could be used on a shoot as a percentage had to be Korean. As I was soon to discover, that rule wasn't actually adhered to at all, although I was told to keep my passport hidden under my pillow just in case.

I wasn't concerned. I just wanted to go on another trip. So in September 1994, three months before my twentieth birthday, I found myself at Charles De Gaulle Airport waiting for a connecting flight to Seoul.

I wrote a diary about this trip. Its brown corrugated cardboard cover has 'KOREA' doodled on the front. The entries show that I'm wowed by the cultural shift I experience before I even board the plane – I wax lyrical like only a nineteen-year-old can about the different ethnicities of the passengers of the Air Korea flight, including 'an Indian man sat cross-legged as if to charm a snake' – a real Adrian Mole moment.

I don't mention the smoking area of the plane in this diary but I remember it well and the memory is enough to bring on a bronchial burst of coughing.

The smoking area used to be at the back, so all the smokers would congregate in the rear aisles of the 747 and stand puffing on their cigs as if at a bus stop or outside a bar. The thought of smoking on a plane is so ludicrous now that I can't quite believe it used to be OK – lots of people holding objects in their hands that burned and released deadly fumes while cruising at 35,000 feet.

*　　*　　*

After the eleven-hour flight and having had scant sleep I was collected from the airport at 10 a.m. by Mr Back, the Classy agency driver. I was fully expecting to be taken to the accommodation to unpack, shower, change traveller's cheques, ring home and generally recover from such a tiring flight (I even wrote 'Face pack here I come!' in my diary.)

What actually happened was I got driven straight to Seoul, where on the street outside the agency I met Mr Kim, Classy's owner, a stern, stocky man in his fifties who smiled so little that it looked like his face might crack when he attempted it.

Mr Kim demanded to see my book, which, as advised, I'd stashed deep in my suitcase, so I had to dig it out, bras, toiletries, shoes all tumbling out onto the dusty pavement. He flicked through it in seconds, grimacing in the manner of an elderly lady who's just found a fresh cat poo in her teapot, and thrust it back into my hands with a grunt.

I was then taken to six castings by Mr Back – one of the good things about South Korea was we weren't expected to find our way around. Either Mr Back or 'Young Mr Kim' would drive us to our castings every day. Negotiating the roads and public transport of Seoul would've been impossible as the Korean language when written is similar to Japanese, only with more squared-off shapes and symbols.

The roads in Seoul felt safe, although that feeling was undermined by a huge electronic sign above the Seongsu Bridge showing an up-to-date tally of the number of road collision casualties so far that year.

Mr Back drove a white sporty Hyundai, which wasn't the most practical car for squeezing four tall girls in, but we were grateful. Not only was South Korea an assault on the senses – huge skyscrapers lit by flashing colourful billboards, a rainbow of neon and LED plastered up the side of buildings – but the area where we were to live, Itaewon, was very complicated to navigate. It was made up of several bustling streets behind which was a huge hill, which was the residential area. To negotiate these without a driver, especially in those first few days, would've been impossible. The steep sides of the hill were smothered in thousands of apartment blocks, like dust on a dropped strawberry. The narrow one-car-width roads that were tangled tightly around the hill were so confusing – every blind corner looked the same as the last.

The blocks of flats weren't high-rise, just two storeys and higgledy-piggledy, balanced haphazardly like a toddler's tower of building blocks, many with extensions added on as an afterthought. They all clung to the sides of the hill, washing lines strung around, zig-zagging across the tiny concrete squares outside each apartment door that were occasionally dotted with bedraggled, thirsty-looking potted trees, the floor seasoned with discarded cigarette butts. People lived piled up on top of one another. The shadows of visitors descending from the apartment upstairs were accompanied by loud post-dinner belches ringing out as they made their way past our lounge window. In South Korea it is taken as a compliment to your cooking skills if guests burp after consuming your food and judging from the gassy cacophony coming from the stairwell outside, our neighbour must've been the Korean equivalent of Delia Smith.

On my first day Mr Back and I didn't arrive at my new home until nearly 9 p.m. – I was almost delirious with tiredness but was chuffed that one of the castings had resulted in a job the following day. I made my way up the steep concrete steps outside, following my new flatmate Diane who'd been on the castings too, dragging my massive case behind me.

On the second floor, through a dark wooden door we found a lounge with a lino floor, open-plan kitchen area to the rear and two bedrooms and a bathroom leading off it.

My two flatmates were both from Hawaii: Caroline was twenty with long dark hair and a perfect smile, an all-American sweetheart who looked like she'd just toppled out of a passing cheerleading troupe. Diane had pale skin and light auburn hair and seemed to me to be more worldly wise. It soon became clear why – despite being only eighteen she was married. This of course was a source of fascination and wonder for me and the other girls; that and the fact that she was a Mormon.

I had my own room, a tiny box room with a single mattress on the floor and a little wicker shelf unit next to it. No bed as such, but I would've happily slept on a pallet I was so knackered.

The apartment was harshly lit and furnished with a big boxy telly and the sort of ripped leather sofa usually found in the waiting area of a minicab office.

The fridge was filthy and on one of my first days there I write in my diary about cleaning it out and removing all the rotting food and cockroach carcasses.

Cockroaches became quite a feature of my time in South Korea. They were everywhere. There'd be an occasional one so big I didn't know whether to squash it or pop a saddle on it and ride it to the shops.

I generally have Buddhist leanings when it comes to killing things, but I had no qualms about bumping off roaches. I'm afraid they're in the same category as wasps on the squishlist – 'entirely killable'.

I became quite brave at squashing them. The bigger ones seemed to scream when I dealt them a deathly blow, especially the ones in the bathroom, where they thrived, forever scuttling up the shower wall. They were shiny and brown like polished mahogany and the size of one half of a Kinder Egg, with long antennae poking out the front like a couple of dressage whips.

They often met their end on the side of my Herbal Essences shampoo bottle.

The roaches I couldn't get to were the smaller ones I'd see in the fridge. This fridge was ancient and there were hiding places for them in various cracks in the back of it and in the rubber seal.

I didn't wage war on these ones, I just made sure my milk carton top was on properly. I used to quickly open and close the fridge door three times before looking in there, giving them time to scuttle off to their hiding places. Out of sight, out of mind. I didn't keep much fresh food in the fridge. I mainly lived on eggs on toast and, randomly, tinned pineapple.

Life in South Korea was pretty full on. The photoshoots would be long days. The best jobs to get were the catalogue shoots because they were the best money, but they were also the hardest graft. I know, I know, calling having your picture taken 'graft' sounds ridiculous, but the big shoots were genuinely knackering, often starting at 10 a.m. and finishing at 3 a.m.

There'd often be around twenty-five outfits to get through, so twenty-five outfit changes, twenty-five changes of background if you were outdoors or, if you were indoors, twenty-five tweaks of the lighting while you held a weird pose. The outfit changes were stressful. The Korean stylists often had a couple of assistants and getting undressed was a bit like being attacked by angry seagulls, each woman pulling at a different bit of clothing.

The styling itself was pretty random. We were shooting winter designs but September in Seoul was still a humid 25 degrees, so we were constantly boiling, in strange long leather macs and tweed baker boy caps. The trousers were always way too short, designed with a Korean body size in mind, not long-legged western models. To hide this fact, the stylists on nearly every shoot simply tucked the bottom of the trews into socks or boots.

We'd then see teenage girls wearing trousers tucked into their boots as if off on an impromptu hike; they'd religiously copied the look from the fashion pages. We'd cringe – this wasn't a new trend to follow, it was just because the trousers didn't fit!

Not speaking the same language as ninety-five per cent of people on the shoot was often stressful too. On a shoot the only word a photographer would know was 'Posey!' – this is what they'd holler again and again from behind their camera, meaning 'Pose'.

Every shoot it was 'Posey!' CLICK. 'Posey!' CLICK. 'Posey!' CLICK.

This took a little getting used to and on my second day on my very first job I struggled. In the UK and Milan the poses were much more relaxed and required just subtle changes every few clicks, usually instructed by the photographer – a chin lowered here, a gaze moved to the left, a hand on the hip. Usually while pouting.

I'd been a nymph wrapped in a dust sheet, of course, but this was something else. Being photographed in South Korea was a mixture of poorly improvised sign language and a weird Christmas parlour game.

I soon learned a variety of poses, and the dafter the better. The trick with posing was to lose all self-consciousness; the more wacky my expression and flamboyant my moves the more they clicked – and occasionally cheered.

A favourite of mine was cowgirl sharpshooter-inspired, making my fingers into pistols because then I could do so much stuff with them: hold 'pistols' by my face – CLICK! – 'shoot' at the camera – CLICK! – blow the invisible smoke from the end of my pistols – CLICK! – put the pistols in my invisible holster while winking to camera – CLICK! CLICK! CLICK!

I have to confess that when working with fellow Brits, for

our own amusement and if the photographer was particularly grumpy we'd joyfully flick the Vs at the camera as one of the poses while doing a big cheesy grin or a faux shocked face, mouth rounded in a big O.

Somewhere in some South Korean catalogues from the mid-nineties are pictures of a couple of young girls performing the British hand signal for Piss Off.

My first few days flew by, busy with either being on shoots or being driven around, and after a shaky start I started to settle into life in South Korea. I could find my way to the little local convenience store and perfected my chopstick technique by eating my breakfast cereal with them. I was still a little nervous at the agency. According to some other Classy models we saw on castings, Mr Kim was in the Korean mafia and had been in jail a lot. There was no basis for these rumours, but take a clutch of gossipy young girls with a penchant for melodrama plus Mr Kim's sour expression, then add to that the way his staff pandered to him, and it was easy to believe he was some sort of under-world big shot.

We heard from the agency that there was a new girl arriving from Manchester called Clare. I was really excited; I liked Diane and Caroline but I didn't have many shared reference points with the two girls from Hawaii, one of whom was a Mormon.

Clare arrived, exhausted after her long flight, and fell into bed immediately. Caroline, Diana and I arrived home from shopping like the three bears finding Goldilocks, except this Goldilocks hadn't eaten our porridge and had cropped brown hair and skin the colour of roast beef thanks to an enthusiastic application of fake tan.

I love that in my diary I talk about this mystery girl called Clare arriving from Manchester (she came from about four miles up the road from me), not knowing that she would go on to reveal herself as the most magnificent woman, my very best friend and soulmate to this day, and godmother to my children.

Of course I had no way of knowing all this as I peeped into Caroline's room and saw Clare passed out face down on a mattress, clothes and high heels spewing from her open suitcase.

We laugh now that the very first conversation we ever had, once she'd woken up, went like this:

Me: 'Hiya, d'ya smoke and drink coffee?'

Clare: 'Yeah.'

Me: 'Right, well there's some fags in me room if you want one and my coffee's over there, we can share it.'

Clare: 'Fanks.'

I liked Clare right away; she was a breath of fresh air in the apartment. She liked to party and, having been surrounded until now by non-drinking, all-American nice girls, it felt great to hook up with someone a bit naughty.

The more I chatted to Clare the more I warmed to her; we spoke about our backgrounds and school. I was wide-eyed when she informed me coolly that she'd been suspended from school eight times, once for setting off a firework in the girls' changing rooms.

Straight away we joined forces to lead Diane from the path of teetotal, married Mormonism and onto the road to Funsville. In fact, Funsville fell at our feet that night in the form of a phone call from England.

The dad of a male model called Ben who used to live in our apartment needed to get hold of him as there'd been a death in the family. Classy had stopped his contract and fired him, so they weren't any use – could we help? We were told that Ben used to drink at the Kettle, a bar about a ten-minute walk up the hill from the apartment.

Clare, Caroline, Diane and I quickly declared it our solemn duty to trace Ben (the fact that our intrepid mission involved discovering a nearby bar had absolutely no bearing on our decision . . .).

We found the Soju Kettle surprisingly easily. It had a small square glass frontage with empty bottle crates stacked outside. The illuminated sign was adorned, randomly, with the MTV logo and a huge dartboard as well as the neon outline of a teapot, guiding us in like the north star.

As we walked through the door two things became immediately apparent.

The first was that the clientele in there were largely male and western; in fact, as we'd soon discover, they were all GIs – American troops, based a couple of miles away at the Yongsan Garrison, the headquarters of the US military in South Korea.

The second was that they were clearly not expecting four young models to come storming into the bar. Indeed, if this had been in the Wild West, the music would've stopped as everyone turned to look in silence at these new visitors, saloon doors still gently swinging behind them.

As it was, the cheesy nineties R 'n' B kept playing but everyone did pause to stare at this walking shampoo advert – four girls with hair of different hues and styles striking a pose in the doorway.

Then the spell was broken and the GIs sprang to life, curious about these alien life forms suddenly inhabiting their orbit.

This colliding of worlds promised to be a lot of fun: young soldiers far from home suddenly in the company of some not-unattractive young women, two of whom bringing with them the familiarity and sunny outlook of Hawaii and the other two slightly feistier and from the UK.

Drinks were bought, introductions were made, accents were

mutually admired and we were treated with lots of respect. The soldiers all had impeccable manners – all that military training had certainly knocked off any rough edges.

We were called 'Ma'am', LIKE WHAT PEOPLE SAY IN THE MOVIES, which we found completely brilliant, and whenever one of us popped a cigarette in our mouths, four or five Zippo lighters would appear from nowhere, eager to be the one to ignite our smokes.

A Korean barmaid brought over a stack of plastic cups and three kettles on a tray – old-fashioned small tin kettles, the sort you'd put on a stove-top.

Inside each was a bright red, orange or yellow liquid that tasted like melted ice lollies – fruity, sickly sweet and slightly cloying. The alcohol in these kettles was soju – a clear spirit, the Korean equivalent of sake.

We soon learned that soju was very strong – it had an almost hallucinatory effect on you and you'd feel fantastically drunk very quickly, as well as slightly woozy. Rumour had it that soju contained loads of trace chemicals including formaldehyde and it was these ingredients that made the drink so potent.

Like only the young can, we absolutely relished the drama of drinking something that was not only considered pretty hardcore but would pickle your organs as you got sozzled on the alcohol content. Soju hangovers were legendary for their brutality, something I was on this first night still blissfully unaware of.

At our Kettle debut we proceeded to get stuck in. It felt magnificent to let loose – we got sloshed, danced, smoked and generally threw ourselves head first into the revelry.

Diane was surprisingly easy to influence and with minimal twisting of arm quietly forgot the whole teetotal Mormon thing and downed a few drinks.

Finally, some time in the early hours, we all linked arms and wound our slightly wobbly way back to the apartment high on life (and formaldehyde).

We never did find Ben that night – to be fair we didn't conduct a fingertip search, but we did ask around, and had no luck. We did, however, find the place that quickly became the very centre of our life in Seoul – all the dramas, drunkenness, fun and forging of lifelong friendships would happen inside those four walls.

The next day we hung limply from the jaws of a vicious hangover as we were driven by Mr Back to our castings. We probably could've fuelled his Hyundai with the soju fumes we were emitting.

Despite being green around the gills, both Clare and I were chosen for the first job of the day and were driven four hours out of Seoul on a coach with all the team – clients, photographer, make-up artists, stylists, numerous assistants and all the clothes. It was boiling on the coach, autumn sunshine streaming through the windows, and everyone slept, apart from myself and Clare. We laughed a lot on this trip – and learned a lot.

We learned that Koreans worked hard and played hard. After a long shoot, no matter how late the wrap, the client would always expect to take out the whole team for food and soju. The downside was that everyone would be roused out of their beds a handful of hours later, and would be expected to be ready to commence work on minimal sleep. This was why the Koreans napped every chance they got. On the coach en route to the next location it was as if someone had pressed the 'standby' button on the whole crew – they all instantly nodded off as the doors swung shut.

Clare and I couldn't sleep; we were too excited to be on a

trip away and by discovering we had so much in common. It was on this trip that Clare and I really bonded. Our belly laughs and non-stop larking about were probably massively irritating for the rest of the crew but we were having a ball. They got their revenge though. After a long day we all went for dinner. Obviously being teenagers a long way from home we were craving our usual Korean dinner – a Wendy's burger – but our options here were kimbap (sushi rolls) and bulgogi – slippery strips of beef and onion in a glossy thick gravy. Clare and I tried our best to be polite and eat it.

Despite the stereotype of age supposedly bringing with it a 'stuck in their ways' mentality, I believe that as you get older you become more, not less, open-minded. Back then I wasn't remotely adventurous with food, but now I'll try anything and in fact currently have in my fridge two jars of kimchi – the Korean fermented cabbage side dish that was my nemesis as a nineteen-year-old in the country of its origin.

After dinner and soju we went to our room, still giggling and overexcited, only to be woken up four hours later at 6 a.m. to set off for the location.

The shoot was exhausting and, probably to everyone's relief, Clare and I finally cottoned on to the snoozing thing and slept on the coach whenever we could.

Mid-morning we arrived at a new location and one of the make-up artists came over and gently woke me. She was wielding what I thought was a toothbrush but turned out to be a small razor blade with a long handle, not unlike a shank – the improvised weapon found hidden down the sock of any self-respecting high-security prison inmate. In broken English she explained she needed to trim my eyebrows.

Usually, of course, tweezers are adequate for this task but in South Korea the make-up artists preferred a razor. I agreed warily, but watched my brows in a hand mirror like a hawk and gently wrestled her away when I thought she'd chipped away at them enough.

She then made her way over to Clare while her assistant coloured in my eyebrows with eye make-up to the extent that it looked like someone had used their thumb to smear Marmite over them.

Clare, who was still sleepy, woke briefly to agree to the eyebrow trim before snoozing back off again. The brows were razored and then covered with the same dark brown streaks as mine.

It was later that evening, back home at the apartment, that Clare discovered the true extent of the trim. Once she'd washed off the eye make-up all that remained was a tiny patch of eyebrow on each side, like quotation marks framing her nose.

She screamed, and we all ran to look and guffawed supportively. They did grow back eventually, but to this day she still blames me for not waking her up.

As the ancient Korean proverb says, 'Snooze, you lose (your eyebrows).'

The shoot itself went well and by the end Clare and I were firm friends. It was a hot day but we were modelling winter fashions and had to wrap up in the usual trousers tucked into boots, long coats and hats.

The photographer had a great idea to recreate a classic snowy scene but we were shooting in a massive disused quarry, surrounded by a red dusty landscape that looked more like Mars than the Alps. They'd planned ahead though, which is how Clare and I found ourselves sitting on a wooden sleigh in sweltering heat as sacks of salt were emptied around us to look like snow.

I knew Clare and I had something special when for one of the shots we were asked to walk away from the camera, chatting.

We got on so well that we didn't hear the photographer yelling at us to stop, he'd got his shot. We were so busy talking and laughing that by the time we turned round the entire crew were dots on the horizon.

My month in Seoul was a real eye-opener. Different models came and went – friendships were intense but fleeting, made in the back of a stuffy Hyundai, on shoots and over kettles. The jobs rolled by and we settled into the rhythm of work, rest and play.

I will be forever indebted to South Korea for landing Clare at my feet and for all the good times we had there – dancing with her to hip-hop at Kings Niteclub up the road from the Kettle, laughing on shoots till our stomachs ached and generally behaving badly. I loved the chilled times too, at the apartment wearing face masks, watching movies like *A Few Good Men*, half-heartedly exercising to a Cindy Crawford workout video, writing our names on our boxes of eggs with biro so the others wouldn't nick them; they came not in a dozen per pack, but ten, so much to our amusement S-A-R-A-H and C-L-A-R-E fit perfectly, a letter per egg.

It was fantastic to earn half-decent money and build up my book as well as seeing a completely different world.

In fact, South Korea was like another planet. Wealth and poverty stood shoulder to shoulder: skyscrapers jostled for space in the city while at ground level tiny shacks smaller than a telephone box sold everything from chewing gum to razors to flip-flops through an envelope-sized hatch, the vendor crammed inside with all their produce closing in around them as if the whole thing had been in the jaws of a car crusher.

All of it was eye-opening, from the indoor market where all us girls ordered made-to-measure traditional dresses in red satin to the food on shoots, which often tested us – one breakfast was small grey crabs floating in what looked, to the untrained eye, like dishwater.

I was unadventurous when it came to the food, and stuck to only a few rice dishes when out with clients. In my own time I preferred a Wendy's or a McDonalds and once Clare and I even tried a vending machine hot dog on a stick wrapped in a smooth, lukewarm deep-fried spongey casing, which was as scrumptious as it sounds.

My struggles with the food also compounded my homesickness: home is where the belly is, that's what the saying should be. I longed for Mum's cooking and I missed her.

I was desperate to get home but at the same time I hoped to get a contract with a New York agency. I kept mithering MMA about it the whole time I was in South Korea. I'm guessing then that my homesickness was only half about missing home and the other half was about being sick of Seoul. Dad didn't call often but that wasn't a surprise – shooting the breeze on the phone wasn't really his thing; he always spoke with such urgency and distraction it sounded like there was a bovine emergency happening mere feet away from him that he was too polite to mention. Chatting on the phone was as awkward to Dad as smiling was to Mr Kim. I'd be all like 'It's mad here, Dad. The food is crazy – the beef isn't as good as British beef' (always on message). I'd prattle on about the friends I'd made: 'There's Clare from – guess where? Whitefield! I know! Up the road!' and downplay any nightlife – 'Oh we went out for a meal but that's about it, nothing too wild' – to which he'd reply, 'Right . . . right . . . well, OK then, love, I'd better go.'

Throughout all my trips Mum was the one steady thing. Whatever side of the world I was on, she'd always call and she'd always be there when I called. Again, Mum. Just there. Quietly by my side even when we were 5,000 miles apart.

This book began as an accidental love letter to my dad. But what's become so clear to me now, as I've continued to remember and rediscover those formative years, the memories and moments bubbling to the surface, is that it's Mum who's always been there, not wanting any credit, no ulterior motive, just driven by a maternal urge to be there for her kid whenever she's needed or wanted. As natural and instinctive to her as breathing. From donkeys to car crashes and long-distance phone calls. Even when I think I've not needed her, I have and she's known it.

Dad was different. It was always the thrill of the chase with him; I needed to work harder for his affection or attention, not because he wanted me to or because he was distant, but because he also had other pressures – his wife, work, the farm. Dad showed me he loved me in subtle ways – a look, a smile, a hug. Mum was much more vocal and in turn needed to know that love and appreciation was returned. It's only in recent years, since I've had kids of my own, that she's felt that love fully returned – and then some.

I love them both, of course. Loads. It's just now I see clearly just how much they each did for me in their own different ways.

The time came for me to leave South Korea and 'cash out' – when the agency would tot up all my earnings and take away the cost of the flights and roach-infested accommodation, leaving me with a decent chunk of money to take home, which I stashed away in a money belt. I waved goodbye to Clare and Korea and headed home, away from the Kettle and bulgogi and back to Bolton and Mum and to contemplate my next adventure.

39. Two-dollar Fried Egg

It was lovely to be home, back in my bed above the Pineapple, the familiar rumble of voices downstairs from the punters punctuated by the odd raucous moment like celebrating a flukey shot on the pool table. It was great to be back with Ani, out with the lads, but as my diary's final entry said, I'd been bitten by the travel bug: I wrote 'New York? Yes please!'

In January 1995 I had a casting at MMA with Wilhelmina's, a reputable agency in Manhattan. They liked my book and wanted me to go over for a three-month contract. I was beside myself. So excited. I'd never been to America, never mind New York.

I packed a woolly hat and my long black coat as I knew the winters were brutal, but it wasn't until we landed at JFK and I stepped outside and the chill slapped my cheek that I realised just how cold it would be. I would probably need to wear all of the contents of my suitcase at once. No matter – I was in New York, and as the yellow taxi hurtled along the highway towards Manhattan the long-flight fug dispersed and I tried to drink it all in.

It was like I'd landed in every movie I'd ever seen – the cars, the big lorries with their boxy long-nosed cabs, the highway itself – this was before I'd even swept over the Brooklyn Bridge and seen Manhattan stood proudly in the distance, that breathtaking skyline that still makes my eyes shine and my fingers tingle.

My time in New York was absolutely one of the biggest adventures of my life. I lived in a two-bedroomed eighth-floor apartment in midtown with six other girls, so it was a pretty chaotic flat, as you can imagine. The rooms each had two sets of bunkbeds, so we all lived and slept literally on top of each other. Girls from France, Holland and the UK thrown together in this amazing city.

The apartment was a complete state – the average age was twenty and the average attitude was 'mañana', so no one could be bothered to wash up. Piles of dirty plates and cereal bowls with solidified matter on them like something from an archaeological dig lay alongside coffee cups and saucers that doubled as ashtrays, with fag butts glued to them with spilt liquid. The work surfaces weren't visible underneath the stacks of crockery and jumbles of empty takeaway, juice and milk cartons that were piled chest-high. As you walked into the apartment the sour smell from the tiny kitchen sprang at your nostrils from its hiding place.

Once past the wall of stink, though, it was a nice big open-plan area with large curtainless windows and two doors leading off to the bedrooms.

These were just for bedtime and the lounge was definitely for lounging: leggy girls in jogging bottoms and face packs, happy to be home from shoots and castings, make-up free at last and slouching about, ringing home, watching Fox, eating takeaway noodles. The apartment was like being at the world's longest slumber party.

There were so many brilliant things about living in New York. From the little things – ringing for takeaway breakfast, a two-dollar fried egg roll and coffee arriving within ten minutes from the Korean-run cafe round the corner – to the bigger stuff like belonging. I just felt like I fitted in. There were times after work when I'd be making my way home from the store with a paper grocery bag in my arms, past all the commuters, sidestepping schoolkids, mums with buggies and the occasional crazy, and I'd feel like a New Yorker. Born and bred. I often fantasised that somehow my spirit had been jumbled up with that of a New York girl and somewhere in Manhattan there was a girl who'd never feel she fully fitted-in until she randomly visited Bolton one time, and strolled past Greggs thinking 'I'm home at last!'

It's easy to feel like you fit in in New York because the second you arrive you can find your way around like a native thanks to the grid system layout of Manhattan's streets. Every morning once I'd scribbled down all the addresses of the castings and worked out my route I'd set off on foot, preferring it to the tube, often walking ninety blocks a day.

Having learned my lesson in Milan that carrying a model portfolio can get you unwanted attention, I used a rucksack to carry my big black book, wrapped up warm in long black military-style coat (it WAS the nineties) and black woolly hat and set off, melting into the crowds going about their business as I went about mine, striding happily under the cold blue skies of New York.

There were test shoots with exciting new young photographers – daring outfits, blaring music, red lips, wind machines on full blast as I swayed and pouted and shouted – unpaid but a benefit for both, great, fun, sexy shots for the photographer and me.

My fellow models were brilliant fun too; there was Melissa

from London with a long straight nose and flashing, playful eyes, poker-straight dark-blond hair falling past her shoulders in a smooth sheet. And April, who was an Amazonian Marilyn Monroe with a Canvey Island accent. Very tall, platinum-blonde wavy bob, hourglass, striking. She couldn't pronounce her Rs and this coupled with her Essex twang meant lots of the Americans couldn't understand her pronunciation of her own name, resulting in her shouting 'January, February, March, APRIL!' until they did.

We'd go out to party only very rarely as we were too young to drink legally, but we had a ball having face packs and noodles back at the flat. I was still going out with Ani so was quite happy to chill at night and, once I'd made my phone calls home, would eat carbs and then crawl into my bunk. Ani and I fell back into the now familiar routine of a long-distance relationship, chatting for ten minutes at a time. Once he came over for a long weekend and I showed him the sights, navigating the city's tourist spots like a local.

The agency was really good, sending us on some great castings, and I got enough work to just about break even and cover the flights and accommodation costs while gaining some lovely pictures for my book.

I still had a few weeks left of my trip when my New York adventure was cut short by a phone call.

I was at a shoot and the agency had tracked me down to tell me I needed to call my mum urgently. My blood ran cold. This was before mobile phones and the very fact that Mum had tried so hard to reach me could only mean terrible news.

My heart was racing as I called home, hands shaking as I dialled the number from a landline in a quiet corner of the studio. I quickly ran through the possibilities – Grandad had died when I was fifteen, so it obviously wasn't him. I was praying it wasn't Dad and that Mum herself was all right. I had always been very close to Nana and she was getting on, so I feared the worst and my chest was thumping as the phone lines crackled and connected.

Mum picked up.

Turns out fearing the worst was pointless as what was to come was beyond even that. In fact, beyond all comprehension. Even now. It wasn't Nana. It was in fact my cousin Tracy. She'd died in the night. Tracy was like a third sister to me – kind and loving, yet hilarious and quick-witted. Sweetest nature, gentlest soul. Stolen from us aged just twenty-seven by what turned out to be natural causes from an undetected heart condition called viral myocarditis, a virus that causes the heart to stop beating.

I don't remember much of the following few hours. The agency was amazing and bundled me onto a flight home, where I drank too much white wine and sobbed until a stewardess came to calm me. Mum met me at the airport and I fell into her arms. Tracy's story isn't mine to tell; it's a story that belongs to the two children, now grown up, she left behind. I just know that the whole family's heart broke when she went, and nothing was really ever the same again. We carried on but it felt like our world had shifted on its axis.

40. Scorch Marks

I threw myself into trying to get away again. I didn't want to be in Bolton. I wanted to escape my home town with its memories round every corner. There was so much sadness. I needed to disappear.

I now had a great modelling book and it wasn't long until a contract came through for a new trip. Maybe I was a coward; this was a time for family to stick together and mourn but I couldn't – it was time to return to Korea.

My second trip to Seoul, in the summer of 1995, was a disaster from the beginning. It quickly became apparent that I wasn't ready to be away from my family and the very thing I thought I needed to escape from was actually the thing I needed to surround myself with: the sorrow, the sadness. The mourning, although hard, is all part of a process.

I was instantly homesick and hated the work. After New York the jobs seemed low-grade. On my first trip the castings and the city itself were an adventure, but this time I resented being schlepped around in the back of a Hyundai.

I tried to think of it as an easy way to get away and make money but it was a struggle. Also Clare wasn't there. She was working over in Thailand, although I was hoping she'd come to visit me during my stay. We'd hung out at home and were confirmed official best friends.

I returned to the Soju Kettle with my new crew and had fun, but it all seemed empty. The work was going well but I was homesick. Every night before drinking at the Kettle I'd stop off at a street stall and cheer myself up by tucking into some bokkeumbap – the most delicious egg fried rice with spring onions and beef – before getting stuck into some soju. If my second trip to Seoul was to be summed up by a montage it would show me wolfing fried rice and then rampaging around the Kettle downing drinks, and comfort-eating boiled eggs in the flat. It would end on a close-up of me on photoshoots struggling to zip up skirts, bloated and miserable, to the tune of the Chemical Brothers' 'Out of Control'.

I was already pushing my luck with the agency, being petulant and tired on castings, but then one night I nearly burned down the apartment and snuffed out my own life in one fell swoop.

I'd been out indulging in a few Kettles and had returned home to my little box room. I fell onto my little mattress on the floor and decided to read by candlelight. Very Jane Austen, except she probably wasn't used to downing ten units of Korean spirit of a night.

I fell fast asleep and the large church candle quickly burned down. In fact, it burned all the way down, and the thing that saved my life was I'd been stupid/clever enough to place the candle on top of a large box of cook's matches. I know that, normally, balancing a candle on top of matches wouldn't be advisable; however, on this occasion it saved me. As the candle burned down it set fire to the massive box of matches and as each match ignited it made a crackling, sharp, hissing noise. The matchbox was full so it sounded like a mini firework display was going off. It was enough to rouse me.

By the time I came round, though, the fire had started to take hold of the little plastic shelving unit underneath the candle. As I came to I screamed, waking the whole flat, and we all rushed around throwing water on the fire. The flames doused, I flopped back onto my mattress and slept again.

The next morning I saw the full extent of the damage – huge black scorch marks stained the wall and floor and the shelf was warped and melted to half its original size.

When Mr Back came to collect us he saw the damage and went crazy. He was furious and said that I would have to pay for the repairs.

I was miserable. The only good thing was that Clare was on her way from Thailand to visit me.

Unfortunately, when she arrived I felt empowered – she has that effect on me – and my attitude towards the agency soured even more; I resented their controlling ways, how they scolded me like a child, how they blew hot and cold with their affections.

We went out and had the night out to end all nights out – well, to end a modelling contract anyway. The next morning I missed the pickup for the castings and that was the final straw for the agency.

The partying, the extra pounds, the attitude. They'd had enough. It was a little like when you don't have the heart or the stomach to finish with someone so you behave badly in the hope they'll dump you first.

I was hauled into the office later that day and they told me it was over. 'You got too fat Sara and you party too much . . . you gotta go home.' I was going home a month early.

I was relieved.

It also turned out to be one of the best things that could've happened to me.

As I strolled through Seoul Incheon International Airport I remember feeling a strange sensation – that for the rest of my life on this earth I would never set foot in that country again. So far that's true but I'd never say never now, my feelings have mellowed over the years. But back in 1995 I'd had enough of Korea – and it'd certainly had enough of me.

41. Nubile Gulls

I felt instantly better as the Korean Air jumbo touched down and although I didn't exactly kiss the tarmac at Manchester International, I silently breathed a long, slow sigh of relief to be home. Mum was there to collect me and as we drove towards home I drank in the familiarity of Bolton – the wet streets, putty-coloured skies and Greggs.

It felt great to be back above the Pineapple. Back with Mum.

I got myself back on track: no boozing, more sleep. It felt fantastic to be home in my own bed; even the sound of the pool cues thudding on the floor of the pub, like some tribal celebration, sounded welcoming.

I was so excited to see my dad, I had butterflies as I bumped Mum's little blue Austin Metro up the farm lane. He was up at the top cattle pens on the tractor, deaf to my shouted greetings because of the roar of the engine, then jumping in shock as I popped up in front of him waving my arms like a drowning man.

Dad put the kettle on for a brew and hugged me to his chest, the fabric of his shirt soft from being washed so often. He smelled faintly of diesel and hay. Everything was familiar and a comfort. I felt like I could relax, like loosening my tie or undoing a top button.

Happy to be home, but I still needed to work out what to do next.

Models have a shelf life for sure, but also a boredom threshold – it had served me well but there's only so long you can find holding your belly in and pouting a worthwhile use of time. I wasn't making big bucks – to cash in I'd need to travel, but for now I just wanted to gather my family and friends to me like a lady gathering her skirts before stepping over a mucky puddle.

I began to ponder the idea of further education – maybe I'd have to bite the bullet and try university. After all, a model's life was pretty much the same as a student's (apart from everyone being a size 8 and having fewer nicked traffic cones knocking about) – I was already used to sharing a chaotic, untidy home with my peers, drinking and smoking too much, sleeping too little, being late, hungover and chased by figures of authority. As a student I'd just be a tad more skint.

In the meantime, I would do what I did best – tread water. I was twenty years old; there was no mad rush. The world may have been my oyster but for now I was content with scraps from the chippy while I pondered my next move.

A fortnight after arriving home I got a call from MMA – a good casting had come in. They said I should make the effort. The total train fare to Manchester was about £3.50 but I was completely skint. Could I be bothered? I asked Mum her opinion and she said 'Sweetheart, you've got to go in for it, what if it's something big? You'd be crackers not to,' and so, for the umpteenth time, she handed over the cash and the confidence I needed.

I set off for Manchester. A bus and two trains later I arrived at the agency to find an absolutely packed back room – a sea of skinny, a wall of waifs – and an accompanying racket of dozens of models squawking excitedly, the occasional slightly hysterical shriek ringing out, like nubile gulls.

There was a real buzz. Forget scrapping for fifty-quid fashion shoots for teen mags. This was a casting for television.

I imagined that if the producers were scouting round the model agencies they weren't after hard-nosed political broadcasters. Less Paxman, more pouting. What kind of show would it be?

I let the other girls go out before me, hoping to suss out the situation from their feedback.

One at a time they stepped out onto the mini catwalk and returned a few minutes later a little confused – they weren't being judged on their looks or on their walk but on their TALK!

Well hello. Talking was what I was good at. I'd barely drawn breath since birth – at school I was quieter, less confident, my natural propensity to perform and entertain dampened down by the bullies, but at home it was a different story. I'd always had an appreciative crowd in my mum and nana as I ran through my repertoire of impressions and general showing off. I was even used to working difficult crowds – Dad was often tired and busy but I'd always make him crack a smile with my gabbing on, jazz-handing around all over the shop.

At work, though, I was never particularly pushy and so on this day I continued to hang back until there was only me left. It was quite a cunning tactical move, with hindsight. Isn't it always best in any competition to go last so the judges remember you, the dozens who went before all having merged into one (long-limbed) mass?

I walked out on the stage clutching a hairbrush to hold like a microphone. This now makes me feel proud and curls my toes in equal measure. There was a guy and a woman sitting at the end of the catwalk. They looked really cool, very London. The man held a video camera and the woman, who had olive skin, dark wavy hair and denim jacket, started to ask me some questions – silly stuff about relationships like a quiz from *Cosmo* magazine: 'What's the worst date you've been on?' and that kind of thing. I was in my element. Being asked my opinion at a modelling casting was unheard of but joyful. I rattled off various humorous and wildly exaggerated anecdotes, speaking into my hairbrush mic and leaning in conspiratorially to the camera. Playing it for laughs.

The floodgates had officially opened. I had an audience for my musings, one that was even chuckling at them. I was hooked. Maybe I was imagining it but it seemed the woman's eyes had lit up a little as I talked, laughed and gurned down the lens.

Afterwards I asked the woman her name – she was called Esta (which I thought was, again, very London and arse-achingly cool) – and before leaving I grilled Georgina in the office about the job. It was a company called Rapido TV and they were making something called *The Girlie Show* for Channel 4.

As I strode through St Ann's Square towards Piccadilly Station I felt strangely confident. Kinetic energy was fizzing and building inside me and I felt like something was about to burst free. I allowed myself to dream, and hoped that this would be the beginning of a new chapter, the first domino in a line of thousands that was about to be flicked by the finger of fate, which would knock into a thousand more, revealing a bright, bold pattern – the next few years of my life. Hopefully.

* * *

This story began as a homage to my darling dad Len, for always being the calm at the eye of the storm; as a thank-you note to my siblings for shaping and moulding me and putting up with me, the show-off with the wonky legs; and as a love letter to the farm where we all grew up, where the brickwork of the buildings holds all our childhood memories.

It's still all of those things but as the tale slowly unfurled I found myself writing and thinking about (and finally appreciating) the unlikely hero of the story – my mum. The bespectacled supermum, and now grandma. The 4'11" heroine. Opinionated, funny, loving and complicated, but always there. Because if my twenty-year-old hunch was true and I was to embark on a big, new adventure it would all be thanks to Mum.

What would I do without her? Honestly. Where would I be if it wasn't for my mum? I've asked myself that question a lot, especially since that day. She's been my safety net, ready to catch me as I've spun and tumbled through life. The Post-it note parenting has made an indelible mark on me, for which I'm now so grateful. It, and she, showed me that working hard for what you want is important; to be independent and to provide is crucial; and no one is perfect, because being a grown-up doesn't mean you have all the answers. My own parenting style has ripples of her running through it – loud but loving.

All the times she annoyed and chivvied me, like a collie dog nipping the ankles of the slowest sheep in the flock, I now know were for my own good – she was simply herding me in the right direction so that eventually I'd grow up to be a good, useful, kind person. I think she succeeded, mostly.

So Mum, this story is for you for all your hard work, your support, your encouragement and for always believing in me and always being there, whether I knew I needed you or not.

Acknowledgements

This book would not have happened without the persistent prodding, poking and persuasion from my agent, Melanie Rockcliffe – half woman, half force of nature – who's been by my side through thick and thin. She always said I could write a book, whether I should have is another matter altogether. If it goes tits up Mel, I'm hiding in your basement till it all blows over.

Huge thanks to everyone at Hodder who held my hand during the whole first book thing, especially Hannah Black, who has the patience and skill of a terrorist negotiator and the soothing, encouraging tone of a thousand Mary Berrys.

Despite the tough times I had at secondary school there were some great teachers and lovely friends too. So I would just like to send a high-five to the people not featured in the book but who are still part of my story, who attended Smithills School 1986-1990 and Canon Stade Sixth Form 1990-1992, plus those who in recent times have come back into my life via the magic of social media: the girl who tried and failed to get me into Def Leppard, Rebecca Charnock; Tim Tatlock and Matt Gill.

Big thanks to my first best friend and fellow Brosette Joanne Lloyd for giving me permission to use photos displaying our unique teen hairdos.

Huge thank you to my kickass crew of best friends, the 'Five go French' gang – Emily, Anna, Eleanor and Megan. I'm awestruck by their continued love and support and capacity to consume rosé wine on our annual jaunt to the south of France. Special mention to Megan who came up with the title for this book. We settled on *Till the Cows Come Home* over her other gem: *A Cox and Bull Story*. Genius.

Thank you to Clare for being my soul mate and 'bestest friend and bridesmaid' – from the moment we clapped eyes on each other a spark ignited that'll never go out.

Thank you to my brilliant, kind, funny husband Ben; even though this book pre-dates him, without his love I wouldn't be as happy, settled and confident as I am and things like this book wouldn't ever come to fruition. I'd like to take this opportunity to warn Ben that at least three of my perms are featured in this book.

Love and thanks to my three awesome children who every day make me proud, make me laugh and make me shout a bit. I love you Lola, Isaac and Renee.

A bit like at the Oscars, I've saved the biggest till last – this thank you is the equivalent of Best Picture: my family.

To my siblings for always being there for me. I love you. Our family is big and complicated, but I know wherever I am in the world I have at least four people who'd fight my corner: Robert, David, Dot and Yvonne. A special thank you to Rob for always answering his phone during the writing of this book, even though he knew I was probably asking him a random question about a dog's name from 1976. Without his kindness, impressive memory and slightly obsessive logging of old photographs this book would not have been possible.

Thank you to my mum Jackie and dad Len. For their unwavering support, for their patience and advice, for the best bear hugs (Dad) and the scrambled egg when I was poorly (Mum). For letting me be myself but teaching me to think of others too (I got that one eventually), and for always loving me. I love you.

Do you wish this wasn't the end?

Join us at www.hodder.co.uk, or follow us on
Twitter @hodderbooks to be a part of our community
of people who love the very best in books and reading.

Whether you want to discover more about a book
or an author, watch trailers and interviews, have the
chance to win early limited editions, or simply browse
our expert readers' selection of the very best books,
we think you'll find what you're looking for.

And if you don't,
that's the place to tell us what's missing.

We love what we do, and we'd love you to be part of it.

www.hodder.co.uk

@hodderbooks

HodderBooks

HodderBooks